# A Course in Mathematical Physics 2

Walter Thirring

# A Course in Mathematical Physics

# 2

## Classical Field Theory

Second Edition

Translated by Evans M. Harrell

With 74 Illustrations

Springer Science+Business Media, LLC

Dr. Walter Thirring
Institute for Theoretical Physics
University of Vienna
A-1090 Vienna
Austria

Dr. Evans M. Harrell (*Translator*)
Department of Mathematics
Georgia Institute of Technology
Atlanta, GA 30332
U.S.A.

Translation of Lehrbuch der Mathematischen Physik
Band 2: Klassische Feldtheorie
Wien-New York: Springer-Verlag 1978

Originally published by Springer-Verlag New York Inc. in 1986
Softcover reprint of the hardcover 2nd edition 1978

ISBN 978-1-4612-6463-7 ISBN 978-1-4419-8762-4 (eBook)
DOI 10.1007/978-1-4419-8762-4

Library of Congress Cataloging-in-Publication Data
Thirring, Walter E.
A course in mathematical physics.
Translation of: Lehrbuch der mathematischen Physik.
Bibliography: v. 2, p.
Includes index.
Contents: — 2. Classical field theory.
1. Mathematical physics. I. Title.
QC20.T4513 vol. 2 530.1'5 86-1813

9 8 7 6 5 4 3 2 1

# Preface

In the past decade the language and methods of modern differential geometry have been increasingly used in theoretical physics. What seemed extravagant when this book first appeared 12 years ago, as lecture notes, is now a commonplace. This fact has strengthened my belief that today students of theoretical physics have to learn that language—and the sooner the better. After all, they will be the professors of the twenty-first century and it would be absurd if they were to teach then the mathematics of the nineteenth century. Thus for this new edition I did not change the mathematical language. Apart from correcting some mistakes I have only added a section on gauge theories. In the last decade it has become evident that these theories describe fundamental interactions, and on the classical level their structure is sufficiently clear to qualify them for the minimum amount of knowledge required by a theoretician. It is with much regret that I had to refrain from incorporating the interesting developments in Kaluza–Klein theories and in cosmology, but I felt bound to my promise not to burden the students with theoretical speculations for which there is no experimental evidence.

I am indebted to many people for suggestions concerning this volume. In particular, P. Aichelburg, H. Rumpf and H. Urbantke have contributed generously to corrections and improvements. Finally, I would like to thank Dr. I. Dahl-Jensen for redoing some of the figures on the computer.

Vienna
*December, 1985*

W. Thirring

# Contents

# Symbols Defined in the Text

| | | |
|---|---|---|
| $e^{i_1 i_2 \dots i_p}$ | basis of the *p*-forms | (1.2.3) |
| $E_p(M)$ | linear space of the *p*-forms | (1.2.5; 2) |
| $d$ | exterior differential | (1.2.6) |
| $\omega\|_N$ | restriction of a form | (1.2.7; 3) |
| $E_m^0(U)$ | space of *m*-forms with compact support | (1.2.9) |
| $\langle e^i(x)\|e^k(x)\rangle$ | scalar product | (1.2.14) |
| $i_v$ | interior product | (1.2.16) |
| $*$ | isomorphism between $E_p$ and $E_{m-p}$ | (1.2.17) |
| $\delta$ | codifferential | (1.2.19) |
| $\Delta$ | Laplace–Beltrami operator | (1.2.20) |
| $L_v$ | Lie derivative | (1.2.23) |
| $\omega^i{}_k$ | affine connection | (1.2.25) |
| $\omega_{ik}$ | affine connection | (1.2.25) |
| $\Theta(x)$ | Heaviside step function | (1.2.31) |
| $\delta(x)$ | Dirac delta function | (1.2.31) |
| $\delta_{\bar{x}}$ | Dirac delta form | (1.2.33) |
| $G_{\bar{x}}$ | Green function | (1.2.35) |
| **E**, **B**, F | electric and magnetic fields | (1.3.1) |
| $A$ | vector potential | (1.3.7) |
| $\Lambda$ | gauge function | (1.3.10; 1) |
| $J$ | current | (1.3.12) |
| $Q$ | total charge | (1.3.18; 2) |
| $T^{\alpha\beta}$ | energy-momentum tensor | (1.3.20) |
| $P^\alpha$ | total energy-momentum | (1.3.21) |
| $\mathcal{T}^\alpha$ | energy-momentum form of the field | (1.3.22) |
| $z(s)$ | world-line | (1.3.25; 2) |
| $t^\alpha$ | energy-momentum form of matter | (1.3.25; 2) |
| $\mathcal{L}$ | Lagrangian | (2.1.1) |
| $W$ | action | (2.1.1) |

| | | |
|---|---|---|
| $\mathscr{S}$ | Poynting's vector | (2.1.13) |
| $D^{\pm}(N)$ | domains of influence | (2.1.15) |
| $D_{\bar{x}}$ | Green function | (2.2.5) |
| $D^{\text{ret}}(x)$ | retarded Green function | (2.2.7) |
| $G_{\bar{x}}^{\text{ret}}$ | retarded Green function (form) | (2.2.7) |
| $F^{\text{ret}}$ | retarded field strength | (2.2.9) |
| $F^{\text{in}}$ | incoming field strength | (2.2.15) |
| $F^{\text{out}}$ | outgoing field strength | (2.2.15) |
| $F^{\text{rad}}$ | radiation field | (2.2.21) |
| $D(x)$ | $D$-function | (2.2.22) |
| $\delta E$ | energy loss per period | (2.4.4; 2) |
| $j$ | specified current | (3.1.7) |
| $\varepsilon(k)$ | dielectric constant | (3.1.19; 3) |
| $S$ | superpotential | (3.1.21; 1) |
| $F(z)$ | Fresnel's integral | (3.3.10) |
| $\langle \| \rangle$ | scalar product | (4.1.3) |
| $S_p$ | Sections | (4.1.9) |
| $D$ | exterior covariant derivative | (4.1.10) |
| $D_X$ | covariant derivative | (4.1.15) |
| $\Omega$ | curvature form | (4.1.19) |
| $R$ | curvature in space-time | (4.1.20; 1) |
| $\Gamma_{ijk}$ | Christoffel symbol | (4.1.36) |
| $R_{ijkm}$ | Riemann-Christoffel tensor | (4.1.44; 2) |
| $C_{jk}$ | Weyl forms | (4.1.44; 3) |
| $r_0$ | Schwarzschild radius | (4.4.41) |
| $K$ | curvature parameter | (4.4.42) |
| $c$ | rate of convergence | (4.6.8) |
| $J^+(x)$ | future of $x$ | (4.6.18(a)) |
| $J^-(x)$ | past of $x$ | (4.6.18(a)) |
| $C(x, S)$ | set of causal curves | (4.6.18(b)) |
| $C^1(x, S)$ | set of differentiable causal curves | (4.6.18(b)) |
| $d(\lambda)$ | length of $\lambda$ | (4.6.18(c)) |

# Introduction 1

## 1.1 Physical Aspects of Field Dynamics

*Electric and magnetic fields are dynamically interconnected in such a way that an electromagnetic disturbance propagates with a universal velocity in empty space. By studying this phenomenon we gain a qualitative understanding of field radiation and are led to expect analogous gravitational behavior.*

The unification of the theories of electric and magnetic phenomena was one of the great scientific events of the nineteenth century. Whereas stationary electric fields **E** have sources at the positions of the charges but are irrotational ($\nabla \times \mathbf{E} = \mathbf{0}$), changing magnetic fields produce circulating electromotive forces. In contrast, magnetic fields **B** are always sourceless and circulate around currents and places where there is a time-dependent electric field. The dynamical interrelation of the two fields is described by Maxwell's equations: If we consider empty space (no sources or currents), then in units where $c = 1$ they require that

$$\oint_{\partial N} d\mathbf{s} \cdot \mathbf{E} = -\int_N d\mathbf{S} \cdot \dot{\mathbf{B}}, \qquad \oint_{\partial N} d\mathbf{s} \cdot \mathbf{B} = \int_N d\mathbf{S} \cdot \dot{\mathbf{E}}, \tag{1.1.1}$$

for integrals over arbitrary surfaces $N$ with boundaries $\partial N$; and if the surface is closed, then

$$\oint_N d\mathbf{S} \cdot \mathbf{E} = \oint_N d\mathbf{S} \cdot \mathbf{B} = 0 \tag{1.1.2}$$

We shall later recognize these apparently independent relationships as different aspects of a single fact, that the field-strength form and its dual form are closed. Before going more fully into this geometrical interpretation, let us try to come to an intuitive understanding of the physical consequences of these equations.

**Electromagnetic Waves** (1.1.3)

At a fixed time, a field

$$E_y = B_z = \cos(\omega(x - t)), \text{ all other components } 0, \tag{1.1.4}$$

looks as follows:

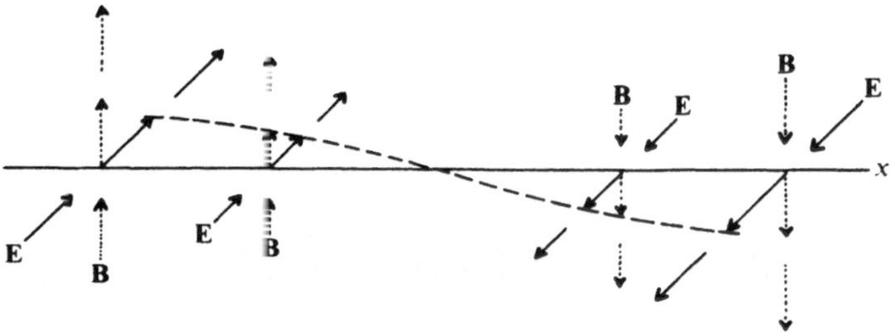


Figure 1 The fields in a plane wave

This is obviously free of sources, and it satisfies (1.1.1), as is easily seen on the surface chosen in Figure 2. Since the wave in Figure 1 moves to the right, $\dot{E}$ and $\dot{B}$ have the same sign as $E$ and $B$ in that region, and so for $t = 0$

$$\oint_{\partial N} \mathbf{E} \cdot d\mathbf{s} = -2L = -L\omega \int_0^{\pi/\omega} dx \sin \omega x = -\int_N \dot{\mathbf{B}} \cdot d\mathbf{S}$$

We see that unlike in the stationary situation, where the electric field of a point source falls off as $1/r^2$, it is dynamically possible for it to travel through space at the speed of light, normalized here to one, without decaying, that is, without the pulse losing intensity.

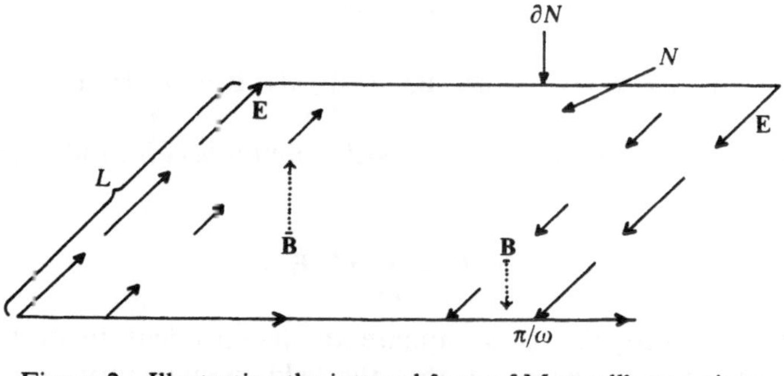


Figure 2 Illustrating the integral form of Maxwell's equations

It can also be seen from the relationships (1.1.1) and (1.1.2), though less directly, that any change in the fields propagates at the speed of light. We shall later study how this follows from the structure of the characteristics of the equivalent differential equation. For the moment let us take this property as given, and use it to investigate how an accelerated charge shakes off some of its Coulomb field and emits it as radiation.

**The Production of Electromagnetic Radiation** (1.1.5)

A charge $e$ moving with a constant velocity $\mathbf{v}$ in the $x$-direction emits no radiation (by Lorentz invariance). However, if it is brought to a stop at the origin during the time $-\tau < t < 0$, then its Coulomb field at some time $t > 0$ looks as follows: At distances $r > t + \tau$ from the origin it is equal to the field of the moving charge, since those parts of space have not yet learned of the braking of the particle. The lines of force out there point at the spot $\mathbf{x} \simeq \mathbf{v}t$ rather than at $\mathbf{x} = \mathbf{0}$, where the charge remains as $t > 0$. Thus the lines of force are displaced by $\mathbf{v}t \sim \dot{\mathbf{v}}\tau t$ compared with the field at $r < t$. At $r < t$ the field is that of a charge at rest at $\mathbf{x} = \mathbf{0}$, as the field has already forgotten that the charge ever moved. In between, in the spherical shell $t < r < t + \tau$, the lines of force progress continuously and without sources. Hence they must bend, and, as shown in Figure 3, as $r$ increases they must get folded more closely together in the parts of the spherical shell that are at most at right angles to $\mathbf{v}$. This causes the field strength to increase by a factor of $r\dot{v}$: The increase in the density of the field lines is proportional to (displacement of the field lines)/(thickness of the spherical shell) $\sim \dot{v}\tau t/\tau \sim r\dot{v}$, because at $r = t \gg \tau$ one would see the field of a charge at $x = t\mathbf{v} \sim t\tau\dot{\mathbf{v}}$,

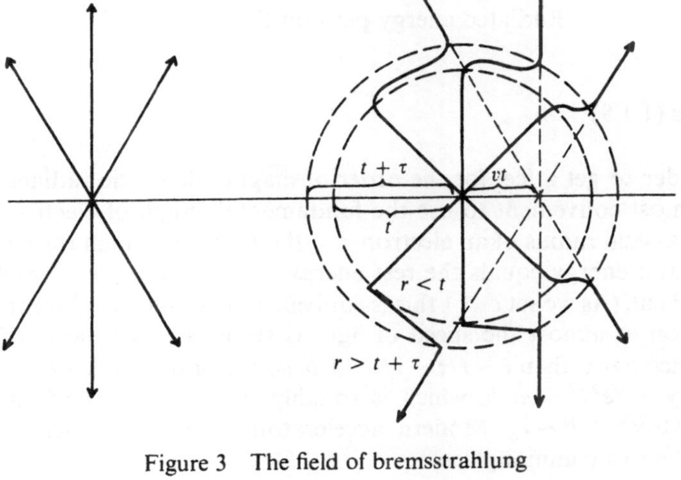


Figure 3 The field of bremsstrahlung

and thus the lines of force are displaced by this amount. The field in the spherical shell is consequently not $|\mathbf{E}| = e/r^2$, but

$$|\mathbf{E}| \sim \frac{e\dot{v}}{r}, \text{ for } t < r < t + \tau, \text{ and } \vartheta \equiv \measuredangle(x, v) \sim \frac{\pi}{2}. \tag{1.1.6}$$

**Remarks** (1.1.7)

1. The compression factor is only significant when $\vartheta$ is near $\pi/2$. If $\vartheta = 0$ or $\pi$, then the lines of force in the spherical shell are not bent at all. A more careful calculation produces an overall factor $\sin^2 \vartheta$ in $|E|^2$.
2. The sign of the field's augmentation is clearly such that its direction is opposite to that of the acceleration.

An electric field $\sim 1/r$ rather than $\sim 1/r^2$ as with the static Coulomb field leads immediately to the radiation of energy. The field energy contained in the spherical shell,

$$\sim \int_{t<r<t+\tau} d^3x |\mathbf{E}|^2 \sim |\mathbf{E}|^2 r^2 \tau \sim e^2 |\dot{\mathbf{v}}|^2 \tau$$

is conserved during the expansion, so that at large $r$ the radiated field still has just as much energy, while the Coulomb energy decreases to zero. The radiated energy is evidently imparted to the field during the braking in the time $\tau$, and then it travels off to infinity at the speed of light within the spherical shell $t < r < t + \tau$. In this way we obtain the basic formula of radiation,

**Larmor's Formula** (1.1.8)

$$\text{Radiated energy per unit time} = \frac{2}{3} \frac{e^2}{4\pi} \dot{v}^2.$$

**Remarks** (1.1.9)

1. In order to get a feel for the order of magnitude of the radiated energy, it is most convenient to use the fundamental length of electrodynamics, the classical radius of an electron, i.e., the radius at which the Coulombic potential energy equals the rest energy of an electron: $r_c = e^2/4\pi mc^2 \sim 10^{-13}$ cm. (As we set $c = 1$ this is equivalent to about $10^{-23}$ seconds.) If an electron at almost the speed of light is stopped in a time $\tau \sim r_c$ over a distance $b \sim \tau$, then $\dot{v} \sim v/\tau \sim 1/\tau \sim 1/b$, so the braking releases a radiative energy $\sim \tau e^2 \dot{v}^2 \sim e^2/b$, which is roughly the rest energy of an electron ($\sim \frac{1}{2}$ MeV) if $b \sim r_c$. Modern accelerators have made such breakneck occurrences commonplace.

2. Electrons are most easily braked with an electric field **E**, and it is natural to ask what the connection between the radiated energy and the energy of the field **E** is. Because $m\dot{\mathbf{v}} = e\mathbf{E}$, an electron radiates $E^2(e^2/m)^2\tau$, which is the fraction of the energy of the field **E** contained in a volume $\tau \times$ (the classical electron radius)$^2$. If the electron is not subjected to a single braking, but is moved periodically, as in a light wave, then we are interested in the cross-section, defined as (the radiated energy per unit time)/(the incident energy per unit time and surface area). Since the energy density of a light wave is $E^2$, the energy incident in a time $\tau$ and a unit of surface area is $E^2\tau$, and the scattering cross-section is about $r_c^2$. This means that the electron is about $10^{-13}$ cm across, in the sense that it blocks a surface area $r_c^2 \sim 10^{-26}$ cm$^2$ from a beam of light. However, quantum effects often make an electron act more as if it were $10^{-11}$ cm across. An electron is best pictured as about $10^{-11}$ cm across but fairly transparent, so that it scatters light only weakly. The explanation for why matter is so often opaque, though the field is predominantly influenced by the rather transparent electrons it contains, will be discussed later.
3. The precise numerical factor given in (1.1.8) comes about because
   (a) We use units in which the Coulomb field is $e/4\pi r^2$, giving an extra $(4\pi)^{-2}$.
   (b) According to (1.1.7; 1) the energy has an angular distribution $\sim \sin^2 \vartheta$. Integrated over a spherical surface, this gives $\frac{2}{3} \cdot 4\pi$.
   (c) The energy density is actually $(|\mathbf{E}|^2 + |\mathbf{B}|^2)/2$, but the contribution from **B** equals that from **E**.

The source of light is ordinarily atoms, in which negative charges orbit positive ones, while the total charge is neutral. If a single charge oscillates, then the pattern of field lines produced is that of a repetition of one-time brakings (Figure 4, left). If two charges oscillate around each other, then this

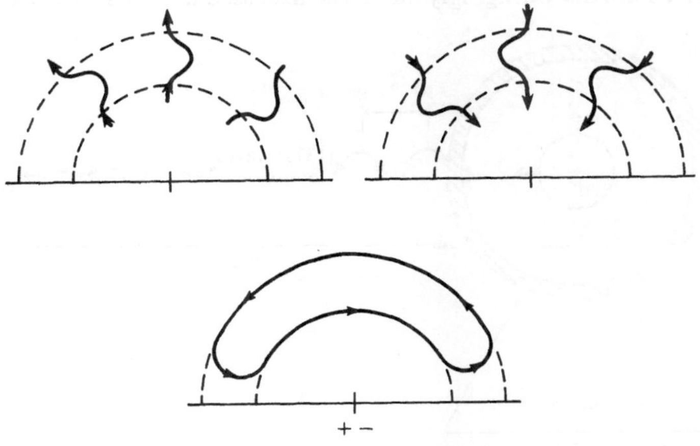


Figure 4 The electric field of an oscillating dipole

field must be superposed on one that is oppositely directed and has a phase-lag (Figure 4). As we see, oscillating dipoles emit circulating electric fields, in which $|\mathbf{E}|$ is again given by (1.1.6). If $\omega$ is the frequency and $L$ the amplitude of oscillation, then $\omega^2 L$ replaces $\dot{v}$ in (1.1.6). This produces the formula for

**Dipole Radiation** (1.1.10)

$$\text{Radiated energy per cycle} \sim e^2 L^2 \omega^3.$$

**Application to Atoms** (1.1.11)

For an atom, one would set $L$ equal to the Bohr radius $r_b \equiv \hbar^2/me^2 = (\hbar c/e^2)^2 r_c = (137)^2 r_c \sim 10^{-8}$ cm. The electron velocity $\omega L$ is roughly $(137)^{-1}$, so the period is $\sim(137)r_b \sim (137)^3 r_c \sim 10^{-15}$ sec. Then from (1.1.10) the energy loss per cycle is $\sim(137)^{-3}e^2/r_b$, and an electron has to orbit $(137)^3$ times to radiate an energy $e^2/r_b$. The available energies of excitation are on the order of $e^2/r_b \sim 10$ eV, so we expect the lifetime of an excited state to be about $(137)^3 \cdot 10^{-15}$ sec. $\sim 10^{-8}$ sec. Of course, a more exact analysis of what goes on requires a quantum-theoretical analysis of the process, but only quantum-mechanical quantity needed for a preliminary orientation is $r_b$. The creation of light may be typically outlined in this way: An atom emits $(137)^3$ waves in $10^{-8}$ seconds. Since the wavelength is about $1/\omega \sim 137 r_b$, the resultant wave-packet is $(137)^4 r_b \sim 10$ cm long (Figure 5). These figures also show up in the widths of the emitted spectral lines and in the coherence length of the radiation.

The original estimate of bremsstrahlung in (1.1.8) used only the electric field's form $\sim\mathbf{x}/r^3$. It is tempting to reason the same way with gravitation, replacing $e^2$ with the formally analogous quantity $\kappa m_p^2$. Before spending too much time on the details of this analogy, one would likely note that the numbers involved are discouragingly large. As discussed in (I: 1.1.1), the coupling

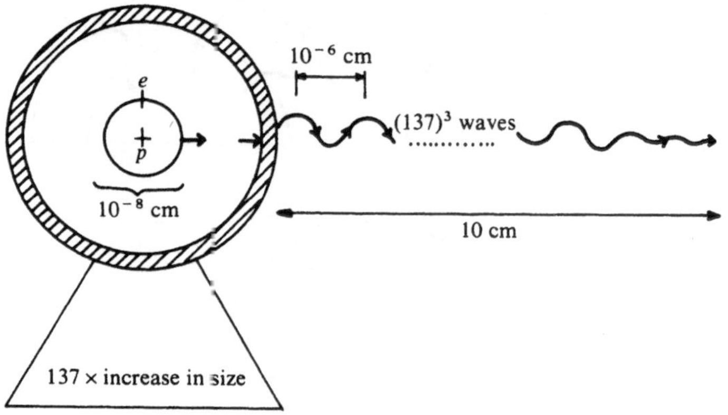


Figure 5 The typical lengths connected with the emission of light by an atom

constants differ by a factor of $10^{36}$. Whereas an atom takes $10^{-8}$ seconds to emit a photon, it would require $10^{36} \cdot 10^{-8}$ sec. $\sim 10^{28}$ sec. to bring a graviton into being. This is $10^{10}$ times the age of the universe, and it seems idle to speculate more about such questions. But masses, unlike charges, all have the same sign, and therefore large bodies benefit from tremendous coherence effects. We shall shortly see that collapsing stars ought to release gigantic energies through gravitational radiation.

**The Production of Gravitational Radiation** (1.1.12)

The points to reconsider in the derivation of (1.1.5) are:

(a) One of the facts used was that $\mathbf{E} = e\mathbf{x}/r^3$. The potential energy $e^2/r$ of two charges $e$ corresponds to the gravitational energy $\kappa M^2/r$ of two masses $M$. Hence the analogue of the Coulomb field is $\sqrt{\kappa}M\mathbf{x}/r^3$, the square of which is the energy density.

(b) The second fact used was that the electric field propagates at the speed of light. We shall later learn that this is also a property of the gravitational field, as a consequence of Einstein's equations. However, the gravitational field in turn influences the speed of light, which complicates the details of the radiation problem, though the orders of magnitude should not be affected.

(c) In the electromagnetic case the center of charge was accelerated, but the analogy fails at this point. Since the gravitational field is coupled to all masses, and the center of mass moves uniformly, this kind of radiation never occurs. This is apparent, for instance, if two masses oscillate around each other, for which two equal fields with a phase shift must be superposed, but this time with the same sign. Then in directions perpendicular to the oscillation, the large component in the direction of $\dot{\mathbf{v}}$ cancels out (see Figure 6), leaving a field $\sim 1/r^2$.

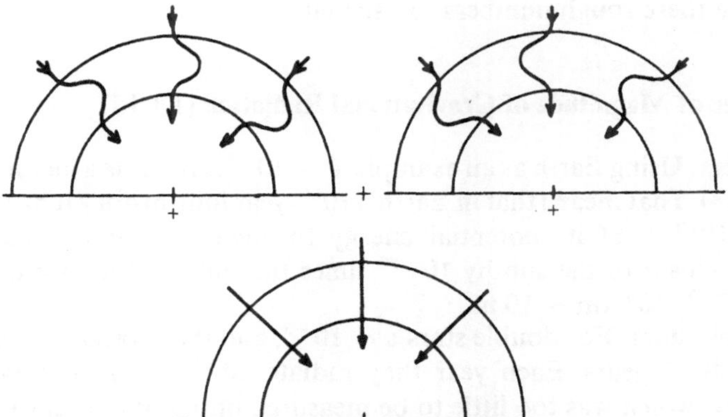


Figure 6 The gravitational field of two oscillating masses

(d) There can be electromagnetic quadrupole radiation even with a stationary center of charge; and in the situation of (c), the asymptotic fields do not quite cancel out at 45°. A more careful calculation shows that they are only reduced by a factor $v \sim \omega L$. The same thing is true for gravitational waves, leading to

**Gravitational Radiation of Rotating Masses** (1.1.13)

$$\text{Radiated energy per unit time} \sim \kappa M^2 \omega^6 L^4 \sim \frac{\kappa M^2}{L} v^5 \omega,$$

or

$$\text{Radiated energy per period} \sim v^5 \cdot \text{the gravitational energy of the system.}$$

**Remarks** (1.1.14)

1. We shall later encounter an apparently different reason for the additional factor $v^2 = \omega^2 L^2$: whereas the electric field is coupled to the current, the gravitational field interacts with the energy-momentum tensor, which is quadratic in the velocity. An additional $v$ in the coupling produces an additional $v^2$ in the radiated energy.
2. By interpreting $\kappa M^2/L$ as the gravitational energy of the system we obtain a relationship in which the miniscule coupling constant no longer appears, or, more precisely, is hidden within the gravitational energy.
3. We have not taken the trouble to puzzle out the exact numerical factor in (1.1.13). Moreover, it only applies in the context of the linear approximation to the Einsteinian theory. No exact solutions, showing how gravitational waves are created, have yet been found.

In (I: §1.1) we have discussed the qualitative features of cosmic phenomena. Let us use these rough numbers to calculate

**The Order of Magnitude of Gravitational Radiation** (1.1.15)

(a) Planets: Using Earth as an example, $v \sim 10^{-4}$, giving us a factor $10^{-20}$ in (1.1.13). That means that in Earth's $10^{10}$-year long history it has lost only one $10^{10}$-th of its potential energy through radiation, causing it to draw closer to the sun by $10^{-10}$ times the radius of its orbit, which is $\sim 10^{-10} \cdot 10^8$ km $\sim 10$ m.
(b) Double stars: For double stars $v \sim 10^{-3}$, and the orbital period is often only $10^{-3}$ years. Each year they radiate $10^{-12}$ of their gravitational energy, which was too little to be measured until quite recently.
(c) Black holes: If a star collapses under its own weight, then its contents are compressed into nuclear matter, and it turns into a neutron star,

with a radius of perhaps 10 km. (cf. §4.5). If further compressed, it produces such a strong gravitational field that the gravitational energy is comparable to the rest energy, and relativistic effects appear. The gravitational analogue of the classical radius of an electron is the Schwarzschild radius $\kappa M$, which is on the order of a kilometer for stars. Kepler orbits at this distance become relativistic, as the potential energy approaches the rest energy. The implosion of a neutron star can lead to a situation where $v \sim 1$ and $\kappa M^2/L \sim M$, and then, by (1.1.13), the radiated energy becomes equal to the rest energy of a star. Even if a more exact calculation would show that the true figure was a few percent, the energies involved are certainly huge.

Unfortunately, not much is known about gravitational radiation experimentally. It is predicted by all sensible theories, however, since our derivation has only posited a finite propagation speed for the gravitational field.

In all of the above discussion the motion of the sources of the fields has been assumed given. In fact the fields also affect the sources, and one really ought to solve the equations for the coupled system. We encountered a simple example of such a problem in the scattering of light, where an electron was accelerated by a light wave, which in turn caused the light impinging on a surface $\sim r_c^2$ to be scattered. That derivation, though, was for a free electron, and certainly does not hold for electrons in matter. Such a small cross-section would render normal matter, with interatomic distances $\sim(137)^2 r_c$, highly transparent. The reason this is not the case is that bound electrons exhibit resonance behavior, increasing $\dot{v}$ compared with (1.1.9; 2).

**The Scattering of Light by Bound Electrons** (1.1.16)

The electrons sit in the electric field $\mathbf{E}$ of the binding force as well as the light wave. Idealizing the binding as a harmonic force with strength $m\omega_0^2$, making the frequency of atomic electron oscillations $\omega_0$, would lead one to expect $m(\ddot{\mathbf{x}} + \omega_0^2\mathbf{x}) = e\mathbf{E}$ for the equation of motion of the electrons. A periodic electric field then gives rise to an acceleration

$$\dot{\mathbf{v}} = \ddot{\mathbf{x}} \sim \frac{e\mathbf{E}}{m}\left(1 - \frac{\omega_0^2}{\omega^2}\right)^{-1},$$

which increases the scattering cross-section by $(1 - \omega_0^2/\omega^2)^{-2}$. Since the frequency $\omega$ of visible light, which is emitted by other atoms, may be near $\omega_0$, the resonance denominator can be very small, and light will not penetrate far. Of course, realistic matter has several different resonance frequencies, which is why it is colorful.

Because of the situation just sketched, it is tempting to simplify the coupled system of matter and an electromagnetic field by replacing the matter with the boundary condition that the field does not penetrate it. This is the approach usually taken in optical problems like diffraction. We shall find that even with this simplification such problems can only be solved with difficulty, if any value is placed on precision.

If one wants to undertake a serious analysis of the coupled system of matter plus field, then it must be clearly understood that the dangers of the electric and gravitational cases lie in opposite quarters: Since like charges repel, but all masses attract, the Coulomb force tends to make a system explode, while gravity tends to make it collapse. These tendencies of the static forces are preserved in the relativistic generalizations of the equations, and can not be abolished by any mathematical sleight of hand. We shall not be able to answer the question of what holds the electronic charge together, and classical electrodynamics will remain an incomplete theory. It is replaced by quantum electrodynamics at short distances, of course, but even that theory can not explain how come the electrical energy of a point charge is finite. On the other hand, the most refined mathematics only confirms the naive expectation that, with gravity, a sufficiently massive object collapses under its own weight. It seems exceedingly difficult to get around these singularity theorems; at the very least, we can say that the prognosis of disaster is likely to be true.

## 1.2 The Mathematical Formalism

*The central concepts of classical field theory are the algebra $\{E_p\}$ of forms and their exterior derivatives. With the *-mapping $E_p \rightarrow E_{m-p}$ defined by the metric, they allow Maxwell's and Einstein's equations to be written in analogous ways.*

The calculus of É. Cartan is especially useful in classical field theory, because it formalizes the mathematical constructions used in the field equations. The most important rules of calculation will be briefly summarized here; the reader is referred to (I, chapter 2) or to the mathematics books cited in the bibliography for more precise definitions of the concepts.

The space of tensor fields over a manifold $M$† has a linear structure‡; every element can be written as a linear combination of certain basis elements, where the coefficients are the tensor components. In the example of the covariant vector fields, any $m$ (= the dimension of $M$ here and henceforth) linearly independent vector fields $e^i(x)$, $i = 1, \ldots, m$, $x \in M$, can be used; any vector field $V$ can be written as

$$V = \sum_{i=1}^{m} V_i(x) e^i(x).$$

† As in the first volume, a sufficiently differentiable manifold with a boundary will be referred to simply as a manifold and differentiability will likewise be implicitly assumed for tensor fields.

‡ More precisely, a module structure. The coefficients are functions of $M$ and not simply real numbers.

**Remarks** (1.2.1)

1. Independence means that the $e^i(x)$ are independent vectors for all $x$.
2. A basis is occasionally called an $m$-frame or tedrad for $m = 4$. For any $i$, $e^i$ is a vector, and not a component of a vector, or any such thing.
3. There does not generally exist a global basis, as a linearly independent set of $e^i$ can not normally be continuously extended to all of $M$. (According to (I: 2.6.17; 6), on the surface of a sphere there does not exist any continuous, nowhere vanishing vector field.) However, on the domain $U$ of a chart there is always a natural basis, as discussed below. Therefore, for local processes like differentiation there is no danger in assuming the existence of a basis.
4. A new basis can always be formed from an old one by $e^i(x) \to \bar{e}^i(x) = A^i{}_j(x)e^j(x)$, where Det $A^i{}_j(x)$ is different from zero for all $x$. No basis is distinguished from the others in the absence of more structure.

With the formation of the tensor product $\otimes$, the $e^i$ also provide a basis for the $p$-fold contravariant tensor fields. On $U$ any tensor field can be written

$$t = \sum_{j_k} t_{j_1 \cdots j_p}(x) e^{j_1} \otimes e^{j_2} \otimes \cdots \otimes e^{j_p}. \tag{1.2.2}$$

Among tensor fields, the totally antisymmetric ones ($\equiv$ $p$-forms) are especially important, because they define $p$-dimensional volume elements. An antisymmetric tensor product $\wedge$ (the wedge, or exterior, product) can be introduced on them. Like the (finite) tensor product, it is associative and distributive, but $e^i \wedge e^j = -e^j \wedge e^i \equiv e^i \otimes e^j - e^j \otimes e^i$. Since we shall frequently use the product bases, we introduce the abbreviation

$$e^{i_1 i_2 \cdots i_p} \equiv e^{i_1} \wedge e^{i_2} \wedge \cdots \wedge e^{i_p}. \tag{1.2.3}$$

With (1.2.3) any $p$-form can be written

$$\omega = \sum_{i_k} \omega_{i_1 \cdots i_p} \frac{e^{i_1 \cdots i_p}}{p!}. \tag{1.2.4}$$

**Remarks** (1.2.5)

1. The basis $e^{i_1 \cdots i_p}$ has only $\binom{m}{p}$ independent elements, due to antisymmetry. If $p = 0$, then $p$-forms are ordinary functions, and if $p > m$ they are defined as zero, as otherwise antisymmetry would be impossible.
2. The $p$-forms on a manifold are a linear space (in fact a module), denoted by $E_p(M)$.
3. The wedge product

$$e^{i_1 \cdots i_p} \wedge e^{i_{p+1} \cdots i_{p+q}} = e^{i_1 \cdots i_{p+q}} = (-1)^{pq} e^{i_{p+1} \cdots i_{p+q}} \wedge e^{i_1 \cdots i_p}$$

creates an obviously associative mapping $E_p \times E_q \xrightarrow{\wedge} E_{p+q}$, and imparts the structure of a (graded) algebra to the space $\bigcup_p E_p$ of forms.

**The Exterior Differential** (1.2.6)

The elementary differential operations, gradient, divergence, and curl, generalize to a linear mapping $d: E_p(M) \to E_{p+1}(M)$, obeying the rules:

(a) $d(\omega_1 + \omega_2) = d\omega_1 + d\omega_2, \qquad \omega_i \in E_p$
(b) $d(\omega_1 \wedge \omega_2) = (d\omega_1) \wedge \omega_2 + (-1)^p \omega_1 \wedge d\omega_2, \qquad \omega_1 \in E_p, \omega_2 \in E_q$
(c) $d(d\omega) = 0, \qquad \omega \in E_p,$

for $p, q = 0, 1, \ldots, m$.

**Remarks** (1.2.7)

1. Rules (a) and (b) describe how $d$ acts with respect to the algebraic operations, by which Leibniz's rule gets an additional $(-1)^p$ from antisymmetry.
2. The coordinates $x^i$, $i = 1, \ldots, m$, of a point of $M$ may be thought of as a mapping $U \to \mathbb{R}$, $M \supset U =$ the domain of the chart. That is, they are an element of $E_0(U)$. The $dx^i$ are thus $m$ 1-forms, known as the **natural basis** of $E_1(U)$. If $p = 0$, then (a) and (b) are the usual rules of differentiation, so the exterior differential of a function $U \to \mathbb{R}: x \to f(x)$ becomes $df = dx^i \, \partial f / \partial x^i$. If $p = 0$, then $d$ is the gradient, and $f_{,i}$ are its components in the natural basis. Because of (c), in the natural basis

$$de^{j_1 \cdots j_p} = 0, \qquad 0 \le p \le m,$$

and the exterior differential of a $p$-form

$$\omega = \omega_{j_1 \cdots j_p} \frac{e^{j_1 \cdots j_p}}{p!}$$

(with the summation convention) becomes

$$d\omega = (d\omega_{j_1 \cdots j_p}) \wedge \frac{e^{j_1 \cdots j_p}}{p!} = \omega_{j_1 \cdots j_p, k} \frac{e^{kj_1 \cdots j_p}}{p!}.$$

3. Local coordinates can be introduced on any $n$-dimensional submanifold $N$ so that $x_{n+1} = x_{n+2} = \cdots = x_m = 0$. The **restriction** $\omega_{|N}$ of a form is defined by setting $dx_{n+1|N} = dx_{n+2|N} = \cdots = dx_{m|N} = 0$, and letting the restriction commute with $+$ and $\wedge$. It is then easy to see that $d\omega_{|N} = d(\omega_{|N})$.
4. Forms with vanishing exterior differentials are called **closed**, and forms that can be written as exterior differentials are called **exact**. Rule (c) means that exact $\Rightarrow$ closed. The opposite implication holds on starlike manifolds but not in general. It always holds locally.

**The Integral** (1.2.8)

Under a coordinate transformation

$$x \to \bar{x}(x), \qquad dx^i = \frac{\partial x^i}{\partial \bar{x}^j}\, d\bar{x}^j,$$

the natural basis of $E_m(U)$ transforms as an $m$-dimensional volume element:

$$e^{1\cdots m} = \operatorname{Det}\left(\frac{\partial x^i}{\partial \bar{x}^j}\right)\bar{e}^{1\cdots m},$$

and it is possible to define a coordinate-independent integral over $\omega$ by

$$\int_U \omega \equiv \int dx^1 \cdots dx^m\, \omega_{1\ldots m}(x), \tag{1.2.9}$$

$$\omega \in E_m^0(U) \equiv \text{the } m\text{-forms with compact support.}$$

As $\omega \in E_m^0(M)$ can be written as a finite sum $\sum_i \omega_i$ of $m$-forms each supported in the domain of a chart, its integral is defined by (1.2.9) and $\int \sum_i \omega_i = \sum_i \int \omega_i$. More generally, the integral of $\omega \in E_n^0(M)$ over an $n$-dimensional submanifold $N$ is defined as the integral of the restriction of $\omega$ to $N$:

$$\int_N \omega \equiv \int_N \omega_{|N}. \tag{1.2.10}$$

Integration is the inverse of differentiation $d$ in the sense that Gauss's and Stokes's theorems generalize to

$$\int_N d\omega = \int_{\partial N} \omega, \qquad \omega \in E_{n-1}^0(M), \tag{1.2.11}$$

where $\partial N$ is the boundary of $N$.†

**Riemannian Structure** (1.2.12)

A symmetric, covariant tensor field of the second degree

$$g = g_{ik} e^i \otimes e^k, \text{ with } \operatorname{Det}(g_{ik}(x)) \neq 0 \quad \forall x \tag{1.2.13}$$

creates an isomorphism between the covariant and contravariant vector fields and lets the spaces of the two types of vector fields be identified. A scalar product is defined by

$$\langle e^i(x) | e^k(x)\rangle = g^{ik}(x), \qquad g^{ij} g_{jk} = \delta^i{}_k. \tag{1.2.14}$$

If the matrix $g_{ik}$ is positive, then the space is said to be **Riemannian**, and otherwise **pseudo-Riemannian**. The $e_i \equiv g_{ik} e^k$ form a dual basis, with

† We shall occasionally use this theorem when $\partial N$ is not actually a manifold, but has corners or some other harmless handicap.

$\langle e^j | e_i \rangle = \delta^j{}_i$; and $v^k \equiv g^k v_i$ are called contravariant components of a vector $v = v_i e^i = v^i e_i$.

The scalar product as map $E_1 \times E_2 \to E_0$ can be generalized in two ways.

**The Scalar Product between *p*-Forms** (1.2.15)

The pseudo-Riemannian structure also induces a bilinear map $E_p \times E_p \overset{<|>}{\to} E_0$. Expressed in components it reads

$$\langle \omega | v \rangle = \frac{1}{p!} \sum_{(i)(j)} \omega_{j_1 \ldots j_p} v_{i_1 \ldots i_p} g^{j_1 i_1} \cdots g^{j_p i_p} = \langle v | \omega \rangle.$$

In a Riemannian space $\langle \omega | \omega \rangle^{1/2}$ measures the $p$-dimensional volume defined by $\omega$. If it is the product of 1-forms.

**The Interior Product of a *q*-Form and a *p*-Form** (1.2.16)

A bilinear map $E_p \times E_q \to E_{p-q}$, $p \geq q$: $(\omega, v) \to i_v \omega$ can be defined by the rules:

(i) $i_v \omega = \langle \omega | v \rangle$ for $p = q$;
(ii) $i_v(\omega_1 \wedge \omega_2) = (i_v \omega_1) + (-)^{p_1} \omega_1 \wedge i_v \omega_2$ for $v \in E_1$, $\omega_i \in E_{p_i}$;
(iii) $i_{v_1 \wedge v_2} = i_{v_2} \circ i_{v_1}$.

By $\langle \, | \, \rangle$ the $E_p$ are identified with their dual spaces and the interior product is the transpose of the exterior product:

$$\langle v \wedge \omega | u \rangle = i_{v \wedge \omega} \mu = i_\omega (i_v \mu) = \langle \omega | i_v \mu \rangle.$$

In an $m$-dimensional manifold the space $E_m$ is one-dimensional and a pseudo-Riemannian structure distinguishes the $\varepsilon \in E_m$ which are normalized: $\langle \varepsilon | \varepsilon \rangle = (-)^s = \operatorname{Det} g / |\operatorname{Det} g|$. If such an $\varepsilon$ exists globally the manifold is called **orientable**. We shall assume from now on that there is globally given an orientation $\varepsilon$. In this case $E_p$ can be identified with $E_{m-p}$.

**The Hodge Duality Map** (1.2.17)

The isomorphism $E_p \overset{*}{\to} E_{m-p}$ is defined by $^*\omega = i_\omega \varepsilon$. It has the properties

(i) $\varepsilon = {}^*1$, $\quad {}^*\varepsilon = (-)^s$;
(ii) $* \circ * = (-)^{p(m-p)+s}$;
(iii) $i_v {}^*\omega = {}^*(\omega \wedge v) = (-)^{pq} i_\omega {}^*v$ for $\omega \in E_p$, $v \in E_q$;
(iv) for $v, \omega \in E_p$ one has

$$v \wedge {}^*\omega = \varepsilon i_v \omega = \omega \wedge {}^*v = \varepsilon (-)^s i^* v^* \omega.$$

One should notice:

ad (i). $E_0$ and $E_m$ are one-dimensional and, at a point, isomorphic to $R$. $\varepsilon$ is the *-image of the number 1.

ad (ii). Except for a sign the map * is its own inverse. Unfortunately, one cannot get rid of these signs by using another $\varepsilon$.

ad (iii). Duality changes the interior product into the exterior product.

ad (iv). The word duality has its root in the fact that $E_{m-p}$ is the dual space for $E_p$ if we define a scalar product $\{\ ,\ \}$ by $v \wedge \omega = \varepsilon\{v, \omega\}$. It is related to $i$ by

$$\{v, \omega\} = (-)^{p(m-p)+s} i_v {}^*\omega.$$

So far the scalar product and other algebraic operations have been understood pointwise, that is $\langle \omega | v \rangle \in E_0 \equiv C(M)$ assigns a number to each point at a manifold. In an orientable Riemannian manifold $\int {}^*\langle \omega | v \rangle$ maps $E_p \times E_p$ into $R$ and is a scalar product in sense of a Hilbert space.

In terms of a basis the definitions imply the

**Rules** (1.2.18)

(a) $${}^*e^{i_1 \dots i_p} = g^{i_1 j_1} \dots g^{i_p j_p} e^{j_{p+1} \dots j_m} \varepsilon_{j_1 \dots j_m} \frac{\sqrt{|\mathrm{Det}\, g|}}{(m-p)!}$$

(b) $$e_j \wedge {}^*e^{j_1 \cdots j_p} = \sum_{r=1}^{p} (-1)^{r+p} \delta^{j_r}{}_j {}^*e^{j_1 \cdots j_{r-1} j_{r+1} \cdots j_p} = (-1)^{p+1} {}^* i_{e_j} e^{j_1 \cdots j_p}$$

(c) $$i_{e_j} {}^*e^{j_1 \cdots j_p} = {}^*e^{j_1 \cdots j_p j}.$$

The **codifferential**

$$\delta \equiv {}^*d{}^*(-1)^{m(p+1)+s}, \qquad d = {}^*\delta{}^*(-1)^{m(p+1)+1+s}, \tag{1.2.19}$$

is a generalization of the divergence. It is a linear mapping $E_p \to E_{p-1}$, and can be coupled with $d$ to construct the generalization $\Delta$ of the Laplace operator for $E_p \to E_p$:

$$\Delta = \delta d + d\delta \qquad \text{(the \textbf{Laplace–Beltrami operator})}. \tag{1.2.20}$$

With a natural basis $e^{j_1 \dots j_p}$ of $E_p(\mathbb{R}^m)$, where $g = dx^i \otimes dx^k \eta_{ik}$, $\eta_{ik} = \pm 1$ if $i = k$ and otherwise 0,

$$\begin{aligned} \delta(fe^{i_1 \cdots i_p}) &= \sum_{j=1}^{p} f_{,}^{\ i_j} e^{i_1 \cdots i_{j-1} i_{j+1} \cdots i_p} (-1)^{j-1}, \qquad f \in E_0(\mathbb{R}^m) \\ \Delta(fe^{i_1 \cdots i_p}) &= f_{,k}{}^{k} e^{i_1 \cdots i_p}, \qquad f_{,}^{\ k} \eta_{ki} \equiv f_{,i}, \end{aligned} \tag{1.2.21}$$

(Problem 6). If, in particular, $p = 1$ and $f_k$ are the components of a vector field, then $\delta$ is the ordinary divergence $f^k{}_{,k}$ and $\Delta$ is the operator $(\eta^{-1})^{ki}\, \partial^2/\partial x^i\, \partial x^k$ as applied to the individual components. (If $m = 3$ and $p = 1$, then $\Delta = -\nabla \times \nabla \times + \nabla\nabla \cdot$).

Problem 7 is to derive the following set of

**Rules** (1.2.22)

(a) $dd = \delta\delta = 0, \quad d\Delta = \Delta d, \quad \delta\Delta = \Delta\delta;$
(b) $\delta * = (-1)^p * d, \quad *\delta = (-1)^{p+1} d*;$
(c) $d\delta * = *\delta d, \quad *d\delta = \delta d*, \quad *\Delta = \Delta *.$

In volume I we learned moreover that if a vector field $v$ is specified, then

**The Lie Derivative** (1.2.23)

$$L_v = i_v \circ d + d \circ i_v : E_p \to E_p:$$
$$L_v(\omega_1 + \omega_2) = L_v\omega_1 + L_v\omega_2,$$
$$L_v(\omega_1 \wedge \omega_2) = (L_v\omega_1) \wedge \omega_2 + \omega_1 \wedge L_v\omega_2$$

gives the rate of change of a form under the action of the flow generated by the contravariant components of $v$. (If $v = v_i\, dx^i$ and $f \in E_0(\mathbb{R}^n)$, then $L_v f = v^i\, \partial f/\partial x^i$.)

To calculate the derivative in a different coordinate system (Problems 1 and 2) it is convenient to use a so-called orthogonal basis. Since $g_{ik}$ is a symmetric matrix, it can be transformed into a diagonal matrix $\eta_{ik}$ with only $\pm 1$ as eigenvalues, by some change of basis. This determines the $e^i$ up to a local Lorentz transformation:

$$\begin{aligned} &e^i(x) \to \Lambda^i{}_k(x) e^k(x), \\ &\eta_{kl} \Lambda^k{}_m(x) \Lambda^l{}_n(x) = \eta_{mn} \quad \forall x. \end{aligned} \tag{1.2.24}$$

An orthogonal basis is not necessarily natural, and the exterior differentials of the orthogonal $e^i$ may not vanish. To handle the general case we define

**The Affine Connections** $\omega^i{}_k \in E_1(M)$ (1.2.25)

$$de^i = -\omega^i{}_k \wedge e^k, \qquad dg_{ik} = \omega_{ik} + \omega_{ki},$$
$$\omega_{ik} = g_{ij}\omega^j{}_k.$$

**Remarks** (1.2.26)

1. In §4.1 we shall show that these properties determine the $\omega^i{}_k$ uniquely.
2. It follows from (1.2.25) for the differentials of the bases in $E_p$ that

$$de^{j_1 \cdots j_p} = -\omega^{j_1}{}_j \wedge e^{j j_2 \cdots j_p} - \omega^{j_2}{}_j \wedge e^{j_1 j \cdots j_p} - \cdots$$
$$\cdots - \omega^{j_p}{}_j \wedge e^{j_1 \cdots j_{p-1} j},$$

and likewise

$$d{*}e^{j_1\cdots j_p} = -\omega^{j_1}{}_j \wedge {*}e^{jj_2\cdots j_p} - \omega^{j_2}{}_j \wedge {*}e^{j_1 j\cdots j_p} - \cdots$$
$$\cdots - \omega^{j_p}{}_j \wedge {*}e^{j_1\cdots j_{p-1}j}$$

(Problem 8).

3. The $\omega$'s generalize the $\Gamma$'s of (I: 1.1.7), and hence play the role of the field strength in gravitational theory. They transform inhomogeneously under a change of basis (Problem 9):

$$\omega^k{}_r \to A^k{}_s \omega^s{}_j (A^{-1})^j{}_r - (A^{-1})^j{}_r \, dA^k{}_j.$$

This means in particular that there is always a basis for which all the $\omega^k{}_r$ vanish at a point (see (I: 5.6.11)).

**Partial Differential Equations** (1.2.27)

All of the equations that will interest us are generally covariant (i.e., chart-independent), and hence the only kind of differentiation they contain is the exterior differential. Orientability and the pseudo-Riemannian structure of the world enter only through the *-mapping. The prototype of this kind of equation for a vector field is the specification of the divergence and curl of the field, or more generally of $\delta F$ and $dF$ for $F \in E_p$.

Partial differential equations determine the derivatives in certain directions, while in others they may allow the fields to vary freely, and even have discontinuities. These directions, known as the **characteristics**, are the key to an understanding of the implications of the equations. To exhibit this fact in its germinal form, we discuss the simplest nontrivial

**Example** (1.2.28)

$M = \mathbb{R}^2$ with the metrics $g_\pm \equiv dx^2 \pm dt^2$. According to (1.2.17; 2), if $p = 1$, then $* \circ * = \mp \mathbf{1}$, and so $*$ is a linear transformation of vectors, and has eigenvalues $\pm i$ for $g_+$ and $\pm 1$ for $g_-$. More specifically, if $\mathbf{F} = \mathbf{E}\, dt + \mathbf{B}\, dx$, then $*\mathbf{F} = -\mathbf{B}\, dt \pm \mathbf{E}\, dx$; for $g_+$, $*$ is a rotation through 90°, and for $g_-$ it is a reflection about the axis $x = -t$ (Figure 7):

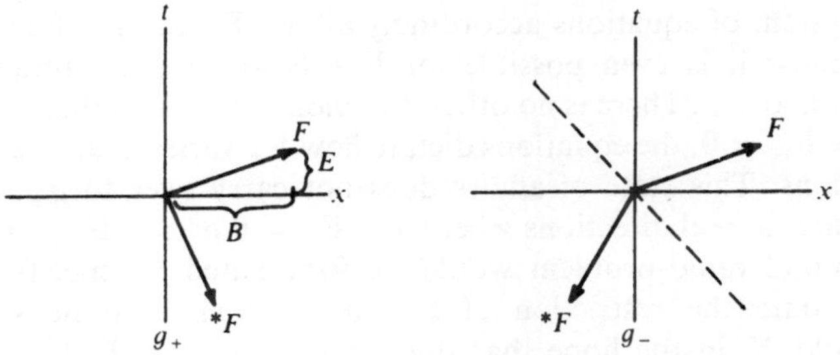


Figure 7 The action of * on $\mathbb{R}^2$

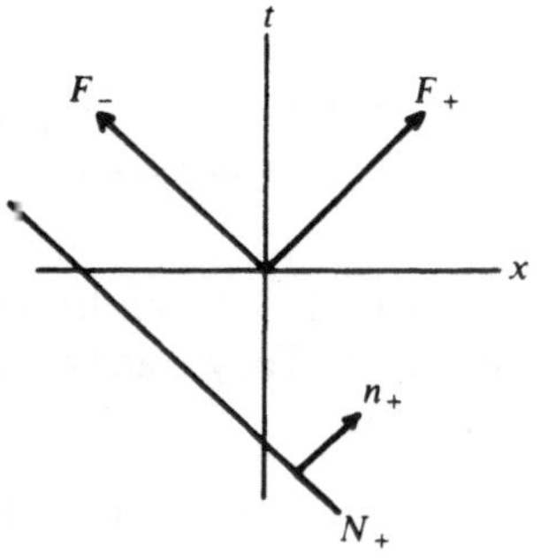


Figure 8 The hypersurface $N_+$ and its normals $n_+$

Let us therefore consider the system of equations for $\mathbf{F}$,

$$d\mathbf{F} = \mathbf{M}, \qquad \delta\mathbf{F} = \mathbf{J},$$

with specified $\mathbf{M}$ and $\mathbf{J}$. For $g_-$ it is possible to decouple the equations using the diagonalization of $*$:

$$\mathbf{F}_\pm \equiv (1 \mp *)\mathbf{F}: d\mathbf{F}_\pm = \mathbf{M} \pm *\mathbf{J}.$$

The $\mathbf{F}_\pm = (\mathbf{E} \pm \mathbf{B})(dt \pm dx)$ are independent of each other in this case, and we need only to work with a single equation. For $g_+$ the same method leads to the equations $\mathbf{F}_\pm = (1 \mp i*)\mathbf{F} = (\mathbf{E} \pm i\mathbf{B})(dt \mp i\,dx)$, in which case $\mathbf{F}_\pm$ are complex conjugates.

If $n_\pm = dx \pm dt$ and $N_\pm$ are lines parallel to $x = \mp t$, so that $n_{\pm|N_\pm} = 0$, then for $g_-$ we get $i_{n_+} n_+ = i_{n_-} n_- = 0$ and therefore

(a) $n_\pm \wedge \mathbf{F}_\pm = \mathbf{0}$;
(b) $\mathbf{F}_{\pm|N_\pm} = 0$;
(c) $i_{n_\pm} \mathbf{F}_\pm = \mathbf{0}$.

**Conclusions** (1.2.29)

(a) If $\mathbf{E} \pm \mathbf{B}$ depends only on $t \pm x$, then $d\mathbf{F}_\pm$ becomes $(\mathbf{E} \pm \mathbf{B})' n_\pm \wedge n_\pm = \mathbf{0}$. The system of equations accordingly allows $\mathbf{F}_\pm$ to vary freely in those directions; it is even possible for $\mathbf{E} + \mathbf{B}$ to be discontinuous in the direction of $n_+$. There is no other direction $n \in E_1$ such that $n \wedge \mathbf{F}_+ = \mathbf{0}$ or $n \wedge \mathbf{F}_- = \mathbf{0}$; the equations dictate how $\mathbf{F}_\pm$ varies in all the remaining directions. This state of affairs does not carry over to $g_+$, for which there are no real directions where $n \wedge \mathbf{F}_+ = \mathbf{0}$ or $n \wedge \mathbf{F}_- = \mathbf{0}$.
(b) The initial-value problem would be formulated by specifying as the initial data the restriction of $\mathbf{F}_\pm$ to some $m - 1$-dimensional submanifold $N$, in the hope that this might determine $\mathbf{F}$. If $n_{\pm|N} = 0$ at

any point of $N$, then $\mathbf{F}_{\pm|N}$ also vanishes at that point, making it impossible to choose the initial data at will. It is possible that they even leave $\mathbf{F}$ indeterminate. By fact (b) there exist $\mathbf{F}_+ \neq \mathbf{0}$ for which $\mathbf{F}_{+|N} = \mathbf{0}$ and $d\mathbf{F}_+ = \mathbf{0}$. If $n_{\pm|N} \neq 0$, the only condition on the initial data is $d(\mathbf{F}_{+|N}) = d\mathbf{F}_{+|N} = (\mathbf{M} - {}^*\mathbf{J})_{|N}$, which automatically vanishes (because there exists no 2-form on a one-dimensional $N$).

(c) By (1.2.23) the Lie derivative $L_{e_k}$ in the direction $x^k$ is given by $i_{e_k} \circ d + d \circ i_{e_k}$ with $e_k = g_{kj}\, dx^j$. Thus $L_{e_\pm}$ with $e_\pm \equiv gn_\pm = n_\mp$ is the Lie derivative in the direction $n_\pm$ (parallel to $x = \pm t$), and because $L_{e_-}\mathbf{F}_+ = i_{e_-}\, d\mathbf{F}_+ + di_{e_-}\mathbf{F}_+ = i_{e_-}\, d\mathbf{F}_+ = i_{e_-}(\mathbf{M} - {}^*\mathbf{J})$, the rate of change of $\mathbf{F}_+$ in the direction $n_-$ is determined. As we shall see later from the explicit solution, $F$ is determined by an arbitrary set of initial data on $N$ when:

   (i) $n_{\pm|N}$ is different from zero at all points of $N$; and
   (ii) every line $t \pm x =$ constant intersects $N$ at exactly one point.

An $m - 1$-dimensional hypersurface $N$ (in other words, in two dimensions, a curve) that satisfies (i) and (ii) is called a **Cauchy surface**. It is nowhere tangent to the characteristics $n_\pm$, and it must be large enough to determine $\mathbf{E} \pm \mathbf{B}$ everywhere. (See Figure 9.)

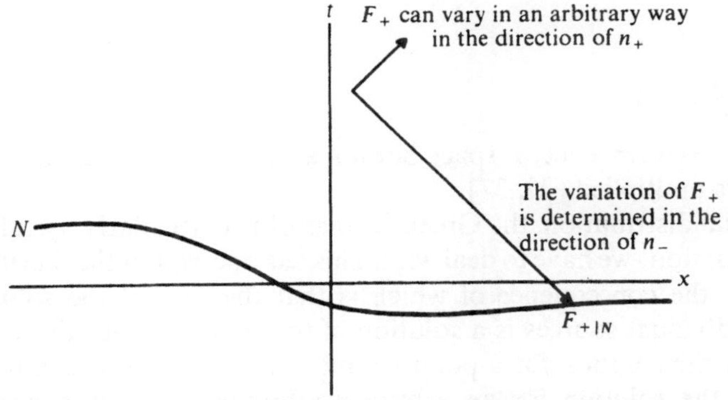


Figure 9 Determination of $F_+$ from its restrictions to the Cauchy surface $N$

**Remarks** (1.2.30)

1. Not all surfaces that are everywhere spacelike and extend to infinity are Cauchy surfaces. (See Figure 10.)
2. Observations like the foregoing can normally answer only local questions. Global questions are more difficult to treat effectively. Some seemingly innocuous manifolds, which are dear to the hearts of cosmologists, have no Cauchy surfaces at all.
3. Since $\mathbf{E} + \mathbf{B}$ can be an arbitrary function of $t + x$, one might want to risk using discontinuous functions, for which $M_\pm : n_\pm|_{M_\pm} = 0$ are surfaces

where $\mathbf{E} \pm \mathbf{B}$ has jump discontinuities. If such functions are admitted, the classical notion of differentiation is fraught with unnecessary difficulties. For instance, the equation $(\partial/\partial u)(\partial/\partial v)\Phi = 0$ has the solution $\Phi = \varphi(u) + \psi(v)$, where $\varphi$ is arbitrary and $\psi$ is differentiable, but this solution does not satisfy $(\partial/\partial v)(\partial/\partial u)\Phi = 0$.

To avoid these inconveniences it is natural to try to use distributions $\varphi$, defined by their integrals against suitable test-functions. It is then always permissible to differentiate, as defined through integration by parts: $\int f\varphi^{(n)} \equiv (-1)^n \int f^{(n)}\varphi$.

A strict formulation of the problem requires the theory of locally convex topological vector spaces. As we shall not need the more profound theorems, we content ourselves with making the

**Definition** (1.2.31)

$$\Theta(x) = \begin{cases} 1, & \text{if } x \geq 0 \\ 0, & \text{if } x < 0 \end{cases} \qquad \text{(Heaviside's step function)}$$

$$\delta(x) = \frac{d\theta(x)}{dx} \qquad \text{(Dirac's delta function)},$$

and the

**Warning** (1.2.32)

Distributions form a vector space but not an algebra; they can be added but not always multiplied [22, 37].

A special distribution, the Green function,† is particularly useful for solving the equations we have to deal with. This lets one exploit the linearity of the equations, the consequence of which is that the sum of the solutions for several individual sources is a solution of the equation with the sum of the sources. If the solution for a point source is known, then it can be used to construct the solution for an arbitrary inhomogeneity by superposition. Putting these ideas into practice as applied to $E_p$ requires a bit more algebra and replacing the sums with integrals.

The first thing to do is to generalize the delta function so that when it is multiplied by a $p$-form and integrated it reproduces the value of the $p$-form at a point $\bar{x} \in M$ (the set of all such values is denoted by $E_{p|\bar{x}}$). Since we are only able to integrate $m$-forms, and the result ought to be in $E_{p|\bar{x}}$, we need a $\delta_{\bar{x}} \in E_{p|\bar{x}} \otimes E_{m-p}(M)$ such that‡

$$\int \delta_{\bar{x}} \wedge F = F(\bar{x}), \qquad \text{for all} \quad F \in E_p(M). \tag{1.2.33}$$

† We follow Jackson's [1] lead and abandon the ungrammatical phrase "the Green's function."

‡ In de Rham's terminology, $\delta_{\bar{x}}$ would be called a $p$-form-valued current. In the older literature one often encounters the term Green's dyadic for the three-dimensional case. We shall write $E_p(M)$ even when the components are distributions rather than differentiable functions.

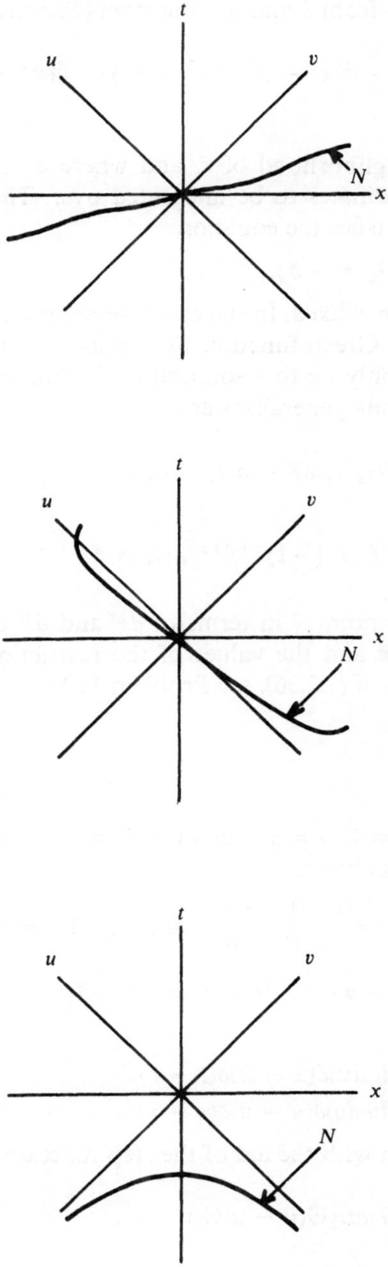


Cauchy surface, $\left.\begin{matrix} v \\ u \end{matrix}\right\} = \dfrac{t \pm x}{\sqrt{2}}$

not a Cauchy surface, because (i) is violated

not a Cauchy surface, because (ii) is violated

Figure 10

The distribution $\delta_{\bar{x}}$ vanishes away from $\bar{x}$ and has the form (Problem 10)

$$\delta_{\bar{x}} = \bar{e}_{i_1 \cdots i_p} \otimes {}^*e^{i_1 \cdots i_p} \frac{(-1)^{p(m-p)}}{p!} \delta(x^1 - \bar{x}^1)\delta(x^2 - \bar{x}^2) \cdots \delta(x^m - \bar{x}^m), \tag{1.2.34}$$

in a natural basis $e^{i_1 \cdots i_p}$ of a neighborhood of $\bar{x}$, and where $\bar{e}_{i_1 \cdots i_p}$ is the basis in $E_{p|\bar{x}}$ and $x^i$ are the coordinates to be integrated over. The Green function $G_{\bar{x}} \in E_{p|\bar{x}} \otimes E_{m-p}(M)$ satisfies the equation

$$\Delta G_{\bar{x}} = -\delta_{\bar{x}}, \tag{1.2.35}$$

where $\Delta = d\delta + \delta d$ acts on $x^i$ with $\bar{x}$ fixed. In the cases we are interested in, we shall verify the existence of a Green function by explicit construction. Equation (1.2.35) determines $G_{\bar{x}}$ only up to a solution of the homogeneous equation. Using $G_{\bar{x}}$, Green's formula generalizes as

$$F(\bar{x}) = (-1)^{p+m} \int_N [dG_{\bar{x}} \wedge \delta F - \delta G_{\bar{x}} \wedge dF]$$

$$- \int_{\partial N} [\delta G_{\bar{x}} \wedge F + (-1)^{m+p+s} {}^*dG_{\bar{x}} \wedge {}^*F], \tag{1.2.36}$$

which expresses $F \in E_p(M)$ at the point $\bar{x}$ in terms of $dF$ and $\delta F$ on some $m$-dimensional submanifold $N \ni \bar{x}$ and the values of the restrictions of $F$ and ${}^*F$ to $\partial N$. (For the derivation of (1.2.36), see Problem 11.)

**Example** (1.2.37)

We return to (1.2.28). $M = \mathbb{R}^2$, $p = 1$, $g = g_-$, and $(-1)^s = -1$. Using the "lightlike" coordinates that diagonalize *,

$$\begin{matrix} v \\ u \end{matrix} = \frac{t \pm x}{\sqrt{2}}, \qquad \text{for which} \quad g_- = \begin{vmatrix} 0 & -1 \\ -1 & 0 \end{vmatrix} \quad \text{and} \quad (-1)^s = -1,$$

$${}^*du = du, \qquad {}^*dv = -dv, \qquad {}^*(du \wedge dv) = -\mathbf{1},$$

and (omitting the $\otimes$ sign)

$$\begin{aligned} \delta_{\bar{x}} &= (d\bar{x}\, dt - d\bar{t}\, dx)\delta(x - \bar{x})\delta(t - \bar{t}) \\ &= (d\bar{v}\, du - d\bar{u}\, dv)\delta(u - \bar{u})\delta(v - \bar{v}). \end{aligned}$$

The Green function can be written with the aid of the step function (1.2.31) as

$$G_{\bar{x}} = (d\bar{v}\, du - d\bar{u}\, dv)\tfrac{1}{2}\Theta(\bar{u} - u)\Theta(\bar{v} - v),$$

because

$$\begin{aligned} dG_{\bar{x}} &= \tfrac{1}{2}[\delta(v - \bar{v})\Theta(\bar{u} - u)d\bar{v} + \Theta(\bar{v} - v)\delta(\bar{u} - u)d\bar{u}]du \wedge dv \\ \delta G_{\bar{x}} &= \tfrac{1}{2}[\delta(v - \bar{v})\Theta(\bar{u} - u)d\bar{v} - \Theta(\bar{v} - v)\delta(\bar{u} - u)d\bar{u}], \end{aligned}$$

with which it is easy to verify that

$$\Delta G_{\bar{x}} = (d\bar{u}\, dv - d\bar{v}\, du)\delta(u - \bar{u})\delta(v - \bar{v}) = -\delta_{\bar{x}}.$$

If $F = \varphi\, du + \psi\, dv$, then the field equations $dF = M$, $\delta F = J$, explicitly read:

$$dF = du \wedge dv(\psi_{,u} - \varphi_{,v}) = M \equiv du \wedge dv\, m(u, v)$$
$$\delta F = \quad -\psi_{,u} - \varphi_{,v} \quad = J(u, v).$$

Now the integral over $N$ of (1.2.36) contributes

$$\begin{aligned} -\int dGJ + \int \delta GM = &-\tfrac{1}{2}\int du \wedge dv(\delta(v - \bar{v})\Theta(\bar{u} - u)d\bar{v} \\ &+ \Theta(\bar{v} - v)\delta(\bar{u} - u)d\bar{u})J(u, v) \\ &+ \tfrac{1}{2}\int du \wedge dv(\delta(v - \bar{v})\Theta(\bar{u} - u)d\bar{v} \\ &- \Theta(\bar{v} - v)\delta(\bar{u} - u)d\bar{u})m(u, v) \\ = &\int_{u(\bar{v})}^{\bar{u}} du\, \tfrac{1}{2}(m(u, \bar{v}) - J(u, \bar{v}))d\bar{v} \\ &- \int_{v(\bar{u})}^{\bar{v}} dv\, \tfrac{1}{2}(m(\bar{u}, v) + J(\bar{u}, v))d\bar{u}, \end{aligned}$$

if the submanifold $\partial N$ is defined as $\{u = u(v)\} = \{v = v(u)\}$:

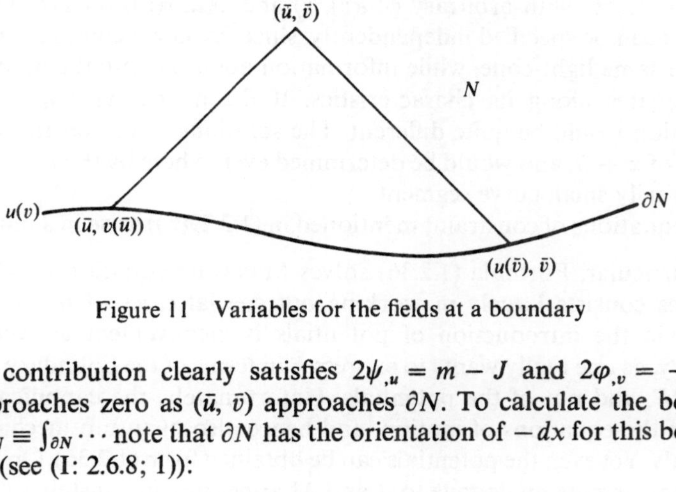


Figure 11 Variables for the fields at a boundary

This contribution clearly satisfies $2\psi_{,u} = m - J$ and $2\varphi_{,v} = -m - J$, and approaches zero as $(\bar{u}, \bar{v})$ approaches $\partial N$. To calculate the boundary term $F_{\partial N} \equiv \int_{\partial N} \cdots$ note that $\partial N$ has the orientation of $-dx$ for this boundary integral (see (I: 2.6.8; 1)):

$$\begin{aligned} \int_{\partial N} \delta G \wedge F = &\frac{d\bar{v}}{2}\int_{\partial N} (\varphi\, du + \psi\, dv)\delta(v - \bar{v})\Theta(\bar{u} - u) \\ &- \frac{d\bar{u}}{2}\int_{\partial N} (\varphi\, du + \psi\, dv)\delta(u - \bar{u})\Theta(\bar{v} - v) \end{aligned}$$

$$\begin{aligned} \int_{\partial N} {}^*dG \wedge {}^*F = &\frac{d\bar{v}}{2}\int_{\partial N} (-\varphi\, du + \psi\, dv)\delta(v - \bar{v})\Theta(\bar{u} - u) \\ &+ \frac{d\bar{u}}{2}\int_{\partial N} (-\varphi\, du + \psi\, dv)\delta(u - \bar{u})\Theta(\bar{v} - v); \end{aligned}$$

since $dv$ contributes positively and $du$ negatively, the net result is that

$$F_{\partial N} = d\bar{v}\, \psi(u(\bar{v}), \bar{v})\Theta(\bar{u} - u(\bar{v})) + d\bar{u}\, \varphi(\bar{u}, v(\bar{u}))\Theta(\bar{v} - v(\bar{u})).$$

As long as $\bar{x} \in N$, so that $\Theta(\bar{u} - u(\bar{v}))\Theta(\bar{v} - v(\bar{u})) = \mathbf{1}$, this solves the homogeneous equation, because $\psi$ is independent of $\bar{u}$ and $\varphi$ is independent of $\bar{v}$. It also clearly satisfies the boundary conditions.

**Remarks** (1.2.38)

1. It was not apparent in (1.2.36) that $\partial N$ was required to be a Cauchy surface. However, the example shows that $dG_{\bar{x}}$ can not be used as a distribution on $\partial N$ if $\partial N$ contains parts of the light-cone of $\bar{x}$, which would contribute $\delta(0)$.
2. The derivation relies on Stokes's theorem, which we only know for forms of compact support. Since the support of $G_{\bar{x}}$ remains on a light-cone, Stokes's theorem may also be used here for non-compact $N$.
3. At this point (1.2.36) appears as an identity for a solution of the field equations. It may be used to construct a solution from its initial data, but gives no information about how arbitrarily the initial data may be chosen. In the example, if $\partial N$ is a Cauchy surface, then the solution meets all requirements with arbitrary $\psi(u(v), v)$ and $\varphi(u, v(u))$ on $\partial N$. The initial values can be specified independently, since $\partial N$ never contains two points of the same light-cone, while information about an initial condition only propagates along its characteristics. If the metric were $g_+$, then the situation would be quite different. The solutions would be analytic functions of $x + it$, and would be determined everywhere by their values on an arbitrarily small curve segment.
4. The equations of constraint mentioned in (1.2.29(b)) arise in 4 dimensions.

In particular, Formula (1.2.36) solves Maxwell's equations, and electrodynamics consists largely in working out special cases of it. The normal detour via the introduction of potentials is inconvenient as well as unnecessary, as one really wants to express $F$ in terms of the boundary values of $F$ and $*F$, and not of the potentials. Unfortunately, the canonical formulation of the equations of motion, and hence also quantum mechanics, use potentials. Yet even the potentials can be obtained from (1.2.36). On the other hand, the currents analogous to $J$ and $M$ appearing in Einstein's equations depend nonlinearly on the fields, so in that case (1.2.36) is useful only in the linear approximation.

**Problems** (1.2.39)

1. The metric $g = d\rho^2 + \rho^2\, d\varphi^2 + dz^2$ in cylindrical coordinates for $\mathbb{R}^3\backslash\{\mathbf{0} \times \mathbb{R}\}$. Calculate $\omega_{ik}$ for the orthogonal basis $e^1 = dz, e^2 = d\rho, e^3 = \rho\, d\varphi$. Write $\nabla \times A$ and

$\nabla \cdot A$ in the components of this basis and of the natural basis. What is the connection between them?

2. The same question for spherical coordinates and $\mathbb{R}^3\backslash\{\mathbf{0} \times \mathbb{R}\}$, for which $g = dr^2 + r^2\, d\vartheta^2 + r^2 \sin^2 \vartheta\, d\varphi^2$.
3. Calculate $\Delta$ from (1.2.20) for $p = 0$ in the natural basis, and specialize to the cases of cylindrical and spherical coordinates on $\mathbb{R}^3\backslash\{\mathbf{0} \times \mathbb{R}\}$.
4. Prove the normalization $* \circ * = (-1)^{p(m-p)+s}1$. Using (1.2.16).
5. Derive the rules (1.2.18):

$$*e^{i_1\ldots i_p} = g^{i_1 j_1} \cdots g^{i_p j_p} e^{j_{p+1}\ldots j_m} \varepsilon_{j_1\ldots j_m} \frac{\sqrt{|g|}}{(m-p)!},$$

$$e_j \wedge *e^{j_1 \cdots j_p} = \sum_r (-1)^{r+p} \delta^{j_r}{}_j *e^{j_1 \cdots j_{r-1} j_{r+1} \cdots j_p}$$

$$i_{e_j} *e^{j_1 \cdots j_p} = *e^{j_1 \cdots j_p j}.$$

6. Verify (1.2.21).
7. Verify (1.2.22).
8. Check (1.2.26; 2).
9. Derive the transformation law for the $\omega$'s (1.2.26; 3).
10. Show that $\delta_{\bar{x}}$ from (1.2.34) has the property (1.2.33).
11. Prove (1.2.36).
12. Find a manifold $M$ and a $J \in E_p(M)$ for which the equation $dF = J$ has no (global) solution, although $dJ = 0$.

## **Solutions** (1.2.40)

1. Orthogonal basis: $\eta_{ik} = \delta_{ik}$, $0 = de^1 = de^2$, $de^3 = e^{23}/\rho$. Hence only $\omega_{32} = -\omega_{23} = e^3/\rho \neq 0$, and $de^{12} = de^{32} = 0$, $de^{31} = -\omega^{32} \wedge e^{21}$. Letting $A = a_i e^i$,

$$dA = (a_{1,2} - a_{2,1})e^{21} + (a_{2,3}/\rho - a_{3,2} - a_3/\rho)e^{32} + (a_{3,1} - a_{1,3}/\rho)e^{13},$$

$$*dA = (a_{3,2} + a_3/\rho - a_{2,3}/\rho)e^1 + (a_{1,3}/\rho - a_{3,1})e^2 + (a_{2,1} - a_{1,2})e^3,$$

$$*A = a_1 e^{23} + a_2 e^{31} + a_3 e^{12},\ *d*A = a_{1,1} + a_{2,2} + a_{3,3}/\rho + a_2/\rho.$$

Natural basis: $\sqrt{|g|} = \rho$. Let $A = A_z\, dz + A_\rho\, d\rho + A_\varphi\, d\varphi$. Then

$$*A = \rho[A_z\, d\rho \wedge d\varphi + A_\rho\, d\varphi \wedge dz + \rho^{-2} A_\varphi\, dz \wedge d\rho],$$

$$*d*A = \rho^{-1}[(\rho A_z)_{,z} + (\rho A_\rho)_{,\rho} + (\rho^{-1} A_\varphi)_{,\varphi}],$$

$$*dA = \rho^{-1}[(A_{\varphi,\rho} - A_{\rho,\varphi})dz + (A_{z,\varphi} - A_{\varphi,z})d\rho + \rho^2(A_{\rho,z} - A_{z,\rho})d\varphi].$$

The connection is that $(A_z, A_\rho, A_\varphi) = (a_1, a_2, \rho a_3)$.

2. Orthogonal basis: $e^1 = dr$, $e^2 = r\,d\vartheta$, $e^3 = r \sin\vartheta\, d\varphi$, $de^1 = 0$, $de^2 = dr \wedge d\vartheta$, $de^3 = r\cos\vartheta\, d\vartheta \wedge d\varphi + \sin\vartheta\, dr \wedge d\varphi \Rightarrow \omega_{21} = e^2/r$, $\omega_{31} = \sin\vartheta\, d\varphi$, $\omega_{32} = \cos\vartheta\, d\varphi \Rightarrow de^{12} = 0$, $de^{13} = -e^{23}/r\tan\vartheta$, $de^{23} = 2e^{23}/r$. Let $A = a_i e^i$. Then

$$*d*A = a_{1,1} + \frac{2}{r}a_1 + \frac{1}{r}a_{2,2} + \frac{a_2}{r\tan\vartheta} + \frac{a_{3,3}}{r\sin\vartheta},$$

$$*dA = \left[\frac{a_{3,2}}{r} + \frac{a_3}{r\tan\vartheta} - \frac{a_{2,3}}{r\sin\vartheta}\right]e^1 + \left[\frac{a_{1,3}}{r\sin\vartheta} - \frac{a_3}{r} - a_{3,1}\right]e^2$$
$$+ \left(a_{2,1} + \frac{a_2}{r} - \frac{a_{1,2}}{r}\right)e^3.$$

Natural basis: $\sqrt{|g|} = r^2\sin\vartheta$. Let $A = A_r\,dr + A_\vartheta\,d\vartheta + A_\varphi\,d\varphi$. Then

$$*A = r^2\sin\vartheta[A_r\,d\vartheta \wedge d\varphi + r^{-2}A_\vartheta\,d\varphi \wedge dr + r^{-2}\sin^{-2}\vartheta A_\varphi\,dr \wedge d\vartheta]$$

$$*dA = \frac{dr}{r^2\sin\vartheta}(A_{\varphi,\vartheta} - A_{\vartheta,\varphi}) + \frac{d\vartheta}{\sin\vartheta}(A_{r,\varphi} - A_{\varphi,r}) + d\varphi\,\sin\vartheta\,(A_{\vartheta,r} - A_{r,\vartheta}),$$

$$*d*A = \frac{1}{r^2\sin\vartheta}\left[(r^2\sin\vartheta A_r)_{,r} + (\sin\vartheta A_\vartheta)_{,\vartheta} + \left(\frac{A_\varphi}{\sin\vartheta}\right)_{,\varphi}\right].$$

The connection is that $(A_r, A_\vartheta, A_\varphi) = (a_1, ra_2, r\sin\vartheta a_3)$.

3.
$$\Delta f = \frac{1}{\sqrt{g}}\frac{\partial}{\partial x^\alpha}(f_{,\beta}g^{\alpha\beta}\sqrt{g})$$

Cylindrical coordinates: $\Delta = \frac{1}{\rho}\left[\rho\frac{\partial^2}{\partial z^2} + \frac{\partial}{\partial\rho}\rho\frac{\partial}{\partial\rho} + \frac{1}{\rho}\frac{\partial^2}{\partial\varphi^2}\right]$

Spherical coordinates: $\Delta = \frac{1}{r^2\sin\vartheta}\left[\frac{\partial}{\partial r}r^2\sin\vartheta\frac{\partial}{\partial r} + \frac{\partial}{\partial\vartheta}\sin\vartheta\frac{\partial}{\partial\vartheta} + \frac{\partial}{\partial\varphi}\frac{1}{\sin\vartheta}\frac{\partial}{\partial\varphi}\right].$

4. From (1.2.16) we infer

$$\langle *\omega | *\omega\rangle = \langle *\omega | i_\omega \varepsilon\rangle = \langle \omega \wedge *\omega | \varepsilon\rangle$$
$$= (-)^{p(m-p)}\langle *\omega \wedge \omega | \varepsilon\rangle = (-)^{p(m-p)}\langle\omega | i_{*\omega}\varepsilon\rangle = (-)^{p(m-p)}\langle\omega | **\omega\rangle.$$

The fact that $**\omega$ can differ from $\omega$ only by a sign is a general property of duality. Therefore $**\omega = (-)^{p(m-p)+s}\omega$ is equivalent to $\langle\omega|\omega\rangle = (-)^s\langle *\omega | *\omega\rangle$. It is obviously enough to show this for a basis

$$e^{i_1\ldots i_p} \equiv \varepsilon_p \in E_p$$

such that $\varepsilon = \varepsilon_p \wedge \varepsilon_{m-p}$ with $i_{\varepsilon_p}\varepsilon_{m-p} = 0$. Then

$$\langle *\varepsilon_p | *\varepsilon_p\rangle = \langle\varepsilon_{m-p}|\varepsilon_{m-p}\rangle\langle\varepsilon_p|\varepsilon_p\rangle\langle\varepsilon_p|\varepsilon_p\rangle$$
$$= \langle\varepsilon|\varepsilon\rangle\langle\varepsilon_p|\varepsilon_p\rangle = (-)^s\langle\varepsilon_p|\varepsilon_p\rangle.$$

5. In a basis we have

$$\varepsilon = \frac{e^{i_1\ldots i_m}}{m!}\varepsilon_{i_1\ldots i_m}\sqrt{|g|},$$

where

$$g = \text{Det } g_{ik} = (\text{Det}\langle e^i | e^k\rangle)^{-1}.$$

Then (a) follows from the definition $*\omega = i_\omega \varepsilon$ and the rules (1.2.16) for $i$. (b) and (c) are a consequence of (1.2.17(iii)).

6.

$$\begin{aligned}\delta(f e^{i_1\cdots i_p}) &= *[(-1)^{m(p+1)+s} f_{,k} e^k \wedge *e^{i_1\cdots i_p}]\\ &= *\left[(-1)^{(m+1)(p+1)+s}\sum_{j=1}^{p} f^{,i_j} *e^{i_1\cdots i_{j-1}i_{j+1}\cdots i_p}(-1)^{j+1}\right]\\ &= \sum_{j=1}^{p} f^{,i_j} e^{i_1\cdots i_{j-1}i_{j+1}\cdots i_p}(-1)^{j+1},\end{aligned}$$

$$d\delta(f e^{i_1\cdots i_p}) = \sum_{j=1}^{p} f^{i_j}_{,\,i_0} e^{i_0 i_1\cdots i_{j-1}i_{j+1}\cdots i_p}(-1)^{j-1}$$

$$\delta d(f e^{i_1\cdots i_p}) = \sum_{j=0}^{p} f^{i_j}_{,\,i_0} e^{i_0 i_1\cdots i_{j-1}i_{j+1}\cdots i_p}(-1)^{j}$$

$$\Rightarrow \Delta(f e^{i_1\cdots i_p}) = \sum_{k=1}^{m} f^{k}_{,\,k} e^{i_1\cdots i_p}.$$

7. (a) $d\Delta = d\delta\, d = \Delta d$, $\delta\Delta = \delta\, d\delta = \Delta\delta$;
(b) $\delta * = *d**(-1)^{m(m-p+1)+s} = (-1)^p * d$, $d* = *\delta**(-1)^{m(m-p)+1+s} = (-1)^{p+1}*\delta$;
(c) $d\delta * = (-1)^p d * d = *\delta d$, $\delta d* = (-1)^{p+1}\delta * \delta = *d\delta$.

8. Letting $\omega_k{}^i = \eta^{ij}\omega_{kj}$, we start with the identity

$$\omega_{k_1}{}^i \varepsilon_{ik_2\cdots k_m} + \omega_{k_2}{}^i \varepsilon_{k_1 i\cdots k_m} + \cdots + \omega_{k_m}{}^i \varepsilon_{k_1\cdots k_{m-1}i} = 0$$

$\forall k_1 = 1, \ldots, m; \ldots; k_m = 1, \ldots, m$. To verify this, consider the three cases:

(i) All $k_i$ are different. Then $i$ must equal the $k$ that is missing in $\varepsilon$, and $\omega_{kj} = -\omega_{jk}$; since $\eta$ is diagonal, $\omega_k{}^i = 0$ if $i = k$.
(ii) Two of the $k_i$ are equal, say $k_1 = k_2$. There remains

$$\omega_{k_1}{}^i \varepsilon_{ik_1\cdots} + \omega_{k_1}{}^i \varepsilon_{k_1 i\cdots} = 0.$$

(iii) Three $k_i$ are equal. Then all the $\varepsilon$'s vanish.

If the identity is multiplied by $e^{k_{p+1}\cdots k_m}$ and the indices are relabeled, then

$$\omega_{k_{p+1}}{}^i \varepsilon_{k_1\cdots k_p i k_{p+2}\cdots m} e^{k_{p+1}\cdots k_m} = -\omega^{k_{p+1}}{}_i \varepsilon_{k_1\cdots k_m} e^{ik_{p+2}\cdots k_m},$$

in the orthogonal basis. In general, because of (1.2.18(a)),

$$\begin{aligned}(m-p)!\,d*e^{j_1\cdots j_p} &= \eta^{j_1k_1}\cdots\eta^{j_pk_p}\varepsilon_{k_1\cdots k_m} de^{k_{p+1}\cdots k_m}\\ &= -\eta^{j_1k_1}\cdots\eta^{j_pk_p}\varepsilon_{k_1\cdots k_m}\{\omega^{k_{p+1}}{}_i e^{ik_{p+2}\cdots k_m} + \cdots + \omega^{k_m}{}_i e^{k_{p+1}\cdots k_{m-1}i}\}\\ &= \eta^{j_1k_1}\cdots\eta^{j_pk_p}\{\omega_{k_{p+1}}{}^i \varepsilon_{k_1\cdots k_p i k_{p+1}\cdots k_m} + \omega_{k_{p+2}}{}^i \varepsilon_{k_1\cdots k_{p+1} i k_{p+3}\cdots k_m} + \cdots\\ &\quad + \omega_{k_m}{}^i \varepsilon_{k_1\cdots k_{m-1}i}\} e^{k_{p+1}k_{p+2}\cdots k_m}\\ &= -\eta^{j_1k_1}\cdots\eta^{j_pk_p}\{\omega_{k_1}{}^i \varepsilon_{ik_2\cdots k_m} + \cdots + \omega_{k_p}{}^i \varepsilon_{k_1\cdots k_{p-1} i k_{p+1}\cdots k_m}\} e^{k_{p+1}\cdots k_m}\\ &= -\omega^{j_1}{}_i *e^{ij_2\cdots j_p} - \cdots - \omega^{j_p}{}_i *e^{j_1j_2\cdots j_{p-1}i}(m-p)!.\end{aligned}$$

9. Let

$$\bar{e}^j = A^j{}_k e^k : d\bar{e} = dA^j{}_r (A^{-1})^r{}_k \bar{e}^k - A^j{}_k \omega^k{}_r (A^{-1})^r{}_s \bar{e}^s = -\bar{\omega}^j{}_k \bar{e}^k$$
$$\Rightarrow \bar{\omega}^j{}_k = A^j{}_s \omega^s{}_r (A^{-1})^r{}_k - (A^{-1})^s{}_k \, dA^j{}_s,$$

because this also satisfies the second defining equation:

$$\bar{g} = A^{-1t} g A^{-1} \Rightarrow d\bar{g} = A^{-1t}(dg)A^{-1} - A^{-1t} g A^{-1}(dA)A^{-1}$$
$$- A^{-1t}(dA^t)A^{-1t} g A^{-1} = \bar{g}\bar{\omega} + (\bar{g}\bar{\omega})^t$$
$$= A^{-1t} g\omega A^{-1} - A^{-1t} g A^{-1}(dA)A^{-1} + A^{-1t}(g\omega)^t A^{-1} - A^{-1t}(gA^{-1}dA)^t A^{-1}.$$

10. According to (1.2.18),

$$\frac{1}{p!} e_{j_1 \cdots j_p} \wedge {}^*e^{i_1 \cdots i_p} \omega^{j_1 \cdots j_p} = \omega^{i_1 \cdots i_p} {}^*\mathbf{1}.$$

Thus

$$\bar{e}_{i_1 \cdots i_p} \otimes {}^*e^{i_1 \cdots i_p} \wedge e_{j_1 \cdots j_p} \omega^{j_1 \cdots j_p} \frac{(-1)^{p(m-p)}}{p!} = \bar{e}^{i_1 \cdots i_p} \omega_{i_1 \cdots i_p} {}^*\mathbf{1}.$$

Integrating this $\int dx_1 \cdots dx_m$ yields

$$\int \bar{c}_{\bar{x}} \wedge e^{i_1 \cdots i_p} \omega_{i_1 \cdots i_p} = \bar{e}^{i_1 \cdots i_p} \omega_{i_1 \cdots i_p}(\bar{x}).$$

11.

$$\delta G_{\bar{x}} \wedge dF = (-1)^{p+m+1}[d(\delta G_{\bar{x}} \wedge F) - (d\delta G_{\bar{x}}) \wedge F],$$
$$dG_{\bar{x}} \wedge \delta F = (-1)^{mp+m+s} d^*F \wedge {}^*dG_{\bar{x}}$$
$$= (-1)^{mp+m+s}[d({}^*F \wedge {}^*dG_{\bar{x}} - (-1)^{m-p}{}^*F \wedge d^*dG_{\bar{x}}]$$
$$= (-1)^{mp+m+s} d({}^*F \wedge {}^*dG_{\bar{x}}) + (-1)^{mp+p+s+1}{}^*d^*dG_{\bar{x}} \wedge F$$
$$= (-1)^{mp+m+s} d({}^*F \wedge {}^*dG_{\bar{x}}) + (-1)^{m+1+p}\delta dG_{\bar{x}} \wedge F$$
$$\Rightarrow (-1)^{p-m}[dG_{\bar{x}} \wedge \delta F - \delta G_{\bar{x}} \wedge dF]$$
$$= \underbrace{-(\delta d + d\delta)G_{\bar{x}}}_{=\delta_{\bar{x}}} \wedge F + d(\delta G_{\bar{x}} \wedge F + (-1)^{mp+p+s}{}^*F \wedge {}^*dG_{\bar{x}}),$$

and when this is integrated over $N$, we get (1.2.36).

12. $M = T^1$ and $J = d\varphi \in E_1(T^1)$, where $\varphi$ is the angle on the torus. If $F \in E_0(T^1)$, then it would be $\varphi +$ constant, which can not be defined continuously over all of $T^1$, even though $d\varphi$ exists on all of $T^1$.

## 1.3 Maxwell's and Einstein's Equations

*Field strengths are described by 2-forms, the exterior differentials of which in the electric (respectively gravitational) case are the 3-forms of charge (resp. energy and momentum). These 3-forms, which function as the sources of the fields, are exact, a fact that implies differential and integral conservation theorems.*

Maxwell wrote Faraday's discoveries in the form of the equations that now bear his name (Table 1), and which describe all electromagnetic phenomena. The way they fuse space and time, and the electric and magnetic fields, is so awe-inspiring that Boltzmann, quoting Goethe, exclaimed, "Was it a god who traced these signs?"† It is regrettable that Cartan's calculus did not exist at that time, as it has perfected the notation in which all the facts of electromagnetism are so concisely formulated.

As discussed in detail in (I: §5.1), the electric and magnetic fields **E** and **B** are combined in a single 2-form $F$, which is written in the natural basis of $\mathbb{R}^4$, with coordinates $x^0 = t$, $\mathbf{x} = (x^1, x^2, x^3)$ and in vector notation, as

$$F = \tfrac{1}{2}F_{\alpha\beta}\, dx^\alpha \wedge dx^\beta = dt \wedge (d\mathbf{x}\cdot\mathbf{E}) - dx^1 \wedge dx^2 B_3 - dx^3 \wedge dx^1 B_2 - dx^2 \wedge dx^3 B_1. \quad (1.3.1)$$

**Table 1**

**Maxwell's Equations in the Course of History**

The constants $c$, $\mu_0$, and $\varepsilon_0$ are set to 1, and modern notation is used for the components.

| The Homogeneous Equation | The Inhomogeneous Equation |
|---|---|
| Earliest Form | |
| $\frac{\partial B_x}{\partial x} + \frac{\partial B_y}{\partial y} + \frac{\partial B_z}{\partial z} = 0$ | $\frac{\partial E_x}{\partial x} + \frac{\partial E_y}{\partial y} + \frac{\partial E_z}{\partial z} = \rho$ |
| $\frac{\partial E_z}{\partial y} - \frac{\partial E_y}{\partial z} = -\dot{B}_x$ | $\frac{\partial B_z}{\partial y} - \frac{\partial B_y}{\partial z} = j_x + \dot{E}_x$ |
| $\frac{\partial E_x}{\partial z} - \frac{\partial E_z}{\partial x} = -\dot{B}_y$ | $\frac{\partial B_x}{\partial z} - \frac{\partial B_z}{\partial x} = j_y + \dot{E}_y$ |
| $\frac{\partial E_y}{\partial x} - \frac{\partial E_x}{\partial y} = -\dot{B}_z$ | $\frac{\partial B_y}{\partial x} - \frac{\partial B_x}{\partial y} = j_z + \dot{E}_z$ |
| At the End of the Last Century | |
| $\nabla\cdot\mathbf{B} = 0$ | $\nabla\cdot\mathbf{E} = \rho$ |
| $\nabla\times\mathbf{E} = -\dot{\mathbf{B}}$ | $\nabla\times\mathbf{B} = \mathbf{j} + \dot{\mathbf{E}}$ |
| At the Beginning of This Century | |
| ${}^*F^{\beta\alpha}{}_{,\alpha} = 0$ | $F^{\beta\alpha}{}_{,\alpha} = j^\beta$ |
| Mid-Twentieth Century | |
| $dF = 0$ | $\delta F = J$ |

† Faust, part I, Dr. Faust's first soliloquy.

**The Homogeneous Maxwell Equations** (1.3.2)

These are the basis-free statement that $F$ is closed:

$$dF = 0. \tag{1.3.3}$$

By (1.2.11) this is equivalent to

$$0 = \int_{\partial N_3} F, \tag{1.3.4}$$

if $F_{|N_3} \in E_2^0(N_3)$, and $N_3$ is three-dimensional. To illustrate this concentrated notation, let us examine some

**Special Cases** (1.3.5)

1. $N_3 = \{(t, \mathbf{x}) \in \mathbb{R}^4 : t = t_0 \in \mathbb{R}, |\mathbf{x}| \leq R\}$ (a spherical ball)
   $\partial N = \{(t, \mathbf{x}) \in \mathbb{R}^4 : t = t_0, |\mathbf{x}| = R\}$ (a spherical surface). (See Figure 12.)

   Then (1.3.4) says that there are no magnetic charges:

   $$\int_{|\mathbf{x}| = R} \mathbf{B} \cdot d\mathbf{S} = 0.$$

   The differential version of this statement, $\nabla \cdot \mathbf{B} = 0$, amounts to $d(F_{|N_3}) = 0$, which follows from (1.3.3) because $d$ commutes with restrictions.

2. $N_3 = \{(t, \mathbf{x}) \in \mathbb{R}^4 : t_0 \leq t \leq t_1, x^3 = 0, |\mathbf{x}| \leq R\}$ (a cylinder).
   $\partial N = \{(t, \mathbf{x}) \in \mathbb{R}^4 : t_0 \leq t \leq t_1, x^3 = 0, |\mathbf{x}| = R\} \cup \{(t, \mathbf{x}) \in \mathbb{R}^4 :$

   $$t = t_0, x^3 = 0, |\mathbf{x}| \leq R\} \cup \{(t, \mathbf{x}) : t = t_1, x^3 = 0, |\mathbf{x}| \leq R\}$$

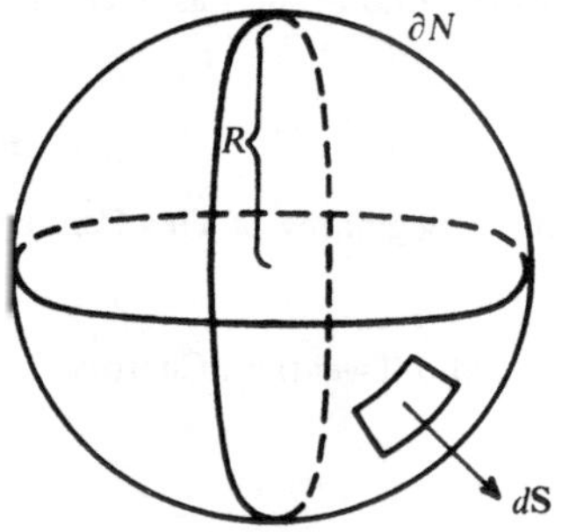


Figure 12 $N_3$ of (1.3.5; 1)

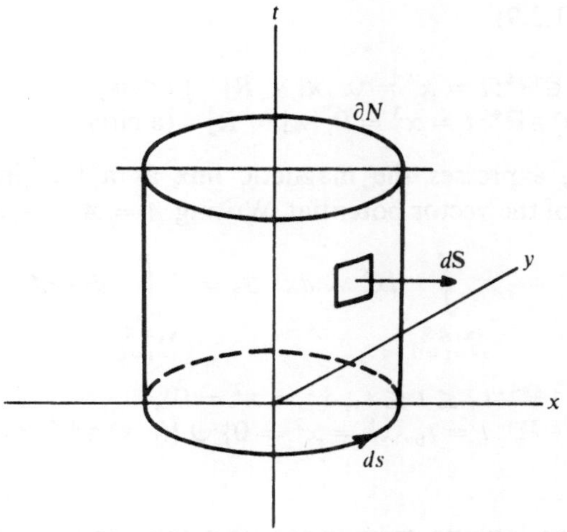


Figure 13 $N_3$ of (1.3.5; 2)

(the outer cylindrical surface plus top and bottom). (See Figure 13.) Then (1.3.4) is the integral form of the law of induction:

$$\int_{t_0}^{t_1} dt \int_{\substack{x^3=0\\|\mathbf{x}|=R}} ds \cdot \mathbf{E} = \int_{\substack{t=t_0,\, x^3=0\\|\mathbf{x}|\leq R}} d\mathbf{S} \cdot \mathbf{B} - \int_{\substack{t=t_1,\, x^3=0\\|\mathbf{x}|\leq R}} d\mathbf{S} \cdot \mathbf{B}.$$

The vanishing of the differential of $F_{|N_3}$ means that

$$0 = d(F_{|N_3}) \Leftrightarrow \dot{B}_3 + (\nabla \times \mathbf{E})_3 = 0.$$

As already mentioned, closure is equivalent to exactness on starlike manifolds, and specifically on $\mathbb{R}^4$. This implies that there exists a

**Vector Potential** $A \in E_1(\mathbb{R}^4)$ (1.3.6)

$$F = dA. \tag{1.3.7}$$

In integrated form,

$$\int_{N_2} F = \int_{\partial N_2} A, \tag{1.3.8}$$

if $F \in E_2^0(N_2)$.

**Special Cases** (1.3.9)

1. $N_2 = \{(t, \mathbf{x}) \in \mathbb{R}^4 : t = x^3 = 0, |\mathbf{x}| \le R\}$ (a disk)
   $\partial N_2 = \{(t, \mathbf{x}) \in \mathbb{R}^4 : t = x^3 = 0, |\mathbf{x}| = R\}$ (a circle).

   Then (1.3.8) expresses the magnetic flux as a line integral over the spatial part of the vector potential. Writing $A = A\,dt - \mathscr{A}\,d\mathbf{x}$,

$$\int_{\substack{|\mathbf{x}| \le R \\ x^3 = t = 0}} dx^1 \wedge dx^2\, B_3 = \int_{\substack{|\mathbf{x}| = R \\ x^3 = t = 0}} d\mathbf{s} \cdot \mathscr{A}.$$

2. $N_2 = \{(t, \mathbf{x}) \in \mathbb{R}^4 : t_0 \le t \le t_1, x^1 = x^2 = 0\}$,
   $\partial N = \{(t, \mathbf{x}) \in \mathbb{R}^4 : t = t_0, x^1 = x^2 = 0\} \cup \{(t, \mathbf{x}) \in \mathbb{R}^4 : t = t_1, x^1 = x^2 = 0\}$.

   We have been looking only at compact submanifolds, thereby trivially fulfilling the requirement of compact support. If $F$ vanishes outside some compact region, or at least goes to zero fast enough at infinity, then (1.3.8) still holds, and implies

$$\int_{t_1}^{t_2} dt \wedge dx^3\, E_3 = \int_{t = t_0} dx^3\, \mathscr{A}_3 - \int_{t = t_1} dx^3\, \mathscr{A}_3.$$

**Remarks** (1.3.10)

1. Equation (1.3.7) determines $A$ only up to a gauge transformation $A \to A + d\Lambda$, where $\Lambda \in E_0(\mathbb{R}^4)$. Because $\partial(\partial N_2) = \varnothing$, there is no contribution from $\Lambda$ to the integral (1.3.8).
2. The statement (1.3.8) is on the whole stronger than (1.3.4). There exist manifolds $N_2$, without boundaries, that are not themselves boundaries of compact submanifolds $N_3$. For instance, the field of a magnetic monopole,

$$\mathbf{E} = \mathbf{0}, \qquad \mathbf{B} = \frac{e'}{4\pi} \frac{\mathbf{x}}{|\mathbf{x}|^3},$$

   satisfies (1.3.3) on $M = \{(t, \mathbf{x}) \in \mathbb{R}^4 : x \ne 0\}$, but gives rise to no vector potential: If $N_2 = \{(t, \mathbf{x}) \in \mathbb{R}^4 : t = 0, |\mathbf{x}| = R\}$, which is compact and has no boundary, then (1.3.2) would lead to the contradiction

$$0 = \int_{N_2} F = -\int_{|\mathbf{x}| = R} \mathbf{B} \cdot d\mathbf{S} = -e'.$$

   $N_2$ is in fact the boundary of $N_3 = \{(t, \mathbf{x}) \in \mathbb{R}^4 : t = 0, 0 < |\mathbf{x}| < R\}$, but since $N_3$ is not compact, one cannot conclude that $0 = \int_{N_3} dF = \int_{N_2} F$: for (1.3.4) to be applicable, $F$ must have compact support on $N_3$. The physical significance of this is that any magnetic sources have been removed from $M$, whereas the field in $M$ is that of a magnetic monopole. There are theories that also claim there are no electric charges either, but that some complicated structure of $M$ makes it appear that there are electrical sources.

3. $M = \mathbb{R}^4 \setminus \{(t, \mathbf{x}) \in \mathbb{R}^4 : x^1 = x^2 = 0,\ x^3 \leq 0\}$ is starlike with respect to every point of the positive $z$-axis, and is contained in the $M$ of Remark 2. Therefore the field of a magnetic monopole satisfies $dF = 0$ on $M$, and there ought to be a vector potential. In fact,

$$\mathscr{A} = \frac{e'}{4\pi}\left(1 - \frac{z}{\sqrt{x^2 + y^2 + z^2}}\right)\frac{x\,dy - y\,dx}{x^2 + y^2}$$

reproduces the field of Remark 2 and is differentiable on $M$ (Problem 6). Equation (1.3.8) does not lead to a contradiction if $N_2 = \{(t, \mathbf{x}) \in \mathbb{R}^4 : t = 0, |\mathbf{x}| = R\} \cap M$, since $N_2$ is not compact. Physically, $A$ is the potential of an infinitely thin solenoid along the negative $z$-axis, which has been removed from $M$. We conclude that the empirical absence of magnetic monopoles indicates that the manifold in which we live has none of the pathologies described here, at least on the small scale.

The inhomogeneous equation describes how $F$ is produced by the sources, which are the charges or, more correctly, the charge densities. Since the charge density ought to yield the charge present in some three-dimensional, region when it is integrated over that region, we describe it as a 3-form $*J$. The corresponding 1-form $J$ consists of a temporal part, the charge density $\rho$, and a spatial part, the current density $\mathbf{J}$: $J = -\rho\,dt + \mathbf{J}\,d\mathbf{x}$. In terms of $J$ we may write down the

**Inhomogeneous Maxwell Equations** (1.3.11)

$$\delta F = J, \quad \text{or} \quad d{*F} = -{*J}; \tag{1.3.12}$$

and in integrated form,

$$-\int_{\partial N_3} {*F} = \int_{N_3} {*J}, \qquad F \in E_2^0(N_3). \tag{1.3.13}$$

**Remarks** (1.3.14)

1. The formulation of these equations requires a manifold with some additional structure, which defines the $*$-mapping. We do not require the concept of a covariant derivative, which will come up later.
2. Since none of these equations make reference to a chart, we are not restricted to the use of Cartesian coordinates. We shall even postulate the same form of Maxwell's equations for use on more general manifolds

than $\mathbb{R}^4$. This will cause the interaction with the gravitational field, and will make the equations describe such things as the bending of light by the sun.
3. The Poincaré transformations of $\mathbb{R}^4$ are the ones that leave the form of the metric $\eta_{ik}$, and consequently also * and Maxwell's equations, invariant.

The equations of (1.3.11) do not have solutions for an arbitrary 1-form $J$. On the contrary, because of (1.2.6(c)), their first corollary is

**Conservation of Charge** (1.3.15)

$$\delta J = 0, \quad \text{or} \quad d{}^*J = 0. \tag{1.3.16}$$

When integrated, this is

$$\int_{\partial N_4} {}^*J = 0, \qquad \text{if } {}^*J \in E_3^0(N_4). \tag{1.3.17}$$

**Special Cases** (1.3.18)

1. $N_4 = \{(t, \mathbf{x}): t_0 \leq t \leq t_1, |\mathbf{x}| \leq R\}$. Then (1.3.17) implies that the charge that flows out through the surface equals the change of the charge contained in the sphere between the time $t_0$ and $t_1$:

$$\int_{\substack{|\mathbf{x}|<R \\ t=t_0}} d^3x\, \rho - \int_{\substack{|\mathbf{x}|<R \\ t=t_1}} d^3x\, \rho = \int_{t_0}^{t_1} dt \int_{|\mathbf{x}|=R} d\mathbf{S} \cdot \mathbf{J}.$$

2. Let $\partial N = \{(t, \mathbf{x}): t = 0\} \cup \{(t, \mathbf{x}): \bar{t} \equiv (t - vx^1)/\sqrt{1 - v^2} = 0\} \equiv A \cup B$. This fails to be the boundary of a submanifold, because it has sharp bends (see Figure 14), though, as already mentioned, the bends could be smoothed out without changing the value of the integral much. It also fails to be compact, so let us assume that $J$ has spatially compact support. Then (recalling that ${}^*dx^0 = -dx^1 \wedge dx^2 \wedge dx^3$, so that $\int_{t=\text{const.}} {}^*J = \int d^3x\, J^0 = \int d^3x\, \rho$),

$$\int_A {}^*J = \int_B {}^*J \equiv Q.$$

If we calculate the integral over $B$ in Lorentz-transformed coordinates,

$$\bar{x}^1 = \frac{x^1 - vt}{\sqrt{1 - v^2}}, \qquad \bar{x}^2 = x^2, \qquad \bar{x}^3 = x^3, \qquad \bar{t} = \frac{t - vx^1}{\sqrt{1 - v^2}},$$

then we find that

$$Q \equiv \int_{t=0} d^3x\, J^0 = \int_{\bar{t}=0} d^3\bar{x}\, \bar{J}^0, \qquad \bar{J}^0 = \frac{J^0 - vJ^1}{\sqrt{1 - v^2}},$$

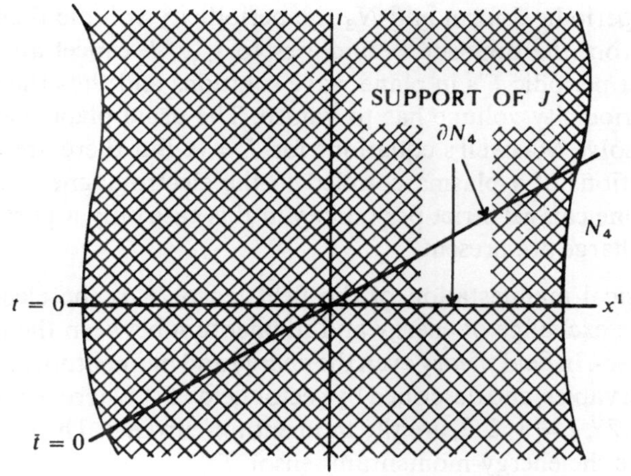


Figure 14 The Lorentz invariance of the total charge

since $(J^0, J^1)$, the components of $J$, transform as $(x^0, x^1)$. The total charge $Q$ is therefore a Lorentz invariant; the increase in the charge density by $\sqrt{1 - v^2}$ is compensated for by the Lorentz contraction of the volume element. One must bear in mind, however, that such concepts as scalar, vector, etc., are not defined for integrated quantities, because the transformation coefficients in general depend on the position.

**Remarks** (1.3.19)

1. The equations (1.3.11) imply more than (1.3.16); conservation of charge necessarily implies the existence of a 2-form $^*F$ such that $^*J = -d^*F$ only if the manifold is starlike, or has some similar property (cf. (1.2.39; 12)). If $t$ is defined globally on $M$, and the submanifold $N_3 = \{t = 0\}$ is compact and has no boundary, then it follows from (1.3.11) that

$$Q = \int_{N_3} {}^*J = \int_{\partial N_3} {}^*F = 0,$$

since $\partial N_3 = \varnothing$. Hence a closed universe must have zero total charge, a conclusion that goes beyond mere charge conservation. The underlying reason is that in this case the lines of force emanating from a charge can not go to infinity, but must terminate in some opposite charge. The fact is that matter appears to be neutral to a remarkable degree. If the charge on the proton were increased relative to the electron charge, then the $1/r^2$ gravitational force would in effect be weakened for protons and strengthened for electrons. Since it is empirically known that the two forces are proportional to the masses and are weaker than the Coulomb force by a factor $10^{-36}$, matter is neutral to at least this accuracy.

2. If $F$ is a periodic 2-form and $N_3$ the periodicity volume then the contributions from the opposite boundaries to $\int_{\partial N_3} {}^*F$ cancel and $Q = 0$. We shall meet periodic $F$'s in plane wave solutions and thus the total charge in the periodicity volume has to vanish. The usual plane wave solutions actually solve Maxwell's equations for $J = 0$. But there are many wave-type solutions of a plasma coupled to the electromagnetic field and we see that one can get strict periodicity in this case only if particles of both signs of charge are present.

The physical manifestation of the field tensor $F$ is the electromagnetic force, i.e., the exchange of energy and momentum between the field and the charge-carriers. In field theory even the energy and momentum are introduced as local observables, and described by vector fields $\mathscr{T}^\alpha$, where $\mathscr{T}^0$ is the energy-current and $\mathscr{T}^j, j = 1, 2, 3$, are the momentum-currents. The components in a basis define the energy-momentum tensor $T^\alpha{}_\beta$:

$$\mathscr{T}^\alpha = T^\alpha{}_\beta e^\beta. \tag{1.3.20}$$

If the corresponding dual 3-forms are integrated over a spacelike† submanifold $N_3$, the result is the energy $P^0$ and the momentum $P^j$ contained in $N_3$:

$$P^\alpha = \int_{N_3} {}^*\mathscr{T}^\alpha. \tag{1.3.21}$$

As in classical mechanics, the Lagrangian formulation of the field equations makes it possible to construct a field energy.

In fact, there will be some close analogy to the expressions we met in Volume 1 except that field theory is richer in components. The coordinate $q$ is replaced by the vector potential $A$ and $\dot{q}$ by $dA = F$. The Lagrangian $\frac{1}{2}\dot{q}^2$ becomes the 4-form $-\frac{1}{2}F \wedge {}^*F$ and the dual $-{}^*F = \partial\mathscr{L}/\partial\, dA$ plays the role of $p = \partial L/\partial\dot{q}$. The sign is dictated by the requirement that the energy density $T_{00}$ be positive. The counterpart to the energy $H = \dot{q}p - L$ is

$$\begin{aligned} {}^*\mathscr{T}^\alpha &= -dA \wedge i^{\alpha *}F + i^\alpha\mathscr{L}, \\ \mathscr{T}^\alpha &= \tfrac{1}{2}{}^*(i^\alpha F) \wedge {}^*F - (i^{\alpha *}F) \wedge F) \\ &= -(F^\alpha{}_\tau F^\tau{}_\beta + \tfrac{1}{4}\delta^\alpha{}_\beta F_{\rho\sigma}F^{\rho\sigma})\, dx^\beta, \end{aligned} \tag{1.3.22}$$

where $i^\alpha$ stands for the interior product (1.2.15) with $dx^\alpha$: $i^\alpha F = F^\alpha{}_\beta\, dx^\beta$. It has the job to pick out the components. We shall derive (1.3.22) in two ways. First by a consideration similar to classical mechanics where the invariance of the Lagrangian under translations produces a conservation law. Later, we shall see that it also equals $\partial\mathscr{L}/\partial e^\alpha$. This shows that change $e^\alpha \to L^\alpha{}_\beta e^\beta$ of the basis produces the homogeneous transformation law $\mathscr{T}^\alpha \to L^\alpha{}_\beta \mathscr{T}^\beta$ for the energy momentum currents of matter and of the

† A submanifold is called **spacelike (timelike, lightlike)** if $g$ is positive (negative, zero) when restricted to it.

electro-magnetic field. Only the Lagrangian of gravity contains *de* which produces an inhomogeneous transformation.

As with charge-conservation, the change in the energy-momentum is determined by the codifferential of the currents, $\delta\mathcal{T}^\alpha$. The rate of change in time of $P^\alpha$ equals a four-dimensional integral over $d{*}\mathcal{T}^\alpha$ and a surface integral over the spatial parts of the energy-momentum currents. For example, if $N_3 = \{(t, \mathbf{x}) : t = \pm T, |\mathbf{x}| \leq R\}$, then

$$P^\alpha(T) - P^\alpha(-T) = \int_{\substack{-T \leq t \leq T \\ |\mathbf{x}| \leq R}} d{*}\mathcal{T}^\alpha - \int_{\substack{-T \leq t \leq T \\ |\mathbf{x}| = R}} {*}\mathcal{T}^\alpha. \tag{1.3.23}$$

If $F$ is coupled to a current, then energy and momentum are exchanged between the field and the current in amounts determined by the codifferential of the energy-momentum currents of the field. This codifferential is determined by Maxwell's equations and is known as the

**Lorentz Force** (1.3.24)

$$\delta\mathcal{T}^\alpha = \langle i^\alpha F | J \rangle$$

or, in components,

$$T^{\alpha\beta}{}_{,\beta} = F^{\alpha\beta} J_\beta.$$

**Derivation of the Equation for the Lorentz Force**

In the natural basis of $\mathbb{R}^n$,

$$L_{e^k}\omega \equiv (i^k \circ d + d \circ i^k)\omega_{j_1 \cdots j_p} e^{j_1 \cdots j_p} = \omega_{j_1 \cdots j_p}{}^{,k} e^{j_1 \cdots j_p},$$

and hence $L_{e^k}{*}\omega = {*}L_{e^k}\omega$. Defining $L^\alpha \equiv L_{e^\alpha}$, this means that

$$\begin{aligned} L^\alpha(F \wedge {*}F) &= (L^\alpha F) \wedge {*}F + F \wedge L^\alpha{*}F \\ &= (L^\alpha F) \wedge {*}F + F \wedge {*}L^\alpha F = 2(L^\alpha F) \wedge {*}F \end{aligned}$$

(cf. (1.2.18)(a) and (1.2.23)). Therefore, using (1.2.6) and (1.3.22),

$$\begin{aligned} d{*}\mathcal{T}^\alpha &= -d[\tfrac{1}{2}i^\alpha(F \wedge {*}F) - (i^\alpha F) \wedge {*}F] \\ &= -\tfrac{1}{2}L^\alpha(F \wedge {*}F) + (L^\alpha F) \wedge {*}F - i^\alpha F \wedge d{*}F = i^\alpha F \wedge {*}J. \end{aligned}$$

According to (1.2.18) (b) and (c), $v \wedge {*}\omega = {*}i_v\omega(-1)^{p+1}$, for $v \in E_1$ and $\omega \in E_p$. Hence

$$\delta\mathcal{T}^\alpha = -{*}d{*}\mathcal{T}^\alpha = -{*}(i^\alpha F \wedge {*}J) = \langle i^\alpha F | J \rangle$$

**Remarks** (1.3.25)

1. It is important that the $i^\alpha$ in (1.3.22) be an interior product with a natural basis $e^\alpha$. The ${*}$-mapping does not commute with $L^\alpha$ in an arbitrary basis,

which spoils the above proof. It is also not possible for (1.3.24) to hold in every basis, because if $e^\alpha \to A^\alpha{}_\beta(x)e^\beta$, then the right side would transform linearly with $A$, while the left side would gain a term containing $dA$. A physical consequence of this is that fictitious forces must be added to the Lorentz force in an accelerated reference system.

2. If $z(s)$ is the world-line of a particle (see (I: §4.1)), then its current is

$$J_\beta(x) = e\int_{-\infty}^{\infty} ds\, \delta^4(x - z(s))\dot{z}_\beta(s),$$

where $\delta^4(x) \equiv \delta(x^0)\delta(x^1)\delta(x^2)\delta(x^3)$ (cf. Problem 7). The Lorentz force with this $J$,

$$e\int_{-\infty}^{\infty} ds\, \delta(x - z(s))F^{\alpha\beta}(z(s))\dot{z}_\beta(s),$$

equals $-\delta t^\alpha$, if we choose the energy-momentum currents of a particle so that the energy becomes $\int {}^*t^0 = m\dot{z}^0 > 0$:

$$t^\alpha = +m\int_{-\infty}^{\infty} ds\, \dot{z}^\alpha\dot{z}_\beta(s)\delta^4(x - z)dx^\beta:$$

$$\begin{aligned}\delta t^\alpha &= +m\int_{-\infty}^{\infty} ds\, \dot{z}^\alpha\dot{z}^\beta\frac{\partial}{\partial x^\beta}\delta^4(x - z(s))\\ &= -m\int_{-\infty}^{\infty} ds\, \dot{z}^\alpha\frac{d}{ds}\delta^4(x - z(s)) = m\int_{-\infty}^{\infty} ds\, \ddot{z}^\alpha\delta^4(x - z(s))\\ &= e\int_{-\infty}^{\infty} ds\, \dot{z}_\beta F^{\beta\alpha}(z(s))\delta^4(x - z(s)).\end{aligned}$$

Thus the total energy and momentum are formally conserved. Yet the singularity of the field of a point-particle at $z(s)$ results in inconsistencies (cf. §2.4), which are resolved only for continuous matter in §3.1. Note that the Lorentz force is concentrated on the world-line of the particle, and that the equation $\delta(\mathcal{T}^\alpha + t^\alpha) = 0$, with the replacement of $\delta^4(x - z(s))$ by $\rho(x - z(s))$ for some continuous function $\rho$, is no longer true. Hence there is no local energy-momentum conservation for an extended charged particle, unless other forces hold it together.

If $\mathcal{T}^\alpha$ is defined in an arbitrary basis by (1.3.22), then it transforms as $e^\alpha$ under a change of basis:

$$e^\alpha \to \bar{e}^\alpha = A^\alpha{}_\beta e^\beta, \qquad \bar{\mathcal{T}}^\alpha = A^\alpha{}_\beta\mathcal{T}^\beta. \tag{1.3.26}$$

The energy and momentum are combined into a vector-valued 1-form. Thus, if $\delta(\mathcal{T}^\alpha + t^\alpha) = 0$, then a global Lorentz transformation treats $P^\alpha$ as a vector (Problem 5). If the $A^\alpha{}_\beta$ are allowed to depend on $x$, then this statement becomes meaningless, however, and the conservation equation $\delta(\bar{\mathcal{T}}^\alpha + \bar{t}^\alpha) = 0$ is false. In fact, in this case

$$d^*(\bar{\mathcal{T}}^\alpha + \bar{t}^\alpha) = dA^\alpha{}_\beta \wedge A^{-1\beta}{}_\gamma{}^*(\bar{\mathcal{T}}^\gamma + \bar{t}^\gamma).$$

Choosing $e^\alpha$ as the Cartesian $dx^\alpha$ of $\mathbb{R}^4$, i.e., setting the $\omega^\alpha{}_\beta$ of (1.2.25) to zero, makes

$$\bar{\omega}^\alpha{}_\gamma = -(dA^\alpha{}_\beta)(A^{-1})^\beta{}_\gamma, \tag{1.3.27}$$

according to (1.2.26; 3), and therefore

$$\delta(\bar{\mathscr{T}}^\alpha + \bar{t}^\alpha) = -\langle \bar{\omega}^\alpha{}_\gamma | \bar{\mathscr{T}}^\gamma + \bar{t}^\gamma \rangle \tag{1.3.28}$$

**Remarks** (1.3.29)

1. It seems odd at first that with general bases there are only nonconservation theorems; they are due to fictitious forces in accelerated reference frames and hence are to be expected on physical grounds. They provide a clue to an understanding of the structure of Einstein's equations.
2. If (1.3.28) is compared with (1.3.24), one sees that $\omega$ plays the same role for fictitious forces as $F$ plays for the Lorentz force. $\omega$ acts on $\mathscr{T} + t$ as $F$ acts on $J$.

**Example** (1.3.30)

A rotating basis. Let

$$\begin{aligned} \bar{e}^0 &= dt \\ \bar{e}^1 &= dx \cos \nu t + dy \sin \nu t \\ \bar{e}^2 &= -dx \sin \nu t + dy \cos \nu t \\ \bar{e}^3 &= dz. \end{aligned}$$

This basis is orthogonal but not natural;

$$d\bar{e}^1 = \nu \, dt \wedge \bar{e}^2, \qquad d\bar{e}^2 = -\nu \, dt \wedge \bar{e}^1 : \bar{\omega}^1{}_2 = -\nu \, dt = -\bar{\omega}^2{}_1.$$

It makes

$$\begin{aligned} d(^*\mathscr{T}_1 + {}^*t_1) &= \nu \, dt \wedge (^*\mathscr{T}_2 + {}^*t_2) \\ d(^*\mathscr{T}_2 + {}^*t_2) &= -\nu \, dt \wedge (^*\mathscr{T}_1 + {}^*t_1); \end{aligned}$$

or, if $P_n(t) \equiv \int_{x^0=t} (^*\mathscr{T}_n + {}^*t_n)$, then it follows from Stokes's theorem that

$$P_1(t_1) - P_1(t_0) = \nu \int_{t_0}^{t_1} dt \, P_2(t)$$

$$P_2(t_1) - P_2(t_0) = -\nu \int_{t_0}^{t_1} dt \, P_1(t).$$

These are just like the equations $\dot{P}_1 = \nu P_2$, $\dot{P}_2 = -\nu P_1$ of the mechanics of point particles (cf. (I: 3.2.15; 2)) in a rotating system.

Einstein's theory introduces gravity in (1.3.28) via the principle of equivalence. The gravitational potential is represented by the metric $g$. In other words, if we use orthogonal bases†, so that $g = e^\alpha \otimes e^\beta \eta_{\alpha\beta}$, then the 1-forms $e^\alpha$ are analogous to the $A$ of (1.3.7). Formula (1.3.28) replaces the Lorentz force (1.3.24), and is assumed to hold not only in Minkowski space, but also in the pseudo-Riemannian space determined by $g$.

**Remarks** (1.3.31)

1. As there are four 1-forms $e^\alpha$ representing gravitational potentials, there is not just one current, but four entering into the force.
2. The $\omega$'s are the new field strength, although they do not transform linearly under a change of basis, as $F$ does, but inhomogeneously according to (1.2.26; 3). As a result, they can always be transformed to zero at a point. A physical realization of such a system would be a freely falling elevator, in which there is no gravitational force.
3. Theories have recently been proposed [16] in which the $\omega$'s are analogous to $A$; however, it would take us too far afield to explore this analogy.

If $\omega$ plays the role of $F$, then

$$de^\alpha = -\omega^\alpha{}_\beta \wedge e^\beta, \qquad \omega_{\alpha\beta} = -\omega_{\beta\alpha},$$

can be viewed as the counterpart of the relation $F = dA$. To construct the inhomogeneous equations, one might try to equate the codifferentials of 2-forms linear in $\omega$ to the energy and momentum currents.

Since now there are four currents they ought to be codifferentials of the same number of 2-forms representing the field strength. They should be linear in the $\omega$'s but cannot be identical because the $\omega$'s are six independent 1-forms. The right combination turns out to be

$$F^\gamma = -\tfrac{1}{2} i_{\omega_{\alpha\beta}} e^{\alpha\beta\gamma}. \tag{1.3.32}$$

Einstein's equations are the counterpart to inhomogeneous Maxwell's equations and assert that $\delta F^\gamma$ equals the energy momentum currents. However, there is the problem that according to (1.3.28) the currents of matter and electromagnetism alone are not conserved in an arbitrary basis and $\delta F^\gamma = \text{const}(\mathcal{T}^\gamma + t^\gamma)$ would be mathematically inconsistent. This non-conservation is due to fictitious gravitational forces and can be compensated by adding to the contributions from matter and electromagnetism the energy momentum currents of gravity $\ell^\alpha$. It turns out that they are constructed following the pattern of (1.3.22) where $de^\alpha$ replaces $dA$ and $F^\alpha$ replaces $F$:

$$*\ell^\alpha = \frac{1}{16\pi\kappa}[de^\beta \wedge i^\alpha F_\alpha - i^\alpha \mathcal{L}_g].$$

† As will be done throughout the rest of this chapter. Then indices are raised and lowered by means of $\eta$: $\omega_{\alpha\beta} = \eta_{\alpha\gamma}\omega^\gamma{}_\beta$, etc., which at most changes some signs.

Here $\kappa$ is the gravitational constant which relates the geometric to the physical quantities. The gravitational Lagrangian

$$\mathscr{L}_g = \frac{1}{16\pi\kappa} {*e^{\beta\gamma}} \wedge \omega_\gamma{}^\alpha \wedge \omega_{\alpha\beta}$$

is such that

$$*F^\alpha = \frac{\partial \mathscr{L}_g}{\partial de_\alpha}.$$

These calculations will be done in detail in §4.2. For the moment we just want to show in analogy to the inhomogeneous Maxwell equation we have

**Einstein's Equations** (1.3.33)

$$\delta F^\gamma = \tfrac{1}{2}\delta(i_{\omega_{\alpha\beta}} e^{\alpha\beta\gamma}) = 8\pi\kappa(\mathscr{T}^\gamma + t^\gamma + \ell^\gamma).$$

**Remarks** (1.3.34)

1. The analogy with (1.3.12) is not quite perfect. Einstein's equations do not contain the codifferentials of $de^\alpha = -{*i_{\omega^\alpha{}_\beta}} {*e^\beta}$, but, instead, the 3-form $*e^\beta$ is replaced with $e^{\alpha\beta\gamma}$. The left side of (1.3.33) equals $\Delta e^\gamma$ only in special bases.
2. In §4.2 (1.3.33) is compared with the form used by Einstein where only the curvature appears.
3. Equation (1.3.28) follows from (1.3.33) in the same way as the Lorentz force follows from Maxwell's equations.
4. Since $\ell^\gamma$ contains $\omega$, it does not transform under a change of basis according to the linear law (1.3.26). Such a law would be inconsistent with the continuity equation

$$\delta(\mathscr{T}^\alpha + t^\alpha + \ell^\alpha) = 0$$

   arising from (1.3.33). In particular, in a basis where all the $\omega$'s vanish at some point, $\ell^\alpha$ also vanishes at that point. Physically speaking, this is a consequence of the principle of equivalence: Since gravity can be transformed away at any point, it can have neither energy nor momentum definitely localized at a point. On the other hand, $\ell^\alpha$ may be nonzero even in the absence of gravity (flat space), if an unnatural basis is used. This is to say that fictitious forces produce fictitious energies and momenta.
5. Under a change of basis, Equations (1.3.33) transform linearly because the inhomogeneous contribution to $\ell^\gamma$ is compensated for in the transformation of the left sides of the equations.

6. There is, of course, not just one $\ell^\gamma$ for which the continuity equation (1.3.28) holds; it is always possible to add a closed form to it. With an orthogonal basis, our $\ell^\gamma$ agrees with the pseudotensor used by Landau and Lifshitz. At this stage, our definition of $\ell^\gamma$ as the source of the term $i_{\omega_{\alpha\beta}} e^{\alpha\beta\gamma}$ may not seem convincing; we shall see later that in some familiar cases it reproduces the gravitational energy of the elementary theory, in a basis that is as Cartesian as possible.

While nonlinearity makes the equations harder to solve, there is no difficulty in drawing general conclusions by using Stokes's theorem and restricting to submanifolds. As with the inhomogeneous Maxwell equations, we obtain some

**Corollaries** (1.3.35)

1. $$\int_{\partial N_4} (*\mathcal{T}^\gamma + *t^\gamma + *\ell^\gamma) = 0.$$

If a timelike coordinate $t$ is specified globally on $M$, and the energy and momentum fall off sufficiently fast at infinity on the submanifold $t =$ constant, then the total energy and momentum are conserved:

$$P^\gamma(t_1) = P^\gamma(t_2),$$

$$P^\gamma(t_1) = \int_{t=t_1} (*\mathcal{T}^\gamma + *t^\gamma + *\ell^\gamma).$$

2. Equations (1.3.33) allow us to make an even stronger statement,

$$\int_{N_3} (*\mathcal{T}^\gamma + *t^\gamma + *\ell^\gamma) = -\frac{1}{16\pi\kappa}\int_{\partial N_3} \omega_{\alpha\beta} \wedge *e^{\alpha\beta\gamma}.$$

More specifically, if the submanifold $t = t_0$ is compact, space-like, and has no boundary, then

$$P^\gamma(t_0) = 0.$$

In analogy with (1.3.19; 1), this means that the total energy and momenta of a closed universe are zero.

**Remarks** (1.3.36)

1. A rough, order-of-magnitude estimate shows that the negative gravitational energy may well balance the rest energy of the matter in the universe. The average distance between galaxies is about $6 \times 10^6$ light-years. Since the universe is some $10^{10}$ years old and about as many light-years across, there are $\sim 10^{11}$ galaxies in all. If each one has about

$10^{11}$ stars at $10^{33}$ grams apiece, then the universe has a mass of perhaps $10^{55}$ grams. Hence the gravitational energy (in cgs units) is on the order of

$$-M\frac{M\kappa}{R} \sim -M \cdot 10^{55} \cdot 10^{-7} \cdot 10^{-28} \sim -M \cdot 10^{20},$$

where the radius of the universe $R \sim 10^{10}$ light-years $\sim 10^{28}$ cm. This falls short of the rest-energy $Mc^2 \sim M \times 10^{21}$ by about a factor of 10, but there are large uncertainties in our calculation, owing to our ignorance about how much of the energy of the universe is detectable. Later we shall again have occasion to convince ourselves that $\not{t}^\gamma$ indeed contributes a gravitational energy $\sim -M^2\kappa/R$.

2. Because of the complicated dependence on the basis, there is no intuitively appealing observable corresponding to $\not{t}^\gamma$. The total energy and momentum $P^\gamma$ are less sensitive to the choice of basis, as they may be expressed as integrals over $\partial N_3$ according to (1.3.35; 2). Changes of basis leave them invariant as long as nothing is changed on the boundary. If the space is asymptotically a Minkowski space, then these observables transform as vectors under any transformation that asymptotically approaches a Lorentz transformation.
3. The question of when the energy-momentum forms can be defined globally even if it requires more than one chart to describe the manifold will be discussed later (4.2.12; 8).

So much for the structure of Maxwell's and Einstein's equations. In the remainder of the book, after this preview, we study them in greater detail, and will not only guess a few particular solutions, but also see what general statements can be made about the solutions of these equations.

**Problems** (1.3.37)

1. What are the consequences of (1.3.13) for $N_3 = \{(t, \mathbf{x}): t = 0, |\mathbf{x}| \leq R\}$, and of (1.3.12) restricted to $N_3$?

2. The same question for $N_3 = \{(t, \mathbf{x}): t_0 \leq t \leq t_1, x^3 = 0, |\mathbf{x}| \leq R\}$.

3. According to (1.3.18; 3), there is a Lorentz-transformed system with a nonzero charge-density $\bar{J}_0$, if $J_0 = 0$ but $J_1 \neq 0$. Explain physically how a neutral system with a current appears charged when seen from a moving reference frame.

4. Show that on $\mathbb{R}^n$ the Lie derivative with respect to the natural basis, $L_k = i_{dx_k} \circ d + d \circ i_{dx_k}$, of $e^{j_1 \cdots j_p}$ simply yields

$$L_k \omega_{j_1 \cdots j_p} e^{j_1 \cdots j_p} = \omega_{j_1 \cdots j_p, k} e^{j_1 \cdots j_p},$$

in the natural basis.

5. Discuss how $P^\alpha \equiv \int_{N_3} {}^*\mathscr{T}^\alpha$ is affected by Lorentz transformations.

6. Calculate $F$ with the potential given in (1.3.10; 3).

7. Verify that $\delta J = 0$ for a current of the form

$$J_\alpha(x) = \int_{-\infty}^{\infty} ds\, \dot{z}_\alpha(s)\rho(x - z(s)), \qquad \rho \in E_0^0(\mathbb{R}^4).$$

**Solutions** (1.3.38)

1.
$$-{}^*F = \begin{pmatrix} 0 & B_1 & B_2 & B_3 \\ -B_1 & 0 & E_3 & -E_2 \\ -B_2 & -E_3 & 0 & E_1 \\ -B_3 & E_2 & -E_1 & 0 \end{pmatrix}$$

Thus (1.3.13) implies that

$$\int_{|\mathbf{x}|=R,\, t=0} d\mathbf{S}\cdot\mathbf{E} = \int_{|\mathbf{x}|\le R,\, t=0} d^3x\, \rho(x); \quad \text{and} \quad d({}^*F_{|N_3}) = -{}^*J_{|N_3}$$

means that $\nabla\cdot\mathbf{E} = \rho$.

2.
$$\int_{t_0}^{t_1} dt \int_{\substack{x^3=0\\|\mathbf{x}|=R}} d\mathbf{s}\cdot\mathbf{B} = \int_{\substack{t=t_1,\, x^3=0\\|\mathbf{x}|\le R}} d\mathbf{S}\cdot\mathbf{E} - \int_{\substack{t=t_0,\, x^3=0\\|\mathbf{x}|\le R}} d\mathbf{S}\cdot\mathbf{E} + \int_{t_0}^{t_1} dt \int_{\substack{x^3=0\\|\mathbf{x}|\le R}} d\mathbf{S}\cdot\mathbf{j},$$

or $\nabla\times\mathbf{B} = \mathbf{j} + \dot{\mathbf{E}}$.

3. A neutral current consists of oppositely charged currents flowing past each other. Because of the differing definitions of simultaneity, the two currents seem to have different charges in the moving frame (see Figure 15).

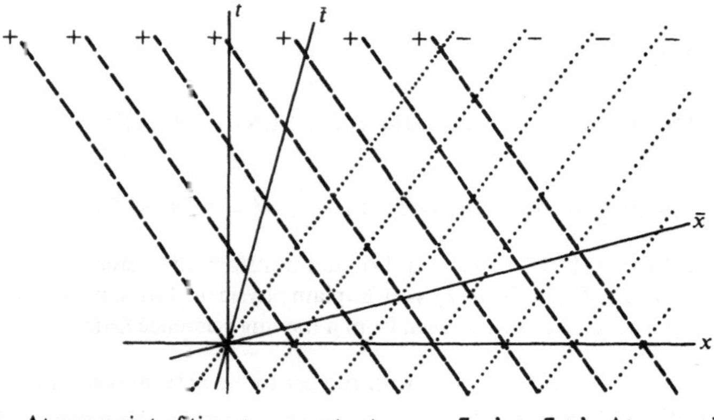


Figure 15 At any point of time $t$ = constant, $x$ sees $7-$'s to $7+$'s. At any point of time $\bar{t}$ = constant, $\bar{x}$ sees $5-$'s to $7+$'s.

4.
$$i^k\, d\omega = \omega_{j_1\cdots j_p,j}\, i^k e^{jj_1\cdots j_p} = \omega_{j_1\cdots j_p,}{}^k e^{j_1\cdots j_p} - p\omega^k{}_{j_2\cdots j_p,j}\, e^{jj_2\cdots j_p}$$
$$di^k\omega = pd(\omega^k{}_{j_2\cdots j_p} e^{j_2\cdots j_p}) = p\omega^k{}_{j_2\cdots j_p,j}\, e^{jj_2\cdots j_p}.$$

5. Since the natural basis transforms according to $d\bar{x}^\alpha = L^\alpha{}_\beta \, dx^\beta$, it is also true that $\bar{\mathcal{T}}^\alpha = L^\alpha{}_\beta \mathcal{T}^\beta$. Now note that $N_3$ looks different in the new coordinates. For instance, if $N_3$ is the hyperplane $t = 0$, then in the new system it is not $\bar{t} = 0$, but $\bar{t} = v\bar{x}$. It is only if $\delta\mathcal{T}^\alpha = 0$ and $\mathcal{T}^\alpha$ vanishes fast enough at infinity that

$$\int_{t=0} {}^*\mathcal{T}^\alpha = \int_{\bar{t}=0} \bar{\mathcal{T}}^\alpha$$

(cf. (1.3.18; 2)).

6. $r^2 \equiv x^2 + y^2 + z^2$: then

$$B_x = -\frac{\partial \mathscr{A}_y}{\partial z} = \frac{e^1}{4\pi} \frac{x}{r} \frac{1}{x^2 + y^2}\left(1 - \frac{z^2}{r^2}\right) = \frac{e^1}{4\pi} \frac{x}{r^3}$$

$$B_y = \frac{\partial \mathscr{A}_x}{nz} = \frac{e^1}{4\pi} \frac{y}{r^3}$$

$$B_z = \frac{\partial \mathscr{A}_y}{\partial x} - \frac{\partial \mathscr{A}_x}{\partial y} = \frac{e^1}{4\pi}\left\{\frac{z(x^2 + y^2)}{r^3(x^2 + y^2)} + \frac{1}{x^2 + y^2} - \frac{2x^2}{x^2 + y^2}\right.$$

$$\left. + \frac{1}{x^2 + y^2} - \frac{2y^2}{x^2 + y^2}\right\} = \frac{e^1}{4\pi} \frac{z}{r^3}.$$

7. $$\frac{\partial}{\partial x^\beta} \int_{-\infty}^{\infty} ds \, \rho(x - z(s))\dot{z}^\beta(s) = -\int_{-\infty}^{\infty} ds \, \dot{z}^\beta(s), \frac{\partial}{\partial z^\beta} \rho(x - z(s))$$

$$= -\int_{-\infty}^{\infty} ds \frac{\partial}{\partial s} \rho(x - z(s)) = 0.$$

No boundary terms appear from the partial integration, if the world-lines are infinite and $\rho \in E_0^0$. The normalization $\langle \dot{z} | \dot{z} \rangle = -1$ is irrelevant: $\delta J = 0$ even if the particles move faster than light.

# 2 The Electromagnetic Field of a Known Charge Distribution

## 2.1 The Stationary-Action Principle and Conservation Theorems

*If the field equations originate from a stationary-action principle, then a conserved current can be constructed for each parameter of an invariance group.*

Field theory may be regarded as a generalization of the mechanics of point particles, in which the dynamical variables $q_i(t)$ are replaced with fields $\Phi(x, t)$, such as $\mathbf{E}(x, t)$ and $\mathbf{B}(x, t)$. The discrete index $i$ goes over to the continuous variable $x$, and, accordingly, the sum $\sum_i$ is replaced with an integral $\int d^3x$. A direct transcription of the formalism of I, §3, leads to infinite-dimensional manifolds, which we would prefer to avoid. Instead, we merely generalize the stationary-action principle (1:2.3.20) in order to find the analogues of the constants arising from the invariance properties. It is clear that in field theory the action $\int dt\, L(q, \dot{q})$ involves an integral over a four-dimensional submanifold $N_4$, and thus requires a 4-form, which allows the construction of a chart-independent integral.

**The Lagrangian Formulation of Field Theory** (2.1.1)

The action is given by

$$W = \int_{N_4} \mathscr{L}(\Phi, d\Phi),$$

where $\mathscr{L} \in E_4$ is the **Lagrangian**. The field equations result from the requirement that $\delta W = 0 \, \forall N_4$ compact and $\forall \delta\Phi$ such that $\delta\Phi_{|\partial N_4} = 0$.†

If we strengthen the homogeneous Maxwell equations to $F = dA$, then in pseudo-Riemannian space, the appropriate

**Electromagnetic Lagrangian** (2.1.2)

*is*

$$\mathscr{L} = -\tfrac{1}{2} dA \wedge {}^*dA - A \wedge {}^*J.$$

**Proof**

Making a variation $A \to A + \delta A$ and using (1.2.18) (a), one finds

$$-\delta W = \int_{N_4} \delta A \wedge [{}^*J + d^*dA] + \int_{\partial N_4} \delta A \wedge {}^*dA,$$

which vanishes if $\delta A_{|\partial N_4} = 0$ and $d^*F = -{}^*J$. □

**Remarks** (2.1.3)

1. The variational formulation offers no guarantee of existence or uniqueness of the solutions of the field equations. Nowhere has it been assumed that $d^*J = 0$, though without this condition it is not possible to satisfy $\delta W = 0 \, \forall \delta A$ such that $\delta A_{|\partial N_4} = 0$. The reason is easy to discover. With the gauge transformation $A \to A + d\Lambda$, where $\Lambda_{|\partial N_4} = 0$, $W$ changes by $\int_{N_4} \Lambda \, d^*J$, and is linear in $\Lambda$ not only for infinitesimal $\Lambda$. As a linear functional, either $W$ has no stationary points, or else, if $d^*J = 0$, it has a plateau. Accordingly, either there are no solutions at all, or else the solution is not uniquely fixed by any boundary condition whatsoever, because there is always the possibility of a gauge transformation.
2. According to (**I**: 5.2.8),
$$-\tfrac{1}{2} F \wedge {}^*F = -\tfrac{1}{4} F_{\sigma\rho} F^{\sigma\rho} {}^*\mathbf{1} = \tfrac{1}{2}(|\mathbf{E}|^2 - |\mathbf{B}|^2) {}^*\mathbf{1}.$$
The sign of $\mathscr{L}$ has been chosen so that the interaction
$$-A \wedge {}^*J = -{}^*i_J A = -J^\alpha A_\alpha {}^*\mathbf{1}$$
of a point particle moving along the world-line $z(s)$ (cf. (1.3.25; 2)) has the same sign
$$-e \int_{-\infty}^{\infty} ds \, \dot{z}^\alpha(s) A_\alpha(z(s)) \delta^4(x - z(s)) {}^*\mathbf{1}$$

† We use the symbol $\delta$ in § 2.1 for variations, rather than for codifferentials.

as in (I: 5.1.8). If a term $\frac{1}{2}m \int ds\, \dot{z}^\alpha(s)\dot{z}_\alpha(s)$ were added to the action, then both the field equations and the equation of motion could be derived from the same stationary-action principle by varying $A(x)$ and $z(s)$. The coupled system of equations suffers from difficulties due to the reaction of a particle on itself, as will be discussed more fully in §2.4.

The advantage of the Lagrangian formulation is that every one-parameter invariance group of $\mathscr{L}$ furnishes a conservation theorem. In field theory the argument goes as follows: If the 4-form $\mathscr{L}$ depends on the $p$-form $\Phi_j$, then a variation of $\mathscr{L}$ (taking $\partial\mathscr{L}/\partial\Phi_j$ with $\delta\Phi$ commuted through to the left) produces

$$\delta\mathscr{L} = \sum_j \delta\Phi_j \wedge \left[\frac{\partial\mathscr{L}}{\partial\Phi_j} - (-1)^p d\frac{\partial\mathscr{L}}{\partial(d\Phi_j)}\right] + d\left(\sum_j \delta\Phi_j \wedge \frac{\partial\mathscr{L}}{\partial(d\Phi_j)}\right), \quad (2.1.4)$$

and the field equations require that the term in the square brackets [ ] is zero. If the variation $\delta$ is that of a Lie derivative $L_v$, $v \in E_1$, then by (1.2.23), $\delta\mathscr{L} = L_v\mathscr{L} = di_v\mathscr{L}$, because the exterior differential of a 4-form vanishes on a four-dimensional manifold. (We think of $\delta$ as a kind of derivative, and not as an infinitesimal quantity.) Hence (2.1.4) says that a certain 3-form is closed and thus, by use of the * mapping, that there is a conserved current.

**Noether's Theorem** (2.1.5)

*Suppose that $v \in E_1$, $\delta\Phi = L_v\Phi$ and $\delta\mathscr{L} = L_v\mathscr{L}$. Then*

$$d\left[\sum_j L_v\Phi_j \wedge \frac{\partial\mathscr{L}}{\partial(d\Phi_j)} - i_v\mathscr{L}\right] \equiv -d{*}\mathscr{T}_v = 0.$$

**Remarks** (2.1.6)

1. The validity of (2.1.5) is premised on the variation $\delta$ including everything that is affected by $L_v$. For instance, even if $J = 0$, the $\mathscr{L}$ of (2.1.2) involves not only $dA$ but also the metric $g$, through the * mapping, and $g$ was not allowed to vary below (2.1.2). Since we supposed that $\delta{*}F = {*}\delta F$, Noether's theorem (2.1.5) applies to (2.1.2) only if $J = 0$ and $L_v{*} = {*}L_v$.
2. If $L_v$ is a translation, then (2.1.5) defines what is known as the **canonical energy-momentum tensor**.
3. The field equations remain unchanged when an exact 4-form is added to $\mathscr{L}$, $\mathscr{L} \to \mathscr{L} + dG$, where $G \in E_3$, since the addition only contributes a boundary integral to $\delta W$. However, the 3-form in the square brackets in (2.1.5) gets a contribution, which, if $G$ depends only on the fields $\Phi$ and not on their derivatives, is $di_vG$ (Problem 1). As an exact 3-form, it does not contribute to integrals over submanifolds without boundaries, but it can affect the conserved observables locally. This difficulty is not

encountered in classical mechanics, which is formally a one-dimensional field theory, and where $G$ would be in $E_0$, and hence $i_v G = 0$. Indeed, an additional $dG$ such that

$$\frac{d}{dt} G(q) = \dot{q}_i \frac{\partial G}{\partial q_i}$$

does not change the Hamiltonian

$$H = \dot{q}_i \frac{\partial \mathscr{L}}{\partial \dot{q}_i} - \mathscr{L}$$

at all.

**Application to the Electromagnetic Field** (2.1.7)

If $J = 0$ and $L_v{}^* = {}^*L_v$, then

$$\begin{aligned} &d[-(L_v A) \wedge {}^*dA + \tfrac{1}{2} i_v(dA \wedge {}^*dA)] \\ &\quad = d[-\tfrac{1}{2}(i_v F) \wedge {}^*F + \tfrac{1}{2} F \wedge i_v{}^*F - (di_v A) \wedge {}^*F] = 0. \end{aligned}$$

**Remarks** (2.1.8)

1. As can be seen above, if $v = e^\alpha$, then in addition to the electromagnetic energy-momentum forms ${}^*\mathscr{T}^\alpha$ given in (1.3.22), there is a new term $(di_v A) \wedge {}^*F$. If $J = 0$, then this is an exact 3-form, $(di_v A) \wedge {}^*F = d(i_v A \wedge {}^*F) + i_v A \wedge {}^*J$, so that (2.1.7) is a special case of (1.3.24). Unfortunately, the new term contains $A$ as well as $F$, and thus depends on the gauge. This does not contradict the gauge-invariance of $\mathscr{L}$ (assuming $J = 0$): Even in the mechanics of point particles the angular momentum $m[\mathbf{x} \times \dot{\mathbf{x}}]$ fails to be translation-invariant, although $\mathscr{L} = m|\dot{\mathbf{x}}|^2/2$ is.
2. The nonuniqueness mentioned in (2.1.6; 3) consists of an additional $d(F \wedge A) = F \wedge F$ in $\mathscr{L}$. Since the corresponding $G$ depends on $dA$ as well as $A$, the extra term in the conserved observable is changed to $2d((i_v A) \wedge F)$ (Problem 5). This is conserved independently of whether ${}^*L_v = L_v{}^*$, because $F \wedge F$ makes no reference to the metric structure of space-time. The expression $d((i_v A) \wedge F)$ not only depends on the gauge, but it also has the wrong reflection property. For example, the energy density would have a term $\sim \mathbf{B} \cdot \nabla V$, which changes sign if $(t, \mathbf{x}) \to (t, -\mathbf{x})$, as $(V, \mathscr{A}) \to (V, -\mathscr{A})$ and $(\mathbf{E}, \mathbf{B}) \to (-\mathbf{E}, \mathbf{B})$. In the so-called gauge theories, similar expressions determine what are known as topological charges $\int F \wedge F$, which characterize the topological structure of the bundle.
3. The local energy is defined by the coupling with gravity, and the formula for it is derived by letting the metric vary in $\mathscr{L}$. This variation has no

contribution from $F \wedge F$, and $F \wedge *F$ gives rise only to the $\mathscr{T}^\alpha$ of (1.3.22), which is gauge-invariant (see (4.2.8)).

4. If $J \neq 0$, then adding $L_v{*J}$ to the variations in (2.1.2) results in the Lorentz force. If $d{*F} = -{*J}$, then from (2.1.2),
$$L_v\mathscr{L} = -d[(L_vA) \wedge {*dA}] - A \wedge L_v{*J}$$
$$= -d[(i_vF) \wedge {*F}] - (di_vA) \wedge {*J} - A \wedge L_v{*J}.$$
On the other hand, due to the rules governing $L_v$,
$$L_v\mathscr{L} = -\tfrac{1}{2}di_v(F \wedge {*F}) - L_v(A \wedge {*J}).$$
Together, these facts imply the formula
$$-d{*\mathscr{T}_v} \equiv \tfrac{1}{2}d[F \wedge i_v{*F} - (i_vF) \wedge {*F}] = -(i_vF) \wedge {*J}$$
for any vector field such that $*L_v = L_v*$.
5. If $d{*J} = 0$ and $\mathscr{L}$ is as in (2.1.2), then $\int_{N_4}\mathscr{L}$ is invariant under $A \to A + d\Lambda$, $\Lambda_{|\partial N_4} = 0$. The gauge group is then a huge (Abelian) invariance group, and is not even locally compact, since it contains arbitrary functions. This leads one to think that there must be an infinite number of conservation theorems. However, they always reduce to trivialities, or rather to identities that hold independently of the field equations. By the way, this is not a characteristic only of field theories, but also occurs in point-particle mechanics (Problem 3). In (2.1.2) the variation $\delta A = d\Lambda$ (and $d{*J} = 0$) produces
$$0 = \delta W = -\int_{N_4}\Lambda d(d{*F}) + \int_{\partial N_4}\Lambda({*J} + d{*F}).$$
Since $\Lambda_{|\partial N_4} = 0$, this means that $d(d{*F}) = 0$, which is true regardless of whether $d{*F} = -{*J}$.

As we have seen, there is a conservation theorem for each $v$ whose Lie derivative does not destroy the structure of $\mathscr{L}$ determined by the metric. Such vector fields are important enough to merit a

**Definition** (2.1.9)

A vector field $v$ satisfying $L_vg = 0$ on a pseudo-Riemannian manifold with the metric $g$ is known as a **Killing vector field**.

**Remarks** (2.1.10)

1. Because $L_xL_y - L_yL_x = L_{[x,y]}$, where [ , ] is the Lie bracket of vector fields (see (I: 2.5.9; 6)), the Killing vector fields form a Lie algebra with [ , ]. (But not a module: $fv$, with $f \in E_0$ is not necessarily a Killing vector field if $v$ is.)

2. If an orthogonal basis $e^i$ ($g = e^i \otimes e^j \eta_{ij}$) is used to decompose the Lie derivative of the $e^i$ as $L_v e^i = A^i{}_j e^j$, $A^i{}_j \in E_0$, then $v$ is a Killing vector field iff $A_{ij} = -A_{ji}$, where $A_{ij} \equiv \eta_{ik} A^k{}_j$.
3. Problem 2 is to show that $*L_v = L_v*$ for Killing vector fields $v$.
4. It is possible for $*L_v\omega$ to equal $L_v*\omega$, $\omega \in E_p$ (for particular values of $p$), even if $v$ is not a Killing vector field. For example, $L_v e^j = fe^j, f \in E_0$, makes $L_v e^{j_1 \cdots j_p} = pfe^{j_1 \cdots j_p}$ and $L_v * e^{j_1 \cdots j_p} = (m-p)f * e^{j_1 \cdots j_p}$, and so $*L_v\omega = L_v*\omega$ for all $\omega \in E_{m/2}$. Yet $L_v$ generates the conformal transformation $L_v g = 2fg$, and $v$ is not a Killing vector field.

**Examples** (2.1.11)

$M$ is taken as $\mathbb{R}^4$ with $e^\alpha = dx^\alpha$ and $g = e^\alpha \otimes e^\beta \eta_{\alpha\beta}$ in these examples.

1. The rigid displacement $v = e^\alpha$ leaves $g$ invariant:

$$L_v e^\gamma = di_{e^\alpha} e^\gamma + i_{e^\alpha} de^\gamma = 0.$$

   The $\mathcal{T}_v$ of (2.1.8; 4) becomes the $\mathcal{T}^\alpha$ of (1.3.22), and (2.1.8; 4) reduces to the Lorentz force (1.3.24).
2. $v = x^\beta e^\alpha - x^\alpha e^\beta$ generates a Lorentz transformation

$$L_v e^\gamma = x^\beta i_{e^\alpha} de^\gamma + d(x^\beta i_{e^\alpha} e^\gamma) - (\alpha \leftrightarrow \beta) = -\eta^{\beta\gamma} dx^\alpha + \eta^{\alpha\gamma} dx^\beta.$$

   The $A$ of (2.1.10; 2) becomes $A^{\gamma\sigma} = \eta^{\alpha\gamma}\eta^{\beta\sigma} - \eta^{\beta\gamma}\eta^{\alpha\sigma}$, and it satisfies the condition of antisymmetry that characterizes Killing vector fields. Because the interior product is linear, the $\mathcal{T}_v$ of (2.1.8; 4) is simply $x^\beta * \mathcal{T}^\alpha - x^\alpha * \mathcal{T}^\beta$. Remark (2.1.8; 4) means that

$$d(x^\beta * \mathcal{T}^\alpha - x^\alpha * \mathcal{T}^\beta) = x^\beta (i_{e^\alpha} F) \wedge *J - x^\alpha (i_{e^\beta} F) \wedge *J,$$

   which implies with (1.3.24) that $dx^\beta \wedge *\mathcal{T}^\alpha - dx^\alpha \wedge *\mathcal{T}^\beta = 0$. Because $dx_\beta \wedge *dx^\gamma = \delta^\gamma{}_\beta *\mathbf{1}$, the energy-momentum tensor must therefore be symmetric; i.e., $T_{\alpha\beta} = T_{\beta\alpha}$. Although (1.3.22) is symmetric, the canonical energy-momentum tensor (2.1.7) is not; to be sure, the canonical energy-momentum tensor is also conserved if $v = x^\beta e^\alpha - x^\alpha e^\beta$, but $x^\alpha$ can not simply be factored out of the gauge-dependent term $di_v A$, which contains $dv$, and the conserved quantity is not $x^\alpha \mathcal{T}^\beta - x^\beta \mathcal{T}^\alpha$.
3. $v = x_\alpha e^\alpha$ generates a dilatation, $L_v e^\gamma = e^\gamma$; and if $F \in E_2(\mathbb{R}^4)$, then according to (2.1.10; 4) $L_v *F = *L_v F$. Then $\mathcal{T}^\alpha$ can be used to formulate a new conservation theorem.

$$d(x_\alpha * \mathcal{T}^\alpha) = x_\alpha i_{e^\alpha} F \wedge *J.$$

   From this and (1.3.24) we conclude that $dx \wedge *\mathcal{T}^\alpha = T_\alpha^\alpha *\mathbf{1} = 0$. We again observe that (1.3.22) satisfies this equation, whereas the canonical energy-momentum tensor (2.1.7) does not.

4. The conformal transformation $L_v g = 2x^\beta g$ is generated by $v = x^\beta x_\alpha e^\alpha - \frac{1}{2}x^\alpha x_\alpha e^\beta$:

$$L_v e^\gamma = x^\beta e^\gamma + x^\gamma e^\beta - \eta^{\gamma\beta} x_\alpha e^\alpha,$$

and the last two terms cancel out in the expression for $L_v g$. According to Remark (2.1.10; 4), it is again true that $L_v{}^*F = {}^*L_v F$ for all $F \in E_2(\mathbb{R}^4)$, and $d^*\mathscr{T}_v$ is

$$d(x^\beta x_\alpha{}^*\mathscr{T}^\alpha - \tfrac{1}{2}x^{2*}\, T^\beta) = ((x^\beta x_\alpha i_{e^\alpha} - \tfrac{1}{2}x^2 i_{e^\beta})F) \wedge {}^*J.$$

The resultant equation

$$0 = x_\alpha\, dx^\beta \wedge {}^*\mathscr{T}^\alpha - x_\alpha\, dx^\alpha \wedge {}^*\mathscr{T}^\beta + x^\beta\, dx_\alpha \wedge {}^*\mathscr{T}^\alpha$$

contains no new information, because the final term vanishes as in Example 3 and the first two vanish as in Example 2.

**Remarks** (2.1.12)

1. The vector fields generating conformal transformations (2.1.11; 4) are not complete (Problem 6); these transformations are not diffeomorphisms of $\mathbb{R}^4$, as they have singularities. This is not serious for the uses we make of them, as we need only the infinitesimal transformations. If a group of diffeomorphisms is desired, then $\mathbb{R}^4$ must be compactified by the addition of points at infinity [17].
2. The Lagrangian $d\Phi \wedge {}^*d\Phi$ for a massless scalar field $\Phi \in E_0$ has no intrinsic length, but even so it fails to be invariant under the dilatation (2.1.11; 3). As a consequence $T^\alpha{}_\alpha \neq 0$ for the energy-momentum tensor of a scalar field even if $m = 0$ (Problem 4).

**The Properties of the Energy-Momentum Tensor** (2.1.13)

Let $e^\alpha$ be an orthogonal basis of a pseudo-Riemannian space ($g = -e^0 \otimes e^0 + \sum_{j=1}^3 e^j \otimes e^j$), and let ${}^*\mathscr{T}^\alpha = \frac{1}{2}(i_{e^\alpha} F \wedge {}^*F - F \wedge i_{e^\alpha}{}^*F) = T^\alpha{}_\beta{}^*e^\beta$. Because of the component representation

$$T_{00} = \tfrac{1}{2}(|\mathbf{E}|^2 + |\mathbf{B}|^2), \qquad T^j{}_0 \equiv \mathscr{S}^j, \qquad j = 1, 2, 3;$$

where

$\mathscr{S} = [\mathbf{E} \times \mathbf{B}]$ is **Poynting's vector,**

we find that for all $x$,

(a) $T_{00}(x) \geq 0$, and $= 0$ only if $F(x) = 0$; and
(b) $\|\mathscr{S}(x)\| \leq T_{00}(x)$.

**Remarks** (2.1.14)

1. If (a) holds in every Lorentz system, then (b) follows. A Lorentz transformation treats $P^\alpha = \int *\mathscr{T}^\alpha$ as a vector (1.3.38; 5), and the equation $P^0 \geq 0$ would be violated by the transformation $P^0 \to (P^0 - \mathbf{v} \cdot \mathbf{P})/\sqrt{1 - v^2}$ unless $\|P\| \leq P^0$.
2. In the orthogonal basis, $\mathscr{S}^j = T^j{}_0 = T_0{}^j$ doubles as a momentum density and as the rate of energy flow: The change in the total energy can be written as

$$\delta\mathscr{T}_0 = \frac{\partial}{\partial t} T_0{}^0 + \frac{\partial}{\partial x^j} T_0{}^j.$$

   The physical interpretation of (b) is that electromagnetic energy can never be transmitted faster than light.
3. The positivity of the energy is expressed mathematically as follows: Let $N_3$ be a spacelike submanifold, and define * with the restriction of $g$ to $N_3$. Then * converts a 3-form into a numerical function, and the positivity of the energy means that $*(T^0{}_{|N_3}) \geq 0$.
4. The signs in (2.1.13) arise from the signature of the metric, and thus depend on the relationship to the standard basis. However, the equations $T^{\alpha\beta} = T^{\beta\alpha}$ and $T^\alpha{}_\alpha = 0$, which follow from (2.1.11; 2) and (2.1.11; 3), hold in any basis on account of the transformation law (1.3.26).

Since the existence of a limiting speed of energy transport follows solely from the structure of the energy-momentum tensor, it is possible to prove the uniqueness of the solution of the Cauchy problem without further analysis of the field equations. Nonuniqueness would contradict the causal propagation of the field at the speed of light. To make this impression mathematically precise, we make some intuitively reasonable

**Definitions** (2.1.15)

(a) A continuous mapping of an interval $I \to M$ is called a **causal curve** iff no two of its points can be connected by a spacelike curve. It is said to be **nonextensible** iff it is not a proper subset of a larger causal curve.

(b) Let $M$ be orientable in time, that is, the forward and backward time directions can be defined smoothly over all of $M$. The **future** (respectively **past) domain of influence** $D^+(N)$ (resp. $D^-(N)$) of a spacelike hypersurface $N$ is the set of all points $p$ of $M$ for which all nonextensible, causal curves through $p$ oriented toward the past (resp. future) intersect

$N$. $D(N) \equiv D^+(N) \cup D^-(N)$ is called the **domain of influence** of $N$, and if $D(N) = M$ then $N$ is a **Cauchy surface**

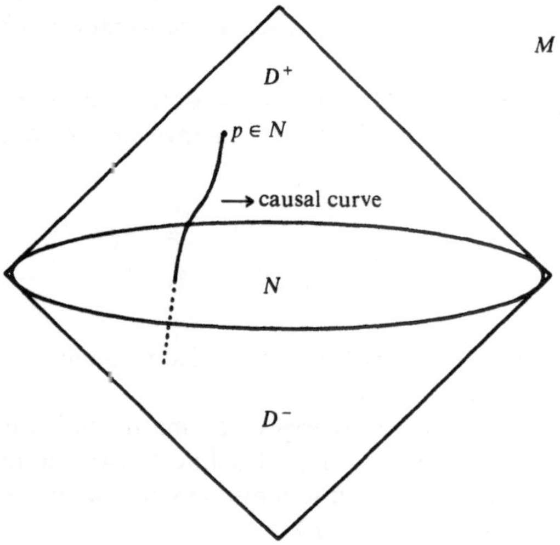


Figure 16 The domain of influence of a hypersurface $N \subset M$

The uniqueness of the Cauchy problem is the statement that $F$ is uniquely determined on $D^{\pm}(N_3)$ by $F_{|N_3}$, $*F_{|N_3}$, and $J$. If $F_{|D^{\pm}(N_3)} \neq 0$ but $J = F_{|N_3} = *F_{|N_3} = 0$, then $F$ would have to propagate from somewhere outside $D^{\pm}(N_3)$ to within $D^{\pm}(N_3)$ without encountering $N_3$, which would be possible only if the propagation speed were greater than the speed of light. Since there exist complicated manifolds $M$ for which the shape of $D^{\pm}$ is confusingly tortuous, let us be contented with the

**Uniqueness of the Cauchy Problem in Minkowski Space** (2.1.16)

*Let $N$ be a three-dimensional, compact, spacelike submanifold of $(\mathbb{R}^4, \eta)$, and suppose that $F_1$ and $F_2$ are two continuous solutions of Maxwell's equations, $dF_1 = dF_2 = 0$, and $\delta F_1 = \delta F_2 = J$. If $F_{1|N} = F_{2|N}$ and $*F_{1|N} = *F_{2|N}$, then $F_1$ and $F_2$ are also equal throughout the interiors of $D^{\pm}(N)$: $F_{1|\text{Int}\, D^{\pm}(N)} = F_{2|\text{Int}\, D^{\pm}(N)}$.*

**Proof**

Let $x \in \text{Int}\, D^+(N)$. Then there exist $\varepsilon > 0$ and $\bar{x} \in D^+(N)$ such that

$$x \in N' \equiv \{y \in \mathbb{R}^4 : |\bar{\mathbf{x}} - \mathbf{y}|^2 - (\bar{x}^0 - y^0)^2 = -\varepsilon^2, y^0 \leq \bar{x}^0\}$$
$$\subset \{y \in \mathbb{R}^4 : |\bar{\mathbf{x}} - \mathbf{y}|^2 \leq (\bar{x}^0 - y^0)^2\}.$$

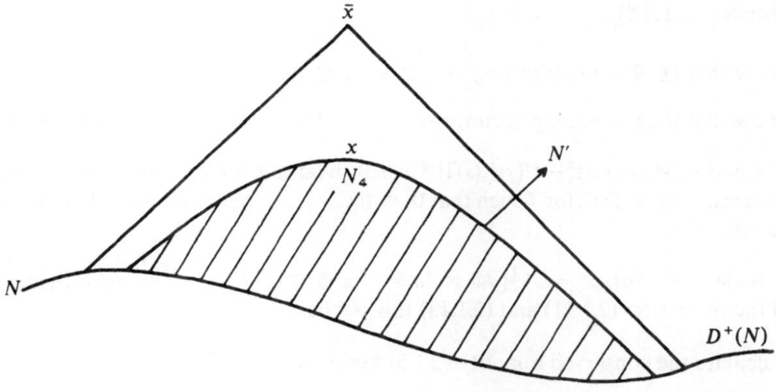


Figure 17 The region of integration used to prove the uniqueness of the Cauchy problem

Hence

$$(N' \cap D^+(N)) \cup \{y \in N : |\bar{\mathbf{x}} - \mathbf{y}|^2 - (\bar{x}^0 - y^0)^2 \leq -\varepsilon^2\} = \partial N_4,$$

where $N_4$ is a compact, four-dimensional submanifold (see Figure 17). If $F = F_1 - F_2$, then $dF = \delta F = 0$ and $F_{|N} = {}^*F_{|N} = 0$. The $\mathscr{T}^\alpha$ formed from $F$ therefore satisfies $\delta \mathscr{T}^\alpha = 0$ and $\mathscr{T}^\alpha{}_{|N} = 0$, and so $0 = \int_{N_4} d{}^*\mathscr{T}^0 = \int_{N' \cap D^+(N)} {}^*\mathscr{T}^0$. Since $N'$ is spacelike, $\mathscr{T}^0$ is a nonnegative measure on $N'$, and the vanishing of the integral implies that $\mathscr{T}^0{}_{|N' \cap D^+(N)}$ is zero almost everywhere. Since it is a continuous function, $\mathscr{T}^0{}_{|N' \cap D^+(N)} = 0$; and because of (2.1.13) (b), $F_{|N' \cap D^+(N)} = 0$, and in particular $F_{|x} = 0$. The proof for $D^-(N)$ is similar. □

**Remarks** (2.1.17)

1. Formula (1.2.36) proves uniqueness by explicit construction, but it assumes the existence of the distributions $G_{\bar{x}}$, $dG_{\bar{x}|N}$, and ${}^*dG_{\bar{x}|N}$. The advantage of the proof given above is that it can be extended to more general manifolds without the necessity of facing the difficult question of the existence of the Green functions.
2. Positivity of the energy is sufficient but not necessary for the uniqueness of the Cauchy problem, which can also be proved (see §3.2), for instance, for the scalar field of Problem 4 with $m^2$ replaced with $-m^2$, although $T^0{}_0$ is then not positive.
3. The electromagnetic current $J$ has no positivity property, and there is no proposition like (2.1.16) for it. Because charges can cancel out, it could happen that $J \equiv 0$ for all $t < t_0$ and $\neq 0$ for all $t > t_0$. More specifically, nothing in Maxwell's equations prevents a charge from moving faster than light.

**Problems** (2.1.18)

1. Show that $(L_v\Phi_j) \wedge (\partial\mathscr{L}/\partial\, d\Phi_j) - i_v\mathscr{L} = di_vG$, if $\mathscr{L} = dG(\Phi)$.

2. Show that if $v$ is a Killing vector, then $L_v* = *L_v$. (Use (1.2.26; 2) and (2.1.10; 2).)

3. The action $W = \int ds[-\dot{z}^\alpha(s)\dot{z}_\alpha(s)]^{1/2}$ is left invariant not only by $s \to s +$ constant, but even by $s \to f(s)$, for $f$ such that $0 < df/ds < \infty$. What is the resultant conserved quantity?

4. Calculate $\mathscr{T}^\alpha$ for $\mathscr{L} = -\frac{1}{2}[d\Phi \wedge *d\Phi + m^2\Phi*\Phi]$, $\Phi \in E_0$, and investigate which of the properties (2.1.11) and (2.1.13) this $\mathscr{T}^\alpha$ has.

5. Calculate the conserved quantity (2.1.5) for $\mathscr{L} = dA \wedge dA$, $A \in E_1$.

6. Integrate the equations $\partial x(t)/\partial t = -a\langle x(t)|x(t)\rangle + 2x(t)\langle a|x(t)\rangle$, which generate the (local) flow of a conformal transformation.

7. Suppose that $\delta(\mathscr{T}^\alpha + t^\alpha) = 0$ and that $\mathscr{T}^\alpha + t^\alpha$ is independent of time in some system and falls off sufficiently in space. Show that in this reference system

$$\int_{0 \le t \le T} dx^j \wedge *(\mathscr{T}^\alpha + t^\alpha) = 0,$$

for $j = 1, 2, 3$, and $\alpha = 0, 1, 2, 3$. (In particular, the "self-stress" terms $\int d^3x\, T_{jj}$ vanish.) Conclude that $\delta(\mathscr{T}^\alpha + t^\alpha)$ can not be zero for the point charge (1.3.25; 2).

**Solutions** (2.1.19)

1. $$\mathscr{L} = d\Phi_j \wedge \frac{\partial G}{\partial \Phi_j}, \qquad (L_v\Phi_j) \wedge \frac{\partial G}{\partial \Phi_j} - i_v\, dG = L_vG - i_v\, dG = di_vG.$$

2. First note that $L_v e^{k_1 \cdots k_p} = \sum_j A^{k_j}{}_k e^{k_1 \cdots k_{j-1} k k_{j+1} \cdots k_p}(-1)^{j+1}$. The identity of (1.2.40; 8) continues to hold when $A^k{}_j$ is substituted for $\omega^k{}_j$, because $A^k{}_j$ has the same antisymmetry. Hence

$$L_v * e^{k_1 \cdots k_p} = \sum_j A^{k_j}{}_k * e^{k_1 \cdots k_j k k_{j+1} \cdots k_p}(-1)^{j+1},$$

and consequently

$$\begin{aligned}*L_v(\omega_{k_1 \cdots k_p} e^{k_1 \cdots k_p}) &= (L_v\omega_{k_1 \cdots k_p}) * e^{k_1 \cdots k_p} + \cdots \\ &\quad + \omega_{k_1 \cdots k_p} * L_v e^{k_1 \cdots k_p} = L_v(\omega_{k_1 \cdots k_p} * e^{k_1 \cdots k_p}).\end{aligned}$$

3. If $s = \bar{s} + f(\bar{s})$, then as $f \to 0$ with $f = 0$ on the boundary,

$$\begin{aligned}W &= \int d\bar{s}(1 + f')\sqrt{-\dot{z}^\alpha(\bar{s})\dot{z}_\alpha(\bar{s}) - 2f\dot{z}^\alpha(\bar{s})\ddot{z}_\alpha(\bar{s})} \\ &= W + \int d\bar{s}\left(f'\sqrt{-\dot{z}^\alpha(\bar{s})\dot{z}_\alpha(\bar{s})} - f\frac{\dot{z}^\alpha(\bar{s})\ddot{z}_\alpha(\bar{s})}{\sqrt{-\dot{z}^\alpha(\bar{s})\dot{z}^\alpha(\bar{s})}}\right) + O(f^2).\end{aligned}$$

Integrating this by parts leads to the identity

$$\frac{d}{ds}\sqrt{-\dot{z}^\alpha(s)\dot{z}_\alpha(s)} = \frac{-\dot{z}^\alpha(s)\ddot{z}_\alpha(s)}{\sqrt{-\dot{z}^\alpha(s)\dot{z}_\alpha(s)}},$$

which also holds if $\ddot{z} \neq 0$.

4. $\mathcal{T}^\alpha = (i_\alpha d\Phi) \wedge {}^*d\Phi - \frac{1}{2}i_\alpha[d\Phi \wedge {}^*d\Phi + m^2\Phi^*\Phi]$. Then $T_{\alpha\beta} = T_{\beta\alpha}$ (Lorentz invariance), but not $T^\alpha{}_\alpha = 0$, even if $m = 0$ (no dilatation invariance). $T_{00} \geq |\dot\Phi\nabla\Phi|$ and $T^0{}_j = \dot\Phi\nabla_j\Phi$, iff $m^2 \geq 0$.

5. $(L_v A) \wedge \dfrac{\partial \mathcal{L}}{\partial\, dA} - i_v\mathcal{L} = 2(L_v A) \wedge F - i_v(F \wedge F) = 2(di_v A) \wedge F = 2d(i_v A \wedge F)$.

6. Letting $x \equiv x(0)$, $x(t) = (x - at\langle x|x\rangle)/(1 - 2t\langle x|a\rangle + \langle x|x\rangle\langle a|a\rangle t^2)$, because

$$\frac{\partial x}{\partial t} = \frac{-a\langle x|x\rangle}{1 - 2t\langle x|a\rangle + \langle x|x\rangle\langle a|a\rangle t^2} + 2\,\frac{(x - at\langle x|x\rangle)(\langle x|a\rangle - \langle x|x\rangle\langle a|a\rangle t)}{(1 - 2t\langle x|a\rangle + \langle x|x\rangle\langle a|a\rangle t^2)^2}$$
$$= -a\langle x(t)|x(t)\rangle + 2x(t)\langle a|x(t)\rangle.$$

For any $t > 0$ there exists $x \in \mathbb{R}^4$, namely the $x$ for which

$$1 - 2t\langle a|x\rangle + t^2\langle a|a\rangle\langle x|x\rangle = 0,$$

which gets sent off to infinity.

7. Integrate $x^j\, d^*(\mathcal{T}^\alpha + t^\alpha) = dx^j \wedge {}^*(\mathcal{T}^\alpha + t^\alpha) + d(x^j{}^*(\mathcal{T}^\alpha + t^\alpha))$ over

$$N_4 \equiv \{(t, \mathbf{x}): 0 \leq t \leq T\}.$$

In the rest-frame of the charge nothing depends on time, and only $t^0 \neq 0$. Since $dx_\alpha \wedge {}^*\mathcal{T}^\alpha = 0$ (2.1.11; 3), we obtain the contradiction

$$0 = \sum_{j=1}^{3} \int_{0 \leq t \leq T} dx^j \wedge {}^*\mathcal{T}^j = \int_{0 \leq t \leq T} dx^0 \wedge {}^*\mathcal{T}^0 > 0.$$

## 2.2 The General Solution

*The characteristics of Maxwell's equations are hypersurfaces with lightlike normals. The Green function is easy to construct in Minkowski space ($\mathbb{R}^4, \eta$), and solves the initial-value problem explicitly.*

If the argument of (1.2.28) is applied to the physically interesting case $m = 4$, $p = 2$, one finds that the Minkowski metric

$$\eta = \begin{vmatrix} -1 & & & \\ & 1 & & \\ & & 1 & \\ & & & 1 \end{vmatrix}$$

contains the $g_+$ of (1.2.28) as its spatial part and also $g_-$ as a spatiotemporal part. Since in the present case $* \circ * = (-)^{p(m-p)+s} = -1$, the combinations of the fields that diagonalize $*$ are $F \pm i^*F$, and when these are written explicitly, both real and complex characteristic directions are found, so there is a combination of both of the cases of (1.2.28) (Problem 1). We have remarked that the characteristics may in general be surfaces of discontinuity for the solutions. If such a hypersurface is specified by $u(x) = 0$, for $u \in E_0(V)$, $V \subset M$, then there should exist solutions that behave locally like $\Theta(u)$. The exterior differential $d\Theta(u) = \delta(u)du$ is then singular at $u = 0$, and if $J$ is a regular function, then the singular contributions to the left sides of the equations $dF = 0$ and $\delta F = J$ must cancel out. These contributions are proportional to $du$, so we are interested in finding solutions for $J = 0$ that depend only on $u$: $F = c_{ij}(u)e^{ij}$. For such solutions the equations say that the exterior and interior products of $F'$ with $du$ must vanish, because $dF = du\, c_{ij}' \wedge e^{ij} \equiv du \wedge F'$, $d{*F} = du \wedge {*F'} = -{*(i_{du}F')}$. This argument leads directly to a

**Condition for the Characteristics** (2.2.1)

*Let $F$ be discontinuous where $u = 0$, but suppose that $J$ is continuous there. Then the equations $dF = 0$ and $\delta F = J$ imply that at $u = 0$*

$$du \wedge F' = 0 \quad \textit{and} \quad i_{du}F' = 0;$$

*which are satisfied only if $\langle du|du\rangle = 0$ or $F' = 0$.*

**Proof**

If $du \wedge F' = 0$, then in a local basis using $du$, $F'$ contains $du$ as a factor, that is, $F' = du \wedge f$, where $f \in E_1$ is independent of $du$. The second equation then requires that $i_{du}\, du \wedge f = \langle du|du\rangle f - \langle du|f\rangle\, du = 0$. Because $f$ and $du$ are independent, we conclude that $\langle du|du\rangle = \langle du|f\rangle = 0$. □

**Remark** (2.2.2)

1. In other words, either the normal to the surface is lightlike, or else there are no discontinuities in $F$.
2. This is only a local statement. Whether $u$ can be defined as a global coordinate depends on the large-scale structure of space-time.
3. A by-product of (2.2.1) is the statement that fields with discontinuities must have a special structure; they are the exterior products of two 1-forms, of which one ($du$) is a null field, and the other ($f$) is orthogonal to it (also in the sense of the metric $\eta$). Both invariants vanish for such fields:

$$*(F \wedge F) = *(du \wedge f \wedge du \wedge f) = 0,$$

and $*(F \wedge *F) = i_{du}i_f F = (\langle du|f\rangle)^2 - \langle du|du\rangle\langle f|f\rangle = 0$. If the space and time parts are separated, then $du = dt + \mathbf{n}\cdot d\mathbf{x}$ and $f = dt + \mathbf{f}\cdot d\mathbf{x}$, which requires that $|\mathbf{n}|^2 = 1 = (\mathbf{n}\cdot\mathbf{f})$. In terms of the field strengths, $\mathbf{E} = \mathbf{n} - \mathbf{f}$ and $\mathbf{B} = [\mathbf{n} \times \mathbf{f}]$, so this means that $(\mathbf{n}\cdot\mathbf{E}) = (\mathbf{n}\cdot\mathbf{B}) = (\mathbf{E}\cdot\mathbf{B}) = 0 = |\mathbf{E}|^2 - |\mathbf{B}|^2$. The field (1.1.4) was of this form with $u = x - t$.

4. We also note that if $\langle du|du\rangle = 0$, then it is not sufficient to specify $F$ and $*F$ at $u = 0$ in order to solve the Cauchy problem: It is possible to choose $F$ not identically zero, such that $f(0) = 0$.
5. Whereas a field satisfying $du \wedge F = i_{du}F = 0$ can vary arbitrarily from one hyperplane $u =$ const. to the next, the Lie derivative $L_{du}$ in the direction $g\,du$ tangential to $u =$ const. is determined by Maxwell's equations:

$$L_{du}F = di_{du}F + i_{du}\,dF = 0$$
$$L_{du}{*F} = di_{du}{*F} + i_{du}\,d{*F} = -i_{du}{*J}.$$

   The higher Lie derivatives can similarly be calculated, which determines $F$ everywhere, if $F$ and $*F$ are specified on some surface that can be translated with $e^{L_{du}}$ so as to cover all of $M$.

After studying these local questions, we will evaluate formula (1.2.36) for Maxwell's equations. This requires the explicit form of the Green function, which can be written down only for the simplest manifolds.

### The Construction of $G_{\bar{x}}$ in Minkowski space $(\mathbb{R}^4, \eta)$ (2.2.3)

If $e^\alpha$ is the natural and orthogonal basis, then according to Equation (1.2.34), $G_{\bar{x}}$ has the form

$$G_{\bar{x}} = \tfrac{1}{2}\bar{e}_{\alpha\beta} \otimes {*e^{\alpha\beta}} D_{\bar{x}}(x),$$

for $p = 2$, and where $D_{\bar{x}} \in E_0(\mathbb{R}^4)$ satisfies the equation

$$-D_{\bar{x},\alpha}{}^{\alpha} = \delta^4(x - \bar{x}).$$

The translation-invariance of Minkowski space allows the partial differential equation to be reduced to an ordinary differential equation, by expanding it in a series in the eigenfunctions of the translation operator, i.e., a Fourier series. For the delta-function this expansion is

$$\delta^4(x - \bar{x}) = (2\pi)^{-4} \int_{-\infty}^{\infty} d^4k\, e^{i\langle k|\bar{x} - x\rangle}, \tag{2.2.4}$$

where $k^i$ and $x^i$ are regarded as the components of vectors, and $\langle\,|\,\rangle$ stands for the Lorentz scalar product (1.2.14) with $g = \eta$. In Fourier-transformed space the Laplacian (i.e.: $_{,\alpha}{}^{\alpha}$) produces a factor $ik_\alpha ik^\alpha = -\langle k|k\rangle \equiv -k^2$, so that, finally, $D_{\bar{x}}$ has the Fourier integral representation

$$D_{\bar{x}}(x) = (2\pi)^{-4} \int_{-\infty}^{\infty} d^4k\, e^{i\langle k|\bar{x} - x\rangle}/k^2. \tag{2.2.5}$$

**Remarks** (2.2.6)

1. The $k$ integrals do not converge in the classical sense, but as distributions—this is the content of Fourier's theorem. We have unscrupulously interchanged integrals by $k$ with derivatives by $x$. Fortunately, distributions are so agreeable that they put up with manipulations that classical analysis considers criminal.
2. Because of the translation-invariance of Minkowski space, $D_{\bar{x}}$ depends only on $x - \bar{x}$.
3. Since $k^2 = (|\mathbf{k}| - k^0)(|\mathbf{k}| + k^0)$, the integrand of (2.2.5) has poles, and it must be decided what to do about them when they are in the integration path. We are not restricted to integrals along the real axis; the analyticity of the integrand of (2.2.4) allows the integration path to be distorted into the complex plane. We use the path denoted as

$$\int_{\frown\frown} d^4k,$$

which passes above the poles in the complex $k$-plane. Other choices of the integration path would produce integrals differing by the contributions of the residues. This nonuniqueness should not be surprising, because the equation in (2.2.3) determines $D_{\bar{x}}$ only up to a solution of the homogeneous equation, and

$$\eta^{\alpha\beta} \frac{\partial^2}{\partial x^\alpha \, \partial x^\beta} e^{i(\mathbf{k}\cdot\mathbf{x} - k^0 t)} = 0,$$

if $k^0 = \pm|\mathbf{k}|$.

The path shown in Figure 18 is chosen on physical grounds, since the Green function it produces corresponds to physically realistic initial conditions. It is denoted $D^{\text{ret}}(\bar{x} - x)$, and an elementary integration (Problem 2) produces the

**Retarded Green Function** (2.2.7)

$$D^{\text{ret}}(x) = \frac{\delta(r - 1)}{4\pi r} = \frac{\delta(x^2)}{2\pi} \Theta(t).$$

$$G_{\bar{x}}^{\text{ret}} = \frac{1}{4\pi} \bar{e}_{\alpha\beta} \otimes {}^*e^{\alpha\beta} \delta((\bar{x} - x)^2) \Theta(\bar{t} - t),$$

$$r \equiv |\mathbf{x}|, \qquad x^2 \equiv \langle x | x \rangle.$$

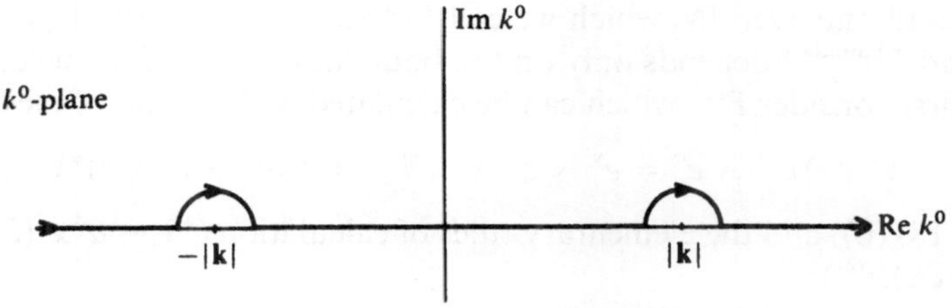


Figure 18 The path of integration for $D^{\mathrm{ret}}$

**Remarks** (2.2.8)

1. For the second form of $D^{\mathrm{ret}}$ we made use of the formula

$$\delta(f(x)) = \sum_i \frac{1}{|f'(x_i)|}\delta(x - x_i), \qquad f(x_i) = 0,$$

to write

$$\delta(x^2)\Theta(t) = \delta((r + t)(r - t))\Theta(t) = \frac{\Theta(t)}{2r}(\delta(r + t) + \delta(r - t))$$

$$= \frac{\delta(r - t)}{2r}\Theta(t).$$

2. The integral (2.2.5) singles out no particular Lorentz system, and thus $D^{\mathrm{ret}}$ depends only on $(x - \bar{x})^2$ and $\Theta(t - \bar{t})$. The preference for one time-direction enters through the choice of the integration path; the path —∪∪— would result in $D^{\mathrm{adv}}(\bar{x} - x) \equiv D^{\mathrm{ret}}(x - \bar{x})$.
3. Because $G^{\mathrm{ret}}_{\bar{x}}$ is supported wholly on the negative light-cone of $\bar{x}$ (see Figure 19), the integration over infinite space-time regions is justified painlessly.

The means by which the explicit forms of the field strengths can be calculated from (1.2.36) are now ready. The field is the sum of two integrals, one

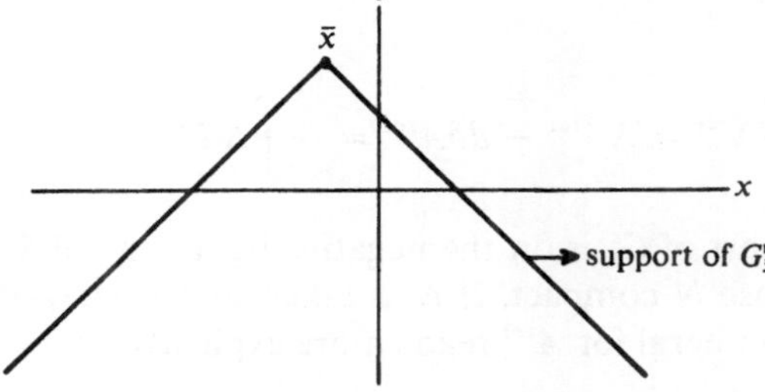


Figure 19 The support property of the retarded Green function

over $N$ and one over $\partial N$, which we call $F^{\text{ret}}$ and $F^{\text{boundary}}$, $F^{\text{ret}}$ depends only on $J$, and $F^{\text{boundary}}$ depends only on the boundary values of the field.

We first consider $F^{\text{ret}}$, which can be calculated with the aid of the formula

$$e^{\gamma} \wedge {}^*e^{\alpha\beta} \wedge e^{\sigma} = e^{\gamma} \wedge e^{\sigma} \wedge {}^*e^{\alpha\beta} = (\eta^{\sigma\beta}\eta^{\gamma\alpha} - \eta^{\sigma\alpha}\eta^{\gamma\beta}){}^*\mathbf{1}$$

(cf. (1.2.18) (b)) and the elementary rules of calculation (${}^*\mathbf{1} \to d^4x$, $(\partial/\partial x)D^{\text{ret}} = -(\partial/\partial \bar{x})D^{\text{ret}}$):

$$F^{\text{ret}}(\bar{x}) = \int_N dG_{\bar{x}}^{\text{ret}} \wedge J = \tfrac{1}{2}\bar{e}_{\alpha\beta} \int_N \frac{\partial}{\partial x^{\gamma}} D^{\text{ret}}(\bar{x} - x) e^{\gamma} \wedge {}^*e^{\alpha\beta} \wedge e^{\sigma} J_{\sigma}(x)$$

$$= \bar{e}^{\alpha\beta} \int_N d^4x\, J_{\alpha}(x) \frac{\partial}{\partial \bar{x}^{\beta}} D^{\text{ret}}(\bar{x} - x) = dA^{\text{ret}}(\bar{x}), \tag{2.2.9}$$

$$A^{\text{ret}}(\bar{x}) \equiv -\bar{e}^{\alpha} \int_N d^4x\, D^{\text{ret}}(\bar{x} - x) J_{\alpha}(x).$$

**Remarks** (2.2.10)

1. $F^{\text{ret}}$ is precisely the exterior differential of a vector potential $A^{\text{ret}}$. It is common to solve for $F$ by first setting $F = dA$, and then using $\delta F = J$ to determine $A$. Equation (1.2.36) with $p = 1$ and $A$ in place of $F$ shows that $A_{|\bar{x}}$ does not depend only on $A_{|\partial N}$, ${}^*dA_{|\partial N}$, and $J_{|N}$, but also on $\delta A_{|N}$. If we rewrite $\int_N \delta G_{\bar{x}} \wedge dA$ as $\int_N G_{\bar{x}} \wedge \delta\, dA + \int_{\partial N} \cdots = \int_N G_{\bar{x}} \wedge J + \int_{\partial N} \ldots$, then there are three contributions to $A(\bar{x})$, viz., $\int_N G_{\bar{x}} \wedge J$ as in (2.2.9), $\int_N dG_{\bar{x}} \wedge \delta A$, and $\int_{\partial N} \ldots$. Gauge invariance makes it impossible to fix $A$ in terms of $J$ and the boundary values of $A$ and $dA$; the equations leave open the possibility of a contribution from $d\Lambda$. The solution of the boundary-value problem is unique if we impose the additional condition that $\delta A = 0$ (the **Lorentz gauge**).
2. The $A^{\text{ret}}$ of (2.2.9) satisfies

$$\delta A^{\text{ret}}(\bar{x}) = -\int_{\partial N} D^{\text{ret}}(\bar{x} - x){}^*J(x) \equiv j(\bar{x}), \qquad j \in E_0(N),$$

(Problem 4), so $F^{\text{ret}}$ satisfies the equations

$$dF^{\text{ret}} = 0,$$

$$\delta F^{\text{ret}} = \delta\, dA^{\text{ret}} = \Delta A^{\text{ret}} - d\delta A^{\text{ret}} = -\int \Delta G^{\text{ret}} \wedge J - dj = J - dj.$$

3. Since the support of $G_{\bar{x}}$ is on the negative light-cone of $\bar{x}$, we are not required to choose $N$ compact. If $N$ is taken as $\{(t, x): t_0 \le t \le t_1\}$, then, by (2.2.7), the integral for $A^{\text{ret}}$ reads more explicitly

$$A_{\alpha}^{\text{ret}}(\bar{x}) = -\int_{|\mathbf{x} - \bar{\mathbf{x}}| < \bar{t} - t_0} \frac{d^3x}{4\pi|\mathbf{x} - \bar{\mathbf{x}}|} J_{\alpha}(\mathbf{x}, \bar{t} - |\mathbf{x} - \bar{\mathbf{x}}|).$$

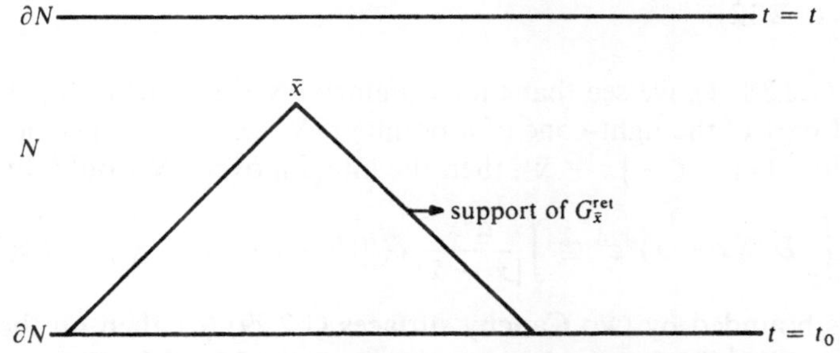


Figure 20 The region of integration for $F^{\text{ret}}(\bar{x})$

This integral always converges for bounded $J_\alpha$, even if $J$ does not have spatially compact support (Figure 20).

4. Although the use of a different Green function, say $D^{\text{adv}}$, does not change $F(\bar{x})$, it does change the integrals $\int_N$ and $\int_{\partial N}$. For example, with the $N$ of Remark 3, $F^{\text{boundary}}$ has contributions only from $t = t_0$, while only the boundary values at $t = t_1$ would show up in $D^{\text{adv}}$. If we want to know $F$ in the half-space later than $t = t_0$, then $D^{\text{ret}}$ is more useful than $D^{\text{adv}}$, because we can let the upper (later) part of $N$ go to infinity. The limit $t_1 \to \infty$ would not necessarily exist for $D^{\text{adv}}$: If we insisted that $F_{|t_0} = 0$, then without the upper boundary term all that would remain of $F$ would be $\int_{t>t_0} dG^{\text{adv}} \wedge J$, which does not vanish at $t = t_0$, and thus requires that there be a contribution to $\int_{\partial N}$ at $t = t_1$. Since the equations are invariant under time-reflection $t \to -t$, the appropriate Green function for the time-reversed question would be $D^{\text{adv}}$.
5. With the aid of Green functions with $p = 3$, we can write (2.2.9) in coordinate-free notation as

$$A^{\text{ret}} = \int_N G_{\bar{x}}^{\text{ret}} \wedge J$$

In order to study $F^{\text{boundary}}$, we use (2.2.7) to write it more explicitly as

$$\begin{aligned} F^{\text{boundary}}(\bar{x}) &= -\int_{\partial N} (\delta G_{\bar{x}} \wedge F - {}^*dG_{\bar{x}} \wedge {}^*F) \\ &= \bar{e}^{\alpha\beta} \int_{\partial N} \left[ \frac{\partial}{\partial \bar{x}^\alpha} D^{\text{ret}}(\bar{x} - x) F_{\gamma\beta}(x) {}^*e^\gamma - \frac{\partial}{\partial \bar{x}^\gamma} D^{\text{ret}}(\bar{x} - x) F_\alpha{}^\gamma(x) {}^*e_\beta \right. \\ &\quad \left. + \frac{1}{2} \frac{\partial}{\partial \bar{x}^\alpha} D^{\text{ret}}(\bar{x} - x) F_{\alpha\beta}(x) {}^*e^\gamma \right] \end{aligned} \tag{2.2.11}$$

(Problem 5).

**Remarks** (2.2.12)

1. As in (1.2.38; 1), we see that the boundary $\partial N$ is not arbitrary. If it contained part of the light-cone of a point $\bar{x} \in N$, i.e., the set specified by the equation $t(x) = \bar{t} - |x - \bar{x}|$, then the integral over $\partial N$ would diverge, as

$$\int_{\partial N} D^{\text{ret}}(\bar{x} - x)^* e^0 \cong \int \frac{d^3x}{|x - \bar{x}|} \delta(t(\mathbf{x}) - \bar{t} - |\mathbf{x} - \bar{\mathbf{x}}|) \sim \int \delta(0).$$

2. If $N$ is bounded by two Cauchy surfaces (1.2.29) (c), then on the earlier surface $F^{\text{boundary}}$ takes on the boundary values of $F$, and $F^{\text{ret}}$ goes to zero. The first of these claims is proved as follows: If $N$ is given in Cartesian coordinates as $t \geq t(x)$, then†

$$^*e^\alpha{}_{|\partial N} = dx^1 \wedge dx^2 \wedge dx^3 \left(1, -\frac{\partial t}{\partial x^1}, \frac{\partial t}{\partial x^2}, -\frac{\partial t}{\partial x^3}\right).$$

The Lorentz system may be chosen so that $\bar{x}$ is the origin and $\partial t(0)/\partial x^j = 0$. Because

$$\lim_{\bar{t} \downarrow 0} \int d^3x\, f(x) \frac{\partial}{\partial \bar{x}^\beta} \frac{\delta(\bar{t} - t - r)}{4\pi r}\bigg|_{t=0} = \delta^0{}_\beta f(0)$$

(see Problem 6),

$$\lim_{\bar{x} \downarrow 0} \int_{\partial N} {}^*e^\alpha \frac{\partial}{\partial \bar{x}^\beta} D^{\text{ret}}(\bar{x} - x) f(x) = \delta^\alpha{}_0 \delta^0{}_\beta f(0).$$

If this is substituted into (2.2.11), the two first terms cancel out as $\bar{x} \downarrow 0$, leaving only $F(0)$. On the other hand

$$\lim_{\bar{x} \downarrow 0} F^{\text{ret}}(\bar{x}) = 0,$$

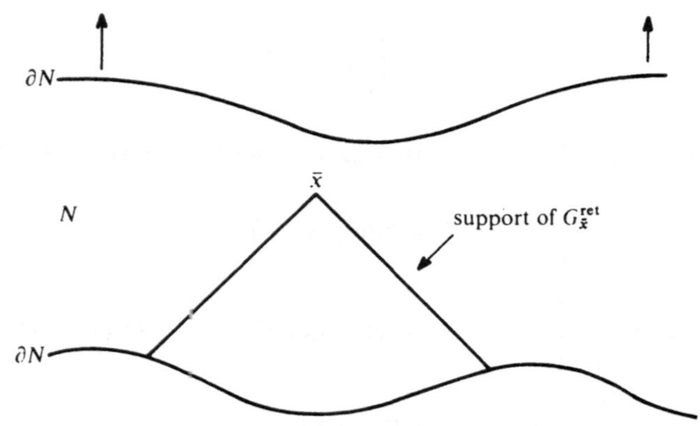


Figure 21 The displacement of the later boundary of $N$

† $^*e^0 = -e^{123}$, but $\partial N$, as the lower side of $N$, has a negative orientation.

because the integration region shrinks to zero (see (2.2.11; 3)), so that if $J$ is bounded, nothing is left over.
3. In Problem 7 it is shown, using the explicit form (2.2.11), that in fact $dF^{\text{boundary}} = 0$ and $\delta F^{\text{boundary}} = dj$, because $d(F_{|\partial N}) = 0$ and $d(*F_{|\partial N}) = -*J$ (cf. (2.2.10; 2)).
4. Although the boundary values normally propagate by $D^{\text{ret}}$ along whole light-cones, particular $F_{|\partial N}$ and $*F_{|\partial N}$ can be found for which the propagation is approximately along a light ray. In this somewhat vague approximation, the problem of how light propagates reduces to the easier problem of the motion of a massless particle. We made use of a similar simplification in (I: 5.7.15) to calculate the bending of light by a gravitational field.
5. By using $D^{\text{ret}}$, one can take $N$ as the whole half-space later than some Cauchy surface, without changing $F^{\text{ret}}$ or $F^{\text{boundary}}$, if the upper boundary is displaced to $t \to +\infty$ (Figure 21).

In our verification that $d(F^{\text{ret}} + F^{\text{boundary}}) = 0$ and $d(*F^{\text{ret}} + *F^{\text{boundary}}) = -*J$, i.e.,

$$\lim_{\bar{x}\downarrow x} F^{\text{ret}}(\bar{x}) = \lim_{\bar{x}\downarrow x} *F^{\text{ret}}(\bar{x}) = 0, \qquad \lim_{\bar{x}\downarrow x} F^{\text{boundary}}(\bar{x}) = F(x),$$

$$\lim_{\bar{x}\downarrow x} *F^{\text{boundary}}(\bar{x}) = *F(x) \quad \forall x \in \partial N,$$

we needed to know only that $d(F_{|\partial N}) = 0$ and $d(*F_{|\partial N}) = -*J_{|\partial N}$. Whereas to use (1.2.36) it is a priori necessary to assume the existence of a solution $F$, the explicit construction demonstrates the

**Existence of the Solution to $dF = 0$, $\delta F = J$, for Given Initial Values** (2.2.13)

Let $N$ be the half-space later than a Cauchy surface $\partial N$ in $(\mathbb{R}^4, \eta)$, and suppose that $H \in E_2(\partial N)$, $K \in E_2(\partial N)$, and $J \in E_1(N)$ satisfy the equations

$$dH = 0, \qquad dK = -*J_{|\partial N}.$$

Then

$$F(\bar{x}) = \int_N dG_{\bar{x}} \wedge J - \int_{\partial N} [\delta G_{\bar{x}} \wedge H - *dG_{\bar{x}} \wedge K]$$

is a solution of $dF = 0$ and $\delta F = J$ on $N$, for which $F$ approaches $H$ and $*F$ approaches $K$ as $x$ approaches $\partial N$.

**Remarks** (2.2.14)

1. On a Cauchy surface $\partial N$, $F_{|\partial N}$ and $*F_{|\partial N}$ are linearly independent, so the boundary values $H$ and $K$ do not depend on each other. If $N$ had characteristic directions, then the above construction would not work, because $\delta G_{\bar{x}}$ and $*dG_{\bar{x}}$ are not distributions on lightlike surfaces.

2. $H$ and $K$ satisfy the restriction of Maxwell's equations to a spacelike hypersurface; $\nabla \cdot \mathbf{H} = 0$ and $\nabla \cdot \mathbf{K} = \rho$. Thus they may be taken as $\mathbf{H} = \nabla \times \mathbf{v}_1$ and $\mathbf{K} = \nabla \times \mathbf{v}_2 - \nabla \int \rho(x)/|\mathbf{x} - \bar{\mathbf{x}}|$, for arbitrary $\mathbf{v}_1$ and $\mathbf{v}_2$.

Often $J$ is sufficiently localized that even $\int_{\mathbb{R}^4} dG_{\bar{x}} \wedge J$ converges. Then the remaining boundary in (1.2.36) can be taken to $t_0 = -\infty$. If $F^{\text{ret}}$ is to converge in this case, $F^{\text{boundary}}$ must also converge, and in fact it approaches a solution of the free equation ($J = 0$), because $F^{\text{ret}}$ then solves Maxwell's equations. This incoming field will be denoted by $F^{\text{in}}$, and in the time-reversed situation, using $D^{\text{adv}}$ from (2.2.8; 2), it will be called $F^{\text{out}}$.

**Definition of the Asymptotic Fields** (2.2.15)

$$F = F^{\text{in}} + F^{\text{ret}} = F^{\text{out}} + F^{\text{adv}},$$

if the integrals

$$F^{\text{ret}}(\bar{x}) = \int_{\mathbb{R}^4} dG_{\bar{x}}^{\text{ret}} \wedge J \quad \text{and} \quad F^{\text{adv}}(\bar{x}) = \int_{\mathbb{R}^4} dG_{\bar{x}}^{\text{adv}} \wedge J$$

exist; in which case

$$dF^{\text{ret}} = dF^{\text{adv}} = dF^{\text{in}} = dF^{\text{out}} = \delta F^{\text{in}} = \delta F^{\text{out}} = 0, \qquad \delta F^{\text{ret}} = \delta F^{\text{adv}} = J.$$

**Remarks** (2.2.16)

1. The existence of $F^{\text{ret}}$ is guaranteed if $J$ is localized well enough in space and does not evolve too nearly along a light-cone. We shall learn from the hyperbolic motion in §2.3 that it is possible for $F^{\text{ret}}$ to converge pointwise while $\delta F^{\text{ret}} \neq J$, because the limit as $t_0 \to -\infty$ does not exist in the appropriate topology.
2. If there exists a $T$ such that $J = 0 \, \forall t < T$, which is, of course, possible only if the total charge $Q = 0$, then $F = F^{\text{in}} \, \forall \, t < T$. Roughly speaking, $F^{\text{in}}$ is the field that existed before $J$ was switched on, and $F^{\text{out}}$ is the field that remains after $J$ has been switched off.

If $N = \mathbb{R}^4$, then $F^{\text{ret}}$ is the differential of the

**Liénard–Wiechert Potentials** (2.2.17)

$$F^{\text{ret}} = dA^{\text{ret}},$$

where

$$A_\alpha^{\text{ret}}(\bar{x}) = -\int \frac{d^3x \, J_\alpha(\bar{t} - |\bar{\mathbf{x}} - \mathbf{x}|, x)}{4\pi |\bar{\mathbf{x}} - \mathbf{x}|}.$$

**The Static Case** (2.2.18)

If $J$ is independent of time, then

$$A_\alpha^{\text{ret}}(\bar{\mathbf{x}}) = -\int \frac{d^3x\, J_\alpha(x)}{4\pi|\bar{\mathbf{x}} - \mathbf{x}|},$$

and, in particular, the Coulomb potential

$$V(\bar{\mathbf{x}}) = A_0^{\text{ret}}(\bar{\mathbf{x}}) = \int \frac{d^3x\, \rho(x)}{4\pi|\bar{\mathbf{x}} - \mathbf{x}|}, \qquad \rho(x) \equiv J^0(x) = -J_0(x).$$

A necessary condition for the existence of this potential is that $\rho$ fall off faster than $1/r^2$.

We close this section with a different application of (2.2.15), to rewrite Equation (1.3.23) for the work expended. To this end we assume that $J$ is sufficiently well localized in space so that $N$ may be set to $\mathbb{R}^4$. Then

$$P^\alpha(T) - P^\alpha(-T) = \int_{-T \le t \le T} (i^\alpha F^{\text{in}} + i^\alpha F^{\text{ret}}) \wedge {}^*J. \tag{2.2.19}$$

The energy and momentum are thus affected partly by the incoming field and partly by the field caused by $J$ alone. The latter contribution is explicitly

$$\int_{-T \le t \le T} i^\alpha F^{\text{ret}} \wedge {}^*J = \int_{-T}^{T} d^4\bar{x} \int_{-\infty}^{\infty} d^4x\, J^\beta(\bar{x})$$
$$\times \left(J^\alpha(x) \frac{\partial}{\partial \bar{x}^\beta} D^{\text{ret}}(\bar{x} - x) - J_\beta(x) \frac{\partial}{\partial \bar{x}_\alpha} D^{\text{ret}}(\bar{x} - x)\right). \tag{2.2.20}$$

If $J$ has compact support ($Q = 0$), then we can take the limit $T \to \infty$. By integration by parts, the first term on the right side of (2.2.20) contributes zero, because $\delta J = 0$. By symmetrization $x \leftrightarrow \bar{x}$, in the second term

$$\frac{\partial}{\partial \bar{x}^\alpha} D^{\text{ret}}(\bar{x} - x)$$

is changed to

$$\frac{1}{2} \frac{\partial}{\partial \bar{x}^\alpha} (D^{\text{ret}}(\bar{x} - x) - D^{\text{adv}}(\bar{x} - x)) \equiv \frac{1}{2} \frac{\partial}{\partial \bar{x}^\alpha} D(\bar{x} - x).$$

Introducing the **radiation field**

$$F^{\text{rad}} = F^{\text{ret}} - F^{\text{adv}} = F^{\text{out}} - F^{\text{in}}$$

allows one to write

$$P_\alpha(\infty) - P_\alpha(-\infty) = \int_{\mathbb{R}^4} (i_\alpha F^{\text{in}} + i_\alpha \tfrac{1}{2} F^{\text{rad}}) \wedge {}^*J,$$

$$-\int i_\alpha F^{\text{rad}} \wedge {}^*J = \int d^4\bar{x}\, d^4x\, J^\beta(\bar{x}) J_\beta(x) \frac{\partial}{\partial \bar{x}^\alpha} D(\bar{x} - x).$$

(2.2.21)

The convolution (2.2.21) becomes a product when Fourier-transformed†:

$$\tilde{J}_\beta(k) = \int d^4x\, e^{i\langle k|x\rangle} J_\beta(x) = \tilde{J}_\beta{}^*(-k).$$

Since $D$ has the Fourier integral representation

$$D(x) = i(2\pi)^{-3} \int d^4k\, \delta(k^2)(\Theta(k^0) - \Theta(-k^0)) e^{i\langle k|x\rangle} \qquad (2.2.22)$$

(Problem 8), there results an expression for the

**Energy and Momentum Lost by the Radiative Reaction of the Field** (2.2.23)

If $F^{\text{in}} = 0$, $J \in E_1^0(\mathbb{R}^4)$, then

$$P^\alpha(\infty) - P^\alpha(-\infty) = (2\pi)^{-3} \int d^4k\, \Theta(k^0)\delta(k^2)(|\tilde{\mathbf{J}}(k)|^2 - |\tilde{\mathbf{J}}^0(k)|^2) k^\alpha.$$

**Remarks** (2.2.24)

1. Because $\tilde{F}^{\text{ret}}_{\alpha\beta}(k) = (ik_\alpha \tilde{J}_\beta(k) - ik_\beta \tilde{J}_\alpha(k))/k^2$, the field reflects the frequency distribution of the source. This allows the integrand to be interpreted as the energy and momentum lost to the field with wave vector $k$. In particular, as the time becomes infinite, the factor $\delta(k^2)$ makes the loss go mainly into the free field, characterized by $|\mathbf{k}|^2 = (k^0)^2$.
2. If the sources are strictly periodic, the assumption that $J \in E_1^0$ is violated. This shows up as a delta function in $\tilde{J}$ and a $\delta^2$ in (2.2.23). The physical significance is that a periodic process radiates an infinite amount of energy in an infinite time (cf. (2.3.29; 7)).
3. In Fourier-transformed space, $\delta J = 0$ reads

$$\tilde{\mathbf{J}} \cdot \mathbf{k} = \tilde{J}^0 k^0.$$

If $|\mathbf{k}| = |k^0|$, then $|\mathbf{J}| \geq |J^0|$. Since with $\alpha = 0$ the rest of the integrand is nonnegative, $P^0(\infty) \geq P^0(-\infty)$. This is a consequence of the positivity

† We shall write * on the right side to denote the complex conjugate.

of the energy. If $F^{\text{in}} = 0$, then $P^0(-\infty) = 0$; thus energy can only be released from the current to the field.

## Problems (2.2.25)

1. Write $F \pm i{}^*F$ out explicitly, and show that $E_1 = B_2 =$ an arbitrary function of $t + z$, and $E_2 = B_1 = E_3 = B_3 = 0$ solves the equations $dF = \delta F = 0$. Show that the field can be written as $du \wedge f$, where $\langle du | du \rangle = \langle du | f \rangle = 0$.
2. Calculate the integral (2.2.5) for $D^{\text{ret}}$, using the integration path (2.2.6; 3).
3. Use $D^{\text{ret}}$ to find the Green function for Laplace's equation in three dimensions.
4. Calculate $\delta A$ for the $A$ of (2.2.9).
5. Calculate $\delta G_{\bar{x}} \wedge F - {}^*dG_{\bar{x}} \wedge {}^*F$ with the $G_{\bar{x}}$ of (2.2.7) and $F = \frac{1}{2}e^{\alpha\beta}F_{\alpha\beta}$.
6. Show that
$$\lim_{t \to 0} \int d^3x\, f(x) \frac{\partial}{\partial x^\alpha} \frac{\delta(t - r)}{4\pi r} = \delta^0{}_\alpha f(0).$$
7. Write (2.2.11) for $\partial N = (0, \mathbf{x})$ explicitly in terms of the components $\mathbf{E}$ and $\mathbf{B}$ of $F^{\text{boundary}}$, and calculate $E_{k,k}$ and $B_{k,k}$.
8. Find the Fourier integral decomposition of the $D$ of (2.2.22).

## Solutions (2.2.26)

1. $$F + i{}^*F = \tfrac{1}{2}(E_1 + B_2)(dt + dx_3)(dx_1 - i\,dx_2) + \tfrac{1}{2}(E_1 - B_2)(dt - dx_3)(dx_1 + i\,dx_2)$$
$$+ \tfrac{1}{2}(E_2 + B_3)(dt + dx_1)(dx_2 - i\,dx_3) + \tfrac{1}{2}(E_2 - B_3)(dt - dx_1)(dx_2 + i\,dx_3)$$
$$+ \tfrac{1}{2}(E_3 + B_1)(dt + dx_2)(dx_3 - i\,dx_1) + \tfrac{1}{2}(E_3 - B_1)(dt - dx_2)(dx_3 + i\,dx_1).$$
The individual terms of this sum are of the required form.
2. If $t \le 0$ (respectively $\ge 0$), then the path of integration can be closed in the upper (resp. lower) complex $k^0$-plane, so that
$$-\frac{1}{2\pi} \int_{\frown\!\smile\!\frown} dk^0 \frac{e^{-ik^0 t}}{(k^0 - k)(k^0 + k)} = \frac{\Theta(t)}{k} \sin kt.$$
There remains
$$(2\pi)^{-3} \int d^3k\, e^{i\mathbf{k}\cdot\mathbf{x}} \frac{\Theta(t)}{k} \sin kt = \frac{\Theta(t)}{2\pi^2 r} \int_0^\infty dk \sin kr \sin kt = \frac{\delta(r - t)}{4\pi r} \Theta(t).$$
3. Because $(\partial^2/\partial t^2) - \Delta)D^{\text{ret}}(x) = \delta^4(x)$, the function $\int_{-\infty}^{\infty} dt\, D^{\text{ret}}(x) = 1/4\pi r$ satisfies the equation $-\Delta(1/4\pi r) = \delta^3(x)$.
4. $$-\delta A^{\text{ret}}(\bar{x}) = \int_N d^4x\, J_\alpha(x) \frac{\partial}{\partial x^\alpha} D^{\text{ret}}(\bar{x} - x) \equiv \int_N d({}^*JD_{\bar{x}}^{\text{ret}}) = \int_{\partial N} {}^*JD_{\bar{x}}^{\text{ret}}.$$

5. Using the abbreviation $D_{,\alpha} = (\partial/\partial x^\alpha)D^{\text{ret}}(\bar{x} - x) = -(\partial/\partial \bar{x}^\alpha)D^{\text{ret}}(\bar{x} - x)$ and Rule (1.2.18)(b),

$$dG_{\bar{x}} = \tfrac{1}{2}\bar{e}_{\alpha\beta}e^\gamma \wedge {}^*e^{\alpha\beta}D_{,\gamma} = -\bar{e}^{\alpha\beta}D_{,\alpha}{}^*e_\beta$$

$${}^*dG_{\bar{x}} \wedge {}^*F = -\tfrac{1}{2}\bar{e}^{\alpha\beta}D_{,\alpha}e_\beta \wedge {}^*e^{\sigma\tau}F_{\sigma\tau} = \bar{e}^{\alpha\beta}D_{,\alpha}F_{\beta\sigma}{}^*e^\sigma$$

$$\delta G_{\bar{x}} = \tfrac{1}{2}\bar{e}^{\alpha\beta}{}^*(dD \wedge e_{\alpha\beta}) = \tfrac{1}{2}\bar{e}^{\alpha\beta}D_,{}^{\gamma}{}^*e_{\gamma\alpha\beta}$$

$$\delta G_{\bar{x}} \wedge F = \tfrac{1}{4}\bar{e}_{\alpha\beta}D_{,\gamma}F_{\sigma\tau}e^{\sigma\tau} \wedge {}^*e^{\gamma\alpha\beta} = -\bar{e}_{\alpha\beta}D_{,\gamma}[F^{\alpha\gamma}{}^*e^\beta - \tfrac{1}{2}F^{\alpha\beta}{}^*e^\gamma].$$

If these equations are combined, then

$$-\delta G_{\bar{x}} \wedge F + {}^*dG_{\bar{x}} \wedge {}^*F = \bar{e}^{\alpha\beta}\left[\frac{\partial}{\partial \bar{x}^\alpha} D(\bar{x} - x)F_{\gamma\beta}(x){}^*e^\gamma - \frac{\partial}{\partial \bar{x}_\gamma} D(\bar{x} - x)F_{\alpha\gamma}(x){}^*e_\beta + \frac{1}{2}\frac{\partial}{\partial \bar{x}_\gamma} D(\bar{x} - x)F_{\alpha\beta}(x){}^*e^\gamma\right].$$

6. $\alpha = j = 1, 2, 3$: $\displaystyle\int_0^\infty dr\, d\Omega(f(0)r^2 + f_{,k}(0)r^2x_k + \cdots)\frac{-\partial}{\partial x_j}\frac{\delta(t - r)}{4\pi r}$

$$= \int_0^\infty dr\, d\Omega(f(0)2x_j + f_{,k}(0)(2x_jx_k + r^2\delta^j{}_k) + \cdots)\frac{\delta(t - r)}{4\pi r} \to 0 \qquad \text{as } t \to 0.$$

$$\alpha = 0: \int_0^\infty dr\, d\Omega(f(0)r^2 + f_{,k}(0)r^2x_k + \cdots)\frac{1}{4\pi r}\frac{-\partial}{\partial r}\delta(t - r) = f(0) + O(t) \to f(0)$$

as $t \to 0$.

7.
$$E_k(\bar{x}) = \int d^3x[\dot{D}^{\text{ret}}(\bar{x} - x)E_k(x) + \varepsilon_{klm}D^{\text{ret}}_{,l}(\bar{x} - x)B_m(x)]$$

$$B_k(\bar{x}) = \int d^3x[\dot{D}^{\text{ret}}(\bar{x} - x)B_k(x) - \varepsilon_{klm}D^{\text{ret}}_{,l}(\bar{x} - x)E_m(x)]$$

$$E_{k,k}(\bar{x}) = \int d^3x\, \dot{D}^{\text{ret}}(\bar{x} - x)E_{k,k}(x) = \frac{\partial}{\partial \bar{t}}\int d^3x\, D^{\text{ret}}(\bar{x} - x)J^0(x)$$

$$B_{k,k}(\bar{x}) = \int d^3x\, \dot{D}^{\text{ret}}(\bar{x} - x)B_{k,k}(x) = 0,$$

since **E** and **B** must satisfy the restriction of Maxwell's equations to $\partial N$.

8. $D(x) = (D^{\text{ret}}(x) - D^{\text{adv}}(x))$

$$= (2\pi)^{-4}\int d^4k\, e^{i\langle k|x\rangle}\lim_{\varepsilon\downarrow 0}\left[\frac{1}{|\mathbf{k}|^2 - (k^0 + i\varepsilon)^2} - \frac{1}{|\mathbf{k}|^2 - (k^0 - i\varepsilon)^2}\right]$$

$$= (2\pi)^{-3}i\int d^4k\, e^{i\langle k|x\rangle}\delta(k^2)[\Theta(k^0) - \Theta(-k^0)].$$

## 2.3 The Field of a Point Charge

*There is an expression in closed form for the field of a point charge undergoing any given motion. It contains all information about the radiation emitted by an accelerating charge.*

Our first application of the formulas derived in the preceding section will be to calculate $A^{\text{ret}}$ and $F^{\text{ret}}$ for the current (1.3.25; 2); $J$ and $D^{\text{ret}}$ will be respectively a four-dimensional and a one-dimensional delta function, which allow the integrals over $d^4x$ and $ds$ to be done. Using the rule $\delta(f(x)) = \sum_i \delta(x - x_i)|f'(x_i)|^{-1}$, where the $x_i$ are the zeroes of $f$, we obtain

$$A_\alpha^{\text{ret}}(\bar{x}) = -\int d^4x\, D^{\text{ret}}(\bar{x} - x)J_\alpha(x) = -e\int_{-\infty}^{\infty} ds\, \dot{z}_\alpha(s)D^{\text{ret}}(\bar{x} - z(s))$$

$$= \frac{-e}{2\pi}\int_{-\infty}^{z^0(s)<\bar{x}^0} ds\, \dot{z}_\alpha(s)\delta((\bar{x} - z(s))^2) = \frac{e}{4\pi}\frac{\dot{z}_\alpha(s_0)}{\langle \dot{z}(s_0)|\bar{x} - z(s_0)\rangle}, \tag{2.3.1}$$

$$(\bar{x} - z(s_0))^2 = 0, \qquad \bar{x}^0 > z^0(s_0).$$

**Remarks** (2.3.2)

1. The negative sign makes (2.3.1) consistent with $\langle \dot{z}(s_0)|\bar{x} - z(s_0)\rangle < 0$.
2. $s_0$ is a function of $\bar{x}$ (see Figure 22).
3. If we consider $\bar{x}_\alpha$, $z_\alpha(s_0(\bar{x}))$, and $\dot{z}_\alpha(s_0(\bar{x}))$ as the components of vector fields $x$, $z$, and $\dot{z}$, then (2.3.1) may be written without indices as

$$A^{\text{ret}} = \frac{e}{4\pi}\frac{\dot{z}}{\langle \dot{z}|x - z\rangle}.$$

After the $x$-integration the coordinate will be called $x$, rather than $\bar{x}$.

The way that $s_0$ depends on $x$ must first be known before $F^{\text{ret}}$ can be calculated from $A^{\text{ret}}$. This dependence can be determined by noting that $x - z \in E_1(\mathbb{R}^4)$ is a null field:

$$0 = \frac{1}{2}\frac{\partial}{\partial x^\beta}(x^\alpha - z^\alpha(s_0(x)))(x_\alpha - z_\alpha(s_0(x))) = x_\beta - z_\beta - \dot{z}^\alpha\frac{\partial s_0}{\partial x^\beta}(x_\alpha - z_\alpha)$$

$$\Rightarrow ds_0 = (x - z)\langle \dot{z}|x - z\rangle^{-1}. \tag{2.3.3}$$

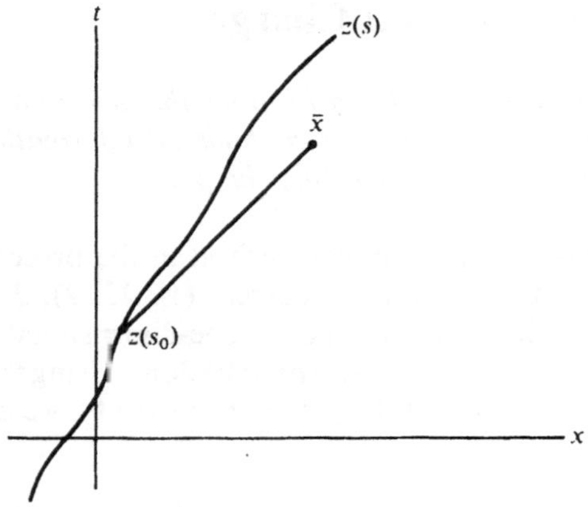


Figure 22 Determination of $z(s_0)$ on the world-line of a charge

The exterior differential of the constituents of $A$ can be calculated from (2.3.3) as

$$
\begin{aligned}
d\dot{z} &= ds_0 \wedge \ddot{z} = (x - z) \wedge \ddot{z}\langle \dot{z} | x - z\rangle^{-1}, \\
\frac{\partial}{\partial x^\beta}\dot{z}^\alpha(x_\alpha - z_\alpha) &= \ddot{z}^\alpha \frac{\partial s_0}{\partial x^\beta}(x_\alpha - z_\alpha) + \dot{z}_\beta - \dot{z}^\alpha \dot{z}_\alpha \frac{\partial s_0}{\partial x^\beta} \\
\Rightarrow d\langle \dot{z} | x - z\rangle &= \dot{z} + (x - z)\frac{\langle \ddot{z} | x - z\rangle + 1}{\langle \dot{z} | x - z\rangle},
\end{aligned}
\tag{2.3.4}
$$

where we have used the normalization $\dot{z}^\alpha \dot{z}_\alpha \equiv \dot{z}^2 \equiv \langle \dot{z} | \dot{z}\rangle = -1$ and have also considered $\dot{z}_\alpha(s_0(x))$ as the components of the vector field $\dot{z}$. Combining the above formulas produces

**The Retarded Field of a Point Charge** (2.3.5)

$$
\begin{aligned}
F^{\text{ret}} &= \frac{e}{4\pi}(\langle \dot{z} | x - z\rangle^{-1}\, d\dot{z} + \langle \dot{z} | x - z\rangle^{-2} \dot{z} \wedge d\langle \dot{z} | x - z\rangle) \\
&= \frac{e}{4\pi}\langle \dot{z} | x - z\rangle^{-2}\left(\dot{z}\frac{\langle \ddot{z} | x - z\rangle + 1}{\langle \dot{z} | x - z\rangle} - \ddot{z}\right) \wedge (x - z).
\end{aligned}
$$

**Remarks** (2.3.6)

1. By using the normalization $\dot{z}^2 = -1$ for the world-line $z(s)$ we assume that the charge never reaches the speed of light.

2. The 2-form $F^{\text{ret}}$ is of a special form, as the exterior product of a null vector field $x - z$ with another vector field $\dot{z}\langle \dot{z}|x - z\rangle^{-1}(\langle \ddot{z}|x - z\rangle + 1) - \ddot{z}$. In contrast to the field in (2.2.2; 3) with discontinuities, the interior product of these fields is not 0 but 1.
3. $F^{\text{ret}}$ is the sum of two fields, one of which, $F^{(\ddot{z})}$, contains the terms proportional to $\ddot{z}$, and the other, $F^{(\dot{z})}$, contains only $\dot{z}$. The two fields have different asymptotic behavior, as expected on dimensional grounds; $F^{(\dot{z})}$, the field of the near zone, falls off as a Coulomb field $(1/r^2)$, whereas $F^{(\ddot{z})}$, the field of the far zone, falls off only as $1/r$.

We note some of the special

**Properties of $F^{\text{ret}}$ for a Point Charge** (2.3.7)

(a) $F^{\text{ret}} \wedge F^{\text{ret}} = 0 = F^{(\dot{z})} \wedge F^{(\dot{z})} = F^{(\ddot{z})} \wedge F^{(\ddot{z})}$
(b) $F^{\text{ret}} \wedge (x - z) = 0 = F^{(\dot{z})} \wedge (x - z) = F^{(\ddot{z})} \wedge (x - z)$
(c) $i_{x-z}F^{(\ddot{z})} = *((x - z) \wedge *F^{(\ddot{z})}) = 0$
(d) $F^{(\ddot{z})} \wedge *F^{(\ddot{z})} = 0 = F^{(\dot{z})} \wedge *F^{(\ddot{z})} = F^{\text{ret}} \wedge *F^{(\ddot{z})}$.

**Proof**

(a) This holds for any element of $E_2$ that is an exterior product of two vectors.
(b) $F^{\text{ret}}$, $F^{(\dot{z})}$, and $F^{(\ddot{z})}$ contain $x - z$ as a factor.
(c) The interior product of the two factors of $F^{(\ddot{z})}$ vanishes, and $x - z$ is a null field.
(d) This follows from (c) because of the factor $x - z$. □

**Remarks** (2.3.8)

1. Since $F \wedge F \sim \mathbf{E} \cdot \mathbf{B}$, Property (a) implies that the electric and magnetic fields are always perpendicular to each other, and this is also a property of $F^{(\dot{z})}$ and $F^{(\ddot{z})}$ considered separately. Therefore it requires more than one charge to produce a magnetic field parallel to an electric field.
2. Because $*((x - z) \wedge F) = i_{x-z}*F$, Property (b) implies that

$$(x - z)^{\beta}(*F^{\text{ret}})_{\alpha\beta} = 0.$$

For $\alpha = 0$, this means that $\mathbf{B}$ is also perpendicular to the 3-vector connecting the reference point (where the field is measured) to the position of the charge at the retarded time $s_0$.
3. If $F^{(\ddot{z})}$ is considered separately, then Remark 2 still applies, and because of Property (c) it applies even if $F$ and $*F$, that is, $\mathbf{B}$ and $\mathbf{E}$, are interchanged. Therefore the electric field stemming from $F^{(\ddot{z})}$, which dominates in the far zone, is also perpendicular to the 3-vector from the reference point to the position of the charge at the retarded time $s_0$.

4. According to Property (d), because $F \wedge {}^*F \sim |\mathbf{E}|^2 - |\mathbf{B}|^2$, in the far zone $|\mathbf{E}^{(\ddot{z})}| = |\mathbf{B}^{(\ddot{z})}|$, and if $z(s_0) = 0$, then

$$\mathbf{E}^{(\ddot{z})} = \left[\mathbf{B}^{(\ddot{z})} \times \frac{\mathbf{x}}{r}\right] \quad \text{and} \quad \mathbf{B}^{(\ddot{z})} = \left[\frac{\mathbf{x}}{r} \times \mathbf{E}^{(\ddot{z})}\right].$$

The Lagrangian $F^{\text{ret}} \wedge {}^*F^{\text{ret}} = F^{(\dot{z})} \wedge {}^*F^{(\dot{z})}$ has contributions only from the near zone.

### The Field in the Rest-Frame of the Particle (2.3.9)

Let us choose the Lorentz system in which $z(s_0) = 0$, $\dot{z}(s_0) = (1, 0, 0, 0)$, and $\ddot{z}(s_0) = (0, \ddot{\mathbf{z}})$. Then for $x$ in the positive light-cone of $z(s_0)$, the field (2.3.5), written in components, is expressed as

$$\mathbf{E} = \frac{e}{4\pi r}\left\{-\ddot{\mathbf{z}} + \frac{\mathbf{x}}{r^2}(1 + (\mathbf{x} \cdot \ddot{\mathbf{z}}))\right\}$$

$$\mathbf{B} = \frac{e}{4\pi r}\left[\ddot{\mathbf{z}} \times \frac{\mathbf{x}}{r}\right].$$

The magnetic lines of force circle the charge, and the electric lines of force leave the charge initially in the radial direction, and later bend to become perpendicular to **B** and to **x** in the far zone (Figure 23)

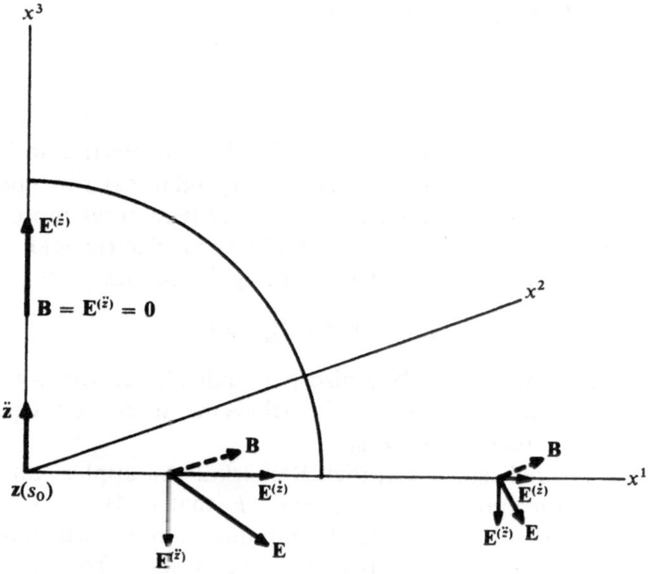


Figure 23 The fields in the near and far zones

**Warning** (2.3.10)

The description in (2.3.9) is not an instantaneous picture of the field, but specifies the field at a given position $\mathbf{x}$ at the time $t = |\mathbf{x}|$. The field in the spacelike section at $t$ = constant does not depend on $\dot{z}$ and $\ddot{z}$ at a single value of $s$, but on a whole segment of the particle's trajectory. It can be written down explicitly only for special kinds of trajectories.

**Examples** (2.3.11)

1. Uniform motion. Let

$$z(s) = \frac{s}{\sqrt{1 - v^2}}(1, v, 0, 0) = \dot{z}s.$$

Then

$$(x - z(s_0))^2 = x^2 - s_0^2 - 2\langle \dot{z} | x \rangle s_0,$$

and thus

$$s_0 = -\langle \dot{z} | z(s_0) \rangle = -\langle \dot{z} | x \rangle - \sqrt{\langle \dot{z} | x \rangle^2 + x^2}.$$

This means that

$$\begin{aligned} A^{\text{ret}} &= \frac{-e}{4\pi} \frac{\dot{z}}{\sqrt{\langle \dot{z} | x \rangle^2 + x^2}} \\ &= \frac{-e(-1, v, 0, 0)}{4\pi \sqrt{1 - v^2}} \left[ \left( \frac{x_1 - vt}{\sqrt{1 - v^2}} \right)^2 + x_2^2 + x_3^2 \right]^{-1/2}, \end{aligned}$$

$$\begin{aligned} F^{\text{ret}} &= \frac{e}{4\pi} \frac{x \wedge \dot{z}}{(\langle \dot{z} | x \rangle^2 + x^2)^{3/2}} \\ &= \frac{e}{4\pi} \begin{vmatrix} 0 & x_1 - vt & x_2 & x_3 \\ -x_1 + vt & 0 & -vx_2 & -vx_3 \\ -x_2 & vx_2 & 0 & 0 \\ -x_3 & vx_3 & 0 & 0 \end{vmatrix} \\ &\quad \times \frac{1 - v^2}{[(x_1 - vt)^2 + (1 - v^2)(x_2^2 + x_3^2)]^{3/2}}. \end{aligned}$$

Note that

(i) If $v = 0$, the above expression reduces to the usual Coulomb field, $A_0 = e/4\pi r$, $\mathbf{A} = \mathbf{0}$; $\mathbf{E} = e\mathbf{x}/4\pi r^3$, $\mathbf{B} = \mathbf{0}$. When $v$ is not zero, the Coulomb field is simply transformed according to the transformation law for $E_2(\mathbb{R}^4)$ (cf. (I: 5.2.7)), which is automatically taken into account by the covariant notation.

(ii) The electric field points at the simultaneous position of the charge, not to its retarded position. This fact was used in (1.1.5).

(iii) The denominator contains the spatial distance to $z(s_0)$ in the rest-frame of the charge, viz.,

$$\langle \dot{z} | x - z \rangle = (\langle \dot{z} | x \rangle^2 + x^2)^{1/2} = \left( \frac{(x_1 - vt)^2}{1 - v^2} + x_2^2 + x_3^2 \right)^{1/2}.$$

This is equal neither to the distance $((x_1 - vt)^2 + x_2^2 + x_3^2)^{1/2}$ from the simultaneous position of the charge, nor to the distance from $z(s_0)$, which would be $\langle dt | x - z \rangle$. On the spacelike section $t = 0$, $\langle \dot{z} | x - z \rangle$ equals $r = |\mathbf{x}|$ for the points perpendicular to the direction of motion (i.e., $x_1 = 0$), and is otherwise greater than $r$. The increase of the denominator is compensated for by the factor $1/\sqrt{1 - v^2}$ of the Lorentz transformation, and the net effect is that the static Coulomb potential is altered as shown in Figure 24; $A^0$ is increased perpendicular to the motion and left unchanged in the $x_1$-direction. The oblately squashed potential produces an electric field increased with respect to $\mathbf{x}/4\pi r^3$ if $\mathbf{x} \perp \mathbf{v}$ and decreased if $\mathbf{x} \parallel \mathbf{v}$. The increased range of the Coulomb field makes a charged particle at nearly the speed of light cause greater ionization.

(iv) Of course, the decrease in the field in the forward direction changes into an increase when the charge gets near the reference point. At the time $t = r$, the denominator $\langle \dot{z} | x - z \rangle$ becomes

$$\frac{(x_1 - vr)^2}{1 - v^2} + x_2^2 + x_3^2 = \frac{(r - x_1 v)^2}{1 - v^2}$$

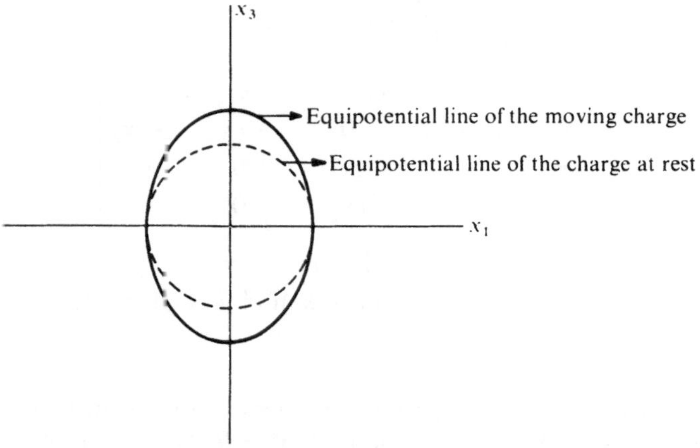


Figure 24 Lorentz contraction of the Coulomb potential

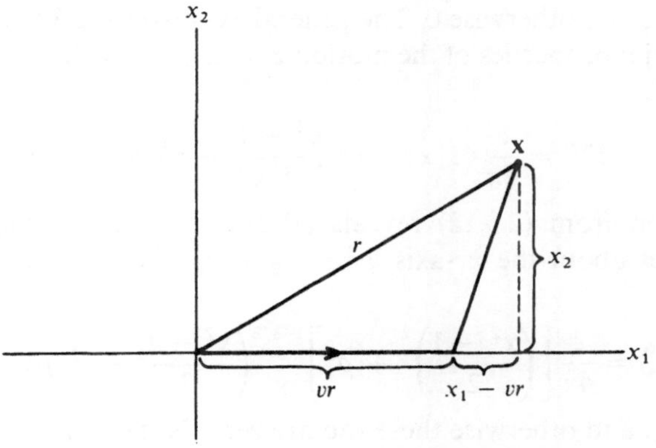


Figure 25 The lengths involved in the field of a point charge

and for $x_2 = x_3 = 0$ is decreased compared with $r^2$ by the factor $(1 - v)/(1 + v)$ (Figure 25). Incidentally, this factor causes the radiation to bunch strongly in the forward direction (Problem 3). The expression $\langle \dot{z} | x - z \rangle$ must be evaluated at the retarded time, and radiation, once emitted, is not affected if the particle flies off afterwards in some different direction, never to come near the reference point.

2. Uniform acceleration. Hyperbolic motion, like that of a charged particle in a constant electric field (cf. I §4.2), is characterized by $z(s)^2$ = constant. Let

$$z(s) = (\text{Sinh } s, \text{Cosh } s, 0, 0) = \ddot{z}, \dot{z} = (\text{Cosh } s, \text{Sinh } s, 0, 0) = \dddot{z}.$$

Then $s_0$ can be calculated from $(x - z(s_0))^2 = x^2 + 1 - 2\langle x | z \rangle = 0$. Although the equation for $s_0$ is transcendental, it is easy to calculate that

$$\dot{z}(s_0) = \left( \frac{t\xi - x_1(1 + x^2)}{2(t^2 - x_1^2)}, \frac{x_1\xi - t(1 + x^2)}{2(t^2 - x_1^2)}, 0, 0 \right),$$

where

$$\xi \equiv -2\langle \dot{z} | x - z \rangle = [(1 + x^2)^2 + 4(t^2 - x_1^2)]^{1/2}. \qquad (2.3.12)$$

Then

$$A^{\text{ret}} = \frac{-e}{4\pi} \frac{1}{\xi(t^2 - x_1^2)} [t\xi - x_1(1 + x^2), x_1\xi - t(1 + x^2), 0, 0]$$

if $t > -x_1$, and otherwise 0. The general expression (2.3.5) simplifies due to the special properties of the motion $\ddot{z} = z$, $\langle \dot{z} | z \rangle = 0$, and $\langle x | z(s_0) \rangle = (x^2 + 1)/2$:

$$F^{\text{ret}} = \frac{\epsilon}{4\pi} \langle \dot{z} | x \rangle^{-2} \left( \dot{z} \frac{x^2 + 1}{2\langle \dot{z} | x \rangle} - z \right) \wedge (x - z).$$

Substitution from (2.3.12) reveals (Problem 1) that, using cylindrical coordinates about the $x_1$-axis, $\rho^2 = x_2^2 + x_3^2$, the only nonzero components are

$$(E_1, E_\rho, B_\varphi) = \frac{e}{4\pi} \left[ \left( \frac{x^2 - 1}{2} \right)^2 + \rho^2 \right]^{-3/2} \left( \frac{x^2 - 1}{2} - \rho^2, \rho x_1, \rho t \right) \quad (2.3.13)$$

if $t > -x_1$, and otherwise these too are zero. Note that

(i) The part of space where $t + x_1 \leq 0$ has no field, because it can not be connected to the world-line by any light-cone (Figure 26):

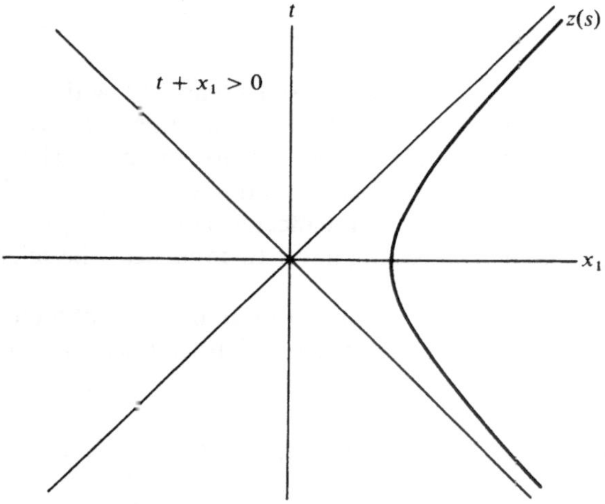


Figure 26 Hyperbolic motion

(ii) As a practical matter, motion is never strictly hyperbolic. If hyperbolic motion is combined with uniform motion at a velocity $-v$ (respectively $v$) at the point $(-v, 1, 0, 0)/\sqrt{1 - v^2}$ (resp. $(v, 1, 0, 0)/\sqrt{1 - v^2}$), then the results of Examples 1 and 2 in the appropriate regions are simply combined (see Figure 27). Interestingly, this field does not converge to (2.3.13) as $v \to 1$, as there remains a contribution from the initial uniform motion:

$$\lim_{v \to 1} \frac{\epsilon}{4\pi} \frac{(x_1 - \sqrt{1 - v^2} + vt, \rho, -v\rho)(1 - v^2)}{[(x_1 - d + vt)^2 + (1 - v^2)\rho^2]^{3/2}}$$

$$= \frac{e}{2\pi} \frac{\delta(x_1 + t)}{1 + \rho^2} (0, \rho, -\rho)$$

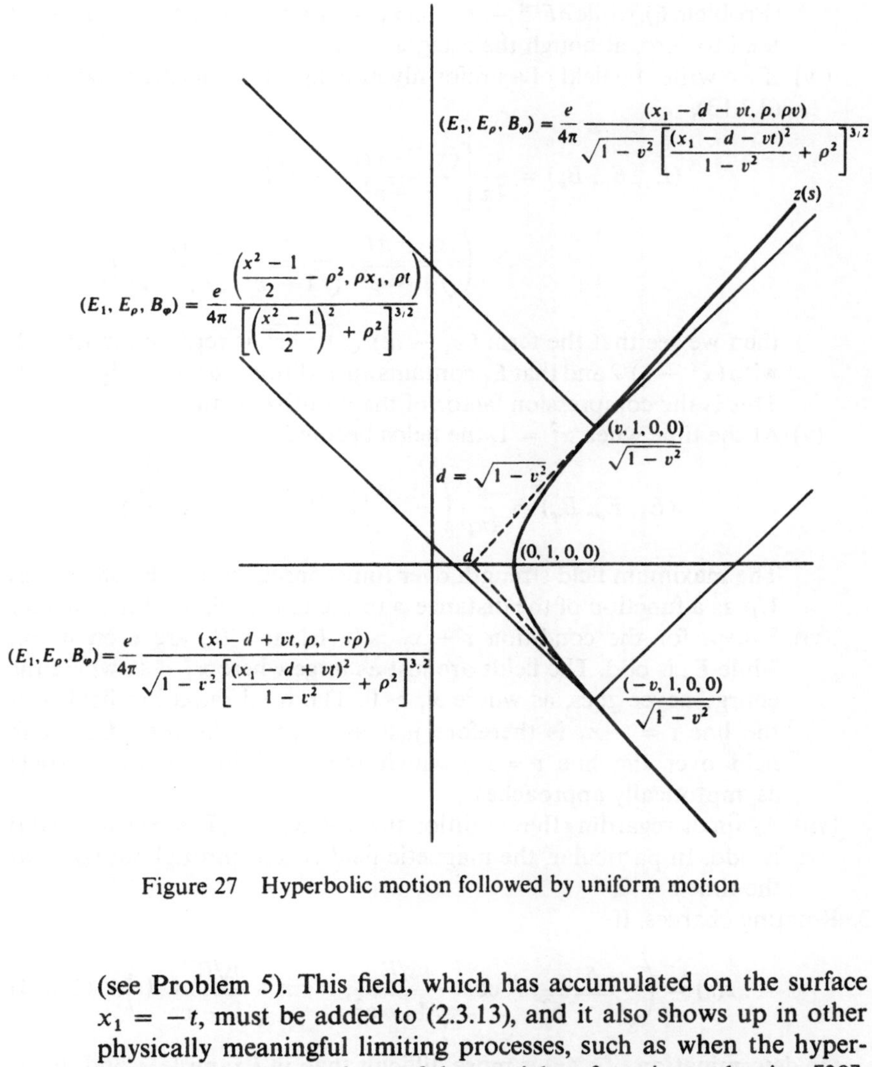


Figure 27 Hyperbolic motion followed by uniform motion

(see Problem 5). This field, which has accumulated on the surface $x_1 = -t$, must be added to (2.3.13), and it also shows up in other physically meaningful limiting processes, such as when the hyperbolic motion is that of one of the particles of a pair-production [38].

(iii) Equation (2.3.13) is not acceptable mathematically, because

$$\delta F^{\text{ret}} - J = \frac{e}{\pi}\frac{\delta(x_1 + t)}{(1 + \rho^2)^2}(1, 1, 0, 0).$$

The apparent surface current vanishes if the field of (ii) is added to $F^{\text{ret}}$. Taken by itself, $F^{\text{ret}}$ is neither the limit of the $F$ of (ii) as $v \to 1$, nor the limit as $T \to \infty$ of an $F_T^{\text{ret}}$ that would result from using the half-space $t \geq -T$ for $N$ in (2.2.9). As noted in (2.2.10; 2), $\delta F_T^{\text{ret}} - J_{|N} = -dj$, but the right side of this equation approaches zero as $T \to \infty$

(Problem 6), while $\delta F^{\text{ret}} - J \sim \delta(x_1 + t)$. Its integral over $x_1$ does not tend to zero, although the integral of $dj$ does.

(iv) If we write the field of a uniformly moving charge in the notation of (2.3.13),

$$(E_1, E_\rho, B_\varphi) = \frac{e}{4\pi}\left[\frac{(x_1 - vt)^2}{1 - v^2} + \rho^2\right]^{-3/2} \times \left(\frac{x_1 - vt}{\sqrt{1 - v^2}}, \frac{\rho}{\sqrt{1 - v^2}}, \frac{v\rho}{\sqrt{1 - v^2}}\right),$$

then we see that the term $(x_1 - vt)/\sqrt{1 - v^2}$ is replaced in (2.3.13) with $(x^2 - 1)/2$ and that $E_1$ contains an additional term $-\rho^2[\ldots]^{-3/2}$. This is the compression factor of the intuitive picture (1.1.5).

(v) At the time when $x^2 = 1$, the fields become

$$(E_1, E_\rho, B_\varphi) = \frac{e}{4\pi\rho}\left(-1, \frac{x_1}{\rho}, \sqrt{1 + (1 + x_1^2)/\rho^2}\right).$$

The maximum field strength over time consequently falls off only as $1/\rho$ as a function of the distance $\rho$ to the line of flight of the charge.

(vi) Except for the condition $t + x_1 > 0$, $E_1$ and $B_\varphi$ are even in $x_1$, while $E_\rho$ is odd. The fields are just as large where $x_1 < 0$, where the charge never goes, as where $x_1 > 0$. The total radiation field over the line $t = -x_1$ is therefore just as large as the total Coulomb field over the line $t = x_1$, which the world-line of the particle asymptotically approaches.

(vii) Again disregarding the condition that $t + x_1 > 0$, **E** is even in $t$ and **B** is odd. In particular, the magnetic field is zero throughout space at the time $t = 0$.

3. Rotating charges. If

$$z(s) = \left(\frac{s}{\sqrt{1 - v^2}}, R\cos\frac{vs/R}{\sqrt{1 - v^2}}, R\sin\frac{vs/R}{\sqrt{1 - v^2}}, 0\right), \quad (2.3.14)$$

the determination of $s_0(x)$ is more difficult than in Example 2, and therefore one usually looks only at the limit $v \to 0$, $R \to 0$, $e \to \infty$, such that $v/R \to \omega$ and $e = 1/R\omega^2$. In order not to be encumbered with the infinite Coulomb field that results, one considers two opposite charges in mirror-image paths about the origin (Hertz's dipole). In this limit, $s_0 = t - r$, $z - x = (r, \mathbf{x})$, $\dot{z} = (1, 0, 0, 0)$, $e\ddot{z} = -(0, \cos\omega(r - t), \sin\omega(r - t), 0)$, and the fields become

$$\mathbf{B} = \frac{e[\ddot{\mathbf{z}} \times \mathbf{x}]}{4\pi r^2}, \qquad \mathbf{E} = \frac{[\mathbf{B} \times \mathbf{x}]}{r},$$
$$e\ddot{\mathbf{z}} = (\cos\omega(r - t), \qquad \sin\omega(r - t), 0). \quad (2.3.15)$$

These examples illustrate $F$ in three representative cases, for free motion and linear and circular acceleration. Often of greater practical interest than the field strengths are the energy and momentum forms created by the charge. These will be sums (1.3.22) of two terms quadratic in $F$. Since each component of $F$ is itself a sum of six fractions, it seems that blind substitution would produce 72 fractions. Fortunately, the special structure of the $F$ of a point charge can be used to reduce the algebraic complexity.

**The Energy-Momentum Forms of the Field of a Point Charge** (2.3.16)

With the rules (1.2.18), $\mathscr{T}_\alpha$ can be rewritten

$$\mathscr{T}_\alpha = *((i_\alpha F) \wedge *F - \tfrac{1}{2} i_\alpha (F \wedge *F)) = \frac{e_\alpha}{2} *(F \wedge *F) - i_{i_\alpha F} F.$$

Our $F$ is of the form

$$F = \frac{e}{4\pi} \langle \dot{z} | x - z \rangle^{-2} v \wedge n, \qquad v = \dot{z} \frac{1 + \langle \ddot{z} | x - z \rangle}{\langle \dot{z} | x - z \rangle} - \ddot{z}, \qquad n = (x - z),$$

and so the invariants are

$$n^2 = 0, \qquad \langle n | v \rangle = 1, \qquad v^2 = \ddot{z}^2 - \left( \frac{1 + \langle \ddot{z} | x - z \rangle}{\langle \dot{z} | x - z \rangle} \right)^2$$

(use $\dot{z}^2 = -1$ and $\langle \dot{z} | z \rangle = 0$). To calculate $\mathscr{T}_\alpha$ we need the equations

$$*(v \wedge n \wedge *(v \wedge n)) = i_v i_n v \wedge n = 1, i_\alpha v \wedge n = v_\alpha n - n_\alpha v,$$

and

$$i_{v_\alpha n - n_\alpha v} v \wedge n = v_\alpha n \langle n | v \rangle - n_\alpha (n v^2 - v \langle n | v \rangle).$$

By substituting for the scalar product, one finds that

$$\mathscr{T}_\alpha = \left( \frac{e}{4\pi} \right)^2 \langle \dot{z} | x - z \rangle^{-4} \Bigg\{ (x - z)(x - z)_\alpha \left( \ddot{z}^2 - \left( \frac{1 + \langle \ddot{z} | x - z \rangle}{\langle \dot{z} | x - z \rangle} \right)^2 \right)$$

$$+ \ddot{z}(x - z)_\alpha + (x - z)\ddot{z}_\alpha - (\dot{z}(x - z)_\alpha + (x - z)\dot{z}_\alpha)$$

$$\times \frac{1 + \langle \ddot{z} | x - z \rangle}{\langle \dot{z} | x - z \rangle} + \frac{e_\alpha}{2} \Bigg\}.$$

**Remarks** (2.3.17)

1. We have only used $F^{\text{ret}}$, which corresponds to the initial condition $F^{\text{in}} = 0$.
2. The terms that contain $\ddot{z}$ quadratically are recognizable as the $\mathscr{T}_\alpha$ of $F^{(\ddot{z})}$. At large distances they would dominate, as they decrease as $1/r^2$. The contribution from $F^{(\dot{z})}$ goes as $1/r^4$, and the mixed term as $1/r^3$.

3. The structure of $\mathcal{T}_\alpha$ shows $T_{\alpha\beta} = T_{\beta\alpha}$ and $T_\alpha^{\ \alpha} = 0$.

Poynting's vector $\mathcal{S}_j = [\mathbf{E} \times \mathbf{B}]_j$ is useful for visualizing how the field energy flows. This can best be understood by returning to the representative

**Examples** (2.3.18)

1. Uniform motion. In cylindrical coordinates, $F$ of the form given in (2.3.11; 2), paragraph (iv), makes Poynting's vector
$$(\mathcal{S}_1, \mathcal{S}_\rho, \mathcal{S}_\varphi) = \left(\frac{e}{4\pi}\right)^2 \frac{\rho v(1 - v^2)}{[(x_1 - vt)^2 + (1 - v^2)\rho^2]^3}(\rho, -x_1 + vt, 0).$$
The streamlines of energy are circles $\rho^2 + (x - vt)^2 = R^2$ around the (simultaneous, not retarded) position of the charge $vt$. The field energy flows toward the future positions of the charge, in other words, along with the charge. (See Figure 28.)
2. Uniform acceleration From (2.3.11; 2),
$$(\mathcal{S}_1, \mathcal{S}_\rho, \mathcal{S}_\varphi) = \left(\frac{e}{4\pi}\right)^2 \frac{\rho t}{\left[\left(\frac{x^2 - 1}{2}\right)^2 + \rho^2\right]^3}\left(x_1\rho, \rho^2 + \frac{1 - x^2}{2}, 0\right),$$
in the notation used above. The streamlines are again circles, $\rho^2 + (x_1 - \sqrt{R^2 + t^2 + 1})^2 = R^2$, but as the radius $R$ increases, the center of the circle moves ahead of the position of the charge. This makes the flow of energy point more and more outward as $\rho$ increases with fixed $x_1$ (Figure 28); this occurs because $\mathbf{E}$ has a stronger component in the direction of the motion.
3. A rotating charge in the dipole limit. With (2.3.15), we get
$$\mathcal{S} = \frac{\mathbf{x}}{(4\pi)^2 r^3}\left(|\ddot{\mathbf{z}}|^2 - \left(\ddot{\mathbf{z}} \cdot \frac{\mathbf{x}}{r}\right)^2\right) = \frac{\mathbf{x}}{(4\pi)^2 r^3}(1 - \sin^2\vartheta\cos^2(\omega(r - t) - \varphi))$$
in polar coordinates. The energy flows radially outward, and the flow is strongest perpendicular to the direction of the acceleration at the retarded time.

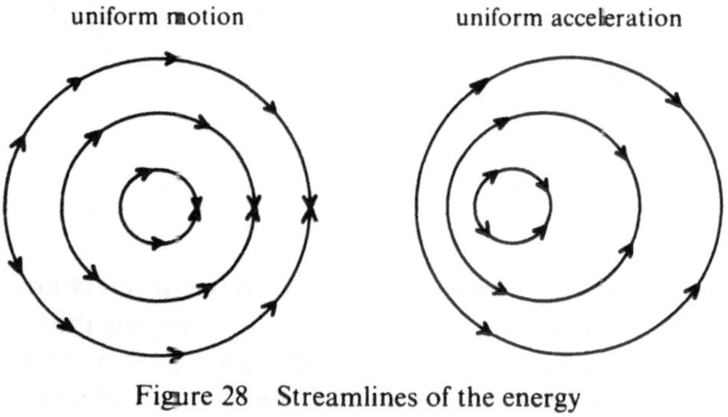


Figure 28 Streamlines of the energy

**Remarks** (2.3.19)

1. If we return to the case of uniform motion at the velocity $\pm v$ joined to hyperbolic motion, then we have to add the streamlines of Examples 1 and 2. In the spherical shell

$$\left(t - \frac{v}{\sqrt{1-v^2}}\right)^2 < \left(x - \frac{1}{\sqrt{1-v^2}}\right)^2 + \rho^2 < \left(t + \frac{v}{\sqrt{1-v^2}}\right)^2$$

we would obtain an increased flow of energy outwards.

2. Opinions differ as to whether a charge in hyperbolic motion emits radiation, owing to the different possible definitions made when posing global questions like that of radiation to infinity. In §3.4 a local definition will be introduced, using the reaction of the radiation on the charge. At this point we summarize the facts supporting the various opinions, and leave the reader to make up his own mind:
   (a) $\mathrm{Max}_t |\mathscr{S}|$ falls off only as $1/\rho^2$.
   (b) Expressions like $\int_K d\mathbf{S} \cdot \mathscr{S}(t = R)$, where $K$ is a sphere of radius $R$ centered on the position of the particle, tend to positive values as $R \to \infty$.
   (c) All the streamlines are closed, and none of them run to infinity.

Since the field equations are linear, the spectral distribution of the sources is inherited by the field. We therefore end this section with a brief discussion of the

**Fourier Decomposition of the Current of a Point Charge** (2.3.20)

$$\tilde{J}_\alpha(k) = \int d^4x\, e^{-i\langle k|x\rangle} J_\alpha(x) = e \int_{-\infty}^{\infty} ds\, \dot{z}_\alpha(s) e^{-i\langle k|z(s)\rangle}.$$

**Examples** (2.3.21)

1. The sudden acceleration of a charge,

$$z(s) = \begin{cases} \dot{\bar{z}} s & \text{if} \quad s > 0, \qquad \dot{\bar{z}} = \dfrac{1}{\sqrt{1-\bar{v}^2}}(1, \bar{\mathbf{v}}) \\[2ex] \dot{z} s & \text{if} \quad s \leq 0, \qquad \dot{z} = \dfrac{1}{\sqrt{1-v^2}}(1, \mathbf{v}). \end{cases}$$

The distribution $\tilde{J}_\alpha(k)$ turns out to be

$$\tilde{J}_\alpha(k) = ie \lim_{\varepsilon \to 0} \left( \frac{\dot{z}_\alpha}{\langle k|\dot{z}\rangle + i\varepsilon} - \frac{\dot{\bar{z}}_\alpha}{\langle k|\dot{\bar{z}}\rangle - i\varepsilon} \right).$$

Consequently,

(i) if $\dot{z} = \bar{z}$, then $\tilde{J}_\alpha(k) = e2\pi\dot{z}_\alpha(\langle k|\dot{z}\rangle)$, and if $v = \bar{v} = 0$, $\tilde{J}_\alpha(k)$ is thus $\sim \delta(k^0)$.

(ii) The use of this $\tilde{J}(k)$ in (2.2.23), although it does not satisfy the assumptions needed to justify that formula, produces

$$P^\alpha(\infty) - P^\alpha(-\infty) = e^2 \int \frac{d^4k}{(2\pi)^3} \delta(k^2) k^\alpha \Theta(k^0) \left| \frac{\dot{z}}{\langle \dot{z}|k\rangle} - \frac{\dot{\bar{z}}}{\langle \dot{\bar{z}}|k\rangle} \right|^2.$$

It might at first be thought that (2.2.19) could not be used for a point charge, since the field at the position of the particle, and hence also $F^{\text{ret}} \wedge {}^*J$, is infinite. As we shall learn in §2.4, however, $F^{\text{rad}} \wedge {}^*J$ remains finite, and so under the right circumstances (2.2.23) is also applicable to point particles. It should furthermore be no surprise that the integral diverges for large $|\mathbf{k}|$, as $\ddot{z}$, and consequently also $F^{(\ddot{z})}$, become infinite for this motion. It can be hoped that if the acceleration were made gentler during a time $\tau$, the integrand could be suppressed for $|\mathbf{k}| > 1/\tau$, allowing the integral to converge, while not essentially changing $\tilde{J}$ for $|\mathbf{k}| < 1/\tau$. With this motive, let us leave the integral as it is, but rewrite it by using the continuity equation for current,

$$k^\alpha \tilde{J}_\alpha = k^0\left(\tilde{J}^0 - \left(\frac{\mathbf{k}}{k^0} \cdot \tilde{\mathbf{J}}\right)\right) = 0.$$

This brings the spatial components $\tilde{\mathbf{J}}_\perp$ of $\tilde{J}$ that are perpendicular to $\mathbf{k}$ into play: If $k^2 = 0$, then

$$|\tilde{J}^\alpha \bar{\tilde{J}}_\alpha| = |\tilde{J}_\perp|^2, \qquad \tilde{\mathbf{J}}_\perp \equiv \tilde{\mathbf{J}} - \frac{\mathbf{k}(\mathbf{k} \cdot \tilde{\mathbf{J}})}{|\mathbf{k}|^2}.$$

Defining $\mathbf{w} \equiv \mathbf{v}/(1 - |\mathbf{v}| \cos \vartheta)$, $\vartheta = \sphericalangle(\mathbf{k}, \mathbf{v})$, and defining $\bar{\mathbf{w}}$ analogously, makes the energy loss

$$P^0(\infty) - P^0(-\infty) = e^2 \int \frac{d^3k}{(2\pi)^3} \frac{|\mathbf{w} - \bar{\mathbf{w}}|_\perp^2}{2|\mathbf{k}|^2}.$$

If $|\mathbf{v}|$ and $|\bar{\mathbf{v}}| \ll 1$, then $|\mathbf{w} - \bar{\mathbf{w}}|_\perp^2 = |\mathbf{v} - \bar{\mathbf{v}}|^2 - \langle \mathbf{k}|\mathbf{v} - \bar{\mathbf{v}}\rangle^2/|\mathbf{k}|^2$, and thus the maximum occurs for $\mathbf{k} \perp \mathbf{v} - \bar{\mathbf{v}}$; for wave-vectors perpendicular to the acceleration there is enhancement. For relativistic motion, the denominators $1 - v \cos \vartheta$ and $1 - \bar{v} \cos \vartheta$ strongly favor the directions of $\mathbf{v}$ and $\bar{\mathbf{v}}$ (see Problem 3). The frequency distribution has the characteristic spectrum $d^3k/|\mathbf{k}|^2 \sim dk$ of bremsstrahlung; the same energy is radiated in every frequency interval.

2. A rotating charge. Whereas the field of a Hertz dipole (2.3.15) is purely harmonic in time, the fields of charges rotating in circles with finite radii also exhibit higher harmonic frequencies. Thus it comes about that highly energetic electrons in a magnetic field emit a characteristic x-ray spectrum. Let us calculate the energy emitted at a given harmonic frequency.

A formula analogous to (2.2.23) can be derived (Problem 7) for the energy lost per period $\omega$ by a current with periodic time-dependence. The only essential change is the definition of $\tilde{J}$, as we now use a Fourier series in time:

$$\tilde{J}^{\alpha}(n\omega, \mathbf{k}) \equiv \frac{\omega}{2\pi}\int_0^{2\pi/\omega} dt \int_{\mathbb{R}^3} d^3x\, e^{-i(\mathbf{k}\cdot\mathbf{x}-n\omega t)} J^{\alpha}(x), \qquad n \in \mathbb{Z}. \quad (2.3.22)$$

There results

$$P^0\left(\frac{2\pi}{\omega}\right) - P^0(0) = \sum_{n\geq 1}\int \frac{d^3k}{(2\pi)^3} 2\pi\delta(|\mathbf{k}|^2 - (n\omega)^2)n\omega \cdot \tilde{J}^{\beta}(n\omega, \mathbf{k})\tilde{J}_{\beta}(-n\omega, -\mathbf{k}). \quad (2.3.23)$$

If the current (2.3.14) is written as

$$\begin{aligned} J^{\alpha}(x) &= e\int_{-\infty}^{\infty} ds\, \dot{z}^{\alpha}(s)\delta^4(x - z(s)) \\ &= e(1, -v\sin\omega t, v\cos\omega t, 0)\delta^3(\mathbf{x} - \mathbf{z}(t)), \end{aligned}$$

then

$$\tilde{J}(n\omega, \mathbf{k}) = e\omega\int_0^{2\pi/\omega}\frac{dt}{2\pi} e^{i(n\omega t - \mathbf{k}\cdot\mathbf{z}(t))}(1, -v\sin\omega t, v\cos\omega t, 0). \quad (2.3.24)$$

Since the radiation is concentrated in the plane of motion within a sector of opening angle $\sim\sqrt{1/v - 1}$ (Problem 3), it is interesting to calculate the frequency distribution for $\mathbf{k}$ in the plane of motion, let us say in the $x_2$-direction. According to the argument in the preceding example, only $J^1$ contributes to the energy loss. The integral (2.3.24) becomes a Bessel function

$$\tilde{J}^1(n\omega, 0, n\omega, 0) = -e\omega v\int_0^{2\pi/\omega}\frac{dt}{2\pi}\sin\omega t e^{in(\omega t - v\sin\omega t)} = -ieR\mathrm{J}'_n(nv). \quad (2.3.25)$$

To discuss this result, it is necessary to distinguish the cases $nv \ll 1$ and $nv \gg 1$. In the former case, one can use the well-known expansion ([20], cf. (3.4.3) and (3.4.4)):

$$\mathrm{J}'_n(nv) = \frac{1}{2(n-1)!}\left(\frac{nv}{2}\right)^{n-1}(1 + O((nv)^2)). \quad (2.3.26)$$

This shows that the frequency distribution of (2.3.23) is $\sim(nv)^{2n}/[(n-1)!]^2$, which has a strong maximum at $n = 1$ if $v \ll 1$. If $v = 1$, the simple formula

$$\mathrm{J}'_n(n) \to \frac{2^{2/3}}{3^{1/3}\Gamma(\frac{1}{3})}n^{-2/3}$$

(see [20], 9.4.43) could be used, and the spectrum would be $n^2|\tilde{J}|^2 \sim n^{2/3}$. Since $v < 1$, however, this formula is not valid for high $n$, and the better asymptotic formula,

$$J_n'(n\iota) \to -\frac{2}{v}\left(\frac{1-v^2}{4\xi}\right)^{1/4} \frac{\mathrm{Ai}'(n^{2/3}\xi)}{n^{2/3}}$$

$$\tfrac{2}{3}\xi^{3/2} = \ln\frac{1+\sqrt{1-v^2}}{v} - \sqrt{1-v^2} \underset{v\to 1}{\to} \frac{(1-v^2)^{3/2}}{3}$$

$$\mathrm{Ai}'(z) \to \begin{cases} -\dfrac{1}{3^{1/3}\Gamma(\frac{1}{3})} & \text{as } z \to 0 \\[2ex] \dfrac{1}{2\sqrt{\pi}} z^{1/4} e^{-(2/3)z^{3/2}} & \text{as } z \to \infty \end{cases} \tag{2.3.27}$$

must be resorted to. This modifies the spectrum to

$$n \exp\left(-\tfrac{2}{3}n(1-v^2)^{3/2}\right)$$

for $n^{2/3}(1-v^2) \gg 1$. Therefore the spectrum has its maximum at $n \sim (1-v^2)^{-3/2} \equiv \gamma^3$, after which it decreases exponentially.

**Remark** (2.3.28)

The reason that the maximum is at $\sim\gamma^3$ can be understood as follows: An observer $O$ first sees the light that is emitted in an angular interval $\sim\gamma^{-1}$. (See Figure 29.) Since the beam of radiation sweeps just barely ahead of the charge, the light arrives at $O$ within a time interval

$$\Delta t = (1-v)\gamma^{-1}R \sim \gamma^{-3}R$$

(if $v \to 1$), and hence

$$\omega_{\max} \sim \frac{1}{\Delta t} \sim \frac{\gamma^3}{R} \sim \omega\gamma^3.$$

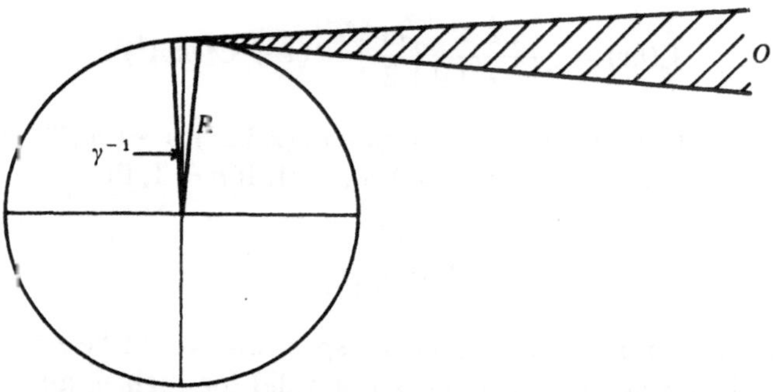


Figure 29 Radiation from relativistic motion in a circle

## Problems (2.3.29)

1. Substitute from (2.3.12) into the expression for $F^{\text{ret}}$.

2. Calculate the field and Poynting's vector $\mathcal{S}$ for a charge moving in a straight line at the speed of light.

3. Calculate the energy distribution $T_{00}$ of the radiation of a fast-moving charge $[\dot{z}(s_0) = 1/\sqrt{1-v^2}\,(1, 0, 0, v), z(s_0) = 0]$ in the far zone at $x^2 = 0$ and
   (a) longitudinal acceleration ($\ddot{z}(s_0) \sim (v, 0, 0, 1)$); and
   (b) transverse acceleration ($\ddot{z}(s_0) \sim (0, 0, 1, 0)$).

   At what angle, in the usual polar coordinates, is the maximum as $v \to 1$?

4. Calculate $\int_{t=0} \mathcal{T}^{\alpha}$ for the field of a uniformly moving point charge (2.3.11; 1). Use $\bar{x} = (x_1/\sqrt{1-v^2}, x_2, x_3)$ as the integration variables, and define the classical radius of the electron by the divergent integral
$$\frac{1}{r_c} = \frac{1}{2}\int \frac{d^3\bar{x}}{|\bar{\mathbf{x}}|^4}.$$
   Show that this destroys the vector transformation property of (1.3.37; 5). How come?

5. Show that
$$\lim_{v\to 1} \frac{ed^2}{4\pi} \frac{(x_1 + vt - d, \rho, -v\rho)\Theta((\rho^2+1)d/2 - x_1 - vt)}{[(x_1 + vt - d)^2 + \rho^2 d^2]^{3/2}}$$
$$= \frac{e}{4\pi} \frac{2\rho}{1+\rho^2} \delta(t + x_1)(0, 1, -1),$$
   where $d \equiv \sqrt{1-v^2}$.

6. Calculate $j(x) = \int_{t=-T} d^3x\, D^{\text{ret}}(\bar{x} - x) J_0(x)$ from (2.2.10; 2) for a point charge, and take the limit as $T \to \infty$ of $j$ for hyperbolic motion.

7. Calculate $P^0(2\pi/\omega) - P^0(0)$ for a $J$ that depends on time periodically, by using the Fourier series (2.3.22).

## Solutions (2.3.30)

1. $((x^2 + 1)/2\langle \dot{z}|x\rangle)\dot{z} - z = -(2/\xi)(x_1, t, 0, 0)$.
$$F^{\text{ret}} = \frac{-8}{\xi^3}(x_1, t, 0, 0)$$
$$\wedge \left(\frac{t(t^2 - x_1^2 + \rho^2 + 1) - x_1\xi}{2(t^2 - x_1^2)}, \frac{x_1(t^2 - x_1^2 + \rho^2 + 1) - t\xi}{2(t^2 - x_1^2)}, x_2, x_3\right)\frac{e}{4\pi}$$
   Using $\mathbf{x} \cdot d\mathbf{x} = x_1\, dx_1 + \rho\, d\rho$, we find (2.3.13).

2. The limit as $v \to 1$ of the fields of (2.3.11; 1) is
$$(E_1, E_\rho, E_\varphi) = \frac{e}{2\pi}\delta(x - t)\frac{1}{\rho}(0, 1, 1);$$

and this satisfies $\delta F = J$ with

$$J = e\delta(x - t)\delta(y)\delta(z)(1, 1, 0, 0) = \lim_{v \to 1} e\delta(x - vt)\delta(y)\delta(z)(1, v, 0, 0).$$

$\mathscr{S}$ points in the direction $x_1$, but because of the factor $\delta^2$, this infinitely Lorentz-contracted field has an infinite energy-momentum density.

3. (a) $T_{00} \sim \dfrac{(1 - v^2)^4 \sin^2 \vartheta}{r^2(1 - v\cos\vartheta)^6}, \qquad \vartheta_{\max} \sim \sqrt{\dfrac{1}{v} - 1} \sim \dfrac{m}{E}.$

(b) $T_{00} \sim \dfrac{(1 - v^2)^2}{r^2(1 - v\cos\vartheta)^6}[(1 - v\cos\vartheta)^2 - (1 - v^2)\sin^2\vartheta\sin^2\varphi].$

4. $$\int \mathscr{T}^0 = \frac{1 + v^3/3}{1 - v^2}\frac{1}{r_c}, \qquad \int \mathscr{T}^j = v^j \frac{4/3}{1 - v^2}\frac{1}{r_c} \neq v^j \int \mathscr{T}^0(v = 0)/\sqrt{1 - v^2}.$$

The condition that $\delta\mathscr{T}^\alpha = 0$, used in the derivation of the transformation law, is violated at the origin. Even adding a $t^\alpha$ for the particle will not avoid the problem, as $P^\alpha$ equals $(m/\sqrt{1 - v^2}, m\mathbf{v}/\sqrt{1 - v^2})$. Thus $\delta(\mathscr{T}^\alpha + t^\alpha)$ is only formally zero (cf. (1.3.25; 2) and (2.1.18; 7)).

5. If $x_1 \neq t$, then the three components approach zero; on the other hand, with $\alpha = (x + vt)/d - 1$,

$$\int_{-\infty}^{\infty} \frac{dt\, d^2\Theta(\ )}{[(x_1 + vt - d)^2 + \rho^2 d^2]^{3/2}} \to \int_{-\infty}^{d(1+\rho^2)/2 - x_1} \frac{dt\, d^2}{[(x_1 + vt - d)^2 + \rho^2 d^2]^{3/2}}$$

$$\to \frac{1}{v}\int_{-\infty}^{(\rho^2 - 1)/2} \frac{d\alpha}{[\alpha^2 + \rho^2]^{3/2}}$$

$$= \frac{1}{v\rho^2}\frac{\alpha}{(\alpha^2 + \rho^2)^{1/2}}\bigg|_{-\infty}^{(\rho^2 - 1)/2}$$

$$= \frac{1}{v\rho^2}\left(1 + \frac{\rho^2 - 1}{\rho^2 + 1}\right) \to \frac{2}{1 + \rho^2}.$$

6. If $z$ is the point at which the world-line $z(s)$ crosses the hypersurface at $t = -T$, then

$$j(x) = \frac{e}{2\pi}\delta((x - z)^2)\Theta(x^0 - z^0).$$

If the motion is hyperbolic,

$$z = \left(\frac{-v}{\sqrt{1 - v^2}}, \frac{1}{\sqrt{1 - v^2}}, 0, 0\right),$$

and as $v \to 1$,

$$\delta(x^2 + 1 - 2\langle x | z\rangle) = \sqrt{1 - v^2}\,\delta(2(x + vt) - (x^2 + 1)\sqrt{1 - v^2}) \to 0.$$

7. Substituting the Fourier series

$$J(x) = \sum_{n \in \mathbb{Z}} \int \frac{d^3k}{(2\pi)^3} \tilde{J}(n\omega, \mathbf{k}) e^{i(\mathbf{k}\cdot\mathbf{x} - n\omega t)}$$

into the integral

$$P^0(2\pi/\omega) - P^0(0) = \int_{0 \le \bar{t} \le 2\pi/\omega} i_0 F^{\mathrm{ret}} \wedge {}^*J$$

$$= \int_{0 \le \bar{t} \le 2\pi/\omega} d^4\bar{x} J^\beta(\bar{x}) \int_{\mathbb{R}^4} d^4x \left( J^0(x) \frac{\partial}{\partial \bar{x}^\beta} D^{\mathrm{ret}}(\bar{x} - x) - J_\beta(x) \frac{\partial}{\partial \bar{x}^0} D^{\mathrm{ret}}(\bar{x} - x) \right)$$

and using (2.2.4) and (2.2.5) yields

$$\int d^4x\, D^{\mathrm{ret}}(\bar{x} - x) J^\alpha(x) = \sum_{n \in \mathbb{Z}} \int \frac{d^3k}{(2\pi)^3} \tilde{J}(n\omega, \mathbf{k}) \frac{e^{i(\mathbf{k}\cdot\bar{\mathbf{x}} - n\omega t)}}{|\mathbf{k}|^2 - (n\omega + i\varepsilon)^2}.$$

Then, finally,

$$P^0\left(\frac{2\pi}{\omega}\right) - P^0(0) = \sum_{n \ge 1} \int \frac{d^3k}{(2\pi)^3} 2\pi\delta\left(|\mathbf{k}|^2 - (n\omega)^2\right) n\omega \tilde{J}^\beta(n\omega, \mathbf{k}) \tilde{J}_\beta(-n\omega, -\mathbf{k}).$$

## 2.4 Radiative Reaction

*The radiation of electromagnetic energy causes a reactive force on a charge. The calculation of this force for point particles is tricky, as it involves divergent integrals.*

The product of $J$ and $F$ appears in the Lorentz force (1.3.24), but it is not well defined for a point particle, since the field is singular at the particle's position. Let us first look into the less problematical matter of the energy-momentum of the emitted radiation, as we slowly work up to the infinities in the equation for the total energy and momentum.

The starting point is Stokes's theorem for the electromagnetic energy-momentum forms:

$$\int_N d^*\mathcal{T}^\alpha = \int_{\partial N} {}^*\mathcal{T}^\alpha. \tag{2.4.1}$$

We choose $N$ as the four-dimensional region bounded by the light-cones

$$L_1 = \{x \in \mathbb{R}^4 : (x - z(s_1))^2 = 0,\ x^0 > z^0(s_1)\}$$
$$L_2 = \{x \in \mathbb{R}^4 : (x - z(s_2))^2 = 0,\ x^0 > z^0(s_2)\}$$

and the cylinder

$$K = \{x \in \mathbb{R}^4 : (x - z(s_1))^2 + \langle \dot{z}(s_1) | x - z(s_1) \rangle^2 = R^2\},$$

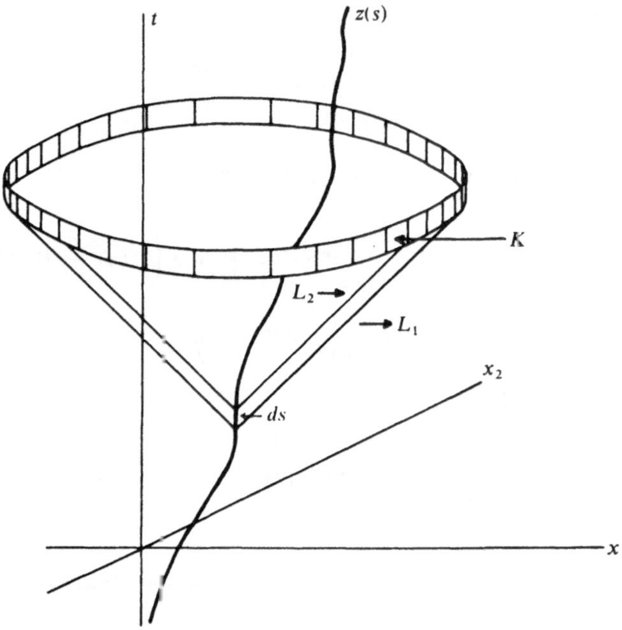


Figure 30 The hypersurface used for $\partial N$ in (2.4.1)

in order to be able to follow the radiation as it goes to infinity (Figure 30). We first calculate the part of (2.4.1) coming from the piece of $\partial N$ contained in $K$. For this purpose $R$ may be increased to $\infty$, and $ds \equiv s_2 - s_1 \to 0$; the result has the interpretation of the amount of energy-momentum that is lost by the charge between $s_1$ and $s_1 + ds$ and escapes to infinity. In this limit $\int_{\partial N \cap K} {}^*\mathscr{T}_\alpha$ consists only of the contribution of $F^{(\ddot{z})}$ to $\mathscr{T}_\alpha$ (cf. (2.3.16))

$$\mathscr{T}_\alpha^{(\ddot{z})} = \left(\frac{e}{4\pi}\right)^2 \langle \dot{z} | x - z \rangle^{-4} (x - z)(x - z)_\alpha \left\{ \ddot{z}^2 - \left( \frac{\langle \ddot{z} | x - z \rangle}{\langle \dot{z} | x - z \rangle} \right)^2 \right\};$$

the integral of this is asymptotically independent of $R$, while the other terms all have higher powers of $R$ in the denominator. The external surface $\partial N \cap K$ has a height $ds$ above $K \cap L_1$, and in the limit as $ds \to 0$, we need to know ${}^*\mathscr{T}_\alpha$ only on $K \cap L_1$. If we write $x - z = R(\dot{z} + n)$, $n \in E_1$, on that surface (see Figure 31), then from $x \in K \cap L_1$:

$$0 = R^{-2}(x - z)^2 = -1 + 2\langle \dot{z} | n \rangle + n^2$$

and

$$1 = R^{-2}\{(x - z)^2 + \langle \dot{z} | x - z \rangle^2\} = (-1 + \langle z | n \rangle)^2$$

it follows that $n^2 = 1$ and $\langle \dot{z} | n \rangle = 0$, and hence that

$$-\langle \dot{z} | x - z \rangle = R \quad \text{and} \quad \langle \ddot{z} | x - z \rangle = R\langle \ddot{z} | n \rangle.$$

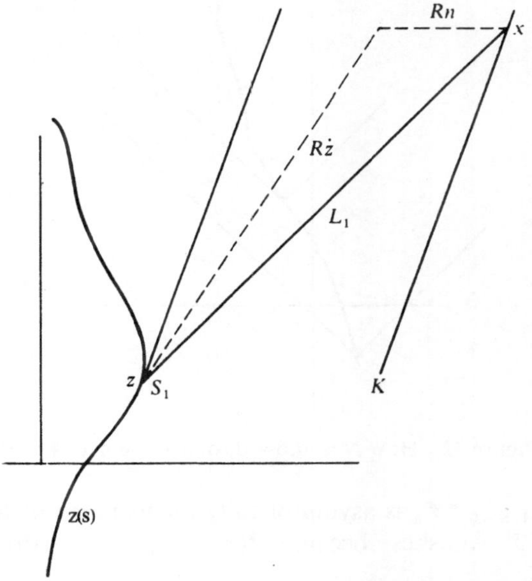


Figure 31 The quantities introduced for the evaluation of (2.4.1)

Consequently,

$$\mathscr{T}^{(\ddot{z})}_{\alpha} = \left(\frac{e}{4\pi}\right)^2 (\dot{z} + n)(\dot{z}_\alpha + n_\alpha)(\ddot{z}^2 - \langle \ddot{z}|n\rangle^2)R^{-2},$$

where $n$ varies over the spatial unit sphere in $K \cap L_{10}$. The 3-form $*(\dot{z} + n)$ acts in the integral $\int *\mathscr{T}_\alpha$ as $R^2\, ds\, d\Omega_n$ ($d\Omega$ is the element of solid angle on the unit sphere). By symmetry, all odd powers of $n$, and hence the part containing $n_\alpha$, drop out of the integral $\int *\mathscr{T}_\alpha$. By taking the average over the unit sphere we simply replace $\langle \ddot{z}|n\rangle^2$ with $\ddot{z}^2/3$, and in this limit we obtain

**Larmor's Formula** (2.4.2)

$$\int_{\partial N \cap K} *\mathscr{T}_\alpha = \frac{2}{3}\frac{e^2}{4\pi}\, ds\, \dot{z}_\alpha \ddot{z}^2.$$

**Remarks** (2.4.3)

1. Formula (2.4.2) is the covariant generalization of (1.1.8) for the loss of energy-momentum.
2. In the rest frame $\dot{z}(s_1) = (1, 0, 0, 0)$ only energy is lost, and

$$\frac{dE}{dt} = \frac{e^2}{4\pi}\frac{2}{3}\ddot{z}^2.$$

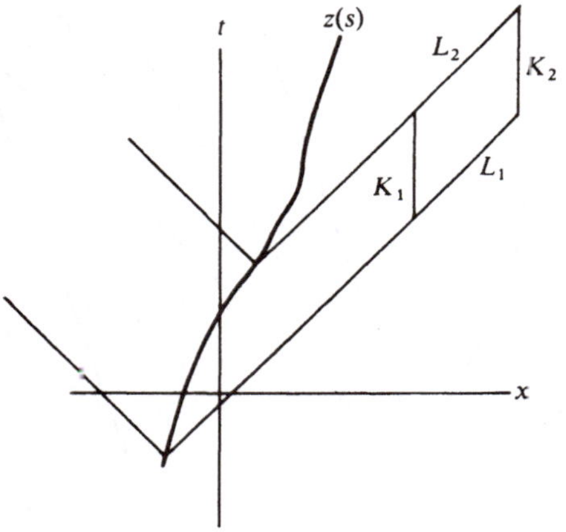


Figure 32 How $N$ is allowed to increase in (2.4.3; 3)

3. The reason $\int_{\partial N \cap K} {}^*\mathcal{T}_\alpha$ is asymptotically independent of $R$ for large $R$ is that $\int_{L_1} {}^*\mathcal{T}_\alpha^{(\dot{z})}$ vanishes, because ${}^*(x - z)_{|L_{1,2}} = 0$ (Problem 1). (See Figure 32.)

   Since the equation $d^*\mathcal{T}_\alpha = 0$ is valid away from the world-line $z(s)$,

$$\int_{K_1} {}^*\mathcal{T}_\alpha^{(\dot{z})} - \int_{K_2} {}^*\mathcal{T}_\alpha^{(\dot{z})} = \int_{L_1} {}^*\mathcal{T}_\alpha^{(\dot{z})} - \int_{L_2} {}^*\mathcal{T}_\alpha^{(z)} = 0.$$

**Examples** (2.4.4)

1. Hyperbolic motion. If

   $z(s) = a^{-1}(\text{Sinh } as, 0, 0, \text{Cosh } as)$ and $\dot{z}(s) = (\text{Cosh } as, 0, 0, \text{Sinh } as)$,

   then

$$\frac{dP^\alpha}{ds} = \frac{3}{2}\frac{e^2}{4\pi}\dot{z}^\alpha a^2.$$

   The energy contained in the radiation field always increases, while momentum is transferred to it in the negative $z$-direction when $s < 0$, and in the positive $z$-direction when $s > 0$. Therefore the charge always radiates forward along its direction of motion.
2. Synchrotron radiation. With the current of a rotating charge (2.3.14),

$$\dot{z} = \frac{1}{\sqrt{1 - v^2}}(1, -v \sin \omega t, v \cos \omega t, 0),$$

$$\ddot{z} = \frac{-v^2}{R(1 - v^2)}(0, \cos \omega t, \sin \omega t, 0),$$

where $\omega = v/R$. As a consequence, $\delta E$, the energy loss per period $\cdot 2\pi$, is

$$\delta E = \frac{2}{3\omega}\frac{e^2}{4\pi}\ddot{z}^2 = \omega\frac{e^2}{6\pi}\frac{v^2}{(1-v^2)^2}.$$

If we use the value $e^2/4\pi\hbar = \frac{1}{137}$, this is $\sim\omega\hbar(v^2/200)(E/m)^4$. So long as $E/m \sim 1$, the charge clearly needs to undergo more than 200 revolutions to give off a quantum at the ground-state frequency. If $v \to 1$, the rate of energy loss increases rapidly; for, e.g., 5 GeV electrons, $(E/m)^4 \sim 10^{16}$. Accordingly, fast electrons moving in a circle lose quite a bit of energy in the form of synchrotron radiation.

Although $\mathscr{T}_\alpha^{(\ddot{z})}$ does not contribute to the integral over the light-cones $L_1$ and $L_2$ in the calculation of the right side of (2.4.1), the other terms in $\mathscr{T}_\alpha$ have infinite integrals! They decrease as $r^{-3}$ and $r^{-4}$, which diverge when integrated over all space. In order to isolate the causes of the problem, we write

$$d{*\mathscr{T}_\alpha} = i_\alpha(F^{\text{in}} + \tfrac{1}{2}F^{\text{rad}} + \tfrac{1}{2}(F^{\text{ret}} + F^{\text{adv}})) \wedge {*J} \tag{2.4.5}$$

(cf. (2.2.21)). The first term is the Lorentz force from the incoming field, and causes no trouble. We discovered that if the current in (2.2.21) had compact support, then aside from $F^{\text{in}}$ only the radiation field $F^{\text{rad}}$ contributes to the loss of energy-momentum in the limit of infinite times. For this reason, we next evaluate the contribution of $F^{\text{rad}}$ to the left side of (2.4.1) for the point charge. It will turn out that this term is finite, and all the difficulties stem from the last term.

Since $J$ is supported on $z(s)$, we must evaluate $F^{\text{rad}}$ on the world-line of the charge. After some simple algebra (Problem 2), we can write the radiation field as

$$F^{\text{rad}}_{\alpha\beta}(x) = e\int_{-\infty}^{\infty} ds\, D(x - z(s))\frac{d}{ds}\left[\frac{\dot{z}_\alpha(x-z)_\beta}{\langle\dot{z}|x-z\rangle} - (\alpha \leftrightarrow \beta)\right]. \tag{2.4.6}$$

We wish now to let $x$ carefully approach some point of the world-line, say $z(0)$. To do this, we expand the integrand of (2.4.6) about $s = 0$:

$$z(s) - z(0) = s\dot{z} + \frac{s^2}{2}\ddot{z} + \frac{s^3}{6}\dddot{z} + \ldots, \qquad \dot{z} \equiv \dot{z}(0), \quad \text{etc.}$$

$$\dot{z}(s) = \dot{z} + s\ddot{z} + \frac{s^2}{2}\dddot{z} + \ldots, \tag{2.4.7}$$

and call $x - z(0) = \lambda$. In order that $x$ stay between the retarded time and the advanced time (Figure 33), let $\langle\dot{z}|\lambda\rangle = 0$ thus $\lambda$ is spacelike as it approaches 0. Since

$$D(x) = \frac{\delta(x^2)}{2\pi}(\Theta(x^0) - \Theta(-x^0))$$

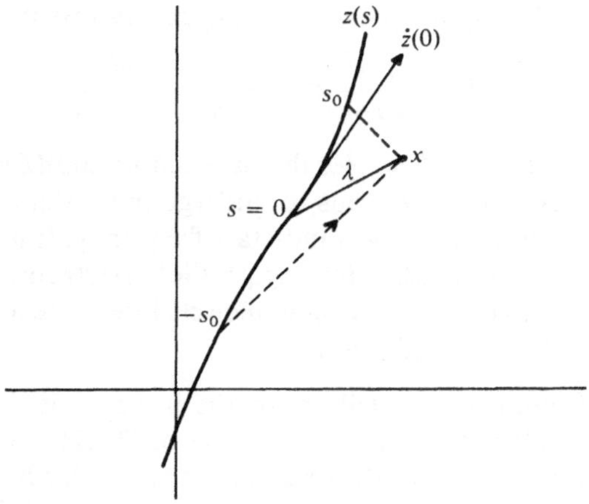


Figure 33 The limit $\lambda \to 0$

and

$$(x - z(s))^2 = \left(\lambda - s\dot{z} - \frac{s^2}{2}\ddot{z}\right)^2 + O(s^3) = \lambda^2 - s^2(1 + \langle\lambda|\ddot{z}\rangle) + O(s^3),$$

if $\lambda \to 0$, then

$$D(x - z(s)) = \frac{\delta(\lambda^2 - s^2)}{2\pi}(\Theta(s) - \Theta(-s)) = \frac{1}{4\pi\lambda}(\delta(s - \lambda) - \delta(s + \lambda)),$$

$$\lambda \equiv \langle\lambda|\lambda\rangle^{1/2} > 0. \tag{2.4.8}$$

If the rest of the integrand of (2.4.6) is also expanded about $s = 0$,

$$\begin{aligned} [\ \ ] &\equiv \frac{N(\lambda)}{s} + A(\lambda) + sB(\lambda) + \frac{s^2}{2}C(\lambda) \\ \frac{d}{ds}[\ \ ] &= -\frac{N(\lambda)}{s^2} + B(\lambda) + sC(\lambda), \end{aligned} \tag{2.4.9}$$

then as $\lambda \to 0$, (2.4.6) becomes equal to

$$\frac{e}{4\pi}\lim_{\lambda\to 0}\frac{1}{\lambda}\left(\frac{d}{ds}[\ \ ]_{|s=\lambda} - \frac{d}{ds}[\ \ ]_{|s=-\lambda}\right) = \frac{e}{4\pi}2C(0).$$

Substitution of the series (2.4.7) reveals that

$$C(0) = \tfrac{2}{3}(\dddot{z}_\alpha \dot{z}_\beta - \dddot{z}_\beta \dot{z}_\alpha),$$

with which we obtain a formula for the

**Radiation Field on the World-Line** (2.4.10)

$$\tfrac{1}{2}F^{\text{rad}}(z(s)) = \frac{e}{3\pi}\dddot{z}(s) \wedge \dot{z}(s).$$

**Remarks** (2.4.11)

1. The radiation field is purely electric in the rest-frame of the particle:

$$\tfrac{1}{2}(\mathbf{E}^{\mathrm{ret}} - \mathbf{E}^{\mathrm{adv}}) = \frac{e}{4\pi}\frac{2}{3}\dddot{\mathbf{z}}, \qquad \tfrac{1}{2}(\mathbf{B}^{\mathrm{ret}} - \mathbf{B}^{\mathrm{adv}}) = 0.$$

2. In the result (2.4.10) it is understood that the term $-\ddot{\mathbf{z}}/r$ in $\mathbf{E}$ of (2.3.9) contributes the limit

$$\lim_{r\to 0} \frac{-\ddot{\mathbf{z}}(s - r) + \ddot{\mathbf{z}}(s + r)}{2r} = \dddot{\mathbf{z}}$$

to $F^{\mathrm{rad}}$.
When averaged over space, $(\mathbf{x}(\mathbf{x}\cdot\ddot{\mathbf{z}}))/r^3$ in like manner contributes $-\ddot{\mathbf{z}}/3$, and the Coulomb field of the near zone and the magnetic field disappear from (2.3.9). Formula (2.4.10) should a priori be averaged over the different possible directions from which $z(s)$ is approached, but we have noted above that it is independent of the direction of $\lambda$ provided that $\langle \dot{z} | \lambda \rangle = 0$.

If we use (2.4.10) in Equation (2.4.5), and conclude from $\langle \dot{z} | \ddot{z} \rangle = 0$ that $\langle \dot{z} | \dddot{z} \rangle = -\ddot{z}^2$, then we get a formula for the

**Energy and Momentum Lost to the Radiation Field** (2.4.12)

$$\frac{dP_\alpha^{\mathrm{rad}}}{ds} \equiv \frac{-d}{2ds}\int_{N_4} {}^*i_J i_\alpha F^{\mathrm{rad}} = \frac{e}{2}\dot{z}^\beta(s)F_{\beta\alpha}^{\mathrm{rad}}(z(s)) = \frac{e^2}{4\pi}\frac{2}{3}(\dot{z}_\alpha \ddot{z}^2 - \dddot{z}_\alpha).$$

**Remarks** (2.4.13)

1. The limit $ds \to 0$ is understood in (2.4.12), and $N_4$ contains the part of $z(s)$ between $s$ and $s + ds$; thus on the right side the derivative of $z$ is taken at the proper time $s$.
2. The first term on the right is precisely the energy-momentum vector that flows to infinity (2.4.2). It would not be possible for this term to be the whole energy-momentum exchange, because $e\dot{z}^\alpha \dot{z}^\beta F_{\alpha\beta}^{\mathrm{rad}}$ must equal zero and not $-(e^2/6\pi)\ddot{z}^2$. The deficiency is made up by the second term.
3. Although the first term has a definite sign because $\ddot{z}^2 \geq 0$, and represents an irretrievable loss of energy, the second is a total differential and contributes nothing to an integral by $ds$, provided that the initial value of $\ddot{z}$ is returned to at the end. This vector represents retrievable energy-momentum stored in the near zone, and for that reason it is not present at infinity.

4. In the rest-frame $\dot{z} = (1, 0, 0, 0)$, $\ddot{z}$ is $(0, \ddot{\mathbf{z}})$, and $\ddot{z}^2 + (\dot{z}|\dddot{z}) = 0 \Rightarrow \dddot{z} = (\ddot{z}^2, \dddot{\mathbf{z}})$. The two terms in the energy loss cancel out, as the particle has no energy to lose. An accelerated particle that is momentarily at rest borrows the energy that it radiates from the field in the near zone.
5. If $\dot{z} = (1/\sqrt{1 - v^2}, \mathbf{v}/\sqrt{1 - v^2})$, then the $\ddot{z}^2$ term of the reaction on the charge acts like a frictional force retarding $\mathbf{v}$, while the term with $\dddot{\mathbf{z}}$ tends to increase the acceleration of the particle. This leads to all sorts of paradoxical consequences, which we shall return to after having discussed the remaining contribution $\sim F^{\text{ret}} + F^{\text{adv}}$.

**Examples** (2.4.14)

1. By taking more derivatives in Example (2.4.4; 1),

$$\ddot{z}(s) = a(\text{Sinh } as, 0, 0, \text{Cosh } as)$$
$$\dddot{z}(s) = a^2(\text{Cosh } as, 0, 0, \text{Sinh } as),$$

we see that the two terms of (2.4.12) exactly cancel, and $dP^{\text{rad}}/ds = 0$. A charge in hyperbolic motion radiates on credit; the energy is not supplied by the particle, but comes from the near zone. Of course, the debt must be repaid later, once the acceleration stops. For example, if the charge is accelerated from rest to the velocity $v = \text{Tanh } as_0$,

$$\dot{z}(s) = \Theta(-s)(1, 0, 0, 0) + \Theta(s)\Theta(s_0 - s)(\text{Cosh } as, 0, 0, \text{Sinh } as)$$
$$+ \Theta(s - s_0)\left(\frac{1}{\sqrt{1 - v^2}}, 0, 0, \frac{v}{\sqrt{1 - v^2}}\right);$$

and then

$$\ddot{z}(s) = \Theta(s)\Theta(s_0 - s)a(\text{Sinh } as, 0, 0, \text{Cosh } as),$$
$$\dddot{z}(s) = \Theta(s)\Theta(s_0 - s)a^2(\text{Cosh } as, 0, 0, \text{Sinh } as) + \delta(s)(0, 0, 0, 1)$$
$$- \delta(s - s_0)\left(\frac{v}{\sqrt{1 - v^2}}, 0, 0, \frac{1}{\sqrt{1 - v^2}}\right).$$

Hence

$$\dot{z}\ddot{z}^2 - \dddot{z} = -\delta(s)(0, 0, 0, 1) + \delta(s - s_0)\left(\frac{v}{\sqrt{1 - v^2}}, 0, 0, \frac{1}{\sqrt{1 - v^2}}\right).$$

The force on the particle is the negative of the rate of change of the energy-momentum of the field. At first the particle feels a jolt in the direction of the acceleration from $F^{\text{rad}}$, and later the radiation force operates in the opposite direction, and $F^{\text{in}}$ has to pay the energy bill.
2. For the rotating charge of (2.4.4; 2),

$$\dddot{\mathbf{z}} = \frac{v^3}{R^2(1 - v^2)^{3/2}}(0, \sin \omega t, -\cos \omega t, 0).$$

The second term of (2.4.12) causes no additional loss of energy in this case, but it does intensify the braking action of the first term, opposing the velocity of the charge:

$$\dot{z}\ddot{z}^2 - \dddot{z} = \frac{v^3}{R^2(1-v^2)^{5/3}}(v, -\sin \omega t, \cos \omega t, 0).$$

The calculation of the last term of (2.4.5) is simply a matter of replacing the difference appearing in (2.4.8) with a sum:

$$D^{\text{ret}}(x - z(s)) + D^{\text{adv}}(x - z(s)) = \frac{1}{4\pi\lambda}(\delta(s - \lambda) + \delta(s + \lambda)). \quad (2.4.15)$$

Hence the expansion (2.4.9) results in a contribution $-N(\lambda)/\lambda^3 + B(\lambda)/\lambda$, where

$$N(\lambda) = \frac{\dot{z}_\alpha \lambda_\beta - \dot{z}_\beta \lambda_\alpha}{1 + \langle \ddot{z} | \lambda \rangle}.$$

And not only is $N(\lambda)/\lambda^3$ divergent as $\lambda \to 0$, but it also depends on the direction of $\lambda$: If no direction has been singled out in Minkowski space, then the net result must be $\sim \dot{z}_\alpha \ddot{z}_\beta - \dot{z}_\beta \ddot{z}_\alpha$ and the coefficient goes as $1/\lambda$ for dimensional reasons. Actually the next term $B(0)$ is $(\dot{z}_\alpha \ddot{z}_\beta - \dot{z}_\beta \ddot{z}_\alpha)/2$ independently of the direction of $\lambda$. Normally only the latter term is retained while $N$ is swept under the rug by some averaging procedure; by this hocus-pocus,

$$\tfrac{1}{2}(F^{\text{ret}} + F^{\text{adv}})(z(s)) = \frac{e}{4\pi} \ddot{z} \wedge \dot{z} \lim_{\lambda \to 0} \frac{c}{\lambda}. \quad (2.4.16)$$

Equation (2.4.16) is not very well defined, but at any rate the numerical factor $c$ is positive when calculated in this way. Then attempts are made to argue away the resulting indeterminacy in the radiative reaction of the field along the following lines: Suppose that the three contributions of (2.4.5) are combined and set equal to minus the rate of change of the energy-momentum of a particle of mass $m_0$ during the time $ds$. Then there results

$$m_0 \ddot{z}_\beta = e\dot{z}^\alpha F^{\text{in}}_{\alpha\beta} - \frac{e^2}{4\pi}\frac{2}{3}(\dot{z}_\beta \ddot{z}^2 - \dddot{z}_\beta) - \delta m \ddot{z}_\beta,$$

$$\delta m = \lim_{\lambda \to 0} \frac{ce^2}{4\pi\lambda}. \quad (2.4.17)$$

Next one calls $m = m_0 + \delta m$ ("mass-renormalization") and smugly solves the

**Renormalized Equation of Motion** (2.4.18)

$$m\ddot{z}_\beta = e\dot{z}^\alpha F^{\text{in}}_{\alpha\beta} - \frac{e^2}{4\pi}\frac{2}{3}(\dot{z}_\beta \ddot{z}^2 - \dddot{z}_\beta).$$

**Remarks** (2.4.19)

1. The mass $m$ is clearly what would be measured by the inertia of the particle in an external field. In (2.4.18) there are no apparent infinities.
2. It may seem peculiar that in a theory that is invariant under reversal of the motion, something so obviously not invariant under motion reversal as the radiative reaction force should occur. It arises because of the use of $F^{\text{in}}$ and $D^{\text{ret}}$; it gets its sign from the initial conditions, and would have the opposite sign if $F^{\text{out}}$ and $D^{\text{adv}}$ were used. By using $(D^{\text{ret}} + D^{\text{adv}})/2$, one can even find stationary solutions of the relativistic two-body problem [23] for which nothing at all is radiated. If the system has a finite energy one finds radiation damping for $t \to \pm\infty$.
3. When discussing (2.2.21) we learned that for currents with compact support the total energy-momentum vector transmitted to the field originates in the Lorentz force with $F^{\text{rad}}$. This result is carried over to point particles in (2.4.16), and it can be seen that $(F^{\text{ret}} + F^{\text{adv}})/2$ simply contributes $\delta m(\dot{z}(\infty) - \dot{z}(-\infty))$, which is the change of the energy-momentum of the self-field attached to the particle.
4. The self-field is eliminated in (2.4.18), and only the particle's coordinates appear. To solve the initial-value problem of the total system, $z$, $\dot{z}$, $F$ and $^*F$ must all be known at some time. One might therefore expect that the elimination of $F$ would render it necessary to take the whole previous history, $z(s)$ for $s < 0$, into account. In fact, the only extra quantity that shows up in the limit of a point particle is $\ddot{z}$, and the solution manifold of the Cauchy problem is only increased by dependence on the three parameters $\ddot{\mathbf{z}}$. But even so, the physically acceptable solution manifold will turn out to be of a lower dimension.
5. The question arises of why we did not circumvent the difficulties connected with point particles by using charges spread over some positive volume. Unfortunately, it is not easy to obtain a theory in this way that has local conservation of energy and momentum (cf. (1.3.25; 2)).
6. The result (2.4.18) can be explained as follows: The field of a point charge has an infinite energy.

$$\frac{1}{2}\int(|\mathbf{E}|^2 + |\mathbf{B}|^2) \to \infty,$$

and since the particle carries this energy along with it, it is subjected to an infinite mass increase $\delta m$ by Einstein's principle that energy is mass. In order that $m$ remain finite, it is necessary to start with an infinite negative "bare" mass $m_0$, obviously a dangerous undertaking. The field $E^{(\ddot{z})} = -e\ddot{z}/r$ causes a reverse acceleration, that is, a braking. This quantity was decomposed above into an infinite term $-\delta m\ddot{z}$ and $+\dddot{z}$. The positive sign of the latter part comes from the use of $D^{\text{ret}}$: The particle feels the field that was produced a short time beforehand, and if the contribution $-\delta m\ddot{z}$ from $(D^{\text{ret}} + D^{\text{adv}})/2$ is subtracted, the net force is in the direction of the positive rate of change of the acceleration. This tends to make the particle fly on ahead, if the braking term is compensated for to a large extent by an infinite negative inertia. If a particle is accelerated so quickly that the "acceleration force" $\sim \dddot{z}$ is as large as the braking terms $\dot{z}\ddot{z}^2$ and $\ddot{z}$, then the particle takes off under its own steam.

Unfortunately, mathematics can not be fooled by such simple tricks; the difficulties that were swept under the rug show up later as all sorts of paradoxical consequences of Equation (2.4.18).

**Examples** (2.4.20)

1. The run-away solution:

$$\dot{z} = (\mathrm{Cosh}[\tau_0 a e^{s/\tau_0}],\ \mathrm{Sinh}[\tau_0 a e^{s/\tau_0}], 0, 0),$$

$$\tau_0 = \frac{e^2}{6\pi m}, \qquad a \text{ arbitrary},$$

solves (2.4.18) with $F^{\text{in}} = 0$ (Problem 3). The charge suddenly begins to run away ($\tau_0 \sim 10^{-23}$ seconds for electrons). Since $\ddot{z}^2 = a^2 e^{2s/\tau_0}$, it radiates a tremendous amount of energy; this is consistent with conservation of energy because on the one hand an acceleration takes energy away from a particle with a negative mass $m_0$, and on the other hand there is always energy to be tapped from the infinite reservoir of self-energy and pumped into the far zone.

2. A well aimed shot can bring the flight of the particle to a stop. If $F^{\text{in}}$ is such that

$$eE_1(z) = ma\tau_0 \,\mathrm{Cosh}[\tau_0 a]\delta(z^0 - z^3),$$

then (2.4.18) has the solution

$$\dot{z}(s) = \Theta(-s)(\mathrm{Cosh}[\tau_0 a e^{s/\tau_0}], \mathrm{Sinh}[\tau_0 a e^{s/\tau_0}], 0, 0) + \Theta(s)(\mathrm{Cosh}[\tau_0 a], \mathrm{Sinh}[\tau_0 a], 0, 0)$$

if $z_3(0) = z_0(0)$, $0 = -\dot{z}^3(0) = \dot{z}^2(0)$ (Problem 3). Such behavior is often felt to be acausal, because the particle starts to accelerate before it is brought to its senses by the pulse from $F^{\text{in}}$ (Figure 34).

**Remark** (2.4.21)

Not all the solutions of (2.4.18) are crazy (see Problem 4). Attempts have been made to separate sense from nonsense by imposing special initial conditions (cf. [24]). It is to be hoped that some day the real solution of the problem of the charge-field interaction will look different, and the equations describing nature will not be so highly unstable that the balancing act can only succeed by having the system correctly prepared ahead of time by a convenient coincidence.

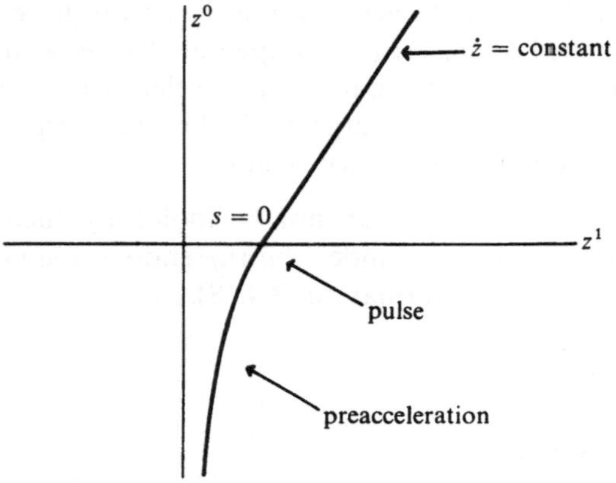


Figure 34 Motion with preacceleration

## **Problems** (2.4.22)

1. Verify the claim made in (2.4.3; 3) that $^*(x - z)_{|L_1} = 0$.
2. Derive (2.4.6).
3. Show that the $\dot{z}$'s of Examples (2.4.20; 1) and (2.4.20; 2) solve (2.4.18).
4. Solve (2.4.18) for $F^{\text{in}}$ a constant electric field in the $x_1$-direction and no magnetic field. Use the ansatz $\dot{z}(s) = (\text{Cosh } \omega(s), \text{Sinh } \omega(s), 0, 0)$. Compare with (I: 5.2.19; 3) and (2.4.20; 1).

## **Solutions** (2.4.23)

1. Let $z(s) = 0: r_{|L_1} = t_{|L_1} \Rightarrow dr_{|L_1} = dt_{|L_1}$. Hence

$$^*x = \mathbf{x} \cdot {}^*d\mathbf{x} - t^*dt = r^*dr - t^*dt = r\, d\Omega \wedge dt - t\, d\Omega \wedge dr: {}^*x_{|L_1} = 0.$$

2. Let $y \equiv (\bar{x} - z(s))^2$. Then

$$A^{\text{rad}}_{\alpha,\beta}(\bar{x}) = -e \int ds\, \dot{z}_\alpha \frac{\partial y}{\partial \bar{x}^\beta} \frac{ds}{dy} \frac{d}{ds} D(y) = -e \int ds\, D(\bar{x} - z(s)) \frac{d}{ds} \frac{\dot{z}_\alpha(s)(\bar{x} - z(s))_\beta}{\langle \dot{z}(s) | \bar{x} - z(s) \rangle}.$$

3. We shall verify Example (2.4.21; 2), and the solution of (2.4.20; 1) will appear as a by-product.

$$\ddot{z} = \Theta(-s) a e^{s/\tau_0} (\text{Sinh}[\ ], \text{Cosh}[\ ], 0, 0)$$

$$\dddot{z} = \Theta(-s) a^2 e^{2s/\tau_0} (\text{Cosh}[\ ], \text{Sinh}[\ ], 0, 0) + \ddot{z}/\tau_0$$
$$- \delta(s) a (\text{Sinh } \tau_0 a, \text{Cosh } \tau_0 a, 0, 0)$$

$$\dot{z}\ddot{z}^2 - \dddot{z} = -\ddot{z}/\tau_0 + \delta(s) a (\text{Sinh } \tau_0 a, \text{Cosh } \tau_0 a, 0, 0).$$

Now,

$$\delta(z^0(s) - z^3(s)) = \frac{1}{\dot{z}^0(0)} \delta(s) = \frac{\delta(s)}{\text{Cosh } \tau_0 a},$$

and hence $e\dot{z}^\alpha F^{\text{in}}_{\alpha\beta} = \delta(s)\, m\, a\, \tau_0(-\text{Sinh } \tau_0\, a, \text{Cosh } \tau_0\, a, 0, 0)$ and $m\ddot{z}_\beta = e\dot{z}^\alpha F^{\text{in}}_{\alpha\beta} - m\tau_0(\dot{z}_\beta \ddot{z}^2 - \dddot{z}_\beta)$.

$$\ddot{z} = \dot{\omega}(\text{Sinh}\ \omega, \text{Cosh}\ \omega, 0, 0), \qquad \ddot{z}^2 = \dot{\omega}^2,$$
$$\dddot{z} = \dot{\omega}^2(\text{Cosh}\ \omega, \text{Sinh}\ \omega, 0, 0) + \ddot{\omega}(\text{Sinh}\ \omega, \text{Cosh}\ \omega, 0, 0),$$
$$-\dot{z}\ddot{z}^2 + \dddot{z} = \ddot{\omega}(\text{Sinh}\ \omega, \text{Cosh}\ \omega, 0, 0).$$

Equation (2.4.18) requires that $\dot{\omega} = E/m + \tau_0 \ddot{\omega}$, which implies that

$$\omega(s) = a + \frac{E}{m} s + c\tau_0 e^{s/\tau_0},$$

where $a$ and $c$ are constants of integration. Only if $c = 0$, that is, for the special initial condition $\ddot{z}(0)^2 = E^2/m^2$, is there no self-acceleration.

# 3 The Field in the Presence of Conductors

## 3.1 The Superconductor

*The superconductor is a simple model of a coupled system of equations for charged matter and an electromagnetic field. As a perfect conductor and diamagnet it excludes all electric and magnetic fields from its interior.*

Realistic situations do not very closely resemble the idealization discussed in the preceding chapter, where the charge distribution is prescribed. The field in turn influences the motion of the charges, so it would be more correct to analyze the coupled system. For a point-particle the analysis is subject to the difficulties encountered in §2.4. Moreover, the charge-carriers in matter, electrons and atomic nuclei, are governed by the laws of quantum mechanics, and their motion is a very complicated many-body problem. Every phenomenological description of matter is of necessity either highly idealized or else so general as to contain little information. Notwithstanding that objection, in order to formulate the ideas of this chapter mathematically, we shall single out one of the many models for a superconductor, which can be cast in a simple mathematical form. It is good enough for our purposes, as we shall always consider an extreme case in the examples, for which the charge-carriers in matter are numerous and move about freely. By responding instantaneously to any applied field, they cause the net field within the material to disappear entirely. Later, when we treat the gravitational interaction, this model will serve as our prototype of charged matter.

**London's Equations** (3.1.1)

Consider the hydrodynamic equations of an incompressible, charged, frictionless fluid in an electromagnetic field, calling the velocity field $\mathbf{v}(\mathbf{x}, t)$:

$$\frac{d\mathbf{v}}{dt} \equiv \frac{\partial \mathbf{v}}{\partial t} + \nabla \frac{v^2}{2} - [\mathbf{v} \times \nabla \times \mathbf{v}] = \frac{e}{m}(\mathbf{E} + [\mathbf{v} \times \mathbf{B}]). \tag{3.1.2}$$

Then from the equation $\nabla \times \mathbf{E} = -\dot{\mathbf{B}}$ results the generalization of Helmholtz's circulation theorem,

$$\dot{\mathbf{w}} = \nabla \times [\mathbf{v} \times \mathbf{w}], \qquad \text{where} \quad \mathbf{w} \equiv \nabla \times \mathbf{v} + \frac{e}{m}\mathbf{B}. \tag{3.1.3}$$

Therefore, if $\mathbf{w}$ is zero at any time, it is always zero. This means that the curl of $\mathbf{v}$ arises only from the vortices created when $\mathbf{B}$ is switched on (cf. I: §5.4), and the equations simplify to

$$\begin{aligned} \frac{\partial \mathbf{v}}{\partial t} + \nabla \frac{v^2}{2} &= \frac{e}{m}\mathbf{E} \\ \nabla \times \mathbf{v} &= -\frac{e}{m}\mathbf{B}. \end{aligned} \tag{3.1.4}$$

If one now writes

$$\mathbf{J} = \frac{e\rho\mathbf{v}}{\sqrt{1 - v^2}}, \qquad J^0 = \frac{e\rho}{\sqrt{1 - v^2}}, \tag{3.1.5}$$

and if $\rho$ is constant, then

$$J_{\beta,\alpha} - J_{\alpha,\beta} = \frac{\rho e^2}{m} F_{\alpha\beta} \tag{3.1.6}$$

to an accuracy of order $v^2 \ll 1$. This equation together with Maxwell's equations will be the foundation of our model. It admits a coordinate-independent formulation,

$$F = \frac{dJm}{\rho e^2}, \qquad \delta F = J + j. \tag{3.1.7}$$

**Remarks** (3.1.8)

1. The current $j$ consists of charges not participating in the superconductive current $J$. We shall take $j$ as given, and assume that $\delta j = 0$; then (3.1.7) implies that $dF = \delta J = 0$.
2. We shall ignore the heuristic derivation to the point of not requiring that $\langle J | J \rangle = -e^2\rho^2$, which follows from (3.1.5).

3. For the present, $\rho$ is regarded as a constant, known as the density of the superconducting electrons. There will later be a discussion of the variable $\rho$.
4. Equation (3.1.7) shows that the manifold is not provided with any additional structure; for instance, there is no distinguished rest-frame.

**The Integral Form of the Equations** (3.1.9)

If $F = dA$, then (3.1.6) is equivalent to

$$\int_{\partial N_2} \left( \mathscr{J} - \frac{\rho e^2}{m} A \right) = 0, \qquad \dim N_2 = 2.$$

Two important special cases are

(i) $N = \{t = z = 0, x^2 + y^2 \leq R^2\}$:

$$\int_{x^2+y^2=R^2} d\mathbf{s} \cdot \mathbf{J} = \frac{\rho e^2}{m} \int_{x^2+y^2=R^2} ds_j A_j = -\frac{\rho e^2}{m} \int_{x^2+y^2<R^2} d\mathbf{S} \cdot \mathbf{B},$$

i.e., the current circulates in proportion to the magnetic flux (cf. Remark (I: 5.1.10; 1), with $A_j$ equal to minus the vector potential).

(ii) $N = \{y = z = 0, t_1 \leq t \leq t_2\}$:

$$\int_{t=t_2} dx\, J_x - \int_{t=t_1} dx\, J_x = \frac{\rho e^2}{m} \int_{t_1}^{t_2} dt\, dx\, E_x.$$

The rate of change of the superconductive current is given by the integral over the electric field.

**The Elimination of the Superconductive Current** (3.1.10)

From (3.1.7) we get a second-order equation for $F$,

$$\left( -\Delta + \frac{\rho e^2}{m} \right) F = -dj. \tag{3.1.11}$$

The solution of (3.1.11) requres a Green function satisfying

$$\left( -\Delta + \frac{\rho e^2}{m} \right) G_{\bar{x}}^{\text{ret}} = \delta_{\bar{x}}, \tag{3.1.12}$$

with which

$$F(\bar{x}) = F^{\text{in}}(\bar{x}) - \int G_{\bar{x}}^{\text{ret}} \wedge dj,$$

$$\left( -\Delta + \frac{\rho e^2}{m} \right) F^{\text{in}} = 0. \tag{3.1.13}$$

We have assumed that $j$ decreases sufficiently fast at infinity, so that the integrals can extend over the whole manifold as in (2.2.15), without any boundary terms.

In Minkowski space $(\mathbb{R}^4, \eta)$ it is easy to construct a Green function satisfying (3.1.12). As in (2.2.3) it has the form

$$G_{\bar{x}}^{\text{ret}} = \tfrac{1}{2}\bar{e}_{\alpha\beta} \otimes {}^*e^{\alpha\beta}\Delta^{\text{ret}}(\bar{x} - x), \tag{3.1.14}$$

where

$$\Delta^{\text{ret}}(x) = (2\pi)^{-4} \int d^4k\, e^{i\langle k|x\rangle}\left(k^2 + \frac{e^2\rho}{m}\right)^{-1}. \tag{3.1.15}$$

The integration path for $k^0$ again passes above the poles at $\pm\sqrt{|\mathbf{k}|^2 + e^2\rho/m}$, as in Figure 18, in order that $\Delta^{\text{ret}}(x) = 0$ for $x^0 < |\mathbf{x}|$. The integral (3.1.15) can be expressed in terms of Hankel functions [22], and if $\Delta^{\text{ret}}$ is integrated over time, the result is a Yukawa potential,

$$\int_{-\infty}^{\infty} dt\, \Delta^{\text{ret}}(x) = \frac{e^{-r(\rho e^2/m)^{1/2}}}{4\pi r} \tag{3.1.16}$$

(Problem 1). If (3.1.13) is written as

$$F = F^{\text{in}} + dA^{\text{ret}}$$
$$A_\alpha^{\text{ret}}(\bar{x}) = -\int d^4x\, \Delta^{\text{ret}}(\bar{x} - x) j_\alpha(x), \tag{3.1.17}$$

then in the static limit,

$$A_\alpha^{\text{ret}}(\bar{\mathbf{x}}) = -\int \frac{d^3x}{4\pi|\mathbf{x} - \bar{\mathbf{x}}|} \exp\left(-|\mathbf{x} - \bar{\mathbf{x}}|\left(\frac{\rho e^2}{m}\right)^{1/2}\right) j_\alpha(x). \tag{3.1.18}$$

**Remarks** (3.1.19)

1. A bounded solution $\sim\exp(i\langle k|x\rangle)$ for $F^{\text{in}}$ exists only for $(k^0)^2 = |\mathbf{k}|^2 + \rho e^2/m \geq \rho e^2/m$. The significance of the **plasma frequency** $\sqrt{\rho e^2/m}$ is evident in the following electrostatic situation: Suppose charges $e$ are arrayed along a line at the points $nL$, $n \in \mathbb{Z}$. If one charge is displaced slightly from equilibrium by $x \ll L$ (see Figure 35), then it feels a force $e^2((L + x)^{-2} - (L - x)^{-2}) \sim -e^2xL^{-3}$ from its two nearest neighbors. If this is set equal to $m\ddot{x}$, the equation is oscillatory at the frequency $(e^2\rho/m)^{1/2}$, where we have identified the density $\rho$ as $L^{-3}$. The oscillations are associated with solutions having $\mathbf{k} = \mathbf{0}$ and $k^0 = (e^2\rho/m)^{1/2}$.

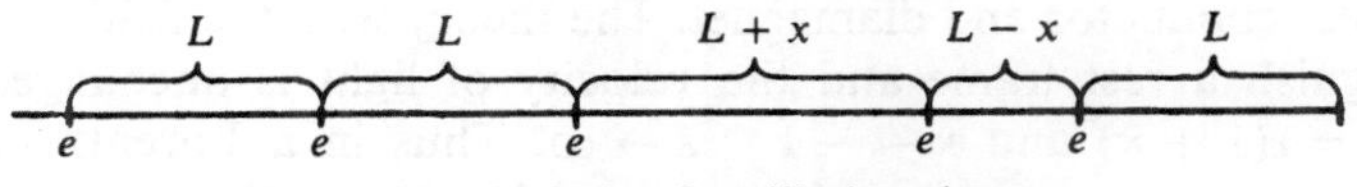


Figure 35 A chain of oscillating charges

2. There are no nontrivial static solutions ($k^0 = 0$) for $F^{\text{in}}$. According to Equation (3.1.18), the field of a static charge does not penetrate a superconductor, but decreases exponentially within a skin-depth also given by $(e^2\rho/m)^{1/2}$. The cause is the induced current $J$, which can be calculated from (3.1.7) with the boundary condition $F^{\text{in}} = 0$ and (3.1.18) as

$$J_\alpha(\bar{\mathbf{x}}) = -\int \frac{d^3x\, e^{-|\mathbf{x}-\bar{\mathbf{x}}|(\rho e^2/m)^{1/2}}}{4\pi|\mathbf{x}-\bar{\mathbf{x}}|} j_\alpha(\mathbf{x}) \frac{\rho e^2}{m}$$

in the static case; it opposes the original current and completely cancels $j$ at large distances:

$$\int d^3x\, j_\alpha(\mathbf{x}) = -\int d^3x\, J_\alpha(\mathbf{x}).$$

3. The connection between the Fourier transforms of the external and induced currents with the appropriate definition of $(\ldots)^{-1}$ is

$$\tilde{J} = -\left(1 + \frac{mk^2}{\rho e^2}\right)^{-1} \tilde{j}.$$

If $k = 0$, they are equal and opposite. The 2-form $G$ with Fourier transform

$$\tilde{G} = \left(1 + \frac{\rho e^2}{mk^2}\right)\tilde{F}$$

satisfies Maxwell's equations with no superconducting current,

$$dG = 0, \qquad \delta G = j.$$

Since $\rho$ is constant, the fields $\mathbf{D}$ and $\mathbf{H}$ of phenomenological electrodynamics satisfy these same equations, where the factor $1 + \rho e^2/mk^2$ corresponds on the one hand to a dielectric constant

$$\varepsilon(k) = 1 + \frac{\rho e^2}{mk^2}, \qquad \tilde{\mathbf{D}} = \varepsilon\tilde{\mathbf{E}};$$

and on the other to a magnetic susceptibility

$$\kappa(k) = \frac{-1}{1 + \dfrac{k^2m}{\rho e^2}}, \qquad \tilde{\mathbf{B}} = (1+\kappa)\tilde{\mathbf{H}}.$$

Both $\varepsilon(k)$ and $\kappa(k)$ are Lorentz-invariant and commute as convolution operators with translations, but they depend on the frequency, because the relationship between $F$ and $G$ is not local. If $k = 0$, then $\varepsilon = \infty$ and $\kappa = -1$; $\mathbf{E}$ and $\mathbf{B}$ become zero within the material, as they are shielded in a perfect conductor and diamagnet. The theory with the field $J$ does not distinguish a rest frame and the velocity of light is unchanged. Thus $c = 1 = \varepsilon(1+\kappa)$ and $\kappa \to -1 \quad \varepsilon \to \infty$. Thus in a Lorentz invariant situation charge screening and expulsion of the magnetic field go together.

4. In our case $G$ is just the field generated by the current $j$. In more realistic modes the connection between $F$ and $G$ is tensorial and nonlocal, even when nonlinear effects are neglected. If

$$G_{ik}(x) = \int d^4x' K_{iklm}(x, x')F^{lm}(x'),$$

then $dG \neq 0$ and $G$ no longer satisfies Maxwell's equations with the external current.

Now that we have a useful system of equations for the field plus the charges, let us take the opportunity to study the conservation theorems for the total system, by using the

**Lagrangian Formulation** (3.1.20)

*If we particularize* (3.1.7) *by setting*

$$F = dA, \qquad J = \frac{e\rho}{m}(dS + eA), \qquad S \in E^0,$$

*then the Lagrangian*

$$\mathscr{L} = -\frac{1}{2}\frac{m}{\rho e^2} J \wedge {}^*J - \frac{1}{2} F \wedge {}^*F$$

*reproduces Equations* (3.1.7) *without* $j$.

**Proof**

Making a variation of $\mathscr{L}$ as in (2.1.2) yields

$$\delta\mathscr{L} = \frac{\delta S\, d^*J}{e} - d\left(\frac{\delta S^*J}{e}\right) - \delta A \wedge [d^*F + {}^*J] - d[\delta A \wedge {}^*F]$$

(where $\delta$ is the variation, not the codifferential). Hence the Euler-Lagrange equations are

$$d^*J = 0, \qquad d^*F = -{}^*J. \qquad \square$$

**Remarks** (3.1.21)

1. Even the "superpotential" $S$ must be changed if a gauge transformation is made; if $F$ and $J$, and consequently $\mathscr{L}$, are to be invariant under $A \to A + d\Lambda$, $S \to S - e\Lambda$.
2. It is possible to express $J \wedge {}^*J$ as $\overline{(d + ieA)\varphi} \wedge {}^*(d + ieA)\varphi$ by use of a complex field $\varphi = \exp(iS)$. The effect of $A$ in this scheme is to make the exterior differential invariant under $\varphi \to \exp(ie\Lambda(x))\varphi$. (See §4.1.)

3. The scalar model discussed so far turned out to be of more importance in elementary particle physics than in condensed matter physics. It showed that in a gauge invariant theory a "mass term" can appear in Maxwell's equations. Exponentially decaying Green functions correspond in quantum theory to particles with a mass and one stumbling block in the unification of weak interactions with electromagnetism was that the particles mediating the weak interaction did not seem to be massless. In fact, they were now found to be exceedingly heavy which is believed to come from their interaction with a scalar field.

**The Energy-Momentum Forms** (3.1.22)

According to Noether's theorem (2.1.5), the 3-forms

$$\frac{(L_v S)^* J}{e} + (L_v A) \wedge {}^*F + i_v \mathscr{L} = \frac{m}{2\rho e^2}(J \wedge i_v{}^*J + (i_v J)^* J)$$
$$+ \tfrac{1}{2}((i_v F) \wedge {}^*F - F \wedge i_v{}^*F) + d((i_v A)^* F)$$

are closed for all Killing vector fields $v$ (cf. Definition (2.1.9)).

**Remarks** (3.1.23)

1. The gauge-dependent term, $d((i_v A)^* F)$, has again shown up. Since it is exact, the rest of the right side of (3.1.22) must be closed. As will be seen in (4.2.9), only that part interacts with the gravitational field, and will be used as the energy-momentum tensor.
2. All the generators of the Poincaré group can function as $v$. However, since $J$ is a 1-form, the presence of matter breaks the conformal invariance (see (2.1.10; 4)); the skin-depth is a distinguished length.
3. If $v$ is the generator $\partial_\alpha$ of a translation, then the 3-form (3.1.22) gets a contribution $t_{\alpha\beta}\, dx^\beta$ from the matter, where

$$t_{\alpha\beta} = \frac{m}{\rho e^2}[J_\alpha J_\beta - \tfrac{1}{2}\eta_{\alpha\beta} J_\gamma J^\gamma].$$

**The Properties of the Energy-Momentum Tensor of Matter** (3.1.24)

(a) $t_{\alpha\beta} = t_{\beta\alpha}$;

(b) $t_{00} = \dfrac{m}{2\rho e^2}[J_0^2 + |\mathbf{J}|^2] \geq 0$, and $= 0$ only if $J = 0$;

(c) $t_{0i} = \dfrac{m}{\rho e^2} J_0 J_i, \qquad t_{00}{}^2 \geq \sum_{i=1}^{3} t_{0i}{}^2.$

**Remarks** (3.1.25)

1. Property (a) follows from Lorentz invariance; but $t_\alpha{}^\alpha \neq 0$, because of the lack of conformal invariance (cf. (2.1.11)).
2. Since there is the same positivity property as in (2.1.13) for the electromagnetic energy-momentum tensor, the argument of (2.1.16) about the uniqueness of the Cauchy problem is still applicable.
3. A relativistic fluid is described phenomenologically by the field $u \in E_1$ of its four-velocity ($\langle u | u \rangle = -1$), its mass-density $\bar{\rho}$, and its pressure $p$. The energy-momentum tensor $T_{\alpha\beta} = (\bar{\rho} + p)u_\alpha u_\beta + pg_{\alpha\beta}$ is by construction diagonal in the rest system ($u = (1, 0, 0, 0)$) and in no other. Its eigenvalues are $(\bar{\rho}, p, p, p)$. By a comparison with (3.1.23; 3) one arrives at the identification
$$u_\alpha = J_\alpha(J_0^2 - |\mathbf{J}|^2)^{-1/2}, \qquad \bar{\rho} = p = \frac{1}{2}\frac{m}{\rho e^2}(J_0^2 - |\mathbf{J}|^2).$$
In normal matter, $p \sim$ the density of kinetic energy $\sim 10^{-10}\bar{\rho}$, while for radiation $p = \bar{\rho}/3$. Thus the pressure is unrealistically high in this model.
4. If $e = 0$, then $S$ satisfies d'Alembert's equation $\Delta S = 0$. The compressional waves, which may be thought of as sound, therefore propagate at the speed of light, as is to be expected from $\partial p/\partial \bar{\rho} = 1$. Thus $S$ describes a fluid that is as incompressible as possible without allowing sound to travel faster than light.
5. The reader may be wondering what happens to these sound waves if $e \neq 0$. In that case $S$ loses its physical significance, and can be made to disappear by a gauge transformation. It turns out that the sound waves then reappear as a longitudinal oscillatory mode of $A$, and if $e = 0$, then $A$ can oscillate only transversally (Problem 3).

In practically important problems, one part of space usually contains the metal and all the rest is a vacuum. In that case $\rho$ is not a constant in the model, but instead it changes discontinuously to zero at the metallic surface. A slowly varying field with frequency $k^0 \ll (\rho e^2/m)^{1/2}$ would consequently not penetrate the metal, but would decrease exponentially at the surface within a depth $(\rho e^2/m - (k^0)^2)^{1/2}$. For simplicity we consider the limit $\rho e^2/m \to \infty$, thus disengaging ourselves from the details of the model. The only essential feature that is preserved is that the field $F$ is excluded from within the metal. It is screened by a surface current $J$, which has a delta-function singularity at the surface, in the limit $\rho e^2/m \to \infty$ (cf. (3.1.19; 2)). The equations (3.1.7) of the model are replaced with the

**Metallic Boundary Conditions** (3.1.26)

*Suppose that the four-dimensional submanifold $CN$ is filled with metal, so that $F = 0$ on it, and let the surface $\partial N$ be given locally by the equation $u = 0$.*

*Then the restriction* (1.2.7; 3) $F_{|\partial N}$ *of the field to* $\partial N$ *must vanish, and* $F$ *is screened by the surface current* $\delta(u)(i_{du}F)_{|\partial N}$.

**Proof**

It is only necessary to extend the arguments of (2.2.1) to allow the current to have a delta-function singularity. If $F = \Theta(u)F'$, where $F'$ is continuous, then $0 = dF = \delta(u)du \wedge F'$ and $-{}^*J = d^*F = \delta(u)du \wedge {}^*F'$, where we consider only the singular parts. Because of the $\delta(u)$, it only matters what $F'$ is at $u = 0$, and all the terms containing a factor $du$ drop out of the exterior product. What is left over is simply the restriction of $F'$ according to (1.2.7; 3), and so we conclude from the first of the two equations above that $F'_{|\partial N} = 0$. If we make use of $du \wedge {}^*F = -{}^*i_{du}F$ in the second equation, the claim made about the surface current follows. □

**Remarks** (3.1.27)

1. If $u = x_1$, making $N = \{x_i \in \mathbb{R}: x_1 > 0\}$, then $dx_{1|\partial N} = 0$ and $F_{|\partial N} = (E_2\, dt \wedge dx_2 + E_3\, dt \wedge dx_3 - B_1\, dx_2 \wedge dx_3)_{|\partial N}$. Therefore $E_2$, $E_3$, and $B_1$ must vanish. The interpretation of this is that surface charges do not produce any discontinuous tangential components of the electric field, and surface currents do not produce discontinuous normal components of the magnetic field.
2. The situation is drawn schematically below:

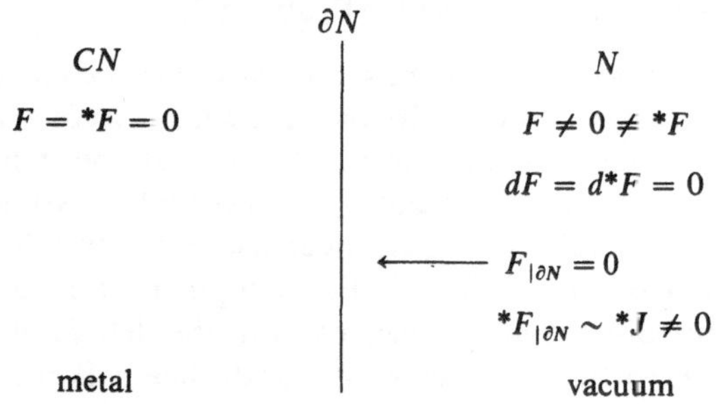


Since the surface current is not specified beforehand, it may be asked how the initial-value problem is to be solved. While the general solution (1.2.36) is always valid, it is not immediately useful, since ${}^*F$ occurs in the surface integral as well as $F$. It would seem to be necessary to know the restrictions of both to the surface, and we only know that $F_{|\partial N} = 0$. If we manage to find a $G_{\bar{x}}$ such that ${}^*dG_{\bar{x}|\partial N} = 0$, however, then there are no unknown surface contributions, and the solution works as in Chapter 2. In other words, the key to the problem is

**The Green Function for Metallic Boundary Conditions** (3.1.28)

Let $M \subset \mathbb{R}^4$ be a part of Minkowski space bounded by spacelike hypersurfaces $\partial M$. Suppose that metal fills $CN \cap M$ and that the current $j$ is known in $N \subset M$ and $\partial_v N \equiv \partial N \backslash \partial N \cap \partial M$ is the vertical boundary of $N$. If $G_{\bar{x}}$ satisfies the equations $-\Delta G_{\bar{x}} = \delta_{\bar{x}}$ and $*dG_{\bar{x}|\partial_v N} = 0\ \forall \bar{x} \in N \backslash \partial N$, then for all $x \in N \backslash \partial N$ the field strength is given by

$$F(\bar{x}) = \int_N dG_{\bar{x}} \wedge j - \int_{\partial M \cap N} [\delta G_{\bar{x}} \wedge F - *dG_{\bar{x}} \wedge *F].$$

**Remarks** (3.1.29)

1. The situation looks as follows:

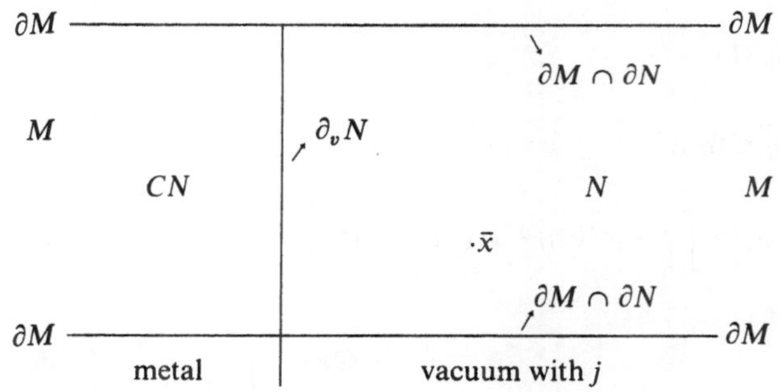


   In (3.1.28) it is only relevant to know $F$ and $*F$ on the initial and final surfaces $\partial M \cap \partial N$, and not on $\partial_v N$. In later examples $G_x$, like $G_x^{\text{ret}}$, will be zero outside the past light-cone of $\bar{x}$, and thus only the initial surface affects the integral. The reason this $G_{\bar{x}}$ is selected is that it expresses $F$ in terms of the initial data, and it automatically takes care of the effect of the currents in the upper surface.
2. Strictly speaking, $N$ is not a manifold with a boundary, because it has a sharp edge. But since the integration by parts used in (1.2.36) can also be justified on regions of the form $\{(x, y) \in \mathbb{R}^2 : x \geq 0, y \geq 0\}$, this presents no real obstacle.
3. The $G_{\bar{x}}$ of (3.1.28) is not uniquely determined. However, as long as we possess some $G_{\bar{x}}$ that vanishes outside the past light-cone, the formula (3.1.28) guarantees the uniqueness of the Cauchy problem.

In the following sections we shall prove the existence of the $G_{\bar{x}}$ used in (3.1.28) by explicit construction, making use of the well-known method of images from electrostatics. This static method is generalized by means of $G_{\bar{x}}$ for charges in arbitrary motion.

**Problems (3.1.30)**

1. Calculate the integral (3.1.16).
2. Show that $t_{\alpha\beta,}{}^{\beta} = J_\beta F^\beta{}_\alpha$, with $t_{\alpha\beta}$ as in (3.1.23; 3).
3. Show that the equations

$$dJ = \frac{e^2\rho}{m} F, \qquad \delta F = J, \qquad \frac{e^2\rho}{m} > 0,$$

have three linearly independent solutions $\sim \exp(i\langle k|x\rangle)$ if

$$k^2 = -\frac{e^2\rho}{m},$$

and otherwise have only the trivial solution $F = 0$.

**Solutions (3.1.31)**

1. If $\mu = \sqrt{e^2\rho/m}$, then

$$\int_{-\infty}^{\infty} dt (2\pi)^{-4} \int d^4k\, e^{i\langle k|x\rangle}(k^2 + \mu^2)^{-1} = (2\pi)^{-3} \int d^3k\, e^{i\mathbf{k}\cdot\mathbf{x}}(|\mathbf{k}|^2 + \mu^2)^{-1}$$

$$= (2\pi)^{-2} \int_0^\infty \frac{k^2\, dk}{k^2 + \mu^2} \int_{-1}^{1} d\eta\, e^{ikr\eta}$$

$$= (2\pi)^{-2} \int_{-\infty}^{\infty} \frac{k\, dk}{k^2 + \mu^2} \frac{e^{ikr}}{ir} = \frac{e^{-\mu r}}{4\pi r}.$$

2. Because of (3.1.6) and $J_{\beta,}{}^{\beta} = 0$,

$$t_{\alpha\beta,}{}^{\beta} = \frac{m}{\rho e^2}[J_{\alpha}{}^{,\beta} J_\beta - J^\beta{}_{,\alpha} J_\beta] = F_{\beta\alpha} J^\beta.$$

3. In the Fourier-transformed space the equations become

$$k_\mu k^\rho \tilde{F}_{\rho\nu} - k_\nu k^\rho \tilde{F}_{\rho\mu} = \frac{e^2\rho}{m} \tilde{F}_{\nu\mu}.$$

Multiplying this by $\cdot\, k^\nu$,

$$-k^2 k^\rho \tilde{F}_{\rho\mu} = \frac{e^2\rho}{m} k^\nu \tilde{F}_{\nu\mu}.$$

Since $\tilde{A}_\mu - k_\mu \tilde{S} \equiv -(m/e^2\rho) k^\rho \tilde{F}_{\rho\mu}$ vanishes iff $F \equiv 0$, either $\tilde{F}_{\nu\mu} = 0$ or else $k^2 = -e^2\rho/m$. Then $\tilde{F}$ is of the form $\tilde{F}_{\nu\mu} = k_\nu \tilde{A}_\mu - k_\mu \tilde{A}_\nu$, which vanishes only if $\tilde{A}_\nu \sim k_\nu$. Thus for the three directions other than $k$ there are nonvanishing solutions.

## 3.2 The Half-Space, the Wave-Guide, and the Resonant Cavity

*The general solution of Maxwell's equations with metallic boundary conditions is easy to construct for simple geometric arrangements of the conductors.*

Classically, the electromagnetic problem in the presence of metallic surfaces is usually conceived as the quest for particular solutions. Here we shall proceed directly to the more general problem, and solve the Cauchy problem by specifying the $G_{\bar{x}}$ of (3.1.28). The interesting question will be what the causal structure of the Green function is. Like the $G_{\bar{x}}^{\text{ret}}$ of (2.2.7), its support will be restricted to the full past light-cone of $\bar{x}$, but unlike $G_x^{\text{ret}}$ not to its surface. This fact is due to waves that reflect from the metallic surfaces and return at some later time. Such echoes may apparently violate causality, as when they give rise to phase velocities greater than the speed of light. In all the problems discussed below we replace the conducting material with a metallic boundary condition. The currents induced on the conductors do not appear, so we may use the symbol $J$ for the externally prescribed current.

We begin with a trivial warm-up exercise, the problem of a plane metallic mirror. The method of solution introduced points the way to the procedure for more complicated problems.

**The Half-Space** (3.2.1)

*In the notation of* (3.1.28), *let*

$$M = \{x^i \in \mathbb{R}^4 : x^0 \geq t^0\},$$
$$N = \{x^i \in M : x^1 \geq 0\}.$$

*The symbol R will stand for the reflection* $(x^0, x^1, x^2, x^3) \to (x^0, -x^1, x^2, x^3)$ *in M and at the same for the induced mapping on the space of tensors* (I: 2.4.19). *Then*

$$G_{\bar{x}} = (1 + R)G_{\bar{x}}^{\text{ret}},$$

*with* $G_{\bar{x}}^{\text{ret}}$ *from* (2.2.7), *is the Green function needed for this problem.*

**Proof**

The rules for manipulating the diffeomorphism $R$ of the space of vectors are such that it can be interchanged with sums, products, and exterior differentiation:

$$R(\omega + \nu) = R\omega + R\nu, \qquad R(\omega \wedge \nu) = R\omega \wedge R\nu, \qquad \omega, \nu \in E,$$
$$R(e^0, e^1, e^2, e^3) = (e^0, -e^1, e^2, e^3), \qquad e^i = dx^i.$$

However, $R$ reverses the orientation so that $R{*}\omega = -{*}R\omega$, $\forall\omega \in E_p$ (cf. (1.2.17; 1)).

We now make use of a

**Lemma**

*If* $\omega \in E_p$ *and* $\omega = -R\omega$, *then* $\omega_{|\partial_v N} = 0$.

**Proof of Lemma**

Let

$$\omega = \sum_{(i)} \omega_{i_1 \cdots i_p}(x) e^{i_1 \cdots i_p}, \qquad e^i = dx^i, \qquad \omega_{i_1 \cdots i_p} \in E_0.$$

If $i_1 \cdots i_p$ contains the index 1, then $e^{i_1 \cdots i_p}{}_{|x^1=0} = 0$, because $dx_{1|x_1=0} = 0$. If $i_1 \cdots i_p$ does not contain the index 1, then $\omega_{i_1 \cdots i_p}(x^0, -x^1, x^2, x^3) = -\omega_{i_1 \cdots i_p}(x^0, x^1, x^2, x^3)$, which must vanish when $x^1 = 0$.

This lemma implies the property ${*}dG_{\bar{x}|\partial_v N} = 0$ required in (3.1.28), because

$$R{*}d(1 + R)G_{\bar{x}}^{\text{ret}} = -{*}dR(1 + R)G_{\bar{x}}^{\text{ret}} = -{*}d(1 + R)G_{\bar{x}}^{\text{ret}}.$$

Observe that none of the operations of this equation affect $\bar{x}$, and that $R^2 = 1$. The property $-\Delta G_{\bar{x}} = \delta_{\bar{x}}\ \forall\bar{x} \in N$ is proved by noting that the factor $\delta((\bar{x} - x)^2)$ has been replaced with $\delta((\bar{x}^0 - x^0)^2 - (\bar{x}^1 + x^1)^2 - (\bar{x}^2 - x^2)^2 - (\bar{x}^3 - x^3)^2)$ in $RG_{\bar{x}}^{\text{ret}}$. When $\Delta$ acts on this it yields zero unless $x = (\bar{x}^0, -\bar{x}^1, \bar{x}^2, \bar{x}^3)$. However, this fails to be in $N \backslash \partial N$ if $\bar{x} \in N \backslash \partial N$. □

**Remarks** (3.2.2)

1. Since $R$ reverses the orientation, $\int_M R\omega = -\int_M \omega\ \forall\omega \in E_p$. If supp $j \subset N$ the integral $\int_N (1 + R) dG^{\text{ret}} \wedge J$ can be taken as $\int_M R$ does not send points out of the integration region, and because

$$\int_M R\, dG^{\text{ret}} \wedge J = \int_M R(dG^{\text{ret}} \wedge RJ) = -\int_M dG^{\text{ret}} \wedge RJ$$

the integral can be written as

$$F(\bar{x}) = \int_M dG_{\bar{x}}^{\text{ret}} \wedge (1 - R)J + \int_{\partial M \cap N} \cdots.$$

The components of $-RJ$ are $(-J_0(Rx), J_1(Rx), -J_2(Rx), -J_3(Rx))$, and thus the field produced by $J$ is as if there were a mirror-image charge of reversed sign at $Rx \in CN$, undergoing the reflected motion. It is easy to see that its field taken together with the field directly produced by $J$ satisfies the metallic boundary conditions (3.1.26) on $\partial_v N$ (Figure 36). In actuality there is an induced charge in the metal, not in the interior of $CN$, but rather, according to (3.1.26), on $\partial_v N$. The surface current $\delta(x^1)(i_{dx^1} F)_{|\partial_v N}$ generates the field in $N$ that would come from $-RJ$.

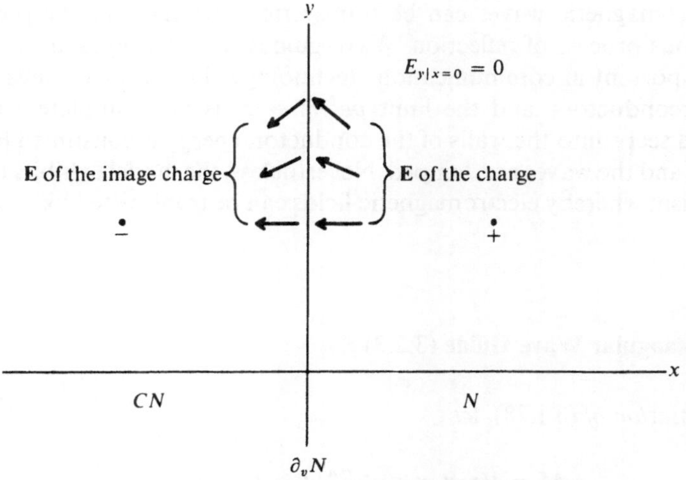


Figure 36 **E** in the presence of a reflecting half-plane

2. It is likewise possible to take $\int_{\partial M \cap N}$ as $\int_{\partial M}$, which involves only $G_{\bar{x}}^{\text{ret}}$, and the appropriate reflected initial data are to be used on $\partial M \cap CN$. If $F$ and $*F_{|\partial M \cap N}$ originate in an incoming wave, then their values in $N$ are the same as if there were no metal present, and the reflected initial values had been specified on $\partial M \cap CN$.
3. The support of $G_{\bar{x}}$ in $N$ is contained in the full past light-cone of $\bar{x}$ (Figure 37). Metallic boundary conditions produce echo effects, but never really violate causality.

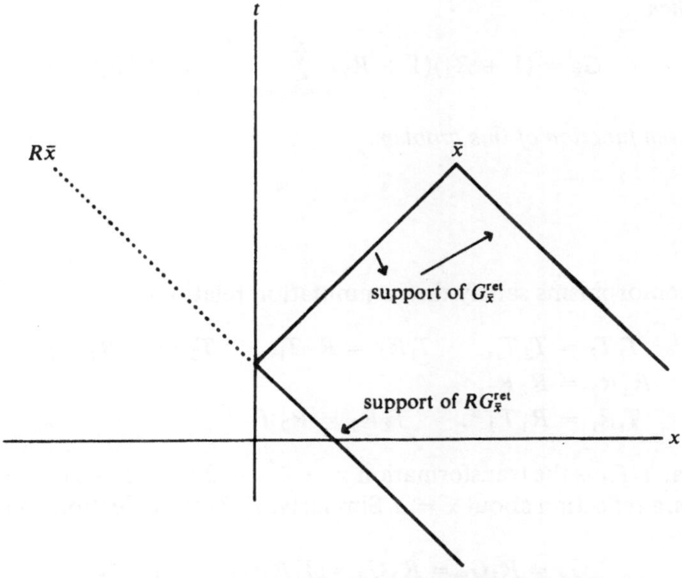


Figure 37 The support of $G_{\bar{x}}$ with a reflecting half-plane

Electromagnetic waves can be transmitted through metallic pipes by a continuous process of reflection. Wave-guides, as these pipes are called, are quite important in communications technology. They are not always made of superconductors, and the limit $\rho e^2/m \to \infty$ is not completely realistic. The field seeps into the walls of the conductor, energy is consumed for Joule heating, and the waves are damped. Nevertheless, the model exhibits the basic mechanism whereby electromagnetic fields can be transmitted like water in a hose.

**The Rectangular Wave-Guide** (3.2.3)

*In the notation of* (3.1.28), *let*

$$M = \{(t, x, y, z) \in \mathbb{R}^4 : t \geq t_0\},$$
$$N = \{x^\alpha \in M : 0 \leq x \leq a, 0 \leq y \leq b\},$$

*and let* $R_1$, $R_2$, $T_1$, *and* $T_2$ *be the transformations on the space of tensors induced by the diffeomorphisms*

$$R_1 : (t, x, y, z) \to (t, -x, y, z)$$
$$R_2 : (t, x, y, z) \to (t, x, -y, z)$$
$$T_1 : (t, x, y, z) \to (t, x + 2a, y, z)$$
$$T_2 : (t, x, y, z) \to (t, x, y + 2b, z)$$

*of* $M$. *Then*

$$G_{\bar{x}} = (1 - R_1)(1 + R_2) \sum_{n=-\infty}^{\infty} \sum_{m=-\infty}^{\infty} T_1^n T_2^m G_{\bar{x}}^{\text{ret}}$$

*is the Green function of this problem.*

**Proof**

The diffeomorphisms satisfy the commutation relations

$$T_1 T_2 = T_2 T_1, \qquad T_1 R_2 = R_2 T_1, \qquad T_2 R_1 = R_1 T_2,$$
$$R_1 R_2 = R_2 R_1,$$
$$T_1 R_1 = R_1 T_1^{-1}, \qquad T_2 R_2 = R_2 T_2^{-1}.$$

Note that $T_1 R_1$ is the transformation $x \to -x + 2a$, or $(x - a) \to -(x - a)$; in words, a reflection about $x = a$. Similarly, $T_2 R_2$ is a reflection about $y = b$. Since

$$G_{\bar{x}} = R_1 G_{\bar{x}} = R_2 G_{\bar{x}} = T_1 R_1 G_{\bar{x}} = T_2 R_2 G_{\bar{x}}$$

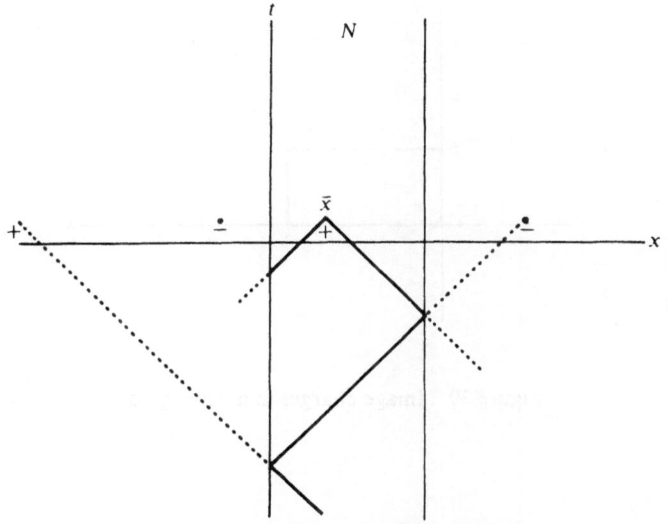


Figure 38 The support of $G_{\bar{x}}$ in a wave-guide

by construction,

$$*dG_{\bar{x}|x=0} = *dG_{\bar{x}|y=0} = *dG_{\bar{x}|x=a} = *dG_{\bar{x}|y=b} = 0,$$

as in the proof of (3.2.1). Moreover, the image of $(0, a)$ under $T_1^n$ equals $(2na, (2n + 1)a)$, and its image under $R_1 T_1^n$ equals $(-2na, -(2n + 1)a)$, and similar statements hold for $T_2$ and $R_2$. As a result, the only part of the sum $\sum_{n,m} (1 + R_1)(1 + R_2)T_1^n T_2^m$ that sends a point of $N \setminus \partial N$ back into $N \setminus \partial N$ is the 1 from the term with $n = m = 0$. Hence that is the only term contributing to $\Delta G_{\bar{x}}$ if $x$ and $\bar{x} \in N$, and therefore $-\Delta G_{\bar{x}} = \delta_x$. □

**Remarks** (3.2.4)

1. Although $N$ is not a manifold, because of its sharp edges, its structure is harmless enough that the integration by parts needed in order to use $G_{\bar{x}}$ is easy to justify.
2. Once again, the support of $G_{\bar{x}}$ in $N$ is contained in the full past light-cone (Figure 38). This is why we were able to choose $M$ as a region extending to $t = +\infty$ as in (2.2.10).

The structure of $G_{\bar{x}}$ results from the infinite number of reflections of the field back and forth between the walls. Consequently, there are infinitely many image charges, as depicted in cross-section in Figure 39. Their periodic configuration gives rise to characteristic normal modes of oscillation.

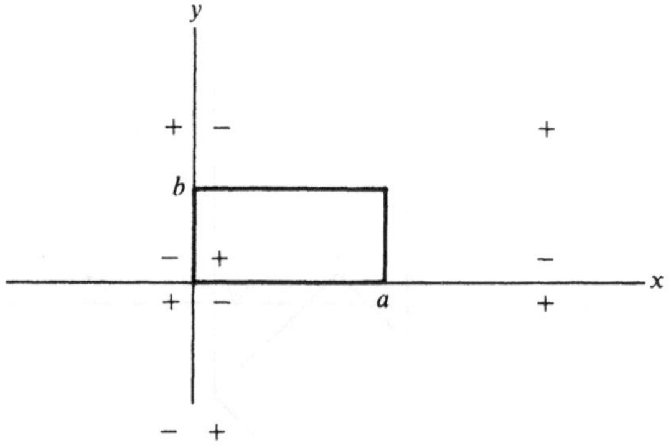


Figure 39 Image charges in a wave-guide

**The Decomposition of $G_{\bar{x}}$ into Normal Modes** (3.2.5)

If we write

$$G_{\bar{x}}^{\text{ret}} = \tfrac{1}{2}\bar{e}_{\alpha\beta} \otimes {}^*e^{\alpha\beta} \int \frac{d^4k}{(2\pi)^4} \frac{e^{i\langle k|\bar{x}-x\rangle}}{k^2} \tag{3.2.6}$$

and use the invariance of $e^\alpha$ under the diffeomorphism

$$T_1^n T_2^m: (t, x, y, z) \to (t, x + 2na, y + 2mb, z)$$

then we find ourselves presented with sums of the form

$$\sum_{n=-\infty}^{\infty} \exp(ik(x + 2na)) = \exp(ikx) \sum_{g \in \mathbb{Z}} \delta\left(k - \frac{\pi g}{a}\right) \frac{\pi}{a}. \tag{3.2.7}$$

Substitution of this into $G_{\bar{y}}^{\text{ret}}$ and summing $\sum_m$ allows the integration $\int dk_x\, dk_y$ to be accomplished with the aid of delta-functions:

$$\sum_{n=-\infty}^{\infty} \sum_{m=-\infty}^{\infty} \int_{-\infty}^{\infty} dk_x\, dk_y\, e^{i(k_x(x+2na)+k_y(y+2mb))}$$

$$= \frac{\pi^2}{ab} \sum_{g_1=-\infty}^{\infty} \sum_{g_2=-\infty}^{\infty} e^{i\pi(g_1 x/a + g_2 y/b)}. \tag{3.2.8}$$

The $G_{\bar{x}}$ of (3.2.3) becomes

$$G_{\bar{x}} = \tfrac{1}{2}\bar{e}_{\alpha\beta}(1 + R_1)(1 + R_2)^* e^{\alpha\beta}$$

$$\cdot \sum_{g_1, g_2} \int \frac{d\omega\, dk}{ab(4\pi)^2} \frac{e^{i((\bar{x}-x)\pi g_1/a + (\bar{y}-y)\pi g_2/b + (\bar{z}-z)k - (\bar{t}-t)\omega)}}{(\pi g_1/a)^2 + (\pi g_2/b)^2 + k^2 - \omega^2}. \tag{3.2.9}$$

**The Field Produced by a Current $J$ (3.2.10)**

For the sake of simplicity we consider $M = \mathbb{R}^4$ with no incoming field. We also suppose that the charges are concentrated within $N$, so that $J_{|\partial N} = 0$. Then, after an integration by parts as in (3.1.13),

$$F(\bar{x}) = -\int G_{\bar{x}} \wedge dJ.$$

The Green function $G_{\bar{x}}$ contains the reflections $(1 + R_1)(1 + R_2)$; if we always combine a reflected term with one having a $g_i$ of opposite sign, then, for example for $(1 + R_1)$, we get

$$\int dx(e^{i(\bar{x}-x)\pi g_1/a} \pm e^{-i(\bar{x}+x)\pi g_1/a})dJ(x) = \begin{matrix}\cos\\ i\sin\end{matrix}\ \bar{x}\pi g_1/a \cdot 2 \int dx\, e^{-ix\pi g_1/a}\, dJ(x).$$

The integration over $\omega$ can be done as in (2.2.25;2), leaving the explicit formula

$$F = \bar{e}_{\alpha\beta} \sum_{g_1, g_2} f^{\alpha\beta} \int \frac{dk}{2\pi ab} \int d^3x\, dt\, \Theta(\bar{t} - t) \frac{\sin \omega(\bar{t} - t)}{-\omega}$$

$$\cdot\, e^{ik(\bar{z}-z)} e^{-i(x\pi g_1/a + y\pi g_2/b)}\, dJ^{\alpha\beta}, \tag{3.2.11}$$

where $\omega = ((\pi g_1/a)^2 + (\pi g_2/b)^2 + k^2)^{1/2}$ and $f^{\alpha\beta}$ are the normal modes

$$f = \begin{pmatrix} 0 & ic_1s_2 & is_1c_2 & -s_1s_2 \\ -ic_1s_2 & 0 & c_1c_2 & ic_1s_2 \\ -is_1c_2 & -c_1c_2 & 0 & is_1c_2 \\ s_1s_2 & -ic_1s_2 & -is_1c_2 & 0 \end{pmatrix} \tag{3.2.12}$$

$$c_1 \equiv \cos \bar{x}g_1\pi/a, \qquad s_1 \equiv \sin \bar{x}g_1\pi/a,$$

$$c_2 \equiv \cos \bar{y}g_2\pi/b, \qquad s_2 \equiv \sin \bar{y}g_2\pi/b.$$

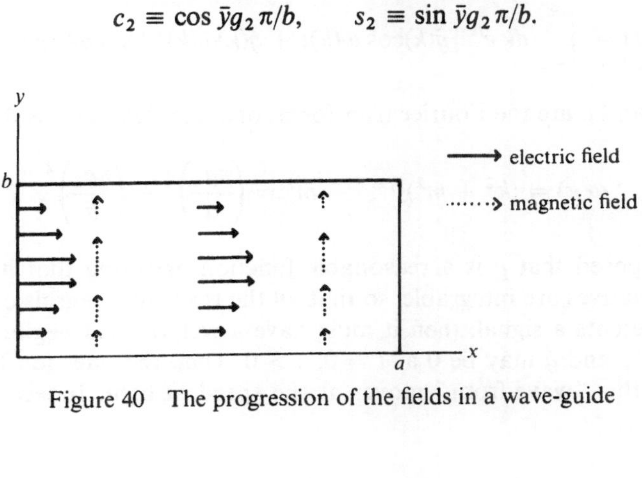


Figure 40 The progression of the fields in a wave-guide

**Remarks** (3.2.13)

1. The boundary conditions (3.1.26) are satisfied because $f_{|\partial N} = 0$. If, say, $\bar{x} = 0$, then they can be checked by deleting the $x^1$ row and column and noting that what is left is $\sim s_1$, which vanishes when $\bar{x} = 0$. It is natural to ask what is meant by the tangential component of **E** or the normal component of **B** at the edges (cf. [2], §7). It turns out that all the components in question are zero at the corners since $s_i = 0$, and only an axial magnetic field $B_3$ is left over.
2. The term with $g_1 = g_2 = 0$ is an axial magnetic field. Electric fields first occur when $g_1 = 0$ and $g_2 = 1$ or vice versa. In the former case the field evolves as shown in Figure 40. The oscillation of the corresponding solution with $J = 0$ (see (3.2.18; 1)) has a minimal frequency $\pi/b$, and the fields depend on $y$. It is clear that a constant electric field can not exist within a metallic pipe, because the boundary conditions would require it to vanish identically. On the other hand, we know that waves of a sufficiently high frequency are able to travel through metal pipes, because we can see through them.
3. If $g_{1,2} \neq 0$ are fixed, then Equation (3.2.11) describes waves moving in the $z$-direction at a phase velocity $((\pi g_1/ka)^2 + (\pi g_2/kb)^2 + 1)^{1/2} > 1$. It is a purely geometrical effect that this velocity is greater than the speed of light. It comes about because the waves do not pass directly through the pipe, but are reflected back and forth at the walls. A wave moving directly in the $z$-direction could not fulfil the boundary conditions; they demand interference from other waves at a certain angle to the $z$-axis. Such a wave $\sim \exp(i(\mathbf{k} \cdot \mathbf{x} - \omega t))$ with $|\mathbf{k}| = \omega$ necessarily has $\omega/k_z > 1$, as the intersection with a plane of constant phase moves along the $z$-axis faster than light, see Figure 41.
4. The group velocity $\partial\omega/\partial k = k/\omega$ is less than 1. Problem 2 establishes its significance as the flow of energy in the $z$-direction per energy.
5. The question of whether the signal velocity is $\leq 1$ is more pertinent. To answer it, consider the wave-packet

$$g(z, t) = \int_{-\infty}^{\infty} dk\, e^{ikz}[\tilde{g}(k)\cos \omega(k)t + \tilde{\dot{g}}(k)\omega(k)^{-1} \sin \omega(k)t], \quad (3.2.14)$$

where $\tilde{g}$ and $\tilde{\dot{g}}$ are the Fourier transforms of $g$ and $\partial g/\partial t$ at $t = 0$, and

$$\omega(k) = (k^2 + m^2)^{1/2}, \qquad m^2 = \left(\frac{\pi g_1}{a}\right)^2 + \left(\frac{\pi g_2}{b}\right)^2.$$

It is supposed that $g$ is a reasonable function, meaning that it and its first derivatives are integrable, so that all the relevant integrals converge. If $g$ represents a signal, then it must have a well-defined beginning; for instance, $g$ and $\dot{g}$ may be 0 at $t = 0$, $z > 0$. Then the question becomes whether this "wave-front" moves at the speed of light. Indeed it does;

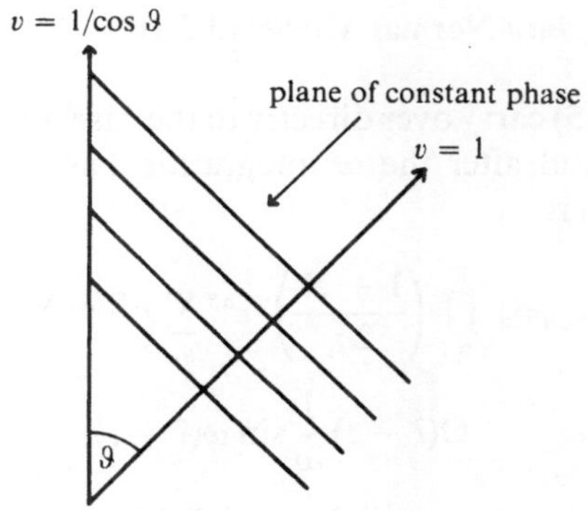


Figure 41 Speeds faster than light in a wave-guide

the above assumptions imply that $g = 0$ for all $z > |t|$ (Problem 5). This does not depend at all on $m^2 > 0$, but holds also if $m^2 < 0$, i.e., for phase velocities less than 1 and group velocities greater than 1 (tachyons) the signal velocity is still 1.

In a wave-guide there are waves with a continuous frequency-spectrum $\omega(k) \geq \omega_{\min}$. If the wave-guide is sealed off in the $z$-direction, then the electromagnetic oscillations have only a discrete spectrum.

**The Resonant Cavity** (3.2.15)

*Let the sets occurring in* (3.1.28) *be*

$$M = \{x^\alpha \in \mathbb{R}^4 : x^0 \geq t_0\},$$
$$N = \{x^\alpha \in M : 0 \leq x^i \leq a_i, i = 1, 2, 3\}.$$

*Let $R_i$ denote the reflections $x^i \to -x^i$, and $T_i$ the translations $x^i \to x^i + 2a^i$, as well as the induced transformations of the tensor spaces. Then the Green function for this problem is*

$$G_{\bar{x}} = \prod_{i=1}^{3} \sum_{n_i=-\infty}^{\infty} (1 + R_i) T_i^{n_i} G_{\bar{x}}^{\text{ret}}.$$

The proof proceeds exactly as for (3.2.3), and will not be repeated here. Since the earlier remarks about the causal structure are still valid, let us immediately make the

**Decomposition of $G_{\bar{x}}$ into Normal Modes** (3.2.16)

The formulas of (3.2.5) carry over directly to the case where $N$ is also bounded in the $z$-direction, and after the $\omega$ integration has been done, we find, in analogy to (3.2.9), that

$$G_{\bar{x}} = \tfrac{1}{2}\bar{e}_{\alpha\beta} \otimes \prod_{i=1}^{3} \left(\frac{1 + R_i}{2a_i}\right)^{*} e^{\alpha\beta} \sum_{g_i} e^{i\sum_{j=1}^{3}(\bar{x}^j - x^j)\pi g_j/a_j}.$$

$$\Theta(\bar{t} - t)\frac{1}{\omega}\sin\omega(\bar{t} - t),$$

$$\omega = \left[\sum_{i=1}^{3}\left(\frac{\pi g_i}{a_i}\right)^2\right]^{1/2}.$$

By making a change of sign in the $g_i$ to replace the reflection of the $x_i$ with a reflection of the $\bar{x}_i$, as in (3.2.10), we can solve for the normal modes:

$$G_{\bar{x}} = \tfrac{1}{2}\bar{e}_{\alpha\beta} \otimes \sum_{g_i} f^{\alpha\beta *} e^{\alpha\beta} e^{-i\sum_{j=1}^{3} x^j \pi g_j/a_j} \Theta(\bar{t} - t)\frac{\sin\omega(\bar{t} - t)}{a_1 a_2 a_3 \omega}.$$

This time,

$$f = \begin{pmatrix} 0 & -c_1 s_2 s_3 & -s_1 c_2 s_3 & -s_1 s_2 c_3 \\ c_1 s_2 s_3 & 0 & ic_1 c_2 s_3 & ic_1 s_2 c_3 \\ s_1 c_2 s_3 & -ic_1 c_2 s_3 & 0 & is_1 c_2 c_3 \\ s_1 s_2 c_3 & -ic_1 s_2 c_3 & -is_1 c_2 c_3 & 0 \end{pmatrix},$$

with the abbreviations

$$\left.\begin{matrix} c_i \\ s_i \end{matrix}\right\} = \left\{\begin{matrix} \cos \\ \sin \end{matrix}\right\} \frac{\bar{x}_i g_i \pi}{a_i}.$$

**Remarks** (3.2.17)

1. The boundary conditions are again satisfied at the edges by the disappearance of the relevant components, and at the corners $f$ is identically zero.
2. There are no static fields with $J = 0$ in a resonant cavity; if $g_i = 0$, $i = 1, 2, 3$, then $f \equiv 0$. Consequently there is a minimal frequency

$$\omega_{\min} = \left(\sum\left(\frac{\pi}{a_i}\right)^2\right)^{1/2}.$$

**Problems** (3.2.18)

1. Provide the $f^{\alpha\beta}$ of (3.2.12) with coefficients for which there exists a solution of the homogeneous Maxwell equations ($J = 0$).

2. For the $f$ of Problem 1, calculate the averages of $T^{00}$ and $T^{03}$, and verify that $\overline{T^{03}}/\overline{T^{00}} = k/\omega$.

3. Find the solutions corresponding to Problem 1 for the cylindrical geometry

$$N = \{x^\alpha \in \mathbb{R}^4 : x^2 + y^2 \leq a^2\}.$$

4. What type of oscillation in the cylinder has the lowest $\omega_{\text{min}}$, and how does $\omega_{\text{min}}^2$ times the cross-sectional area compare with the rectangular case?

5. Suppose that the norms

$$\int_{-\infty}^{\infty} dz\,|g(z, 0)| \equiv \|g\|, \qquad \int_{-\infty}^{\infty} dz \left|\frac{\partial}{\partial t} g(z, 0)\right| \equiv \|\dot{g}\|$$

and

$$\int_{-\infty}^{\infty} dz \left|\frac{\partial}{\partial z} g(z, 0)\right| \equiv \|g'\|$$

are finite for the $g$ of (3.2.14). Show that if $g = \dot{g} = 0$ at $t = 0$, $z > 0$, then $g(z, t) = 0$ for all $z > |t|$.

## **Solutions** (3.2.19)

1.

$$f^{\alpha\beta} = \begin{array}{|c|c|c|c|} \hline & i\dfrac{kg_1\pi}{a}c_1s_2 & i\dfrac{kg_2\pi}{b}s_1c_2 & (\omega^2 - k^2)s_1s_2 \\ \hline & & 0 & -i\dfrac{\omega g_1\pi}{a}c_1s_2 \\ \hline & & & -i\dfrac{\omega g_2\pi}{b}s_1c_2 \\ \hline & & & \\ \hline \end{array} \cdot \exp(i\omega t) \quad \text{or} \quad f \to {}^*f.$$

For this $f$, $B_3 = 0$, and it is known as a $TM$ (transverse magnetic) solution, and hence for ${}^*f$, $E_3 = 0$, which is a $TE$ (transverse electric) solution.

2. Because $\overline{c_i^2} = \overline{s_i^2} = \frac{1}{2}$, etc., we find that

$$\overline{T^{00}} = \frac{\omega^2}{4}(\omega^2 - k^2), \qquad \overline{T^{03}} = \frac{k\omega}{4}(\omega^2 - k^2).$$

3. If $J_n$ denotes the $n$-th Bessel function. then

$$
f^{\mu\nu} = \frac{1}{\rho}
\begin{array}{|cccc|}
t & z & \varphi & \rho \\
 & (\omega^2 - k^2)\rho & -\dfrac{kn}{\rho} & ik\rho\dfrac{\partial}{\partial\rho} \\
(-\omega^2 + k^2)\rho & 0 & -\dfrac{n\omega}{\rho} & i\omega\rho\dfrac{\partial}{\partial\rho} \\
\dfrac{kn}{\rho} & \dfrac{n\omega}{\rho} & 0 & 0 \\
\hline
-ik\rho\dfrac{\partial}{\partial\rho} & -i\omega\rho\dfrac{\partial}{\partial\rho} & 0 & 0
\end{array}
\cdot e^{i(kz + n\varphi - \omega t)} J_n(\rho\sqrt{\omega^2 - k^2}).
$$

$$
f^{*\mu\nu} = \frac{1}{\rho}
\begin{array}{|cccc|}
0 & 0 & i\omega\dfrac{\partial}{\partial\rho} & n\omega \\
0 & 0 & ik\dfrac{\partial}{\partial\rho} & kn \\
-i\omega\dfrac{\partial}{\partial\rho} & -ik\dfrac{\partial}{\partial\rho} & 0 & \omega^2 - k^4 \\
\hline
-n\omega & -kn & k^2 - \omega^2 & 0
\end{array}
\cdot e^{i(kz + n\varphi - \omega t)} J_n(\rho\sqrt{\omega^2 - k^2}).
$$

Solving Maxwell's equations for $f_{|\partial N} = 0$ means that the parts enclosed in the dotted lines must vanish when $\rho = a$. This implies that

$$
J_n(a\sqrt{\omega^2 - k^2}) = 0, \qquad \text{i.e., } \omega = \pm\sqrt{\frac{k^2 + j_{ni}^2}{a^2}},
$$

where $j_{ni}$ is the $i$-th zero of $J_n$. Interchanging $f$ and $^*f$ produces a $TE$ solution, for which it is required that

$$
J_n'(a\sqrt{\omega^2 - k^2}) = 0, \qquad \text{where } J_n'(\rho) \equiv d\, J_n(\rho) d\rho,
$$

i.e., $\omega_{\min} = j_{ni}'(a)$, where $j_{ni}'$ is the $i$-th zero of $J_n'$.

4. Because $j_{01} = 2.40$, $j_{01}' = 3.83$, and $j_{11}' = 1.84$, the $TE$ solution with $n = 1$ has the lowest $\omega_{\min}$, and its $\omega_{\min}^2 \cdot a^2\pi = \pi(1.84)^2$. This is always somewhat larger than the analogous product for the $g_1 = 1, g_2 = 0$ oscillation with a square cross-section:

$$
\left(\frac{\pi}{a}\right)^2 \cdot a^2 = \pi \cdot 3.14 \leq \pi \cdot (1.84)^2 = \pi \cdot 3.38.
$$

5. One can analytically continue $g$ into the upper half-plane, where it goes to zero as $|k|^{-1}$, because

$$|\tilde{g}(u+iv)| = \left|\int_{-\infty}^{\infty} dz\, e^{-iz(u+iv)} g(z,0)\right| = \left|\int_{-\infty}^{0} dz\, e^{zv} e^{-izu} g'(z,0) \frac{i}{u+iv}\right|$$
$$\leq \frac{\|g'\|}{\sqrt{u^2+v^2}} \qquad \forall u \in \mathbb{R},\, v \in \mathbb{R}^+.$$

Similarly, $\tilde{\dot{g}}$ is analytic and bounded by $\|\dot{g}\|$ where $v > 0$. Since $\cos[\omega(k)t]$ and $\sin[\omega(k)t]/\omega(k)$ are entire functions in $k$, the $k$ integral of (3.2.14) can be deformed into the upper half-plane. If we decompose

$$\int_{-\infty}^{\infty} dk \quad \text{into} \quad \int_{-\infty}^{-R} dk + \int_{\pi}^{0} i\, d\varphi\, R \exp(i\varphi) + \int_{R}^{\infty} dk,$$

where $k = R\exp(i\varphi)$ is on the semicircle, then $\int_{-\infty}^{-R}$ and $\int_R^\infty$ go to zero as $R \to \infty$. The remaining integrals are of the form

$$\int_0^\pi d\varphi \exp[(z \pm t\sqrt{1+m^2e^{-2i\varphi}/R^2})R(i\cos\varphi - \sin\varphi)]\left[\frac{e^{i\varphi}R\tilde{g}(Re^{i\varphi})}{(1+m^2e^{-2i\varphi}/R^2)\tilde{\dot{g}}(Re^{i\varphi})}\right].$$

Since $|R\tilde{g}(R\exp(i\varphi))|$ and $|\tilde{\dot{g}}(R\exp(i\varphi))|$ remain bounded as $R \to \infty$, $0 \leq \varphi \leq \pi$, such integrals go to zero if $z > |t|$, by the Riemann–Lebesgue lemma.

## 3.3 Diffraction at a Wedge

*This was the first diffraction problem to succumb to a rigorous treatment. Not only does it confirm the general outlines of one's naive expectations, but it also displays the wave nature of light with a wealth of complex detail.*

The solution of boundary-value problems in somewhat complicated geometrical settings is fraught with difficulties even in the two-dimensional case. For a long time people had to settle for an imprecise theory as handed down primarily from Kirchhoff. The procedure was to substitute into certain exact integral equations the field that would be present if the conductors were removed. By so doing, one obtains a solution of the field equations but violates the boundary conditions. Since the solution involves oscillatory integrals, i.e., the small differences between large positive and negative terms, it is nearly impossible to estimate the errors incurred if one makes a small change in the integrand. Yet the result exhibits the features of experimentally known diffraction patterns, and thus the approximations have remained popular down to the present day among young and old alike. It must nonetheless be counted as a great step forward, that A. Sommerfeld succeeded in solving a nontrivial diffraction problem in 1895; it was only then possible to determine when the approximate theory of diffraction was good, and when it failed.

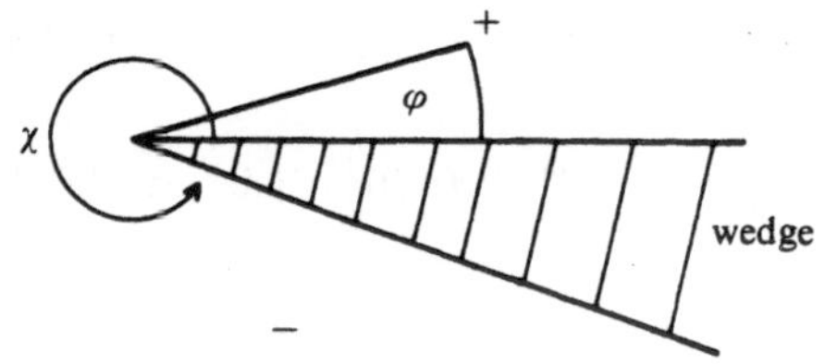


image charges at $-\varphi + 2\chi - 2\pi$

Figure 42 The image charges for a wedge

That problem concerned the diffraction at a wedge, where the metallic surface consisted of the planes where $\varphi = 0$ and where $\varphi = \chi > \pi$. A plausible attempt to solve this problem would proceed by mimicking the solution of the problem of two metallic mirrors at $x = 0$ and $x = a$. Following §3.2. image charges at $x + 2na$ would produce a solution with period $2a$, and then by a reflection at $x = 0$ the boundary conditions on both mirrors could be satisfied. The hitch in trying to transcribe this procedure to the wedge is that if $x$ is the variable, the image charges always remain outside of $0 \leq x \leq a$ when reflected by $R$, but if $\varphi$ is the variable on the torus $T^1$, then because it is periodic with period $2\pi$, the image charges at $-\varphi + 2n\chi$ would at some point enter the region $0 \leq \varphi \leq \chi \pmod{2\pi}$. Sommerfeld's brilliant stroke was to forget about the periodicity with period $2\pi$, and to seek a solution with period $2\chi$. His solution has branches, in the sense that by continuation through the metal $\chi \leq \varphi \leq 2\pi$ the variable would not return to its initial value in $(0, \chi)$, and consequently image charges could appear there. The problem does not require the absence of images at such values of the coordinates; the solution in $0 \leq \varphi \leq \chi$ has no way of knowing that $\varphi$ is actually a variable on $T^1$.

The most convenient way to construct this ramified solution is to write it as a complex integral. The sums $\sum T^n$ that have appeared above can be represented for analytic $f$ as

$$\sum_n T^n f(x) = \sum_n f(x + 2na) = \int_C \frac{dx'}{2a} \frac{-f(x')}{1 - \exp(2\pi i(x' - x)/2a)}, \tag{3.3.1}$$

where the contour $C$ goes around the poles of the integrand at $x' = x + 2na$.

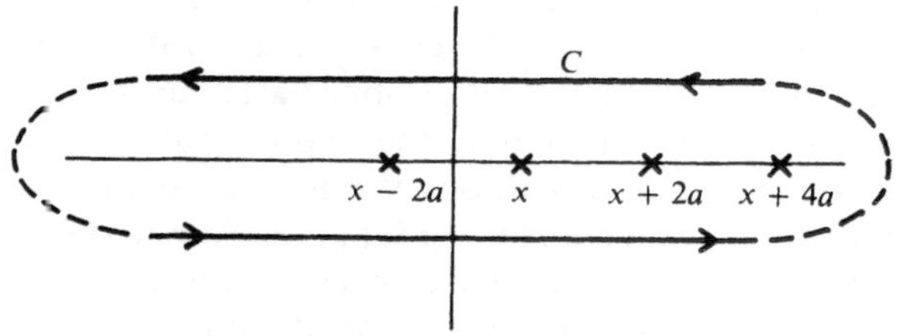


The path of integration in the integral representation of a sum

The expression (3.3.1) always has a period $2a$ in $x$, as long as the contour $C$ is flexible enough to avoid the poles as $x$ increases to $x + 2a$. The best choice of $C$ depends on the analytic properties of the function $f$.

If we try to apply this procedure to the Green function, we soon run up against the difficulty that $D^{\text{ret}}$ is not an analytic function but only a distribution, $\delta(r - t)/4\pi r$. However, it can be approximated by analytic functions, for example by

$$D^{\text{ret}}(x) = \lim_{\varepsilon \downarrow 0} \operatorname{Re} \int_0^\infty d\omega \frac{e^{i(r - t + i\varepsilon)\omega}}{4\pi^2 r}, \tag{3.3.2}$$

where the limit is taken in the sense of distributions. Since the expressions that we require are of the form

$$\int D^{\text{ret}}(\bar{x} - x) f(x) dx,$$

the integrand will be multiplied by the Fourier transform of $f$, and for suitable $f$ the integral $\int d\omega$ will be convergent. The integration becomes rather insensitive to the limit $\varepsilon \downarrow 0$, and therefore we may interchange $\varepsilon \downarrow 0$ with other limiting processes without worry. If we use the above expression and the notation of (3.1.28), we arrive at

**The Green Function for the Wedge** (3.3.3)

*In cylindrical coordinates*

$$x^\alpha = (t, \rho \cos \varphi, \rho \sin \varphi, z),$$

*let*

$$M = \{x^\alpha \in \mathbb{R}^4 : t \geq t_0\}, \qquad N = \{x^\alpha \in M : 0 \leq \varphi \leq \chi\}.$$

*If $R_\varphi$ is the transformation of forms induced by $\varphi \to -\varphi$, then the Green function of this problem is*

$$G_{\bar{x}} = -\tfrac{1}{2}\bar{e}_{\alpha\beta} \otimes (1 + R_\varphi)^* e^{\alpha\beta} \lim_{\varepsilon \uparrow 0} \operatorname{Re} \int_0^\infty d\omega \, e^{-i\omega(\bar{t} - t + i\varepsilon)}$$

$$\int_C \frac{d\varphi'/2\chi}{1 - e^{i\pi(\varphi' - \bar{\varphi} + \varphi)/\chi}} g([(\bar{z} - z)^2 + \bar{\rho}^2 + \rho^2 - 2\bar{\rho}\rho \cos \varphi']^{1/2}),$$

*where*

$$g(r) = \frac{e^{i\omega r}}{4\pi^2 r}.$$

*The integration contour $C$ consists of two curves, one in the upper and one in the lower half-plane, which come in from infinity, enclose the zeros of the term in brackets [ ], avoid the zeroes of the denominator, and then return to infinity (Figure 43).*

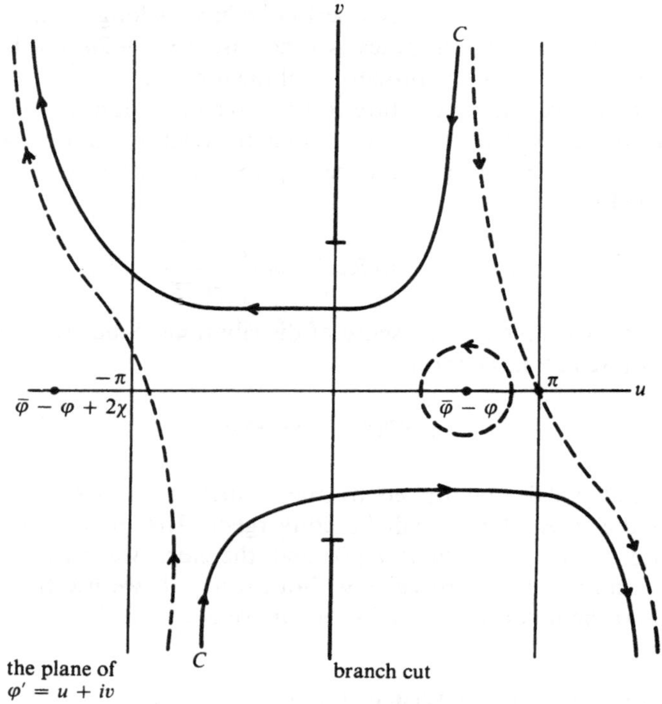


Figure 43 Paths of integration in (3.3.3)

**Proof**

Let $\varphi' = u + iv$, so $\cos \varphi' = \cos u \cosh v - i \sin u \sinh v$. The zeroes of $[\ ] \equiv r$ where $u = 0$ and $\cosh v = ((\bar{z} - z)^2 + \bar{\rho}^2 + \rho^2)/2\bar{\rho}\rho$ require the branch cuts drawn in the figure. Because $\omega > 0$, $\exp(i\omega r)$ decreases exponentially if $\operatorname{Im} r > 0$. If $\varphi'$ is far from the real axis, then $r$ has an imaginary part $\sim \sin u \sinh v$. Therefore $g$ decreases in the upper half-plane if $0 < u < \pi$ or $-2\pi < u < -\pi$, and in the lower half-plane if $-\pi < u < 0$ or $\pi < u < 2\pi$. Since the denominator is small only in the neighborhood of $\varphi' = \bar{\varphi} - \varphi + 2n\chi$, $n \in \mathbb{Z}$, the convergence of the infinite integral is verified if $x \neq \bar{x}$. If $\varphi$ increases by $2\chi$, then the poles move along the real axis, which, however, does not interfere with the integration contour. This immediately implies periodicity in $\varphi$ (and $\bar{\varphi}$) with period $2\chi$, and thus that the boundary conditions are fulfilled exactly as in §3.2.

It remains to verify that $-\Delta G_{\bar{x}} = \delta_{\bar{x}}$. Note that

$$\Delta \frac{e^{i\omega(|\bar{\mathbf{x}} - \mathbf{x}| - \bar{t} + t)}}{|\bar{\mathbf{x}} - \mathbf{x}|} = 0$$

for all $\bar{x} \neq x$, and in particular for complex $\bar{x}$. If we treat $\varphi'$ as a complex part to be added to $\bar{\varphi}$, then we see that the wave equation is satisfied if $x \neq \bar{x}$.

If $x \to \bar{x}$, then the two branch points and the pole move to the origin and pinch the integration contour in two. The integration path is then unable to avoid the point $\bar{\varphi} - \varphi$. In order to see better what the value of the integral is in this limit, let us deform $C$ into the paths drawn with dotted lines in Figure 43 and a circle around $\bar{\varphi} - \varphi$. The integration over the circle just produces $D^{\text{ret}}(\bar{x} - x)$, of course, while the other two parts are not affected by the singularities in the limit $x \to \bar{x}$. This means that $G_{\bar{x}}$ differs from $G_{\bar{x}}^{\text{ret}}$ only by a solution of the homogeneous wave equation, and consequently it satisfies everything required of it. □

**Remark** (3.3.4)

The edge still keeps $N$ from being a manifold. At the edge ($\bar{\rho} \to 0$), $G_{\bar{x}}$ behaves like $(\bar{\rho})^{\pi/\chi}$. Although this approaches zero, if $\chi > \pi$ the derivative, and consequently the nonvanishing components of the field strength, diverge. However, because they approach infinity more slowly than $\rho^{-1}$, the square-integrability is not in danger, and thus the total energy remains finite.

As the first application of (3.3.3) we find out how it accords with the naive idea that the wedge reflects some of the light and casts a shadow. To this end, we assume in the following that $\chi > \pi$.

**Geometric Optics** (3.3.5)

According to geometric optics, if $|\bar{\varphi} - \varphi| < \pi$, then one can see directly from $\bar{x}$ to $x$; if $|\bar{\varphi} + \varphi| < \pi$, then the reflected image from the upper surface is also visible; and if $|\bar{\varphi} - \chi + \varphi - \chi| < \pi$, then the image from the lower surface is visible (Figure 44). These effects can be separated off and the corrections to geometric optics calculated if we make a decomposition of $C$ into the following parts. Two curves, $C_1$ and $C_2$, join $-i\infty - \pi/2$ to $i\infty - 3\pi/2$ and respectively $i\infty + \pi/2$ to $-i\infty + 3\pi/2$, and intersect the real axis at $-\pi$ and respectively $\pi$; and if there is a pole in $(-\pi, \pi)$, then there is also a loop around it (Figure 45).

The loop integral yields $G_{\bar{x}}^{\text{ret}}$, and if $|\bar{\varphi} - \varphi| < \pi$, then we see the light as if there were no wedge, because in the physical region, where $\bar{\varphi}$ and $\varphi \in (0, \chi)$, the poles at $\bar{\varphi} - \varphi + 2n\chi$, $n \neq 0$, are never in $(-\pi, \pi)$. In the reflected part (the one with the $R$ in (3.3.3)), the poles at $\bar{\varphi} + \varphi$ and $\bar{\varphi} + \varphi - 2\chi$ may lie within $(-\pi, \pi)$, which causes light to be reflected by the upper, or respectively lower, surface. Therefore the integrals over $C_1$ and $C_2$ are precisely the corrections to geometric optics. It is apparent that if $\omega\rho \gg 1$ and $\omega\bar{\rho} \gg 1$, that is, at several wavelengths distance from the edge, geometric optics dominates. Because of the factor $\exp(i\omega r)$, the low-lying values of the integrand, in the shaded parts of Figure 45, become deep valleys in this limit. The trail leading up to the pass at $\pm\pi$, where $\exp(i\omega r)$ has absolute value 1, becomes steeper and steeper, and the contribution from going over the top, though dominating

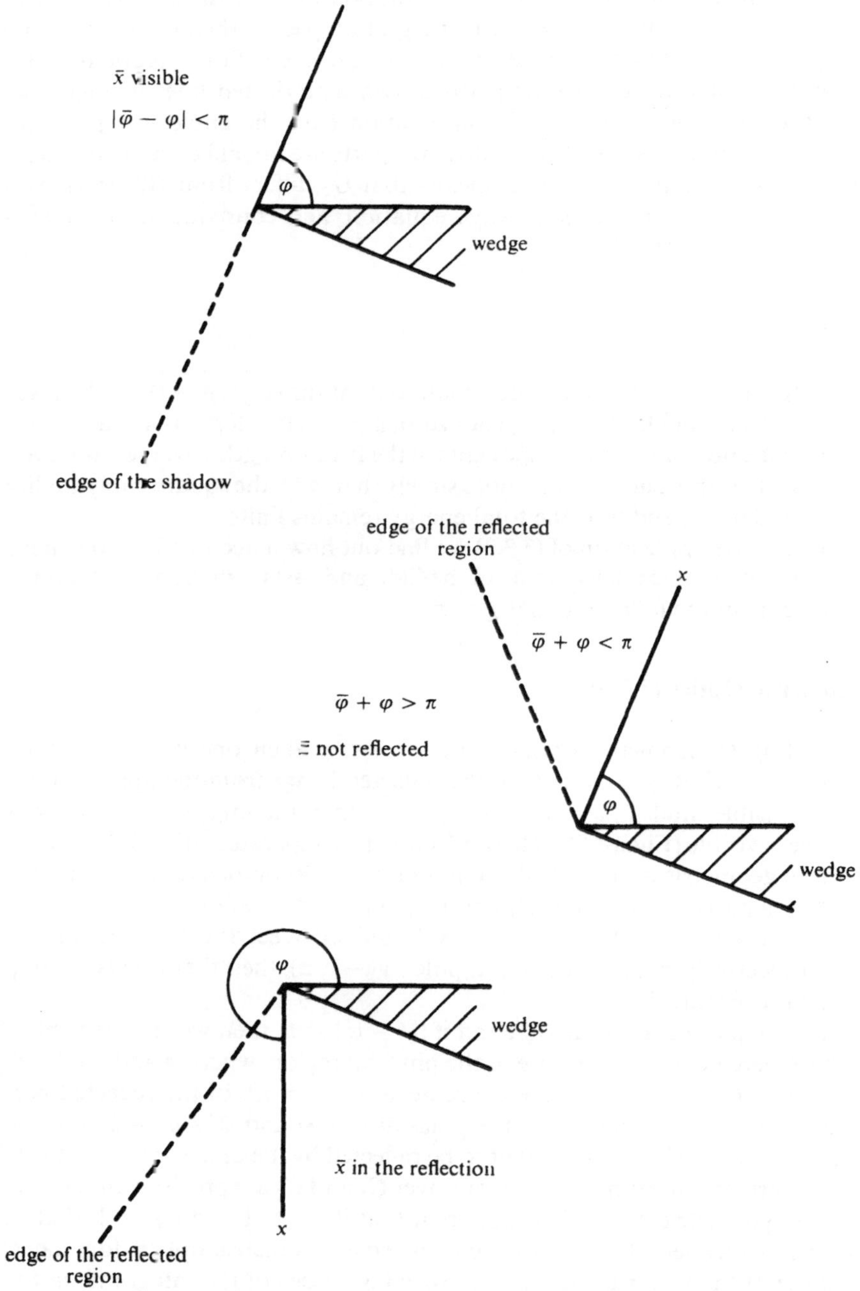

$\bar{x}$ visible
$|\bar{\varphi} - \varphi| < \pi$
$\varphi$
wedge
edge of the shadow
edge of the reflected region
$x$
$\bar{\varphi} + \varphi < \pi$
$\bar{\varphi} + \varphi > \pi$
$\bar{x}$ not reflected
$\varphi$
wedge
$\varphi$
wedge
$\bar{x}$ in the reflection
$x$
edge of the reflected region


Figure 44

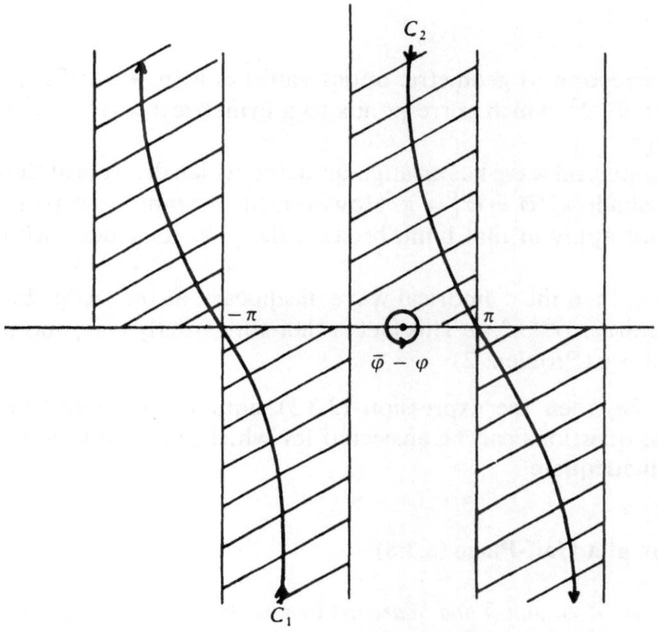


Figure 45 Paths of integration in the hatched regions, where the integrand decreases

the integral, becomes negligible. In order to make these ideas quantitative, let us evaluate

$$\int_{C_1, C_2} \frac{d\varphi'}{2\chi} \frac{e^{i\omega r}}{4\pi r} (1 - e^{i\pi(\varphi' - \bar{\varphi} + \varphi)/\chi})^{-1}$$

by the saddle-point method (= the method of steepest descent). We start by expanding the rapidly varying exponent in the vicinity of the saddle points,

$$r = ((\bar{z} - z)^2 + \bar{\rho}^2 + \rho^2 - 2\bar{\rho}\rho\cos\varphi')^{1/2} \sim R - \frac{\rho\bar{\rho}}{2R}\tilde{\varphi}^2,$$

$$R = ((\bar{z} - z)^2 + (\bar{\rho} + \rho)^2)^{1/2}, \qquad \tilde{\varphi} = \varphi' \pm \pi,$$

and replacing the rest of the integrand by its value at the saddle point. There remains a Gaussian integral over $\tilde{\varphi}$, from which we obtain the asymptotically exact formula:

$$\begin{aligned}
&\int_{C_1 \cup C_2} \frac{d\varphi'}{2\chi} \frac{e^{i\omega r}}{4\pi r} (1 - e^{i\pi(\varphi' - \bar{\varphi} + \varphi)/\chi})^{-1} \\
&\quad \cong \frac{1}{4\chi}\left(\frac{1}{1 - e^{i\pi(\varphi - \bar{\varphi} - \pi)/\chi}} - \frac{1}{1 - e^{i\pi(\varphi - \bar{\varphi} + \pi)/\chi}}\right) \cdot \frac{e^{i\omega R - i\pi/4}}{\sqrt{2\pi\omega\bar{\rho}\rho R}} \\
&\quad = -\frac{1}{4\chi} \frac{\sin(\pi^2/\chi)}{\cos\pi(\bar{\varphi} - \varphi)/\chi - \cos\pi^2/\chi} \frac{e^{i\omega R + i\pi/4}}{\sqrt{2\pi\omega\bar{\rho}\rho R}}
\end{aligned} \tag{3.3.6}$$

(cf. Problem 1).

**Remarks** (3.3.7)

1. The correction to geometric optics vanishes as $\bar{\rho} \to \infty$. If $\rho \gg \bar{\rho}$, then it goes as $\bar{\rho}^{-1/2}$, which corresponds to a cylindrical wave emanating from the edge.
2. The cylindrical wave has an angular factor, which diverges at the boundary of the shadow, $|\bar{\varphi} - \varphi| = \pi$. However, the asymptotic expansion (3.3.6) does not apply in that limit, because the pole coincides with the saddle point.
3. If $\chi = \pi$, then the cylindrical wave disappears, as the wedge becomes the plane mirror of (3.2.1) This fact is therefore strictly true, and not merely asymptotic (Problem 2).

If $\chi = 2\pi$, then the expression (3.3.3) simplifies so greatly that many additional questions can be answered for which the asymptotic expression (3.3.6) is inadequate.

**Diffraction at a Half-Plane** (3.3.8)

*If $\chi = 2\pi$, $\rho \to \infty$, and $\bar{\rho}$ and $\bar{\varphi}$ are arbitrary, then $G_{\bar{x}}$ is asymptotically*

$$G_{\bar{x}} = (1 + R_{\varphi})G_{\bar{x}}^{\text{ret}} \frac{e^{-i\pi/4}}{\sqrt{2}} \int_{-\infty}^{v} dv' \, e^{i\pi v'^2/2},$$

$$v = \sqrt{\frac{\bar{\rho}\rho\omega}{\pi R}} \, 2\cos\left(\frac{\bar{\varphi} - \varphi}{2}\right), \qquad R = ((\bar{z} - z)^2 + \rho^2)^{1/2}.$$

**Remarks** (3.3.9)

1. The $\omega$ in $v$ should be chosen so that $G_{\bar{x}}^{\text{ret}}$ is represented as the integral of (3.3.2).
2. For (3.3.6) it was necessary to assume that $\omega\bar{\rho}\rho((\bar{z} - z)^2 + \rho^2 + \bar{\rho}^2)^{-1/2} \gg 1$. This is not the case if $\rho \to \infty$ and $\bar{\rho} < 1/\omega$; in this sense (3.3.8) is more general. The boundary of the shadow is also described by (3.3.8), as long as the source of the light is sufficiently far from the edge of the wedge.

**Proof of** (3.3.8)

First observe that

$$G_{\bar{x}}(\varphi) + G_{\bar{x}}(\varphi + 2\pi) = G_{\bar{x}}^{\text{ret}}(\varphi),$$

because

$$\frac{1}{1 - e^{i(\varphi' - \bar{\varphi} + \varphi)/2}} - \frac{1}{1 - e^{i(\varphi' + 2\pi - \bar{\varphi} + \varphi)/2}} = \frac{2}{1 - e^{i(\varphi' - \bar{\varphi} + \varphi)}}.$$

Since this sum has a periodicity $2\pi$, the contributions from the paths $C_1$ and $C_2$ of Figure 45 cancel out, and only the loop around $\bar{\varphi} - \varphi$ remains. On the other hand, the difference of the same terms is

$$\frac{1}{1 - e^{i(\varphi' - \bar{\varphi} + \varphi)/2}} - \frac{1}{1 + e^{i(\varphi' - \bar{\varphi} + \varphi)/2}} = \frac{i}{\sin(\varphi' - \bar{\varphi} + \varphi)/2},$$

and we shall find a similar factor in the exponent in the limit $\rho \to \infty$. In (3.3.3) in this limit, let us write

$$r \cong R - \frac{\bar{\rho}\rho}{R}\cos\varphi', \qquad R = ((\bar{z} - z)^2 + \rho^2)^{1/2},$$

and compare it with the quantity

$$|\bar{\mathbf{x}} - \mathbf{x}| \cong R - \frac{\bar{\rho}\rho}{R}\cos(\bar{\varphi} - \varphi)$$

occurring in $G_{\bar{\mathbf{x}}}^{\text{ret}}$ in place of $r$; then in the difference there occurs

$$\cos(\bar{\varphi} - \varphi) - \cos\varphi' = 2\cos^2\frac{(\bar{\varphi} - \varphi)}{2} - 2\cos^2\frac{\varphi'^2}{2}$$
$$= 2\sin\frac{(\varphi' + \bar{\varphi} - \varphi)}{2}\sin\frac{(\varphi' - \bar{\varphi} + \varphi)}{2}$$

If the factor $\exp(i\omega|\bar{\mathbf{x}} - \mathbf{x}|)$ is removed, then in the limit $\rho \to \infty$ we are left with the integral

$$I \equiv \frac{1}{8\pi i}\int_C \frac{d\varphi'\, e^{2i\omega(\bar{\rho}\rho/R)\sin(\varphi' - \bar{\varphi} + \varphi)/2\sin(\varphi' + \bar{\varphi} - \varphi)/2}}{\sin(\varphi' - \bar{\varphi} + \varphi)/2}$$

in $G_{\bar{x}}(\varphi) - G_{\bar{x}}(\varphi + 2\pi)$. The denominator can be disposed of by differentiating by $\bar{\rho}$;

$$\frac{\partial I}{\partial \bar{\rho}} = \frac{\omega\rho}{4\pi R}\int_C d\varphi' \sin\frac{(\varphi' + \bar{\varphi} - \varphi)}{2}\, e^{(2i\omega\bar{\rho}\rho/R)(\cos^2(\bar{\varphi} - \varphi)/2 - \cos^2\varphi'/2)}.$$

Using

$$\sin\frac{(\varphi' + \bar{\varphi} - \varphi)}{2} = \sin\frac{\varphi'}{2}\cos\frac{\bar{\varphi} - \varphi}{2} + \cos\frac{\varphi'}{2}\sin\frac{\bar{\varphi} - \varphi}{2},$$

we see that the second term contributes nothing, because its part of the integrand is even in $\varphi'$, and the integral $\int_C d\varphi'$ is taken in the opposite direction when $\varphi' \to -\varphi'$. We may choose $\cos\varphi'/2$ as a new integration variable in the first term. Then there is no pole, and the two curves $C_1$ and $C_2$ have the same contribution, since the integral changes its sign under $\varphi' \to \varphi' + 2\pi$:

$$\frac{\partial I}{\partial \bar{\rho}} = \frac{\omega\rho}{\pi R}\int_{-i\infty}^{i\infty} e^{-2i(\omega\bar{\rho}\rho/R)(t^2 - \cos^2(\bar{\varphi} - \varphi)/2)}\, dt \cos\frac{\bar{\varphi} - \varphi}{2}$$
$$= \frac{\omega\rho}{\pi R} e^{-i\pi/4} e^{2i(\omega\bar{\rho}\rho/R)\cos^2(\bar{\varphi} - \varphi)/2}\sqrt{\frac{2\pi R}{\omega\bar{\rho}\rho}}\cos\frac{\bar{\varphi} - \varphi}{2}.$$

In order to do the $\bar{\rho}$ integration, we introduce the variable

$$v = \sqrt{\frac{\bar{\rho}\rho\omega}{\pi R}}\, 2\cos\frac{(\bar{\varphi} - \varphi)}{2},$$

$$dv = d\bar{\rho}\sqrt{\frac{\rho\omega}{\pi\bar{\rho}R}}\cos\frac{(\bar{\varphi} - \varphi)}{2}.$$

Since $I = 0$ for $\bar{\rho} = 0$ where $C$ can be shifted away

$$I = \sqrt{2}e^{-i\pi/4}\int_0^v dv'\, e^{i\pi v'^2/2}.$$

Combining this with the sum $G_{\bar{x}}(\varphi) + G_{\bar{x}}(\varphi + 2\pi)$ and using

$$\sqrt{2}e^{-i\pi/4}\int_{-\infty}^0 dv'\, e^{i\pi v'^2/2} = 1,$$

we obtain (3.3.8). □

To discuss the problem in more detail, we need some

**Properties of Fresnel's Integral** (3.3.10)

Let

$$F(z) = \int_{-\infty}^{z} dv\, e^{i\pi v^2/2}.$$

Then $F(\infty) = 2F(0) = 1 - i$, and $F$ behaves asymptotically as

$$z \to 0: F(z) = F(0) + z + i\frac{\pi}{6}z^3 + O(z^5)$$

$$z \to \infty: F(z) = F(\infty) + \frac{e^{i\pi z^2/2}}{i\pi z} + O(z^{-3}) \qquad (3.3.11)$$

$$z \to -\infty: F(z) = \frac{e^{i\pi z^2/2}}{i\pi z} + O(z^{-3}).$$

If $F(x)$ is graphed as a curve in the complex plane depending on $x \in \mathbb{R}$ as a parameter, it has the form of Cornu's spiral. When $x \leq x_0 > 0$, $|F(x)|$

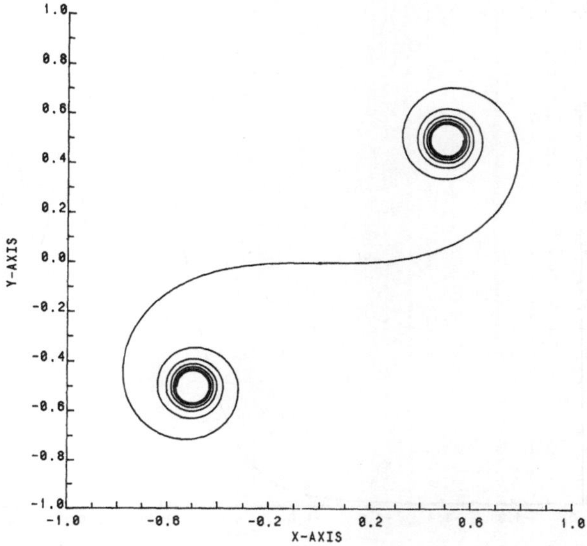


Figure 46 Cornu's spiral

increases monotonically, and after that point it oscillates about the limiting value $\sqrt{2}$.

We next investigate the regions not covered by (3.3.6):

**The Boundary of the Shadow** (3.3.12)

Let $\bar{\varphi} - \varphi = \pi + \delta$, $\delta \ll 1$. The time average of the intensity equals the absolute square of the field, and its ratio to the unscreened intensity $i^{\text{ret}}$ becomes

$$\frac{i}{i^{\text{ret}}} = \frac{1}{2}\left|F\left(\delta\sqrt{\frac{\bar{\rho}\omega}{\pi}}\right)\right|^2$$

as $\rho \to \infty$ with $\rho/R \to 1$. We have disregarded the reflected light; the interference with the term containing $R\varphi$ produces a correction $O(\bar{\rho}^{-1/2})$. During the transition from light to shadow, the intensity thus first oscillates about the value $i^{\text{ret}}$ and then decreases monotonically. The width $\sim\sqrt{\bar{\rho}/\omega}$ of the transition zone, where the shadow is hazy, increases with $\bar{\rho}$, while its angle $1/\sqrt{\bar{\rho}\omega}$ from the edge of the wedge approaches zero as $\bar{\rho}$ increases.

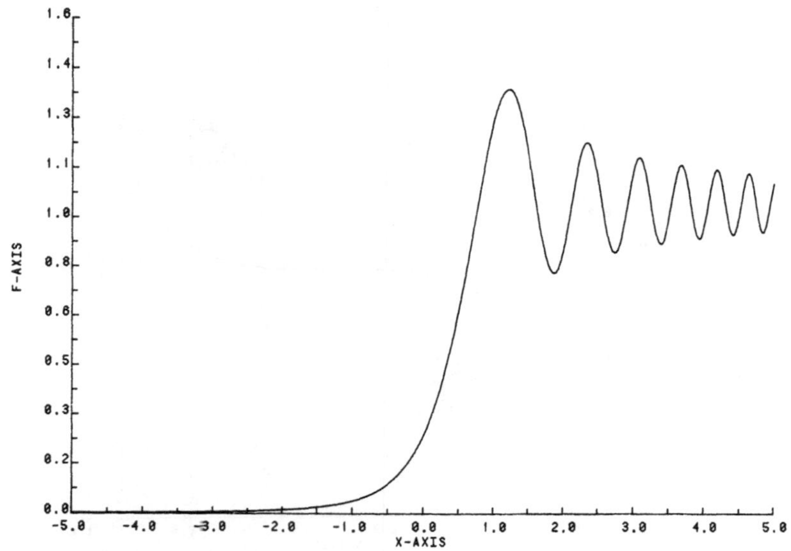


Figure 47 The intensity at the edge of a shadow

**The Edge** (3.3.13)

If $\bar{\rho} \to 0$ and $\rho/R \to 1$, and we consider such a small neighborhood of the edge that the contribution from $G_{\bar{x}}^{\text{ret}}$ is effectively constant and hence unaffected by a reflection, then the components that go to zero at the edge are

$$F\left(\sqrt{\frac{\bar{\rho}\omega}{\pi}}\,2\cos\left(\frac{\bar{\varphi}-\varphi}{2}\right)\right) - F\left(\sqrt{\frac{\bar{\rho}\omega}{\pi}}\,2\cos\left(\frac{\bar{\varphi}+\varphi}{2}\right)\right) \sim \sqrt{\frac{\bar{\rho}\omega}{2}}\,2\sin\frac{\bar{\varphi}}{2}\cdot\sin\frac{\varphi}{2}.$$

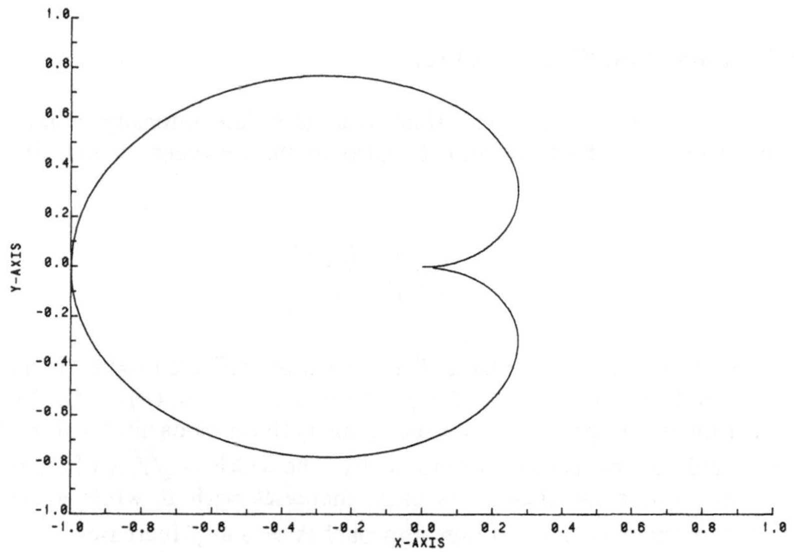


Figure 48 Polar graph of $i(\varphi)$

Thus the intensity dies down in directions along the half-plane, and the angular distribution in $\bar{\varphi}$ with $\varphi$ fixed has no preference for the incident direction.

**The Plane of the Screen** (3.3.14)

If $\bar{\varphi} = \pi$, and the factor $G_{\bar{x}}^{\text{ret}}$ is still effectively constant, then the components that vanish at the screen are diminished by

$$F\left(2\sqrt{\frac{\bar{\rho}\omega}{\pi}}\sin\frac{\varphi}{2}\right) - F\left(-2\sqrt{\frac{\bar{\rho}\omega}{\pi}}\sin\frac{\varphi}{2}\right)$$

compared with $G_{\bar{x}}^{\text{ret}}$. It takes a few wavelengths for the field to increase from zero at $\bar{\rho} = 0$ to the value it would have were it not for the screening, and it takes longer to do so if the light is incident at a smaller angle $\varphi$

**Remark** (3.3.15)

In the integral, Kirchhoff's theory of diffraction replaces the field within the screening plane with the value it would have without screening. This is a very good approximation at many wavelengths away from the screen, but it is poor in the vicinity of the screen. It has the unsettling consequence that the field strength calculated in this manner becomes singular at the edge as $1/\rho$, and is thus not square-integrable, so the field energy is infinite (cf. [2], §25).

In conclusion, geometric optics gives the right answers if one can take a bird's-eye view, from which the wavelength is not perceptible. But up close there is such a complicated pattern of interference that any simple approximation is doomed to failure.

**Problems** (3.3.16)

1. Show how and where (3.3.6) is an asymptotic expansion.
2. Check that $G_{\bar{x}} = (1 + R)G_{\bar{x}}^{\text{ret}}$ if $\chi = \pi$.
3. See whether (3.3.6) agrees with (3.3.8).
4. How large is the error in (3.3.11)?

**Solutions** (3.3.17)

1. Use the saddle-point method discussed, e.g., in Dieudonné, *Infinitesimal Calculus* IX, §1.
2. If $\chi = \pi$, then the function of period $2\chi$ reduces to the function $G_{\bar{x}}^{\text{ret}}$ of period $2\pi$.
3. Applying (3.3.11) to (3.3.8), we get

$$\frac{e^{i\omega|\bar{x}-x|}}{4\pi|\bar{x}-x|}\frac{e^{-i\pi/4}}{\sqrt{2}}\frac{e^{i\pi v^2/2}}{i\pi v} \cong -\frac{e^{i\omega R}ie^{-i\pi/4}e^{i\omega\bar{\rho}\rho/R}}{8\pi\sqrt{2\pi\omega\bar{\rho}\rho R}\cos(\bar{\varphi}-\varphi)/2}$$

for the correction to geometric optics, in agreement with (3.3.6) (where the $R$ is $\sim R + \bar{\rho}\rho/R$).

4. By integration by parts with $z \in \mathbb{R}^+$,

$$-\int_z^\infty dv\, e^{i\pi v^2/2} = -\int_z^\infty \frac{dv}{i\pi v}\frac{\partial}{\partial v} e^{i\pi v^2/2} = \frac{e^{i\pi v^2/2}}{i\pi v} - \int_z^\infty \frac{dv}{i\pi v^2} e^{i\pi v^2/2},$$

$$\left|\int_z^\infty \frac{dv}{i\pi v^2} e^{i\pi v^2/2}\right| = \left|\int_z^\infty \frac{dv}{i\pi^2 v^3}\frac{\partial}{\partial v} e^{i\pi v^2}\right| \le \frac{1}{\pi^2 z^3} + \frac{1}{\pi^2 z^3}.$$

Therefore

$$\left|F(z) - F(\infty) - \frac{e^{i\pi z^2/2}}{i\pi z}\right| \le \frac{2}{\pi^2 z^3} \qquad \forall z \in \mathbb{R}^+.$$

## 3.4 Diffraction at a Cylinder

*This problem is so tractable mathematically that all the various phenomena of geometric and wave optics can be worked out.*

Despite their being complementary problems to the wave-guide and the resonant cavity, it is an elaborate job to analyze diffraction at a cylinder or at a sphere. Even if it is not required to write down the complete Green function, but is enough to solve for the individual waves that $G_{\bar{x}}$ is constructed from, there are still infinite sums and complex integrals to cope with. These expressions reduce to elementary functions only in particular limits, in which the relevant properties are well displayed. Depending on the ratio of the wavelength to the size of the metallic body, it may be possible for the wave to pass around the obstruction. This phenomenon is somewhat simpler for a cylinder than for a sphere, for which reason we restrict our discussion to cylinders. We also consider only scalar solutions $u$ to the wave equation $-\Box u = 0$, with either Dirichlet or Neumann boundary conditions, $u_{|\partial N} = 0$ or $du_{|\partial N} = 0$. Although it is possible to construct $G_{\bar{x}}$ with the aid of these solutions, the expressions that result are too involved to be illuminating. We are again primarily interested in the limit where the source of the waves is at infinity; therefore we utilize $u$'s incident in a plane and scattering into cylindrical waves. In cylindrical coordinates they would have the form

$$\left(e^{-iqx} + \frac{e^{iq\rho}}{\sqrt{\rho}} f(\varphi)\right) e^{i(kz-\omega t)}.$$

as $\rho \to \infty$. If the first term is expanded in a Fourier series,

$$e^{iqx} = \sum_{n=-\infty}^{\infty} e^{in\varphi} \frac{1}{2\pi}\int_0^{2\pi} d\varphi'\, e^{-i(q\rho\cos\varphi' + n\varphi')} = \sum_{n=-\infty}^{\infty} e^{in(\varphi-\pi/2)} J_n(q\rho), \quad (3.4.1)$$

then we can represent the solution $u(\rho, \varphi)$ that satisfies the boundary conditions at $\rho = a$ and $\rho \to \infty$ as a

**Fourier Series** (3.4.2)

$$u = \sum_{n=-\infty}^{\infty} e^{in(\varphi - \pi/2)}\left[ J_n(q\rho) - \frac{J_n(qa)}{H_n^{(1)}(qa)} H_n^{(1)}(q\rho) \right].$$

In order to evaluate the sum, recall the

**Asymptotic Behavior of** $H_\nu^{(1)}(x)$ (3.4.3)

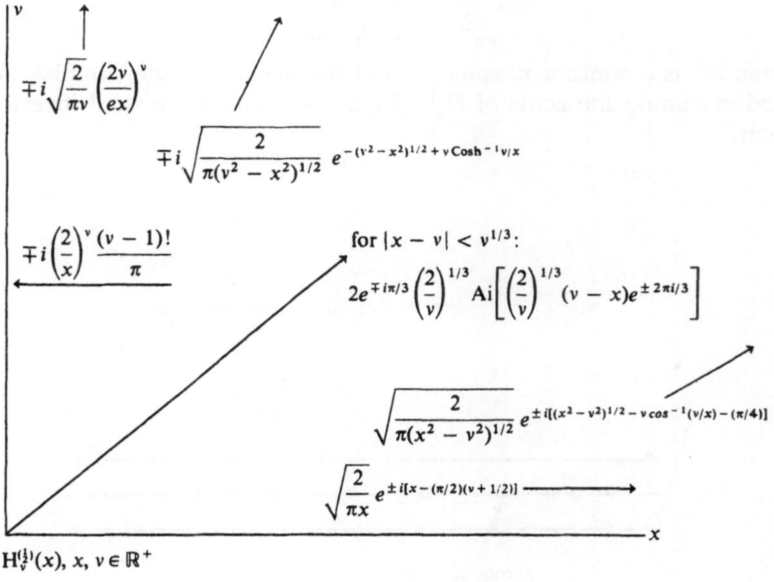


**Remarks** (3.4.4)

1. The asymptotic behavior of the other solution follows from

$$H_\nu^{(2)}(x) = (H_{\nu^*}^{(1)}(x^*))^*, \qquad H_{-\nu}^{(1)}_{(2)}(x) = e^{\pm i\pi\nu} H_\nu^{(1)}_{(2)}(x),$$

$$J_\nu = \tfrac{1}{2}(H_\nu^{(1)} + H_\nu^{(2)}),$$

$$J_\nu \overset{\nu \to \infty}{\simeq} \frac{1}{\sqrt{2\pi\nu}}\left(\frac{ex}{2\nu}\right)^\nu, \qquad \overset{x \to 0}{\simeq} \left(\frac{x}{2}\right)^\nu \frac{1}{\nu!}.$$

2. Consequently, if $n$ is large, then the $n$-th term in (3.4.2) approaches
$$\frac{e^{in(\varphi-\pi/2)}}{\sqrt{2\pi n}}\left(\frac{e}{2n}\right)^n\left[(q\rho)^n-\frac{(qa)^{2n}}{(q\rho)^n}\right],$$
and the sum converges uniformly on compact subsets of the $x$-$y$-plane. Even so, the first term in the sum dominates the scattered wave only if $qa \ll 1$. In that case,
$$\frac{\mathrm{J}_n(qa)}{\mathrm{H}_n^{(1)}(qa)} \to \pi i\left(\left(\frac{qa}{2}\right)^{|n|}\frac{1}{|n|!}\right)^2 n \quad \text{for } |n| \geq 1,$$
and
$$u = e^{-iqx} + \frac{i\pi}{2\ln qa}\mathrm{H}_0^{(1)}(q\rho) + O((qa)^2).$$
3. If $qa \gg 1$, then we make use of the fact that $\mathrm{H}_\nu^{\left({1 \atop 2}\right)}$ are analytic functions in $\nu$. This allows the sum in (3.4.2) to be turned into an integral over $d\nu$,
$$\sum_{n=-\infty}^{\infty} = \frac{1}{2i}\int_C \frac{d\nu\, e^{-i\pi\nu}}{\sin \pi\nu}, \tag{3.4.5}$$
where $C$ is a contour passing around the integers in the complex plane and excluding the zeros of $\mathrm{H}_\nu^{(1)}$ (Figure 49), which we shall investigate later.

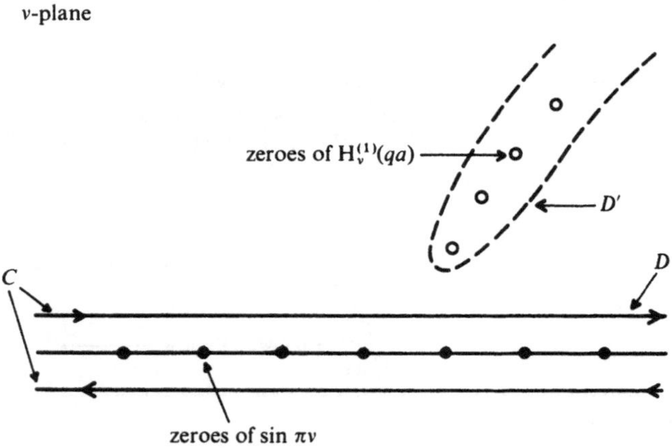

Figure 49 Paths of integration for (3.4.6)

This integral can be further transformed into an integral along the path $D$ strictly above the real axis, by taking account of (3.4.4; 1). The result of this is the

**Representation as a Fourier Integral** (3.4.6)
$$u = \frac{1}{4i}\int_D \frac{d\nu\, e^{-i\nu\pi/2}}{\sin \pi\nu} 2\cos \nu(\varphi - \pi)$$
$$\cdot\frac{\mathrm{H}_\nu^{(1)}(qa)\mathrm{H}_\nu^{(2)}(q\rho) - \mathrm{H}_\nu^{(2)}(qa)\mathrm{H}_\nu^{(1)}(q\rho)}{\mathrm{H}_\nu^{(1)}(qa)}.$$

To evaluate the above integral by means of the residue theorem, we need to also know the

**Asymptotic Behavior of the Bessel Functions in the Upper ν-Plane** (3.4.7)

Suppose that $x$ is real, and define

$$A \equiv \sqrt{\frac{2}{\pi}}(\nu^2 - x^2)^{-1/4} \to \sqrt{\frac{2}{\pi\nu}},$$

$$e^{\alpha} \equiv \exp(\nu^2 - x^2)^{1/2}\left[\frac{\nu}{x} + \frac{\sqrt{\nu^2 - x^2}}{x}\right]^{-\nu} \to \left(\frac{ex}{2\nu}\right)^{\nu}.$$

Then for large $x$ and $|\nu|$, $H^{(\frac{1}{2})}$ and J behave as:

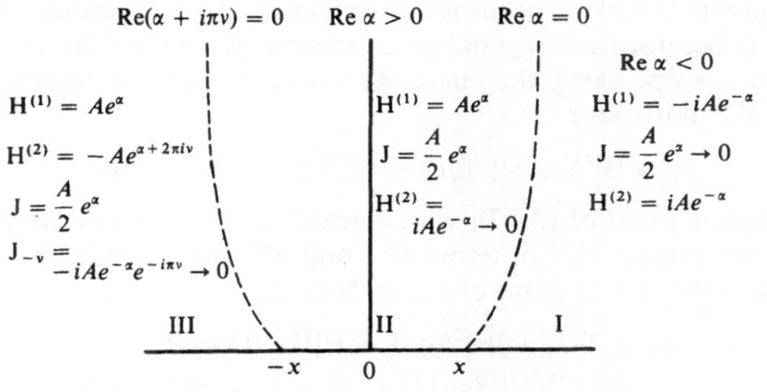


ν-plane

**Remarks** (3.4.8)

1. The function J is always of the form $(A/2)e^{\alpha}$, whereas $H^{(1)}$ (respectively $H^{(2)}$) changes its asymptotic behavior along the curve $\text{Re}\,\alpha = 0$ (respectively $\text{Re}(\alpha + i\pi\nu) = 0$), on which its zeroes are situated. These curves intersect the real axis at $\pm x$ at an angle of $\pm\pi/3$. Far away from the curve $\text{Re}\,\alpha = 0$, $H^{(1)}$ increases rapidly in the upper half-plane, going as $|\nu|^{\text{Re}\,\nu}$.
2. The zeroes $\nu_m$, $m = 1, 2, 3$, of $H_{\nu}^{(1)}(qa)$ closest to the real axis are near $\nu = qa$, where the asymptotic representation in terms of Airy functions is applicable; if $qa \gg 1$, then

$$H_{\nu_m}^{(1)}(qa) = 0, \qquad \nu_m \cong qa + c_m\left(\frac{qa}{2}\right)^{1/3} e^{i\pi/3},$$

$$\text{Ai}(-c_m) = 0, \qquad c_m = (2.3, 4.1, 5.5, 6.7, 7.9, \ldots).$$

3. Farther from the real axis along the curve $\text{Re}\,\alpha = 0$, $H^{(1)}$ is given simply as the sum of the asymptotic forms of regions I and II, and becomes zero if

$$\text{Sinh}\left(\alpha - \frac{i\pi}{4}\right) = 0.$$

This implies that

$$\nu_m \sim \frac{i\pi m}{\ln \dfrac{2\pi m}{eqa}} e^{-(i\pi/2)\ln(2\pi m/eqa)}, \qquad m \gg 1.$$

These facts allow us to discover expressions for $u$ in different regions of the $\rho$-$\varphi$-plane when $qa \gg 1$.

**The Shadow** (3.4.9)

The integral in (3.4.6) can be turned into an integral over $D'$ around the zeroes of $H_\nu^{(1)}(qa)$, because the integrand goes to zero exponentially fast in the upper half-plane, except along the curve $\operatorname{Re}\alpha = 0$. To see the reason for this, consider the numerator

$$g_\nu \equiv H_\nu^{(1)}(qa)H_\nu^{(2)}(q\rho) - H_\nu^{(2)}(qa)H_\nu^{(1)}(q\rho) = g_{-\nu}. \tag{3.4.10}$$

In the region $I \cup II$ of (3.4.7), within which the asymptotic form of $H_\nu^{(1)}$ changes, we express $H_\nu^{(1)}$ in terms of J and $H_\nu^{(2)}$. Meanwhile, for the same reason we write $H_\nu^{(2)}$ in terms of $J_{-\nu}$ in $II \cup III$:

$$\begin{aligned} g_\nu &= 2[J_\nu(qa)H_\nu^{(2)}(q\rho) - H_\nu^{(2)}(qa)J_\nu(q\rho)] \\ &= 2e^{i\pi\nu}[H_\nu^{(1)}(qa)J_{-\nu}(q\rho) - J_{-\nu}(qa)H_\nu^{(1)}(q\rho)]. \end{aligned} \tag{3.4.11}$$

In either case (3.4.7) shows us the asymptotic expression

$$g_\nu \sim \frac{2i}{\pi\nu}\left[\left(\frac{a}{\rho}\right)^\nu - \left(\frac{\rho}{a}\right)^\nu\right], \tag{3.4.12}$$

which thus holds uniformly in the upper half-plane. Hence $g_\nu$ is essentially $c^{\pm\nu}$, whereas the denominator $H_\nu^{(1)}(qa)$ is asymptotically dominated by $\nu^{\pm\nu}$, which wins out over $c^{\pm\nu}$, as can be seen if we let $\nu = t\exp(i\psi)$, $t$ and $\psi \in \mathbb{R}^+$; then for $t \to \infty$,

$$\begin{aligned} |\nu^{\pm\nu}| &= e^{t|\cos\psi \ln t - \psi \sin\psi|}, \\ |c^{\pm\nu}| &= e^{t|\cos\psi \ln c|}. \end{aligned} \tag{3.4.13}$$

Because all the other factors in (3.4.6) grow at the fastest as $c^{\pm\nu}$, the integrand decreases exponentially in the upper half-plane for all $\rho$ and $\varphi$. An exception is made, of course, for the curve $\operatorname{Re}\alpha = 0$ (asymptotically, $\ln t = \psi\tan\psi$), because the denominator can vanish, and the integral $\int_{D'}$ can be rewritten as the sum of the residues:

$$u = \pi \sum_{m=1}^{\infty} \frac{e^{-i\nu_m\pi/2}}{\sin \pi\nu_m} \cos(\varphi - \pi)\nu_m \frac{H_{\nu_m}^{(2)}(qa)}{\dfrac{\partial}{\partial\nu_m} H_{\nu_m}^{(1)}(qa)} H_{\nu_m}^{(1)}(q\rho). \tag{3.4.14}$$

This equation is only useful if the term with $m = 1$ is so much larger than the rest of the sum that it is essentially equal to $u$. To discover when this happens, we refer again to (3.4.3) to calculate

$$r_m \equiv \frac{H^{(2)}_{\nu_m}(qa)}{\frac{\partial}{\partial \nu_m} H^{(1)}_{\nu_m}(qa)} \tag{3.4.15}$$

for small $m$, with $\nu_m$ as in (3.4.8; 2). We learn that the order of magnitude of this factor is the same for the first few $m$'s.

If $\rho - a \gg a(qa)^{-2/3}$, then the other asymptotic form can be used for $H^{(1)}_{\nu_m}(q\rho)$,

$$H^{(1)}_{\nu_m}(q\rho) = \sqrt{\frac{2}{\pi}}(q^2\rho^2 - \nu_m^2)^{-1/4} \cdot \exp\left\{i\left[(q^2\rho^2 - \nu_m^2)^{1/2} - \nu_m \arccos\frac{\nu_m}{q\rho} - \frac{\pi}{4}\right]\right\}. \tag{3.4.16}$$

Whether the summands of (3.4.14) decrease rapidly with $m$ depends mainly on whether the coefficient of $i\nu_m$ in the exponent is positive. Since

$$\operatorname{Im} \nu_m = \left(\frac{qa}{2}\right)^{1/3} c_m \sqrt{\frac{2}{3}} \gg 1,$$

we see that

$$\sin \pi\nu_m \cong \frac{e^{-i\pi\nu_m}}{2i};$$

and if $|\varphi - \pi| \gg (qa)^{-1/3}$, then also

$$\cos(\varphi - \pi)\nu_m \cong \tfrac{1}{2}e^{-i\nu_m|\varphi - \pi|}.$$

All together, there is an overall factor (up to terms $O((qa)^{-1/2})$) of

$$\exp\left\{i\nu_m\left[\frac{\pi}{2} - |\varphi - \pi| - \arccos\frac{a}{\rho}\right]\right\},$$

making the condition for the first term of (3.4.14) to dominate that

$$|\varphi - \pi| + \arccos\frac{a}{\rho} - \frac{\pi}{2} < -(qa)^{-1/3}. \tag{3.4.17}$$

This amounts to the geometric condition that there is a shadow with a boundary fuzzy within an angle $\sim(qa)^{-1/3}$ (Figure 50). With the numerical

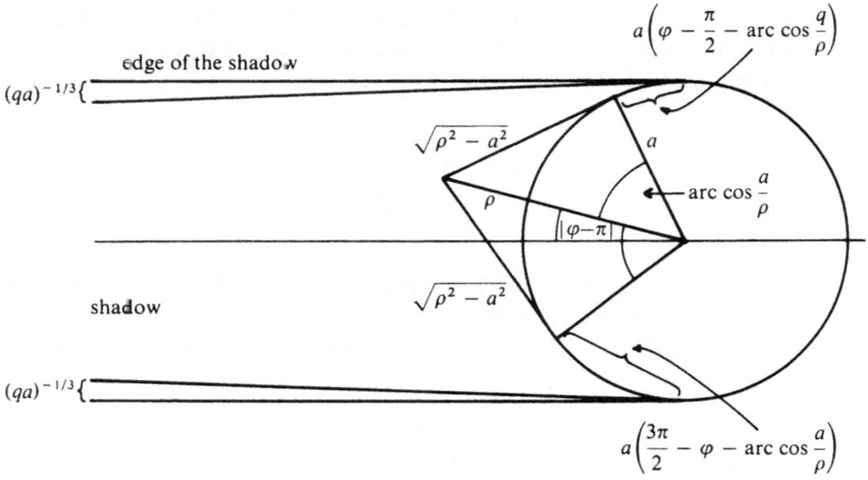


Figure 50 The shadow according to geometric optics

value of $c_1$, the resulting expression for $u$, up to corrections $\sim(qa)^{-1/3}$, is

$$u \cong \sqrt{\frac{\pi}{2q}}\exp\left(\frac{i\pi}{4}\right)r_1\frac{\exp(iq\sqrt{\rho^2 - a^2})}{(\rho^2 - a^2)^{1/4}}$$

$$\cdot\left[\exp(iqa - 1.5(qa)^{1/3})\left(\varphi - \frac{\pi}{2} - \text{arc cos}\,\frac{a}{\rho}\right)\right.$$

$$\left. + \exp(iqa - 1.5(qa)^{1/3})\left(-\varphi + \frac{3\pi}{2} - \text{arc cos}\,\frac{a}{\rho}\right)\right]. \quad (3.4.18)$$

The interpretation of this is that waves impinging on $(x, y) = (0, \pm a)$ are initially attenuated as they propagate along the surface to the points from which they can emanate in a straight line to the reference position. The damping factor of such "creeping waves" is correlated with the fuzziness of the shadow. During the linear propagation through the distance $\sqrt{\rho^2 - a^2}$ there is the usual amplitude factor $(\rho^2 - a^2)^{-1/4}$ for cylindrical waves, along with the phase factor $\exp(iq\sqrt{\rho^2 - a^2})$.

**The Illuminated Region at** $|\varphi| > \pi/2$ (3.4.19)

Since (3.4.14) is no good outside the shadow, we rewrite the $\cos\nu(\varphi - \pi)$ of (3.4.6) as $\exp(i\nu\pi)\cos\nu\varphi - i\exp(i\nu\varphi)\sin\nu\pi$. Taking the first term of this expression changes (3.4.17) to

$$|\varphi| + \text{arc cos}\,\frac{a}{\rho} \le \frac{3\pi}{2} - (qa)^{-1/3}, \quad (3.4.20)$$

and this is always true except exactly in the forward direction. Hence this part produces only the damped waves that manage to slip by the cylinder, and can be neglected in comparison with the remainder, $u_2$. The sin $\nu\pi$ cancels out of $u_2$, so the integration contour can be deformed at will across the real axis, and the integral can be calculated by the saddle-point method. We start by extending the integral

$$u_2 \equiv -\frac{1}{2}\int d\nu\, e^{i\nu(\varphi-\pi/2)}\,\frac{H_\nu^{(2)}(q\rho)H_\nu^{(1)}(qa) - H_\nu^{(2)}(qa)H_\nu^{(1)}(q\rho)}{H_\nu^{(1)}(qa)} \tag{3.4.21}$$

along a contour $D'$ that passes around the curve Re $\alpha = 0$ so as to stay to the right of $\operatorname{Re}(\alpha + i\pi\nu) = 0$ $(\alpha = \sqrt{\nu^2 - (qa)^2} - \nu \ln(\nu + \sqrt{\nu^2 - (qa)^2})/qa)$, and which then follows a curve containing the zeroes of $H_\nu^{(1)}(q\rho)$:

$$\operatorname{Re}\left[\sqrt{\nu^2 - (q\rho)^2} - \nu\ln\nu + \frac{\sqrt{\nu^2 - (q\rho)^2}}{q\rho}\right] = 0. \tag{3.4.22}$$

According to (3.4.7), in this region $H_\nu^{(2)}(q\rho)$ decreases exponentially, and the first term in the numerator of (3.4.21) contributes nothing, because it is an entire function in $\nu$. As for the second term, on the left-side portion of $D'$ the factor $H_\nu^{(2)}(qa)$ is exponentially decreasing, whereas on the right-side portion the variable can follow a path through the zeroes of $H_\nu^{(1)}(q\rho)$, which are on (3.4.22). On this portion of the contour the factor $\exp(i\nu(\varphi - \pi/2))$ takes care

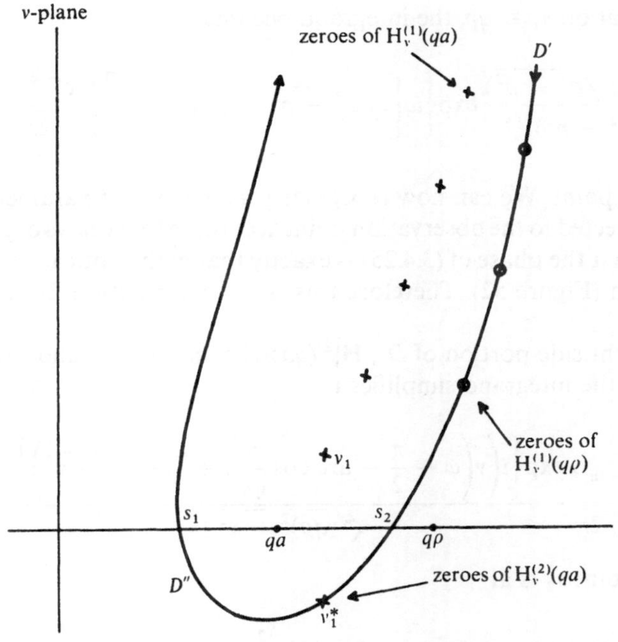


Figure 51 Path of integration for calculating geometric optics

of the exponential decrease, and this part of the contour can be connected to a path $D''$ that crosses the real axis at a saddle point $s_1 < qa$, proceeds to the first zero $\nu^*$ of $H_\nu^{(2)}(qa)$, and then climbs over another saddle point $s_2 > qa$ to reach the valley containing the zeroes of $H_\nu^{(1)}(q\rho)$. We next determine $s_i$ in the case $|s_i - qa| > (qa)^{1/3}$; the other case will be considered in (3.4.34). The integrand of $u_2$ on the left-side portion of $D''$ looks like

$$\frac{1}{\sqrt{2\pi}} e^{i\nu(\varphi-\pi/2)}((q\rho)^2 - \nu^2)^{-1/4} e^{i\pi/4}$$

$$\cdot \exp\left\{i\left[-2(q^2a^2 - \nu^2)^{1/2} + (q^2\rho^2 - \nu^2)^{1/2}\right.\right.$$

$$\left.\left. + 2\nu \arccos\frac{\nu}{aq} - \nu \arccos\frac{\nu}{q\rho}\right]\right\}, \tag{3.4.23}$$

because of the asymptotic properties of the Hankel functions. The $\nu$ dependence is dominated by the exponent, and the saddle point $s_1$ is at the position where the derivative by $\nu$ vanishes,

$$\varphi - \frac{\pi}{2} + 2 \arccos\frac{s_1}{qa} - \arccos\frac{s_1}{q\rho} = 0. \tag{3.4.24}$$

With the notation $s_1 \equiv qp$, the integrand becomes

$$\frac{\exp(-iq\sqrt{a^2 - p^2})}{\sqrt{q}(\rho^2 - p^2)^{1/4}} \exp\left\{iq\left[\sqrt{\rho^2 - p^2} - \sqrt{a^2 - p^2}\right]\right\} \frac{e^{i\pi/4}}{\sqrt{2\pi}} \tag{3.4.25}$$

at the saddle point. We can now recognize $p$ as the impact parameter of the ray that is reflected to the observation point according to the laws of geometric optics, and that the phase of (3.4.25) is exactly that of the optical path having this reflection (Figure 52). Therefore this is the contribution from reflected light.

On the right side portion of $D''$, $H_\nu^{(2)}(qa)/H_\nu^{(1)}(qa) \sim -1$, and the asymptotic form of the integrand simplifies to

$$-\frac{e^{-i\pi/4}}{\sqrt{2\pi}} \frac{\exp\left\{i\left(\nu\left(\varphi - \frac{\pi}{2} - \arccos\frac{\nu}{q\rho}\right) + \sqrt{\rho^2q^2 - \nu^2}\right)\right\}}{\sqrt[4]{(q\rho)^2 - \nu^2}}. \tag{3.4.26}$$

The saddle point $s_2$ is at

$$\varphi - \frac{\pi}{2} - \arccos\frac{s_2}{q\rho} = 0, \tag{3.4.27}$$

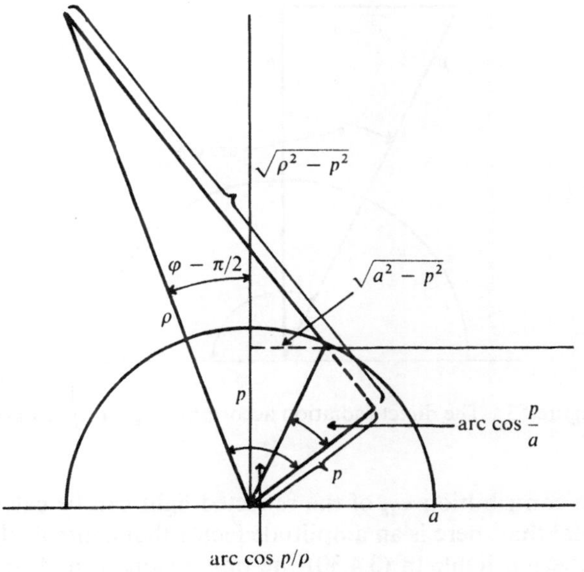


Figure 52 The reflected light according to geometric optics

and, with $p = v/q$, the integrand has the value

$$-\frac{e^{-i\pi/4}}{\sqrt{2\pi}} \frac{\exp(iq\sqrt{\rho^2 - p^2})}{\sqrt[4]{(q\rho)^2 - v^2}} \tag{3.4.28}$$

at the saddle point. Consequently $p$ is the impact parameter of the incident ray, and the phase of (2.4.28) is simply $\exp(-iqx)$ (Figure 53).

To calculate the incoming wave precisely, it is necessary to go through the usual manipulations of the method of steepest descent:

$$\int dv\, A(v)e^{i\alpha(v)} \cong A(s_2)e^{i\alpha(s_2)} \int_{-\infty}^{\infty} dv \exp\left\{i\alpha''(s_2)\frac{(v - s_2)^2}{2}\right\}$$

$$= A(s_2)\sqrt{\frac{-2\pi}{i\alpha''(s_2)}}\, e^{i\alpha(s_2)}; \tag{3.4.29}$$

since

$$A(s_2) = \frac{-e^{-i\pi/4}}{\sqrt{2\pi}((q\rho)^2 - s_2)^{1/4}},$$

$$\alpha''(s_2) = ((q\rho)^2 - s_2)^{-1/2}, \tag{3.4.30}$$

we find exactly $\exp(iq\sqrt{\rho^2 - p^2}) = \exp(-iqx)$. The corrections to the asymptotic expression can be calculated systematically, and are $O(\rho^{-2})$.

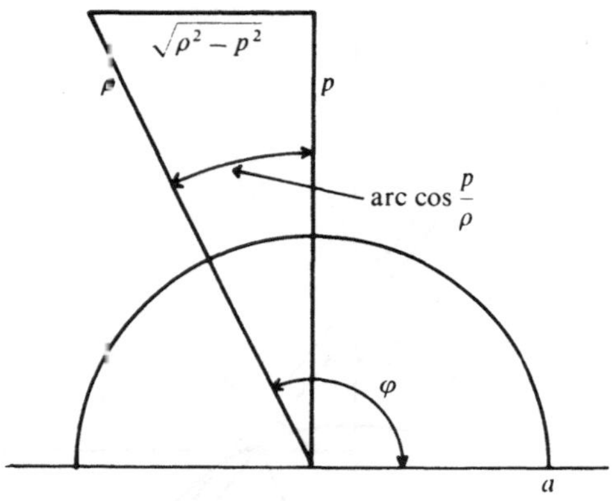


Figure 53 The direct radiation according to geometric optics

Similarly, the contribution $u_{2r}$ of the reflected light can be calculated accurately, to reveal that there is an amplitude factor that controls the spreading out of the reflected light. In (3.4.30) the only changes in $A$ are signs, but, because of (3.4.24),

$$\alpha''(v) = 2(a^2q^2 - v^2)^{-1/2} - (\rho^2q^2 - v^2)^{-1/2}. \tag{3.4.31}$$

Together with $u_2 \cong \exp(-iqx) + u_{2r}$, this makes

$$u_{2r} = \left(\frac{a^2 - p^2}{\rho^2 - p^2}\right)^{1/4}\left(2 - \left(\frac{a^2 - p^2}{\rho^2 - p^2}\right)^{1/2}\right)^{-1/2} \cdot \exp\{iq[\sqrt{\rho^2 - p^2} - 2\sqrt{a^2 - p^2}]\}. \tag{3.4.32}$$

If $\rho \gg a$, then $\sqrt{a^2 - p^2} = a\cos\varphi/2 \equiv a\sin\vartheta/2$, where $\vartheta = \pi - \varphi$ is the scattering angle. Then

$$u_{2r} \cong \sqrt{\frac{a\sin\vartheta/2}{2\rho}}\exp\left\{i\left(q\rho - 2qa\sin\frac{\vartheta}{2}\right)\right\}. \tag{3.4.33}$$

The angular distribution $\rho|u_{2r}|^2\,d\vartheta = dp$ is exactly what the geometric law of reflection predicts. Although our derivation works only if $\varphi > \pi/2$, the same result can be obtained for $\varphi < \pi/2$ by making minor changes, and by symmetry $u(\varphi) = u(-\varphi)$.

**The Boundary of the Shadow** (3.4.34)

The validity of the result of the previous paragraph relied on

(a) $\varphi > \pi/2$, so that the exponential factor decreases in the upper half-plane;
(b) staying away from the region where $\varphi \to \pi$ and $\rho \to \infty$, in which the creeping waves are insufficiently attenuated; and
(c) $|p - a| > a(qa)^{-2/3}$, so that the saddle point is sufficiently far from $qa$.

At the boundary of the shadow, assumption (c) fails, and we would have to assume that $\rho$ is not too large, in order to preserve (b). Granting that, Equation (3.4.21) could still be used, although both saddle points now move toward $qa$. The integration contour is divided into $D_1$, reaching from $aq + i\infty$ to $aq$; $D_2$, from $aq$ to $q\rho_1 < q\rho$; and $D_3$, from $q\rho_1$ to infinity along (3.4.22). The integrand decreases exponentially fast along $D_1$ and $D_3(\varphi > \pi/2)$, and on $D_2$ the ratio $H^{(2)}/H^{(1)} = -1 + 2J/H^{(1)}$ can be replaced with $-1$. The other contributions are straightforward to estimate as being $O((qa)^{1/3}(\varphi - \varphi_0))$, where $\varphi_0$ is the angle such that $(\rho, \varphi_0)$ is on the geometric boundary of the shadow, and we are interested in the values $|\varphi - \varphi_0| \sim (qa)^{-1/2}$.

Under these circumstances we are thus led back to

$$u_2 = \frac{e^{-i\pi/4}}{\sqrt{2\pi}} \int_{qa}^{q\rho_1} d\nu \frac{\exp\left\{i\nu\left(\varphi - \frac{\pi}{2} - \arccos\frac{\nu}{q\rho}\right) + i\sqrt{\rho^2 q^2 - \nu^2}\right\}}{[(q\rho)^2 - \nu^2]^{1/4}} \tag{3.4.35}$$

(cf. (3.4.26)), since the dominant part of the integral ought to come from the vicinity of $qa$, and we have used the proper asymptotic form of $H_\nu^{(1)}(q\rho)$ for that region. By the substitution

$$\nu \equiv q\rho \sin\omega, \qquad \varphi_0 > \omega > \varphi_1 > \frac{\pi}{2} \tag{3.4.36}$$

$(a = \rho\sin\varphi_0, \rho_1 = \rho\sin\varphi_1)$, (3.4.35) becomes

$$u_2 = \frac{e^{-i\pi/4}}{\sqrt{2\pi}} \int_{\varphi_1}^{\varphi_0} d\omega(|q\rho\cos\omega|)^{1/2} e^{iq\rho[(\varphi-\omega)\sin\omega - \cos\omega]}. \tag{3.4.37}$$

The function in the brackets [ ] in the exponent has its minimum at $\omega = \varphi$, and its leading term as $q\rho \to \infty$ can be approximated by a parabolic function,

$$[\ \ ] \cong -\cos\varphi\left(1 + \frac{(\omega - \varphi)^2}{2}\right).$$

If we write

$$\rho\cos\varphi = x, \qquad -\frac{iqx(\varphi - \omega)^2}{2} = \frac{i\pi\tau^2}{2}, \tag{3.4.38}$$

then we discover that

$$u_2 = e^{-iqx} \frac{e^{-i\pi/4}}{\sqrt{2|x|q}} \int_{\sqrt{q|x|/\pi}(\varphi-\varphi_0)}^{\sqrt{q|x|/\pi}(\varphi-\varphi_1)} d\tau\, e^{i\pi\tau^2/2}$$

$$\cdot\left[q|x|\cos\tau\sqrt{\frac{\pi}{q|x|}} + q\rho\sin\varphi\sin\tau\sqrt{\frac{\pi}{q|x|}}\right]^{1/2}. \tag{3.4.39}$$

As $q|x| \to \infty$, the upper limit of integration goes to infinity and $[\quad]^{1/2}$ approaches $\sqrt{q|x|}$; but unlike what we had earlier, the lower limit of integration remains finite if we are in the region $|\varphi - \varphi_0| \sim (q|x|)^{-1/2}$. With Fresnel's integral

$$F(x) = \int_{-\infty}^{x} d\tau\, e^{i\pi\tau^2/2} \tag{3.4.40}$$

we can simply write

$$u_2 = e^{-iqx} \frac{e^{-i\pi/4}}{\sqrt{2}} \left( F(\infty) - F\left( (\varphi - \varphi_0) \sqrt{\frac{q|x|}{\pi}} \right) \right). \tag{3.4.41}$$

In other words, $F$ specifies how the incident wave $\exp(-iqx)$ dies down as it passes from light ($\varphi - \varphi_0 < 0$, $F(\infty) = \sqrt{2}\exp(i\pi/4)$) to shadow ($\varphi - \varphi_0 > 0$, $u_2 \to 0$). At the geometric edge of the shadow ($\varphi = \varphi_0$) it has only half the original amplitude, because $F(0) = F(\infty)/2$, and as it passes from that point into the light, it oscillates according to Cornu's spiral around the original amplitude (cf. (3.3.10)).

**The Fraunhofer Region** (3.4.42)

Finally, we undertake a study of the region $\rho > (qa)a$, where the shadow has already faded out. If $|\varphi - \pi| > (qa)^{-1/3}$, then we know from (3.4.19) that the wave is simply composed of the incident wave plus a wave reflected according to geometric optics. If $|\varphi - \pi| < (qa)^{-1/3}$, then the facts discussed in (3.4.19) do not suffice to cause attenuation of the creeping waves, and we are forced to take drastic steps. We shall rely on the uniformly convergent expansion in the upper $\nu$-plane,

$$\frac{e^{-i\nu\pi/2}}{\sin \nu\pi} = -2ie^{i\nu\pi/2} \sum_{m=0}^{\infty} e^{2i\pi m\nu}. \tag{3.4.43}$$

Note that the condition for the contributions from the residues to decrease sufficiently fast is now

$$|\varphi - \pi| + \arccos \frac{a}{\rho} < \pi\left(2m + \frac{1}{2}\right) - (aq)^{-1/3} \tag{3.4.44}$$

(cf. (3.4.17)), which is always true for $m \geq 1$, so we only need to worry about the contribution $u_0$ with $m = 0$ in the sum. To take care of that, we again make the partition $D' = D_1 \cup D_2 \cup D_3$ of the earlier paragraph, and see that the main contribution again comes from $D_2$ with $H^{(2)}/H^{(1)} \to -1$. We shall only calculate the main contribution, as an estimate of the others shows that they become negligible in comparison with it as $\rho \to \infty$. Choosing $\rho_1 = \rho/2$, we can write the $m = 0$ term of the integral over $D_2$ of the asymptotic form of the integrand (3.4.6) as

$$u_{20} = -\frac{e^{-i\pi/4}}{\sqrt{2\pi}}\int_{qa}^{q\rho/2}\frac{dv}{((q\rho)^2 - v^2)^{1/4}}$$

$$\exp\left\{iv\left(\vartheta + \frac{\pi}{2} - \arccos\frac{v}{q\rho}\right) + i\sqrt{(q\rho)^2 - v^2}\right\} + (\vartheta \to -\vartheta)$$

$$= -e^{i(q\rho - \pi/4)}\sqrt{\frac{q\rho}{2\pi}}\int_{a/\rho}^{1/2} d\tau(1-\tau^2)^{-1/4}e^{iq\rho\alpha(\tau)} + (\vartheta \to -\vartheta),$$

$$\alpha(\tau) = \vartheta\tau + \tau \arcsin\tau + \sqrt{1-\tau^2} - 1. \tag{3.4.45}$$

We have introduced the scattering angle $\vartheta = \pi - \varphi > 0$ and substituted from $\tau \equiv v/q\rho$. Since the derivative of the exponent $\alpha'(\tau) = \vartheta + \arcsin\tau$ never vanishes, we may make the usual integration by parts,

$$\int_{a/\rho}^{1/2} d\tau\, A(\tau)e^{iq\rho\alpha(\tau)} = \int_{a/\rho}^{1/2} d\tau \frac{A(\tau)}{iq\rho\alpha'(\tau)}\frac{\partial}{\partial\tau}e^{iq\rho\alpha(\tau)}$$

$$= \frac{A(\tau)e^{iq\rho\alpha(\tau)}}{iq\rho\alpha'(\tau)}\Bigg|_{a/\rho}^{1/2} + O((q\rho)^{-2}) \tag{3.4.46}$$

to determine the asymptotic properties. For values $\vartheta \sim (qa)^{-1}$ but $\vartheta \gg a/\rho$, i.e., $\rho \gg aaq$, we thereby find that

$$u_{20} \rightleftarrows \frac{e^{iq\rho}}{\sqrt{\rho}}e^{-i\pi/4}\sqrt{\frac{2}{\pi q}}\frac{\sin qa\vartheta}{\vartheta}. \tag{3.4.47}$$

This is the same as the diffraction pattern at a slit or cylinder as calculated with the older theory, in which Formula (1.2.36) is used with $u$ in the lighted regions replaced with the incident wave. However, the older procedure gives no indication of the errors incurred, whereas the above theory is able to specify corrections to the asymptotic formulas.

If $\rho \gg aqa$, then $u$ is of the form

$$u \cong e^{-iqx} + \frac{e^{iq\rho}}{\sqrt{\rho}}f(\vartheta) \equiv u_{\text{in}} + u_{\text{scat}}. \tag{3.4.48}$$

If the wave is normally incident ($k = 0$ in (3.2.19; 3) and $q = \omega$), $f$ determines the

**Scattering Cross-Section** (3.4.49)

$$\sigma(\vartheta) \equiv \lim_{\rho\to\infty}\rho\frac{(T^{0\rho})_{\text{scat}}}{(T^{00})_{\text{in}}} = |f(\vartheta)|^2,$$

where the energy-momentum tensor is to be calculated respectively with $u_{\text{scat}}$ or $u_{\text{in}}$.

**Remarks** (3.4.50)

1. The cross-section $\sigma$ is the radiated energy per unit angle divided by the incident energy per unit length. It has the dimensions of a length. We are using the energy current from (3.1.24) for the scalar field $u$ (see Problem 1); the electromagnetic problem has a somewhat more complicated formulation.
2. In the limit $qa \gg 1$ and $\rho \to \infty$, we find that

$$|f(\vartheta)|^2 = \frac{a}{2}\sin\frac{\vartheta}{2} \qquad \text{if } \vartheta > (qa)^{-1/3},$$

$$|f(\vartheta)|^2 = \frac{2}{\pi q}\left|\frac{\sin qa\,\vartheta}{\vartheta}\right| \text{ if } \vartheta \sim (qa)^{-1};$$

in words, in addition to the geometrically scattered light there is an extra forward maximum, originating in the waves that permeate the shadow. Interestingly enough, both components have the geometric cross-section $2a$; but there is hardly any contribution from $(qa)^{-1} \ll q < (qa)^{-1/3}$:

$$2\int_{(qa)^{-1/3}}^{\pi} d\vartheta\,\frac{a}{2}\sin\frac{\vartheta}{2} = 2a(1 + O((qa)^{-2/3}))$$

$$2\int_0^{(qa)^{-1/3}} d\vartheta\,\frac{2}{\pi q}\left|\frac{\sin qa\vartheta}{\vartheta}\right|^2 = 2a(1 + O((qa)^{-1/3})).$$

Hence the total cross-section is twice the geometric cross-section. The significance is that not only is the directly incident light reflected, but also the light passing by at a distance $a$ is somewhat deflected.
3. In the limit $qa \ll 1$, the scattering cross-section is strongly dependent on the polarization. Until now we have always assumed that $u_{|\partial N} = 0$, for which (3.4.4; 2) makes the scattering cross-section

$$\sigma(\vartheta) = \frac{\pi a}{2qa|\ln qa|^2}.$$

This can exceed the geometric cross-section by an arbitrary amount, even though it obviously vanishes if $q$ is fixed and $a \to 0$.
4. If $du_{|\partial N} = 0$, then the Fourier coefficients in (3.4.2) are replaced with $J_n'(qa)/H_n^{(1)\prime}(qa)$, and if $qa \to 0$, then the terms with $n = 0, \pm 1$ contribute:

$$\sigma = \frac{\pi(qa)^3}{8}a(1 - 2\cos\vartheta)^2.$$

The total cross-section $a(3\pi^2/4)(qa)^3$ is much smaller than the geometric cross-section.

5. In electromagnetism Remark 3 applies to a field **E** parallel to the $z$-axis, and Remark 4 to **E** perpendicular to the $z$-axis. The physical reason for the greater cross-section of Remark 3 is that it is easier to move charges along the axis of the cylinder than transversely.

The example we have discussed has shown how complicated diffraction phenomena can be. Even Cartan's formalism, which makes the algebraic complications trivial, is not very effective for these analytical problems. However, the integral representation (3.4.6) is versatile enough that the whole range of diffraction phenomena can be derived from it, like rabbits pulled from a magician's hat. In general interference of waves renders the solution chaotic. Only in some limits the simple laws of geometrical optics emerge.

**Problems** (3.4.51)

1. Derive (3.4.49) for the energy-momentum forms (3.1.24) with

$$J_\alpha = u_{,\alpha}, \qquad u_{\text{in}} = e^{-iq(x+t)}, \qquad u_{\text{scat}} \cong \frac{e^{iq(\rho-t)}}{\sqrt{\rho}} f(\varphi).$$

2. Derive the optical theorem

$$\sigma_{\text{tot}} = \int_{-\pi}^{\pi} d\vartheta\, \sigma(\vartheta) = -2\sqrt{\frac{2\pi}{q}}\, Re(e^{i\pi/4} f(0)),$$

where $f$, as in (3.4.48) is defined by the asymptotic form of

$$u = \sum_{n=-\infty}^{\infty} e^{in((\pi/2)-\vartheta)}\left[\mathrm{J}_n(q\rho) - \frac{\mathrm{J}_n(qa)}{\mathrm{H}_n^{(1)}(qa)}\mathrm{H}_n^{(1)}(q\rho)\right] \xrightarrow{\rho\to\infty} e^{-iqx} + f(\vartheta)\frac{e^{iq\rho}}{\sqrt{\rho}}.$$

Use the formula

$$\frac{\mathrm{J}_n}{\mathrm{H}_n^{(1)}} = \frac{1}{2}\frac{\mathrm{H}_n^{(1)} + \mathrm{H}_n^{(2)}}{\mathrm{H}_n^{(1)}} \equiv \frac{1}{2}(1 - e^{2i\delta_n}), \qquad \delta_n \text{ real}.$$

3. Calculate the scattering cross-sections given in (3.4.50; 3) and (3.4.50; 4).

**Solutions** (3.4.52)

1. $$\frac{(T^{0\rho})}{(T^{00})} = \frac{\dot{u}u_{,\rho}}{\frac{1}{2}(\dot{u}^2 + |\nabla u|^2)}, \frac{(T^{0\rho})_{\text{scat}}}{(T^{00})_{\text{in}}} = \rho\frac{q^2|f(\varphi)|^2/\rho}{q^2} = |f(\varphi)|^2.$$

Since $|u(t)|^2 = (\text{Re}\, u(t))^2 + (\text{Im}\, u(t))^2 = (\text{Re}\, u(t))^2 + (\text{Re}\, u(t + \pi/2\omega))^2$, the use of the absolute value is equivalent to a time-average.

2. $$\frac{f(\vartheta)}{\sqrt{\rho}} e^{iq\rho} = \frac{1}{2}\sum_{n=-\infty}^{\infty} e^{in((\pi/2)-\vartheta)}(e^{2i\delta_n} - 1)\mathrm{H}_n^{(1)}(q\rho).$$

From the asymptotic expression

$$\mathrm{H}_n^{(1)}(q\rho) \cong \sqrt{\frac{2}{\pi q \rho}}\, e^{i(q\rho - (n\pi/2) - (\pi/4))}$$

one finds that

$$f(\vartheta) = \frac{e^{-i\pi/4}}{\sqrt{2\pi q}} \sum_{n=-\infty}^{\infty} e^{-in\vartheta}(e^{2i\delta_n} - 1), \qquad \sigma = \int_{-\pi}^{\pi} |f(\vartheta)|\, d\vartheta = \frac{4}{q} \sum_{n=-\infty}^{\infty} \sin^2 \delta_n,$$

$$\mathrm{Re}(e^{i\pi/4} f(0)) = \frac{-1}{\sqrt{2\pi q}} \sum_{n=-\infty}^{\infty} (1 - \cos^2 \delta_n + \sin^2 \delta_n).$$

3. The cross-section of Remark 3 follows from (3.4.4; 2). As for Remark 4,

$$u = \sum_{n=-\infty}^{\infty} e^{in((\pi/2) - \vartheta)} \left( \mathrm{J}_n(q\rho) - \frac{\mathrm{J}_n'(qa)}{\mathrm{H}_n^{(1)\prime}(qa)} \mathrm{H}_n^{(1)}(q\rho) \right),$$

$$f(\vartheta) = \sqrt{\frac{2}{\pi q}}\, e^{-i\pi/4} \left( \frac{\mathrm{J}_0'(qa)}{\mathrm{H}_0^{(1)\prime}(qa)} + 2 \cos \vartheta \frac{\mathrm{J}_1'(qa)}{\mathrm{H}_1^{(1)\prime}(qa)} + \cdots \right)$$

$$\cong \sqrt{\frac{2}{\pi q}}\, e^{-i\pi/4} \left( \frac{qa}{2} \right)^2 (1 - 2 \cos \vartheta) i\pi.$$

# Gravitation 4

## 4.1 Covariant Differentiation and the Curvature of Space

*The covariant derivative defines the rate of change of a tensor field in the direction of a vector. Covariant derivatives in two different directions do not in general commute; their commutator determines the curvature of space.*

In field theory one has to deal with derivatives of vector fields and in modern theories there appear quantities which are vectors not in space-time but in an internal space. In both cases one deals with vector bundles where vectors at different points are not canonically oriented towards each other. A chart independent notion of a derivative requires an additional structure, the so-called connection. It will be the subject of this chapter. As one hopes that eventually space, time and internal space will turn out to be only different directions in a unifying entity we start with some definitions which allow us to treat both cases in the same way.

**Definition** (4.1.1)

A **section** of a vector bundle $V$ with basis $B$ and projection $\Pi$ is a map $\Phi$: $B \to V$ with $\Pi \circ \Phi = 1_B$. The set of sections is denoted by $S_0(V)$.

**Examples** (4.1.2)

1. The charged scalar field (3.1.21; 2) $\Phi$ has two components $\varphi^1$, $\varphi^2$ and can be considered as a section of the bundle $B \times \mathbb{R}^2$ where $B$ is space-time.

It associates to $x \in B$ the point $(x; b_1\varphi^1(x) + b_2\varphi^2(x)) \in V$ if $b_{1,2}$ are a basis of $\mathbb{R}^2$ which is the fibre in this case.

2. Besides the charged pions $\pi^\pm$ there exists the neutral pion $\pi^0$ with about the same mass. Their description requires a 3-component pseudo-scalar field $\varphi^i$, $i = 1, 2, 3$ where $\varphi^1 \pm i\varphi^2$ describe $\pi^\pm$ and $\varphi^3$ describes $\pi^0$. Thus the pion field is a section in a vector bundle with fibre $\mathbb{R}^3$, the so-called isospin space. There are many examples of similar internal degrees of freedom in particle physics.
3. Vector fields (resp. 1-forms) are sections in the bundles $T(B)$ (resp. $T^*(B)$), $S_0(T^*(E)) = \mathscr{T}^0_1(B)$, etc. In our case $B$ will be space-time, $\dim B = 4$. Thus the fibres in $T(B)$ will always be $\mathbb{R}^4$. Similarly tensor fields are sections of the tensor bundles.

We shall encounter only situations where the fibres have in addition to their structure of a vector space another important property which deserves a name.

**Definition** (4.1.3)

A nondegenerate bilinear map (or sequilinear form for complex bundles) $S_0(V) \times S_C(V) \to C(B)$: $(\Phi, \Psi) \to \langle\Phi|\Psi\rangle$ is called **fibre metric**.

**Remark** (4.1.4)

$S_0(V)$ is a module over $C(B)$ and bilinearity implies $\langle f\Phi|g\Psi\rangle = fg\langle\Phi|\Psi\rangle\ \forall f, g \in C(B)$. By nondegenerate we mean $\langle\Phi|\Psi\rangle = 0\ \forall\Phi \in S_0 \Rightarrow \Psi = 0$ at every point of $B$.

**Examples** (4.1.5)

1. In the Lagrangian for the charged scalar field or the pion field appears the combination $\sum_i \varphi^{i2}(x)$. It corresponds to the usual metric in $\mathbb{R}^n$ and will be taken as fibre metric for this bundle.
2. On a Riemannian space $M$ the metric defines a fibre metric in the bundle $T(M)$. We have seen how it can be extended to $T^*(M)$ and the tensor bundles.
3. On any manifold $M$ the canonical 2-form on $T^*(M)$ defines a fibre metric in the bundle $T(T^*(M))$. Note that the fibre metric need not be positive.

Every element of a vector space can be expanded in terms of a basis $\{b_i\}$, but for a vector bundle a basis for the sections may not exist globally but only locally. In any case the metric is locally determined by its action on a basis and for a symmetric metric it is expedient to use a basis where the metric assumes the normal form

$$\langle b_i|b_j\rangle = \eta_{ij} \qquad \text{with} \qquad \eta_{ij} = \begin{cases} \pm 1 & \text{for } i = j, \\ 0 & \text{for } i \neq j, \end{cases}$$

("orthonormal" or simply "orthogonal basis"). This requirement does not fix the $b_i$ uniquely, $b_i = b_k L^k{}_i$ is also an orthogonal basis if $L^t \eta L = \eta$. Even though the $\eta_{ik}$ do not depend on $x \in B$ the $L$ may do so.

**Definition** (4.1.6)

Let $\{b_i\}$ be a local basis of sections in a vector bundle with fibre metric $\langle b_i | b_j \rangle = \eta_{ij}$. The transformations $b_i \to b_k L^k{}_i$ with $L^t \eta L = \eta$ are called **gauge transformations**. The $L^k{}_i(x)$ form the same group for all $x \in B$, the so-called **gauge group** $G$ of the bundle. If the bundle is trivializable and $L: B \to G$ is a constant map one speaks of global gauge transformations.

**Examples** (4.1.7)

1. For the fields $\Phi = \sum b_i \varphi^i$ with fibre metric $\langle \Phi | \Phi \rangle = \sum_{i=1}^n \varphi^{i^2}$ the gauge group $G$ is $O(n)$.
2. In a pseudo-Riemannian space with signature $(n, m)$ ($\eta$ has $n$ positive and $m$ negative eigenvalues) the gauge group is $O(n, m)$.
3. In $T(T^*(M))$ with the canonical 2-form as fibre metric the gauge group is the symplectic group.

**Remark** (4.1.8)

There are groups which cannot be characterized by the invariance of a quadratic form. To treat them as a gauge group one has to ascend to the next level of abstraction and define principal fibre bundles where the fibres are the group manifold itself. In our cases the gauge group will always come from the invariance of a fibre metric and we shall not need this construction.

For an exterior derivative of a section we still need a notion which combines sections and $p$-forms.

**Definition** (4.1.9)

The sections over the vector bundle $B \times V \otimes_\pi \Lambda_p T^*(B)$ are called **$p$-form valued sections** (or vector valued $p$-forms), their set is denoted by $S_p$.† They are a linear space and the exterior product $\wedge$ maps $E_p \times S_q$ into $S_{p+q}$. Thus $\Phi \in S_p$ can be written $\Phi = \sum_i b_i v^i$ with $v^i \in E_p$ and $b_i$ the basis in $S_0$. Given a vector field $u \in \mathscr{T}^1_0(B)$ we define an inner product $i_u: S_p \to S_{p-1}$ by

$$i_u \Phi = \sum_j b_j i_u v^j.$$

† $\otimes_\Pi$ denotes a product of bundles with the same basis. In it the fibres are the tensor product of individual fibres and the basis is the common basis. $\Lambda_p T^*(B)$ is the $p$-fold antisymmetric product of the $T^*(B)$.

So far there is no chart-independent notion of parallelism between fibres at different points. To define the derivative of a section $\Phi$ we need a prescription how to parallel transport $\Phi(x)$ to $x + dx$ such that $\Phi(x + dx) - \Phi(x)$ can be defined. This prescription will be expressed on the infinitesimal level by the connection. We shall wonder later whether and how this parallelism can be extended to the local or global levels. First we list some formal desiderata for the derivative and then successively narrow it down by requiring that it conserve further structures. To start with we want a derivative $D_u$ of a section in direction of a vector field $u$ to be linear in $u$ so that one ought to be able to write it $i_u D$, $D: S_0 \to S_1$, hence we concentrate on $D$.

**Definition** (4.1.10)

A **covariant exterior derivative** $D$ is a map $S_p \to S_{p+1}$ with the properties:

(i) $D(\Phi_1 + \Phi_2) = D\Phi_1 - D\Phi_2, \qquad \Phi_i \in S_p;$
(ii) $D(\Phi \wedge v) = (D\Phi) \wedge v + (-)^p \Phi \wedge dv, \qquad \Phi \in S_p, \qquad v \in E_q.$

The 1-forms ${\omega^k}_i$ appearing in the expansion of the covariant exterior derivative of a local basis $b_i \in S_0$

$$Db_i = b_k {\omega^k}_i, \qquad {\omega^k}_i \in E_1,$$

are called the **linear connection or gauge potential**.

**Remarks** (4.1.11)

1. Through the rules (i) and (ii) $D$ is completely determined by the $\omega$'s. For instance, with the natural basis $dx^\alpha$ in $E_1$ we can write

$$\Phi = \frac{1}{p!} \sum_{i,(\alpha)} b_i \varphi^{i(\alpha)} \, dx^{\alpha_1} \wedge \cdots \wedge dx^{\alpha_p}$$

and $D\Phi$ becomes with the rules (4.1.10)

$$D\Phi = \frac{1}{p!} \sum_{i,(\alpha)} b_i (d\varphi^{i(\alpha)} + {\omega^i}_k \varphi^{k(\alpha)}) \wedge dx^{\alpha_1} \wedge \cdots \wedge dx^{\alpha_p}.$$

2. Since the $b_i$ need not be defined globally the $\omega$ will not be either, even though $D$ exists globally.
3. $\omega$ can be considered as element of $S_1(L(F))$, a 1-form valued section of the bundle over the domain of the $b_i$ with the linear transformations of the fibres $F$ as fibres. In this sense one might use the shorthand $D = d + \omega \wedge$. However, one has to keep in mind that under a change of the basis in $F$ the $\omega$'s do not transform the way elements of $S_1(L(F))$ do.

**Transformation of the Gauge Potential under Gauge Transformations** (4.1.12)

Under a change of the local basis $b \to bL$, $L \in S_0(L(F))$ the $\omega$ transform as

$$\omega \to L^{-1}\omega L + L^{-1}\, dL.$$

**Remarks** (4.1.13)

1. $dL \in S_1(L(F))$ is to be understood as the matrix with elements $dL^i{}_k$ if the change of basis is $b_k \to \bar{b}_k = b_i L^i{}_k$. Thus, with index notation
$$\bar{\omega}^i{}_k = (L^{-1})^i{}_j\, \omega^j{}_l L^l{}_k + (L^{-1})^i{}_j\, dL^j{}_k.$$
2. The transformation law follows from the rules (4.1.10)
$$\bar{b} = bL \Rightarrow D\bar{b} = (Db)L + bdL = b\omega L + b\, dL = \bar{b}(L^{-1}\omega L + L^{-1}\, dL).$$
3. Looked at from the point of view of the components $\varphi^i$ of $\Phi = \sum b_i \varphi^i \in S_0$ one can say that the components $d\varphi^i + \omega^i{}_k \varphi^k$ of $D\Phi$ transform like the $\varphi^i$ under the gauge transformation $\varphi^i \to \bar{\varphi}^i = (L^{-1})^i{}_k \varphi^k$. The inhomogeneous term $(dL^{-1})\Phi$ is cancelled by the corresponding term in the transformed $\omega$ since $L^{-1}L = 1 \Rightarrow dL^{-1}L = -L^{-1}\, dL$.

**Examples** (4.1.14)

1. It may happen that one basis $b_i$ is covariant constant, $\omega^i{}_k = 0$. In another basis $\bar{b}_k = b_i L^i{}_k$ we have then
$$D\bar{b}_k = \bar{b}_j (L^{-1})^j{}_i\, dL^i{}_k.$$
If there is no global basis it is therefore not clear whether there is a $D$ with $\omega = 0$ on all of $B$.
2. The model (3.1.21; 2) of a charged field can be written in terms of two real fields $\varphi^1 = \sqrt{\rho} \cos S$, $\varphi^2 = \sqrt{\rho} \sin S$ or one complex field $\Phi = \varphi^1 + i\varphi^2 = \sqrt{\rho}\, e^{iS}$. To get a vector bundle we have to allow $\rho$ to vary over $\mathbb{R}^+$. In this case we have the fibres $F$ equal to $\mathbb{R}^2$ or $C$ and the bundle $V = \mathbb{R}^4 \times F$. In the Lagrangian (3.1.20) occurred the combination $d\varphi + iA\varphi$ (or $d\varphi^i + \omega^i{}_k \varphi^k$, $\omega^i{}_k = \begin{pmatrix} 0 & 1 \\ -1 & 0 \end{pmatrix} A(x)$) which is a covariant exterior derivative in this bundle. $S \to S + \Lambda$ corresponds to a change of basis in which case $A$ undergoes the gauge transformation $A \to A + d\Lambda$. Here the gauge transformations and the connection involve only the matrix $\begin{pmatrix} 0 & 1 \\ -1 & 0 \end{pmatrix}$ and therefore the $L$ cancel out in $L^{-1}\omega L$.

So far our motivation for $D$ was more formal. Its geometrical significance becomes more apparent when one defines the covariant derivative in the direction of a vector field over $B$.

**Definition** (4.1.15)

Given a vector field $u \in \mathscr{T}^1_0(B)$ the **covariant Lie derivative** is defined by

$$\mathscr{L}_u = i_u \circ D + D \circ i_u.$$

It maps $S_p$ into $S_p$. The particular case $p = 0$ is called **covariant derivative** and is denoted by $D_u \equiv i_u \circ D$.

**Remarks** (4.1.16)

1. Remember how the inner product $i_u$ was defined. We have more explicitly for $\varphi = \sum_i b_i \varphi^i \in S_0$ the expression
$$\mathscr{L}_u \Phi = D_u \Phi = \sum_i b_i((d\varphi^i | u) + (\omega^i{}_k | u)\varphi^k).$$
2. For $\mathscr{L}_u$ like for $L_u$ the Leibniz rule, $\mathscr{L}_u(v \wedge \Phi) = (\mathscr{L}_u v) \wedge \Phi + v \wedge \mathscr{L}_u \Phi$ holds. But generally $\mathscr{L}_{fu} \neq f \mathscr{L}_u$, only for $D_u$ we have $\forall f \in E_0$: $D_{fu} = f D_u$.
3. Following the construction of the tensor bundles $\mathscr{T}^r_s(M)$ out of $T(M)$ one can associate to $V$ its dual bundle $V^*$ and the tensor products
$$V^r_s = \overbrace{V \underset{\pi}{\otimes} V \underset{\pi}{\otimes} \cdots}^{r} \overbrace{V^* \underset{\pi}{\otimes} V^* \underset{\pi}{\otimes} \cdots V^*}^{s}.$$
The differential processes $D$ and thus $\mathscr{L}_u$ can be carried over to the sections $S_p(V^r_s)$ by postulating Leibniz' rule for $p = 0$,
$$D(\Phi_1 \otimes \Phi_2) = (D\Phi_1) \otimes \Phi_2 + \Phi_1 \otimes D\Phi_2 \quad \text{for} \quad \Phi_i \in S_0(V),$$
and the invariance of the map
$$(S_0(V^*) \times S_0(V)) \to E_0(B):$$
$$D(\Psi | \Phi) = (D\Psi | \Phi) + (\Psi | D\Phi), \qquad \Psi \in S_0(V^*), \qquad \Phi \in S_0(V).$$
$D(\Psi | \Phi) = d(\Psi | \Phi)$ because $D = d$ on the scalars. In particular, for the dual basis $b^{*j}$, $(b^{*j} | b_k) = \delta^j{}_k$ we infer from $d\delta^j{}_k = 0$ that $Db^{*j} = -\omega^j{}_k b^{*k}$. One typical example will be $S(L(F))$ where $L(F)$ is identified with $F^* \otimes F$: If $L(F) \ni L = \sum v_i w_i$, $v_i \in F$, $w_i \in F^*$ then $L \cdot u = \sum_i v_i(w_i | u)$. Again $\mathscr{L}_u$ depends only on the direction of $u$ and not on its derivative when applied to $S_0(V^r_s)$ where it is denoted by $D_u$.

**Geometrical Significance of the Covariant Derivative** (4.1.17)

A section $\Phi \in S_0(V)$ as map $B \to V$ also maps $T(B) \xrightarrow{T(\Phi)} T(V)$. With reference to a basis this can be written: $x \to (x, \varphi^i(x))$,

$$T(\Phi): (x, v) \to \left(x, \varphi^i(x); v, v^\alpha \frac{\partial \varphi^i}{\partial x^\alpha}\right).$$

If we identify $F$ and $T(F)$ this looks schematically as follows.

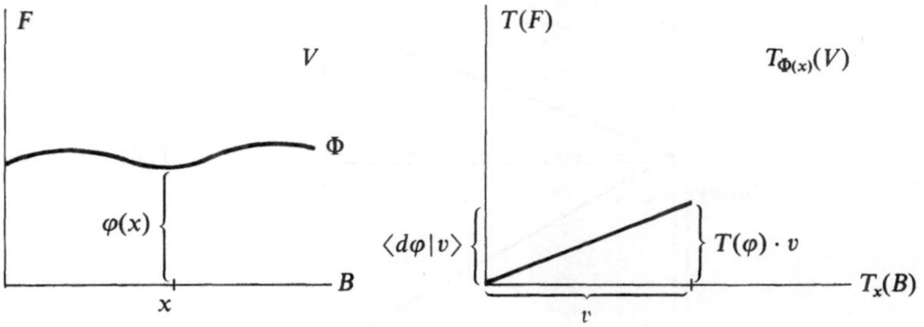


What we now want for $D_v(\Phi)$ is the change of the vertical part of $\Phi$ (in the direction of $F$) as one goes in the direction $v$ of $B$. However, in $T(V)$ there is no preferred horizontal direction, only the vertical direction is defined by $T(\Pi)$. For the components in the vertical direction we need a preferred horizontal direction. The latter is defined by the 1-forms $-\ \omega^i{}_k$ which can be considered as map†

$$(x, v) \to (x, \varphi^i(x);\ v,\ -(\omega^i{}_k | v)\varphi^k).$$

$D_v(\Phi)$ as difference between the two maps measures the change $T(\Phi)v$ relative to the horizontal direction given by $\omega$ (see the following illustration). Thus $D_v\Phi$ should be looked at as the change of $\Phi_x \in F$ as one proceeds in direction of $v$. It comes about because the components change by $(d\varphi^i | v)$ and the basis is rotated by $(\omega^i{}_k | v)$.

A section whose vertical component does not change will be one with $D_v\Phi = 0\ \forall v$, that is $D\Phi = 0$. One says that such a section is covariant constant or $\Phi_x \in F$ is parallel transported as one moves in $B$. If one has a covariant constant basis the $\omega$'s are zero and the covariant constant vectors are the ones with constant components. In this case the notion of parallelity given infinitesimally by $D$ can be extended to the domain of this basis. One can see on $T(S^2)$ that in general the extension even to the local level might be impossible. There the standard prescription for parallel transport along great circles is to keep the angle with them fixed. If one now takes a vector at the north pole and transports it to the south pole one arrives at vectors pointing in different directions if one takes different great circles to get there. Even for points which are a small but finite distance apart the above description does not lead to a unique result. Obviously, a necessary condition for $D\Phi = 0$ is $DD\Phi = 0$. Unlike $d$ the square of $D$ is not identically zero. It

† Analogously to $\times_\Pi$ the sum $+_\Pi$ of vector bundles is defined such that the fibres are the sum of the fibres and the basis is the common basis.

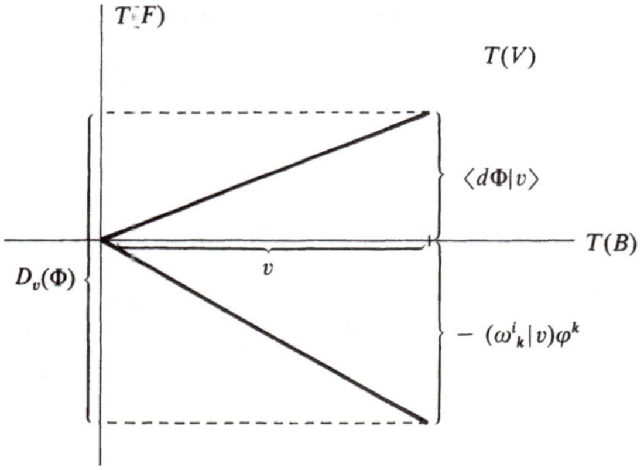


The geometrical meaning of $D_v$

turns out that $D \circ D$ does not depend on the derivatives of the section one applies it to, our rules (4.1.8) say

$$\begin{aligned} DD(f \wedge \Phi) &= D(df \wedge \Phi + (-)^q f \wedge D\Phi) \\ &= ddf \wedge \Phi + (-)^{q+1}\, df \wedge D\Phi + (-)^q\, df \wedge D\Phi \\ &\quad + f \wedge DD\Phi = f \wedge DD\Phi \qquad \forall f \in E_q, \quad \Phi \in S_p. \end{aligned} \tag{4.1.18}$$

Thus $DD\Phi \in S_{p+2}$ can be linearly expressed by $\Phi$ and thus defines a 2-form with values in $L(F)$.

**Definition** (4.1.19)

The **curvature form** $\Omega \in S_2(L(F))$ is defined by

$$DD\Phi = \Omega \wedge \Phi, \qquad \Phi \in S_p.$$

**Remarks** (4.1.20)

1. $\Omega$ is determined by its action on a basis $\Omega b_i \equiv b_k \Omega^k{}_i$ since (4.1.18) shows $DDb_i\varphi^i = b_k\Omega^k{}_i\varphi^i$. Now

$$DDb_i = D(b_k\omega^k{}_i) = b_k\, d\omega^k{}_i + b_k\omega^k{}_j \wedge \omega^j{}_i$$

and we arrive at "Cartan's second structure equation"

$$\Omega^i{}_k = d\omega^i{}_k + \omega^i{}_j \wedge \omega^j{}_k.$$

(In the tangent bundle of space-time we shall use the letter $R$ for $\Omega$.) Considering $\omega \in S_1(L(F))$ one may write this more compactly $\Omega = d\omega + \omega \wedge \omega$. Since matrix multiplication is not commutative $\omega \wedge \nu \neq -\nu \wedge \omega$ for matrix valued 1-forms and $\omega \wedge \omega$ need not vanish.

2. For the connection (4.1.13) in $V = \mathbb{R}^4 \times \mathbb{R}^2$ with the abelian gauge group $\mathscr{A}(2)$ the term $(\omega \wedge \omega)^i{}_k = A \wedge A\varepsilon^i{}_j\varepsilon^j{}_k$ vanishes since $A \wedge A = 0$

and $\Omega^i{}_k = dA\varepsilon^i{}_k = F\varepsilon^i{}_k$ $(\varepsilon^i{}_k = (\begin{smallmatrix} 0 & 1 \\ -1 & 0 \end{smallmatrix}))$. Thus the electromagnetic field $F = dA$ is the curvature of this connection.

3. By a change of basis $b_i \to \bar{b}_i = b_k L^k{}_i$ we have because of (4.1.18)
$$DD\bar{b}_i = b_j \Omega^j{}_k L^k{}_i = \bar{b}_k \Omega^k{}_i,$$
which shows that
$$\bar{\Omega}^i{}_k = (L^{-1})^i{}_j \Omega^j{}_m L^m{}_k.$$
Thus $\Omega$, in contradistinction to $\omega$, transforms under a change of basis like an element of $S_2(L(F))$ ought to transform. One can verify (Problem 4) that the inhomogeneous terms in the transformation law (4.1.12) do cancel out in the combination $d\omega + \omega \wedge \omega$.

4. To see the significance of $\Omega$ for the directional derivative we have to use the identity for $\Phi \in S_0$
$$(i_v i_u DD)\Phi = i_u D i_v D\Phi - i_v D i_u D\Phi - D_{[u,v]}\Phi \tag{4.1.21}$$
(Problem 7). It tells us that if we have two vector fields with $[u, v] = 0$ so that they form two-dimensional surfaces then $(D_u D_v - D_v D_u)\Phi = (i_v i_u \Omega)\Phi$. We see that $\Omega$ measures the difference between the covariant changes of $\Phi$ by first proceeding in the direction of $u$ and then in the direction of $v$ and secondly by coming to the same (infinitesimally adjacent) point by proceeding in the opposite order.

We return to the question of covariant constant sections. $D\Phi = 0$ means for the components of $\Phi$
$$d\varphi^i + \omega^i{}_k \varphi^k = 0. \tag{4.1.22}$$
This implies
$$0 = d(\omega^i{}_k \varphi^k) = (d\omega^i{}_k + \omega^i{}_j \wedge \omega^j{}_k)\varphi^k = \Omega^i{}_k \varphi^k. \tag{4.1.23}$$
Thus a covariant constant section has to be at each point $x \in B$ an eigenvector of the matrix $\Omega^i{}_k(x)$ with eigenvalue zero. If we have a covariant constant basis $\Omega$ has to be zero since its action on a basis gives zero. This also follows from the fact that in such a basis $\omega = 0$ and therefore $\Omega$ is zero and $\Omega$ transforms homogeneously under a change of basis. Conversely, $\Omega = 0$ implies $d\omega^i{}_k + \omega^i{}_j \wedge \omega^j{}_k = 0$ which are the integrability condition for the system (4.1.22). Frobenius' theorem [22, Vol. I] tells us that they are also sufficient for the local existence of solutions of (4.1.22) with arbitrary initial conditions. This situation deserves a name.

**Definition** (4.1.24)

A connection of a vector bundle is called locally **flat** if one of the two equivalent conditions hold:

(i) $\Omega = 0$;
(ii) there exists (locally) a covariant constant basis.

**Remarks** (4.1.25)

1. If the bundle is parallelizable, that is, there is a global basis, then we get a flat connection by declaring this basis as covariant constant. However, flatness is a local property and a flat bundle like the Möbius strip need not be a product. But even if the tangent space of a manifold is a flat product bundle the manifold need not be an open set of $\mathbb{R}^n$ but can have different global properties as is the case for $T^n$.
2. Flatness is an internal property of the manifold, $\Omega$ does not reproduce the curvature of a surface as imbedded in, say, $\mathbb{R}^3$. $\Omega$ vanishes for a cylinder because it could be rolled out flat on a plane without affecting its metric and thereby its internal structure.

The covariant exterior derivative $D$ can be extended in a natural way to $S_p(L(F))$ (see (4.1.16; 3) and one finds when applied to $\Omega$.

**Bianchi's Identity** (4.1.26)

$$D\Omega = 0.$$

**Remarks** (4.1.27)

1. For the components (4.1.20) of $\Omega = b_k \Omega^k{}_i b^{*i}$ we see from (4.1.16; 3)
$$0 = D\Omega = b_r(\omega^r{}_k \Omega^k{}_j + d\Omega^r{}_j - \Omega^r{}_i \omega^i{}_j) b^{*j}$$
or
$$d\Omega^i{}_j = -\omega^i{}_k \wedge \Omega^k{}_j + \Omega^i{}_k \wedge \omega^k{}_j.$$
This relation is proved in Problem 6.
2. One might think $\Omega = D\omega$ and thus $D^2 = 0$ on $\omega$. However, it has to be noted that though $\Omega$ is a covariant exterior derivative of $\omega$ in the sense that $\Omega$ transforms like $L(F)$ under a change of basis this $D$ is not the one used in (4.1.26) since $\omega$ does not transform like $S_1(L(F))$. Written in components as above we would get
$$(D\omega)^i{}_k = d\omega^i{}_k + 2\omega^i{}_l \wedge \omega^l{}_k$$
and differs from $\Omega^i{}_k$ by the factor 2 in the last term.

Our general considerations so far do not constrain $\omega$ any further. We shall now impose restrictions on the covariant derivative which finally determine it uniquely from the fibre metric.

The fibre metric defines a scalar product in $F$ and thus maps $S_p(V) \times S_0(V)$ into $E_p(B)$. We shall require that the connection conserves this structure in the sense that the derivative of the scalar product contains just the derivatives of the sections involved.

**Compatibility of the Connection with the Fibre Metric** (4.1.28)

In a vector bundle $V$ with fibre metric $\langle\,|\,\rangle$ we shall require

$$D\langle\Psi|\Phi\rangle = d\langle\Psi|\Phi\rangle = \langle D\Psi|\Phi\rangle + (-)^p\langle\Psi|D\Phi\rangle,$$
$$\Psi \in S_p(V), \qquad \Phi \in S_0(V).$$

Such connections will be called metric connections.

**Remark** (4.1.29)

For the $\omega$'s this imposes the condition

$$\begin{aligned} 0 = d\eta_{ik} = d\langle e_i|e_k\rangle &= \langle De_i|e_k\rangle + \langle e_i|De_k\rangle \\ &= \omega^j{}_i\langle e_j|e_k\rangle + \langle e_i|e_j\rangle\omega^j{}_k = \omega^j{}_i\eta_{jk} + \omega^j{}_k\eta_{ij}. \end{aligned}$$

Thus if the $\eta$ are symmetric $\omega_{ik} \equiv \omega^j{}_k\eta_{ij}$ has to be antisymmetric

$$\omega_{ik} = -\omega_{ki}. \tag{4.1.30}$$

This means that $\omega_{ik}(x)$ cannot be an arbitrary element of $L(F)$, but must belong to the Lie algebra of the gauge group. In our cases $G$ will be for the inner gauge symmetries an $O(n)$ or $U(n)$ group and for the space-time symmetry we have $O(3, 1)$. For an arbitrary basis with $\langle e_i|e_k\rangle = g_{ik}$ (4.1.30) generalizes to $dg_{ik} = \omega_{ik} + \omega_{ki}$.

**Examples** (4.1.31)

1. The Lie algebra of the one-dimensional group $O(2)$ are the multiples of $\begin{pmatrix} 0 & 1 \\ -1 & 0 \end{pmatrix}$ and the connection (4.1.14; 2) conserves the $O(2)$-invariant fibre metric $\langle\Phi|\Phi\rangle = (\varphi^1)^2 + (\varphi^2)^2$ since $\omega^i{}_k = \omega_{ik} = -\omega_{ki}$.
2. In the tangent bundle of a Riemannian space $\eta_{ik} = \delta_{ik}$ and thus the $\omega^i{}_k = -\omega^k{}_i$. For the pseudo-Riemannian space-time we have

$$\eta_{ik} = \begin{cases} -1 & \text{if } i = k = 0, \\ 1 & \text{if } i = k = 1, 2, 3, \\ 0 & \text{otherwise.} \end{cases}$$

Hence $\omega^i{}_k = -\omega^k{}_i$ for $i, k = 1, 2, 3$, $\omega^i{}_0 = \omega^0{}_i$, and $\omega^i{}_i = 0$.

We shall finally study special properties of the tangent bundle of a Riemannian space. To indicate that we are considering this special case we shall denote the curvature in $T(M)$ by the traditional letter $R$. First, we use the fact that in this case the sections $S_0(T(M))$ can be naturally identified with the vector fields $\mathscr{T}^1_0(M)$. Since vector fields $\mathscr{T}^1_0(M)$ map $E_1$ into $E_0$ they also map $S_1(T(M))$ into $S_0(T(M)) \equiv \mathscr{T}^1_0(M)$. The vector valued 1-form which produces in this way the identity map has a special name.

**Definition** (4.1.32)

If the $b_i$ form a local basis in $\mathscr{T}^1_0(M)$ and $e^i$ the dual basis, $(e^i|b_j) = \delta^i{}_j$, then

$$\theta \equiv \sum_i b_i \otimes e^i \in S_1(T(M))$$

is called **soldering form** and $T \equiv D\theta$ **torsion**. Metric connections with $T = 0$ are called torsion-free or Levi-Civita connections.

**Remarks** (4.1.33)

1. $D\theta = \sum b_i(de^i + \omega^i{}_k \wedge e^k)$ and for a torsion-free connection we have $de^i = -\omega^i{}_k \wedge e^k$. Unless specified otherwise we will only consider torsion-free metric connections.
2. In Problem 11 it is verified that $T = 0$ is equivalent to $D_X Y - D_Y X = [X, Y]$, $X, Y \in \mathscr{T}^1_0$ or to $(D_X u|Y) - (D_Y u|X) = (du|X \otimes Y)$, $u \in \mathscr{T}^0_1$.

**Examples** (4.1.34)

1. In a flat Riemannian space we have in the covariant constant basis $\omega^i{}_k = 0$. If, in addition, $T = 0$ we see from (4.1.31; 1) that $de^i = 0$ or locally $e^i = dx^i$. Therefore $R = T = 0$ imply the local existence of a natural covariant constant basis in $E_1$.
2. Consider polar coordinates in $\mathbb{R}^2\setminus\{0\}$. The metric is $g = dr^2 + r^2\,d\varphi^2 = e^1 \otimes e^1 + e^2 \otimes e^2$ in the orthogonal basis $e^1 = dr$, $e^2 = r\,d\varphi$. From $de^1 = 0$, $de^2 = dr \wedge d\varphi$ we conclude for a torsion-free connection which preserves $g$ that $\omega_{12} = -\omega_{21} = -d\varphi = -e^2/r$, $\omega_{11} = \omega_{22} = 0$. Thus $R_{ik} = d\omega_{ik} + \omega_i{}^l \wedge \omega_{lk} = 0$. With a connection for which the orthogonal

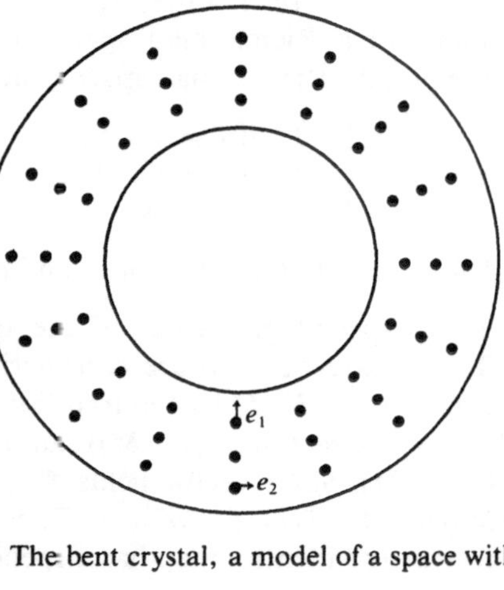


Figure 54 The bent crystal, a model of a space with torsion

basis is covariant constant we get the torsion $b_2\,de^2 = (1/r)\,\partial\varphi \otimes dr \wedge d\varphi$ since the dual basis is $b_1 = \partial r$, $b_2 = (1/r)\,\partial\varphi$. One verifies that in this case the relation $D_X Y - D_Y X = [X, Y]$ from (4.1.31; 2) is violated. If we take $X = b_1$, $Y = b_2$ the left-hand side vanishes since the basis is covariant constant but $[b_1, b_2] = [\partial r, 1/r]\,\partial\varphi = -(1/r^2)\,\partial\varphi \neq 0$. This definition of the connection could be made concrete by bending a rectangular crystal into a ring so that the atoms sit at the points $r = nr_0$, $\varphi = m\varphi_0$ for $n$, $m \in \mathcal{N}$ (Figure 54). If we lived in such a crystal and defined parallel transport by the lattice of atoms as usually, then our space would have torsion.

3. The sphere $S^2$ with the metric $g = R^2(d\theta^2 + \sin^2\theta\, d\varphi^2)$. In the orthogonal basis $e^1 = R\,d\theta$, $e^2 = R\sin\theta\, d\varphi$ we have $de^1 = 0$, $de^2 = R\cos\theta\, d\theta \wedge d\varphi$ and therefore $\omega_{12} = -\omega_{21} = -\cos\theta\, d\varphi$. This gives a curvature

$$R^1{}_2 = d\omega^1{}_2 = \sin\theta\, d\theta \wedge d\varphi = \frac{1}{R^2}\, e^1 \wedge e^2.$$

4. The Lorentz hyperboloid $H = \{(x, y, z) \in \mathbb{R}^3 : x^2 + y^2 - z^2 = R^2\}$ and $g =$ restriction of $dx^2 + dy^2 - dz^2$ to $H$. In cylindrical coordinates $(x, y) = \rho(\cos\varphi, \sin\varphi)$ we find that $e^1 = R\,d\rho/\sqrt{\rho^2 - R^2}$ and $e^2 = \rho\, d\varphi$ are an orthogonal basis on $H$: $g_{|H} = -e^1 \times e^1 + e^2 \times e^2$. $de^1 = 0$, $de^2 = d\rho \wedge d\varphi \Rightarrow \omega^1{}_2 = -\omega^2{}_1 = -\sqrt{\rho^2/R^2 - 1}\, d\varphi \Rightarrow R^1{}_2 = d\omega^1{}_2 = -\rho R^{-2}/\sqrt{\rho^2/R^2 - 1}\, d\rho \wedge d\varphi = -R^{-2} e^1 \wedge e^2$.

**Remark** (4.1.35)

In Example 2, those $\omega$'s imply for the covariant derivative that

$$D_X e^2 = -(e^2|X)\frac{e^1}{r}, \qquad D_X e^1 = (e^2|X)\frac{e^2}{r}.$$

If $X$ has only the 1-component, then the $D_X e^i$ vanish. The covariant derivative of $e^i$ in the 2-direction is a rotation, which is faster for smaller $r$. This accords exactly with the following intuitive picture of the rotation of the local basis in polar coordinates.

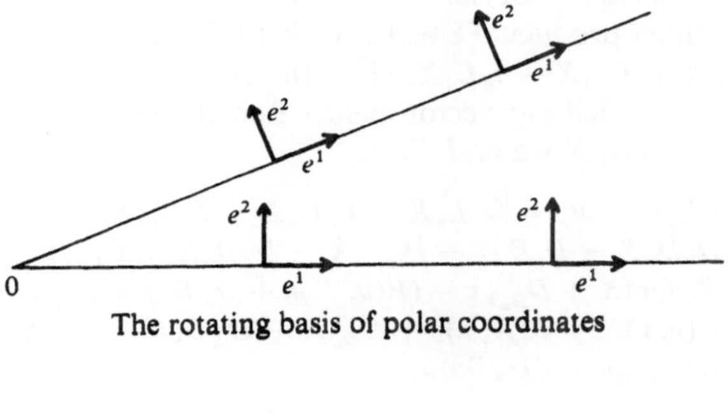

The rotating basis of polar coordinates

We have seen that the two conditions of invariance of $g$ under parallel transport and absence of torsion imply the relations (1.2.25). There it was stated that they determined the connection $\omega$. We shall now verify this claim by an explicit expression of $\omega$ in terms of the derivatives of $g$.

**Calculation of the Affine Connections** (4.1.36)

(a) In the natural basis, $e^j = dq^j$: The so-called **Christoffel symbols** $\Gamma_{ijk} \in \mathscr{T}^0_0$ occur in the decomposition of $\omega_{ij}$ in the basis, $\omega_{ij} = \Gamma_{ijk}\, dq^k$.
Since in the present case $de^j = 0 = \Gamma_{ijk}\, dq^j \wedge dq^k$, they are symmetric in $j$ and $k$. In this basis, $dg_{ik} = g_{ik,l}\, dq^l$, so the $\Gamma$'s are determined by the facts

$$g_{jk,l} = \Gamma_{jkl} + \Gamma_{kjl}$$

and

$$\Gamma_{ijk} = \Gamma_{ikj},$$

which are solved by

$$\Gamma_{jkl} = \tfrac{1}{2}(g_{jk,l} + g_{lj,k} - g_{kl,j}) \tag{4.1.37}$$

Since this solution is unique, we see that our axioms fix $D$ uniquely (Problem 5).

(b) In an orthogonal basis: From $dg_{ik} = d\eta_{ik} = 0 = \omega_{ik} + \omega_{ki}$ together with

$$(de^j | e_k \otimes e_i) = -(\omega^j{}_l \wedge e^l | e_k \otimes e_i) = (\omega^j{}_k | e_i) - (\omega^j{}_i | e_k),$$

one can derive

$$(\omega_{kj} | e_i) = \tfrac{1}{2}[(de_j\; e_i \otimes e_k) + (de_k | e_j \otimes e_i) - (de_i | e_k \otimes e_j)] \tag{4.1.38}$$

(Problem 5)

Next we study the special features which arise because a flow in the base manifold induces in a natural way a flow in all the tensor bundles. Thus the Lie derivative of a section with respect to a vector field in $B$ makes sense and we can study the

**Commutator of the Covariant Derivative and the Lie Derivative** (4.1.39)

We noted in (4.1.21) that for vector fields $u$, $v$, $X$ the commutator $(D_u D_v - D_v D_u - D_{[u,v]})X = (i_v i_u R)X \equiv R(u, v)X$ does not depend on the derivative of $X$ but defines the curvature. One can derive a similar relation for $DL_v - L_v D$ using (4.1.21) and the fact that the Lie derivative $L_v$ (like $D_v$) applied to the inner product of a vector with a 1-form just takes the derivative of both: $L_v i_u X = i_{L_v u} X + i_u L_v X$. (For the inner product of forms this is true only if $v$ is a Killing vector field.) Remembering $[X, Y] \equiv L_X Y = -L_Y X = D_X Y - D_Y X$ we find

$$\begin{aligned}
\langle (DL_v - L_v D)X | u\rangle &= D\; L_v X - L_v D_u X + D_{[v,u]} X \\
&= D_u D_v X - D_u D_X v - D_{[u,v]} X - D_v D_u X + D_{D_u X} v \\
&= R(u, v)X + D_{D_u X} v - (R(u, X)v + D_X D_u v + D_{[u,X]} v) \\
&= R(u, v)X + R(X, u)v + D_{D_u X} v - D_X D_u v - D_{D_u X} v + D_{D_X u} v \\
&= R(X, v)u - \langle D_X D v | u\rangle.
\end{aligned}$$

In the last step we used a cyclic property of $R$ derived in (4.1.44; 2(c)) and $D_X D_u v = D_X i_u Dv = i_{D_X u} Dv + i_u D_X Dv = D_{D_X u} v + \langle D_X Dv | u \rangle$. Thus we see that the commutator does not contain any derivatives of $X$ but second derivatives of $v$.

**Commutativity of $D$ and * (4.1.40)**

In (1.2.16) we widened the scalar product $\langle\,|\,\rangle$ to the interior product $i_v$. (4.1.28) generalizes for $i_v$ to $D_X i_v = i_{D_X v} + i_v D_X$ as used above. In particular, we see $D_X \varepsilon = 0$ since $0 = D_X 1 = D_X \langle \varepsilon | \varepsilon \rangle = 2 \langle D_X \varepsilon | \varepsilon \rangle$ and $\varepsilon$ is not degenerate. Therefore $D_X i_v \varepsilon = i_{D_X v} \varepsilon$ and thus $D_X {*v} = {*D_X v}$. Here $E_p$ is identified with $S_0(\bigwedge_p T^*(B))$ and not with $S_p(T^*(B))$. Thus $D_X$ on $E_p$ cannot be written as $i_X D$ where $i_X$ is the map $E_p \to E_{p-1}$ in which sense it is to be understood here.

In a pseudo-Riemannian space vector fields define directions in the vector bundles. Thus the covariant derivative $D_X$ distinguishes the vector fields that are consistent with it in the sense that a vector translated parallelly in its own direction from some point equals the vector at the neighboring point:

**Definition** (4.1.41)

A vector field $X$ such that $D_X X = 0$ is called a **geodesic vector field**.

**Remarks** (4.1.42)

1. The connection with the geodesic lines: Let $z(s)$ be a streamline of a geodesic vector field $X$, that is, $\dot{z}(s) = X(z(s))$. Then the components of $z$ in the natural basis satisfy the equation $\ddot{z}^i = -\Gamma^i{}_{jk} \dot{z}^j \dot{z}^k$, because with Remark (4.1.16; 3),
$$0 = D_X(X^i \partial_i) = (X^i{}_{,k} X^k + (\omega^i{}_j | \partial_k) X^j X^k) \partial_i$$
and
$$\ddot{z}^i(s) = \frac{d}{ds} X^i(z(s)) = X^i{}_{,k} \dot{z}^k.$$
From (4.1.36(a)), $(\omega^i{}_j | \partial_k)$ is just $\Gamma^i{}_{jk}$, and we obtain exactly the geodesic equations of motion (I: 1.1.6). The shortest lines are thus the straightest, in the sense of the tangent vector to a curve being transformed into itself under parallel transport along them.
2. The connection with Killing vector fields: Remember that for a Killing vector field $X$ we had $L_X g = 0$, whereas $L_X X = 0$ for any vector field. In contradistinction for the covariant derivative $D_X X = 0$ holds only for geodesic vector fields and $D_X g = 0$ holds for any vector field. In general "Killing" does not imply "geodesic" nor vice versa. But there is the

following relation: Let $v$ be a Killing vector field (2.1.9), so that $L_v\langle X|X\rangle = 2\langle L_vX|X\rangle$, and let $X$ be a geodesic vector field. Then

$$L_X\langle v|X\rangle = D_X\langle v|X\rangle = \langle D_Xv|X\rangle = \langle D_vX|X\rangle - \langle L_vX|X\rangle$$
$$= \tfrac{1}{2}(D_v\langle X|X\rangle - L_v\langle X|X\rangle) = 0.$$

Consequently, the component of a geodesic vector field in the direction of a Killing vector field is constant along a geodesic line. This reveals a new significance of a Killing vector field $v$, that it provides a constant of the motion of a particle in a gravitational field, namely the component of the four-momentum $\dot{z}$ of the particle in the direction of $v$.

3. If one vector field in an orthogonal basis is also natural it is geodesic. To see this, let $X$ be this vector field and $v$ be either $X$ (then $\langle v|X\rangle = \pm 1$) or a member of a natural basis orthogonal to $X$. In any case $\langle v|X\rangle$ is constant and with (4.1.33; 2)

$$0 = L_X\langle v|X\rangle = D_X\langle v|X\rangle = \langle D_Xv|X\rangle + \langle v|D_XX\rangle$$
$$= D_v\tfrac{1}{2}\langle X|X\rangle - \langle L_vX|X\rangle + \langle v|D_XX\rangle.$$

Now $\langle X|X\rangle = \pm 1$ so the first term vanishes and $L_vX = 0$ since the Lie derivatives between members of a natural basis vanish. Therefore all components of $D_XX$ are zero.

In a pseudo-Riemannian space the curvature forms are not arbitrary, but are subject to certain algebraic conditions which we now enumerate:

**Algebraic Identities Satisfied by the Curvature Forms** (4.1.43)

(a) $R_{ij} = -R_{ji}$;
(b) $R_{ij} \wedge e^i = 0$.

**Proof**

(a) This fact is independent of the basis. If the basis is changed by $\bar{e} = Ae$ (Problem 4), then $R \to ARA^{-1}$ and $g \to A^{t-1}gA^{-1}$, and thus $R_{ij} = g_{ik}R^k{}_j \equiv (gR)_{ij} \to (A^{t-1}gRA^{-1})_{ij}$, which preserves the antisymmetry. And Fact (a) is obviously true in an orthogonal basis, since $\omega_{ij} = -\omega_{ji} \Rightarrow \omega_{ik} \wedge \omega^k{}_j = -\omega_{kj} \wedge \omega_i{}^k = -\omega_{jk} \wedge \omega^k{}_i$.
(b) $0 = d\,de^i = -d(\omega^i{}_k \wedge e^k) = -R^i{}_k \wedge e^k$. □

**Consequences** (4.1.44)

1. The $m$ 3-forms (b) we get from the $\binom{m}{2}$ 2-forms vanish. In the absence of other algebraic conditions there are consequently

$$\binom{m}{2}^2 - m\binom{m}{3} = \frac{m^2(m^2-1)}{12}$$

independent components of the curvature. In one dimension, there is no curvature 2-form, in two dimensions there is one, and in four dimensions there are twenty. However, many of the components can be made to vanish by a suitable choice of bases and coordinate systems. Let us take an orthogonal basis, to fix the $g_{ik}$ as $\eta_{ik}$, and try to find how many invariants can be constructed from the $m^2(m^2-1)/12$ components of $R_{ik}$ and the $m^2$ components of the $e^i$. There are available $m^2$ functions $\partial \bar{x}^i/\partial x^j$ from a possible change of coordinate system and $m(m-1)/2$ components of $A$ (since $A\eta A^t = \eta$), with which to make $m^2 + m(m-1)/2$ of the total of $m^2 + m^2(m^2-1)/12$ components vanish. There remain $m^2(m^2-1)/12 - m(m-1)/2 = m(m-1)(m-2)(m+3)/12$ invariants. This simple argument does not work in two dimensions, however. Moreover, we have only imposed the algebraic conditions, and there may be some differential dependence among the invariants as well.

2. If $R^i{}_j$ are decomposed in a basis $R^i{}_j = \frac{1}{2}R^i{}_{jkm}e^{km}$, then the **Riemann–Christoffel tensor** $R_{ijkm} = g_{in}R^n{}_{jkm}$ obeys the equations (Problem 8)

(a) $R_{ijkm} = -R_{ijmk}$;
(b) $R_{ijkm} = -R_{jikm}$;
(c) $R_{ijkm} + R_{ikmj} + R_{imjk} = 0$;
(d) $R_{ijkm} = R_{kmij}$.

3. If $R_{jk}$ is written in terms of the contractions $i_{e_j}R^j{}_k = R^j{}_{kjm}e^m \equiv R_k \in E_1$ and $i_{e_k}R^k = R^{jk}{}_{jk} \equiv \mathrm{R} \in E_0$ in such a way that the contractions of the remainders vanish, then those remainders define the **Weyl forms** $C_{jk}$:

$$R_{jk} \equiv \frac{-R}{(m-2)(m-1)} e_j \wedge e_k + \frac{1}{m-2}(e_j \wedge R_k - e_k \wedge R_j) + C_{jk}.$$

It is clear that the $C_{ij}$ also satisfy (4.1.23), and by construction $i_{e_j}C^j{}_k \equiv C_k = 0$ (Problem 9). Because $i_{e_j}R_k = i_{e_k}R_j$, the equations $C_k = 0$ pose only $m(m+1)/2$ independent conditions, which leaves $m^2(m^2-1)/12 - m(m+1)/2 = m(m+1)(m+2)(m-3)/12$ components for the $C_{ij}$. In three dimensions all $C_{ij}$ are zero; they first occur in four dimensions, with ten components. The $C^i{}_j$ are important because they remain invariant under conformal transformations $g \to fg, f \in E_0$ (Problem 10), and in particular they vanish on all conformally flat spaces. Conversely, if $C = 0$, then $g = f\eta$ [25].

We conclude the section by collecting the formulas that connect the metric and the curvature, to lay the foundation for later calculations:

$$\begin{aligned} g &= g_{ik}e^i \otimes e^k, \\ de^i &= -\omega^i{}_j \wedge e^j, \\ dg_{ik} &= g_{ij}\omega^j{}_k + \omega_k{}^j g_{ji}, \\ R^i{}_k &= d\omega^i{}_k + \omega^i{}_j \wedge \omega^j{}_k. \end{aligned} \tag{4.1.45}$$

**Problems** (4.1.46)

1. Find a $u$ such that $(du|X \otimes Y) = (D_X u|Y) - (D_Y u|X)$ is violated in Example (4.1.34; 2) with torsion.
2. Use (1.2.26; 3) to calculate the $\omega$'s for the plane with polar coordinates. (The $\omega$'s vanish for $dx^{1,2}$, so set $\bar{e}^i = A^i{}_k\, dx^k$, and compare with (4.1.34; 2).)
3. With the help of (1.2.26; 3), find a necessary and sufficient condition to make the $\omega$'s vanish by a change of basis.
4. Show that for $\bar{b} = bL$, we have $\Omega = L^{-1}\Omega L$.
5. Derive (4.1.37) and (4.1.38).
6. Prove (4.1.27; 1).
7. Show (4.1.21) by verifying the more general relation for $\omega \in S_1$:

$$i_v i_u D\omega = i_u D(\omega|v) - i_v D(\omega|u) - i_{[u,v]}\omega.$$

*Hint*: $\omega$ can be written $\sum_i b_i v^i, b_i \in S_0$, $v^i \in E_1$ and because of linearity it suffices to show the relation for one term,. Use

$$i_v i_u dv = i_u d(v|v) - i_v d(v|u) - i_{[u,v]} v \qquad \text{for } v \in E_1.$$

8. Derive the equations (4.1.44; 2).
9. Show that $i_{e_j} C^j{}_k \equiv C_k = 0$.
10. Suppose that $\bar{e}^i = fe^i$ but $\bar{g}_{ik} = g_{ik}$ such that $g = f^2 g$. Show that $\bar{C}_{ik} = C_{ik}$.
11. Show $de^j = -\omega^j{}_k e^k \overset{a}{\Leftrightarrow} (D_X u|Y) - (D_Y u|X) = (du|X \otimes Y) \overset{b}{\Leftrightarrow} D_X Y - D_Y X = [X, Y]$, $X, Y \in \mathcal{T}^1_0$, $u \in \mathcal{T}^0_1$ and $e^j$ a basis in $\mathcal{T}^0_1$.

**Solutions** (4.1.47)

1. $u = e^2$; $de^2 = dr \wedge d\varphi \neq 0$, but $D_X e^2$ ought to vanish for all $X \in T^1_0$.
2. $A = \begin{vmatrix} \cos\varphi & \sin\varphi \\ -\sin\varphi & \cos\varphi \end{vmatrix}$, $dA = d\varphi \begin{vmatrix} -\sin\varphi & \cos\varphi \\ -\cos\varphi & -\sin\varphi \end{vmatrix}$, $-dA \cdot A^{-1} = d\varphi \begin{vmatrix} 0 & -1 \\ 1 & 0 \end{vmatrix}$.
3. Choose $A$ so that $dA^i{}_k = \omega^i{}_l A^l{}_k$. As remarked in (4.1.20; 3), it is both necessary and sufficient for solubility that $d(\omega^i{}_l A^l{}_k) = R^i{}_l A^l{}_k = 0$, and therefore $R^i{}_k = 0 \Leftrightarrow \omega^i{}_k = 0$.
4. $\bar{\omega} = L^{-1}\omega L + L^{-1}\, dL$,

$$\begin{aligned} d\bar{\omega} + \bar{\omega} \wedge \bar{\omega} &= (dL^{-1}) \wedge \omega L + L^{-1}\, d\omega L - L^{-1}\omega\, dL + (dL^{-1}) \wedge dL \\ &\quad + (L^{-1}\omega L + L^{-1}\, dL) \wedge (L^{-1}\omega L + L^{-1}\, dL) \\ &= L^{-1}(d\omega + \omega \wedge \omega)L. \end{aligned}$$

5.

$$\begin{aligned} \Gamma_{kjl} + \Gamma_{jkl} &= g_{jk,l} \\ \Gamma_{jlk} + \Gamma_{ljk} &= g_{lj,k} \\ -\Gamma_{lkj} - \Gamma_{klj} &= -g_{kl,j} \\ \hline 2\Gamma_{jkl} &= g_{jk,l} + g_{lj,k} - g_{kl,j}. \end{aligned}$$

Let $(de_j | e_i \otimes e_k) = (de_j)_{ik}$, etc. Then

$$\begin{aligned}(de_j)_{ik} &= (\omega_{ji})_k - (\omega_{jk})_i \\ (de_k)_{ji} &= (\omega_{kj})_i - (\omega_{ki})_j \\ -(de_i)_{kj} &= -(\omega_{ik})_j + (\omega_{ij})_k \\ \hline (de_j)_{ik} + (de_k)_{ji} - (de_i)_{kj} &= 2(\omega_{kj})_i.\end{aligned}$$

6. $d(d\omega^i{}_j + \omega^i{}_s \wedge \omega^s{}_j) = d\omega^i{}_s \wedge \omega^s{}_j - \omega^i{}_s \wedge d\omega^s{}_j$

$$= \Omega^i{}_s \wedge \omega^s{}_j - \omega^i{}_s \wedge \Omega^s{}_j - \omega^i{}_k \wedge \omega^k{}_s \wedge \omega^s{}_j + \omega^i{}_s \wedge \omega^s{}_k \wedge \omega^k{}_j$$
$$= \Omega^i{}_s \wedge \omega^s{}_j - \omega^i{}_s \wedge \Omega^s{}_j.$$

7. Insert $\omega = bv$, $b \in S_0$, $v \in E_1$

$$\begin{aligned}i_v i_u D\omega = (i_u Db)(i_v v) - (i_v Db)(i_u v) \\ + b(i_u d(i_v v) - i_v d(i_u v) - i_{[u,v]} v).\end{aligned}$$

This equals

$$i_u D(\omega | v) - i_v D(\omega | u) - i_{[u,v]}\omega = i_u D(b i_v v) - i_v D(b i_u v) - b i_{[u,v]} v.$$

8. (a) holds because both sides are components of a 2-form;
(b) follows from (4.1.23(a));
(c) $0 = R^i{}_j \wedge e^j = R^i{}_{jkm} e^{jkm}$. Because of (a), only the cyclic permutation remains in the sum;
(d) follows from (a) – (c):

$$\begin{aligned}R_{ijkm} &= -R_{ikmj} - R_{imjk} = R_{kimj} + R_{mijk} \\ &= -R_{kmji} - R_{kjim} - R_{mjki} - R_{mkij} \\ &= 2R_{kmij} + R_{jkim} + R_{jmki} = 2R_{kmij} - R_{jimk} \Rightarrow 2R_{ijkm} = 2R_{kmij}.\end{aligned}$$

9. Because $i_{e_j} e^j = m$ and $e^j i_{e_j} \omega = p\omega$ for all $\omega \in E_p$, we find

$$R_k = i_{e_j} R^j{}_k = \frac{-R(m-1)e_k}{(m-2)(m-1)} + \frac{1}{m-2}[Re_k - R_k - R_k + mR_k] + C_k \Rightarrow C_k = 0.$$

10. It suffices to show that $\bar{R}_{ik} = R_{ik} + v_i \wedge e_k - v_k \wedge e_i$ for some $v_i \in E_1$. Because $d\bar{e}^i = df \wedge e^i + f\, de^i = -\bar{\omega}^i{}_k \wedge \bar{e}^k$ and $\bar{\omega}_{ik} + \bar{\omega}_{ki} = \omega_{ik} + \omega_{ki}$,

$$\bar{\omega}_{ik} = \omega_{ik} + (df | e_i)e_k - (df | e_k)e_i.$$

In a natural basis, $e_s \wedge \omega^s{}_n = 0$, and therefore

$$((df | e_i)e_s - (df | e_s)e_i) \wedge \omega^s{}_k = v_k \wedge e_i, \qquad v_k = (df | e_s)\omega^s{}_k,$$

and

$$d\bar{\omega}_{ik} - d\omega_{ik} = v_i \wedge e_k - v_k \wedge e_i, \qquad v_i = d(df | e_i).$$

11. (a) $\Leftarrow$: $(de^j | X \otimes Y) = (D_X e^j | Y) - (D_Y e^j | X)$

$$= -(\omega^j{}_k | X)(e^k | Y) + (\omega^j{}_k | Y)(e^k | X) = -(\omega^j{}_k \wedge e^k | X \otimes Y)$$

$\Rightarrow$ If $u = fe, f \in E_0$ then $du = df \wedge e + f\, de$ and $D_X u = i_X(df \wedge e + f\, De)$.

Thus the first term also satisfies the equation.

(b) Remember $(du | X \otimes Y) = L_X(u | Y) - L_Y(u | X) - (u | [X, Y])$ and $L_X(u | Y) = D_X(u | Y) = (D_X u | Y) + (u | D_X Y)$.

## 4.2 Gauge Theories and Gravitation

*Energy and momentum imprint a structure on space and time through Einstein's equations, which equate the energy and momentum forms to quantities constructed from $R^\alpha{}_\beta$.*

To derive the field equations with the Lagrangian formalism one needs a 4-form as Lagrangian density. However, to get fields equations which are consistent one has to make sure that the action functional has a stationary point. We have seen in §2.1 that in Maxwell's theory this is only the case if the current is conserved. In §3.1 the current was derived from a scalar field $S$ and the field equations for $S$ were just the requirement that the current be conserved. We shall now study this phenomenon more generally with a scalar field $\Phi = \sqrt{\rho}\, e^{iS}$ where the density $\rho$ is no longer constant. Instead of the complex field $\varphi$ we shall use two real scalar fields $\varphi^1$, $\varphi^2$ with $\Phi^1 = \varphi^1 + i\varphi^2$. Expressed in terms of these fields the Lagrangian $\overline{(d + ieA)\varphi} \wedge *(d + ieA)\varphi$ of (3.1.21; 2) becomes in the notation (4.1.14; 2)

$$-\frac{1}{2}\sum_{i=1}^{2}(D\varphi)^i \wedge *(D\varphi)^i, \qquad (D\varphi)^i = d\varphi^i + e\tau^i{}_k A\varphi^k, \qquad \varepsilon = \begin{pmatrix} 0 & 1 \\ -1 & 0 \end{pmatrix}. \tag{4.2.1}$$

Another term invariant under the local gauge transformation would be a "mass-term" $-\frac{1}{2}m^2 \sum_i \varphi^i \wedge *\varphi^i$. The general total Lagrangian

$$\mathscr{L} = -\tfrac{1}{2}\, dA \wedge *dA - \tfrac{1}{2}\sum_i ((D\varphi)^i \wedge *(D\varphi)^i + m^2\varphi^i \wedge *\varphi^i) \tag{4.2.2}$$

is invariant under $A \to A + d\Lambda$, $\varphi \to e^{-e\varepsilon\Lambda}\varphi$, $\Lambda \in C(\mathbb{R}^4)$ and the linear dependence of the action $W = \int \mathscr{L}$ on $\Lambda$ which prevented $W$ in (2.1.3; 1) to have stationary points is absent. Thus there is hope that the Euler equations which result from (4.2.2)

$$d{*F} = -{*J}, \qquad D{*D\Phi} = m^2\, {*\Phi}, \tag{4.2.3}$$

with

$$*J = -\frac{\partial \mathscr{L}}{\partial A} = e\varepsilon^j{}_k \varphi^k \wedge *(D\Phi)^j$$

are consistent. Indeed, $d{*J} = 0$ is a consequence of the equations for $\Phi$. To see this most directly consider the variation of $\mathscr{L}$ with $\Phi$

$$\delta\mathscr{L} = \sum_i \delta\varphi^i\left[\frac{\partial\mathscr{L}}{\partial\varphi^i} - d\frac{\partial\mathscr{L}}{\partial\, d\varphi^i}\right] + d\left(\sum_i \delta\varphi^i \frac{\partial\mathscr{L}}{\partial\, d\varphi^i}\right). \tag{4.2.4}$$

The Euler equations require [ ] = 0 and for $\delta\Phi$ the change $\Phi \to e^{-e\varepsilon\Lambda}\Phi$ to first order in $\Lambda$ gives $\delta\mathscr{L} = ed(\Lambda\varepsilon^j{}_k\varphi^k \wedge *(D\Phi)^j)$. If we choose a global gauge transformation, ($d\Lambda = 0$), the invariance of $\mathscr{L}$ guarantees $\delta\mathscr{L} = 0 = -d^*J$.

Next we shall extend these considerations to the non-abelian gauge group $O(3)$. In this case $\Phi$ has three components and one might think that in (4.2.2) one just has to take $\sum_i$ from one to three. However, now one also has three gauge potentials and $dA$ is not invariant under the gauge transformation $A \to L^{-1}AL + L^{-1}\,dL$ (see 4.1.12). This suggests replacing $dA$ with $DA = \Omega$ (see 4.1.20; 1) which transforms as $(DA) \to L^{-1}AL$. Thus the bilinear expression $\operatorname{tr}\Omega \wedge {}^*\Omega = \Omega^i{}_k \wedge {}^*\Omega^k_i$ is gauge invariant and seems to be a good candidate for the Lagrangian of the gauge field. Such a Lagrangian has first been proposed by Yang and Mills and has later turned out to describe the strong and electroweak interactions if further fields are added.

**The Yang-Mills Lagrangian** (4.2.5)

$$\mathscr{L} = \tfrac{1}{2}\operatorname{tr} DA \wedge {}^*DA - \tfrac{1}{2}\langle D\varphi \wedge {}^*D\varphi\rangle - \tfrac{1}{2}m^2\langle \Phi \wedge {}^*\Phi\rangle.$$

**Remarks** (4.2.6)

1. We used the notation $\langle \Phi \wedge \Psi\rangle = \sum_i \varphi^i \wedge \psi^i$ for the scalar product in the fibres.
2. Although we were thinking of $O(3)$ the Lagrangian (4.2.5) has the same appearance for all $O(n)$.

The Euler equations derived from (4.2.5) are similar to (4.2.3) except that now the Maxwell field $F$ is replaced by the curvature $\Omega$. Here we just state the results as we shall go in (4.2.15) through the details of the variational procedure.

**The Yang-Mills Equations** (4.2.7)

$$d^*\Omega = -{}^*J \equiv \frac{\partial\mathscr{L}}{\partial A}, \qquad D^*D\Phi = m^2{}^*\Phi.$$

**Remarks** (4.2.8)

1. The derivation of (4.2.7) follows the lines of (2.1.7), one just notes ${}^*\Omega = \partial\mathscr{L}/\partial\, dA$ since both factors in $\Omega \wedge {}^*\Omega = {}^*\Omega \wedge \Omega$ give the same contribution.
2. To get more explicit expressions one might use a basis $b_i$ for the Lie-algebra = ($3 \times 3$ antisymmetric matrices) which diagonalizes the trace. If $\operatorname{tr} b_\alpha b_\beta = -\delta_{\alpha\beta}$ and $A = \sum b_\alpha A^\alpha$, $A^\alpha \in E_1$, then the curvature $\Omega = \sum b_\alpha F^\alpha$ has according to (4.1.20; 1) the components

$$F^\alpha = dA^\alpha + \tfrac{1}{2}c_{\beta\gamma}{}^\alpha A^\beta \wedge A^\gamma.$$

For $O(n)$ $\alpha$ goes from 1 to $n(n-1)/2$ whereas $i$ for $\varphi^i$ goes from 1 to $n$. $c_{\beta\gamma}{}^{\alpha}$ are the structure constants of $O(3)$ of this basis: $b_\beta b_\gamma - b_\gamma b_\beta = c_{\beta\gamma}{}^{\alpha} b_\alpha$. With this decomposition the Lagrangian of the gauge field becomes $\mathscr{L}^{\text{gf}} = -\frac{1}{2}F^\alpha \wedge {}^*F^\alpha$.

3. Since the $F^\alpha$ do not depend on $dA$ only, but also on $A$, there is a contribution $-\partial\mathscr{L}^{\text{gf}}/\partial A^\alpha = c_{\alpha\beta}{}^{\gamma} A^\beta \wedge {}^*F^\gamma$ to the current ${}^*J$. This does not appear in Maxwell's theory and means that the Yang–Mills equations are nonlinear even in the absence of the scalar field. On the other hand, the contribution of the scalar field to the current $(b^\alpha)_{kl}\varphi^k \wedge {}^*(D\Phi)^l$ has the same structure as previously.
4. The consistency of the Yang–Mills equations requires $d{}^*J = 0$. This follows indeed from the variational principle as can be seen by considerations similar to the preceding ones. A gauge transformation $L = 1 + b_\alpha \Lambda^\alpha$ with $d\Lambda^\alpha = 0$ and $\Lambda^\alpha \to 0$ generates $\delta A^\alpha = c_{\beta\gamma}{}^{\alpha} A^\beta \Lambda^\gamma$ and since $\delta\mathscr{L} = 0$ the generalization of (4.2.4) tells us that

$$\Lambda^\alpha d(c_{\beta\alpha}{}^{\gamma} A^\beta \wedge {}^*F^\gamma + (b^\alpha)_{mn}\varphi^m \wedge {}^*(D\varphi)^n) = 0.$$

Thus the general conclusions which we drew from Maxwell's equations like the vanishing of the total charge in a closed universe or a plane wave still hold.
5. There is an essential difference to Maxwell's theory where the current was gauge invariant. Here the current does not transform like $\Omega$ under a gauge transformation, $J \not\to L^{-1}JL$ but transforms inhomogeneously. Indeed, $J \to L^{-1}JL$ would be incompatible with (4.2.7). Although ${}^*\Omega$ transforms like ${}^*\Omega \to L^{-1}{}^*\Omega L$, $d{}^*\Omega$ does not. Therefore it has no gauge invariant meaning to say that one has a certain amount of currents at a point $x$ in space, one can always find a gauge such that $J(x) = 0$. The total charges $Q_\alpha = \int_{N_3} {}^*J_\alpha = -\int_{\partial N_3} {}^*F_\alpha$ are more robust. Since they can be expressed as boundary integrals they transform homogeneously under all gauge transformations which have $dL = 0$ on $\partial N_3$.

The culprit responsible for the inhomogeneous transformation law of $J$ is the $A$ contained in the contribution from the gauge field. The contribution ${}^*J(\Phi) = b\Phi \wedge {}^*(D\Phi)$ from the scalar field transforms homogeneously. Indeed, if we put the current from the gauge field to the other side of (4.2.7) we obtain on the left-hand side the combination $d{}^*F^\alpha + c_{\beta\gamma}{}^{\alpha} A_\beta \wedge {}^*F^\gamma$ since the structure constants are totally antisymmetric, $c_{\beta\gamma}{}^{\alpha} = c_{\alpha\beta}{}^{\gamma}$. The extension (4.16.3) of the covariant derivative to $S_p(L(F))$ shows (compare Problem 7) that these are just the components of the covariant exterior derivative $D$ which is constructed such that it transforms with $L^{-1}D{}^*\Omega L$. Thus the Yang–Mills equation can also be written $D{}^*\Omega = -{}^*J(\Phi)$. ${}^*J(\Phi)$ alone is not conserved, however, one finds that $D$ applied twice to ${}^*\Omega$ gives zero (Problem 7) and thus $d{}^*J \neq 0$ but $D{}^*J(\Phi) = 0$. Together with (4.1.26) we find a form for (4.2.7) which looks like Maxwell's equation only that $d$ is replaced by $D$.

**Covariant Form of the Yang–Mills Equations** (4.2.9)

$$D\Omega = 0, \qquad D^*\Omega = -{}^*J(\Phi) \Rightarrow D^*J(\Phi) = 0.$$

Since the gauge theories discussed so far seem to be the appropriate description for the electroweak and strong interactions, it is tempting to think that a theory of gravitation should follow the same pattern. According to the equivalence principle discussed in §1.3 gravity is related to the metric structure of space and time $M$. This suggests taking as fibre bundle $T(M)$, as fibre metric the pseudo-Riemannian metric, as gauge group the local Lorentz transformation and as gauge potential the connection $\omega$. Thus a possible Lagrangian would be $-\frac{1}{2}R^{\alpha}{}_{\beta} \wedge {}^*R^{\beta}{}_{\alpha}$ + the matter Lagrangian (3.1.20). However, this does not give a theory of gravitation because the connection $\omega$ does not appear in (3.1.20), everything being expressible in terms of $*$ and $d$. Thus the current ${}^*J = \delta\mathscr{L}/\delta\omega$ does not get a contribution from the Maxwell or scalar fields. Only if we had a spinor field its spin density [21] would contribute to the current. However, one knows that all matter acts as source of the gravitational potential and so this theory does not seem to work. More specifically one believes that all energy and momentum currents are the source of gravitation. Since they are the generators of translations one might think that it is more the translation part of the Poincaré group and not the rotations which are relevant for gravity. If this is so we lose the strict analogy with the previous development where the gauge groups were orthogonal (or unitary) groups. Therefore the construction of the gravitational Lagrangian will involve some guess-work. In any case the energy and momentum currents of matter turn out to be $\partial\mathscr{L}^{\text{matter}}/\partial e^{\alpha}$. Here the tetrads $e^{\alpha} \in E_1$ are a basis and therefore it is reasonable to look at them as the gauge potentials replacing the $A$'s. In this case the $\omega$'s play the role of a field strength and not a potential, which they actually do in the equations (I: 1.1.6) for the motion of a particle in a gravitational field. If we use an orthogonal basis they contain all the information of the metric $g = \sum \eta_{\alpha\beta} e^{\alpha} \otimes e^{\beta}$ and its influence on matter comes only through the $*$-map contained in $\mathscr{L}^{\text{matter}}$. We shall represent matter by a scalar and the electromagnetic field and take the Lagrangian from (3.1.20) (in units $\rho e^2/m = 1$)

$$\mathscr{L}^{\text{matter}} = -\tfrac{1}{2}J \wedge {}^*J - \tfrac{1}{2}F \wedge {}^*F. \tag{4.2.10}$$

When choosing the gravitational Lagrangian $\mathscr{L}^{\text{gf}}$ some points should be born in mind:

**Remarks** (4.2.11)

1. At the time of the birth of gravitational theory, the requirement of general covariance provided some relief from the labor pains, but later on it was more often a source of confusion. The concept of a manifold incorporates

it automatically when the definition uses equivalence classes of atlases, and hence only chart-independent statements are regarded as meaningful. This program is by no means unique to gravitational theory—we have also followed it in classical mechanics and electrodynamics. The big difference is that now the metric $g$ on $M$ is not determined *a priori*.

2. There are some coordinate systems in which Einstein's equations simplify, just as Maxwell's equations are easier to work with in the Lorentz gauge $\delta A = 0$. For example, some formulas are shorter (cf. (4.2.20)) when written in the popular "harmonic coordinates," which satisfy $\delta\, dx = 0$ [26]. It is a matter of opinion, which we leave to the reader, whether this fact is of fundamental importance.
3. Even when the chart is fixed, there still remains the choice of a basis $e^i$. Orthogonal bases are special in that they standardize $g_{ik}$ as $\eta_{ik}$. They still leave open the possibility of a Lorentz transformation $e^\alpha \to L^\alpha{}_\beta(x)e^\beta$, $L^t(x)\eta L(x) = \eta$ for all $x$. $\mathscr{L}^{\text{matter}}$ in (4.2.10) is invariant under this transformation, which means that matter does not define a "teleparallelism." Looking just at matter one cannot observe the orientation of the local Lorentz systems relative to each other. It is thus reasonable to postulate that gravity does not prefer one frame either and therefore $\mathscr{L}^{\text{gf}}$ should be invariant under this transformation too.

We now have to select $\mathscr{L}^{\text{gf}} \in E_4$ from the material gathered in §4.1 for $T(M)$ and pose the requirements that it:

(a) is invariant under a change of basis; and
(b) is quadratic in the derivative of the $e^\alpha$.

This leads uniquely, up to a factor, to (for the last way of writing see Problem 8)

$$\mathscr{L}^{\text{g}} \simeq {}^*R = R_{\alpha\beta} \wedge {}^*e^{\alpha\beta} = 2d(e^\alpha \wedge {}^*de_\alpha) - (de^\alpha \wedge e^\beta) \wedge {}^*(de_\beta \wedge e_\alpha) + \tfrac{1}{2}(de^\alpha \wedge e_\alpha) \wedge {}^*(de^\beta \wedge e_\beta). \tag{4.2.12}$$

Thus the total Lagrangian will have three contributions. One from a scalar field $\mathscr{L}^s = -\frac{1}{2}J \wedge {}^*J$ which represents matter. For simplicity we choose units where $\rho e^2/m = 1$. Secondly, the electromagnetic field adds $\mathscr{L}^e = -\frac{1}{2}F \wedge {}^*F$. Finally, comes $\mathscr{L}^{\text{g}}$ equipped with a factor to be adjusted later.

**Lagrangian for the Total System** (4.2.13)

$$\mathscr{L} = -\tfrac{1}{2}J \wedge {}^*J - \tfrac{1}{2}F \wedge {}^*F + \frac{1}{16\pi\kappa} R_{\alpha\beta} \wedge {}^*e^{\alpha\beta},$$

$$F = dA, \qquad J = dS + eA.$$

**Remarks** (4.2.14)

1. If we consider the $e^\alpha$ as counterpart to $A$ the simplest analog to $dA \wedge *dA$ which is invariant under global Lorentz transformations would be $de^\alpha \wedge *de_\alpha$. This leads to a theory with teleparallelisms and contradicts our information gained from experiments on the bending of light rays by the sun. The somewhat different form of (4.2.13) insures invariance under local Lorentz transformations. Because of this invariance we may use an orthonormal basis for the derivation of Euler's equation. The variation of the metric $g$ results from our making the variation of $e$ without imposing any orthogonality constraints. Since the metric $\eta$ is constant we may freely pull indices up and down under the derivative and use the notation $\omega_{\alpha\beta} = \eta_{\alpha\gamma}\varepsilon^\gamma{}_\beta$ etc. It amounts only to a change of sign for each subscript zero.
2. Another quantity invariant under local gauge transformations would be simply $*1$. It contains no derivatives and by itself cannot lead to a differential equation. In an early stage of the theory Einstein added it in as a "cosmological term." Later it fell into disfavour and there is no empirical evidence for it. Yet its absence is still mysterious since there are many possible sources which could contribute such a term. We shall occasionally resurrect it for the purpose of comparison.
3. In an orthogonal basis we find that

$$\begin{aligned} *e^{\alpha\beta} \wedge d\omega_{\alpha\beta} - d(*e^{\alpha\beta} \wedge \omega_{\alpha\beta}) &= -(d*e^{\alpha\beta}) \wedge \omega_{\alpha\beta} \\ &= -\omega^\alpha{}_\gamma \wedge *e^{\gamma\beta} \wedge \omega_{\alpha\beta} + \omega^\beta{}_\gamma \wedge *e^{\alpha\gamma} \wedge \omega_{\alpha\beta} \\ &= -2*e^{\beta\gamma} \wedge \omega_\gamma{}^\alpha \wedge \omega_{\alpha\beta}, \end{aligned}$$

and so we could simply use

$$\mathscr{L}' = \frac{-1}{16\pi\kappa} *e^{\beta\gamma} \wedge \omega_\gamma{}^\alpha \wedge \omega_{\alpha\beta}$$

for gravitation. This equivalent $\mathscr{L}'$, when taken by itself, is altered by a change of basis. It is useful for showing why we get an equation of second order: it contains only $\omega$, i.e., derivatives of $e$, but not $d\omega$.

**Derivation of the Euler–Lagrange Equations** (4.2.15)

(a) Variation† of the Lagrangian of $S$: From $e^\alpha \wedge *J = J \wedge *e^\alpha$ (see (1.2.18(a))), we conclude that

$$e^\alpha \wedge \delta*J = \delta J \wedge *e^\alpha + J \wedge \delta*e^\alpha - (\delta e^\alpha) \wedge *J.$$

If we successively vary the $e$'s occurring in $*e^\alpha$, we discover that

$$\delta*e^\alpha = \delta e^\beta \wedge (i_\beta*e^\alpha), \qquad i_\beta \equiv i_{e_\beta}.$$

† Here $\delta$ denotes the variation, not the codifferential.

Using this in the above formula, multiplying by $J_\alpha$, and summing:

$$J \wedge \delta{*J} = \delta J \wedge {*J} - \delta e^\alpha \wedge [J \wedge i_\alpha{*J} + (i_\alpha J)\cdot{*J}] \Rightarrow$$
$$\delta(-\tfrac{1}{2}J \wedge {*J}) = -\delta J \wedge {*J} + \tfrac{1}{2}\delta e^\alpha \wedge [J \wedge i_\alpha{*J} + (i_\alpha J)\cdot{*J}].$$

(b) Variation of the electromagnetic part of $\mathscr{L}$: As in part (a), we conclude from $e^{\alpha\beta} \wedge {*F} = F \wedge {*e^{\alpha\beta}}$ that

$$e^{\alpha\beta} \wedge \delta{*F} = (\delta F) \wedge {*e^{\alpha\beta}} + F \wedge \delta{*e^{\alpha\beta}} - (\delta e^{\alpha\beta}) \wedge {*F}.$$

As before, multiplication by $F_{\alpha\beta}$ yields

$$\delta(-\tfrac{1}{2}F \wedge {*F}) = -\delta F \wedge {*F} + \tfrac{1}{2}\delta e^\alpha \wedge [(i_\alpha F) \wedge {*F} - F \wedge i_\alpha{*F}].$$

(c) Variation of the gravitational part: First note that by (1.2.18(c)),

$$\delta{*e^{\alpha\beta}} = \delta e^\gamma \wedge i_\gamma{*e^{\alpha\beta}} = \delta e^\gamma \wedge {*e^{\alpha\beta}}_\gamma, \qquad e^{\alpha\beta}{}_\gamma \equiv \eta_{\gamma\sigma}e^{\alpha\beta\sigma} \text{ etc.}$$

Although $R_{\alpha\beta}$ is itself constructed from the $\omega$'s and a variation of $e$ induces a variation of $\omega$, we need not calculate these variations, because

$$*e^{\alpha\beta} \wedge \delta R_{\alpha\beta} = d(*e^{\alpha\beta} \wedge \delta\omega_{\alpha\beta}) \tag{4.2.16}$$

(Problem 1), and hence the variation of $R$ does not affect the Euler–Lagrange equations.

The combination of these three results produces

$$\delta\mathscr{L} = -\delta J \wedge {*J} - \delta F \wedge {*F} + \delta e^\alpha$$
$$\wedge \left[{*t_\alpha} + {*\mathscr{T}_\alpha} + \frac{1}{16\pi\kappa}{*e_{\alpha\beta\gamma}} \wedge R^{\beta\gamma}\right] + \frac{1}{16\pi\kappa}d(*e^{\alpha\beta} \wedge \delta\omega_{\alpha\beta})$$
$$*t_\alpha = \tfrac{1}{2}((i_\alpha J){*J} + J \wedge i_\alpha{*J}),$$
$$*\mathscr{T}_\alpha = \tfrac{1}{2}((i_\alpha F) \wedge {*F} - F \wedge i_\alpha{*F})$$

or, writing $J = dS + eA$ and $F = dA$,

$$\delta\mathscr{L} = \delta S\, d{*J} - \delta A \wedge (d{*F} + e{*J}) + \delta e^\alpha$$
$$\wedge \left[{*t_\alpha} + {*\mathscr{T}_\alpha} + \frac{1}{16\pi\kappa}{*e_{\alpha\beta\gamma}} \wedge R^{\beta\gamma}\right] \tag{4.2.17}$$
$$- d\left(\delta S{*J} + \delta A \wedge {*F} - \frac{1}{16\pi\kappa}{*e^{\alpha\beta}} \wedge \delta\omega_{\alpha\beta}\right).$$

The requirement that $\delta \int \mathscr{L} = 0$ therefore results in the

**Field Equations of the Total System** (4.2.18)

(a) $d{*J} = 0$,
(b) $d{*F} = -e{*J}$,
(c) $-\tfrac{1}{2}{*e_{\alpha\beta\gamma}} \wedge R^{\beta\gamma} = 8\pi\kappa({*t_\alpha} + {*\mathscr{T}_\alpha})$.

**Remarks** (4.2.19)

1. These calculations generalize those of §3.1, because the metric has also been varied. The effect is to produce Einstein's equations (c) for the gravitational potential, in addition to the equations we got before.
2. The right sides of (c) are the energy-momentum currents of a scalar field and an electromagnetic field. Note that the gauge-dependent contribution (2.1.8; 1) of the canonical energy-momentum tensor does not appear, and that the Maxwellian contribution is coupled with gravity. These currents have the structure of the Hamiltonian $p\dot{q} - L$ in mechanics as anticipated in (1.3.22). They can be written
$$*t_\alpha = (i_\alpha {*J}) \wedge dS - i_\alpha \mathscr{L}^s, \qquad *\mathscr{T}_\alpha = i_\alpha {*F} \wedge dA - i_\alpha \mathscr{L}^e.$$
Now $*J = -\partial\mathscr{L}^s/\partial\, dS$ (resp. $*F = -\partial\mathscr{L}^e/\partial\, dA$) correspond to $p$; $dS$ (resp. $dA$) correspond to $\dot{q}$ and the inner product picks out the right component.
3. Although we have used an orthogonal basis to derive (4.2.17), (c) has the same form in any basis, because of the homogeneous transformation law for $R_{\alpha\beta}$ (4.1.20; 3). Equation (c) seems to have a different structure than (a) and (b), it does not say that the derivative of a field strength is a current. (c) is the analog to the covariant form of the Yang–Mills equations (4.2.9). We shall shortly cast it into the form (4.2.7) which is like (a) and (b) but thereby the two sides of the equation lose the property of transforming homogeneously under gauge transformations.

**Different Versions of Einstein's Equations** (4.2.20)

(a) The classical version. The 3-forms of energy, momentum, and electric current all occur in (4.2.8). In §3.1 we wrote down the equations for the corresponding 1-forms, and we can similarly rewrite (c). By Rule (1.2.18(b)), for all $\omega \in E_2$,
$$-{*e_{\alpha\beta\gamma}} \wedge \omega = ({*e_\alpha} i_\beta i_\gamma + {*e_\beta} i_\gamma i_\alpha + {*e_\gamma} i_\alpha i_\beta)\omega.$$
If $\omega = R^{\beta\gamma}$ and we observe that $i_\alpha i_\beta R^\beta{}_\gamma = i_\alpha R_\gamma = i_\gamma R_\alpha$ and ${*e^\gamma} i_\gamma R_\alpha = *R_\alpha$ (cf. (4.1.24; 3)), then we see that
$$R_\alpha - \tfrac{1}{2} e_\alpha R = 8\pi\kappa(t_\alpha + \mathscr{T}_\alpha),$$
i.e.,
$$R_\alpha = 8\pi\kappa(t_\alpha + \mathscr{T}_\alpha - \tfrac{1}{2} e_\alpha(t + \mathscr{T})), \qquad t = i_\alpha t^\alpha, \text{ etc.}$$
For the components of the Riemann-Christoffel tensor (4.1.24; 2) and the energy-momentum tensor $T_{\alpha\beta}$ such that $t_\alpha + \mathscr{T}_\alpha = T_{\alpha\beta} e^\beta$, this means that
$$R^\gamma{}_{\alpha\gamma\beta} - \tfrac{1}{2} g_{\alpha\beta} R^\gamma{}_{\sigma\gamma}{}^\sigma = 8\pi\kappa T_{\alpha\beta}.$$

(b) As a Yang–Mills type equation. $\mathscr{L}'$ of (4.2.14; 3) can be written (see (4.2.12))

$$\mathscr{L}' = -\tfrac{1}{2}(de^\alpha \wedge e^\beta) \wedge *(de_\beta \wedge e_\alpha) + \tfrac{1}{4}(de^\alpha \wedge e_\alpha) \wedge *(de^\beta \wedge e_\beta)$$

if we momentarily put $8\pi\kappa = 1$. Performing the variational procedure (4.2.17) with the $e^\alpha$ we get another version of (4.2.18(c))

$$d{*}F_\alpha = -{*}J_\alpha$$

$${*}F_\alpha = \frac{\partial \mathscr{L}}{\partial de^\alpha} = e^\beta \wedge *(de_\beta \wedge e_\alpha) - \tfrac{1}{2}e_\alpha \wedge *(de^\beta \wedge e_\beta) \tag{4.2.21}$$

$${*}J_\alpha = \frac{\partial \mathscr{L}}{\partial e_\alpha} = i_\alpha{*}J \wedge dS + i_\alpha{*}F \wedge dA + i_\alpha{*}F_\beta \wedge de^\beta - i_\alpha \mathscr{L}'.$$

They are now in the form of the exterior derivative of a field strength yielding the dual of a current and have the following features in common with the Yang–Mills equations:

(i) There is a contribution ${*}t_\alpha = i_\alpha{*}F_\beta \wedge de^\beta - i_\alpha\mathscr{L}'$ from the $e^\alpha$ to the currents. One can interpret it as the energy-momentum currents of the gravitational field. It has exactly the same structure $p\dot{q} - L$ as the contributions from $S$ and $A$.

(ii) Under $e^\alpha \to L^\alpha{}_\beta(x)e^\beta$ both $d{*}F_\alpha$ and ${*}J_\alpha$ transform inhomogeneously. From the three contributions to ${*}J_\alpha$ it is only $t_\alpha$ which does not transform homogeneously. For any given point $x$ we can find a frame such that $t_\alpha(x) = 0$. Conversely, on a flat space with non-Cartesian coordinates $t^\alpha \neq 0$. As discussed in §1.3 these facts reflect how the balance of energy-momentum is affected in accelerated reference frames. The total currents are conserved, $d{*}J_\alpha = 0$ and thus we draw again the conclusions that the total energy momentum of a closed universe or a periodic field is zero. If space is asymptotically flat and we restrict ourselves to transformations which are asymptotically constant Lorentz transformations then the total energy and momentum transform like a vector.

To see the relation to the classical version we have to re-express $de$ in terms of the $\omega$'s. This is done most easily if one remembers that the exterior product is the dual of the inner product

$$\begin{aligned} F^\alpha &= i_\beta(de^\beta \wedge e^\alpha - \tfrac{1}{2}\eta^{\alpha\beta}de^\gamma \wedge e_\gamma) \\ &= -i_\beta(\omega^\beta{}_\gamma \wedge e^{\gamma\alpha} - \tfrac{1}{2}\eta^{\alpha\beta}\omega^\gamma{}_\sigma \wedge e^\sigma{}_\gamma) \\ &= -\langle\omega_{\beta\gamma}|e^\beta\rangle e^{\gamma\alpha} + \tfrac{1}{2}\langle\omega_{\gamma\sigma}|e^\alpha\rangle e^{\sigma\gamma} \\ &= -\tfrac{1}{2}i_{\omega_{\beta\gamma}}e^{\alpha\beta\gamma} = -\tfrac{1}{2}{*}(\omega_{\beta\gamma} \wedge {*}e^{\alpha\beta\gamma}). \end{aligned} \tag{4.2.22}$$

Thus $d{*F^\alpha} = \frac{1}{2}d(\omega_{\beta\gamma} \wedge {*e^{\alpha\beta\gamma}})$ which has some similarity to the left side of (4.2.18(c)), the difference being ${*t^\alpha}$. To get the latter in terms of the $\omega$'s we use ${*e^{\alpha\beta\gamma}} = \varepsilon^{\alpha\beta\gamma\delta}e_\delta$,

$$d(\omega_{\beta\gamma} \wedge e_\delta) = d\omega_{\beta\gamma} \wedge e_\delta - \omega_{\beta\gamma} \wedge \omega_{\sigma\delta} \wedge e^\sigma$$

and

$$R_{\beta\gamma} = d\omega_{\beta\gamma} - \omega_{\sigma\beta} \wedge \omega^\sigma{}_\gamma.$$

Comparing (4.2.21) with (4.2.18(c)) we see

$$*t^\alpha = -\frac{\varepsilon^{\alpha\beta\gamma\delta}}{16\pi\kappa}(\omega_{\sigma\beta} \wedge \omega^\sigma{}_\gamma \wedge e_\delta - \omega_{\beta\gamma} \wedge \omega_{\sigma\delta} \wedge e^\sigma).$$

This form of the currents has been put forward by Landau and Lifshitz so that we shall call (4.2.21) the Landau–Lifshitz form of Einstein's equation. They did not use an orthogonal but a natural basis. In the natural basis $e^\alpha = dx^\alpha$, $\omega_{\alpha\beta} \equiv \Gamma_{\alpha\beta\mu}dx^\mu$ with $\Gamma_{\sigma\beta\mu} = \Gamma_{\sigma\mu\beta}$. Then the energy-momentum tensor of gravity, that is the components of $t^\alpha$ in this basis, are symmetric

$$dx^\rho \wedge {*t^\alpha} = -\frac{\sqrt{|g|}}{16\pi\kappa}(\Gamma_{\sigma\beta\mu}\Gamma^\sigma{}_{\gamma\nu}\varepsilon^{\rho\mu\nu}{}_\delta + \Gamma_{\beta\gamma\mu}\Gamma_{\sigma\delta\nu}\varepsilon^{\rho\mu\nu\sigma})\cdot\varepsilon^{\alpha\beta\gamma\delta}\cdot {*1}$$
$$= dx^\alpha \wedge {*t^\rho}.$$

This symmetry ensures local angular momentum conservation: $d({*t^\beta} + {*\mathcal{T}^\beta} + {*t^\beta}) = 0$ implies $d(x^\alpha({*t^\beta} + {*\mathcal{T}^\beta} + {*t^\beta}) - x^\beta({*t^\alpha} + {*\mathcal{T}^\alpha} + {*t^\alpha})) = 0$ only if it is true for $t$ as well as for $t$ and $\mathcal{T}$ (cf. (2.1.11; 2)) that $dx^\alpha \wedge {*t^\beta} - dx^\beta \wedge {*t^\alpha} = 0$. Here $x^\alpha$ is a local coordinate, so at this stage the theorem of conservation of angular momentum is formulated strictly on the domain of a single chart. The appropriate 3-form is defined globally only on special manifolds.

From (2.2.22) one deduces immediately (Problem 8) the so-called ADM-expression for the total energy in an asymptotically flat space. It was a major discovery [41] when it was shown that it is positive provided the external energies $\mathcal{T}^0$ and $t^0$ have this property. Thus the negative energy of gravitation can never exceed the energy of the sources if the space is to stay flat in the large.

**Remarks** (4.2.23)

1. The versions (4.2.20) do not exhaust the possibilities of writing Einstein's equations in terms of the exterior differentials of 2-forms. Numerous other variants have been proposed [10][29][39]. We just wanted to exhibit the analogy to non-abelian Yang–Mills theories. The field equations are nonlinear because the gravitational field carries energy and momentum. However these quantities evade localization since they transform inhomogeneously under local Lorentz transformations.

2. It is not the curvature forms $R_{\alpha\beta}$ but rather their contractions $R_\alpha$ that are locally determined by Einstein's equations. However, if the Weyl tensor is known, which happens, for instance, if the space is conformally flat ($C_{\alpha\beta} = 0$), then (4.2.11(c)) does determine $R_{\alpha\beta}$. In vacuo ($\mathcal{T}_\alpha = t_\alpha = 0$), $R_{\alpha\beta}$ and $C_{\alpha\beta}$ are the same and conformally flat solutions are flat. In two and three dimensions where the Weyl-tensor vanishes all solutions of Einstein's equations in vacuo are flat.
3. If one follows Cartan's suggestion and retains the torsion, generalizing the foregoing argument, then, like $R_\alpha$, it is determined locally, by the spin density of the matter present [21]. In the absence of spin, the space becomes torsionless, and the theory reduces to the one presented above. Since we know of no objects with sufficiently high spin densities, this variant of the theory agrees with experiment as well as Einstein's.
4. The geometrical significance of the 1-forms $R_\alpha$ determined by (4.2.20(a)) is brought out through the following heuristic argument: Suppose that $v$ points in the time direction $e^0$. Then if it is translated parallelly around the infinitesimal loop formed by $e^1 \wedge e^0$, it changes by $(\delta v)^1 = v^0 R^1{}_{010} \times$ the surface area, or, summing over the three spatial components,

$$(\delta v)^1 + (\delta v)^2 + (\delta v)^3 = v^0 R^\alpha{}_{0\alpha 0} \times \text{surface area}$$
$$= v^0 8\pi\kappa \int [T_{00} - \tfrac{1}{2} g_{00} T^\alpha{}_\alpha].$$

If $e^0$ is a geodesic vector field, then $v$ remains in the tangential direction of the geodesic vectors when translated parallelly along $e^0$. The positivity of the right side of the above equation indicates that the geodesic lines converge (Figure 55). This reflects the attractive character of gravity, and contains the seeds of the destruction of space and time that will be discussed in §4.6.

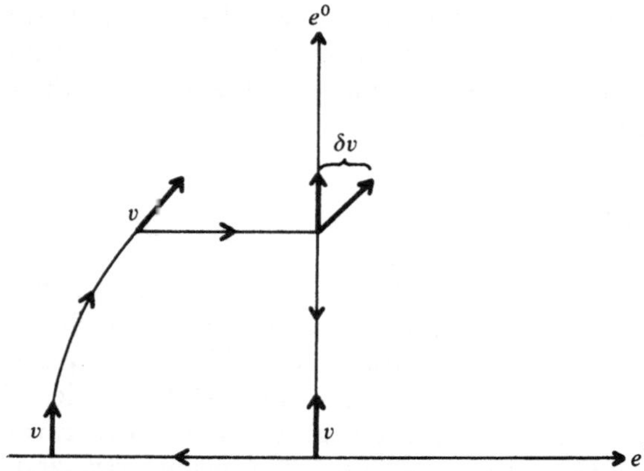


Figure 55 Parallel displacement of $v$ along the streamlines of $e^1$ and $e^0$

5. The $e^i$, and thus also the $t^\alpha$, can not in general be defined globally, though they can if there exists a compact, spatially orientable submanifold $N$ without a boundary. It is known that such a manifold is parallelizable, and hence $e^1$, $e^2$, and $e^3$ exist globally on $N$. If $e^0$ is taken as the timelike normal vector field, then $t^\alpha$ can be defined globally on $N$. Therefore the statement (1.3.35; 2) that an orientable, closed universe has zero total energy-momentum is always valid.

Now that the observables introduced in §1.3 have been identified, we are ready to derive the finite speed of propagation of the gravitational field, which, as anticipated in §1.1, equals the speed of light. Speaking mathematically, it is a matter of finding the characteristics of the equations (4.2.11(c)). In §1.3, the characteristics were defined as the possible

**Surfaces of Discontinuity of the Solutions** (4.2.24)

*Let $e^\alpha$ have possibly discontinuous second derivatives with respect to a local coordinate $u$, and let $t^\alpha$ and $\mathscr{T}^\alpha$ be continuous. Then either $R_{\alpha\beta}$ is continuous, or else $du = n_\alpha e^\alpha$ in an orthogonal basis $e^\alpha$, where $n^2 \equiv n_\alpha n_\beta \eta^{\alpha\beta} = \langle du | du \rangle = 0$.*

**Remarks** (4.2.25)

1. If the $e^\alpha$ are allowed not to be $C^\infty$, then it is of course possible to choose them with discontinuous second derivatives even in flat space, a fact which reflects only the choice of basis. Any genuine discontinuity would have to show up in the $R_{\alpha\beta}$, and (4.2.13) states that such discontinuities can only occur along surfaces with lightlike normals.
2. The equations $\delta\, dA = J$ also allow arbitrary discontinuities in $A$ even if $J$ is continuous; after all, they can be contained in a gauge potential $A = d\Lambda$. The analogous alternatives are that either $F = dA$ is continuous, or else $n^2 = 0$ (cf. (2.2.1)).

**Proof of** (4.2.24)

The part of $de^\alpha$ that contains discontinuous first derivatives must be proportional to $du$:

$$de^\alpha = (A^\alpha{}_\beta + S^\alpha{}_\beta) du \wedge e^\beta,$$
$$A_{\alpha\beta} \equiv \eta_{\alpha\gamma} A^\gamma{}_\beta = -A_{\beta\alpha}, \qquad S_{\alpha\beta} = S_{\beta\alpha}.$$

We have separated the coefficients into a symmetric and an antisymmetric part, because they act differently in $\omega_{\alpha\beta} = -\omega_{\beta\alpha}$. If we accept the following equations modulo continuous terms, then

$$\omega_{\alpha\beta} = -A_{\alpha\beta}\, du + S_\alpha n_\beta - S_\beta n_\alpha, \qquad S_\alpha = S_{\alpha\gamma} e^\gamma,$$

as can be verified by substitution into $de^{\alpha} = -\omega^{\alpha}{}_{\beta} \wedge e^{\beta}$. By assumption, any possible discontinuity in the curvature originates with $d\omega_{\alpha\beta}$, hence either with $S'_{\alpha}$ such that $dS_{\alpha} = du \wedge S'_{\alpha}$ or with $dA_{\alpha\beta} = A'_{\alpha\beta}\, du$. The latter possibility does not contribute to $d\omega_{\alpha\beta}$, so the discontinuous part of the curvature becomes

$$R_{\alpha\beta} = du \wedge (S'_{\alpha} n_{\beta} - S'_{\beta} n_{\alpha}),$$

and thus

$$R_{\beta} \wedge du = i_{\alpha} R^{\alpha}{}_{\beta} \wedge du = (n^{\alpha} S'_{\alpha} n_{\beta} - S'_{\beta} n^2) \wedge du.$$

But in that case,

$$(R_{\beta} n_{\gamma} - R_{\gamma} n_{\beta}) \wedge du = n^2 (S'_{\gamma} n_{\beta} - S'_{\beta} n_{\gamma}) \wedge du = n^2 R_{\gamma\beta}.$$

Since $R_{\beta}$ must be continuous according to Einstein's equations (4.2.11(c)), either $n^2 = 0$ or else $R_{\gamma\beta}$ stays continuous. □

**The Linear Approximation** (4.2.26)

The analogy we have just discovered between the characteristics of Einstein's and Maxwell's equations extends in the case of weak fields to a simple wave equation for the gravitational field. Even though we shall not show here when the contributions we drop are actually negligible, the approximation is of value as a first orientation to the problem, especially as the space around us is quite weakly curved.

Let

$$e^{\alpha} = dx^{\alpha} + \varphi^{\alpha}{}_{\beta}\, dx^{\beta}, \qquad \varphi_{\alpha\beta} \equiv \eta_{\alpha\sigma} \varphi^{\sigma}{}_{\beta} = \varphi_{\beta\alpha},$$

be an orthogonal basis, (the symmetry of $\varphi$ will be justified later) and suppose $|\varphi^{\alpha}{}_{\beta}(x)| \ll 1$ for all $x$. Then from

$$de^{\alpha} = \varphi^{\alpha}{}_{\beta,\gamma}\, dx^{\gamma} \wedge dx^{\beta}$$

it follows to first order in $\varphi$ that

$$\omega^{\alpha}{}_{\beta} = (\varphi^{\alpha}{}_{\gamma,\beta} - \varphi_{\beta\gamma,}{}^{\alpha}) dx^{\gamma},$$

since (1.2.25) is satisfied to this order. In that case,

$$d\omega^{\alpha}{}_{\beta} = (\varphi^{\alpha}{}_{\gamma,\beta\rho} - \varphi_{\beta\gamma,}{}^{\alpha}{}_{\rho}) dx^{\rho} \wedge dx^{\gamma},$$
$$i_{\alpha}\, d\omega^{\alpha}{}_{\beta} = (\varphi^{\alpha}{}_{\gamma,\beta\alpha} - \varphi_{\beta\gamma,}{}^{\alpha}{}_{\alpha} - \varphi^{\alpha}{}_{\alpha,\beta\gamma} + \varphi_{\beta\alpha,}{}^{\alpha}{}_{\gamma}) dx^{\gamma}.$$

In harmonic coordinates, where $\delta\, dx^{\alpha} = 0$,

$$\varphi_{\beta\alpha,}{}^{\alpha} = \tfrac{1}{2} \varphi^{\alpha}{}_{\alpha,\beta} \tag{4.2.27}$$

(Problem 2), and Einstein's equations (4.2.11(a)) become

$$-\varphi_{\beta\gamma,}{}^{\alpha}{}_{\alpha}\, dx^{\gamma} = 8\pi\kappa (T_{\beta\gamma} - \tfrac{1}{2} \eta_{\beta\gamma} T^{\alpha}{}_{\alpha}) dx^{\gamma} \tag{4.2.28}$$

in this approximation. These can be solved with the Green function (2.2.3):

$$\varphi_{\alpha\beta}(\bar{x}) = \varphi_{\alpha\beta}^{\text{in}}(\bar{x}) + 8\pi\kappa \int d^4x\, D^{\text{ret}}(\bar{x} - x)(T_{\alpha\beta}(x) - \tfrac{1}{2}\eta_{\alpha\beta}T^{\rho}{}_{\rho}(x))$$

$$\varphi^{\text{in}}_{\alpha\beta,\rho}{}^{\rho} = 0, \qquad \varphi^{\text{in}\ \alpha}_{\beta\alpha,} = \tfrac{1}{2}\varphi^{\text{in}\ \alpha}_{\alpha}{}_{,\beta}. \tag{4.2.29}$$

**Remarks** (4.2.30)

1. The $T_{\alpha\beta}\,dx^{\beta}$ are the energy-momentum forms without the gravitational contribution. It is not inconsistent to neglect gravity in this approximation, because $\kappa T_{\beta\gamma,}{}^{\gamma}$ vanishes to zeroth order. This makes (4.2.29) and (4.2.27) consistent to first order.
2. In the static limit, $\int dt\, D^{\text{ret}}(x - \bar{x}) = \dfrac{1}{4\pi|\mathbf{x} - \bar{\mathbf{x}}|}$, $T_{\alpha\beta} = Mj_{\alpha}j_{\beta}$, where $j = (-1, \mathbf{v})\delta^3(\mathbf{x})$, $\varphi_{\alpha\beta}$ becomes

$$\varphi_{\alpha\beta}(\bar{x}) = \frac{2M\kappa}{r}(j_{\alpha}j_{\beta} + \tfrac{1}{2}\eta_{\alpha\beta});$$

   i.e., the same result as stated in (I: 5.6.2), because $g = (\eta_{\alpha\beta} + 2\varphi_{\alpha\beta})dx^{\alpha} \otimes dx^{\beta}$. This justifies the choice of the factor $8\pi\kappa > 0$. The sign is not dictated by the geometry, but is only found empirically.
3. The analogy with electrodynamics should not make us overlook that, as discussed in volume I, chapter 6, the metric as measured is $g$ and not $\eta_{\alpha\beta}\,dx^{\alpha} \otimes dx^{\beta}$, although the difference is not great if the fields are weak.
4. The symmetry of $T_{\alpha\beta}$ justifies our ansatz $\varphi_{\alpha\beta} = \varphi_{\beta\alpha}$ a posteriori.

We close this section by investigating whether the generalization of the calculations of §2.1 and §3.1 connected with conservation laws brings new insights to the case of a gravitational field. Since the metric is not fixed a priori, there are now more invariance properties, and one would expect to find additional conserved quantities.

Returning to Equation (4.2.17) for the variation of $\mathscr{L}$, we start by looking at the new contribution from gravitation. If the variation comes from the Lie derivative $L_X$ in the direction of the vector field $X$, then Equation (4.2.17) implies that

$$\begin{aligned} &L_X(*e^{\alpha\beta} \wedge R_{\alpha\beta}) - d(*e^{\alpha\beta} \wedge L_X\omega_{\alpha\beta}) \\ &\quad = (L_X e_{\alpha}) \wedge *T^{\alpha} \\ &\quad = (i_X de_{\alpha} + di_X e_{\alpha}) \wedge *T^{\alpha} \\ &\quad = -(i_X\omega_{\alpha\sigma})e^{\sigma} \wedge *T^{\alpha} + (i_X e^{\sigma})\omega_{\alpha\sigma} \wedge *T^{\alpha} - (i_X e_{\alpha})d{*T^{\alpha}} \\ &\qquad + d((i_X e_{\alpha}){*T^{\alpha}}), \end{aligned} \tag{4.2.31}$$

where $*T^{\alpha} \equiv *e^{\alpha\beta\gamma} \wedge R_{\beta\gamma} \in E_3$ and $e^{\alpha}$ is an orthogonal basis.

If (4.2.31) is integrated over a four-dimensional manifold $N$ without a boundary, and $X$ has compact support in $N$, then from $\int L_X\omega = 0$ for all

$\omega \in E_m$, the invariance of the integral under Lie differentiation (I: 2.6.11), and from $(i_X \omega_{\alpha\sigma})e^\sigma \wedge {}^*T^\alpha = 0$ (because $e^\sigma \wedge {}^*T^\alpha = e^\alpha \wedge {}^*T^\sigma$) we infer that

$$\int_N x_\alpha(d^*T^\alpha + \omega^\alpha{}_\sigma \wedge {}^*T^\sigma) = 0, \qquad x_\alpha \equiv i_X e_\alpha.$$

Since this must be true for all vector fields $X$ of compact support in $N$—no invariance properties have been assumed of $X$—we obtain the

**Contracted Bianchi Identity** (4.2.32)

$$d^*T^\alpha = -\omega^\alpha{}_\sigma \wedge {}^*T^\sigma.$$

**Remarks** (4.2.33)

1. This fact follows from the more general equation (4.1.26) (Problem 3).
2. The invariance of the integral under general coordinate transformations is expressed by $\int L_X \omega = 0$. No new conservation theorems result from this general covariance, but only identities that hold independently of any field equation (cf. (2.1.8; 5) and (2.1.18; 3)).
3. Although (4.2.21) was derived for orthogonal bases, it has the same form in all bases, since the ${}^*T^\alpha$ transform as ${}^*e^\alpha$.
4. If Einstein's equations $T^\alpha = 16\pi\kappa(\mathscr{T}^\alpha + t^\alpha)$ hold, then (4.2.21) implies Equation (1.3.28):
$$d({}^*\mathscr{T}^\alpha + {}^*t^\alpha) = -\omega^\alpha{}_\beta \wedge ({}^*\mathscr{T}^\beta + {}^*t^\beta).$$
5. Because of the symmetry $i_\beta T_\alpha = i_\alpha T_\beta$, the 1-forms $T_\alpha$ have 10 linearly independent components, among which (4.2.21) creates 4 differential identities. Thus only 6 of Einstein's equations are independent of one another. This is felt to be the correct number of equations for the 10 components $g_{\alpha\beta} = g_{\beta\alpha}$ of $g = g_{\alpha\beta}\, dx^\alpha \otimes dx^\beta$: the equations ought not to fix the coordinate system, and so there must remain 4 arbitrary functions $\bar{x}^\alpha(x)$ to play with.
6. It is part of the relativity folklore that Einstein's theory differs from other field theories inasmuch as Einstein's equations also determine the equations of motion of matter. In particular (4.2.32) is supposed to imply that particles move on geodesics provided no other forces act on them. Whereas the proof of this claim for point particles encounters the difficulty that they generate singular gravitational fields we shall find a simple proof for continuous matter in (4.6.30; 4). There we will show that for an ideal fluid without pressure Bianchi's identity demands that the velocity field be geodesic. Similarly, if one has only one scalar field (4.2.32) implies the field equation. With two scalar fields the condition (4.2.32) for the

sum of their energy momentum tensors can not imply the field equation because they may interact without changing the conservation of their total energy-momentum. Thus the situation is not too different from Maxwell's equations where for one real field describing charged matter $d^*J = 0$ is already the field equation $d^*(dS + eA) = 0$. For one complex field $\sqrt{\rho}\, e^{iS}$ the current is $\rho(dS + eA)$ and its conservation does not imply the field equation.

**Problems** (4.2.34)

1. Show that ${}^*e^{\rho\tau} \wedge \delta R_{\rho\tau} = d({}^*e^{\rho\tau} \wedge \delta\omega_{\rho\tau})$.
2. Show that $\delta\, dx^\alpha = 0$ implies that $\varphi_{\alpha\beta.}{}^{\beta} = \frac{1}{2}\varphi^\beta{}_{\beta.\alpha}$ for weak fields.
3. Derive (4.2.32) from (4.1.26).
4. What effect would a term $\Lambda^*\mathbf{1}$ in $\mathscr{L}$ have on Einstein's equations?
5. Calculate ${}^*t_0$ from (4.2.18(c)) on flat space in the orthogonal basis of polar coordinates.
6. Show that $R^\alpha{}_\beta \wedge R^\beta{}_\alpha$ is exact.
7. Use the natural extension of $D$ for $V \in \mathrm{Sp}(L(F))$: $DV = dV + \omega \wedge V + (-)^p V \wedge \omega$ to show $DD^*\Omega = 0$.
8. Show

$$\frac{1}{2}R_{\alpha\beta} \wedge {}^*e^{\alpha\beta} = d(e^\alpha \wedge {}^*de_\alpha) - \frac{1}{2}(de^\alpha \wedge e^\beta) \wedge {}^*(de_\beta \wedge e_\alpha) + \frac{1}{4}(de^\alpha \wedge e_\alpha) \wedge {}^*(de^\beta \wedge e_\beta).$$

*Hint*: Use (4.2.22) and (4.2.14; 3).

9. Show that the total energy ("ADM-energy")

$$\frac{-1}{8\pi\kappa}\int_N d^*F^0 = \frac{-1}{8\pi\kappa}\int_{\partial N} {}^*F^0,$$

where $\partial N = \{x \in \mathbb{R}^4 : |\vec{\mathbf{x}}| = R,\ x_0 = 0\}$ is an asymptotically flat space is given by

$$\lim_{R\to\infty}\frac{1}{16\pi\kappa}\int_{\partial N} {}^*e^{jo}(g_{jk,k} - g_{kk,j}) \qquad \text{with} \quad j, k = 1, \ldots, 3.$$

(Use a natural basis $e^k = dx^k$, (4.1.37) for $\omega$ and that $g_{ik} \to \eta_{ik}$ for $R \to \infty$.)

**Solutions** (4.2.35)

1.

$$d({}^*e^{\rho\tau} \wedge \delta\omega_{\rho\tau}) = -2\omega^\rho{}_\sigma \wedge {}^*e^{\sigma\tau} \wedge \delta\omega_{\rho\tau} + {}^*e^{\rho\tau} \wedge d\delta\omega_{\rho\tau}$$
$$= {}^*e^{\rho\tau} \wedge (d\delta\omega_{\rho\tau} + 2\omega_\rho{}^\sigma \wedge \delta\omega_{\sigma\tau}) = {}^*e^{\rho\tau} \wedge \delta R_{\rho\tau}.$$

2. To first order, $dx^\alpha = e^\alpha - \varphi^\alpha{}_\beta e^\beta$; so

$$0 = {}^*\delta\, dx^\alpha = d({}^*e^\alpha - \varphi^\alpha{}_\beta {}^*e^\beta) = -\omega^\alpha{}_\beta \wedge {}^*e^\beta - \varphi^\alpha{}_{\beta,\gamma} e^\gamma \wedge {}^*e^\beta$$
$$= (\varphi_{\beta\gamma,}{}^\alpha - \varphi^\alpha{}_{\gamma,\beta} - \varphi^\alpha{}_{\beta,\gamma}) e^\gamma \wedge {}^*e^\beta.$$

By (1.2.28), however, $e^\gamma \wedge {}^*e^\beta = \eta^{\gamma\beta}{}^*\mathbf{1}$.

3. $$d{}^*T^i = d({}^*e^{imk} \wedge R_{mk}) = -\omega^i{}_j \wedge {}^*e^{jmk} \wedge R_{mk} - 2\omega^m{}_s \wedge {}^*e^{iskh} \wedge R_{mk}$$
$$+ 2{}^*e^{imk} \wedge \omega_m{}^s \wedge R_{sk} = -\omega^i{}_j \wedge {}^*T^j.$$

4. $\delta(\Lambda{}^*\mathbf{1}) = \Lambda\delta e^j \wedge i_j{}^*\mathbf{1} = \Lambda\delta e^j \wedge {}^*e_j$ (see (1.2.18(c)). Thus there is an additional term $\sim{}^*e_j$ in ${}^*\mathcal{T}_j$.

5. $e^\alpha = (dt, dr, r\, d\vartheta, r \sin\vartheta\, d\varphi)$, $\omega_{12} = -d\vartheta$, $\omega_{23} = -\cos\vartheta\, d\varphi$, $\omega_{31} = \sin\vartheta\, d\varphi$,
$$\tfrac{1}{2}\varepsilon_{0\beta\gamma\delta} d(\omega_{\beta\gamma} \wedge e_\delta) = d(-\cos\vartheta\, d\varphi \wedge dr + 2r \sin\vartheta\, d\varphi \wedge d\vartheta)$$
$$= -\sin\vartheta\, dr \wedge d\vartheta \wedge d\varphi.$$

6. $R^\alpha{}_\beta \wedge R^\beta{}_\alpha = d[\omega^\alpha{}_\gamma \wedge d\omega^\gamma{}_\alpha + \tfrac{2}{3}\omega^\alpha{}_\beta \wedge \omega^\beta{}_\gamma \wedge \omega^\gamma{}_\alpha]$.

7. $D{}^*\Omega = d{}^*\Omega + \omega \wedge {}^*\Omega - {}^*\Omega \wedge \omega$
$$DD{}^*\Omega = d\omega \wedge {}^*\Omega - {}^*\Omega \wedge d\omega - \omega \wedge d{}^*\Omega - d{}^*\Omega \wedge \omega$$
$$+ \omega \wedge (d{}^*\Omega + \omega \wedge {}^*\Omega - {}^*\Omega \wedge \omega) + (d{}^*\Omega + \omega \wedge {}^*\Omega - {}^*\Omega \wedge \omega) \wedge \omega$$
$$= \Omega \wedge {}^*\Omega - {}^*\Omega \wedge \Omega = 0$$
according to (1.2.17).

8. According to (4.2.14; 3) ${}^*e^{\alpha\beta} \wedge R_{\alpha\beta} = d({}^*e^{\alpha\beta} \wedge \omega_{\alpha\beta}) + {}^*e^{\beta\gamma}\omega_\gamma{}^\alpha \wedge \omega_{\alpha\beta}$. Now always using (1.2.17),

$$e_\alpha \wedge {}^*de^\alpha = {}^*e^\beta \langle \omega^\alpha{}_\beta | e_\alpha \rangle$$

and

$$({}^*e^{\alpha\beta}) \wedge \omega_{\alpha\beta} = {}^*i_{\omega_{\alpha\beta}} e^{\alpha\beta} = 2{}^*e^\beta \langle \omega^\alpha{}_\beta | e_\alpha \rangle \Rightarrow$$
$$\tfrac{1}{2}{}^*e^{\alpha\beta} \wedge \omega_{\alpha\beta} = e^\alpha \wedge {}^*de^\alpha.$$

Next with (4.2.22)

$$-de^\alpha \wedge {}^*F_\alpha = \tfrac{1}{2}\omega_{\alpha\sigma} \wedge e^\sigma \wedge \omega_{\beta\gamma} \wedge {}^*e^{\alpha\beta\gamma}$$
$$= \tfrac{1}{2}\omega_{\alpha\sigma} \wedge \omega_{\beta\gamma}[\eta^{\sigma\alpha}{}^*e^{\beta\gamma} - \eta^{\sigma\beta}{}^*e^{\alpha\gamma} + \eta^{\sigma\gamma}{}^*e^{\alpha\beta}]$$
$$= {}^*e^{\gamma\alpha} \wedge \omega_\gamma{}^\beta \wedge \omega_{\alpha\beta}.$$

Inserting the expression (4.2.21) for ${}^*F_\alpha$ completes the demonstration of the equality.

9. With (4.2.22) we have

$$-{}^*F^0 = -\tfrac{1}{2}(g_{ik,j} - g_{ij,k} - g_{jk,i})\tfrac{1}{2}e^j \wedge {}^*e^{ik0} = \tfrac{1}{2}(-g_{ii,k} + g_{ik,i}){}^*e^{0k}$$

since by symmetry the first term cancels out and the two others give the same contribution.

# 4.3 Maximally Symmetric Spaces

*The spaces with the simplest structure, after flat spaces, are those of constant curvature. They are a generalization of the spherical surface and, though simple, have some physically interesting aspects.*

Killing vector fields generate isometries (i.e., diffeomorphisms that leave $g$ invariant) of the space and are bijectively related to the constants of motion and conserved currents. Yet the fields need not be complete; it is possible for their flow to lead out of the manifold (see (I: 2.3.7)). However, they imprint a local structure on the space even when they do not generate one-parameter groups of isometries. Generally there are none but if there are enough of them around, Einstein's equations become more tractable, and explicit calculations are possible.

The prototype of a Killing vector field is a rotation of $\mathbb{R}^m$; for some pair of indices $(i, k)$, $v^i = x^k$ and $v^k = -x^i$, and the other components are zero. Note that

(a) $v^i{}_{,k} + v^k{}_{,i} = 0$, and
(b) $v^l{}_{,jk} = 0$.

The generalizations of these facts to pseudo-Riemannian spaces are

### The Relationships among the Covariant Derivatives of Killing Vector Fields $v$ (4.3.1)

*We use the natural basis $e_\alpha = \partial_\alpha$ and the notation*

$$\langle D_{e_\beta} v | e_\alpha \rangle \equiv v_{\alpha;\beta}, \qquad \langle D_{e_\gamma} D_{e_\beta} v | e_\alpha \rangle \equiv v_{\alpha;\beta;\gamma}.$$

*Then, with the $R^\lambda{}_{\sigma\rho\mu}$ of* (4.1.24; 2),

(a) $v_{\alpha;\beta} + v_{\beta;\alpha} = 0$, *and*
(b) $v_{\mu;\rho;\sigma} = R^\lambda{}_{\sigma\mu\rho} v_\lambda$.

### Proof

(a) Killing vector fields leave the scalar product invariant, $L_v\langle X | Y\rangle = \langle L_v X | Y\rangle + \langle X | L_v Y\rangle$. On the other hand, with Axioms (4.1.6(f)) and (g′),

$$\begin{aligned} L_v\langle X | Y\rangle = D_v\langle X | Y\rangle &= \langle D_v X | Y\rangle + \langle X | D_v Y\rangle \\ &= \langle D_X v | Y\rangle + \langle X | D_Y v\rangle + \langle L_v X | Y\rangle + \langle X | L_v Y\rangle. \end{aligned}$$

Combined, these make $\langle D_X v | Y\rangle + \langle X | D_Y v\rangle = 0$; i.e., (a), if $X$ and $Y$ are taken as the basis fields.

(b) Since a Killing vector field $v$ conserves the metric structure and a torsion-free connection is uniquely determined by the metric $L_v$ commutes with the covariant exterior derivative $D$. To demonstrate this formally one has to consider the one parameter (local) group $\Phi_t$ of isometries generated by $v$ and consider

$$\frac{\partial}{\partial t} D\Phi_t \bigg|_{t=0} = \frac{\partial}{\partial t} \Phi_t D \bigg|_{t=0}$$

Thus we infer from (4.1.39) that $R(X, v)u = \langle D_X Dv | u \rangle$. Using the properties (4.1.44; 2) of $R$ this becomes (b) when written in index notation.

**Discussion** (4.3.2)

(b) allows the second derivative of $v$ to be written linearly in $v$. By carrying the procedure further, one can reduce all the higher derivatives to $v$ and its first derivatives. If we assume analyticity, we can express $v$ locally in terms of $v$ and its first derivative at a single point, say 0:

$$v_\rho(x) = A_\rho{}^\lambda(x) v_\lambda(0) + B_\rho{}^{\lambda\sigma}(x) v_{\lambda;\sigma}(0).$$

This greatly restricts the number of possible Killing vector fields. In an $m$-dimensional space, $v_{\lambda;\sigma}(0) = -v_{\sigma;\lambda}(0)$ can assume $m(m-1)/2$ values, and $v_\lambda(0)$ can assume $m$ values. Hence there are at most $m + m(m-1)/2 = m(m+1)/2$ independent Killing vector fields.

**Remarks** (4.3.3)

1. "Independent" means that they have no linear relationships with constant coefficients. They may satisfy equations with variable coefficients as rotations $x_j \partial_i - x_i \partial_j$ are expressed by translations $\partial_i$. Since the Killing vector fields do not form a module (2.1.10; 1), this does not mean dependence.
2. On flat space the Euclidean group is the largest group of isometries (I: 4.1.13; 4), and it has exactly $m(m+1)/2$ parameters. The statement that the group can be at most this large if space is curved is therefore quite plausible.

To classify the symmetric spaces, we begin with a

**Definition** (4.3.4)

(a) A space is **maximally symmetric** iff it possesses $m(m+1)/2$ independent Killing vector fields.

(b) A space is **isotropic about the point** $x$ iff it has $m(m-1)/2$ Killing vector fields, for which $x$ is a fixed point of the flow, and the $A^i{}_k$ of $(L_v e^i)(x) = A^i{}_k(x)e^k(x)$, where $e^k$ are an orthogonal basis, generate the total Lorentz group of $(\mathbb{R}^m, \eta)$.
(c) A space is **isotropic** iff it is isotropic about all of its points.
(d) A space is **homogeneous** if it has a transitive† group of isometries.
(e) A space is **stationary** if it has a timelike Killing vector field. If the latter is orthogonal to a family of spacelike hypersurfaces the space is called **static**.

**Remarks** (4.3.5)

1. Definition (b) means that there are $m(m-1)/2$ Killing vector fields $v^i$ such that $v^i{}_\alpha(x) = 0$, and the $v^i{}_{\alpha;\beta}(x)$ form a basis for the space of $m \times m$ antisymmetric matrices, where $i$ runs from 1 to $m(m-1)/2$.
2. On a maximally symmetric space there are Killing vector fields as in (b) at every point. Therefore it is isotropic.
3. To appreciate the distinction made in (e) we remark that in general for a vector field $X$ there is locally a family of transversal hypersurfaces $N_\alpha$ such that $X$ and the vectors in $N_\alpha$ span the tangent space. However, this hypersurfaces will not be orthogonal to $X$ in the sense of a given metric and there will be no orthogonal family of hypersurfaces. In contradistinction to the above for a given spacelike hypersurface one can find, at least locally, an orthogonal geodesic vector field.

The curvature must be the same in every direction on an isotropic space. Because of this, there is an extremely simple

**Structure of the Curvature Forms of an Isotropic Space** (4.3.6)

*On an isotropic space,*

$$R^{ik} = Ke^{ik}, \textit{for some constant } K.$$

**Proof**

Let $\Phi$ be an isometry and $\Phi_*$ the translation it induces on tensors. We noted that $\Phi_*$ commutes with $D$ and from $DD\Phi_* = \Phi_* DD$ we conclude $R\Phi_* = \Phi_* R$. Let $\Phi$ be an isometry with $x$ as fixed point such that $(\Phi_* e^i)(x) = L^i{}_k e^k(x)$ with $L^t \eta L = \eta$. If $R$ is decomposed in this basis then its invariance under $\Phi_*$ means that the Riemann–Christoffel-tensor (4.1.42; 2)

† Transitive means that any point can be reached from every point by a transformation from the group.

is invariant under $R_{ijkl} \to R_{mnop}L^m{}_iL^n{}_jL^o{}_kL^p{}_l$, it transforms as a matrix in the antisymmetric tensor product space. This representation is irreducible (except for $m = 4$) and the matrix group elements must be proportional to the unit matrix: $R_{ij}{}^{lm} = K(\delta_i^l\delta_j^m - \delta_i^m\delta_j^l)$ or for the curvature forms $R_{ij}(x) = K(x)e_{ij}(x)$. For $m = 4$ there would be the possibility $R_{ij} = K{*e_{ij}}$ which is excluded by $R_{ij} \wedge e^j = 0$. In order to see why $K$ has to be constant, consider the equation $dR_{ik} = dK \wedge e_{ik} + K\,de_{ik}$. By using Bianchi's identity (4.1.26), we discover that $dK = 0$, and hence $K$ is independent of $x$. (For $m = 2$ another argument works.) □

**Remarks** (4.3.7)

1. If the curvature is independent of the direction, then it is also independent of the position. For that reason, one says simply that such spaces have constant curvature.
2. By (4.1.44; 3) and (4.3.6), isotropic spaces have vanishing Weyl forms, and hence they are conformally flat if $m > 3$.
3. It is even true that the existence of more than $m(m - 1)/2 + 2$ Killing vector fields implies that $R_{ik} = Ke_{ik}$ (cf. [4]).

**Construction of Isotropic Spaces with** $m \geq 3$ (4.3.8)

Since such spaces are conformally flat, there is an orthogonal basis of the form $e^a = dx^c/\psi$, $\psi \in E_0$. Therefore $de^a = \psi_{,b}e^{ba}$, and thus $\omega_{ab} = \psi_{,a}e_b - \psi_{,b}e_a$. With the help of (4.1.45) again, this leads to the curvature forms

$$R_{ab} = \psi(\psi_{,ac}e^c{}_b - \psi_{,bc}e^c{}_a) - \psi_{,c}\psi^{,c}e_{ab}.$$

If this is to equal $Ke_{ab}$, then $\psi_{,ab}$ must be zero for all $a \neq b$, and hence

$$\psi = \sum_{i=1}^m f^i(x^i).$$

If $f_a = \eta_{ai}f^i$, then (4.3.6) implies that

$$f_b'' + f_a'' = \psi^{-1}(K + f_c'f^{c\prime}).$$

Since the left side depends only on $x^a$ and $x^b$, while the right side is the same for all $a$ and $b$, both sides are in fact constant. This makes $f$ a quadratic function, and therefore $\psi$ can be put into the form

$$\psi = 1 + \frac{K}{4}x^ax^b\eta_{ab}.$$

Therefore, locally a space of constant curvature always has coordinates for which

$$g = \frac{dx^i\, dx^k\, \eta_{ik}}{(1 + Kx^2/4)^2}, \qquad e^i = \frac{dx^i}{1 + Kx^2/4},$$

$$\omega^{ik} = \frac{K}{2}(x^i e^k - x^k e^i) \tag{4.3.9}$$

(where $x^2 \equiv x^a x^b \eta_{ab}$).

**The Killing Vector Fields of Isotropic Spaces** (4.3.10)

Because of the isotropy about the origin, the generators of rotations are Killing vector fields: Let $v = x_j\,\partial_k - x_k\,\partial_j$, where $x_j = \eta_{jk}x^k$ and $\partial_k$ is the dual basis to $dx^k (i_{\partial_k} dx^i = \delta^i{}_k)$, and fix a pair of indices $(j, k)$; then

$$\begin{aligned}
L_v e^m &= di_v\, e^m + i_v\, de^m \\
&= d\left(x_j\left(1 + \frac{Kx^2}{4}\right)^{-1} i_{\partial_k} dx^m\right) \\
&\quad - x_j\frac{K}{2}\left(1 + \frac{Kx^2}{4}\right)^{-2} x_l\, i_{\partial_k} dx^l \wedge dx^m - (j \leftrightarrow k) \\
&= d\left[x_j \delta^m{}_k\left(1 + \frac{Kx^2}{4}\right)^{-1}\right] \\
&\quad + \frac{K}{2}x_j\left(1 + \frac{Kx^2}{4}\right)^{-2}(\delta^m{}_k x_l\, dx^l - x_k\, dx^m) - (j \leftrightarrow k) \\
&= dx_r\left(1 + \frac{Kx^2}{4}\right)^{-1}(\delta^m{}_k \delta^r{}_j - \delta^m{}_j \delta^r{}_k).
\end{aligned}$$

Since $L_v e^m = A^{mr} e_r$ and $A^{mr} = \delta^m{}_k \delta^r{}_j - \delta^m{}_j \delta^r{}_k = -A^{rm}$, rotations about the origin are Killing vector fields. (See (2.1.10; 2).)

Moreover, there are generalizations of translations: If $k$ is fixed, then $v = \partial^k(1 - x^2K/4) + (K/2)x^k x_j\,\partial^j$ is a Killing vector field. This can be verified as in (4.3.10), or, what is easier in this case, one can verify the equation of Problem 1 for the components $v^i$ and $g_{lm}$ of $v$ and $g$ in the natural basis,

$$\begin{aligned}
0 &\overset{?}{=} v^i g_{lm,i} + g_{im} v^i{}_{,l} + g_{il} v^i{}_{,m} \\
&= \left(1 + \frac{Kx^2}{4}\right)^{-3}\left\{-\eta_{lm}\left[Kx^k\left(1 - \frac{x^2K}{4}\right) + \frac{x^k x^2 K^2}{2}\right]\right. \\
&\quad + \left(1 + \frac{Kx^2}{4}\right)\left[-\frac{K}{2}\delta^k{}_m x_l + \frac{K}{2}(x^k\eta_{ml} + x_m\delta^k{}_l) - \frac{K}{2}\delta^k{}_l x_m\right. \\
&\quad \left.\left. + \frac{K}{2}(x^k\eta_{ml} + x_l\delta^k{}_m)\right]\right\} \\
&= 0.
\end{aligned}$$

These $v$'s form a basis for small $x$, and so the group they generate in a neighborhood of the origin acts transitively: any point can be sent to any other. We can assemble our discoveries in a

**Proposition** (4.3.11)

*For a pseudo-Riemannian manifold $M$, the following properties are equivalent:*

(a) *$M$ is maximally symmetric;*
(b) *$M$ has constant curvature;*
(c) *$M$ is isotropic;*
(d) *$M$ is homogeneous and isotropic about some point.*

**Remark** (4.3.12)

Our arguments have been strictly local, so nothing can be concluded about the global behavior—The Killing vector fields need not even be complete. Unions and pieces of spherical surfaces are also isotropic spaces. However, isotropic spaces with the same $K$ are locally isometric.

Every $m$-dimensional manifold can be imbedded as a submanifold in $\mathbb{R}^{2m+1}$. Isotropic spaces can be imbedded in $\mathbb{R}^{m+1}$, whereby the metric is the restriction of a pseudo-Euclidean metric on $\mathbb{R}^{m+1}$. (Restricting the metric means that the scalar product determining it is restricted to the vectors in the tangent space of the submanifold.)

**The Geometrical Imbedding of Isotropic Spaces** (4.3.13)

*Let $\eta_{\alpha\beta}$, $\alpha, \beta = 0, 1, \ldots, m-1$, fix the sign of the metric of an isotropic space, and choose the curvature $|K|$ as the unit of length, so that $K = \pm 1$. Let the $\eta_{ik}$ on $\mathbb{R}^{m+1}$ equal $\eta_{\alpha\beta}$ for $i, k = 0, 1, \ldots, m-1$, and let $\eta_{mm} = K$. Then the isotropic space is locally isometric to the submanifold $H = \{\bar{x} \in \mathbb{R}^{m+1} : \bar{x}^i\bar{x}^k\eta_{ik} = K\}$, where $g$ is the restriction of $d\bar{x}^i \otimes d\bar{x}^k\, \eta_{ik}$ to $H$.*

**Proof**

The equation for $H$ can be written as ($m = \dim H$)

$$K\bar{x}^m\bar{x}^m = K - \eta_{\alpha\beta}\bar{x}^\alpha\bar{x}^\beta \equiv K - \bar{r}^2.$$

Introduce the coordinates $x^\alpha \in \mathbb{R}^m$ on $H$ by

$$\bar{x}^\alpha = \frac{x^\alpha}{1 + Kr^2/4}, \qquad \bar{x}^m = \frac{1 - Kr^2/4}{1 + Kr^2/4},$$

where $r^2 = x^\alpha x^\beta \eta_{\alpha\beta}$; then

$$d\bar{x}^\alpha = \frac{dx^\alpha(1 + Kr^2/4) - Kx^\alpha r\, dr/2}{(1 + Kr^2/4)^2},$$

$$d\bar{x}^m = \frac{-Kr\, dr}{(1 + Kr^2/4)^2},$$

and the restriction of the metric

$$d\bar{x}^i\, d\bar{x}^k\, \eta_{ik} = d\bar{x}^\alpha\, d\bar{x}^\beta\, \eta_{\alpha\beta} + K\, d\bar{x}^m\, d\bar{x}^m = dx^\alpha\, dx^\beta\, \eta_{\alpha\beta}\left(1 + \frac{Kr^2}{4}\right)^{-2}$$

takes on the form (4.3.9). □

**Remarks** (4.3.14)

1. The Killing vector fields of (4.3.10) are simply the restrictions of the generators $\bar{x}^k\, \bar{\partial}^j - \bar{x}^j\, \bar{\partial}^k$ of the Lorentz group of $\mathbb{R}^{m+1}$ to $H$. Their flow leaves $H$ and the metric $\eta_{ik}\, d\bar{x}^i\, d\bar{x}^k$ invariant, and hence so does their restriction to $H$. The group of isometries of spaces of maximal symmetry is therefore isomorphic to some Lorentz group $O(n, m + 1 - n)$.
2. The geodesics are the intersections of $H$ with hyperplanes passing through the origin. They are the solutions of a variational problem $\delta \int ds\, \dot{\bar{x}}^i \dot{\bar{x}}^k \eta_{ik} = 0$ with the constraint $\bar{x}^i \bar{x}^k \eta_{ik} = K$. The introduction of a Lagrange multiplier $\lambda$ leads to the equations $\ddot{\bar{x}}^i = \lambda \bar{x}^i$, making the

$$L^{ik} = \bar{x}^i \dot{\bar{x}}^k - \bar{x}^k \dot{\bar{x}}^i$$

constant. Thus $\bar{x}$ lies on the plane of $\bar{x}(0)$, $\bar{x}(0)$ since

$$0 = \varepsilon_{ike} \bar{x}^i L^{ke} = \varepsilon_{ihe} \bar{x}^i \bar{x}^k x^l.$$

**The Physical Significance of Isotropic Spaces** (4.3.15)

Spaces of a high degree of symmetry gratify the esthetic feelings of physicists, and are therefore popular as models of the world. Since the space around us is isotropic and homogeneous as far as we can see, its isotropy about every point is often elevated to a cosmological principle. Some theorists went beyond this and require a maximal symmetry for space and time, the "perfect cosmological principle." But aside from such cosmological speculation, maximally symmetric spaces are also important as solutions of Einstein's equations when the energy-momentum distribution is sufficiently symmetric. It is necessary to distinguish between the cases of positive and negative $K$ ($K = 0$ is Minkowski space):

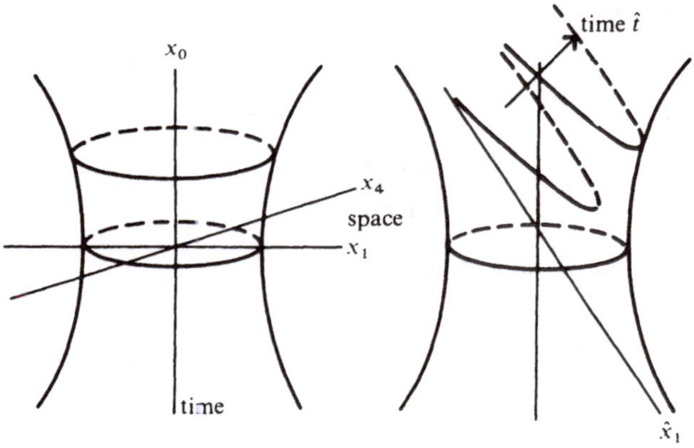


Figure 56 Various cross-sections of the de Sitter universe

$K = 1$ **The de Sitter Universe** (4.3.16)

This can be represented as the hyperboloid

$$-x_0^2 + x_1^2 + x_2^2 + x_3^2 + x_4^2 = 1$$

in $\mathbb{R}^5$ with $g = -dx_0^2 + dx_1^2 + dx_2^2 + dx_3^2 + dx_4^2$. When reduced to $\mathbb{R}^3$, it looks as in Figure 56. The intersections with planes containing the $x_0$-axis are timelike geodesics. If these geodesics are introduced as coordinate lines with the proper time as a new coordinate $t$ (these are known as **synchronous**, or **comoving**, coordinates; see Problem 5),

$$x_0 = \sinh t, \qquad x_1 = \cosh t \sin \chi \sin \vartheta \cos \varphi,$$
$$x_2 = \cosh t \sin \chi \sin \vartheta \sin \varphi, \qquad x_3 = \cosh t \sin \chi \cos \vartheta,$$
$$x_4 = \cosh t \cos \chi,$$

then the metric takes on the form

$$g = -dt^2 + \cosh^2 t\{d\chi^2 + \sin^2 \chi(d\vartheta^2 + \sin^2 \vartheta \, d\varphi^2)\} \qquad (4.3.17)$$

(see Figure 56a). The sections where $t = \text{constant}$ are Riemannian spaces with constant positive curvature and radius $\cosh t$; the universe first contracts, and then expands again. The geodesic vector field $dt$, however, is not unique; if intersections are taken with surfaces at 45° to the $x_0$-axis, then half of the hyperboloid is covered by the coordinates

$$\hat{t} = \ln(x_0 + x_4), \qquad \hat{x}_j = \frac{x_j}{x_0 + x_4}, \qquad j = 1, 2, 3$$

in which the metric has the form

$$g = -d\hat{t}^2 + e^{2\hat{t}}(d\hat{x}_1^2 + d\hat{x}_2^2 + d\hat{x}_3^2) \qquad (4.3.18)$$

(see Figure 56b). The intersections where $\hat{t}$ = constant are expanding Euclidean spaces, in the sense that the geodesics $\hat{x}_j$ = constant, $j = 1, 2, 3$, grow steadily farther apart.† There are a great many facets of the de Sitter universe; we shall even be able to find coordinates in which $g_{ik}$ does not depend on time at all (4.4.42) will show that it is even a static space. In order to survey the causal relationships better, it is convenient to map the whole space into a compact set, in what is known as a **Penrose diagram**. To this end, we write the $t$ of (4.3.17) as

$$t' = 2 \arctan(\exp t) - \frac{\pi}{2},$$

which makes the metric

$$\begin{gathered} g = \cosh^2(t)\,(-dt'^2 + d\chi^2 + \sin^2\chi\, d\Omega^2), \\ d\Omega^2 = d\vartheta^2 + \sin^2\vartheta\, d\varphi^2, \qquad 0 < \chi < \pi, \qquad -\frac{\pi}{2} < t' < \frac{\pi}{2}. \end{gathered} \tag{4.3.19}$$

The conformal equivalence to Minkowski space is again evident if the coordinates

$$t + r = \tan\frac{t' + \chi}{2}, \qquad t - r = \tan\frac{t' - \chi}{2},$$

$$0 < \chi < \pi, \qquad -\pi + \chi < t' < \pi - \chi,$$

are used to turn the Minkowski metric into

$$g = \left[2\cos\frac{t' + \chi}{2}\cos\frac{t' - \chi}{2}\right]^{-2}(-dt'^2 + d\chi^2 + \sin^2\chi\, d\Omega^2). \tag{4.3.20}$$

Both the de Sitter universe and Minkowski space are mapped into a relatively compact part of $\mathbb{R} \times S^3$. The difference between the causal structures of the two spaces comes about because they cover different parts of the $(t', \chi)$ plane (Figure 57). In de Sitter space, timelike geodesics begin on the lines $t' = -\pi/2$, $0 < \chi < \pi$, and end at $t' = \pi/2$, $0 < \chi < \pi$. There are some that do not intersect the past of a given point $p$, and so an observer at $p$ would be unaware of them—they are "beyond the particle horizon." Conversely, the union of the past light-cones of a particle's trajectory does not fill the whole space; there are points that an observer on a geodesic would never see, "beyond the event horizon." This does not occur in Minkowski space, where a timelike geodesic begins at a point $(t', \chi) = (-\pi, 0)$ and ends at $(\pi, 0)$. Only an accelerated particle could emerge from within the line $(t, r) = (-\infty, \infty)$ of Figure 57, and in that case it is possible for two accelerated observers never to see each other (recall (I: 6.4.10; 2)).

† This is observable, since, for example, the proton and electron in a hydrogen atom do not move geodesically, but are electrically bound together, which keeps the Bohr radius from expanding with the geodesics.

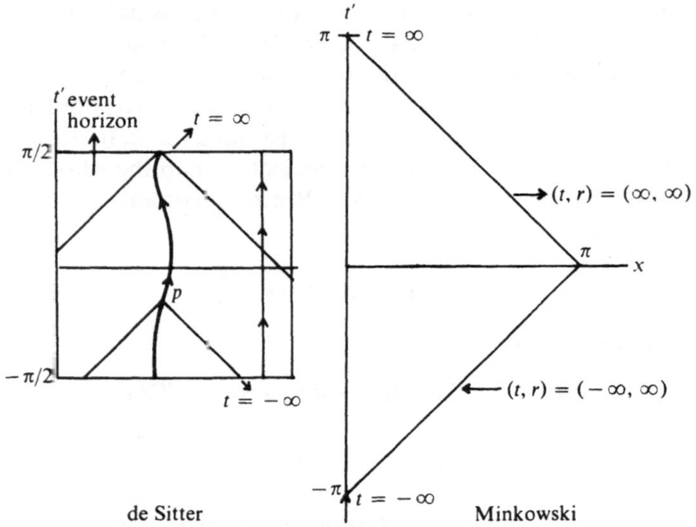


Figure 57 Penrose diagrams for flat and curved spaces

$K = -1$. **The Anti-de Sitter Universe** (4.3.21)

This can be represented as the hyperboloid

$$-x_0^2 + x_1^2 + x_2^2 + x_3^2 - x_4^2 = -1$$

in $\mathbb{R}^5$ with the metric

$$g = -dx_0^2 + dx_1^2 + dx_2^2 + dx_3^2 - dx_4^2$$

(Figure 58). We now observe that the intersection with $x_1 = x_2 = x_3 = 0$ is a closed, timelike geodesic. If the causal structure (cf. (I: 6.4.7)) is to be saved, it is necessary to pass to a covering surface, which can be mapped

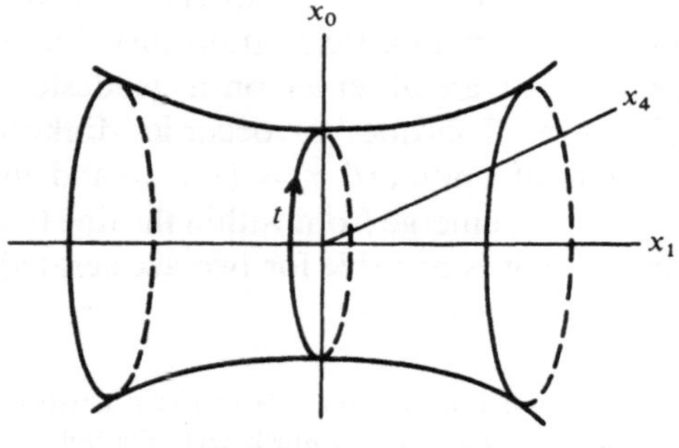


Figure 58 The anti-de Sitter universe

onto the region $\mathbb{R} \times \mathbb{R}^+ \times S^2$ for the variables $(t', \eta, \vartheta, \varphi)$ by setting

$$r = \sqrt{x_1^2 + x_2^2 + x_3^2} = \sinh \eta, \qquad x_0 = \cosh \eta \cos t',$$
$$x_4 = \cosh \eta \sin t'.$$

The metric is turned into

$$g = -\cosh^2 \eta \, dt'^2 + d\eta^2 + \sinh^2 \eta \, d\Omega^2. \tag{4.3.22}$$

Now the intersection where $t' = 0 = x_4$ is a Riemannian space of constant negative curvature. A sphere of radius $R$ in this space has the surface area $4\pi \sinh^2 R$, and proper time $ds = dt' \cosh R$ elapses much faster at large distances. To be able to compare this physical system with the earlier ones, we change to radial variables $\chi = 2 \arctan(\exp \eta) - \pi/2$. This makes the metric

$$g = \cosh^2 \eta(-dt'^2 + d\chi^2 + \sin^2 \chi \, d\Omega^2),$$
$$0 < \chi < \frac{\pi}{2}, \qquad -\infty < t' < \infty, \tag{4.3.23}$$

which is again of the form (4.3.19), except that the overall factor depends on $\chi$ rather than $t'$. The Penrose diagram is an infinite strip, which shows that in

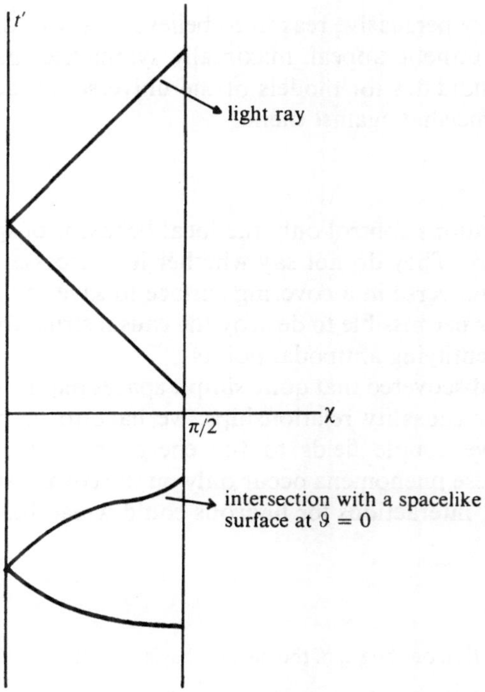


Figure 59 Penrose diagram for the covering surface of the anti-de Sitter universe

anti-de Sitter space there are no Cauchy surfaces at all. If we take the intersection with a spacelike surface at $\vartheta = 0$, then it must lie at an angle of less than 45° in the Penrose diagram. It is always possible to find a light ray along $\vartheta = 0$ which will never intersect this surface (Figure 59). Thus, in the covering surface, which has infinitely many sheets, time has the unusual property that for any infinite spacelike surface it is possible to find an event at a much later time having no causal connection with it.

**Einstein's Equations for Isotropic Spaces** (4.3.24)

If $R_{\alpha\beta} = Ke_{\alpha\beta}$, then $R_\beta = i_\alpha R^\alpha{}_\beta = 3Ke_\beta$, $R = i_\alpha R^\alpha = 12K$, and $R_\alpha - e_\alpha R/2 = -3Ke_\alpha$, so the energy-momentum tensor is $T_{\alpha\beta} = -\eta_{\alpha\beta}3K/8\pi\kappa$. In the phenomenological description (3.1.25; 3), this corresponds to a fluid at rest with energy density $= -$pressure $= 3K/8\pi\kappa$. A negative pressure is necessary to maintain the de Sitter universe ($K > 0$), and a negative energy is needed for the anti-de Sitter universe ($K < 0$), and of course the distributions must be homogeneous. Such unphysical $T_{\alpha\beta}$'s could be caused by

(a) a contribution $\Lambda{*}\mathbf{1}$ in $\mathscr{L}$ (cf. (4.2.14; 2));
(b) the vacuum expectation value of the energy-momentum tensor of fields, which can be $\sim \simeq \eta_{\alpha\beta}$ for reasons of invariance [40];
(c) extra terms in Einstein's equations [31].

There is no very persuasive reason to believe in any of these suggestions, so, despite their esthetic appeal, maximally symmetric spaces are not the front-running candidates for models of the universe; in fact, the empirical evidence goes somewhat against them.

**Remarks** (4.3.25)

1. Einstein's equations control only the local behavior of space, and not its global structure. They do not say whether it is necessary to enclose the anti-de Sitter universe in a covering surface to save the causal structure, or whether it is permissible to destroy the causal structure of the de Sitter universe by identifying antipodal points.
2. Since we have discovered that quite simple spaces require drastic revisions of our familiar causality relationships, we have to be prepared for the worst when we couple fields to the energy-momentum tensor. With gravitation, these phenomena occur only on the cosmic scale, but similar, though strong, interactions for hadrons could wreak havoc on the small scale.

**Problems** (4.3.26)

1. What is the condition on the $g_{lm}$ of the natural basis so that $v$ is a Killing vector field?
2. Show that the Poisson bracket of $p_\alpha v^\alpha(x)$ with the Hamiltonian $g^{\alpha\beta}(x)p_\alpha p_\beta$ vanishes iff $v$ is a Killing vector field (cf. (I: 5.1.10; 2)).

3. Suppose there exists a Killing vector field v. Find a coordinate system in which the $g_{lm}$ of the natural basis are independent of one of the coordinates.

4. It is always true in two dimensions that $R_{ik} = K(x)e_{ik}(x)$ (cf. (4.1.43)). Why is $K$ not necessarily constant?

5. Let $e^0 = dt$ in an orthogonal basis. Show that $e^0$ is a geodesic vector field (see 4.1.41.3).

6. Calculate $d{*F^\alpha}$ in the basis (4.3.9). What is the sign of the gravitational energy?

7. Calculate the integral $\int *1$ over a Riemannian space of constant positive curvature ($K = 1$).

## Solutions (4.3.27)

1. It is that $L_v\, dx^m = d(i_v\, dx^m) = dv^m = v^m{}_{,i}\, dx^i$ (cf. (I: 2.5.12; 5)); thus

$$0 = L_v g_{lm}\, dx^l\, dx^m = v^i g_{lm,i}\, dx^l\, dx^m + 2g_{lm} v^l{}_{,i}\, dx^i\, dx^m$$
$$\Rightarrow v^i g_{lm,i} + g_{im} v^i{}_{,l} + g_{il} v^i{}_{,m} = 0.$$

2. $\{p_\alpha v^\alpha(x),\ g^{\beta\gamma}(x)p_\beta p_\gamma\} = -2p_\alpha v^\alpha{}_{,\gamma} g^{\gamma\beta} p_\beta + v^\alpha g^{\beta\gamma}{}_{,\alpha} p_\beta p_\gamma = 0$ for Killing vector fields: $g_{\alpha\beta} g^{\beta\gamma} = \delta_\alpha{}^\gamma \Rightarrow g^{\beta\gamma}{}_{,\alpha} = -g^{\beta\sigma} g_{\sigma\rho,\alpha} g^{\rho\gamma}$, which brings us to the condition of Problem 1.

3. By Theorem (I: 2.3.12), it is always possible to find local coordinates for which the 1-component of v (in the natural basis) equals 1, and the others vanish. Then by Problem 1, $g_{lm,1} = 0$. If $v$ is timelike, this coordinate can be treated as time, making the metric constant and $v = \partial_t$.

4. There are no 3-forms in two dimensions, and hence there is no Bianchi identity.

5. $0 = de^0 = \omega^0{}_k \wedge e^k$ implies that $\omega^0{}_k \sim e^k$, for $k = 1, 2, 3$. Hence (cf. (4.1.11) and (4.1.17)),

$$D_{e_0} e^0 = -e^k(\omega^0{}_k | e) = 0.$$

6. The basis is orthogonal, hence $\varepsilon_{\alpha\beta\gamma\delta} = -\varepsilon^{\alpha\beta\gamma\delta}$, and

$$-\tfrac{1}{2}\varepsilon^{\alpha\beta\gamma\delta}\omega_{\beta\gamma} \wedge e_\delta = -\frac{K}{2}\varepsilon^{\alpha\beta\gamma\delta} x_\beta e_{\gamma\delta};$$

$$-\tfrac{1}{2}\varepsilon^{\alpha\beta\gamma\delta}\, d(\omega_{\beta\gamma} \wedge e_\delta) = -\frac{K}{2}\varepsilon^{\alpha\beta\gamma\delta}\left[e_{\beta\gamma\delta}\left(1 + \frac{Kr^2}{4}\right) - Kx_\beta x^\sigma e_{\sigma\gamma\delta}\right]$$
$$= -\frac{K}{2}\left[3!{*e^\alpha}\left(1 + \frac{Kr^2}{4}\right) - Kx_\beta x^\sigma 2!\, e_\sigma \wedge {*e^{\alpha\beta}}\right]$$
$$= -3K{*e^\alpha} + \frac{K^2 r^2}{4}{*e^\alpha} - K^2 x^\alpha x_\beta {*e^\beta} = 8\pi\kappa({*t^\alpha} + {*\mathcal{T}^\alpha}).$$

The term linear in $K$ is the same as the right side of Einstein's version, and the parts $\sim K^2$ are the gravitational contribution $\sim \omega \wedge \omega$, which is always negative for the energy density $T_{00}$: $-K^2((|\mathbf{x}|^2 + 3t^2)/4$. It is necessarily negative if $K > 0$, since the integral of the energy density over a compact space must be zero (1.3.35; 2), and the part $\sim K$ is positive.

7. $\int *1 = \int d^m x/(1 + r^2/4)^m = S_m \int_0^\infty dr\, r^{m-1}/(1 + r^2/4)^m$, where $S_m = 2\pi^{m/2}/\Gamma(m/2)$ is the surface area of the $m$-sphere. If $\beta = (1 + r^2/4)^{-1}$, then, recalling that $\Gamma(m)/\Gamma(m/2) = (2^{m-1}/\sqrt{\pi})\Gamma(\frac{1}{2}(m + 1))$,

$$\int *1 = S_m 2^{m-1} \int_0^\infty d\beta\, \beta^{m-2}\left(\frac{1}{\beta} - 1\right)^{(m-2)/2} = S_m 2^{m-1} \frac{\Gamma(m^2/2)}{\Gamma(m)}$$

$$= \frac{2\pi^{(m+1)/2}}{\Gamma\left(\frac{m+1}{2}\right)} = S_{m+1}.$$

Since the space is isometric to the $m + 1$-sphere, the calculation of the volume is correct.

## 4.4 Spaces with Maximally Symmetric Submanifolds

*The nonlinearity complicates Einstein's equations so much that the general solution lies beyond human capabilities. Explicit solutions can be written down only if the space is of sufficiently high symmetry.*

If the symmetry of a maximally symmetric space is reduced, the variety of possible curvature forms becomes great enough to conceivably correspond to physically acceptable energy-momentum currents. It would, however, lead too far afield if we tried to classify all the possibilities exhaustively, so instead our plan will be to investigate the physically relevant metrics that come up when the symmetry is reduced in successive stages.

**Spaces with Six Killing Vector Fields** (4.4.1)

The interesting case is the **Friedmann universe**, with six spacelike Killing vector fields, generating a group isomorphic to O(4). The trajectories of a point under the action of the group form a spacelike submanifold with six Killing vector fields, i.e., a three-dimensional Riemannian space of constant curvature. It is convenient to choose co-moving coordinates, for which the geodesic vector field furnishes the coordinate lines $x$ = constant perpendicular to this space, and the proper time on these geodesic lines is the time-coordinate $t$ (cf. (4.3.26; 5)). Writing $r^2 = |\mathbf{x}|^2$, the metric $g$ is of the type of a

**Robertson–Walker Metric** (4.4.2)

$$g = -dt^2 + R(t)^2 \frac{|d\mathbf{x}|^2}{(1 + Kr^2/4)^2}.$$

**Remarks** (4.4.3)

1. $R(t)$ is an as yet unspecified function of time, something like $\cosh(t)$ in de Sitter space (see (4.3.17), and note that $|d\mathbf{x}|^2(1 + Kr^2/4)^{-1}$ is $d\chi^2 + \sin^2 \chi\, d\Omega^2$ in the coordinates used there), or like $\exp(t)$ for $K = 0$ (see (4.3.18)). If $K > 0$, then the submanifold $t =$ constant has the finite volume $2\pi^2 R^3(t)/K^{3/2}$ (see (4.3.26; 7)).
2. If a new time-variable $t'$ such that $dt'/dt = 1/R(t)$ is introduced as in (4.3.17), then $g$ becomes conformally equivalent to that of de Sitter space, and consequently of Minkowski space,

$$g = R(t)^2\left(-dt'^2 + \frac{|d\mathbf{x}|^2}{(1 + Kr^2/4)^2}\right)$$

(cf. (4.3.19)). It frequently happens that $t'$ takes values only in a finite interval $t_0 < t < t_1$, as in de Sitter space. In that case the causality relationships turn out to be similar to those discussed in (4.3.16), and in particular there are particle and event horizons.

**The Curvature Forms of the Friedmann Universe** (4.4.4)

Let Greek indices run from 0 to 3 and Roman ones from 1 to 3. If we write the orthogonal basis as

$$e^\alpha = \left(dt, \frac{R(t)\, dx^a}{1 + Kr^2/4}\right),$$

then

$$de^\alpha = \left(0, \frac{\dot{R}}{R} e^{0a} - \frac{K}{2R} x_b e^{ba}\right),$$

and therefore

$$\omega^a{}_0 = \omega^0{}_a = \frac{\dot{R}}{R} e^a, \qquad \omega_{ab} = \frac{K}{2R}(x_a e_b - x_b e_a). \tag{4.4.5}$$

Consequently,

$$d\omega^{0a} = \frac{\ddot{R}}{R} e^{0a} - \frac{K\dot{R}}{2R^2} x_b e^{ba},$$

$$d\omega^{ab} = \frac{K}{R^2} e^{ab}\left(1 + \frac{Kr^2}{4}\right) + \frac{K^2}{4R^2}(x^a x_c e^{bc} - x^b x_c e^{ac})$$

and a similar calculation to that of (4.3.8) leads to

$$R^{0a} = \frac{\ddot{R}}{R} e^{0a}, \qquad R^{ab} = \frac{K + \dot{R}^2}{R^2} e^{ab}. \tag{4.4.6}$$

The contracted quantities become

$$R^0 = 3\frac{\ddot{R}}{R}e^0, \qquad R^a = \left(\frac{\ddot{R}}{R} + 2\frac{K + \dot{R}^2}{R^2}\right)e^a,$$
$$R = 6\left(\frac{K + \dot{R}^2(t)}{R^2(t)} + \frac{\ddot{R}(t)}{R(t)}\right). \tag{4.4.7}$$

**Remarks** (4.4.8)

1. If $R(t) =$ constant, then the curvature is constant only in spatial directions,

$$R^{ab} = \frac{K}{R^2}e^{ab}, \qquad R^{0a} = 0.$$

The time-independence of $R$ gives rise to an $R^{0a}$ and contributes to $R^{ab}$.
2. A comparison of (4.4.7) and (4.1.44; 3) reveals that the Weyl forms are now zero, as required by the conformal equivalence to Minkowski space.

**Einstein's Equations in the Classical Form** (4.4.9)

In order to satisfy (4.2.20(a)), the energy-momentum forms of matter must be $e^\alpha \times$ (some function of $t$), because of (4.4.7). Therefore the energy-momentum tensor of matter is necessarily diagonal, and in the spirit of the phenomenological description (3.1.25; 3) we set $T_{00} = \rho =$ energy density and $T_{jj} = p =$ pressure. Einstein's equations then imply that

$$3\frac{\dot{R}^2 + K}{R^2} = 8\pi\kappa\rho$$
$$-\frac{2\ddot{R}}{R} - \frac{\dot{R}^2 + K}{R^2} = 8\pi\kappa p. \tag{4.4.10}$$

**Remarks** (4.4.11)

1. The Bianchi identity (4.2.32) relates $\rho$ and $p$ to $R$, and of course the same relationship follows from (4.4.10). It implies that

$$d^*\mathscr{T}_0 = d(\rho^*e^0) = d\rho \wedge {}^*e^0 + \rho\, d^*e^0$$
$$= -\omega^0{}_j \wedge {}^*\mathscr{T}^j = -p\omega^0{}_j \wedge {}^*e^j,$$

and therefore

$$d\rho \wedge {}^*e^0 = (\rho - p)\omega^0{}_j \wedge {}^*e^j, \text{ i.e., } \dot{\rho} = 3\frac{\dot{R}}{R}(\rho - p).$$

In the form

$$p = -\frac{\frac{d}{dt}(\rho R^3)}{\frac{d}{dt}R^3}$$

this has the interpretation that pressure = −(rate of change of energy)/(rate of change of volume). It is noteworthy that gravity does not appear in the total energy in co-moving coordinates.

2. The static situation $\dot{R} = 0$ requires a negative pressure if $K > 0$, as in de Sitter space, and a negative energy if $K < 0$. This originally induced Einstein to include the cosmological term $\Lambda^*\mathbf{1}$ in his action principle. Friedmann later discovered the solution bearing his name, and the modern tendency is to accept the dynamical equations as is. In order to illustrate their significance better, let us examine

**Einstein's Equations in Landau and Lifshitz's Form** (4.4.12)

In (4.2.20(a)) and (4.2.21(b)) the energy-momentum forms for matter and gravitation were represented as exterior differentials of 2-forms. Since it is not yet apparent what the gravitational energy is, let us track it down. To this end we calculate the restriction of the exterior differential of the 2-form

$$-\tfrac{1}{2}\varepsilon^{0bcd}\omega_{bc} \wedge e_d = -\frac{K}{2R}\varepsilon^{0bcd}x_b e_{cd},$$

$$-\tfrac{1}{2}\varepsilon^{0bcd}\, d(\omega_{bc} \wedge e_d)\Big|_{t=\mathrm{const}} = -\frac{K}{2R^2}\varepsilon^{0bcd}\left[e_{bcd}\left(1 + \frac{Kr^2}{4}\right) - Kx_b x_m e_{mcd}\right]\Bigg|_{t=\mathrm{const}}$$

$$= \frac{3K - K^2r^2/4}{R^2}\,{}^*e^0\Big|_{t=\mathrm{const}} = 8\pi\kappa({}^*\mathscr{T}^0 + {}^*t^0) \qquad (4.4.13)$$

to $t$ = constant.

**Remarks** (4.4.14)

1. Equation (4.4.13) states that

$$8\pi\kappa \cdot \text{the total energy density} = \frac{3K - K^2r^2/4}{R^2}$$

so $8\pi\kappa \cdot$ the gravitational energy density equals $8\pi\kappa \cdot$(total energy density $- \rho) = -(K^2r^2 + 12\dot{R}^2)/4R^2$. In the static situation ($\dot{R} = 0$) this is the energy density of a homogeneous mass distribution $\rho$ according to Newton's theory, constructed just as in electrodynamics, but with the other sign and $4\pi\kappa \to e^2$. If we chose units where $R = 1$, then

$$\Delta V = e\rho = \frac{3K}{2e}, \qquad V = \frac{Kr^2}{4e},$$

$$\nabla V = \frac{\mathbf{x}K}{2e}, \qquad -\tfrac{1}{2}|\nabla V|^2 = -\frac{1}{2e^2}\frac{K^2r^2}{4},$$

and thus, as promised in §1.3, $-K^2r^2/4R^2$ corresponds exactly to $8\pi\kappa \cdot$ the Newtonian gravitational energy in the most nearly Cartesian coordinates. In the dynamical case, there is also a contribution $\dot{R}^2$.

2. The gravitational energy is exactly large enough so that if $K > 0$, the integral of the total energy over the whole space is zero:

$$3K \int_0^\infty \frac{r^2\,dr}{(1 + Kr^2/4)^3} = \frac{K^2}{4} \int_0^\infty \frac{r^4\,dr}{(1 + Kr^2/4)^3}.$$

3. If we write the first of Equations (4.4.10) as

$$\frac{\dot{R}^2}{2} - 4\pi\kappa\rho\frac{R^2}{3} = -\frac{K}{2} = \text{constant},$$

then it has the form of the conservation equation for the energy of a (nonrelativistic) particle with coordinate $R$ and speed $\dot{R}$. The kinetic energy plus the potential energy is constant, where the latter is taken as the potential energy on the surface of a ball of radius $R$ and homogeneous density $\rho$:

$$-V(r) = \frac{4\pi R^3\rho}{3r}\Theta(r - R) + \frac{4\pi R^2}{2}\left(-1 + \frac{r^2}{3R^2}\right)\Theta(R - r).$$

Note that $V(0) - V(R)$ is smaller than $V(R) - V(\infty)$ by a factor of 2.

In the following section, we shall return to (4.4.10) when we study the collapse of stars, and we shall solve them for selected pressure-density ratios. A detailed discussion of their importance for cosmology may be found in [8], [9], [10].

**Spaces with Five Killing Vector Fields** (4.4.15)

In (I: §5.8) we learned that there are in general 5 constants of the motion in the field of a gravitational wave, which are linear in the momentum components. This motion accordingly allows 5 Killing vector fields, and the invariance group they generate is isomorphic to the invariance group of a gravitational wave, and consequently (I: 5.8.4; 1) to that of an electromagnetic wave (I: 5.5.3). The metric given in (I: 5.8.1) is a special case of what we examined in (4.2.13), where it depended on only one coordinate $u$. The interesting situation is where discontinuities are allowed, in which $du = n_\alpha e^\alpha$, $0 = n_\alpha n^\alpha \equiv n^2$. If the coordinates are chosen so that

$$n = (-1, 1, 0, 0), \qquad u = x - t, \tag{4.4.16}$$

then the metric (I: 5.8.1), with the required invariance structure, is of the form

$$g = -dt^2 + dx^2 + p(u)^2\,dy^2 + q(u)^2\,dz^2. \tag{4.4.17}$$

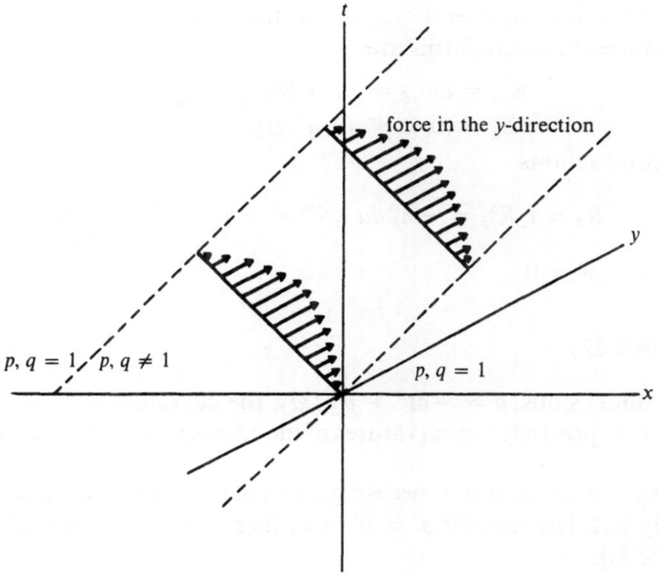


Figure 60 Schematic drawing of a gravitational pulse

**Remarks** (4.4.18)

1. The form of $g$ is that of a metric in a co-moving coordinate system (cf. (4.3.26; 5)). Therefore the coordinate lines $\mathbf{x}$ = constant are possible particle trajectories.
2. The gravitational field described by $g$ is a kind of transverse wave, which alters the distance perpendicular to the direction of propagation between particle trajectories. If $p - 1$ and $q - 1$ have compact support, then the pulse looks schematically as shown in Figure 60. For example, in the solution given in (I: 5.8.7) $p \leq 1$ and $q \leq 1$, so the trajectories draw coser together in the $y$-direction and spread apart in the $z$-direction, as with a quadrupole field. This effect would not be measured by measuring rods, as they would be stretched in the same way. However, the deformation would be observable by measuring the time-delay of a reflected light signal.

**The Curvature Forms** (4.4.19)

If the orthogonal basis

$$e^\alpha = (dt, dx, p\, dy, q\, dz),$$
$$de^\alpha = (0, 0, p'\, du \wedge dy, q'\, du \wedge dz)$$

is used, then the affine connections become

$$\omega_{\alpha\beta} = S_\alpha n_\beta - S_\beta n_\alpha, \qquad S_\alpha = (0, 0, p'\, dy, q'\, dz),$$

as in (4.2.13). Since $n_\alpha S^\alpha = 0 = n_\alpha n^\alpha$, we find that $\omega_{\alpha\beta} \wedge \omega^\beta{}_\gamma = 0$ for all $\alpha$ and $\beta$, and the curvature forms are

$$R_{\alpha\beta} = d\omega_{\alpha\beta} = du \wedge (S'_\alpha n_\beta - S'_\beta n_\alpha),$$
$$S'_\alpha = (0, 0, p''\, dy, q''\, dz), \tag{4.4.20}$$

with the contractions

$$R_\beta = i_\alpha R^\alpha{}_\beta = -n_\beta\, du\, i_\alpha S'^\alpha = -n_\beta\, du\left(\frac{p''}{p} + \frac{q''}{q}\right),$$
$$R = 0. \tag{4.4.21}$$

**Remarks** (4.4.22)

1. In two dimensions, $g = -dt^2 + p^2(t)dy$, the curvature is $p''\, dt \wedge dy$, and thus $R_{\alpha\beta}$ are precisely the curvatures of the corresponding two-dimensional surfaces.
2. The Weyl forms are not of necessity zero, and the space need not be conformally flat. However, if $p'' = q'' = 0$, then it is always flat (cf. Remark (I: 5.8.8; 2)).

**Einstein's Equations in the Classical Form** (4.4.23)

By Equations (4.4.21),

$$R_\alpha - \tfrac{1}{2}e_\alpha R = -n_\alpha\, du\left(\frac{p''}{p} + \frac{q''}{q}\right),$$

and thus an energy current $\sim du$ could well be a source of the gravitational pulse, as long as it is accompanied by an equally large current of the 1-component of the momentum while the other components vanish. Such an energy-momentum current could be produced by an electromagnetic wave. If

$$F = f(u) \wedge du \quad \text{with} \quad \langle f \mid du\rangle = \langle du \mid du\rangle = 0,$$

then $F \wedge {}^*F = 0$, and the energy-momentum forms (1.3.22) are

$$\begin{aligned}\mathscr{T}_\alpha &= {}^*((i_\alpha F) \wedge {}^*F) = -i_{i_\alpha F}F = -(f_\alpha i_{du} - n_\alpha i_f) f \wedge du \\ &= n_\alpha\, du \langle f \mid f\rangle.\end{aligned}$$

Therefore Einstein's equations imply that

$$\frac{p''}{p} + \frac{q''}{q} + 8\pi\kappa\langle f \mid f\rangle = 0.$$

To solve this equation, set $p = L\exp(\beta)$ and $q = L\exp(-\beta)$. Then

$$\frac{L''}{L} + \beta'^2 + 4\pi\kappa\langle f \mid f\rangle = 0, \tag{4.4.24}$$

$$\beta(u) = \int_0^u du'\left(-\frac{L''(u')}{L(u')} - 4\pi\kappa\langle f(u') \mid f(u')\rangle\right)^{1/2} \tag{4.4.25}$$

**Remarks** (4.4.26)

1. In the approximation linear in $\kappa$,

$$L(u) = 1 - 4\pi\kappa \int_0^u du' \int_0^{u'} du'' \langle f(u'') | f(u'') \rangle,$$

while $\beta$ remains arbitrary to first order. If the equations are homogeneous ($f = 0$), this provides us with a solution that could be used for $\varphi^{\text{in}}$ in (4.2.18), because to first order, $\beta = \varphi_{22} = -\varphi_{33}$, and the other $\varphi$'s are zero, so that $0 = \varphi_\alpha{}^\alpha = \varphi_{\beta\alpha,}{}^\alpha = \varphi_{\alpha\alpha,\rho}{}^\rho$.
2. If $L > 0$, then it is a concave function, because $\langle f | f \rangle > 0$, and $f$ and $\beta'$ contribute in similar ways to the curvature $L''$ of the function $L(u)$. In this situation, the trajectories of particles are focused in the $y - z$-plane. This is an effect of the gravitational field produced by the electromagnetic or gravitational wave.
3. If $L'$ is ever negative, then $L$ must sooner or later have a zero. This singularity in the metric might not be a genuine one, but may only indicate that the gravitational wave has disrupted the coordinate system. The space might appear as Minkowski space in some other chart (cf. (I: 5.8.8; 2)), as soon as the wave has passed.

**Einstein's Equations in Landau and Lifshitz's Form** (4.4.27)

It remains to find out how well the interpretation (4.4.26; 2) of $\beta'^2$ as an energy density accords with the formulations (4.2.21(b)). In the latter formulation gravity contributes

$$\ell_\alpha = \frac{-1}{8\pi\kappa}\frac{1}{2}\omega_{\beta\gamma} \wedge \omega_{\alpha\rho} \wedge {}^*e^{\beta\gamma\rho}$$

to the energy, and the other summand is zero. Substitution from (4.4.19) yields

$$\begin{aligned} 8\pi\kappa\ell_\alpha &= {}^*(S_\beta n_\gamma \wedge S_\rho n_\alpha \wedge {}^*e^{\rho\beta\gamma}) \\ &= -n_\alpha i_{S_\beta} i_{S_\rho}(e^{\rho\beta} \wedge du) = -n_\alpha\, du\, 2\frac{p'q'}{pq} \qquad (4.4.28) \\ &= 2n_\alpha\, du\left(\beta'^2 - \left(\frac{L'}{L}\right)^2\right), \end{aligned}$$

so that an additional negative term $-(L'/L)^2$ occurs along with $\beta'^2$.

**Remarks** (4.4.29)

1. In the linear approximation (4.4.26; 1), $L' = 0$, $2\dot{\beta} = \dot{g}_{22} = -\dot{g}_{33}$, and the result is that

$$\text{energy density} = \text{momentum density} = \frac{(\dot{g}_{22} - \dot{g}_{33})^2}{64\pi\kappa}.$$

2. Provided that $L$ has a slowly varying amplitude in comparison with $\beta$, (4.2.21(b)) states that $\dot{\beta}^2$ creates gravity like any other kind of energy, and that the energy of a gravitational wave is positive.
3. Fictitious energies associated with fictitious forces also appear in (4.4.27); their origin is that the $t_\alpha$ do not vanish even in flat space ($p'' = q'' = 0$).
4. The speeding-up of a double star with a short period seems to be consistent with the energy loss due to gravitational radiation as calculated with this formula.

**Spaces with 4 Killing Vector Fields** (4.4.30)

We shall consider the spaces that are the counterpart to the problem of a central force in mechanics. The energy and angular momentum will correspond to the operations of time-displacement and rotations that leave $g$ invariant. In the polar coordinates for $M = \mathbb{R} \times \mathbb{R}^+ \times S^2$, the $g_{\alpha\beta}$ depend only on $r = |\mathbf{x}|$, and the metric can be written as

$$g = -dt^2 \exp(2a(r)) + dr^2 \exp(2b(r)) + r^2(d\vartheta^2 + \sin^2 \vartheta\, d\varphi^2).$$

**Remarks** (4.4.31)

1. If Einstein's equations in vacuo hold, then it can be shown that the existence of the timelike Killing vector field follows simply from the spherical symmetry (Birkhoff's theorem, Problem 5).
2. In co-moving coordinates, the $g_{\alpha\beta}$ are in general time-dependent.

**The Curvature Forms** (4.4.32)

In the orthogonal basis

$$e^\alpha = (e^a\, dt, e^b\, dr, r\, d\vartheta, r \sin \vartheta\, d\varphi),$$

$$de^\alpha = (a' e^a\, dr \wedge dt, 0, dr \wedge d\vartheta, \sin \vartheta\, dr \wedge d\varphi + r \cos \vartheta\, d\vartheta \wedge d\varphi)$$

the affine connections turn out to be

$$\omega^{\alpha}{}_{\beta} = \begin{array}{|c|c|c|} \hline e^{a-b}a'\,dt & 0 & 0 \\ \hline & -e^{-b}\,d\vartheta & -e^{-b}\sin\vartheta\,d\varphi \\ \hline & & -\cos\vartheta\,d\varphi \\ \hline \end{array} \tag{4.4.33}$$

(since $\omega_{\alpha\beta} = -\omega_{\beta\alpha}$, we write them only for $\alpha < \beta$). They make

$$d\omega^{\alpha}{}_{\beta} =$$

$$\begin{array}{|c|c|c|} \hline e^{a-b}(a'' + a'(a' - b'))dr \wedge dt & 0 & 0 \\ \hline & e^{-b}b'\,dr \wedge d\vartheta & e^{-b}(-\cos\vartheta\,d\vartheta + \sin\vartheta\,b'\,dr) \wedge d\varphi \\ \hline & & \sin\vartheta\,d\vartheta \wedge d\varphi \\ \hline \end{array}$$

$$\omega^{\alpha}{}_{\gamma} \wedge \omega^{\gamma}{}_{\beta} = \begin{array}{|c|c|c|} \hline 0 & -e^{a-2b}a'\,dt \wedge d\vartheta & -e^{a-2b}a'\sin\vartheta\,dt \wedge d\vartheta \\ \hline & 0 & e^{-b}\cos\vartheta\,d\vartheta \wedge d\varphi \\ \hline & & e^{-2b}\sin\vartheta\,d\vartheta \wedge d\varphi \\ \hline \end{array}$$

The term $\sim d\vartheta \wedge d\varphi$ cancels out of $d\omega^{1}{}_{3}$, and the $R^{\alpha}{}_{\beta}$ become proportional to $e^{\alpha}{}_{\beta}$:

$$R^{\alpha}{}_{\beta} = \begin{array}{|c|c|c|} \hline (a'b' - a'' - a'^2)e^{-2b}e^{0}{}_{1} & -\dfrac{a'e^{-2b}}{r}e^{0}{}_{2} & -\dfrac{a'e^{-2b}}{r}e^{0}{}_{3} \\ \hline & \dfrac{b'e^{-2b}}{r}e^{1}{}_{2} & \dfrac{b'e^{-2b}}{r}e^{1}{}_{3} \\ \hline & & \dfrac{(1 - e^{-2b})}{r^2}e^{2}{}_{3} \\ \hline \end{array} \tag{4.4.34}$$

**Einstein's Equations** (4.4.35)

Since the remaining symmetry still suffices to make $R^\alpha{}_\beta$ of the form $R^\alpha{}_\beta = K^\alpha{}_\beta e^\alpha{}_\beta$ (no sum), the energy-momentum forms $\mathscr{T}^\alpha$ (letting this embrace everything coupled to gravity) must likewise be $\sim e^\alpha$. The coefficients are $-\sum_{\substack{\beta<\gamma \\ \beta\neq\alpha,\,\gamma\neq\alpha}} K_{\beta\gamma}$, and depend only on $r$:

$$\begin{aligned}
\left(-\frac{1-e^{-2b}}{r^2} - 2b'\frac{e^{-2b}}{r}\right)e^0 &= 8\pi\kappa\mathscr{T}^0, \\
\left(-\frac{1-e^{-2b}}{r^2} + 2a'\frac{e^{-2b}}{r}\right)e^1 &= 8\pi\kappa\mathscr{T}^1, \\
\left(a'' + a'^2 - a'b' + \frac{a'-b'}{r}\right)e^{-2b}e^2 &= 8\pi\kappa\mathscr{T}^2, \\
\left(a'' + a'^2 - a'b' + \frac{a'-b'}{r}\right)e^{-2b}e^3 &= 8\pi\kappa\mathscr{T}^3,
\end{aligned} \qquad (4.4.36)$$

**Remarks** (4.4.37)

1. Because of the spherical symmetry, there is an invariance under $2 \leftrightarrow 3$, and more specifically $\mathscr{T}^2$ and $\mathscr{T}^3$ have the same factor in front of $e^2$ and respectively $e^3$.
2. $\mathscr{T}^\alpha$ is of the form $c^\alpha e^\alpha$ (no sum), and the contracted Bianchi identity (4.2.32) subjects the coefficients $c^\alpha$ to the equation

$$dc^\alpha \wedge {*e^\alpha} = \sum_\beta \omega^\alpha{}_\beta \wedge {*e^\beta}(c^\alpha - c^\beta).$$

3. The $\mathscr{T}^\alpha$ are written with the basis of (4.4.32), and thus the $\mathscr{T}^j, j = 1, 2, 3$, are obtained from Cartesian energy-momentum forms by local rotations (cf. (1.3.26)). If, for instance, $\mathscr{T}^j = pe^j$ in the Cartesian basis, then it is also true in this basis.

**Special Cases** (4.4.38)

1. If we make the phenomenological assumption that

$$\mathscr{T}_\alpha = (\rho e^0, pe^j)$$

(cf. (3.1.25; 3)), where $\rho$ and $p$ are not too singular and decrease sufficiently fast as $r \to \infty$, then the first of Equations (4.4.36) is solved by

$$e^{-2b} = 1 - \frac{8\pi\kappa}{r}\int_0^r dr'\, r'^2\rho(r') \equiv 1 - \frac{2\kappa M(r)}{r}, \qquad (4.4.39)$$

and the second one determines $a$ once $b$ is known:

$$a = -b - 4\pi\kappa \int_r^\infty dr'\, r' e^{2b(r')}(\rho(r') + p(r')). \tag{4.4.40}$$

The last two identical equations relate $\rho$ to $p$, and are equivalent to requiring the contracted Bianchi identity. This subject will be pursued in the following section, for the $\mathscr{T}^\alpha$ treated here it is satisfied.

If it happens that $\rho(r) = p(r) = 0$ for all $r > r_1$, then the metric in the region where $r > r_1$ is the

**Schwarzschild Metric** (4.4.41)

$$g = -\left(1 - \frac{r_0}{r}\right)dt^2 + \frac{dr^2}{1 - \frac{r_0}{r}} + r^2\, d\Omega^2, \qquad r_0 = 8\pi\kappa \int_0^\infty r^2\, dr\, \rho(r).$$

On the other hand, if $\rho = \text{constant} = -p$, then we return to the situation of (4.3.23) and obtain (cf. Problem 4) a

**Static Form of the de Sitter Metric** (4.4.42)

$$g = -(1 - Kr^2)dt^2 + \frac{dr^2}{1 - Kr^2} + r^2\, d\Omega^2, \qquad K = \frac{8\pi\kappa}{3}\rho.$$

2. Equations (4.4.36) allow the pressure in the radial direction to differ from the pressure in the $\vartheta$ and $\varphi$ directions. This could happen for the Coulomb field of a point particle, for which the energy-momentum forms can be calculated as

$$\mathscr{T}^\alpha = \frac{e^2}{2r^4}[e^0, -e^1, e^2, e^3]. \tag{4.4.43}$$

If we set $\exp(2a) = \exp(-2b) = \psi(r)/r$ in (4.4.36), then Einstein's equations read:

$$\left(\frac{1 - \psi'}{r^2} e^0, \frac{-1 + \psi'}{r^2} e^1, \frac{\psi''}{2r} e^2, \frac{\psi''}{2r} e^3\right) = 8\pi\kappa\mathscr{T}_\alpha.$$

If $\psi = r - r_0 + 4\pi\kappa e^2/r$, then this simply reproduces (4.4.43), and the resulting metric is called the

**Reissner–Nordstrøm Metric** (4.4.44)

$$g = -\left(1 - \frac{r_0}{r} + \frac{4\pi\kappa e^2}{r^2}\right)dt^2 + \frac{dr^2}{1 - \frac{r_0}{r} + \frac{4\pi\kappa e^2}{r^2}} + r^2\, d\Omega^2.$$

**Remarks** (4.4.45)

1. In the linear approximation, and with $|p| \ll \rho$, Equation (4.4.40) becomes

$$a = -\frac{g_{00} - 1}{2} = -4\pi\kappa\left[\frac{1}{r}\int_0^r dr'\,\rho(r')r'^2 + \int_r^\infty dr'\,\rho(r')r'\right],$$

which is the Newtonian potential of the spherically symmetric energy density $\rho$, as it must be on account of (4.2.30; 2).

2. The pressure contributes to (4.4.40) as the density contributes to $g_{00}$. Hence the negative pressure $p = -\rho$ of the de Sitter universe in fact makes $a = -$the Newtonian potential, because for constant densities,

$$\frac{1}{r}\int_0^r r'^2\,dr'\,\rho = -\frac{1}{2}\int_r^R r'\,dr'\,\rho + \text{constant}.$$

3. If $\rho$ is more singular than $r^{-3}$ at $r = 0$, then one can write $\int_r^\infty$ instead of $-\int_0^r$ in (4.4.39). This is the case with the Reissner solution (4.4.44), so the positive field energy contributes with a reversed sign to the gravitational potential. The interpretation is that $M = M(\infty)$ represents the total energy, and the potential

$$-\frac{M\kappa}{r} + \frac{4\pi e^2\kappa}{2r^2}$$

shows that as one approaches the origin, part of the energy density is left behind, and the potential is effectively decreased compared with its asymptotic value, $-M\kappa/r$. If $M < \infty$, then it follows that the "naked mass" at the origin must be $-\infty$, since the electromagnetic mass

$$\frac{4\pi}{2}\int_0^\infty \frac{dr}{r^4}$$

is divergent. This has the paradoxical consequence that gravity is repulsive at short distances. Once again, the infinite electrostatic self-energy of a point charge is causing trouble.

4. The basis of (4.4.32) is less suitable for a discussion of gravitational energy using the version (4.2.21(b)) of Einstein's equations, since it is possible to simulate a gravitational $t_\alpha$ even in flat space, with polar coordinates. In Problem 6 the gravitational energy is discussed in the maximally Cartesian coordinates (see (I: 5.7.17; 4)). It turns out that as long as there is asymptotically a Schwarzschild metric, the total energy including the gravitational energy is $M$. Note that

$$\int_{t=0} {}^*\mathscr{T}^0 = 4\pi\int_0^\infty dr\,r^2 e^{b(r)}\rho(r) \neq M = 4\pi\int_0^\infty dr\,r^2\rho(r).$$

5. In Problem 7 the energy density of gravitation is calculated as $-\kappa M^2/8\pi r^4$ in these coordinates. This is equal to the negative of the energy density

$(e/4\pi r^2)^2/2$ of a Coulomb field, where $e^2$ is replaced with $4\pi\kappa M^2$, analogously to Remark (4.4.14; 1). Since $M$ is the integral of the total energy density, there can be an everywhere regular solution with $M > 0$ only if $\rho > 0$ counterbalances the negative gravitational energy. The increase of the Schwarzschild potential compared with the Newtonian potential at small $r$ can be interpreted as the field produced by the negative gravitational energy.

**Properties of Spherically Symmetric Fields** (4.4.46)

1. The Geometric Interpretation of the Spatial Metric. The restriction of the metric to a plane passing through the origin,

$$g\Big|_{\substack{t=\text{const}\\ \varphi=\text{const}}} = e^{2b}\,dr^2 + r^2\,d\vartheta^2,$$

is the metric on a surface of rotation in $\mathbb{R}^3$. If it is written in cylindrical coordinates as $z(r)$, then

$$dz^2 + dr^2 + r^2\,d\vartheta^2 = dr^2(1 + z'^2) + r^2\,d\vartheta^2 = e^{2b}\,dr^2 + r^2\,d\vartheta^2,$$

or, using (4.4.39),

$$z' = \sqrt{1 - e^{2b}} = \sqrt{\frac{2\kappa M(r)}{r - 2\kappa M(r)}},$$

$$z(r) = \int_0^r dr' \sqrt{\frac{2\kappa M(r')}{r' - 2\kappa M(r')}}.$$

As a consequence, the Schwarzschild metric $M(r)$ = constant gives the intersection $\varphi$ = constant the geometry of a paraboloid of revolution,

$$z(r) = \sqrt{4r_0}\sqrt{r - r_0}$$

(Figure 61). The metric is singular at $r = 2\kappa M(r)$. The paraboloid of the Schwarzschild metric can be extended beyond that point, but, if so, $r$ is no longer a monotonic function of $z$.

2. The Causal Structure. The Schwarzschild metric was extended beyond $r = r_0$ in (I: 5.7.2; 5), with the aid of the coordinates

$$u = \sqrt{\frac{r}{r_0} - 1}\,\exp\left(\frac{r}{2r_0}\right)\cosh\left(\frac{t}{2r_0}\right)$$
$$v = \sqrt{\frac{r}{r_0} - 1}\,\exp\left(\frac{r}{2r_0}\right)\sinh\left(\frac{t}{2r_0}\right), \qquad u^2 - v^2 > -1, \tag{4.4.47}$$

in which

$$g = 4r_0^3\exp\left(-\frac{r}{r_0}\right)\frac{du^2 - dv^2}{r} + r^2\,d\Omega^2. \tag{4.4.48}$$

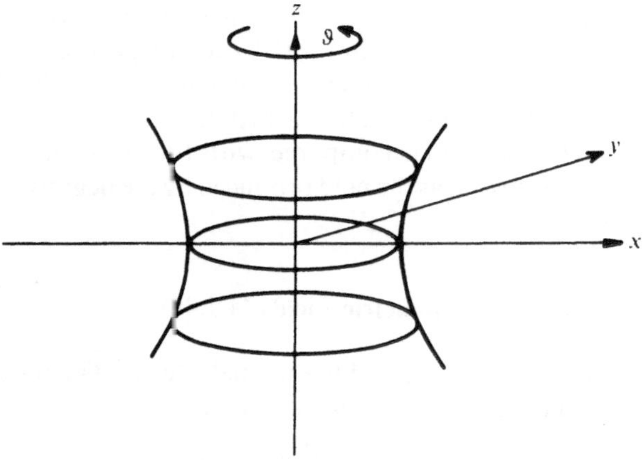


Figure 61 A choice of the metric on the surface $t$ = constant, $\varphi$ = constant

There remains a singularity at $r = 0$, which is now a spacelike hypersurface $u^2 - v^2 = -1$. The quickest way to understand the causal relationships is to draw the Penrose diagram that results from using the coordinates

$$v + u = \tan\left(\frac{\psi + \xi}{2}\right), \qquad v - u = \tan\left(\frac{\psi - \xi}{2}\right),$$
$$-\pi < \psi \pm \xi < \pi, \qquad -\frac{\pi}{2} < \psi < \frac{\pi}{2}. \tag{4.4.49}$$

The metric

$$g = A^2(-d\psi^2 + d\xi^2 + R^2\,d\Omega^2),$$
$$A = \frac{r_0}{\sqrt{r}} e^{-r/2r_0} \cos^{-1}\tfrac{1}{2}(\psi + \xi)\cos^{-1}\tfrac{1}{2}(\psi + \xi),$$
$$R = \frac{r}{A},$$

reveals that timelike lines run in the $\psi$-$\xi$-plane at angles of at least 45°, and radial light rays run at 45°. The region covered by the new time and radial coordinates looks as shown below. Since the boundary contains the spacelike piece where $r = 0$, there is a horizon. Although $r = r_0$ is not a singularity, it is the event horizon for all trajectories that remain in Region I, where $r > r_0$. Regions II and III are invisible from Region I, which is itself invisible from Regions III and IV. Although nothing exceptional happens locally at $r = r_0$, the surface $r = r_0$ has a global significance.

The Reissner metric (4.4.44) becomes for some $r \in (0, \infty)$ singular if $4\pi\kappa e^2 < M^2\kappa^2$. For hadrons this inequality is far from being satisfied since in natural units $e^2 = \frac{1}{137} \geq (\kappa/4\pi)M^2 \equiv$ (Planck length/Compton wave length)$^2 = (10^{-33}\,\text{cm}/10^{-14}\,\text{cm})^2$. Thus no horizon prevents people from

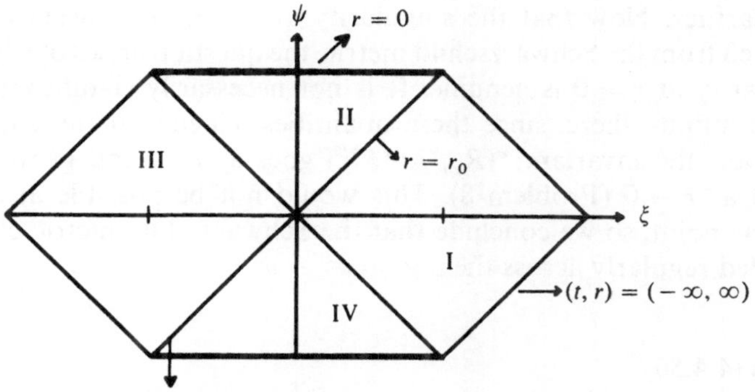


Figure 62 Penrose diagram for the Schwarzschild solution

starting at the "naked singularity" at $r = 0$. The "cosmic censorship hypothesis" conjectures that naked singularities do not develop in reality. Indeed $4\pi e^2 \geq M^2\kappa$ means that the Coulomb repulsion of charged matter would be stronger than the gravitational attraction thus preventing the collapse to a singularity. If $M^2\kappa^2 > 4\pi\kappa e^2$, then the singularity of (4.4.44) at small $r$ lies only in the choice of coordinates; with other coordinates it would be possible to continue to $r = 0$. In that case, the repulsive nature of gravity makes $r = 0$ a time-like line. The appropriate Penrose diagram. Figure 63, thus extends in the timelike direction to infinity, as with the anti-de Sitter universe ([9], p. 921). Hence there are again no global Cauchy surfaces, but instead there is a bizarre possibility that one might crawl through the wormhole bounded by $r = 0$ into another universe just like ours (I′ in Figure 63).

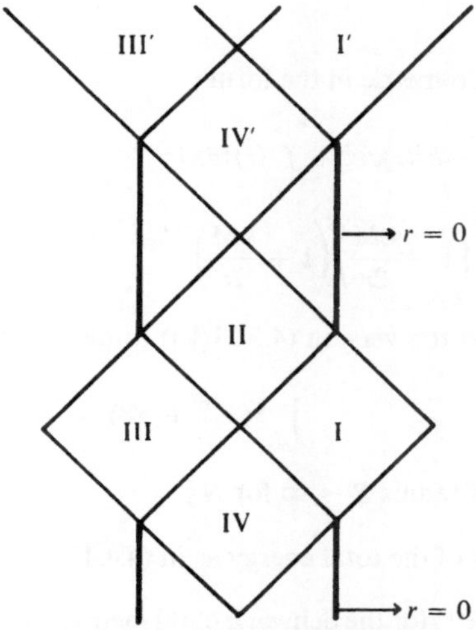


Figure 63 Penrose diagram for the Reissner solution

3. Singularities. Now that the singularity at $r = r_0$ has successfully been removed from the Schwarzschild metric, the question arises of whether the singularity at $r = 0$ is genuine. It is not necessarily significant that the $g_{ik}$ are infinite there, since these quantities depend on the coordinates. However, the invariant $*(R_{\alpha\beta} \wedge *R^{\alpha\beta})$ goes as $r^{-6}$, and grows without bound as $r \to 0$ (Problem 8). This would not be possible at a regular singular point, so we conclude that the Schwarzschild metric can not be extended regularly across the region $u^2 - v^2 > -1$.

**Problems** (4.4.50)

1. Construct the five Killing vector fields of the metric (4.4.17) (cf. (I: 5.8.3)).

2. In the linear approximation, the metric at large distances

$$g_{\alpha\beta} = \eta_{\alpha\beta} + \frac{4\kappa}{r}\int d^3x'(T_{\alpha\beta} - \tfrac{1}{2}T\eta_{\alpha\beta})_{t-|x-x'|}$$

looks like the field of a plane wave. Using (4.4.29; 1), calculate the energy radiated in the 1-direction, and express it in terms of the quadrupole tensor

$$D_{ab} = \int d^3x\, T_{00}(3x_a x_b - \delta_{ab}|\mathbf{x}|^2).$$

3. Calculate the $\mathscr{T}_\alpha$ of (4.4.43) for $F = (e/r^2)e^{01}$. Verify that $i_{e^\alpha}\mathscr{T}_\alpha = 0$.

4. Reexpress (4.4.42) in the form (4.3.9).

5. Prove Birkhoff's theorem: If the $a$ and $b$ of (4.4.30) also depend on time, and Einstein's equations hold with $\mathscr{T}_\alpha = 0$, then there exists a time-coordinate $t'$ such that $g$ is of the form (4.4.30) with $a$ and $b$ independent of time. The metric $g$ must then be of the form (4.4.41).

6. Use the Schwarzschild metric in the form

$$g = -h^2(r)dt^2 + f^2(r)|d\mathbf{x}|^2, \qquad f = \left(1 + \frac{\kappa M}{2r}\right)^2,$$

$$h = \left(1 - \frac{\kappa M}{2r}\right)\left(1 + \frac{\kappa M}{2r}\right)^{-1},$$

(cf. (I: 5.7.17; 4)) and the version (4.2.21(b)) to identify the total energy

$$\int_{N_3} *(\mathscr{T}^0 + \ell^0)$$

with $M$. Use a ball of radius $R \to \infty$ for $N_3$.

7. Calculate the density of the total energy as in (4.4.45; 4) with the $\omega$'s of Problem 6.

8. Calculate $*(R_{\alpha\beta} \wedge *R^{\alpha\beta})$ for the Schwarzschild metric, and check that it is unbounded at $r = 0$.

**Solutions** (4.4.51)

1. The fields with components

$$v^i = (0, 0, 1, 0), (0, 0, 0, 1), (1, 1, 0, 0), \left(z, z, 0, \int_0^{t-x} \frac{du}{q^2(u)}\right), \left(y, y, \int_0^{t-x} \frac{du}{p^2(u)}, 0\right)$$

satisfy

$$v^i g_{lm,i} + g_{im} v^i{}_{,l} + g_{il} v^i{}_{,m} = 0,$$

and are therefore Killing vector fields according to (4.3.27; 1).

2. From the continuity equation of the linear approximation, $T^{\alpha\beta}{}_{,\beta} = 0$, it follows that

$$\frac{\partial^2}{\partial t^2} T_{00} = \nabla_a \nabla_b T_{ab}, \quad \text{i.e.,} \int d^3x\, T_{ab} = \frac{1}{2} \frac{\partial^2}{\partial t^2} \int d^3x\, x_a x_b T_{00}.$$

Therefore

$$\dot{g}_{ab} = \frac{2\kappa}{r} \frac{\partial^3}{\partial t^3} \int d^3x\, x_a x_b T_{00} + \eta_{ab} c.$$

The contribution $\sim \eta_{ab}$ is irrelevant, because we require only the square of the difference of the eigenvalues of the 2-3 subspace $(\dot{g}_{33} - \dot{g}_{22})^2 + 4\dot{g}_{32}^2$. Substitution of $D_{ab}$ yields

$$\mathscr{T}_{10} = \left(\frac{\kappa}{144\pi r^2}\right)[(\dddot{D}_{33} - \dddot{D}_{22})^2 + 4\dddot{D}_{23}^2]$$

(cf. (1.1.13) and [32, §104]).

3. $$*F = -\frac{e}{r^2} e^{23}, \qquad \mathscr{T}_\alpha = \tfrac{1}{2}[-i_{i_\alpha *F} {*F} - i_{i_\alpha F} F],$$

$$\mathscr{T}_0 = -\tfrac{1}{2}\frac{e}{r^2} i_{e_1} \frac{e}{r^2} e^{01} = \frac{e^2}{r^4} e^0,$$

$$\mathscr{T}_1 = \tfrac{1}{2}\frac{e}{r^2} i_{e_0} \frac{e}{r^2} e^{01} = -\frac{e^2}{r^4} e^1,$$

$$\mathscr{T}_2 = -\tfrac{1}{2}\frac{e}{r^2} i_{e_3} \frac{e}{r^2} e^{23} = \frac{e^2}{r^4} e^2,$$

$$\mathscr{T}_3 = \tfrac{1}{2}\frac{e}{r^3} i_{e_2} \frac{e}{r^2} e^{23} = \frac{e^2}{r^4} e^3.$$

4. Suppose $K = 1$, and introduce the coordinates

$$x_4 = (1 - r^2)^{1/2} \cosh t \quad \text{and} \quad x_5 = (1 - r^2)^{1/2} \sinh t$$

on the surface where $x_1^2 + x_2^2 + x_3^2 + x_4^2 - x_5^2 = 1$. Then

$$g = |d\mathbf{x}|^2 + dx_4^2 - dx_5^2 = |d\mathbf{x}|^2 - (1 - r^2)dt^2 + \frac{r^2\, dr^2}{1 - r^2}.$$

If $K = -1$, then take the coordinates

$$x_4 = (1 + r^2)\cos t \quad \text{and} \quad x_5 = (1 + r^2)\sin t$$

on the surface where $-|\mathbf{x}|^2 + x_4^2 + x_5^2 = 1$.

5. If $a$ and $b$ depend on time, then the only immediate change is $\omega^0{}_1$, by $\exp(b - a)\dot{b}\,dr$. This produces the following extra terms in the $8\pi\kappa\mathscr{T}^\alpha$ of (4.4.35):

$$\alpha = 0{:}\ 2\dot{b}\,\frac{e^{-a-b}}{r}\,e^1$$

$$\alpha = 1{:}\ -2\dot{b}\,\frac{e^{-a-b}}{r}\,e^0$$

$$\alpha = 2, 3{:}\ -e^{-2a}(\ddot{b} + \dot{b}^2 - \dot{a}\dot{b})e^\alpha.$$

If $\alpha = 0$, then we also find that $\dot{b} = 0$ and hence $\exp(-2b) = 1 - r_0/r$. And if $\alpha = 1$, then $a' = -b'$, and thus $\exp(2a) = (1 - r_0/r)f^2(t)$. With the variables $dt' = f(t)dt$, we have the Schwarzschild metric. No new conditions result if $\alpha = 2$ or 3.

6. $$e^\alpha = (h\,dt,\ f\,d\mathbf{x}), \qquad \omega^{0j} = \frac{h'}{f}\frac{x^j}{r}\,dt, \qquad \omega^{jk} = \frac{f'}{fr}(x^k\,dx^j - x^j\,dx^k),$$

$$8\pi\kappa\int_{N_3}({}^*\mathscr{T}^0 + {}^*\ell^0) = -\frac{1}{2}\varepsilon^{0bcd}\int_{\partial N_3}\omega_{bc}\wedge e_d$$

$$= -\varepsilon^{0bcd}\int_{\partial N_3}\frac{f'}{r}x^c\,dx^b\wedge dx^d$$

$$= -\lim_{R\to\infty}\int_{r=R} d\Omega\,2f'r^2 = 8\pi\kappa M.$$

In the above equations, $d\Omega$ is the element of solid angle, and we have recalled that $-\varepsilon^{0123} = \varepsilon_{0123} = 1$.

7. $8\pi\kappa({}^*\mathscr{T}^0 + {}^*\ell^0) = -\frac{1}{2}\varepsilon^{0bcd}\,d(\omega_{bc}\wedge e_d) = -\varepsilon^{0bcd}$.

$$d\left(\frac{f'}{r}x^c\right)dx^b\wedge dx^d = -dx^1\wedge dx^2\wedge dx^3\left[6\frac{f'}{r} + 2r\left(\frac{f'}{r}\right)'\right]$$

$$= -\frac{(\kappa M)^2}{r^4}\,dx^1\wedge dx^2\wedge dx^3.$$

8. $R^{\alpha\beta} = c^{\alpha\beta}e^{\alpha\beta}$ (no sum);

$$c^{\alpha\beta} = \frac{\kappa M}{r^3}\ \begin{array}{|c|c|c|} \hline 2 & -1 & -1 \\ \hline & -1 & -1 \\ \hline & & 2 \\ \hline \end{array}\ ,$$

$${}^*(R_{\alpha\beta}\wedge{}^*R^{\alpha\beta}) = \sum_{\alpha,\beta} c_{\alpha\beta}^2 = \frac{24(\kappa M)^2}{r^6}.$$

## 4.5 The Life and Death of Stars

*Gravity differs from other interactions by having a very small coupling constant, and by being universal. For cosmic bodies, the latter property makes the action of gravity sum constructively to such an extent that it dominates all other interactions.*

**The Orders of Magnitude** (4.5.1)

The gravitational energy of $N$ protons (mass $m$) in a volume $V$ is on the order of

$$E_G \sim -\frac{\kappa(Nm)^2}{V^{1/3}} = -\kappa m^2 N^{2/3} N \rho^{1/3}, \qquad \rho = \frac{N}{V}.$$

Although the Coulomb interaction is unimaginably stronger than this, $e^2 \sim 10^{36}\, \kappa m^2$, it is neutralized in normal matter, so that the electrical energy per particle is $\sim -e^2$/(the distance between nearest neighbors). This distance is $\sim \rho^{-1/3}$, so the total electrostatic energy is

$$E_e \sim -e^2 N \rho^{1/3} = \frac{e^2}{\kappa m^2 N^{2/3}} E_G. \tag{4.5.2}$$

We see that if $N \sim (e^2/\kappa m^2)^{3/2} \sim 10^{54}$, then gravity starts to dominate the electrical forces. The mass of Jupiter is about that of $10^{54}$ protons, which is the point at which the Newtonian potential supplants the Coulomb potential as the determiner of the structure. In a larger body, gravity crushes the atoms together, and the matter turns into a highly compressed plasma.

The Fermi energy, which is the origin of the solidity of matter, is (the number of electrons) × (the nearest-neighbor distance)$^{-2}$ × (the electron mass)$^{-1}$, in natural units ($\hbar = c = 1$):

$$E_F \sim \frac{N\rho^{2/3}}{m_e}. \tag{4.5.3}$$

The density $\rho$ of an object adjusts so as to minimize the total energy. Whereas for the Coulombic energy (4.5.2) this makes the density independent of $N$,

$$\rho \sim (e^2 m_e)^3 = (\text{Bohr radius})^{-3}, \tag{4.5.4}$$

in the case of gravitation objects containing more particles are smaller:

$$\rho^{1/3} \sim m^2 \kappa N^{2/3} m_e, \qquad V \sim N^{-1} m_e^{-3} (\kappa m^2)^{-3}. \tag{4.5.5}$$

However, as soon as the separation between nearest neighbors is on the order of magnitude of the Compton wavelength, $\rho^{1/3} \sim m_e$, the relativistic energy $|p|$ is to be used in calculating $E_F$ instead of $|p|^2/2m_e$, and (4.5.3) is replaced by

$$E_F \sim N\rho^{1/3}. \tag{4.5.6}$$

The gravitational energy consequently dominates the Fermi energy when $\kappa m^2 N^{2/3} > 1 \Rightarrow N > (\kappa m^2)^{-3/2} \sim 10^{57}$, i.e., when the mass is somewhat greater than the mass of the sun; and the minimum of the total energy is attained when $\rho = \infty$ and $v = 0$. After that point, there is a process in nature that dramatically controls what happens. The rate of energy loss from stars is normally rather slow—one photon takes several million years to escape from the interior of the sun—but sufficiently energetic electrons can create neutrinos by inverse beta decay $e^- + p \to \nu + n$, which, as they feel no strong interaction, leave the star immediately. This makes the transition to states of lower energy proceed at a much higher rate, and in a matter of seconds the star collapses to a neutron star, of nuclear density. Hence the energy released is on the order of the kinetic energy of neutrons at this density, about 10 MeV per particle, and thus as much energy is emitted as in the normal thermonuclear reactions, but much more rapidly. That is why it is assumed that the catastrophe just described is what takes place in a supernova, for which a single star may radiate with the brilliance of a whole galaxy for a week. The energy released would be the same, because a galaxy has typically $10^{10}$ stars, and normally a star takes $10^9$ years $\sim 10^{10}$ weeks to burn up all its nuclear fuel.

This line of reasoning makes use of a naive, Newtonian picture of gravity, and it is interesting to see how it changes in Einstein's theory, with the help of the material developed in the preceding section. It might be hoped that a sufficiently great pressure could counteract the gravitational attraction and render the stars stable. This is not necessarily the case, however, because in the relativistic theory pressure can also produce gravity, which can aggravate the situation.

Recall, in the spirit of the phenomenological description of the energy and momentum of matter,

$$\mathscr{T}_0 = -\rho e_0, \qquad \mathscr{T}_j = p e_j, \qquad j = 1, 2, 3, \tag{4.5.7}$$

that the energy density $\rho$ and pressure $p$ can not be chosen completely arbitrarily, because of the contracted Bianchi identity (4.2.32) connecting them; for the special form (4.5.7) it requires that

$$dp \wedge {}^*e^1 = \omega^1{}_0 \wedge {}^*e^0(p + \rho) \tag{4.5.8}$$

(see (4.4.37; 2) with $\alpha = 1$ and $\omega^1{}_\beta$ only nonzero if $\beta = 0$).

**Remarks** (4.5.9)

1. We shall later be primarily interested in the static, spherically symmetric case. Then both sides of the equation in (4.4.37; 2) vanish for $\alpha \neq 1$, with the $e$'s and $\omega$'s of (4.4.30); hence (4.5.8) contains all the information of (4.2.32).
2. Since we have earlier expressed the metric in terms of $\rho$ and $p$, Equation (4.5.8) creates a relationship between $\rho$ and $p$, which must be satisfied in

order to have static equilibrium. If an equation of state is known for $\rho$ and $p$, then there can be a static state only at the density distribution for which (4.5.8) agrees with the equation of state.

Taking the $\omega$ of (4.4.33),

$$dp \wedge {}^*e^1 = \exp(-b)\left(\frac{\partial p}{\partial r}\right){}^*\mathbf{1} = -\exp(-b)a'(\rho + p){}^*\mathbf{1}, \qquad (4.5.10)$$

and according to (4.4.39) and (4.4.40) (with $\kappa M(r) \to M(r)$),

$$\begin{aligned} a' &= -b' + 4\pi\kappa r(\rho + p)e^{2b} \\ &= \left(1 - \frac{2M(r)}{r}\right)^{-1} \cdot \left[-4\pi\kappa r\rho + \frac{M(r)}{r^2} + 4\pi\kappa r(\rho + p)\right] \end{aligned}$$

When this is substituted into (4.5.10), there results the

**Tolman–Oppenheimer–Volkoff Equation** (4.5.11)

$$-\frac{dp}{dr} = \frac{(\rho + p)[M(r) + 4\pi\kappa p r^3]}{r(r - 2M(r))}.$$

**Remarks** (4.5.12)

1. Of course, this also follows from (4.4.36), but the Bianchi identity does the trick without the extraneous information of (4.4.36).
2. Equation (4.5.11) generalizes the nonrelativistic fact that

$$-\frac{\partial p}{\partial r} = \frac{\rho M(r)}{r^2}.$$

The increase of the pressure for decreasing $r$ is intensified by the following relativistic effects:

(a) There is an additional term $\sim p$ in $M(r)$, since pressure also produces gravity;
(b) It is necessary to add $p$ to $\rho$, since the gravitational force also acts on $p$;
(c) Gravity increases faster than $\sim 1/r^2$ as $r \to 0$.

We saw at the outset that large, gravitating masses lose their stability in the special theory of relativity, because a relativistic electron gas is not as stiff as a nonrelativistic one, and does not stand firm against gravity. The general relativistic situation is even more precarious, because the solidity of matter also fails to help. In order to see this, we integrate (4.5.11) for the most extreme equation of state, viz., that of incompressible matter, which can not be squashed to arbitrarily high density. If $\rho =$ constant and we require the boundary condition $p(R) = 0$, where $R$ is the radius of the star, then in dimensionless variables we find

$$x = r\sqrt{\frac{8\pi\kappa\rho}{3}}, \qquad x_0 = R\sqrt{\frac{8\pi\kappa\rho}{3}} = \sqrt{\frac{r_0}{R}},$$

$$p(x) = \rho\frac{\sqrt{1 - x^2} - \sqrt{1 - x_0^2}}{3\sqrt{1 - x_0^2} - \sqrt{1 - x^2}} \tag{4.5.13}$$

(Problem 1). As a consequence, we can read off the

**Maximal Pressure in Homogeneous Stars** (4.5.14)

$$p(0) = \rho\frac{1 - \sqrt{1 - r_0/R}}{3\sqrt{1 - r_0/R} - 1}.$$

**Consequences** (4.5.15)

1. Whereas $p(0)$ goes as $\rho r_0/4R$ for stars of homogeneous densities whose radii are much larger than the Schwarzschild radius, and thus $p(0)$ is normally much less than $\rho$, if $R \to r_0$ it increases rapidly and becomes infinite at $R = 9r_0/8$.
2. The pressure in matter comes from the electrons, while the protons give rise to the energy density. The relative orders of magnitude are that $p/\rho \sim$ (electron speed $v)^2 \times$ (electron mass)/(proton mass) $\sim v^2 \cdot 10^{-3}$, so that the pressure in the center of a star like the sun, with $R \sim 10^5 r_0$, requires electrons to be moving at $\sim\frac{1}{10}$ the speed of light. The electrons must be relativistic in stars of the same mass but hundreds of times smaller (white and black dwarfs), and the situation becomes critical.

The next question to answer is how Einstein's theory affects the naive expression (4.5.1) for the gravitational energy. In §4.4 we saw that

$$\frac{r_0}{2\kappa} = 4\pi\int_0^\infty r^2\,dr\,\rho(r) \equiv M$$

is the total energy of the system, while

$$\int \rho{*}e^0 = 4\pi\int_0^\infty r^2\,dr\,\rho(r)\left(1 - \frac{2M(r)}{r}\right)^{-1/2} \tag{4.5.16}$$

equals the total energy of the matter alone. If $\rho =$ constant, then Equation

(4.5.16) can be evaluated easily, and there results the

**Gravitational Energy of a Homogeneous Star** (4.5.17)

$$E_G = M - \int \rho^* e^0 = M\left(1 - \frac{3}{2}\left[\arcsin\sqrt{\frac{r_0}{R}} - \sqrt{\frac{r_0}{R}}\sqrt{1 - \frac{r_0}{R}}\right]\left(\frac{R}{r_0}\right)^{3/2}\right);$$

if

$$R \gg r_0 : E_G = -\frac{3}{5}\frac{\kappa M^2}{R}$$

$$R = r_0 : E_G = -M\left(\frac{3\pi}{4} - 1\right).$$

When the density is small, this reduces to the Newtonian self-energy of a ball of uniform density, and as $R \to r_0$ this formula is of the same order of magnitude, but its numerical factor is somewhat different. At the limit of stability $R = 9r_0/8$, it reads $E_G = -0.37M$.

If the pressure called for in (4.5.11) can not be provided, then it is not possible to have static equilibrium, and the star collapses. In order to pursue this drama analytically, let us consider only stars of uniform pressure and density distributions. The Friedmann solution (4.4.2) applies in the interior, while in the exterior the free Einstein equations hold, for which, according to Birkhoff's theorem (4.4.50; 5) the only available solution with spherical symmetry is the Schwarzschild solution. The problem of matching the solutions will be discussed later, after we study the dynamics in the interior.

The point of departure is Equations (4.4.10), which will be used in the form

$$p = -\frac{\frac{d}{dt}(\rho R^3)}{\frac{d}{dt}R^3} \tag{4.5.18}$$

$$\frac{\dot{R}^2}{2} - \kappa\frac{4\pi}{3}\frac{\rho R^3}{R} = -\frac{K}{2}. \tag{4.5.19}$$

We solve them first for the extremal equation of state, $p = 0$. In normal matter, $p$ is always much less than $\rho$; the greatest pressure is that of massless particles, $p = \rho/3$. When either $p = \rho/3$ or $p = \rho$, analytic solutions can be written down (Problem 3), and we shall later figure out the qualitative behavior for all $p \geq 0$, which is generally similar.

**Solutions with $p = 0$** (4.5.20)

Equation (4.5.18) implies that $M = 4\pi\rho R^3/3 = \text{constant}$, making (4.5.19) of the form of the energy of the radial Kepler motion, with no angular momentum. This equation was integrated in (I: §4.2), and the solution is most conveniently written in the form of Kepler's equation (I: 4.2.24; 7).

We can identify the variables used there as

$$m = 1, \qquad E = -\frac{K}{2}, \qquad \alpha = -\kappa M \Rightarrow a = \frac{\kappa M}{K}, \qquad \varepsilon = 1,$$

and must distinguish three cases:

(a) $K > 0$

$$R = \frac{\kappa M}{K}(1 - \cos u), \qquad t - t_0 = \frac{\kappa M}{K^{3/2}}(u - \sin u) \tag{4.5.21}$$

(b) $K = 0$

$$R = (t - t_0)^{2/3}\left(\frac{9\kappa M}{2}\right)^{1/3} \tag{4.5.22}$$

(c) $K < 0$

$$R = \frac{\kappa M}{|K|}(\cosh u - 1), \qquad t - t_0 = \frac{\kappa M}{|K|^{3/2}}(\sinh u - u). \tag{4.5.23}$$

**Remarks** (4.5.24)

1. The case $K > 0$ in (4.5.19) corresponds to a negative energy in the Kepler problem. Accordingly, $R$ equals zero when $u = 0$ ($t = t_0$) and when $u = 2\pi$ ($t = t_0 + 2\pi\kappa M/K^{3/2}$). In this case Kepler's equations (4.5.21) are the parametric representation of a cycloid, specifying how time elapses during a free fall into the center:

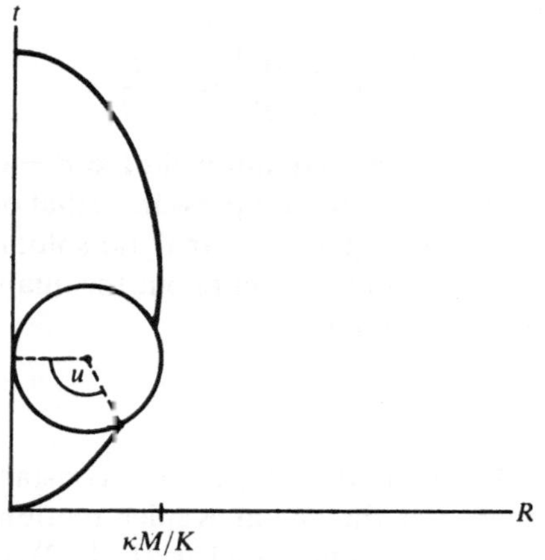


Figure 64 Cycloid for $R(t)$

2. Case (a) describes a big bang at $t = t_0$, which was so weak that the particles flying along the geodesics $x =$ constant are eventually stopped by gravity, turn around, and all will eventually crash together again. In case (b), on the other hand, the initial velocity is high enough to send the particles to infinity. The space $t =$ constant is not compact, but in fact simply $\mathbb{R}^3$.
3. In case (c) the particles retain some kinetic energy when they reach infinity, and the space $t =$ constant has negative curvature.
4. If $t = t_0$, then $R = 0$, and we learn from (4.4.7) that the metric has a genuine singularity at that point, because the curvature scalar approaches infinity.

These results are probably not too surprising, since matter without pressure or angular momentum would be expected to fall into the center unless it has a large enough initial outward radial velocity. A positive pressure changes nothing, because the extra gravity it produces actually favors the collapse. The reason can be seen formally in (4.4.10), by which a positive $p$ contributes negatively to $\ddot{R}$, thus increasing the concavity of the function $t \to R(t)$. In that case, $R(t)$ must approach zero if the initial slope is too small. To be mathematically precise, let us state these thoughts as a

**Criterion for Collapse of the Friedmann Universe** (4.5.25)

*Let* $\dot{R}(0)^2 < K > 0$ *and* $p \geq 0$. *Then* $R(t)$ *vanishes for some*

$$t < \frac{2R(0)}{\sqrt{K} - \dot{R}(0)},$$

*in which* $R(0)$ *is connected with* $\dot{R}(0)$ *by* $R(0)^2 = 3(K + \dot{R}(0))/8\pi\kappa\rho(0)$, *according to* (4.4.10).

**Proof**

Let us write the second equation of (4.4.10) as

$$\ddot{R}(t) = -\frac{\dot{R}(t)^2}{2R(t)} - \frac{K}{2R(t)} - \frac{R(t)8\pi\kappa p(t)}{2} \leq -\frac{K}{2R(t)}.$$

Then

$$R(t) = R(0) + t\dot{R}(0) + \int_0^t dt' \int_0^{t'} dt''\, \ddot{R}(t'')$$

$$\leq R(0) + t\dot{R}(0) - \frac{Kt^2}{4a} = R(0) + \frac{\dot{R}(0)^2 a}{K} - \frac{K}{4a}\left(t - \frac{2\dot{R}(0)a}{K}\right)^2,$$

$$\text{where } a \equiv \sup_t R(t).$$

If $\dot{R}(0)^2 < K$, then $a \leq R(0) + a\dot{R}(0)^2/K$ implies

$$a \leq R(0)\left(1 - \frac{\dot{R}(0)^2}{K}\right)^{-1},$$

and the zero for $R(t)$ happens before

$$t_0 = \frac{2\dot{R}(0)a}{K} + \sqrt{\frac{4aR(0)}{K} + \frac{4a^2\dot{R}(0)^2}{K^2}}, \qquad \text{as} \quad R(0) + t_0\dot{R}(0) = \frac{Kt_0^2}{4a},$$

The bound on $a$ implies (4.5.25) □

**Remarks** (4.5.26)

1. If $p = 0$, then the condition $\dot{R}(0)^2 < K$ corresponds to the statement for the equivalent Kepler problem that the kinetic energy is less than minus the total energy. This obvious criterion preventing escape is valid for all $p \geq 0$.
2. The time $t$ is that of a co-moving coordinate system, and thus the space collapses to a point within a finite proper time for freely falling observers.

Finally, we construct a solution of Einstein's equations, which describes gravitational collapse. The physical picture of what takes place is as follows: If, after having exhausted its nuclear fuel, a star has shrunk down so far that the Fermi energy of the electrons has risen above the threshold for inverse beta decay $e^- + p \to \nu + n$, then the greater part of the matter is turned into neutrons. Since the star is supported against collapse mainly by the Fermi pressure of the electrons, it suddenly gives way. Thus the model would be that of a star in static equilibrium, whose pressure at some time is suddenly reduced to zero. The solution of Einstein's equations before that time is as in (4.4.30). Afterwards, the solution in the interior is (4.4.2), and in the exterior it is the Schwarzschild metric. We now need to show that the solutions can be joined smoothly at the surface to satisfy Einstein's equations with $\rho$ = constant inside and 0 outside, and $p = 0$. Since the surface of the star falls freely, its radius in the co-falling coordinates (4.4.2) is $r = a$ = constant. For simplicity we use units in which $r_0 = 8\pi\kappa a^3\rho/3 = 1$ and consider the case $K = 0$. This makes the motion parabolic, with the surface of the star infinitely large at the beginning. Similarly, the solution with $K > 0$ is a Friedmann space in the interior, matched to a Schwarzschild metric. If $p > 0$, the calculation becomes much more complicated, because it can not be constant inside the star, as otherwise there would be an infinite pressure gradient at the surface. However, the essential features are not greatly altered if $p > 0$ [36].

In order to join (4.4.30) to (4.4.2), we have to express the two metrics in the same coordinates. For this reason, we write the

**Schwarzschild Metric in Co-Falling Coordinates** (4.5.27)

It is convenient to introduce the coordinates $(\tau, \tilde{r})$ in place of $(t, r)$, where $\tau$ is the proper time for radial parabolic motion, and $\tilde{r}$ is $r$ at the time $t = 0$. Since the speed approaches zero asymptotically,

$$p^0 = \left(\frac{dt}{d\tau}\right)\frac{r-1}{r} = 1,$$

and thus

$$-1 = \frac{r}{r-1}\left(-1 + \left(\frac{dr}{d\tau}\right)^2\right) \Rightarrow \frac{dr}{d\tau} = -\frac{1}{\sqrt{r}} \Rightarrow \tau = \frac{2}{3}(\tilde{r}^{3/2} - r^{3/2}).$$

Consequently,

$$\frac{dt}{dr} = \frac{d\tau}{dr}\left(1 + \frac{1}{r-1}\right) \Rightarrow t = \tau - 2\sqrt{r} + \ln\frac{\sqrt{r}+1}{\sqrt{r}-1}.$$

This puts the metric in the normal form $g = -d\tau^2 + g_{ij}\,dx^i\,dx^j$, because

$$d\tau = \sqrt{\tilde{r}}\,d\tilde{r} - \sqrt{r}\,dr,$$

$$dt = d\tau - dr\frac{\sqrt{r}}{r-1} = d\tau\frac{r}{r-1} - d\tilde{r}\frac{\sqrt{\tilde{r}}}{r-1}$$

leads to

$$\begin{aligned} g &= -dt^2\frac{r-1}{r} + dr^2\frac{r}{r-1} + r^2\,d\Omega^2 \\ &= -d\tau^2 + (1 - \tfrac{3}{2}\tau\tilde{r}^{-3/2})^{-2/3}\,d\tilde{r}^2 + (1 - \tfrac{3}{2}\tau\tilde{r}^{-3/2})^{4/3}\tilde{r}^2\,d\Omega^2. \end{aligned} \tag{4.5.28}$$

**Remarks** (4.5.29)

1. This chart can be used for $3\tau/2 < \tilde{r}^{3/2}$, though at $3\tau/2 = \tilde{r}^{3/2}$, which corresponds to $r = 0$, it becomes singular. Equation (4.4.28) therefore extends the Schwarzschild metric beyond $r = r_0 = 1$, but it is not the maximal extension (4.4.48).
2. A particle falling freely from infinity travels from $r = \tilde{r}$ to $r = 0$ in proper time $2\tilde{r}^{3/2}/3$, in units where $r_0 = 1$.

To discover the proper Friedmann solution, note that Equation (4.5.19), with $a = R(0)$, $r_0 = 8\pi\kappa\rho(0)a^3/3 = 1$, and $K = 0$, implies that

$$\dot{R} = -\frac{1}{\sqrt{R}},$$

and hence that

$$R(t) = a(1 - \tfrac{3}{2}ta^{-3/2})^{2/3}.$$

By redefining the coordinates $t \to \tau$, $ar \to \tilde{r}$, we come up with the

**Oppenheimer–Snyder Solution** (4.5.30)

*The metric*

$$g = -d\tau^2 + \begin{cases} (1 - \frac{3}{2}\tau a^{-3/2})^{4/3}(d\tilde{r}^2 + \tilde{r}^2\, d\Omega^2) & \text{if } \tilde{r} \le a \\ (1 - \frac{3}{2}\tau\tilde{r}^{-3/2})^{-2/3}\, d\tilde{r}^2 + (1 - \frac{3}{2}\tau\tilde{r}^{-3/2})^{4/3}\tilde{r}^2\, d\Omega^2 & \text{if } \tilde{r} \ge a, \end{cases}$$

*satisfies Einstein's equations with* $p = 0$,

$$\rho = \begin{cases} 3(a^{3/2} - 3\tau/2)^{-2}/8\pi\kappa & \text{if } \tilde{r} \le a \\ 0 & \text{if } \tilde{r} > a. \end{cases}$$

**Proof**

Einstein's equations are satisfied for $\tilde{r} > a$ and $\tilde{r} < a$ by construction, and, as the curvature ought to be discontinuous at $\tilde{r} = a$ (cf. (4.2.13)), it is only necessary to check that it has no delta-function singularity there. Let us write the orthogonal basis for $g$,

$$e^\alpha = (d\tau, e^\nu \nu'\, d\tilde{r}, e^\nu\, d\vartheta, e^\nu \sin\vartheta\, d\varphi)$$

$$e^{\nu(\tau,\tilde{r})} \equiv \begin{cases} \tilde{r}(1 - \frac{3}{2}\tau a^{-3/2})^{2/3} & \text{if } \tilde{r} \le a \\ \tilde{r}(1 - \frac{3}{2}\tau\tilde{r}^{-3/2})^{2/3} & \text{if } \tilde{r} \ge a, \end{cases} \tag{4.5.31}$$

where $\nu' \equiv \partial\nu/\partial\tilde{r}$ and $\dot{\nu} \equiv \partial\nu/\partial\tau$; $\nu'$ is discontinuous and $\dot{\nu}$ is continuous with a discontinuous first derivative. When restricted to $r = a$ ($d\tilde{r}_{|\tilde{r}=a} = 0$), only the continuous parts of $e^\alpha$ remain, and even the $de^\alpha$ are continuous at $\tilde{r} = a$:

$$de^\alpha = (0, e^\nu(\dot{\nu}' + \dot{\nu}\nu')d\tau \wedge d\tilde{r}, e^\nu(\dot{\nu}\, d\tau + \nu'\, d\tilde{r}) \wedge d\vartheta,$$
$$e^\nu((\dot{\nu}\, d\tau + \nu'\, d\tilde{r})\sin\vartheta + \cos\vartheta\, d\vartheta) \wedge d\varphi).$$

From this formula we get the affine connections,

$\omega^\alpha{}_\beta =$ (4.5.32)

| $\tau$ | $\tilde{r}$ | $\vartheta$ | $\varphi$ |
|---|---|---|---|
| 0 | $e^\nu(\dot{\nu}' + \dot{\nu}\nu')d\tilde{r}$ | $e^\nu\dot{\nu}\, d\vartheta$ | $e^\nu\dot{\nu}\sin\vartheta\, d\varphi$ |
| | 0 | $-d\vartheta$ | $-\sin\vartheta\, d\varphi$ |
| | | 0 | $-\cos\vartheta\, d\varphi$ |
| | | | 0 |

Observe that the discontinuous functions $\nu'$ and $\dot{\nu}'$ are multiplied by $d\tilde{r}$, and so no $\nu''$ shows up in $d\omega^\alpha{}_\beta$ (see Problem 4). Hence, while $R^\alpha{}_\beta$ is discontinuous, it does not contain a delta function. □

**Remarks** (4.5.33)

1. As Einstein's equations do not allow delta-function singularities in the contractions $R_\alpha$, the question arises of whether they can occur in $R^\alpha{}_\beta$. The answer is that they can not, because the surface of discontinuity has a spacelike normal $d\tilde{r}$, and by (4.2.24), the regularity of $R_\alpha$ implies that of $R^\alpha{}_\beta$ in this case.
2. If the basis given in the proof is supplied with a more general function $v(\tau, r)$, then it is easy to find a solution of Einstein's equations that describes the gravitational collapse of a $C^\infty$ density distribution. The discontinuous solution (4.5.30) can be considered as the limiting case of a $C^\infty$ solution (Problem 4).

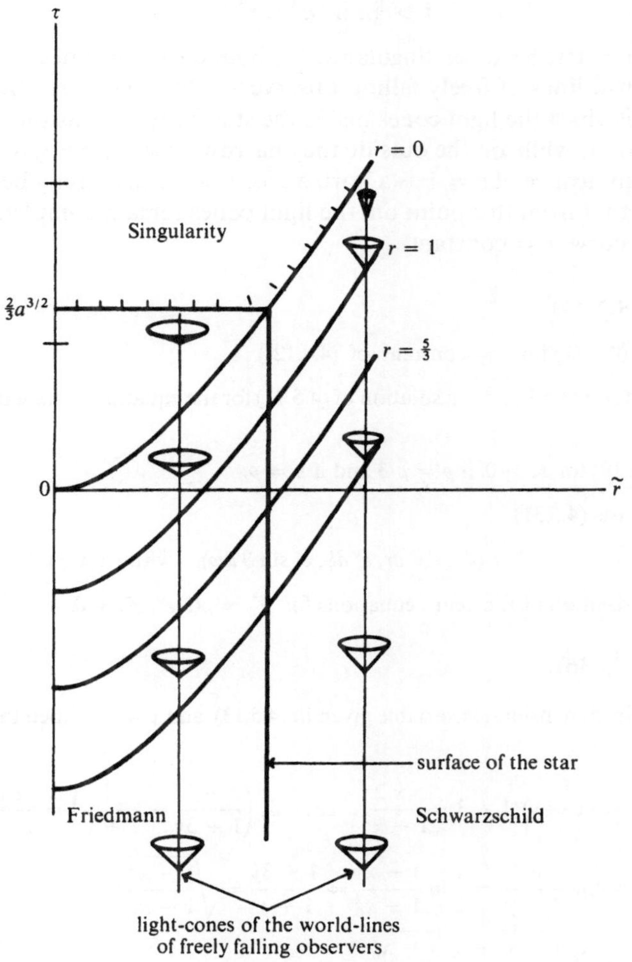


Figure 65

3. In order that the solutions join without a seam, the Schwarzschild radius of the outer solution must be

$$8\pi\kappa \int_0^a d\tilde{r}\, \tilde{r}^2 \rho(\tilde{r}, \tau)_{|\tau=0}.$$

This is to be expected, because when $\tau = 0$, $dr = d\tilde{r}$. In the static coordinates $(t, r)$, the dramatic action in the interior is not detectable from outside. The Schwarzschild solution is as silent as the grave about what it hides within.

**The Geometric Significance of the Oppenheimer–Snyder Solution** (4.5.34)

The chart of (4.5.30) is workable if

$$\tau > \tfrac{2}{3}\min[a^{3/2}, \tilde{r}^{3/2}],$$

though the metric becomes singular at the boundary. The lines $\tilde{r}$ = constant are the world-lines of freely falling observers, and $\tau$ measures their proper time. In this chart the light cones inside the star flatten out as they approach the singularity, while on the outside they narrow down (see Figure 65). Note that the significance of $r = 1$ as a horizon for $\tilde{r} > a$ can thereby be expressed as the fact that from this point on, the light cones remain completely on one side of the curve $r$ = constant.

**Problems** (4.5.35)

1. Integrate (4.5.11) for $\rho$ = constant (cf. (4.4.12)).
2. Show that $\rho = 3/56\pi\kappa r^2$ is a solution of (4.5.11) for the equation of state of radiation, $p = \rho/3$.
3. Solve (4.4.10) for $K > 0$, if $p = \rho/3$ and if $p = \rho$.
4. Use the basis (4.5.31)

$$e^\alpha = (d\tau, e^\nu \nu' \, d\tilde{r}, e^\nu \, d\vartheta, e^\nu \sin\vartheta \, d\varphi), \quad \text{with} \quad \nu(\tau, \tilde{r}),$$

to find a solution of Einstein's equations for $\mathscr{T}_0 = \rho(\tilde{r})e^0$, $\mathscr{T}_j = 0$.

**Solutions** (4.5.36)

1. Let $x$ be the dimensionless variable given in (4.5.13), and $y = p/\rho$; then this equation becomes

$$-\frac{dy}{dx} = (1 + y)(1 + 3y)\frac{x/2}{1 - x^2} \quad \text{i.e.,} \quad dy\left(\frac{-3}{1 + 3y} + \frac{1}{1 + y}\right) = \frac{dx\, x}{1 - x^2}$$

$$\Rightarrow \ln\frac{1 + 3y}{1 + y} = \frac{1}{2}\ln\frac{1 - x^2}{1 - x_0^2} \Rightarrow \frac{1 + 3y}{1 + y} = \sqrt{\frac{1 - x^2}{1 - x_0^2}}$$

$$\Rightarrow \frac{\sqrt{1 - x^2} - \sqrt{1 - x_0^2}}{3\sqrt{1 - x_0^2} - \sqrt{1 - x^2}} = y.$$

2. $p = 1/56\pi\kappa r^2$, $\quad M(r) = 4\pi\kappa \int_0^r dr'\, r'^2\, \rho(r') = 3r/14$,

$$-\frac{\partial p}{\partial r} = \frac{1}{28\pi\kappa r^3} = \frac{4}{56\pi\kappa r^3}\,\frac{\frac{3}{14}r + r/14}{r(1 - \frac{3}{7})} = \frac{\rho + p}{r}\,\frac{M(r) + 4\pi\kappa p r^3}{r - 2M(r)}.$$

3. $p = \dfrac{\rho}{3}$: $\quad \dfrac{K + R^2}{R^2} = -\dfrac{2R\dot{R} + \dot{R}^2 + K}{R^2} \Rightarrow \dot{R}^2 + R\ddot{R} + K = 0 \Rightarrow \dfrac{d^2}{dt^2}R^2 = -2K$

$$\Rightarrow R = (ct - Kt^2)^{1/2}.$$

$p = \rho$: Let $d\tau/dt = 1/R$. Then $4K + 4\dot{R}^2 + 2R\ddot{R} = 0$ becomes the oscillator equation,

$$\frac{d^2R^2}{d\tau^2} = -4KR^2 \Rightarrow R = R_{\max}[\sin 2\sqrt{K}\,\tau]^{1/2};$$

consequently $t$ is given by

$$R_{\max}\int_0^\tau d\tau'[\sin 2\sqrt{K}\,\tau']^{1/2}.$$

In both cases we have chosen $R(0) = 0$, and we observe that $R$ decreases again to zero after a finite time.

4. By (4.1.28) the affine connections (4.5.32) lead to the curvature forms

$$\begin{aligned}
R^0{}_{\tilde r} &= e^\nu(2\dot\nu\nu' + \dot\nu^2\nu' + \ddot\nu' + \dot\nu\nu')d\tau \wedge d\tilde r\\
R^0{}_{\vartheta} &= e^\nu(\dot\nu^2 + \ddot\nu)d\tau \wedge d\vartheta\\
R^0{}_{\varphi} &= e^\nu(\dot\nu^2 + \ddot\nu)\sin\vartheta\, d\tau \wedge d\varphi\\
R^{\tilde r}{}_{\vartheta} &= e^{2\nu}(\dot\nu' + \dot\nu\nu')\dot\nu\, d\tilde r \wedge d\vartheta\\
R^{\tilde r}{}_{\varphi} &= e^{2\nu}(\dot\nu' + \dot\nu\nu')\dot\nu \sin\vartheta\, d\tilde r \wedge d\varphi\\
R^{\vartheta}{}_{\varphi} &= e^{2\nu}\dot\nu^2 \sin\vartheta\, d\vartheta \wedge d\varphi.
\end{aligned}$$

Once again, $R_{\alpha\beta} \sim e_{\alpha\beta}$, so $T_{\alpha\beta}$ is diagonal. Einstein's equations require that

$$\begin{aligned}
8\pi\kappa\rho &= 3\dot\nu^2 + 2\frac{\dot\nu'\dot\nu}{\nu'} = 8\pi\kappa T_{00}\\
0 &= e^\nu(3\dot\nu^2 + 2\ddot\nu) = T_{\tilde r\tilde r}\\
0 &= \left(1 + \frac{1}{2\nu'}\frac{d}{d\tilde r}\right)(3\dot\nu^2 + 2\ddot\nu) = T_{\vartheta\vartheta} = T_{\varphi\varphi}.
\end{aligned}$$

The last two of these equations are solved by $e^\nu = (F(\tilde r)\tau + G(\tilde r))^{2/3}$. Since the basis is invariant under a change of charts $\tilde r \to \bar r(\tilde r)$, we may set $G = \tilde r^{3/2}$, and are left with only one function, $F(\tilde r)$. If $\tau = 0$, then the first of Einstein's equations becomes

$$FF' = 9\pi\kappa\tilde r^2\rho(\tilde r, 0) \Rightarrow F(\tilde r) = -\left[18\pi\kappa\int_0^{\tilde r} dr\, r^2\rho(r, 0)\right]^{1/2}$$

In the case $\rho(r, 0) = 3/8\pi\kappa a^3$, we revert to (4.5.30).

## 4.6 The Existence of Singularities

*The solutions of nonlinear differential equations have a tendency to develop singularities; in particular this is true of Einstein's equations, where the attractive nature of gravity reveals its physical origins.*

The models we have considered of gravitational collapse, in which a singularity develops, were all radially symmetric. It is not at all surprising that a fall directed right at the center will end in a catastrophe. The one new feature of Einstein's theory is that the catastrophe can not be prevented by any pressure, no matter how strong, because the pressure itself produces more gravity. There is a question, however, whether the situation is qualitatively changed by a perturbation of the radial symmetry, just as the angular momentum in the Kepler problem prevents the plunge into the center. In the relativistic Kepler problem, the effective gravitational potential goes as $-1/r^3$, which is stronger than the centrifugal potential (cf. (I: §5.7)), but it is conceivable that other mechanisms might impede the growth of a singularity. It is often claimed ([32], §1.14), on account of this observation, that normally the solutions are free of singularities, which are pathologies afflicting spaces of high symmetry. It was the accomplishment of R. Penrose and others of the school of D. Sciama to disprove this claim: as long as energy and pressure are positive in some reasonable sense, and at some instant there exists the kind of geometry set up by a large mass, then the formation of a singularity is unavoidable, regardless of any symmetry.

Let us agree at this stage what we mean by a singular space. Regularity is incorporated in the concept of a manifold, and any singular points are removed. It might be suggested that unbounded growth of $R^\alpha{}_\beta$ could be taken as a sign of a singularity in the vicinity. We shall see shortly that the $R^\alpha{}_\beta$ describe the tidal force, and consequently this conjectured indicium has a direct physical significance, as it can be observed as bodily discomfort. Unfortunately, it is difficult to express this mathematically, since the components of $R^\alpha{}_\beta$ depend on the basis, and could also become infinite in the absence of a genuine singularity. Conversely, it is possible for all 14 of the invariants that can be constructed from $R^\alpha{}_\beta$ to vanish without $R^\alpha{}_\beta$ itself vanishing. For example, this happens for plane gravitational waves, and is analogous to a nonzero vector in Minkowski space having zero length.

Hence we resort to a different feature of the solutions we have discussed as the criterion for a singularity, viz., that an observer falls into the singularity in a finite proper time, thus leaving the manifold. There is, of course, the trivial possibility that the manifold has simply been chosen too small—if the manifold were only a piece of Minkowski space, then one could leave it in a finite time, although there is not necessarily any singularity outside the piece. In order to exclude such cases, we make a

**Definition** (4.6.1)

A pseudo-Riemannian manifold $M$ is **extensible** iff it is a proper subset of a larger manifold $M'$, i.e., its metric is the restriction to $M$ of the metric on $M'$.

**Remarks** (4.6.2)

1. $M'$ is not, of course, uniquely determined by $M$, so our criterion can not involve examining an extensible manifold to see where there are singularities. For instance, the Schwarzschild metric for $r > 5r_0$ can be extended either to the regular solution with a continuous mass distribution for $r < 5r_0$ or to the singular solution.
2. When confronted with an extensible manifold, one gets the feeling that something has been intentionally left out. Therefore we postulate that the physical space-time continuum is nonextensible.
3. There are examples ([33], p. 58) of nonextensible manifolds that can be escaped from, so it is actually necessary to postulate a more refined property, local nonextensibility. However, the examples seem rather artificial, so we shall content ourselves with the primitive definition.

The next step is to decide what observers we will grant an unlimited stay in the manifold.

**Definition** (4.6.3)

A pseudo-Riemannian manifold is said to be **geodesically complete** in timelike directions iff every timelike geodesic can be extended to an arbitrarily long proper-time parameter.

**Remarks** (4.6.4)

1. A positive metric $g$ defines a metric for the topology of a Riemannian space $M$, and then geodesic completeness means the same thing as completeness in the sense of a metric topological space.
2. An affine parameter could also be defined on lightlike geodesic lines, and one can speak of lightlike and spacelike geodesic completeness. These conditions are not equivalent (Problem 1); but at any rate (4.6.3) must be required on physical grounds.
3. Geodesic incompleteness puts an observer who can stay in the manifold for only a finite time into a predicament, but is not necessarily evidence

of any kind of infinity. This is shown by the example

$$g = -dt^2\left(1 - \frac{h}{2}\right) + dx^2\left(1 + \frac{h}{2}\right) + dx\,dt\,h,$$

$$h = \frac{\lambda}{2}(\cos^4(t - x) - 1), \qquad \lambda \in (0, 2),$$

on $\mathbb{R}^2$ (if desired, $dy^2 - dz^2$ can be added in). If $\lambda$ is small, this is only a weak gravitational wave that spreads throughout the flat space, but nonthetheless the space fails to be geodesically complete in timelike directions, even for arbitrarily small $\lambda$ (Problem 2). The reason is that a particle of the right initial velocity rides the crests of the waves, as in a linear accelerator, and reaches nearly the speed of light. Its proper time runs ever more slowly and never exceeds some finite value. There is no singularity, and the only $R^i{}_j$ that does not vanish is $R^0{}_1 = e^0{}_1 h''/2$. Since $h$ is periodic in $u = t - x$, $g$ can be used as a pseudometric on $T^2$, in which case even this compact set is geodesically incomplete, although it is certainly not a piece of a larger connected manifold. Despite that, we follow common usage and refer to the space as singular.

4. Even in Minkowski space it is possible to reach the end of the manifold after a finite proper time on certain timelike lines. If, for instance, $x = t + 1/t^2$ for $t > 1$, then

$$\int ds = \int \frac{ds}{dt}\,dt = \int_1^\infty dt\sqrt{1 - \left(\frac{dx}{dt}\right)^2} < 2\int_1^\infty dt\, t^{-3/2} = 4.$$

It is only the choice of coordinates that makes the end at $x = \infty$, and it can be transformed to any finite point, just as the end lay at $-\infty$ in the Schwarzschild metric with the variable $\ln r$.

5. One might require that timelike lines with bounded acceleration $\ddot{z}^\alpha \ddot{z}^\beta g_{\alpha\beta}$ can be continued to arbitrarily long proper times. If this were not so, then the crew of a rocket with a finite supply of fuel could conceivably find themselves at the edge of the universe, and would not know what to do. Yet geodesic completeness leaves this possibility open [34].

Geodesic lines are the world-lines of freely falling observers (cf. Problem 4). The nonrelativistic analogue of a geodesic vector field is the velocity field $v_i$ of an ideal fluid with no pressure in a gravitational potential $\Phi$. For stationary fluid flow, the equations of hydrodynamics require that $v_k v_{i,k} = -\Phi_{,i}$. Let $n$ be the distance-vector field between nearby fluid particles, which is carried along with the stream. Its Lie derivative with respect to $v$ vanishes, so

$$v_k n_{i,k} = n_k v_{i,k} \tag{4.6.5}$$

(recall (I: 2.5.12; 5)), which makes the second derivative along the stream-lines

$$v_k \frac{\partial}{\partial x^k}\left(v_j \frac{\partial}{\partial x^j} n_i\right) = -n_k \Phi_{,ik}. \tag{4.6.6}$$

Thus the gradient of the field $\Phi_{,i}$ affects the distance between two particles, and in fact the effect of the second derivative of $\Phi$ is to focus them together: Since $\Phi$ satisfies the equation

$$\Phi_{,jj} = \rho \geq 0, \tag{4.6.7}$$

the net effect of the gravitational field, when averaged over all directions, is to focus particles. For irrotational fluid flow, $v_{i,k} = v_{k,i}$, this can be expressed as an increase in the rate of convergence $c = -v_{i,i}$ of the flow along the streamlines:

$$v_i \frac{\partial}{\partial x_i} c = -v_i v_{k,ik} = v_{k,i} v_{i,k} + \Phi_{,kk} \geq \frac{c^2}{3}. \tag{4.6.8}$$

This equation used (4.6.7) and the irrotationality, which entered through the trace inequality for symmetric $n \times n$ matrices

$$(\operatorname{Tr} M)^2 \leq n \operatorname{Tr}(M^2) \tag{4.6.9}$$

(Problem 3). If $c$ is positive at some point, then it increases so rapidly by (4.6.8) that it soon reaches infinity, and the streamlines meet. If $s$ is the parameter on a streamline, given as $x(s)$, $v_i(x) = dx_i/ds$ (cf. (4.1.42; 1)), then (4.6.8) implies that

$$\frac{dc}{ds} \geq \frac{c^2}{3} \Rightarrow c(s) \geq \frac{c(0)}{1 - sc(0)/3}, \tag{4.6.10}$$

and thus $c$ gets arbitrarily large before $s = 3/c(0)$. This elementary property of gravity contains the essential features of the relativistic theory discussed below.

The relativistic generalizations of (4.6.6) are

**The Equations of Geodesic Deviation** (4.6.11)

*Let $v = v^\alpha e_\alpha$ be a geodesic vector field and $n$ a vector field such that $L_v n = 0$. Then*

$$D_v D_v n = -e_\alpha (R^\alpha{}_\beta | n \otimes v) v^\beta.$$

**Proof**

By (4.1.7(g′)), $D_v n = D_n v$, and because of (4.1.19), (4.1.33; 2), and the equation $D_v v = 0$,

$$\begin{aligned} 0 = D_n D_v v &= D_v D_v n + (D_n D_v - D_v D_n) v \\ &= D_v D_v n + e_\alpha (R^\alpha{}_\beta | n \otimes v) v^\beta. \end{aligned}$$

□

**Example** (4.6.12)

Consider the Friedmann universe (4.4.2). Let the fields $v$ and $n$ be the natural contravariant bases $\partial_t$ and $\partial_x$. Their Lie brackets with each other vanish, and $\partial_t$ is geodesic, since we are using co-falling coordinates. The contravariant components of the metric (4.4.2) are

$$g^{00} = -1, \qquad g^{jj} = \left(\frac{1 + Kr^2/4}{R}\right)^2,$$

so $v$ and $n$ can be written in the orthogonal basis as

$$v = \partial_0 = e_0, \qquad n = \partial_1 = \frac{e_1}{\sqrt{g^{11}}} = e_1 \frac{R}{1 + Kr^2/4}.$$

With the affine connections of (4.4.5), we see that

$$D_v n = \frac{\dot{R}}{1 + Kr^2/4} e_1 + \frac{R}{1 + Kr^2/4} D_v e_1 = \frac{\dot{R}}{1 + Kr^2/4} e_1,$$

$$D_v D_v n = \left(D_v \frac{\dot{R}}{1 + Kr^2/4}\right) e_1 = \frac{\ddot{R}}{1 + Kr^2/4} e_1 = \frac{\ddot{R}}{R} n,$$

which, because of (4.4.6), is precisely

$$-e_1(R^1{}_0 | n \otimes v).$$

**Remarks** (4.6.13)

1. Since $R(t)$ describes how the distance between neighboring world-lines $x =$ constant varies, we perceive that $D_v D_v n$ has the significance of a relative acceleration.
2. Proposition (4.6.11) shows that from the physical point of view it is $R^\alpha{}_\beta$ rather than $\omega^\alpha{}_\beta$ that takes over the role of the electric field strength. Because of the principle of equivalence (I: 5.6.11), there is no trace of the $\omega$'s; freely falling observers can only notice the gradient of the field $R$, specifying the corrections to the principle of equivalence, which holds only in the infinitesimal limit.

The curvature forms may have either sign, either focusing or defocusing. The contractions $R_\alpha$ are immediately determined by the energy and momentum, from which they inherit the positivity (2.1.13). As with (4.6.8), this leads to an

**Increase in the Rate of Convergence of Geodesic Vector Fields** (4.6.14)

*Let $v$ be a timelike geodesic vector field perpendicular to a hyperplane $t = 0$, and assume $i_v R_0 \geq 0$. Then the rate of convergence $c = -\sum_\alpha \langle e^\alpha | D_{e_\alpha} v \rangle$ (cf. Problem 4) satisfies the differential inequality*

$$D_v c \geq \frac{c^2}{3}.$$

**Proof**

We work in the natural basis of a co-falling coordinate system, so that $v = \partial_t$ and $g = -dt^2 + g_{ab}\,dx^a \otimes dx^b$. As in (4.6.11), $D_v v = 0$, and $D_v \partial_a = D_{\partial_a} v$. By also recalling that

$$0 = D_v \delta^\alpha{}_\beta = \langle D_v\, dx^\alpha | \partial_\beta \rangle + \langle dx^\alpha | D_v \partial_\beta \rangle,$$

we find

$$\begin{aligned} D_v c &= -D_v \langle dx^\alpha | D_{\partial_\alpha} v\rangle = -\langle D_v\, dx^\alpha | D_v \partial_\alpha \rangle - \langle dx^\alpha | D_v D_{\partial_\alpha} v\rangle \\ &= -\langle D_v\, dx^\alpha | \partial_\beta \rangle \langle dx^\beta | D_v \partial_\alpha \rangle + \langle R^\alpha{}_\beta | \partial_\alpha \otimes v \rangle v^\beta \\ &= \langle dx^\alpha | D_v \partial_\beta \rangle \langle dx^\beta | D_v \partial_\alpha \rangle + i_v R_0 \geq \frac{c^2}{3}, \end{aligned}$$

since the trace inequality (4.6.9) is again applicable:

$$M_{\alpha\beta} \equiv \langle \partial_\alpha | D_v \partial_\beta \rangle = \langle \partial_\alpha | D_{\partial_\beta} v \rangle = \Gamma^0_{\alpha\beta}$$

is symmetric in $\alpha$ and $\beta$ and vanishes when $\alpha$ or $\beta$ is zero, because of $D_v v = 0$. In the space orthogonal to $v$, $g$ is positive and $c = \mathrm{Tr}(Mg) = \mathrm{Tr}(\sqrt{g}M\sqrt{g})$, while $\mathrm{Tr}(\sqrt{g}M\sqrt{g}\sqrt{g}M\sqrt{g})$ occurs in the above equation. □

**Example** (4.6.15)

Let us take another look at the Friedman universe, for which

$$D_{\partial_a} v = D_v \partial_a = \frac{\dot{R}}{R}\partial_a,$$

and thus $c = -3\dot{R}/R$. Proposition (4.6.14):

$$D_v c = -\frac{\partial}{\partial t}\frac{3\dot{R}}{R} = \frac{3\dot{R}^2}{R^2} - \frac{3\ddot{R}}{R} \geq \frac{c^2}{3} = \frac{3\dot{R}^2}{R^2}$$

holds if $\ddot{R} \leq 0$, which amounts to the condition that $i_v R_0 \geq 0$. From Equations (4.4.10), this condition is met if $\rho + 3p \geq 0$.

**Remarks** (4.6.16)

1. According to Einstein's equations (4.2.20(a)), the condition $i_v R_0 \geq 0$ $\forall v$: $\langle v | v \rangle < 0$ implies for the total energy-momentum tensor that $T_{00} + T_{11} + T_{22} + T_{33} \geq 0$ ("positivity of the energy"). This is true for all sensible models of matter. The reason for the positivity condition is that negative energy produces a repulsive gravitational force, which could prevent the convergence of the geodesics.

2. Generalizing the nonrelativistic result (4.6.10), we see that if $c$ is ever positive, then it must become infinite after a finite time under the circumstances of (4.6.14). Let $s$ be the proper-time parameter (cf. (4.1.42; 1)) on the geodesics of $\iota$, $v^i(x(s)) \equiv \dot{x}^i(s)$, and $c(s) \equiv c(x(s))$. If $c(0) > 0$, then $c$ becomes infinite for some $s$ such that $0 \leq s \leq 3/c(0)$.
3. If $N_{ab} \equiv \langle e_a | e_b \rangle$, then $(\text{Det } N)^{1/2}$ is the volume spanned by the spacelike basis vectors. Since

$$c = -\langle e^a | D_v e_a \rangle = -\tfrac{1}{2} D_v \ln(\text{Det } N),$$

$c \to \infty$ would mean that the volume approaches zero, and neighboring geodesic lines meet. Hence there exists a basis field $n$, $L_v n = 0$, which becomes zero.

It is not necessarily a sign of a disaster if $c$ does not remain bounded; (4.6.14) made no use of the strict positivity of $i_v R_0$, which means that the same conclusions could be reached in flat space, even though the metric is everywhere regular.

**Example** (4.6.17)

Let us introduce the coordinates $\tau = -\sqrt{t^2 - x^2}$ and $u = x/t$ on $M = \{(t, x) \in \mathbb{R}^2 : t < 0, x < 0, t^2 > x^2; g = -dt^2 + dx^2\}$, and let the vector fields

$$v = \partial_\tau = -\frac{t}{\sqrt{t^2 - x^2}} \partial_t - \frac{x}{\sqrt{t^2 - x^2}} \partial_x,$$

$$n = \partial_u = \frac{xt^2}{t^2 - x^2} \partial_t + \frac{t^3}{t^2 - x^2} \partial x,$$

serve as the basis; it is easy to verify that $D_v v = 0$, $D_v \tau = 1$, and $D_v n = D_n v = -n/\tau$. The streamlines of $v$ are geodesic since they are straight lines through the origin, and the streamlines of $n$ are normal to them in the sense of the metric (Figure 66).

$$c = -\langle v | D_v v \rangle - \frac{\langle n | D_v n \rangle}{\langle n | n \rangle} = \frac{1}{\tau},$$

and in fact

$$D_v c = -\frac{\partial}{\partial \tau} \frac{1}{\tau} = \frac{1}{\tau^2} = c^2,$$

and in one space dimension (4.6.14) can be strengthened to $D_v c \geq c^2$. At the origin $c$ becomes infinite, although the space is not singular; it is only that the chart $(\tau, u)$ is unsuitable there.

When geodesics cross, they lose the property of being extremal, which can bring about some contradictions. This will lead to the conclusion that one

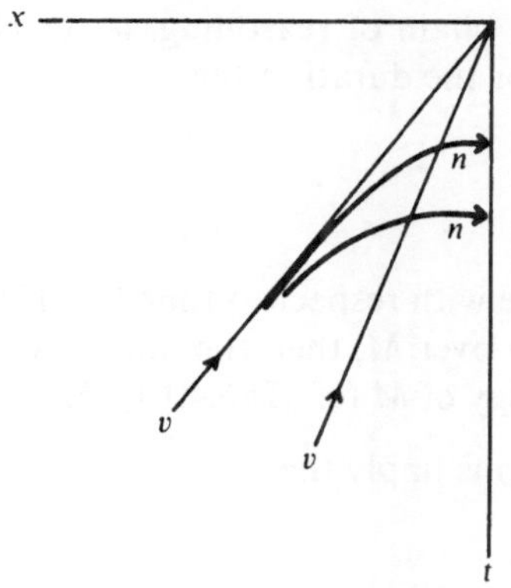


Figure 66 The convergence of a geodesic vector field on flat space

can not make the assumption that geodesics are extensible past the points where they cross. Let us distinguish the different possible states of affairs by making

**Definition** (4.6.18)

(a) The **future** $J^+(x)$ (respectively **past** $J^-(x)$) of $x \in M$ consists of the points of $M$ that can be connected to $x$ by causal curves (see (2.1.15)) directed toward the past (respectively future).
(b) Let $S$ be a spacelike hypersurface and $x \in D^+(S)$. Then the set of causal (respectively, differentiable causal) curves from $x$ to $S$ is denoted by $C(x, S)$ (resp., $C^1(x, S)$). The set $C(x, S)$ is topologized as follows: A basis of neighborhoods of the curve $\lambda$ consists of all curves that stay in a neighborhood of $\lambda$, in the sense of the topology of $M$. We let $C^1$ have the topology induced by $C$.
(c) The **length**, or **proper time**, of $\lambda \in C^1(x, S)$ is defined by

$$d(\lambda) = \int_{s_0}^{s_1} ds \sqrt{-g_{\alpha\beta}\dot{\lambda}^\alpha \dot{\lambda}^\beta}, \qquad \lambda(s_0) \in S, \qquad \lambda(s_1) = x.$$

**Remarks** (4.6.19)

1. The topology on $C(x, S)$ is that of uniform convergence. It is metrizable, since the topology of $M$ is metrizable: the Hausdorff distance function between two subsets of $M$ is defined in terms of the topology of $M$, and the distance function of two causal curves produces the metric on $C(x, S)$. Consequently, compactness becomes synonymous with sequential compactness.
2. $C^1$ is dense in $C$, and so $d(\lambda)$ can be extended to $C$ (Problem 5).

To simplify the next chain of reasoning, let us eliminate pathologies at the outset by making for the duration the

**Assumptions** (4.6.20)

(a) that $M$ is orientable with respect to time (2.1.15); and
(b) that if $x$ and $y$ vary over $M$, then the interiors of $J^-(x) \cap J^+(x)$ form a basis for the topology of $M$ (cf. (I: 6.4.10; 3)).

The above assumptions imply the

**Propositions** (4.6.21)

*For all $x$ in the interior of $D^+(S)$,*
(a) *$J^-(x) \cap D^+(S)$ is compact;*
(b) *$C(x, S)$ is compact; and*
(c) *$d: C^1(x, S) \to \mathbb{R}^+$ is upper semicontinuous.*

**Remarks** (4.6.22)

1. Neither time direction is distinguished, either here or below. Where appropriate, $D^-$(and $J^-$) can be substituted for $D^+$(and $J^+$).
2. Proposition (b) is a variant of Ascoli's theorem, according to which any family of equicontinuous curves on compact sets is relatively compact in the topology of uniform convergence. However, a set of curves of arbitrary gradient is not compact: for example, $x = \sin nt, n = 1, 2, \ldots.$. is not uniformly convergent to anything. The requirement that a curve never gets out of a light-cone prevents this from happening in (4.6.21).
3. Proposition (a) is a necessary condition for (b); for example, $x = \sin t/n$, $n = 1, 2, \ldots$, does not converge uniformly to zero on $-\infty < t \leq 0$.
4. The function $d$ is not continuous, because in any neighborhood of $\lambda \in C(x, S)$ it is possible to reflect lightlike curves back and forth to make $d$ vanish.
5. The extension of $d$ to $C(x, S)$ is likewise upper semi-continuous (Problem 5). We let it define the proper time on a nondifferentiable causal curve.

The proofs of these propositions are rather technical, and are left for Problem 6. An important consequence of them is

**Theorem** (4.6.23)

*Let $S$ be a spacelike hypersurface and $p$ be in the interior of $D^+(S)$. Then there is a curve of greatest proper time from $p$ to $S$, and it is the geodesic through $p$ that is orthogonal to $S$ in the sense of the pseudometric $g$.*

**Proof**

It follows from the compactness of $C$ and the upper semicontinuity of $d$ that $d$ achieves its supremum ([22], 12.7.9). The maximal curve must be geodesic, because otherwise one could find a nearby curve of greater proper time to the same point of intersection with $S$. The orthogonality follows from the requirement that geodesics to other nearby points of $S$ take less time: according to (I: 3.2.18; 6), the change in the time taken on a geodesic line having endpoint $x$ is

$$\delta x^\alpha \frac{\partial}{\partial \dot{x}^\alpha}\sqrt{-\dot{x}^\beta \dot{x}^\gamma g_{\beta\gamma}} = \frac{\delta x^\alpha \dot{x}^\beta g_{\alpha\beta}}{\sqrt{-\dot{x}^\beta \dot{x}^\gamma g_{\beta\gamma}}}.$$

If $\dot{x}$ were not perpendicular to all the tangent vectors of $S$, then one could find a way to increase the time. □

**Example** (4.6.24)

As in Example (4.6.17), let $g = -dt^2 + dx^2$, $p = (t_0, 0)$, and $S = \{(t, x) \in \mathbb{R}^2: t^2 - x^2 = 1, t \leq -1\}$. The straight lines through the origin are the geodesics perpendicular to $S$. The straight lines through $p$, $x = v(t - t_0)$, intersect $S$ where $x^2 = v^2(t - t_0)^2 = t^2 - 1$, so the distance along them to $S$ is $(t_0 - t)\sqrt{1 - v^2} = (1 + t_0^2 - 2t_0 t)^{1/2}$ (see Figure 67). If $t_0 \leq 0$, the maximum is achieved at $t = -1$, since $t \leq -1$, and the geodesic orthogonal to $S$ is then the longest line. If $t_0 > 0$, then the distance grows without bound

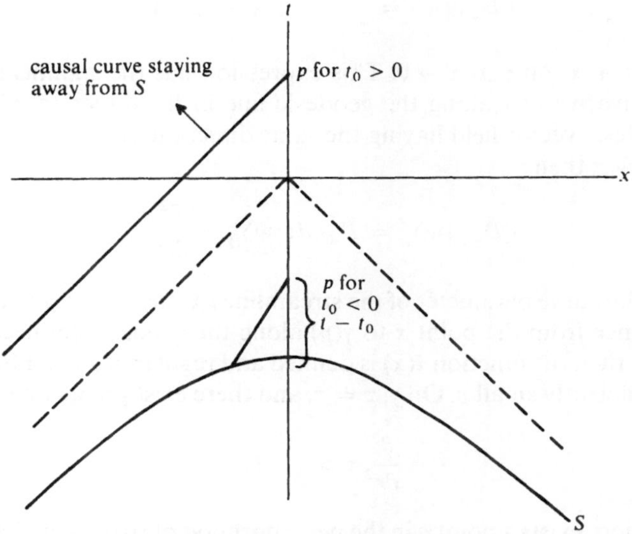


Figure 67 Geodesic lines perpendicular to $S$ in (4.6.24)

as $t \to -\infty$; there exists no maximum, and the line through $(-1, 0)$ is the shortest. This does not contradict (4.6.23), since in this case $p \notin D^+(S)$.

It is intuitively reasonable that one could obtain a more nearly extremal curve from two crossing geodesics, by rounding them off near the intersection. This expectation is confirmed by the

**Theorem** (4.6.25)

*Let $v: D_v v = 0$ and $\langle v | v \rangle = -1$ be the geodesic vector field perpendicular to a hypersurface $S$, let $\gamma(\tau)$ be a streamline of $v$ and suppose $n$ with the properties $L_v n = 0$ and $\langle n | v \rangle = 0$ vanishes at $\gamma(0)$ but not on all of $\gamma$. Then for all $p > 0$, $\gamma$ is not the curve of greatest proper time from $\gamma(p)$ to $S$.*

**Proof**

We choose co-moving coordinates $g = -d\tau^2 + g_{ik}\, dx^i\, dx^k$, and let $\tau$ be the proper time along $-v = \partial_\tau$. Since $n$ satisfies (4.6.11) and vanishes at $\tau = 0$, it is impossible for $D_v n$ to vanish at that point, as it would otherwise be zero on $\gamma$. Therefore, letting $w \equiv n/\tau$, $\lim_{\tau \to 0} \langle w | w \rangle$ is positive (i.e., $n$ is spacelike). Because

$$D_w v = \frac{1}{\tau} D_n v = \frac{1}{\tau} D_v \tau w = D_v w + \frac{w}{\tau}$$

the quantity

$$\langle D_w v\ w \rangle = \frac{1}{\tau} \langle w | w \rangle + \langle D_v w | w \rangle$$

approaches $-\infty$ on $\gamma$ as $\tau \to 0$. This expression has the significance of the second derivative of $\tau$ along the geodesic line in the direction of $w$: Let $\bar{w}$ be the geodesic vector field having the same direction as $w$ on $\gamma$, so $D_{\bar{w}} \bar{w} = 0$ and $w_{|\gamma} = \bar{w}_{|\gamma}$; then

$$\langle D_w v | w \rangle_{|\gamma} = D_{\bar{w}} \langle d\tau | \bar{w} \rangle_{|\gamma} = \frac{\partial^2}{\partial r^2} \tau,$$

where $r$ is the curve parameter of the streamline $dx^\alpha/dr = \bar{w}^\alpha(x(r))$. If $p - \bar{\tau}(x)$ is the distance from the point $x$ to $\gamma(p)$ along the geodesic connecting these two points, then the function $\bar{\tau}(x)$ is defined and regular in a neighborhood of $\gamma(0)$ for sufficiently small $p$. On $\gamma$, $\tau = \bar{\tau}$, and there exist points on $\gamma$ such that

$$\frac{\partial^2}{\partial r^2} \tau > \frac{\partial^2}{\partial r^2} \bar{\tau};$$

therefore there exists a point $q$ in the neighborhood of $\gamma(0)$, for which $\tau_1 > \bar{\tau}_1$. If $\gamma$ intersects $S$ at $\tau = \tau_0 < 0$, then the distance from $\gamma(p)$ to $S$ along $\gamma$ equals

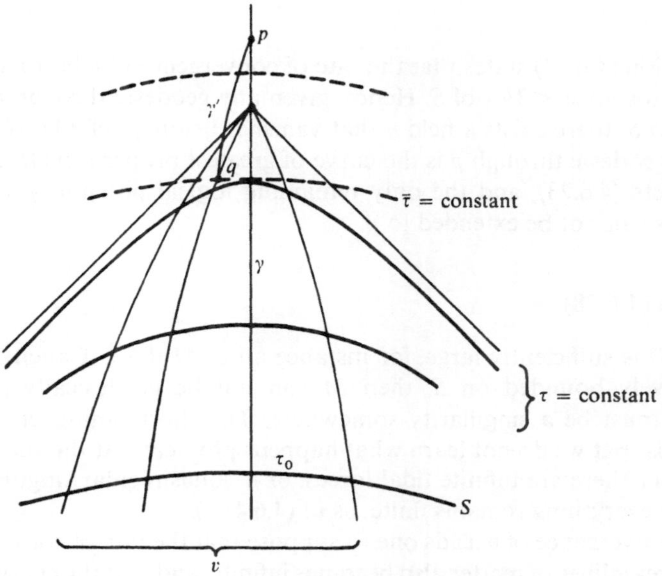


Figure 68 A curve $\gamma'$ having longer proper time than the geodesic $\gamma$

$-\tau_0 + p$. The distance along the geodesics from $p$ to $q$ is $p - \bar{\tau}_1$, and the distance from there to $S$ is $\tau_1 - \tau_0$, so that the total proper time along this path $\gamma'$ is $p - \bar{\tau}_1 + \tau_1 - \tau_0 > p - \tau_0$ (Figure 68). □

**Example** (4.6.26)

In the case of straight lines (4.6.17) and (4.6.24) with $S = \{(t, x) \in \mathbb{R}^2 : t^2 - x^2 = 1, t \leq -1\}$, $\gamma \mathrel{\mathop{=}\limits_{\smash{\bar{}}}}$ the $t$-axis, $\tau = \sqrt{t^2 - x^2}$, and $p = (t_0, 0)$ for $t_0 > 0$, we have $\bar{w} = \partial_x$, and thus $r = x$ and

$$\frac{\partial^2}{\partial r^2} \tau_{|\gamma} = \frac{1}{|t|} > \frac{\partial^2}{\partial r^2} \bar{\tau} = \frac{\partial^2}{\partial x^2} \sqrt{(t - t_0)^2 - x^2}_{|x=0} = \frac{1}{|t - t_0|}$$

for all $t < 0$. The explicit calculation of (4.6.24) confirms the conclusions reached earlier.

Finally, let us collect our results in a

**Theorem** (4.6.27)

*Let $(M, g)$ be orientable with respect to time, $i_v R_0 \geq 0$ for timelike vectors $v$, and let $S \subset M$ be a spacelike hypersurface on which the rate of convergence of the orthogonal geodesic vector field $v$ is always $\geq c_0 > 0$. Then there can not exist a point $p$ in $D(S)$ at a distance greater than $3/c_0$ from $S$.*

**Proof**

Proposition (4.6.14) states that the rate of convergence of $v$ becomes infinite within a distance $\leq 3/c_0$ of $S$. Hence, given any geodesic through $p$ perpendicular to $S$, there exists a field $n$ that vanishes before $p$ (cf. (4.6.16; 3)), and thus no geodesic through $p$ is the curve of greatest proper time from $S$. This contradicts (4.6.23), and the only remaining logical possibility is that the geodesics can not be extended to $p$. □

**Remarks** (4.6.28)

1. If $D(S)$ is sufficiently large, for instance all of $M$ if $S$ is Cauchy, and $c$ is positively bounded on $S$, then $M$ can not be geodesically complete; there must be a singularity somewhere. The shortcoming of this statement is that we do not learn what happens physically at the singularity—whether there are infinite tidal forces, or a "quasiregular singularity" for which everything remains finite, as in (4.6.4; 3).
2. The convergence of $v$ leads one to suppose that the rate of convergence of the streamlines of matter also becomes infinite, and that the energy density is divergent at some point. The difficulty in proving this is that there might exist an earlier, quasiregular singularity, and the time-evolution might stop before reaching an infinite density.
3. In order to draw conclusions from (4.6.23) about the existence of singularities, it is necessary to know something about the size of $D(S)$. If there were a Cauchy surface $S$ with $c_0 > 0$, then singularities would be unavoidable.
4. There are numerous variations and refinements of this theorem [33], yet the precise physical nature of what happens at the singularity is still unclear.

**Examples** (4.6.29)

1. The Friedmann universe (4.6.15) with $S$ equal to the hypersurface at $t =$ constant. This is a Cauchy surface, so every later point lies in $D(S)$. Hence, if $c = -3\dot{R}/R > 0$, then no pressure, however great, can prevent the formation of a singularity.
2. Let $(M, g)$ be $(\mathbb{R}^2, -dt^2 + dx^2)$ and $S = \{(t, x): t^2 - x^2 = 1, t \leq -1, |x| \leq r\}$. Then it can be calculated that

$$D(S) = \{(t, x): t^2 - x^2 > 1, t < x + r - \sqrt{r^2 + 1},$$
$$t < -x - r - \sqrt{r^2 + 1}\}$$

(see Figure 69). In this case, $c_0 = 1$ (cf. (4.6.17)), and the geodesics perpendicular to $S$ leave $D(S)$ at the latest at $r + 1 - \sqrt{1 + r^2} < 1 < 3/c_0$, so (4.6.27) predicts no singularity.

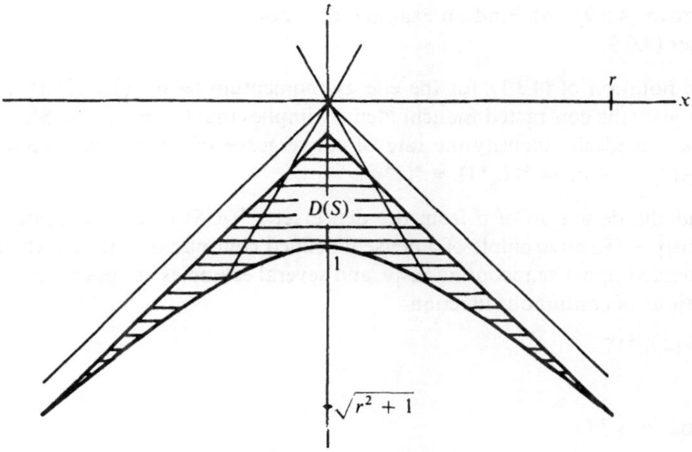


Figure 69 Intersection of geodesics in flat space

3. The Oppenheimer-Snyder Solution (4.5.30). With $S$ such that $\tau = 0$ and $v = \partial_0$, we calculate that

$$c = *di_v*\mathbf{1} = -\left(3\dot{v} + \frac{\dot{v}'}{v'}\right) = \begin{cases} \dfrac{3}{a^{3/2} - 3\tau/2} & \text{if } \tilde{r} < a \\[2ex] \dfrac{3/2}{\tilde{r}^{3/2} - 3\tau/2} & \text{if } \tilde{r} > a. \end{cases}$$

Therefore $c$ is again infinite at the singularity at $r^{3/2} = \tilde{r}^{3/2} - 3\tau/2 = 0$. In the static coordinates, $c \sim r^{-3/2}$, because the radial speed

$$v_r \sim |\text{potential energy}|^{1/2} \sim r^{-1/2}$$

means that $v_{r,r} \sim r^{-3/2}$.

The great interest in singularities is clearly because what is at stake is whether most stars end up as black holes or black holes exist only in peculiar circumstances, and whether the universe originated in a point and will some day return to a point. The question of singularities reveals how incomplete our understanding of natural laws is—are they ever superseded by something beyond human understanding?

**Problems** (4.6.30)

1. Construct an example of a pseudo-Riemannian manifold that is geodesically complete in spacelike and lightlike directions, but not in timelike directions. Do this by choosing $g = \Omega(t, x)(dx^2 - dt^2)$ on $\mathbb{R}^2$, with a suitable function $\Omega$.
2. Find timelike geodesics for the metric of (4.6.4; 3), on which $x + t$ becomes infinite after a finite proper time.

3. (a) Prove (4.6.9). (b) Find an example of a nonsymmetric, real $2 \times 2$ matrix that violates (4.6.9).

4. In the notation of (4.3.1), for the energy-momentum tensor (3.1.25; 3) with $p = 0$, $\mathcal{T}^{\alpha} = \rho v v^{\alpha}$; the contracted Bianchi identity implies that $(\rho v^{\beta} v^{\alpha})_{;\beta} = 0$. Show that this makes $v$ geodesic. Identify the rate of convergence of an arbitrary vector field as $c = -v^{\alpha}{}_{;\alpha} = -\delta v = {}^*(L_v {}^*1) = {}^*(d^*v)$.

5. Extend the definition of $c$ from the dense set $C^1(p, S)$ to $C(p, S)$ upper semicontinuously. Give an example of a densely defined continuous function which can not be extended upper semicontinuously, and several examples of upper semicontinuous extensions of continuous functions.

6. Prove (4.6.21).

## Solutions (4.6.31)

1. Let $\Omega = 1$ for $|x| \geq 1$, $\Omega_{,x}(t, 0) = 0$, and $\lim_{t \to \infty} |t|^{2+\varepsilon}\Omega(t, 0) = 0$ for some $\varepsilon > 0$. Then the time axis is geodesic, and the proper time on it is

$$\int_{-\infty}^{\infty} ds = \int_{-\infty}^{\infty} dt \sqrt{\Omega(t, 0)} < \infty.$$

However, light rays and spacelike lines leave the strip $|x| < 1$ and continue on as in Minkowski space.

2. Let $u = t - x$ and $v = t + x$; then the Lagrangian for the motion becomes

$$\mathcal{L} \equiv \frac{h}{2}\dot{v}^2 - \dot{u}\dot{v}.$$

Consequently

$$\frac{h}{2}\dot{v}^2 - \dot{u}\dot{v} = -1 \quad \text{and} \quad P \equiv \frac{\partial \mathcal{L}}{\partial \dot{v}} = h\dot{v} - \dot{u} = \text{constant}.$$

Therefore we must integrate

$$\dot{u} = \sqrt{P^2 + 2h}, \qquad \dot{v} = \frac{1}{h}(P + \dot{u}) = \frac{2}{\dot{u} - P}.$$

If $P^2 \equiv \lambda$, then

$$\dot{u} = P \cos^2 u \Rightarrow Ps = \tan u \Rightarrow \dot{u} = \frac{P}{P^2 s^2 + 1}$$

$$\Rightarrow \dot{v} = -2\,\frac{1 + 1/P^2 s^2}{P} \Rightarrow v = -\frac{2s}{P} + \frac{2}{sP^3} + \text{constant}.$$

3. (a) If $M = TmT^{-1}$, where $m$ is diagonal, with eigenvalues $m_i$, then (4.6.9) is the Cauchy–Schwarz inequality

$$\left(\sum_i m_i\right)^2 \leq \sum_i 1 \cdot \sum_i m_i^2.$$

(b) If $M = \begin{pmatrix} 1 & 1 \\ -1 & 0 \end{pmatrix}$, then $(\operatorname{Tr} M)^2 = 1$, but $M^2 = \begin{pmatrix} 0 & 1 \\ -1 & -1 \end{pmatrix}$, and $\operatorname{Tr}(M^2) = -1$.

4. Multiply $0 = (\rho v^\beta)_{;\beta} v^\alpha + \rho v^\beta (v^\alpha{}_{;\beta})$ by $v_\alpha$. From $\langle v | v \rangle = -1$ it follows that $v_\alpha (v^\alpha{}_{;\beta}) = 0$, and we conclude that $0 = (\rho v^\beta)_{;\beta}$. In that case, $v^\beta (v^\alpha{}_{;\beta}) = \langle e^\alpha | D_v v \rangle = 0$. The equivalence of the expressions for $c$ follows from

$$L_v {*1} = di_v {*1} = d{*v},$$

$$d({*e^\alpha} v_\alpha) = -v_\alpha \omega^\alpha{}_\beta \wedge {*e^\beta} + dv_\beta \wedge {*e^\beta} = {*\langle e^\beta | dv_\beta - v_\alpha \omega^\alpha{}_\beta \rangle}$$
$$= {*\langle e^\beta | D_{e_\beta} v \rangle} = {*(v^\beta{}_{;\beta})}.$$

5. Let $d(\lambda) = \inf_{C^1 \supset U \ni \lambda} \sup_{\bar{\lambda} \in U} d(\bar{\lambda})$. This is upper semicontinuous and workable as long as the supremum is finite for sufficiently small $U$, which is the case as a corollary of the proof of (4.6.21) (see Problem 6(c)). If $f: \mathbb{R}\backslash\{0\} \to \mathbb{R}$ sends $x \to |x|^{-1}$, then this is not the case at $\{0\}$, and this function can not be extended upper semicontinuously to a function $f: \mathbb{R} \to \mathbb{R}$. Incidentally, the above extension is maximally continuous; for example, to $f: \mathbb{R}\backslash\{0\} \to \mathbb{R}$, $x \to |x|$ it ascribes the value $f(0) = 0$, whereas $f(0) = a > 0$ would make the extension only upper semicontinuous.

6. (a) If $J^-(x) \cap D^+(S)$ were not compact, then there would exist an infinite, locally finite covering with relatively compact neighborhoods $U_i$ with $a_i \in U_i$ for $\{a_i\}$ without a point of accumulation. Let $x \in U_1$ and $\gamma_i$ be a family of causal curves from $x$ to $a_i$. Then $\gamma_i \cap \partial U_1$ has a point of accumulation $h_1$. If $c_1$ is a causal curve from $x$ to $h_1$, then $c_1$ contains a point $x_1$ that lies not only in $U_1$ but also in another set $U_2$. Since $J^-(x_1)$ contains a neighborhood of $h_1$ by Assumption (4.6.20(b)), it also contains an infinite subfamily $\{\gamma_{1_i}\}$ of the $\{\gamma_i\}$, and consequently infinitely many $a_{1_i}$. There is a point of accumulation $h_2$ for $\{\gamma_{1_i}\} \cap \partial U_2 \cap J^-(x_1)$, and there exists a causal curve $c_2$ from $x_1$ to $h_2$, and so on (Figure 70). This procedure yields a causal line connecting $x, x_1, x_2, \ldots$, which can not be extended farther downwards, since the $a_i$ have no point of accumulation. However, for the same reason, it can not intersect $S$, as otherwise one of the relatively compact $U_i$ would contain infinitely many $a_j$. The existence of a nonextensible causal curve not meeting $S$ contradicts the definition of $D^+(S)$, and therefore $J^-(x) \cap D^+(S)$ must be compact.

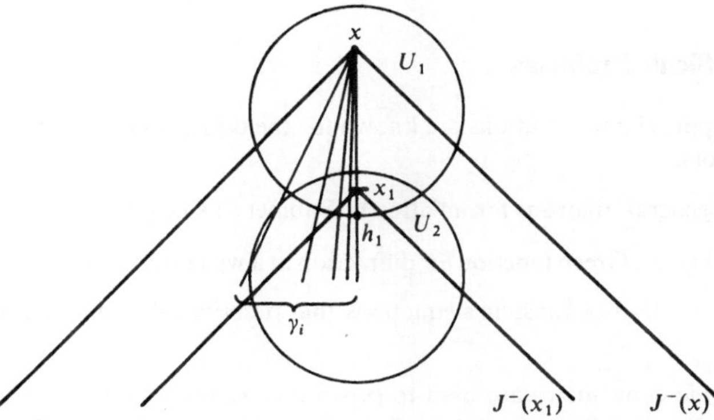


Figure 70 Construction of a nonextensible, causal curve that does not intersect $S$

(b) $C(x, S)$ is compact as a metric space if it is complete and precompact. A uniform limit of causal curves is continuous [(22], 7.2.1), and, because of (4.6.20(b)), causal. Therefore completeness follows immediately. Precompact means that for all $\varepsilon$ there exists a finite covering of $C(x, S)$ with neighborhoods of diameter $\varepsilon$.

Such a neighborhood of a curve $\gamma$ is the set of curves

$$\{\gamma' : \sup_{y \in \gamma'} \inf_{x \in \gamma} \rho(x, y) < \varepsilon\},$$

where $\rho$ is a distance function for the metric on $M$. Since $J^-(x) \cap D^+(S)$ is compact, we can cover it with finitely many $A_i$ = the interior of $J^-(a_i) \cap J^+(a_i')$, $i = 1, \ldots, n$, with diameter $< \varepsilon$. Let $x$ be in $A_1$, and form all unions

$$B_j = \bigcup_{k=1}^{n_j} A_{j_k},$$

with

$$A_{j_1} = A_1, \qquad A_{j_{n_j}} \cap S \neq \emptyset,$$
$$A_{j_k} \cap A_{j_{k+1}} \neq \emptyset, \qquad J^-(A_{j_k}) \cap A_{j_{k+2}} \neq \emptyset.$$

The $B_j' = \{\gamma \in C(x, S) : \gamma \subset B_j\}$ are a covering for $C(x, S)$, since every causal curve of $x$ must be in some such union. The number of the $B_j$ is finite, and their diameter $< \varepsilon$.

(c) We need to show that for all $\varepsilon$ there exists a neighborhood $U$ of $\lambda$ such that $d(\bar{\lambda}) < d(\lambda) + \varepsilon$ for all $\bar{\lambda} \in U$. For this purpose, we use co-moving coordinates moving with $\lambda$, so that $\lambda$ is one of the time axes $x^j$ = constant, $j = 1, 2, 3$, and the $g_{0j}$ vanish. Since $d$ does not depend on the choice of the curve parameter, we can take this parameter as $x^0$; we then have to compare

$$d(\lambda) = \int_\lambda dx^0 \sqrt{-g_{00}}$$

with

$$d(\bar{\lambda}) = \int_{\bar{\lambda}} dx^0 \sqrt{-g_{00} - g_{jk}\dot{\bar{\lambda}}^j \dot{\bar{\lambda}}^k}.$$

But since $g_{jk}\dot{\bar{\lambda}}^j \dot{\bar{\lambda}}^k > 0$ and $g_{00}$ is uniformly continuous, as a continuous function on the compact set $J^-(x) \cap D^+(S)$, we can make $d(\bar{\lambda}) < d(\lambda) + \varepsilon$ by taking a small enough neighborhood.

## Some Difficult Problems

1. Only approximate solutions are known for the diffraction at a slit. Find bounds for the errors.
2. Give a general, rigorous formulation of Babinet's principle [2].
3. Show that the Green function for diffraction at a wedge has causal support properties.
4. Find a solution of Einstein's equations that describes the emission of gravitational waves.
5. Harmonic coordinates are used to prove that Einstein's equations are hyperbolic. Give a purely geometric formulation of this state of affairs, without reference to particular coordinates.
6. Discover singularity theorems that show that $M$ is not only geodesically incomplete, but that the curvature invariants are in fact unbounded, in the right circumstances.
7. Solve the general relativistic two-body problem.

# Bibliography

## Works Cited in the Text


[1] J. D. Jackson. Classical Electrodynamics. New York: Wiley, 1975.

[2] W. Franz. Theorie der Beugung elektromagnetischer Wellen, Ergebnisse der angew. Math., vol. 4. Berlin: Springer, 1957.

[3] T. Fulton and F. Rohrlich. Classical Radiation from a Uniformly Accelerated Charge. *Ann. Phys.* **9**, 499–517, 1960.

[4] K. Yano. The Theory of Lie Derivatives and Its Applications. Amsterdam: North-Holland, 1955.

[5] H. M. Nussenzveig. High Frequency Scattering by an Impenetrable Sphere. *Ann. Phys.* **34**, 23–95, 1965.

[6] F. London. Superfluids, vol. I: Macroscopic Theory of Superconductivity. New York: Wiley, 1950.

[7] J. L. Anderson. Principles of Relativity Physics. New York: Academic Press, 1967.

[8] R. U. Sexl, H. K. Urbantke. Gravitation und Kosmologie, BI-Hochschultaschenbuch, Mannheim: BI-Wissenschaftsverlag, 1974.

[9] C. W. Misner, K. S. Thorne, J. A. Wheeler. Gravitation. San Francisco: Freeman, 1973.

[10] S. Weinberg. Gravitation and Cosmology. New York: Wiley, 1972.

[11] J. M. Souriau. Géometrie et Rélativité. Paris: Hermann, 1964.

[12] M. D. Kruskal. Maximal Extension of Schwarzschild Metric. *Phys. Rev.* **119**, 1743–1745, 1960.

[13] J. C. Graves and D. R. Brill. Oscillatory Character of Reissner–Nordström Metric for an Ideal Charged Wormhole. *Phys. Rev.* **120**, 1507–1513, 1960.

[14] B. Carter. The Complete Analytic Extension of the Reissner–Nordström Metric in the Special Case $e^2 = m^2$. *Phys. Letters* **21**, 423–424, 1966.

[15] A. Trautman. Theory of Gravitation. In: The Physicist's Conception of Nature, J. Mehra, ed. Boston: D. Reidel, 1973.

[16] C. N. Yang. Integral Formalism for Gauge Fields. *Phys. Rev. Lett.* **33**, 445–447, 1974.
[17] W. Rühl. Finite Conformal Transformations in Local Quantum Field Theory, in: Electromagnetic Interactions and Field Theory, *Acta Phys. Austriaca Suppl.* **XIV**, 643–646, 1975.
[18] M. Schönberg. *Revista Brasileira de Fisica 1*, 91, 1971.
[19] A. Uhlmann. *Wissenschaftliche Zeitschrift der Friedrich-Schiller-Universität* **8**, 31, 1958.
[20] M. Abramowitz and I. E. Stegun, eds. Handbook of Mathematical Functions, Applied Mathematics Series 55. Washington: National Bureau of Standards, 1964.
[21] F. Hehl, P. von der Heyde, and G. D. Kerlick. General Relativity with Spin and Torsion: Foundations and Prospects. *Rev. Mod. Phys.* **48**, 393–416, 1976.
[22] J. Dieudonné. Foundations of Modern Analysis, vols. III and IV. New York: Academic Press, 1972 and 1974.
[23] A. Schild. Electromagnetic Two-Body Problem. *Phys. Rev.* **131**, 2762–2766, 1962.
[24] F. Rohrlich. Classical Charged Particles: Foundations of Their Theory. Reading, Mass.: Addison-Wesley, 1965.
[25] L. P. Eisenhart. Riemannian Geometry. Princeton: Princeton Univ. Press, 1949.
[26] V. Fock. The Theory of Space, Time and Gravitation. New York: Macmillan, 1964.
[27] A. Lichnerowicz. Théories Rélativistes de la Gravitation et de l'Electromagnetisme: Rélativité Générale et Théories Unitaires. Paris: Masson, 1955.
[28] J. Wess and B. Zumino. Superspace Formulation of Supergravity. *Phys. Letters* **66B**, 361–364, 1977.
[29] A. Trautman. Conservation Laws in General Relativity. In: Gravitation, an Introduction to Current Research, L. Witten, ed. New York: Wiley, 1962.
[30] E. Pechlaner and R. Sexl. On Quadratic Lagrangians in General Relativity. *Comm. Math. Phys.* **2**, 165–175, 1966.
[31] F. Hoyle and J. V. Narlikar. Cosmological Models in Conformally Invariant Gravitational Theory—II. A New Model. *Mon. Not. Roy. Astr. Soc.* **155**, 323–335, 1972.
[32] L. Landau and E. Lifshitz. The Classical Theory of Fields. Reading, Mass.: Addison-Wesley, 1977.
[33] G. F. R. Ellis and S. W. Hawking. The Large Scale Structure of Space-Time. Cambridge: At the University Press, 1973.
[34] R. P. Geroch. What is a Singularity in General Relativity? *Ann. Phys.* **48**, 526–540, 1968.
[35] M. Fierz and R. Jost. Affine Vollständigkeit und kompakte Lorentz'sche Mannigfaltigkeiten. *Helv. Phys. Acta* **38**, 137–141, 1965.
[36] F. Hoyle et al. In: Quasi-stellar Sources and Gravitational Collapse, I. Robinson, ed. Chicago: The Univ. of Chicago Press, 1965.
[37] M. Reed and B. Simon. Methods of Modern Mathematical Physics, in four volumes. New York: Academic Press, 1974–1979.
[38] H. Bondi and T. Gold. The Field of a Uniformly Accelerated Charge, with Special Reference to the Problem of Gravitational Acceleration. *Proc. Roy. Soc. London* **A229** 416–424, 1955.
[39] W. Thirring and R. Wallner. The Use of Exterior Forms in Einstein's Gravitation Theory. *Revista Brasileira de Fisica*, to appear.

## Further Reading

*1. Alternating Differential Forms*


H. Cartan. Differential Calculus. Paris: Hermann, 1971.

H. Cartan. Differential Forms. Paris: Hermann, 1970.

G. A. Deschamps. Exterior Differential Forms. In: Mathematics Applied to Physics, E. Roubine, ed. New York: Springer, 1970.

H. Flanders. Differential Forms with Applications to the Physical Sciences. New York: Academic Press, 1963.

S. J. Goldberg. Curvature and Homology. New York: Academic Press, 1962.

W. Greub, S. Halperin, and R. Vanstone. Connections, Curvature, and Cohomology. New York: Academic Press, 1972.

H. Holmann and H. Rummler. Alternierende Differentialformen. Mannheim: BI-Wissenschaftsverlag, 1972.


*2. Tensor Analysis and Geometry of Manifolds*


L. Auslander and R. E. MacKenzie. Introduction to Differential Manifolds. New York: McGraw-Hill, 1963.

R. L. Bishop and R. J. Crittenden. Geometry on Manifolds. New York: Academic Press, 1964.

R. L. Bishop and S. I. Goldberg. Tensor Analysis on Manifolds. New York: Macmillan, 1968.

F. Brickell and R. S. Clark. Differentiable Manifolds. New York: Van Nostrand-Reinhold, 1970.

T. Bröcker and K. Jänich. Einführung in die Differentialtopologie, Heidelberger Taschenbuch 143. Heidelberg: Springer, 1968.

Y. Choquet–Bruhat, C. DeWitt–Morette, and M. Dillard–Bleick. Analysis, Manifolds, and Physics. Amsterdam: North-Holland, 1978.

D. Gromoll, W. Klingenberg, and W. Meyer. Riemannsche Geometrie im Großen, Lecture Notes in Mathematics, 55. New York: Springer, 1968.

N. J. Hicks. Notes on Differential Geometry. New York: Van Nostrand-Reinhold, 1971.

S. Kobayashi and K. Nomizu. Foundations of Differential Geometry, vols. I and II. New York: Interscience, 1963 and 1969.

A. Lichnerowicz. Elements of Tensor Analysis. New York: Wiley, 1962.

C. W. Misner. Differential Geometry. In: Relativity, Groups, and Topology, C. DeWitt and B. S. DeWitt, eds. New York: Gordon and Breach, 1964.

E. Nelson. Tensor Analysis. Princeton: Princeton Univ. Press, 1967.

S. Sternberg. Lectures on Differential Geometry. Englewood Cliffs, N.J.: Prentice-Hall, 1964.

T. J. Willmore. An Introduction to Differential Geometry. Oxford: Oxford Univ. Press, 1959.

J. A. Wolf. Spaces of Constant Curvature. New York: McGraw-Hill, 1967.


*3. General Relativity*


R. Adler, M. Bazin, and M. Schiffer. Introduction to General Relativity. New York: McGraw-Hill, 1965.

A. Einstein. The Meaning of Relativity. Princeton: Princeton Univ. Press, 1955.
W. Pauli. Theory of Relativity. New York: Pergamon, 1958.
W. Rindler. Essential Relativity. New York: Springer, 1977.
R. U. Sexl and H. K. Urbantke. Gravitation und Kosmologie, BI-Hochschultachsenbuch. Mannheim: BI-Wissenschaftsverlag, 1975.
J. L. Synge. Relativity, the General Theory. Amsterdam: North-Holland, 1965.
A. Trautman, F. Pirani, and H. Bondi. Lectures on General Relativity. Englewood Cliffs, N.J.: Prentice-Hall, 1972.

*4. Global Analysis*

Y. Choquet-Bruhat and R. Geroch. Global Aspects of the Cauchy Problem in General Relativity. *Comm. Math. Phys.* **14**, 329–335, 1969.
G. F. R. Ellis and D. W. Sciama. Global and Nonglobal Problems in Cosmology. In: General Relativity, Papers in Honor of J. L. Synge, L. O'Raifeartaigh, ed. Oxford: The Clarendon Press, 1972.
D. Farnsworth, J. Fink, J. Porter, and A. Thomson, eds. Methods of Local and Global Differential Geometry in General Relativity, Lecture Notes in Physics 14. New York: Springer, 1972.
R. P. Geroch. Topology in General Relativity. *J. Math. Phys.* **8**, 782–786, 1967.
R. P. Geroch. Domain of Dependence. *J. Math. Phys.* **11**, 437–449, 1970.
R. P. Geroch. Space-Time Structure from a Global Point of View. In: General Relativity and Cosmology, R. K. Sachs, ed. New York: Academic Press, 1971.
ICTP, Global Analysis and its Applications, vols. I, II, and III. Lectures Presented at an International Seminar Course at Trieste from 4 July to 25 August, 1972. New York: Unipub, 1975.
W. Kundt. Global Theory of Spacetime. In: Differential Topology, Differential Geometry and Applications, J. R. Vanstone, ed. Montreal: Canadian Mathematical Congress, 1972.
A. Lichnerowicz. Topics on Space-Time. In: Batelle Rencontres: 1967 Lectures in Mathematics and Physics. C. DeWitt and J. A. Wheeler, eds. New York: Benjamin, 1968.
R. Penrose. Structure of Space-Time. *Ibid.*

*5. Proceedings, Summer Schools, and Collected Papers*

P. G. Bergmann, E. J. Fenyves, and L. Motz, eds. Seventh Texas Symposium on Relativistic Astrophysics. *Annals of the New York Acad. of Sci.* **262**, 1975.
M. Carmeli, S. Fickler, and L. Witten, eds. Relativity. New York: Plenum, 1970.
H-Y. Chiu and W. F. Hoffman, eds. Gravitation and Relativity. New York: Benjamin, 1964.
C. DeWitt and J. A. Wheeler, eds. Batelle Rencontres: 1967 Lectures in Mathematics and Breach, 1964.
C. DeWitt and J. A. Wheeler, eds. Batelle Rencontres: 1967 Lectures in Mathematics and Physics. New York: Benjamin, 1968.
Editorial Committee. Recent Developments in General Relativity. New York: Macmillan, 1962.
J. Ehlers, ed. Relativity Theory and Astrophysics. Providence: Amer. Math. Soc., 1967.
W. Israel, ed. Relativity, Astrophysics, and Cosmology. Boston: D. Reidel, 1973.

C. W. Kilmister, ed. General Theory of Relativity, Selected Readings in Physics. New York: Pergamon, 1973.

C. G. Kuper and A. Peres, eds. Relativity and Gravitation. New York: Gordon and Breach, 1971.

L. O'Raifeartaigh, ed. General Relativity, Papers in Honor of J. L. Synge. Oxford: The Clarendon Press, 1972.

R. K. Sachs, ed. General Relativity and Cosmology, Proceedings of Course 47 of the International School of Physics "Enrico Fermi." New York: Academic Press, 1971.

G. Shaviv and J. Rosen, eds. General Relativity and Gravitation. New York: Wiley, 1975.

P. Suppes, ed. Space, Time, and Geometry. Boston: D. Reidel, 1973.

J. R. Vanstone, ed. Differential Topology, Differential Geometry and Applications, Proceedings of the Thirteenth Biennial Seminar of the Canadian Mathematical Congress. Montreal: Canadian Mathematical Congress, 1972.

L. Witten, ed. Gravitation, an Introduction to Current Research. New York: Wiley, 1962.

*Section 4.1*

R. L. Bishop and S. I. Goldberg. Tensor Analysis on Manifolds. New York: Macmillan, 1968.

N. J. Hicks. Notes on Differential Geometry. New York: Van Nostrand-Reinhold, 1971.

S. Kobayashi and K. Nomizu. Foundations of Differential Geometry, vols. I and II. New York: Interscience, 1963 and 1969.

B. Schmidt. Differential Geometry from a Modern Standpoint. In: Relativity, Astrophysics, and Cosmology, W. Israel, ed. Boston: D. Reidel, 1973.

*Section 4.2*

J. Gel'fand and S. Fomin. Calculus of Variations. Englewood Cliffs, N.J.: Prentice-Hall, 1963.

P. Havas. On Theories of Gravitation with Higher Order Field Equations, *Gen. Rel. Grav.* **8**, 631, 1977.

D. Lovelock and H. Rund. Variational Principles in the General Theory of Relativity. *Jahresbericht der Deutschen Mathematiker-Vereinigung*, **74**, No. 1/2, 1972.

A. Trautman. Conservation Laws in General Relativity. In: Gravitation, an Introduction to Current Research, L. Witten, ed. New York: Wiley, 1962.

*Sections 4.3 and 4.4*

S. Helgason. Lie Groups and Symmetric Spaces. In: Batelle Rencontres: 1967 Lectures in Mathematics and Physics, C. DeWitt and J. A. Wheeler, eds. New York: Benjamin, 1968.

*Section 4.5*

B. K. Harrison, K. S. Thorne, M. Wakano, and J. A. Wheeler. Gravitational Theory and Gravitational Collapse. Chicago: The Univ. of Chicago Press, 1965.

H. Scheffler and H. Elsässer, Physik der Sterne und der Sonne. Mannheim: BI-Wissenschaftsverlag, 1974.

Ya. B. Zel'dovich and I. D. Novikov. Relativistic Astrophysics, vols. I and II. Chicago: The Univ. of Chicago Press, 1971.

*Section 4.6*

C. J. S. Clarke. The Classification of Singularities. *Gen. Rel. Grav.* **6**, 35–40, 1975.

C. J. S. Clarke. Space-Time Singularities. *Comm. Math. Phys.* **49**, 17–23, 1976.

G. F. R. Ellis and B. Schmidt. Singular Space-Times. *Gen. Rel. Grav.* **8**, 915–953, 1977.

R. P. Geroch. Singularities in the Spacetime of General Relativity, Their Definition, Existence, and Local Characterization. Dissertation, Princeton University, 1967.

R. P. Geroch, What is a Singularity in General Relativity? *Ann. Phys.* **48**, 526–540, 1968.

S. W. Hawking. Singularities and the Geometry of Spacetime. Essay submitted for the Adams Prize, Cambridge, 1966.

S. W. Hawking. The Occurrence of Singularities in Cosmology I, II, III. *Proc. Roy. Soc. London* **294A**, 511–521, 1966. *Ibid.* **295A**, 490–493, 1966. *Ibid.* **300A**, 187–201, 1967.

W. Kundt. Recent Progress in Cosmology, Springer Tracts in Modern Physics, 47. New York: Springer, 1968.

R. Penrose. Gravitational Collapse and Space-Time Singularities. *Phys. Rev. Lett.* **14**, 57–59, 1965.

# Index

# A Course in Mathematical Physics 4

# A Course in Mathematical Physics

**Volume 1**

## Classical Dynamical Systems

**By W. Thirring**

**Translated by E.M. Harrell**

This textbook for physics, mathematics, and applied mathematics students is a unique combination of mathematical rigor and realistic physical approach. Starts with the development of the concept of a manifold and proceeds to a discussion of Hamiltonian systems, canonical transformations, and constants of motion. Among specific problems discussed in detail are nonrelativistic motion of particles and systems, relatilistic motion in electromagnetic and gravitational fields, and the structure of black holes. Numerous helpful examples are included.

1978. ISBN 0-387-**81496**-5

**Volume 2**

## Classical Field Theory

**By W. Thirring**

**Translated by E.M. Harrell**

This is a modern and extensive treatment of the classical theories of electromagnetism and gravitation. Among the intriguing areas covered are the mathematical and practical aspects of space time and black hole physics. Other topics include the electromagnetic field of a known charge distribution; the field in the presence of conductors; and gravitation.

1979. ISBN 0-387-**81532**-5

**Volume 3**

## Quantum Mechanics of Atoms and Molecules

**By W. Thirring**

**Translated by E.M. Harrell**

Applying the same mathematical rigor as the first two books, this unique and comprehensive volume explores simple systems in quantum mechanics. The author employs the most modern mathematical methods to produce concrete results comparable to experimental fact. Not only does the text introduce basic axioms, from which quantum mechanics are subsequently derived, but it also demonstrates relevant applications in an in-depth and self-contained fashion.

1981. ISBN 0-387-**81620**-8

**Volume 4**

## Quantum Mechanics of Large Systems

**By W. Thirring**

**Translated by E.M. Harrell**

*A Course in Mathematical Physics* concludes with an innovative development of statistical quantum mechanics from a distinctly modern perspective. Much of the material in this volume has never before appeared in book form. Discounting the traditional perturbation-theoretical calculations, Dr. Thirring concentrates on a number of successful treatments of fundamental issues. Topics explored include: properties of entropy, noncommutative ergodic theory, the proof of the existence for thermodynamic functions, and a mathematical analysis of Thomas-Fermi theory.

1983. ISBN 0-387-**87101**-8

Walter Thirring

# A Course in Mathematical Physics

4

## Quantum Mechanics of Large Systems

Translated by Evans M. Harrell

Springer-Verlag Wien GmbH

Dr. Walter Thirring
Institute for Theoretical Physics
University of Vienna
Austria

Dr. Evans M. Harrell
The Johns Hopkins University
Baltimore, Maryland
U.S.A.

Translation of Lehrbuch der Mathematischen Physik
Band 4: Quantenmechanik grosser Systeme
Wien · New York: Springer-Verlag 1980



ISBN 978-3-7091-7528-6 ISBN 978-3-7091-7526-2 (eBook)
DOI 10.1007/978-3-7091-7526-2

Library of Congress Cataloging in Publication Data

Thirring, Walter E., 1927
Quantum mechanics of large systems.

(A course in mathematical physics; 4)
Translation of: Quantenmechanik grosser Systeme.
Bibliography: p.
Includes index.
1. Statistical thermodynamics. 2. Statistical mechanics I. Title. II. Series: Thirring, Walter E., 1927– Lehrbuch der mathematischen Physik. English; 4.
QC20.T4513 vol. 4 [QC311.5] 530.1′5s [530.1′33] 82-19159

With 39 Figures


Originally published by Springer-Verlag Wien New York in 1983
Softcover reprint of the hardcover 1st edition 1983



Typeset by Composition House Ltd., Salisbury, England.

9 8 7 6 5 4 3 2 1

ISBN 978-3-7091-7528-6

# Preface

In this final volume I have tried to present the subject of statistical mechanics in accordance with the basic principles of the series. The effort again entailed following Gustav Mahler's maxim, "Tradition = Schlamperei" (i.e., filth) and clearing away a large portion of this tradition-laden area. The result is a book with little in common with most other books on the subject.

The ordinary perturbation–theoretic calculations are not very useful in this field. Those methods have never led to propositions of much substance. Even when perturbation series, which for the most part never converge, can be given some asymptotic meaning, it cannot be determined how close the $n$th order approximation comes to the exact result. Since analytic solutions of nontrivial problems are beyond human capabilities, for better or worse we must settle for sharp bounds on the quantities of interest, and can at most strive to make the degree of accuracy satisfactory.

The last two decades have seen successful and beautiful treatments of many fundamental issues—I have in mind the ordering of the states (2.1), properties of the entropy (2.2), noncommutative ergodic theory (3.1), the proof of the existence of the thermodynamic functions (4.3), and the mathematical analysis of Thomas–Fermi theory (4.1.2), which provides an understanding of the stability of matter. The day is surely not far off when most of the remaining holes in the conceptual structure of quantum statistical mechanics will have been filled in and the questions that are not satisfactorily answered today will be added to the list of achievements.

The successful completion of this course of mathematical physics in a reasonable time required the fortunate conjunction of several circumstances. As with volume III, I had active support from several collaborators, and in particular I am greatly obliged to B. Baumgartner, H. Narnhofer, A. Pflug, and A. Wehrl. Countless other colleagues have helped indirectly by coping with other time-consuming duties for me. The English edition has again greatly benefited from the critical reading of B. Simon. The working conditions at the University of Vienna were invaluable for the completion of this project. Last but not least, the frictionless collaboration of Springer-Verlag in Vienna and my secretary and calligrapher F. Wagner enabled the books to appear quickly and at a reasonable price.

I am aware that the uncompromising way of mathematical physics is not the easiest. Yet I feel that it has been one of the greatest intellectual accomplishments of our era to cast the laws of Nature in a clear mathematical form with rigorously deducible consequences. No amount of labor is too high a price to have paid for this. Let me conclude by also acknowledging and expressing my thanks to the reader who has borne with me to the end of the course.

Walter Thirring

# Contents

# Symbols Defined in the Text

| | | |
|---|---|---|
| $\sigma, \sigma^z, \sigma^\pm$ | spin components | (1.1.1) |
| $\mathscr{H}_F$ | Fock space | (1.3.1) |
| Ⓢ, $\Lambda$ | symmetric and antisymmetric tensor product | (1.3.1) |
| $a^*(f)$, $a(f)$ | creation and annihilation operators | (1.3.2) |
| $[\ \ ]_+$ | anticommutator | (1.3.3; 2) |
| $\mathscr{A}_B$ | $C^*$-algebra for bosons | (1.3.3; 3) |
| $\mathscr{A}_F$ | $C^*$-algebra for fermions | (1.3.3; 4) |
| $a(\mathbf{x})$ | annihilation operator | (1.3.3; 7) |
| $\Omega$ | cyclic vector | (1.3.5) |
| $\mathscr{A}_G, \mathscr{A}_E$ | even and gauge-invariant algebras | (1.3.8) |
| Tr | trace | (1.4.10) |
| I, II, III | factors of type I, II, III | (1.4.16) |
| $\rho(n)$ | sum of the first $n$ eigenvalues | (2.1.9) |
| $\preceq$ | ordering of density matrices | (2.1.10; 1) |
| $S_\alpha$ | $\alpha$-entropy | (2.2.2) |
| $S(\rho)$ | von Neumann entropy | (2.2.4) |
| $d\Omega_z^N$ | Liouville measure on phase space | (2.2.7) |
| $\rho_{qu}$ | quantum-theoretic phase-space density | (2.2.7) |
| $\rho_{cl}$ | classical phase-space density | (2.2.7) |
| $S(\sigma \mid \rho)$ | relative entropy | (2.2.22) |
| $\varepsilon, \sigma, \rho$ | energy, entropy and particle densities | (2.3.8) |
| $T$ | temperature | (2.3.16) |
| $C_V, c_V$ | total and specific heat capacity at constant volume | (2.3.17; 3) |
| $P$ | pressure | (2.3.21) |
| $\kappa$ | compressibility | (2.3.22; 3) |
| $\mu$ | chemical potential | (2.3.27) |
| $v_u$ | smeared potential | (2.4.9) |
| $\varphi(T, \rho)$ | Legendre transform of $\varepsilon$ | (2.4.14) |
| $\mathscr{L}$ | Legendre transformation | (2.4.15; 2) |

| | | |
|---|---|---|
| $z$ | fugacity | (2.5.9) |
| tr | trace on the one-particle space | (2.5.10) |
| $v^u$ | unsmeared potential | (2.5.17) |
| $F_\sigma(z)$ | generalized $\zeta$ function | (2.5.20) |
| m | magnetization per volume | (2.5.37) |
| $\mathscr{R}$ | covariance algebra | (3.1) |
| $U$ | unitary operators | (3.1) |
| $B_{\text{eff}}$ | effective magnetic field | (3.1.1; 4) |
| $\tau_t$ | time-automorphism | (3.1.2) |
| $\tau_t^*$ | dual time-evolution | (3.1.2) |
| $\eta, \eta(a), \eta(\sigma)$ | invariant mean | (3.1.14), (3.1.15) |
| $E_0$ | projection onto eigenvectors of $H$ with eigenvalue 0 | (3.1.16; 1) |
| $a_t$ | transformed operator | (3.2.16; 2) |
| $J$ | conjugate-linear operator | (3.2.1) |
| $\pi'$ | conjugate-linear representation | (3.2.1) |
| $\tilde{A}$ | algebra of analytic elements | (3.2.6; (v)) |
| $\sim$ | Fourier transform | (3.2.6; (v)) |
| $\tau_t^h$ | perturbed time-automorphism | (3.3.4) |
| $R_h$ | corresponding operator | (3.3.2) |
| $F_{ab}, G_{ab}$ | correlation function | (3.3.14) |
| $\mathbf{X}_k, Z_k$ | coordinates and charges of the nuclei | (4.1.3; 1) |
| $W(\mathbf{x})$ | wall potential | (4.1.3; 4) |
| $H_n$ | Hamiltonian with an effective field | (4.1.6) |
| $C_n$ | correlation correction | (4.1.6) |
| $\Xi(H)$ | grand canonical partition function | (4.1.8) |
| $\\|n\\|_c$ | $c$-norm | (4.1.10) |
| $h_n$ | one-particle Hamiltonian | (4.1.17) |
| $v_s$ | singular part of the potential | (4.1.18) |
| $a_{p,q}, \rho_{p,q}$ | annihilation and density operators | (4.1.25) |
| $K, A, R$ | contributions to the energy | (4.1.33) |
| $\Phi(\mathbf{x})$ | potential | (4.1.36) |

## Symbols Defined in Earlier Volumes

| | | |
|---|---|---|
| $\mathscr{B}(\mathscr{H})$ | bounded operators | (III: 2.1.24) |
| $\mathscr{C}_1(\mathscr{H})$ | trace-class operators | (III: 2.3.21) |
| $\mathscr{C}_2(\mathscr{H})$ | Hilbert–Schmidt operators | (III: 2.3.21) |
| $\otimes$ | tensor product | (I: 2.4.5) |
| $\oplus$ | direct sum of Hilbert spaces | (III: 2.1.9; 2) |
| $\mathscr{A}'$ | commutant of $\mathscr{A}$ | (III: 2.3.4) |
| $\\|a\\|_p$ | trace norm | (III: 2.3.21) |

# Systems with Many Particles 1

## 1.1 Equilibrium and Irreversibility

*Macroscopic bodies act in an irreversible and deterministic manner in contrast with the reversible and indeterministic character of the underlying laws of quantum physics. How can the apparent contradiction be understood?*

We have learned to describe systems of finitely many particles with an algebra $\mathscr{A}$ of observables, and information about the systems with a state $w$ on the algebra (cf. (III: 2.2.32)). As our main goal is the study of everyday matter, our framework will be that of nonrelativistic quantum theory. For the purposes of contrast, or of aiding intuition, we shall also have occasion to call upon classical mechanics, where states are measures on phase space, and extremal states are point measures. In either framework time-evolution can be represented as an automorphism $a \to a_t$ for $a \in \mathscr{A}$ in the Heisenberg picture. If desired, time-dependence can alternatively, in the Schrödinger picture, be put upon the state: $w \to w_t$, such that $w_t(a) = w(a_t)$. If the algebra is Abelian (classical mechanics), then the point of an extremal state moves along a classical trajectory in phase-space.

In our earlier experience, systems of $N$ particle are so complex for large $N$ that it becomes impossible to reach precise, quantitative conclusions. It turns out, however, that the theoretical analysis again simplifies in the limit $N \to \infty$. Many properties become independent of the exact number of particles and other detailed characteristics of the physical system, somewhat in analogy to what happens in the central limit theorem of probability theory. This may seem peculiar at first; we have always had $\mathscr{A} = \mathscr{B}(\mathscr{H})$, $\mathscr{H}$ a separable Hilbert space, and time-evolution was given by a unitary group on $\mathscr{H}$. What, then, appears so special about a many-particle system? Just that the information contained in a pure state about a many-particle system is so overwhelming that it would be too ambitious to employ the whole of $\mathscr{B}(\mathscr{H})$ for the observables. Actual measurements could never be made on more than a few observables, so $\mathscr{B}(\mathscr{H})$ has to be cut down to size. For instance, suppose that a device is only equipped to observe one particle at a time, and is unable to detect correlations between particles. Then, rather than taking the entire tensor product of the individual particles as the algebra of observables, it is reasonable to regard $\mathscr{A}$ as a single factor. Accordingly, many states differing on $\mathscr{B}(\mathscr{H})$ reduce to the same state when restricted to $\mathscr{A}$. (The classical situation is similar; the restriction of

$$w(\mathbf{q}_1, \ldots, \mathbf{p}_N)$$

is

$$\int d^3q_2 \ldots d^3q_N \, d^3p_2 \ldots d^3p_N \, w(\mathbf{x}_1, \ldots, \mathbf{p}_N),$$

so whole cylindrical regions of phase-space reduce to a single restricted state.) As a consequence large portions of the space of states on $\mathscr{B}(\mathscr{H})$ are quite similar from the point of view of the reduced algebra $\mathscr{A}$. If, in the Schrödinger picture, the state $w_t$ travels throughout the space of states, then its restriction takes on a certain value with a very high probability, unless prevented by some constants of the motion. This most probable state is called the **equilibrium state** over $\mathscr{A}$.

The irreversible tendency toward equilibrium has always aroused wonder, especially as the basic equations of dynamics are invariant under reversal of the motion (III: 3.3.18). We have even seen in classical mechanics that the trajectory of any point on a compact energy surface returns arbitrarily close to its initial position (I: 2.6.13). In quantum theory the Hamiltonian $H$ of a system confined to a finite volume has purely discrete spectrum. If $\varepsilon_j$ and $|j\rangle$ denote the eigenvalues and eigenvectors of $H$, then the time-dependence of an observable $a$ is given by

$$w_t(a) = \sum_{j,k} c_j^* c_k \exp(it(\varepsilon_j - \varepsilon_k))\langle j|a|k\rangle,$$

where the state $w$ is represented by the vector $\sum_j c_j|j\rangle$. The state $w_t(a)$ is now an almost-periodic function of $t$; if the sum is finite, and the $\varepsilon_j$ are rationally dependent, then it is actually strictly periodic. At any rate, to arbitrarily good accuracy, $w_t(a)$ again becomes nearly $w(a)$ after some sufficiently long delay. The trouble is that the recurrence times are so unimaginably long that they have no physical relevance. Suppose, for instance, that there are $N$ distinct energy differences $\omega_j$. The recurrence time can then be estimated as follows. The factors $\exp(i\omega_j t)$ can be pictured as $N$ clocks with hands moving at $N$ different rates. The question is how long it takes for a certain configuration of clock faces to reappear to within some angular accuracy $\Delta\varphi$. The configuration in the space of angles has measure $(\Delta\varphi/2\pi)^N$, so the recurrence time is on the order of $(\Delta\varphi/2\pi)^{-N}/\omega$, where the reciprocal angular velocity $1/\omega$ is an average of the $1/\omega_j$. Even for just $N = 10$, $1/\omega = 1$ sec., and $(\Delta\varphi/2\pi) = 1/100$, so that $w_t$ returns to $w$ to within 1 % accuracy, the recurrence time is $10^{20}$ sec., which is much longer than the age of the universe.

The approach to equilibrium is connected to a loss of information; to be more precise, information does not get lost, but only less accessible. We have seen that when the wave-packet of a free particle spreads (III: 3.3.3), $\Delta x$ grows linearly with time, although the state remains pure and thus has maximal information content. The observable with least deviation from the mean is, however, not $x(t)$ but $x(0) \equiv x(t) - pt$.

This behavior can be seen even in classical motion if a minimal spread of the support of the probability distribution function in phase space is hypothesized to account for quantum effects. If, say, the initial probability density $\rho(p, q)$ is concentrated on a part of the energy shell $\{(q; p) | p_1 \leq p \leq p_2\}$ and is not pointlike, and it moves freely on a torus, then it eventually fills the energy shell densely with a "fuzzy" distribution. Faster particles overtake the slower ones, as bicycles racing in a stadium start packed closely together but later draw apart and eventually spread around the whole track (see Figure 1).

The ergodic hypothesis has figured importantly in the history of statistical mechanics; it is the assumption that the trajectory of almost every point winds densely around the energy shell in phase space, so that the time average can be replaced with the average over the energy shell. On the one hand this requires more than is necessary, since it suffices to fill a sufficiently typical part of the energy shell, the average on which equals the average on the whole shell for the reduced algebra of observables. On the other hand, although macroscopic measurements last much longer than the collision time, they last much less than the recurrence time, so one does not wait for the whole energy shell to be sampled. We shall discuss examples in which the equilibrium state is actually attained by the state in a reasonable time after reduction to one particle.

A pictorial description of the situation is as follows. The information about a subsystem (i.e., the opposite of the entropy, to be defined later) as a function on the space of states of the total system consists mainly of a plain with few hills and still fewer mountains. The larger the total system, the further apart the prominences. Even if a path begins on a peak, it soon descends to the plain, and there is only the slightest probability that it will ascend another mountain in any conceivable time. The time of descent to the plain and the recurrence time are of completely different orders of magnitude. It takes only the time corresponding physically to a few collisions to descend to a level near that of the plain, whereas the other mountains lie in the unfathomable distance. This means that equilibrium is reached long before the immense recurrence time required to wind throughout the space of states;

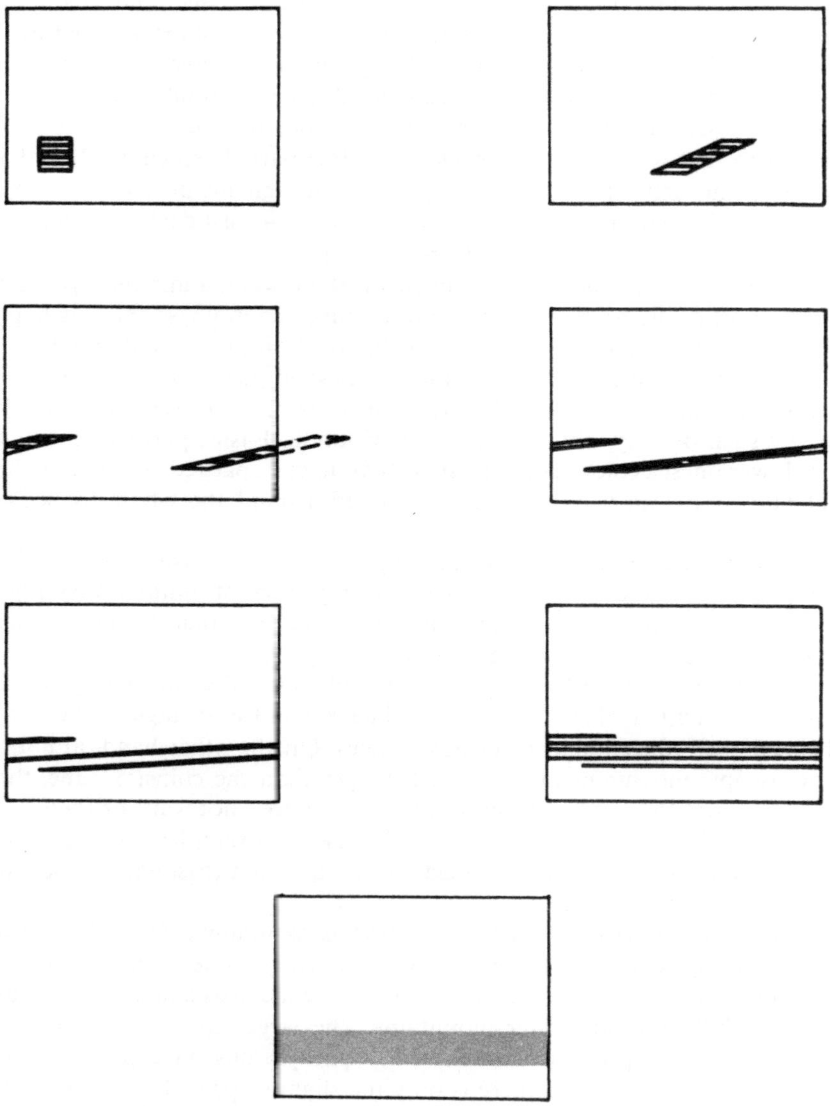

Figure 1 The motion of the density in phase space for a free particle on a torus.

generally, a path soon reaches states that can not be distinguished from equilibrium because of the limits of our measuring abilities. Of course, there is still the question of how one happened, at the beginning, to be at the top of the mountain, but that brings up the one of how the current state of the universe came about and is outside the scope of this book.

Another puzzle is the apparent causal behavior that classical thermodynamics prescribes for macroscopic bodies. According to the arguments that have been advanced, one would rather suspect that the fluctuations of the observables are increased by the loss of information. This is actually true for microscopic variables like the positions and momenta of individual particles. However, if only the so-called macroscopic observables are considered, that is, roughly what was accessible to the more primitive experimental arts of an earlier epoch, then deterministic features arise. Their origin is simply that statistically independent quantities are being averaged: if $a = (1/N)\sum_{j=1}^{N} a_j$, where $w(a_i a_j) = w(a_i)w(a_j)$ for $i \neq j$, then

$$(\Delta a)^2 = \frac{1}{N^2}\left[w\left(\sum_{j,k}(a_j a_k)\right) - \sum_{j,k} w(a_j)w(a_k)\right] = \frac{1}{N^2}\sum_{j=1}^{N}(\Delta a_j)^2.$$

Thus $\Delta a \sim N^{-1/2}$, and for sufficiently large $N$ the deviations from the average are negligible. We shall learn that in the quantum-theoretical formalism such an $a$ approaches a multiple of the identity operator as $N \to \infty$. The limiting coefficient depends on the representation of the algebra.

Let us verify the phenomena described above in two explicitly soluble models. Of necessity they will lack some of the complications arising in reality, but they exhibit the important features. They are embryonic forms of systems of fermions and bosons.

### **The Chain of Spins** (1.1.1)

Let the algebra of observables of the total system be generated by $\boldsymbol{\sigma}_j$, $j = 1, \ldots, N$, where each $\boldsymbol{\sigma}_j$ is a copy of the usual Pauli matrices $\boldsymbol{\sigma}$. Instead of Cartesian components we use $\sigma \equiv \sigma^z$ and $\sigma^{\pm} \equiv (\sigma^x \pm i\sigma^y)/2$, which satisfy the commutation relations

$$[\sigma_j, \sigma_k^{\pm}] = \pm\,\delta_{jk} 2\sigma_k^{\pm},$$
$$[\sigma_j^{+}, \sigma_k^{-}] = \delta_{jk}\sigma_k. \tag{1.1.2}$$

The chain is closed by the identification of $\boldsymbol{\sigma}_{j+N}$ with $\boldsymbol{\sigma}_j$, and the Hamiltonian that determines the time-evolution will be assumed to be of the form

$$H = B\sum_{j=1}^{N}\mu_j\sigma_j + \sum_{n=1}^{N-1}\sum_{j=1}^{N}\sigma_j\sigma_{j+n}\varepsilon(n). \tag{1.1.3}$$

The physical meaning of this is that the spins are coupled with magnetic moments $\mu_j$ to an external magnetic field $B$, and in addition there is an Ising-like spin–spin interaction with the $n$th neighbor. The strength $\varepsilon(n)$ of this interaction is a function that can be specified later, and the periodicity allows us to assume $\varepsilon(n) = 0$ for $n > N/2$. If the contributions to $H$ are denoted as in

$$H \equiv H_0 + \sum_n H_n, \tag{1.1.4}$$

then the $H_k$ commute with one another and with the $\sigma_j$. They are therefore constant in time, and the time-evolution of $\sigma^+$ and $\sigma^- = (\sigma^+)^*$ can be calculated easily from the relationship

$$f(\sigma)\sigma^+ = \sigma^+ f(\sigma + 2), \tag{1.1.5}$$

which follows from (1.1.2). We find

$$\begin{aligned}\sigma_k^+(t) = (\sigma_k^-(t))^* &= \sigma_k^+(0)\exp\left\{2it\left[B\mu_k + \sum_n \varepsilon(n)(\sigma_{k+n} + \sigma_{k-n})\right]\right\}\\ &= \sigma_k^+(0)\exp(2itB\mu_k)\prod_n(\cos 2t\varepsilon(n) + i\sigma_{k+n}\sin 2t\varepsilon(n))(\cos 2t\varepsilon(n)\\ &\quad + i\sigma_{k-n}\sin 2t\varepsilon(n)),\end{aligned} \tag{1.1.6}$$

where $a(t) = \exp(iHt)a\exp(-iHt)$.

The time-evolution consists of Larmor precession in the external field and a kind of diffusion along the chain due to the spin–spin interaction. Suppose that the state at $t = 0$ is pure and has the form of a product, where the spins have a 3-component $s$ and $\sigma_k^+$ has phase $\alpha_k$:

$$\langle\sigma_k(0)\rangle = s, \qquad \langle\sigma_k^+(0)\rangle = \tfrac{1}{2}\sqrt{1 - s^2}\exp(i\alpha_k), \qquad \left\langle\prod_j \boldsymbol{\sigma}_j\right\rangle = \prod_j\langle\boldsymbol{\sigma}_j\rangle. \tag{1.1.7}$$

Then

$$\begin{aligned}\langle\sigma_k^+(t)\rangle &= \tfrac{1}{2}\sqrt{1 - s^2}\exp\{i(\alpha_k + 2tB\mu_k)\}f^2(t),\\ f(t) &= \prod_{n=1}^{N/2}(\cos 2t\varepsilon(n) + is\sin 2t\varepsilon(n)).\end{aligned} \tag{1.1.8}$$

If $N$ is finite, then $f$ is almost periodic, and if $N = \infty$, then $f(t)$ will generally tend to zero as $t \to \infty$ (supposing that $\varepsilon(n)$ tends to zero in such a way that the infinite product makes sense). To make this more explicit, let us consider the special case $s = 0$ and $\varepsilon(n) = 2^{-n-1}$. If $N = \infty$, then $f$ satisfies the equation

$$f(t) = \prod_{n=1}^{\infty}\cos 2^{-n}t = \frac{f(2t)}{\cos t}. \tag{1.1.9}$$

Since $f$ is an entire function, this functional equation and the condition $f(0) = 1$ determine $f$ uniquely—differentiate (1.1.9) to get the Taylor series of $f$. Since the function $(\sin t)/t$ satisfies (1.1.9), it equals $f$. Hence, as $N \to \infty$, the expectation value of $\sigma^{\pm}$ approaches zero. For finite $N$ it follows from (1.1.9) that

$$f_N(t) = \prod_{n=1}^{N/2} \cos 2^{-n}t = \frac{\sin t}{t}\left[\frac{\sin t2^{-N/2}}{t2^{-N/2}}\right]^{-1}. \tag{1.1.10}$$

Therefore, as discussed earlier, the recurrence time $2^{N/2}/\pi$ grows exponentially with $N$, while the time it takes to reach equilibrium is independent of $N$.

To summarize, we have ascertained that for $N = \infty$ the initially pure state of the algebra reduced to one spin tends as $t \to \infty$ to $\langle\sigma\rangle = s, \langle\sigma^{\pm}\rangle = 0$, which corresponds to a mixture:

$$\langle\boldsymbol{\sigma}\rangle = \mathrm{Tr}(\rho\boldsymbol{\sigma}), \qquad \rho = \frac{\exp(-\eta\sigma)}{\mathrm{Tr}\exp(-\eta\sigma)}, \qquad \tanh\eta = s. \tag{1.1.11}$$

Even though the expectation values of the $\sigma_k^{\pm}$ go to zero, their fluctuations remain nonzero, since $\sigma_k^{+}\sigma_k^{-} = (1 + \sigma_k)/2$ is constant. The average magnetization

$$\mathbf{M}_N(t) = \frac{1}{N}\sum_k \boldsymbol{\sigma}_k(t) \tag{1.1.12}$$

works differently. In the state (1.1.7) of our example, $\langle M_N^z\rangle = s$, whereas $\langle M_N^{\pm}\rangle$ is $O(N^{-1/2})$, provided either that the initial phases are disordered or that the $\sigma_k^{\pm}$ get out of phase after a while because the $\mu_k$ differ. The latter situation can in fact be undone by a sudden reversal of $B$, in the spin–echo effect. If $N = \infty$, the diffusion caused by suitable $\varepsilon(n)$ is irreversible, and $\lim_{t\to\infty}\langle M_{\infty}^{\pm}(t)\rangle = 0$. At $t = 0$ the fluctuations are $O(N^{-1/2})$ and remain at this magnitude for all time: If $\sigma_k^{+}(t)\sigma_{k'}^{-}(t)$ is calculated by multiplying together two expressions of the form (1.1.6), then it should be recalled that $\sigma^2 = 1$. However, if the function $\varepsilon(n)$ falls off sufficiently rapidly with $n$, then the $\sigma^2$ terms make little difference for large $k - k'$, and the argument given earlier for the deviations of statistically independent quantities remains valid.

**Chain of Oscillators** (1.1.13)

Now represent the total system by positions and momenta $q_1, \ldots, q_N$, $p_1, \ldots, p_N$, such that $[q_j, p_k] = i\delta_{jk}$, and let the time-evolution be determined by

$$H = \sum_{j=1}^{N} \tfrac{1}{2}(p_j^2 + (q_j - q_{j+1})^2). \tag{1.1.14}$$

This Hamiltonian contains interactions only between nearest neighbors, and the chain can be closed by the condition of periodicity $q_{j+N} = q_j, p_{j+N} = p_j$. The masses and force constants have been set to 1, which amounts to measuring the time in units of the natural period of oscillation. The equations of motion are

$$\dot{q}_j = p_j, \qquad \dot{p}_j = q_{j+1} + q_{j-1} - 2q_j. \tag{1.1.15}$$

With a periodic extension of the variables, $\xi_1, \ldots, \xi_{2N}$, such that

$$\xi_{2n} = p_n, \qquad \xi_{2n+1} = q_{n+1} - q_n, \tag{1.1.16}$$

they are put into the form

$$\dot{\xi}_j = \xi_{j+1} - \xi_{j-1}. \tag{1.1.17}$$

The variables $\xi_n$ satisfy

$$\xi_{n+2N} = \xi_n, \qquad \sum_n \xi_{2n+1} = 0.$$

Recall that the Bessel functions satisfy the recursion formula $\dot{J}_n = (J_{n-1} - J_{n+1})/2$; as a consequence we see that the solution of the initial-value problem is

$$\xi_n(t) = \sum_{k=-\infty}^{\infty} \xi_k(0) J_{k-n}(2t). \tag{1.1.18}$$

**Remarks** (1.1.19)

1. Since $|J_\nu(z)| \sim |z/\nu|^{|\nu}$ as $|\nu| \to \infty$, the sum over $k$ in (1.1.18) converges for, say, bounded $\{\xi_k(0)\}$.
2. If $N < \infty$, then (1.1.18) still holds provided that $\xi_{k+2N}(0) = \xi_k(0)$.
3. Since the equations of motion are linear, the classical and quantum time-automorphisms are identical.
4. There are still $N$ constants of motion with the variables $\xi$:

$$I_k = \sum_{j=1}^{2N} \xi_j \xi_{j+n}, \qquad k = 1, \ldots, N.$$

With the auxiliary condition that $\sum_n \xi_{2n+1} = 0$, only $N - 1$ of the constants are independent, and we find that $\sum_n I_{2n+1} = 0$. If $N = \infty$, then $I_k$ remains significant classically, provided that $\{\xi_k\} \in l^2$.

In order to have a useful framework for discussing the questions that will arise as in these two examples, it is convenient for technical reasons to make use of the Weyl algebra (cf. (III, §3.1)). With one particle, the Weyl algebra consists of the operators $W(r + is) = \exp(i(pr + qs))$, $r, s \in \mathbb{R}$, along with their linear combinations and norm-limits. A state on the Weyl algebra is uniquely characterized by the function $E(r, s) \equiv \langle \exp(i(pr + qs)) \rangle$. We shall only concern ourselves with coherent states (III: 3.1.13), which are of the form $W(z')\, u\rangle$, where $|u\rangle$ is a Gaussian function, the width of which determines the ratio between $\Delta p$ and $\Delta q$. Since

$$\langle u | W(r + is) | u \rangle = \exp\left[-\frac{1}{4}\left(\omega r^2 + \frac{s^2}{\omega}\right)\right],$$

it follows that

$$(\Delta p)^2 = -\frac{d^2}{dr^2} \ln E_{|r,s=0} = \frac{\omega}{2}, \qquad (\Delta q)^2 = -\frac{d^2}{ds^2} \ln E_{|r,s=0} = \frac{1}{2\omega}.$$

The expectation value in the more general state $W(z')|u\rangle$ can be calculated according to (III: 3.1.2; 1) as

$$\begin{aligned}\langle W(z')u|W(z)|W(z')u\rangle &= \langle u|W(-z')W(z)W(z')|u\rangle \\ &= \langle u|W(z)|u\rangle \exp\left[\frac{i}{2}\operatorname{Im}(z^*z' - z^{*\prime}z)\right] \\ &= \exp\left[-\frac{1}{4}\left(\omega r^2 + \frac{s^2}{\omega}\right) + i(rs' - r's)\right]. \qquad (1.1.20)\end{aligned}$$

Thus, the quantities $\Delta p$ and $\Delta q$ are the same as with $|u\rangle$, but the expectation values of $p$ and $q$ are now $s'$ and $-r'$.

Let us return to the issue of how the restriction of the many-particle state to a subsystem evolves in time. The operators $\exp[i(r\xi_0(t) + s\xi_1(t))]$, which describe the momentum of a single particle and its position relative to its neighbor, are useful to this end. Since $[\xi_0(t), \xi_1(t)] = i$, they form a Weyl system. A state characterized by

$$\left\langle \exp\left[i \sum_{n=-\infty}^{\infty} (\xi_{2n}r_n + \xi_{2n+1}s_n)\right]\right\rangle = \exp\left[-\frac{1}{4}\sum_{n=-\infty}^{\infty}\left(\omega r_n^2 + \frac{s_n^2}{\omega}\right) + i(r_n s_n' - r_n' s_n)\right] \qquad (1.1.21)$$

can be regarded as the generalization of (1.1.20).

**Remarks** (1.1.22)

1. The exponent on the left is a linear combination of $p_k$ and $q_k$, as appropriate for a Weyl system for several particles, yet the variables $\xi_{2n}$ and $\xi_{2n+1}$ are not pairs of canonically conjugate variables, since $[\xi_{2n}, \xi_{2n-1}] \neq 0$. Thus (1.1.21) is not simply the tensor product of coherent states of a tensor product of Weyl systems.
2. The significance of (1.1.21) is once again that the variables $\xi_{2n}$ (resp. $\xi_{2n+1}$) all have deviation $\omega$ and expectation values $s_n'$ (resp. $1/\omega$ and $-r_n'$).

With (1.1.21), the desired state on the one-particle system turns out to be

$$\begin{aligned}E(r, s) &\equiv \langle \exp(i(r\xi_0(t) + s\xi_1(t)))\rangle \\ &= \left\langle \exp\left(i \sum_{n=-\infty}^{\infty} [\xi_{2n}(0)(rJ_{2n} + sJ_{2n-1}) + \xi_{2n+1}(0)(rJ_{2n+1} + sJ_{2n})]\right)\right\rangle \\ &= \exp \sum_{n=-\infty}^{\infty} \left\{-\frac{1}{4}\left(\omega(rJ_{2n} + sJ_{2n-1})^2 + (rJ_{2n+1} + sJ_{2n})^2 \frac{1}{\omega}\right)\right. \\ &\qquad \left. + is_n'(rJ_{2n} + sJ_{2n-1}) - ir'(rJ_{2n+1} + sJ_{2n})\right\}. \qquad (1.1.23)\end{aligned}$$

The sums can be evaluated by recourse to the formulas

$$\sum_{n=-\infty}^{\infty} J_{2n}(2t)J_{2n+j}(2t) = \tfrac{1}{2}(\delta_{0j} + J_j(4t)), \qquad j \in \mathbb{Z},$$

$$\sum_{n=-\infty}^{\infty} J_{2n+1}(2t)J_{2n+1+j}(2t) = \tfrac{1}{2}(\delta_{0j} - J_j(4t)), \tag{1.1.24}$$

which are derived in Problem 2. As $t \to \infty$, only the terms with $j = 0$ remain. Moreover, it can be seen from the integral representations and the Riemann–Lebesgue lemma that the contributions linear in the $J_k$ go to zero as $t \to \infty$. In all, we get

$$\lim_{t\to\infty} E(r, s) = \exp\left[-\frac{1}{4}\left(\omega + \frac{1}{\omega}\right)(r^2 + s^2)\right]. \tag{1.1.25}$$

**Remarks** (1.1.26)

1. The limiting state corresponds to the mixture $E = \mathrm{Tr}\, \rho W(z)$, $\rho = \exp[-\eta(p_1^2 + q_1^2)]/\mathrm{Tr} \exp[-\eta(p_1^2 + q_1^2)]$, $\coth \eta = (\omega + 1/\omega)/2$ (Problem 3). As $\omega \to 1$, that is, for minimal mean-square deviation, $\eta \to \infty$, and the state becomes pure. With larger mean-square deviations, $\omega \neq 1$, $(\omega + 1/\omega)/2 > 1$, the limiting state is a mixture.
2. Whereas at $t = 0$ the ratio of $\Delta p$ to $\Delta q$ is $\omega^2$, they become equal as $t \to \infty$, i.e., their ratio, 1, becomes the one defined by $H$. This corresponds to equal amounts of kinetic and potential energy.
3. The reason that the existence of the constants (1.1.19; 4) does not prevent the onset of equilibrium is again the choice of the initial state. Of course, equilibrium can not occur if the system starts off in an eigenstate of a normal mode of oscillation.

These few remarks will serve as our first orientation to irreversible phenomena. We have already studied an example of an irreversible phenomenon in volume II, the emission of light. It is always important to take the limit $N \to \infty$ before $t \to \infty$, as in a finite volume the light returns to the point of emission, and the behavior is almost periodic rather than irreversible. The next section will deal with how the energy is affected by the first limiting process.

**Problems** (1.1.27)

1. Calculate the entropy $S(t) = -\mathrm{Tr}\, \rho(t) \ln \rho(t)$ for one spin, where $f$ is given by (1.1.9).
2. Calculate $\sum_{n=-\infty}^{\infty} J_{2n}(x)J_{2n+j}(x)$ and $\sum_{n=-\infty}^{\infty} J_{2n+1}(x)J_{2n+1+j}(x)$.
3. Show that the density matrix $\rho$ has the property stated in (1.1.26; 1).

**Solutions** (1.1.28)

1. Since Tr $\rho(t) = 1$, the density matrix is of the form $\rho(t) = \frac{1}{2} + \mathbf{c}(t)\cdot\boldsymbol{\sigma}$. Let $c(t) = |\mathbf{c}(t)|$, which $\leq \frac{1}{2}$. The eigenvalues of $\rho(t)$ are $\frac{1}{2} \pm c(t)$, so

$$S(t) = -\left[\frac{1+c(t)}{2}\ln\frac{1+c(t)}{2} + \frac{1-c(t)}{2}\ln\frac{1-c(t)}{2}\right].$$

Because $\operatorname{Tr}\sigma_i\sigma_j = 2\delta_{ij}$, we find $\mathbf{c}(t) = \frac{1}{2}\langle\boldsymbol{\sigma}\rangle$, and therefore $c(t) = (s^2 + (1-s^2)f^4(t))^{1/2}$. Observe that $f$ is not monotonic, and hence that $S$ does not increase monotonically from 0 to its equilibrium value,

$$-\left[\frac{1+s}{2}\ln\left(\frac{1+s}{2}\right) + \frac{1-s}{2}\ln\left(\frac{1-s}{2}\right)\right].$$

2.
$$\exp\left[\frac{z}{2}\left(t - \frac{1}{t}\right)\right] = \sum_{n=-\infty}^{\infty} t^n J_n(z).$$

Putting $z = x + y$ yields

$$\sum_{j=-\infty}^{\infty} t^j J_j(x+y) = \exp\left[\frac{x}{2}\left(t-\frac{1}{t}\right)\right]\exp\left[\frac{y}{2}\left(t-\frac{1}{t}\right)\right]$$
$$= \left(\sum_k t^k J_k(x)\right)\left(\sum_l t^l J_l(y)\right) = \sum_{j=-\infty}^{\infty} t^j \sum_{n=-\infty}^{\infty} J_n(x)J_{j-n}(y),$$

so $J_j(x+y) = \sum_{n=-\infty}^{\infty} J_n(x)J_{j-n}(y)$, which is the addition theorem of Schläfli and Neumann. Putting $y = -x$ and changing $j$ to $-j$ then yields $\sum_n J_n(x)J_{n+j}(x) = \delta_{j0}$, and with $y = x$, there results

$$\sum_n J_n(x)J_{-j-n}(x) = \sum_n (-1)^{n+j}J_n(x)J_{n+j}(x) = J_{-j}(2x) = (-1)^j J_j(2x),$$

from which formulas (1.1.24) follow.

3.
$$\operatorname{Tr}\exp[-\eta(p_1^2 + q_1^2)] = \sum_{n=0}^{\infty}\exp[-\eta(1+2n)]$$

and

$$\langle p^2 + q^2\rangle = \left(-\frac{\partial^2}{\partial r^2} - \frac{\partial^2}{\partial s^2}\right)E(r,s) = -\frac{\frac{\partial}{\partial\eta}\operatorname{Tr}\exp[-\eta(p^2+q^2)]}{\operatorname{Tr}\exp[-\eta(p^2+q^2)]}$$

lead to the result.

## 1.2 The Limit of an Infinite Number of Particles

*The first issues to confront for large systems are what happens to macroscopic properties like energy and volume as $N \to \infty$.*

The models examined in § 1.1 were only caricatures of reality. We shall now determine the physical properties of large bodies. The first question is how the volume $V$ has to vary as $N \to \infty$, in order to ensure that the potential and kinetic energies will be comparable in magnitude and that the interaction between the particles is correctly accounted for. In particular, when are $E$ and $V$ normal, extensive quantities proportional to $N$? In order to fix our ideas, we shall pay particular attention to certain special cases, large atoms and macroscopic or cosmic objects. The dominant force is then electrostatic, except that in cosmic matter gravity also has a decisive effect. Heuristic arguments will sometimes be adduced in this section for guidance in finding which quantities have limits as $N \to \infty$ in these systems.

**Free Particles** (1.2.1)

We begin with a consideration of noninteracting particles confined to a box of side $R$. The energy consists of the quantum-mechanical zero-point energy plus a thermal component proportional to the temperature $T$. As we are only interested in the dependence on $N$ for large $N$, we set $\hbar = k = m = 1$. As explained in (III: 1.2.11) the zero-point energy of a system of fermions is $\sim (\Delta p)^2 \sim (\Delta x)^{-2}$, where $\Delta x$ is about $RN^{-1/3}$, since the volume available per fermion is only $R^3/N$. We arrive at

$$E = \frac{N^{5/3}}{2R^2} + \tfrac{3}{2}NT. \tag{1.2.2}$$

If the two contributions are to remain comparable as $N \to \infty$, and if $T$ goes as $N^t$ for some power $t$, then $R$ must be $\sim N^{1/3-t/2}$, and $EN^{-1-t}$ will tend to a limiting value. The type of interaction will determine the value of $t$ at which the limit is nontrivial and thus of physical interest. For this to happen the kinetic and potential energies have to remain of the same order of magnitude.

Bosons do not have the solitary temperament, so $\Delta x$ may be set equal to $R$. The energy is then on the order of

$$E = \frac{N}{2R^2} + \tfrac{3}{2}NT. \tag{1.2.3}$$

If the two contributions are to have the same dependence on $N$ and we make $T \sim N^t$, then $R \sim N^{-t/2}$ and $E \sim N^{t+1}$. If it is insisted that $T$ remain constant and $R \sim N^{1/3}$, then $E \sim N$, but the zero-point energy drops below the thermal energy. The exact calculation for free bosons in fact reveals that, with a fixed particle density and below a critical temperature, a certain fraction $\lambda(T) > 0$ of the particles are to be found in the ground state with $E_0 \sim N^{1/3}$, and thus $N$ may be replaced with $(1 - \lambda(T))N$. This makes this usual limit also nontrivial.

**Large Atoms** (1.2.4)

The Hamiltonian of a large atom (with $e^2 = 1$) is

$$H = \sum_{i=1}^{N} \left( \frac{|\mathbf{p}_i|^2}{2} - Z|\mathbf{x}_i|^{-1} \right) + \sum_{i>j} |\mathbf{x}_i - \mathbf{x}_j|^{-1}, \tag{1.2.5}$$

which can, if one wishes, be confined in a box. Recall that in volume III we figured out that if $T = 0$ and $Z = N$, the energy is about $N^{5/3}/2R^2 - N^2e^2/R$, which has a minimum about $-\frac{1}{2}N^{7/3}$ for $R \sim N^{-1/3}$. Therefore, in the limit $N \to \infty$ we should expect to set $t = \frac{4}{3}$. In §4.1 it will not only be proved that these limits converge, but even that the Thomas–Fermi theory becomes exact in that limit. The problem can thus be solved in the limit $N \to \infty$, though the solution is not suitable for a direct numerical comparison of theory and experiment. Since there are corrections of about $N^{-1/3}$, 10% accuracy can not be expected for $N \lesssim 10^3$. On the other hand, relativistic effects become significant when $N \sim 10^2$. The kinetic energy is then $\sim N^{4/3}/R$ and if $Ze^2 > 1$ the energy is no longer bounded below. Hence the picture that emerges of a large atom is only an idealization, but at least one with many instructive aspects.

Systems of bosons depend on $N$ in a different way. They all settle into the ground state, and with $Z \sim N$ the radius goes as $N^{-1}$ and the energy as $N^3$. The limits of $EN^{-3}$ and $N^3\rho(xN)$ would be expected to exist, where $\rho$ is the one-particle density distribution. For thermal effects to remain significant, $T$ must be chosen $\sim N^2$. This problem is mostly of academic interest, and the convergence of the quantities mentioned above has not yet been proved.

**Jellium** (1.2.6)

Like an atom, jellium consists of particles repelling one another with a Coulomb force and immersed in the field of an external charge distribution. The difference is that the charge distribution is not concentrated at a point, but rather homogeneously spread with density $\xi$ through a box $\Lambda$ ($\Lambda$ will also sometimes denote the volume of $\Lambda$). It can be regarded as a model of highly compressed matter, with the homogeneous background charge coming from fast-moving electrons, and the particles with explicit coordinates being the nuclei. It is nevertheless often used to describe electrons in a metal, although it is rather far-fetched to speak of the assemblage of ions as a homogeneous background. The Hamiltonian is

$$H = \sum_{i=1}^{N} \frac{|\mathbf{p}_i|^2}{2} + \sum_{i>j} |\mathbf{x}_i - \mathbf{x}_j|^{-1} - \sum_{i=1}^{N} U(\mathbf{x}_i) + \frac{\xi}{2} \int_\Lambda d^3x U(\mathbf{x}), \tag{1.2.7}$$

where $U(\mathbf{x}) = \xi \int_\Lambda d^3x'/|\mathbf{x} - \mathbf{x}'|$. For the system to be neutral, $\xi \int_\Lambda d^3x = N$. The electrostatic energy of the background has been added in so that the potential energy will remain bounded below by $N(RN^{-1/3})^{-1}$, where $R$ is the linear dimension of $\Lambda$. The proof of this relies on the well-known fact of electrostatics that the Coulomb repulsion of two homogeneously charged spheres is less than or equal to that of two point charges at their centers—the inequality occurs when they overlap. Now imagine blowing the charged particles up to homogeneously charged spheres of radius $a$, and let

$$\left(\frac{4\pi a^3}{3}\right)^{-2} \int_{\substack{|\mathbf{x}-\mathbf{x}_i|\leq a \\ |\mathbf{x}'-\mathbf{x}_j|\leq a}} \frac{d^3x\, d^3x'}{|\mathbf{x}-\mathbf{x}'|} = U_{ij}(a),$$

$$\left(\frac{4\pi a^3}{3}\right)^{-1} \int_{|\mathbf{x}-\mathbf{x}_i|\leq a} d^3x U(\mathbf{x}) = U_i(a). \tag{1.2.8}$$

Then $H$ may be written in the form

$$H = \sum_{i=1}^{N} \frac{|\mathbf{p}_i|^2}{2} + \overbrace{\frac{1}{2}\sum_{i,j=1}^{N} U_{ij}(a) - \sum_{i=1}^{N} U_i(a) + \frac{\xi}{2}\int d^3xU(\mathbf{x})}^{\alpha}$$

$$+ \overbrace{\sum_{i=1}^{N}(U_i(a) - U(\mathbf{x}_i))}^{\beta} \overbrace{- \frac{1}{2}\sum_i U_{ii}(a)}^{\gamma} + \overbrace{\sum_{i<j}(|\mathbf{x}_i - \mathbf{x}_j|^{-1} - U_{ij}(a))}^{\delta}. \tag{1.2.9}$$

Contribution $\alpha$ is positive, since it is of the form

$$\int \frac{dx\, dx'}{|\mathbf{x}-\mathbf{x}'|} \rho(\mathbf{x})\rho(\mathbf{x}'),$$

and $1/\sigma$ has a positive Fourier transform. It is easy to show (Problem 1) that $\beta \geq -(2\pi/5)\xi a^2 N$, equality holding provided that all the spheres lie within $\Lambda$, and $\gamma = (N/2)(6/5a)$, the self-energy of homogeneously charged spheres. As discussed earlier, $\delta \geq 0$. The lower bound $-N((2\pi/5)\xi a^2 + (3/5a))$ is optimized at $a = (3/4\pi\xi)^{1/3} \equiv r_s$, which is precisely the radius at which the sum of the volumes of the spheres equals that of $\Lambda$. This computation leads to the

**Lower Bound for the Energy** (1.2.10)

$$H \geq \sum_{i=1}^{N} \frac{|\mathbf{p}_i|^2}{2} - \frac{9}{10}\frac{N}{r_s}.$$

**Remarks** (1.2.11)

1. Nothing has yet been assumed about the shape of $\Lambda$ or the statistics of the particles. In particular, if $\Lambda$ is spherical, then by Problem 2,

$$-\sum_{i=1}^{N} U(\mathbf{x}_i) + \frac{\xi}{2}\int_\Lambda d^3xU(\mathbf{x}) \leq \frac{N}{2R^3}\sum_{i=1}^{N}|\mathbf{x}_i|^2 - \frac{9}{10}\frac{N^2}{R},$$

where equality holds if $x_i \in \Lambda$ for all $i$.

2. Despite its great generality, the numerical accuracy of the bound (1.2.10) is surprisingly good. If $\mathbf{x}_i$ are the sites of a simple, face-centered, or body-centered cubic lattice, computer studies have been made of the limit as $N \to \infty$ of the potential energy over $Nr_s^{-1}$, yielding respectively the values $-0.880$, $-0.895$, and $-0.896$ [3].

Lower bounds for $H$ depending on the particle statistics may be derived from (1.2.10). The energy of free fermions is, as seen earlier, $\sim N^{5/3}/R^2 \sim Nr_s^{-2}$, and with the aid of the more precise proportionality factor,

$$H \geq N(1.1r_s^{-2} - 0.9r_s^{-1}) \geq -\frac{0.81}{4.4}N \quad \text{for all } r_s \in \mathbb{R}^+ \tag{1.2.12}$$

for spin-$\frac{1}{2}$ particles. Even if the volume and consequently $r_s$ are treated as variables, the resultant lower bound is $\sim N$. We shall discover later that with no more than first-order perturbation theory we can obtain an upper bound not much different from (1.2.12): the Pauli exclusion principle makes the electrons stay at a distance $r_s$ apart, and this correlation imitates the energetically favorable configurations of (1.2.11; 2). Since the minimizing radius $r_s$ does not depend on $N$, in this model $E \sim N$ and $R \sim N^{1/3}$, so the exponent $t$ of (1.2.1) equals zero.

A very different picture emerges of bosons. With the kinetic energy (1.2.3) we find, ignoring precise coefficients, that

$$H \geq \frac{N^{1/3}}{r_s^2} - \frac{N}{r_s}. \tag{1.2.13}$$

The minimizing $r_s$ is $\sim N^{-2/3}$, and so $E \sim N^{5/3}$.

**Remarks** (1.2.14)

1. It is uncertain whether the lower bound $\sim N^{5/3}$ displays the correct dependence on $N$. Upper bounds obtained with trial functions include more kinetic energy since the particles have to be correlated in order to attain a sufficiently negative potential energy. Until recently it was only possible to show that $E < -cN^{7/5}$ [1].
2. If the background charge is concentrated at discrete points of a lattice, then trial functions can be thought up that show $E < -cN^{5/3}$, and thus in this case the energy in fact goes as $N^{5/3}$ [2].
3. So far only the electrostatic energy has been accommodated in the background, and minimized according to the density $\xi$. If the background consists of electrons, then its zero-point energy must also be calculated. In a jellium of deuterium atoms, which are bosons, the energy turns out to be $\sim N$: The background density prevents them from collapsing, and for fixed $r_s$ (1.2.13) is on the order of $N$.

**Real Matter** (1.2.15)

Real matter consists of positive and negative point-particles interacting with a Coulomb force, so

$$H = \sum_{i=1}^{N} \frac{|\mathbf{p}_i|^2}{2m_i} + \sum_{i>j} \frac{e_i e_j}{|\mathbf{x}_i - \mathbf{x}_j|} \tag{1.2.16}$$

for particles confined to a box of volume $\Lambda \sim R^3$. We shall often particularize to the situation wherein all negative particles are identical with $m = |e| = 1$ and all positive particles are identical with mass $M$ and charge $Z$. Provided that $Z$ is not so large that relativistic effects become significant, (1.2.16) gives a reasonably accurate description of ordinary matter. We therefore expect to find that $E \sim -N$ for $R \sim N^{1/3}$.

The proof of this fact, known as the "stability of matter," has to be deferred to §4.3. At this point we shall make do with several

**Remarks** (1.2.17)

1. Roughly speaking, the difficulty is that the double sum for the kinetic energy contains $\sim N^2$ terms, so many cancellations are needed for the result to be only $\sim N$. If, as in the gravitating system to be described shortly (1.2.19), all the contributions are of like sign, then cancellations certainly do not occur. Similarly, if the total charge $Q \equiv \sum_i e_i$ is $\sim N^{2/3+\varepsilon}$ and the system is restricted to a region of linear dimension $R \sim N^{1/3}$, the energy fails to be extensive. The electrostatic energy $Q^2/R$ is $\leq N$ only if $Q \leq N^{2/3}$.
2. Even requiring that $Q = 0$ will not guarantee that $|E| \sim N$ if all the particles are bosons. To prove this, rewrite (1.2.16) (with $M = Z = 1$) as

$$H = \sum_{i=1}^{N^-} \frac{|\mathbf{p}_i^-|^2}{2} + \sum_{\alpha=1}^{N^+} \frac{|\mathbf{p}_\alpha^+|^2}{2} + \sum_{i>j} |\mathbf{x}_i^- - \mathbf{x}_j^-|^{-1} + \sum_{\alpha>\beta} |\mathbf{x}_\alpha^+ - \mathbf{x}_\beta^+|^{-1}$$
$$- \sum_{i,\alpha} |\mathbf{x}_i^- - \mathbf{x}_\alpha^+|^{-1}, \tag{1.2.18}$$

where $N^+ = N^-$ for a neutral system. Now take the expectation value in a state with $\Psi^+ \otimes \Psi^-$, where $\Psi^\pm$ are the trial functions that led to $E \sim -N^{7/5}$ for Bose-jellium. Although the particles are correlated, the charge density is homogeneous, as for instance

$$\left\langle \Psi^+ \middle| -\sum_{i,\alpha} |\mathbf{x}_i^- - \mathbf{x}_\alpha^+|^{-1} \middle| \Psi^+ \right\rangle = -\xi \sum_i \int_\Lambda \frac{d^3x}{|\mathbf{x}_i^- - \mathbf{x}|}.$$

The last term in (1.2.28) is therefore equivalent to $-\sum_i U(\mathbf{x}_i^-) - \sum_\alpha U(\mathbf{x}_\alpha^+) + 2(\xi/2)\int d^3x U(\mathbf{x})$, and there results the sum of the energies of the positive and negative Bose-jellia. The expectation value is consequently about $-N^{7/5}$, which is an upper bound to the energy by the min–max principle

(III: 3.5.21). This "instability," which corresponds to the ground-state energy being nonextensive and the spatial contraction of many-particle aggregates of charged bosons, does not imply that individual atoms consisting of oppositely charged bosons would be unstable. A single, nonrelativistic atom of $He^4$ with its electrons subjected to Bose statistics (but with their original mass and charge) would have the same ground-state energy as real $He^4$, since the two-particle ground-state wave-function is symmetric in the spatial coordinates. The lesson here is that experience with two-electron molecules is not a trustworthy guide to the problem of the stability of matter: Since the Pauli exclusion principle makes no difference, the two electrons might just as well be bosons, but a system of many bosons would be unstable, whereas a many-fermion system is stable.

3. Since $He^3$ is just as stable as $He^4$, stability is not a matter of the type of statistics of one of the kinds of charge-carrier. Moreover, the relevant energy is always measured in Rydbergs, using the electronic mass, so matter should remain stable even in the limit of infinite nuclear masses.
4. It could be argued heuristically that the potential energy should go as $-N^{4/3}R^{-1}$, since each charge sees an opposite charge at a distance $RN^{-1/3}$, while charges further away should be screened. If this is added to the kinetic energy $N^{5/3}R^{-2}$ of fermions or $NR^{-2}$ of bosons, the minimum is respectively $\sim -N$ at $R \sim N^{1/3}$ or $\sim -N^{5/3}$ at $R \sim N^{-1/3}$.
5. In relativistic dynamics the kinetic energy is $\sim |\mathbf{p}| \sim 1/\Delta x$, so the system is softer. The heuristic arguments would evaluate the total energy of bosons as $\sim N/R - e^2 N^{4/3}/R$, which is unbounded below when $N$ is sufficiently large. Whereas nonrelativistic energies are always semibounded for any fixed $N$, it may happen that the relativistic energy goes to $-\infty$ for sufficiently large, but still finite, values of $N$.
6. The instability of a Coulomb system of bosons has nothing to do with the long range of the $1/r$ potential, but comes from its short-range features. If the singularity is chopped off by changing the potential to $V(x) = (1 - \exp(-\mu r))/r$, the system of bosons also becomes stable: Since the Fourier transform of $V$ is

$$\tilde{V}(\mathbf{k}) = \frac{4\pi\mu^2}{|\mathbf{k}|^2(|\mathbf{k}|^2 + \mu^2)} > 0,$$

with $|e_i| = e$, we find that

$$V \equiv \sum_{i>j} e_i e_j V(\mathbf{x}_i - \mathbf{x}_j) = \frac{1}{2}\int \frac{d^3k}{(2\pi)^3}\, \tilde{V}(\mathbf{k}) \left|\sum_j \exp(i\mathbf{k}\cdot\mathbf{x}_j)e_j\right|^2 - \frac{1}{2}\sum_{i=1}^{N} e_i^2 V(\mathbf{0})$$

$$> -\frac{N}{2}e^2 V(0) = -\frac{N}{2}e^2\mu,$$

so $H$ is bounded below by $-cN$. It could be argued that nuclei have a form factor, and that if $\mu$ is taken as the reciprocal of the nuclear radius, then $V$ would be a more realistic potential than $1/r$. This would lead to a simple proof of stability, but it misses the real point. Since the Rydberg, which is measured in electronvolts (eV), is determined by the mass of the electron, it is the kinetic energy of the electrons rather than the size of the nuclei that matters most for stability. The lower bound from the size of the nuclei alone would be $\sim -N$ MeV.

**Cosmic Bodies** (1.2.19)

The $1/r$ potentials in an object with gravitationally interacting particles are all attractive, so the situation is drastically different. The ground state of the Hamiltonian

$$H_G = \sum_{i=1}^{N} \frac{|\mathbf{p}_i|^2}{2} - \kappa \sum_{i>j} |\mathbf{x}_i - \mathbf{x}_j|^{-1} \tag{1.2.20}$$

goes as $-N^{7/3}$ for fermions. By the now familiar argument, $E \sim N^{5/3}/R^2 - N^2/R$, which has its minimum value $\sim -N^{-7/3}$ for $R \sim N^{-1/3}$. This can easily be translated into an exact upper bound by the use of trial functions localized in $\mathbb{R}^3$. Lower bounds are harder to come by, since energetically more favorable possibilities have to be ruled out. In this case there is an easier way: Write

$$H_G = \sum_{i=1}^{N} \sum_{j \neq i} \left( \frac{|\mathbf{p}_j|^2}{2(N-1)} - \frac{\kappa}{2} |\mathbf{x}_i - \mathbf{x}_j|^{-1} \right) \equiv \sum_{i=1}^{N} h_i, \tag{1.2.21}$$

so that each $h_i$ is the Hamiltonian of an atom with electrons having no Coulomb repulsion. Particle number $i$ stands for the atomic nucleus, as it has no kinetic energy, and the others are electrons, with mass $N - 1$ and potential $-|\mathbf{x}_i - \mathbf{x}_j|^{-1}/2$. According to (III: 4.5.15) it follows that $h_i \geq -cN^{4/3}$, and indeed the result is a

**Bound for the Energy of Gravitating Fermions** (1.2.22)

$$H_G > -cN^{7/3}, \qquad c = O(1).$$

**Remarks** (1.2.23)

1. Fermi statistics were not fully taken into account, since we have only antisymmetrized with respect to $N - 1$ particles when filling the energy levels. Since complete antisymmetrization restricts the set of admissible functions further, (1.2.22) is at any rate a lower bound.
2. The limit as $N \to \infty$ in this case exists with the scaling behavior $t = \frac{4}{3}$ of (1.2.1), as in (1.2.4). This does not mean that the limit with $t = \frac{4}{3}$ fails to exist for ordinary matter, but only that it is trivial. The potential energy goes to zero and the particles remain free.
3. If the particles are bosons, then they can all be put into the ground state, and $E \sim -N^3$. The radius of the ground state then goes as $N^{-1}$.
4. The Hamiltonian (1.2.20) was for the discussion of electrically neutral particles; if they are instead charged, then $\kappa$ must be replaced with

$\kappa - e_i e_j$. If we bear normal matter in mind, the gravitational force comes from the protonic mass, and in units where the mass of the proton is 1, $\kappa/e^2 \sim 10^{-36}$. Inequality (1.2.22) then *a fortiori* provides a lower bound, since

$$\frac{1}{2}\sum_i \frac{|\mathbf{p}_i|^2}{2} + \sum_{i>j} \frac{e_i e_j}{|\mathbf{x}_i - \mathbf{x}_j|} + \frac{1}{2}\sum_i \frac{|\mathbf{p}_i|^2}{2} - \sum_{i>j} \frac{\kappa}{|\mathbf{x}_i - \mathbf{x}_j|}$$
$$\geq -2c_e e^4 N - 2c\kappa^2 N^{7/3}.$$

The number of particles determines which $N$-dependence dominates. Gravity begins to win out when $N \sim (e^2/\kappa)^{3/2} \sim 10^{54}$, which is about the mass of Jupiter, and the energies of larger heavenly bodies are controlled mainly by gravitation. A concrete consequence is that the atoms get squashed and turn into a plasma of nuclei and electrons. This inequality provides a more rigorous foundation for the heuristic considerations of (II: 4.5.1).

We shall see in §4.2 that the system (1.2.20) can be solved in the limit $N \to \infty$, as the Thomas–Fermi theory becomes exact. Thomas–Fermi theory provides an idealization of stars, various corrections again being needed to make it realistic. In particular, if $N \sim 10^{57}$ relativistic effects become important. As with atoms with $Z > 137$, the Hamiltonian is unbounded below, which leads to a catastrophe. Nonetheless, Thomas–Fermi theory reflects the thermodynamic properties of stars rather well.

This section concludes with Table 1 displaying the many possibilities:

Table 1 The $N$-dependence of the kinetic energy $K$ and the potential energy $V$ when $N$ is large.

| | | | $K$ | $V$ | $R_{\min}$ | $E(R_{\min})$ |
|---|---|---|---|---|---|---|
| Nonrelativistic | electric | Bose | $N/R^2$ | $-N^{4/3}/R$ | $N^{-1/3}$ | $-N^{5/3}$ |
| | | Fermi | $N^{5/3}/R^2$ | $-N^{4/3}/R$ | $N^{1/3}$ | $-N$ |
| | gravitational | Bose | $N/R^2$ | $-N^2/R$ | $N^{-1}$ | $-N^3$ |
| | | Fermi | $N^{5/3}/R^2$ | $-N^2/R$ | $N^{-1/3}$ | $-N^{2/3}$ |
| Relativistic | electric | Bose | $N/R$ | $-N^{4/3}/R$ | 0 | $-\infty$ |
| | | Fermi | $N^{4/3}/R$ | $-N^{4/3}/R$ | 0 | $-\infty$ |
| | | † | | | or $\infty$ | or 0 |
| | gravitational | Bose | $N/R$ | $-N^2/R$ | 0 | $-\infty$ |
| | | Fermi | $N^{4/3}/R$ | $-N^2/R$ | 0 | $-\infty$ |

† If $R_{\min}$ tends to $+\infty$ more rapidly than $N^{1/3}$, then the kinetic energy per particle, $N^{1/3}/R$, becomes arbitrarily small, eventually $\ll m$, and the system is nonrelativistic. Hence $R_{\min}$ certainly can not increase faster than $N^{1/3}$. Which energy breaks the stalemate depends on the strength of the charge. If $Z < 137$, the kinetic energy wins out, and if $Z > 137$, the potential energy wins out.

**Problems** (1.2.24)

1. Calculate the $\beta$ and $\gamma$ of (1.2.9).
2. Verify (1.2.11; 1).

**Solutions** (1.2.25)

1.

$$\gamma: \int_{\substack{|\mathbf{x}|\leq a\\|\mathbf{x}'|\leq a}} \frac{d^3x\, d^3x'}{|\mathbf{x}-\mathbf{x}'|} = \int r^2\, dr\, d\Omega r'^2\, dr'\, d\Omega' \sum_{n,m} \left[\frac{r^n}{r'^{n+1}}\Theta(r'-r) + \frac{r'^n}{r^{n+1}}\Theta(r-r')\right]\frac{4\pi}{2n+1}$$

$$\cdot\, Y_n^m(\Omega)Y_n^{m*}(\Omega') = \int_0^a\int_0^a r^2\, dr\, r'^2\, dr'\left(\frac{\Theta(r'-r)}{r'} + \frac{\Theta(r-r')}{r}\right)(4\pi)^2$$

$$= \frac{2a^5}{15}(4\pi)^2.$$

$$\beta: \int_{\substack{|\mathbf{x}|\leq a\\ \mathbf{x}'\in\Lambda}} d^3x\, d^3x'\left(\frac{1}{|\mathbf{x}-\mathbf{x}'|} - \frac{1}{|\mathbf{x}'|}\right) = \int_{|\mathbf{x}|,|\mathbf{x}'|\leq a} \cdots + \int_{\substack{|\mathbf{x}|\leq a\\|\mathbf{x}'|\geq a}} \cdots.$$

The second integral equals 0, as can be seen by expanding $|\mathbf{x}-\mathbf{x}'|^{-1}$ in spherical harmonics. The first integral equals $-(2\pi a^2/5)(4\pi a^3/3)$ if $\{\mathbf{x}': |\mathbf{x}'|\leq a\}\subset\Lambda$, and is otherwise greater than or equal to this.

2. $U(\mathbf{x}_i) \leq -(3N/2R) + (N/2R)(|\mathbf{x}_i|^2/R^2)$, equality holding for $|x_i| < R$. The self-energy of the background charge is $3N^2/5R$.

## 1.3 Arbitrary Numbers of Particles in Fock Space

*The properties of large systems should not depend on the exact number of particles, so it is convenient to use a representation with a variable number of particles.*

We are used to dealing with atomic systems on $\mathscr{H}_n$, the $n$-particle Hilbert space. As it is impossible to count the particles in a large system, it is convenient to regard the number $N$ of particles as an observable capable of assuming various values. Accordingly, we shall study **Fock space**

$$\mathscr{H}_F = \bigoplus_{n=0}^{\infty} \mathscr{H}_n, \qquad N_{|\mathscr{H}_n} = n, \tag{1.3.1}$$

as the foundation for later analysis. The space $\mathscr{H}_0$ is one-dimensional and spanned by the **vacuum vector** $|0\rangle$. If the particles under consideration are either all bosons or all fermions, then $\mathscr{H}_n$ is either the $n$-fold symmetric or totally antisymmetric tensor product of $\mathscr{H}_1 = L^2(\mathbb{R}^3, d^3x)$ with itself, which will be denoted $\mathscr{H}_1 \circledS \mathscr{H}_1 \circledS \cdots \circledS \mathscr{H}_1$ or $\mathscr{H}_1 \wedge \mathscr{H}_1 \wedge \cdots \wedge \mathscr{H}_1$. If $f_j$, $j = 1, 2, \ldots$, is a complete orthonormal set of functions on $\mathscr{H}_1$, then the vectors $|f_{j_1} \circledS f_{j_2} \circledS \cdots \circledS f_{j_n}\rangle$ or respectively $|f_{j_1} \wedge f_{j_2} \wedge \cdots \wedge f_{j_n}\rangle$ are a basis for $\mathscr{H}_n$. In the latter case all the $j_k$ are to be taken different. For bosons the same $f$'s can be collected together and written as $|f_{j_1}^{n_1}, \ldots, f_{j_k}^{n_k}\rangle$, with $\sum_k n_k = N$. The $C^*$ algebra generated on the individual $\mathscr{H}_n$ of the boson Fock space by the symmetrized Weyl operators

$$\sum_{\pi} \exp\left[i \sum_j (r_{\pi_j} x_j + s_{\pi_j} p_j)\right],$$

where $(\pi_1, \ldots, \pi_n)$ is a permutation of $(1, \ldots, n)$, will be called the **Weyl algebra**, and is represented reducibly on $\mathscr{H}_F$—all bounded functions of $N$ alone belong to the commutant of the representation.

The irreducible **field algebra** on $\mathscr{H}_F$ turns out to be invaluable for the many-body problem:

**Definition** (1.3.2)

Let $|f_1, f_2, \ldots\rangle \equiv |f_1 \circledS f_2 \ldots\rangle$, and define the **creation and annihilation operators** $a^*(f)$ and $a(f)$ by linear extension of

$$\begin{aligned} a(f_m)|f_{j_1}^{n_1}, \ldots, f_{j_k}^{n_k}\rangle &= \delta_{mj_1}\sqrt{n_1}\,|f_{j_1}^{n_1-1}, f_{j_2}^{n_2}, \ldots, f_{j_k}^{n_k}\rangle \\ &\quad + \delta_{mj_2}\sqrt{n_2}\,|f_{j_1}^{n_1}, f_{j_2}^{n_2-1}, \ldots, f_{j_k}^{n_k}\rangle + \cdots \\ &\quad + \delta_{mj_k}\sqrt{n_k}\,|f_{j_1}^{n_1}, f_{j_2}^{n_2}, \ldots, f_{j_k}^{n_k-1}\rangle \qquad \text{(for bosons)}, \end{aligned}$$

$$\begin{aligned} a(f_m)|f_{j_1} \wedge \cdots \wedge f_{j_n}\rangle &= \delta_{mj_1}|f_{j_2} \wedge \cdots \wedge f_{j_n}\rangle - \delta_{mj_2}|f_{j_1} \wedge f_{j_3} \wedge \cdots \wedge f_{j_n}\rangle \\ &\quad + \cdots + (-1)^{n+1}|f_{j_1}, \ldots, f_{j_{n-1}}\rangle \qquad \text{(for fermions)}, \end{aligned}$$

$$\begin{aligned} a^*(f_m)|f_{j_1}^{n_1}, \ldots, f_{j_k}^{n_k}\rangle &= \delta_{mj_1}\sqrt{n_1 + 1}\,|f_{j_1}^{n_1+1}, f_{j_2}^{n_2}, \ldots, f_{j_k}^{n_k}\rangle \\ &\quad + \delta_{mj_2}\sqrt{n_2 + 1}\,|f_{j_1}^{n_1}, f_{j_2}^{n_2+1}, \ldots, f_{j_k}^{n_k}\rangle + \cdots \\ &\quad + \delta_{mj_k}\sqrt{n_k + 1}\,|f_{j_1}^{n_1}, \ldots, f_{j_k}^{n_k+1}\rangle \\ &\quad + \left(1 - \sum_{l=1}^{k} \delta_{mj_l}\right)|f_m f_{j_1}^{n_1}, \ldots, f_{j_k}^{n_k}\rangle \qquad \text{(for bosons)}, \end{aligned}$$

$$a^*(f_m)|f_{j_1} \wedge \cdots \wedge f_{j_n}\rangle = |f_m \wedge f_{j_1} \wedge \cdots \wedge f_{j_n}\rangle \qquad \text{(for fermions)},$$

and $a(\alpha f + \beta g) = \alpha a(f) + \beta a(g)$ for $f$ and $g \in \mathscr{H}_1$.

**Remarks** (1.3.3)

1. The prototypes of the $a$'s for bosons are the $a$ and $a^*$ of a harmonic oscillator (III: 3.3.5; 2), and for fermions they are the matrices $\sigma^{\pm}$ of (1.1.2). The formal analogy is not just superficial; the operators $a(f)$ show up when one quantizes coupled oscillators and then passes to a continuous limit, in the procedure known as **field quantization**, or **second quantization**.

2. Formally, the $a$'s satisfy the commutation or anticommutation relations:

$$[a(f), a^*(g)] = (f|g) \quad \text{(the scalar product on } \mathscr{H}_1\text{)},$$

$$[a(f), a(g)] = 0 \quad \text{for bosons,}$$

$$a(f)a^*(g) + a^*(g)a(f) \equiv [a(f), a^*(g)]_+ = (f|g),$$

$$[a(f), a(g)]_+ = 0 \quad \text{for fermions.}$$

Conversely, (1.3.2) can be derived from the commutation relations and $a(f)|0\rangle = 0$. The commutation relations are invariant under unitary transformations of the $f_j$, so (1.3.2) is independent of the choice of the basis. In the spirit of the GNS Construction, vector states may be identified with operators:

$$|f_{j_1}^{n_1}, \ldots, f_{j_k}^{n_k}\rangle = (n_1! \ldots n_k!)^{-1/2} a^*(f_{j_1})^{n_1} \ldots a^*(f_{j_k})^{n_k}|0\rangle,$$

or

$$|f_{j_1} \wedge \cdots \wedge f_{j_k}\rangle = a^*(f_{j_1}) \ldots a^*(f_{j_k})|0\rangle.$$

3. As in (III: 3.1.10; 2) the commutation relations reveal that the operators $a(f)$ are unbounded. To get a $C^*$ algebra, it is necessary to use the bounded operators $\exp[i(\alpha a(f) + \alpha^* a^*(f))]$; the algebra they generate is called $\mathscr{A}_B$.
4. The anticommutation relations for fermion fields are the same as those of $\sigma^\pm$, for which reason their $a(f)$ are bounded: $\|a(f)\Psi\|^2 + \|a^*(f)\Psi\|^2 = \langle\Psi|(a^*(f)a(f) + a(f)a^*(f))\Psi\rangle = (f|f)\|\Psi\|^2$, so $\|a(f)\| \leq \|f\|$. Because $\langle 0|a(f)a^*(f)|0\rangle = \|f\|^2$, this means $\|a(f)\| = \|a^*(f)\| = \|f\|$. The operators $a(f)$ generate a $C^*$ algebra $\mathscr{A}_F$, which is the norm-closure of the polynomials in $a$ and $a^*$.
5. It follows from Remark 4 that the mapping $f \to a^*(f)$ is an isometric homomorphism of the Banach-space structure of $\mathscr{H}_1$ to that of $\mathscr{A}_F$. (The mapping $f \to a(f)$ is continuous but antilinear, that is, $a(\lambda f + \mu g) = \lambda^* a(f) + \mu^* a(g)$.) For every unitary transformation $U \in \mathscr{B}(\mathscr{H}_1)$ there is a linear transformation $a(f) \to a(Uf)$, which can be extended to an automorphism $u$:

$$u(a(f_1) \ldots a(f_k)a^*(g_1) \ldots a^*(g_j)) = \\ = a(Uf_1) \ldots a(Uf_k)a^*(Ug_1) \ldots a^*(Ug_j). \tag{1.3.4}$$

In particular, for every strongly continuous unitary group $U(t)$ there is a norm-continuous group of automorphisms $u_t$ on $\mathscr{A}_F$ (i.e., the mapping $t \to u_t(a)$ from $\mathbb{R}$ to $\mathscr{B}(\mathscr{H}_F)$ is continuous in norm for all $a$). Therein lies a difference from the Weyl algebra, for which, although the free time-evolution $\exp[i(rp + sx)] \to \exp[i(rp + s(x + pt))]$ is strongly continuous in $t$, it is not continuous in norm. The time-evolution on $\mathscr{A}_B$ is also not continuous in norm, so the property of continuity can not be expressed without reference to a representation. In this regard the field algebra of fermions is much the nicer, owing ultimately to its being modeled on the matrices

$$\begin{pmatrix} 0 & 1 \\ 0 & 0 \end{pmatrix}, \quad \begin{pmatrix} 0 & 0 \\ 1 & 0 \end{pmatrix}.$$

Fermion fields will consequently be preferred when investigating more problematic cases.

6. The algebras $\mathscr{A}_F$ and $\mathscr{A}_B$ may be thought of as constructed from local algebras $\mathscr{A}_\Lambda$, containing only those $a(f)$ and $a^*(f)$ for which supp $f \subset \Lambda$. Clearly, $\mathscr{A}_\Lambda \subset \mathscr{A}_{\Lambda'}$ when $\Lambda \subset \Lambda'$. Since $\mathscr{H}_1$ is the norm-closure of $\bigcup_{\Lambda \subset \mathbb{R}^3} L^2(\Lambda, d^3x)$, $\mathscr{A}_F$ equals the norm-closure of $\bigcup_{\Lambda \subset \mathbb{R}^3} \mathscr{A}_\Lambda$.
7. It is common for annihilation operators to be introduced at single points, for which formally $[a(\mathbf{x}), a^*(\mathbf{x}')] = \delta^3(\mathbf{x} - \mathbf{x}')$, $a(f^*) = \int d^3x a(\mathbf{x}) f(\mathbf{x})$, $a^*(f) = \int d^3x' a^*(\mathbf{x}') f(\mathbf{x}')$. Although $a(\mathbf{x})$ is densely defined as an operator, it is not closeable, so $a^*(\mathbf{x})$ exists only in the sense of a quadratic form and not as an operator (Problem 8). The object $a^*(\mathbf{x})$ is called an operator-valued distribution.
8. Since $a$ annihilates a particle and $a^*$ creates one, the spaces $\mathscr{H}_n$ are not invariant subspaces of Fock space. It can in fact be shown that $\mathscr{A}_F$ and $\mathscr{A}_B$ are irreducibly represented on $\mathscr{H}_F$ (Problem 1). The algebra $\mathscr{A}_F$ is said to be **quasilocal**.

Remark (1.3.3; 5) implies that such things as translations and free time-evolution correspond to norm-continuous one-parameter groups of automorphisms on $\mathscr{A}_F$. The question arises as to whether they can be presented as strongly continuous, one-parameter unitary groups on $\mathscr{H}_F$. If the representation called for is just like the GNS representation of (III: 2.3.9) with the vacuum $|0\rangle$ as a cyclic, and also invariant, vector, then the answer is yes (however, see Problems 6 and 7):

**The Unitary Representability of the Automorphism** (1.3.5)

*Let $u_g$ be a group of automorphisms of a $C^*$ algebra $\mathscr{A}$, $w$ be an invariant state (i.e., $w(u_g(a)) = w(a)$ for all $g$), and $\pi_w$ be the representation constructed with $w$. Then the group of automorphisms has a unique unitary representation $U_g$ on the Hilbert space $\mathscr{H}_F$, such that*

$$\pi_w(u_g(a)) = U_g \pi_w(a) U_g^{-1}, \qquad U_g \Omega = \Omega, \tag{1.3.6}$$

*where $\Omega$ is the cyclic vector.*

**Proof**

If we let $U_g \pi_w(a)\Omega = \pi_w(u_g(a))\Omega$, then the $U_g$ thereby defined satisfies the stated requirements. It is unique, since if there existed another $\tilde{U}_g$ with the same properties, then it would follow that $(\tilde{U}_g U_g^{-1} - \mathbf{1})\Omega = 0$, $\tilde{U}_g U_g^{-1} \in \pi(\mathscr{A})'$. Now, because $\Omega$ is cyclic for $\pi(\mathscr{A})$, it separates $\pi(a)'$, and therefore

$\tilde{U}_g U_g^{-1} = \mathbf{1}$ (cf. Problem 5). (**Separating** means that for $a' \in \pi(\mathscr{A})', a'|\Omega\rangle = 0$ implies $a' = 0$.) □

**Remarks** (1.3.7)

1. If the group is topological and the realization as a group of automorphisms is weakly continuous, then $U_g$ is strongly continuous,
$$\|(U_g - \mathbf{1})\pi_w(a)\Omega\|^2 = 2w(a^*a) - w(a^*u_g(a)) - w(u_g(a^*)a) \to 0$$
as $g$ approaches the identity.
2. Our representation of $\mathscr{A}_F$ (1.3.2) is a $\pi_w$ such that $w(a) = \langle 0|a|0\rangle$ for $a \in \mathscr{A}_F$. Therefore $\Omega$ is the vacuum vector $|0\rangle$ ,and is invariant under the transformations brought up in (1.3.3; 5). It follows that the Euclidean group and free time-evolution can be represented by strongly continuous unitary groups of operators on Fock space. They consequently have self-adjoint generators (Problem 2), which are, however, not bounded. Even the operators $U_g$ do not belong to $\mathscr{A}_F$. To prove this fact we shall make use of

**Definition** (1.3.8)

The $C^*$ algebra obtained by closing the even polynomials in $a$ and $a^*$ in norm is denoted $\mathscr{A}_G$. The norm-closure of the polynomials having the same number of $a$'s as $a^*$'s in each summand is $\mathscr{A}_E$.

**Remarks** (1.3.9)

1. $\mathscr{A}_F \supset \mathscr{A}_G \supset \mathscr{A}_E$. In the Fock representation, $\mathscr{A}_E = \{N\}' \cap \mathscr{A}_F$.
2. Because $[ab, c] = a[b, c]_+ - [a, c]_+ b = a[b, c] + [a, c]b$, if $d \in \mathscr{A}_{\Lambda G}$ and $c \in \mathscr{A}_{\bar{\Lambda}}$, $\bar{\Lambda} \cap \Lambda = \emptyset$, then $[d, c] = 0$.

**Asymptotic Commutativity** (1.3.10)

*Let $V(t) \in \mathscr{B}(L^2\mathbb{R}^3))$ be a one-parameter, unitary group of operators with absolutely continuous spectrum, such that $V(t) \rightharpoonup 0$ as $t \to \infty$, and let $u_t(a(f)) \equiv a(V(t)f)$. Then $\lim_{t\to\infty} \|[a, u_t(b)]\| = 0$ for all $a \in \mathscr{A}_G$ and $b \in \mathscr{A}_F$; this state of affairs is described by saying that $\mathscr{A}_G$ is* **asymptotically Abelian** *with respect to $u_t$.*

**Proof**

First note that $\|[a(f), u_t(a^*(g))]_+\| = \|[a^*(f), u_t(a(g))]_+ = |(V(t)g|f)| \to 0$ as $t \to \infty$. If $d$ is an even polynomial and $c$ is any polynomial in $a(f)$ and $a^*(g)$, then with Remark (1.3.9; 2) it follows that the commutator vanishes asymptotically. Because the algebraic operations are continuous in norm, this extends to $\mathscr{A}_G$ and $\mathscr{A}_F$. □

**Corollaries** (1.3.11)

1. Since the generators of the spatial translation group and the free time-evolution have purely continuous spectrum, for them $V(t) \rightharpoonup 0$, and the appropriate commutators involving them go to zero.
2. The corresponding one-parameter groups of unitary operators on Fock space, $U_t \in \mathscr{B}(\mathscr{H}_F)$, can not belong to $\mathscr{A}_F$. Since every $U_t$ commutes with $N$, it must belong to $\mathscr{A}_E$, and hence $\|[U_t, u_{t'}(a)]\| < \varepsilon$ for all $\varepsilon \in \mathbb{R}^+$, $a \in \mathscr{A}_F$, and sufficiently large $t'$. Note that $\|U_t U_{t'} a U_{t'}^{-1} - U_{t'} a U_{t'}^{-1} U_t\| = \|U_t a U_t^{-1} - a\|$ which obviously can not be arbitrarily small for all $t$. It is even true that $\mathscr{A}_F \cap \bigcup_t U_t = U_0$.
3. Since $\mathscr{A}_F$ is irreducible, $\mathscr{A}_F'' = \mathscr{B}(\mathscr{H}_F)$ (III: 2.3.4), so $U_t$ is certainly attainable as the strong limit of elements of $\mathscr{A}_F$, or even $\mathscr{A}_E$.

**Remarks** (1.3.12)

1. Since commuting observables are jointly diagonable, and hence can be measured simultaneously, if $V$ is a group of translations, this implies that measurements separated by a large spatial distance do not interfere with each other. The local character of the algebra is important for this, and it does not apply to the Weyl operators, as $\exp[i(rp + sx)]$ and $\lim_{a \to \infty} \exp[i(r'p + s'(x + a))]$ do not commute. Even the bicommutant $\mathscr{A}_F''$ in the Fock representation is not asymptotically Abelian—for instance, the generators of the Euclidean group belong to the strong closure of $\mathscr{A}_F$ and are constant with respect to the free time-evolution but do not commute. Therefore $A_F''$ is not asymptotically Abelian with respect to free time-evolution.
2. The point of (1.3.10) for the time-evolution is that as time passes the disturbance due to a measurement diffuses so widely that local observables are not affected at much later times. This does not apply to the observables $x$ and $p$, as $p$ and $x + pt$ fail to commute even at large $t$. Observe that we have as yet proved commutativity only for free time-evolution; the question of whether it also holds for more realistic time-evolutions remains open.
3. This phenomenon does not occur for compact groups like the rotations; for them $U$ is a sum of finite-dimensional representations, for which it is impossible that $U \rightharpoonup 0$.

**Global Observables** (1.3.13)

The particle-number operator $N$ was defined in (1.3.1). It is unbounded and thus $\notin \mathscr{B}(\mathscr{H}_F)$, which $\supset \mathscr{A}_F$. Its domain of self-adjointness is

$$D_N = \left\{\psi_0 \oplus \psi_1 \oplus \cdots \oplus \psi_n \oplus \cdots \in \mathscr{H}_F : \sum_{n=1}^{\infty} n^2 \|\psi_n\|^2 < \infty\right\}.$$

Moreover, unitary gauge transformations $U(\alpha) = \exp(iN\alpha) \in \mathscr{B}(\mathscr{H}_F)$ also do not belong to $\mathscr{A}_F$, but can be attained as strong limits of elements of $\mathscr{A}_E$. In the Fock representation,

$$U(\alpha) = \underset{M \to \infty}{s\text{-lim}} \exp\left(i\alpha \sum_{j=1}^{M} a^*(f_j)a(f_j)\right),$$

where $\{f_j\}$ is an orthonormal basis. Although $U(\alpha)$ does not depend on the basis, it can only be defined in certain representations.

**Remark** (1.3.14)

Since $N$ is conserved in all of the systems treated here, it is not physically possible to measure the relative phase of states of different $N$. This means that $N$ creates a superselection rule in the sense of (III: 2.3.6; 7), and the algebra of observables should, properly speaking, be $\{N\}' = \mathscr{A}_E''$. The representation of this algebra on $\mathscr{H}_F$ is reducible, as its commutant is $\{N\}'' \neq \{\lambda \cdot \mathbf{1}\}$.

**Observables at a Point** (1.3.15)

One frequently considers the particle density and current at a point,

$$\rho(\mathbf{x}) = a^*(\mathbf{x})a(\mathbf{x}) = \sum_{j,k} a^*(f_j)a(f_k)f_j^*(\mathbf{x})f_k(\mathbf{x}),$$

$$\mathbf{j}(\mathbf{x}) = -\frac{1}{2mi}(a^*(\mathbf{x})\nabla a(\mathbf{x}) - (\nabla a^*(\mathbf{x}))a(\mathbf{x}))$$

$$= \sum_{j,k} a^*(f_j)a(f_k)\left(\frac{1}{2mi}(f_j^*(\mathbf{x})\nabla f_k(\mathbf{x}) - (\nabla f_j^*(\mathbf{x}))f_k(\mathbf{x}))\right).$$

The $f_k$ in these formulas must be chosen as an orthonormal basis of $C^1$ functions, in which case these observables are densely defined as quadratic forms. They are not, however, closeable: Their restrictions to $\mathscr{H}_1$ are the quadratic forms of

$$\psi^*(\mathbf{x})\psi(\mathbf{x}) \quad \text{and} \quad \frac{1}{2mi}(\psi^*(\mathbf{x})\nabla\psi(\mathbf{x}) - (\nabla\psi^*(\mathbf{x}))\psi(\mathbf{x})),$$

the former of which is recognizable as the prototype of this phenomenon as encountered in (III: 2.5.18; 3). Matrix elements with, say, $\rho(\mathbf{x})$ may be understood as distributional limits of matrix elements of the bounded operators $a^*(f)a(f)$ as $f \to \delta^3(\mathbf{x})$. Similarly, the continuity equation $\dot{\rho} + \nabla \cdot \mathbf{j} = 0$ holds at least for matrix elements $if$, evolving freely in time, $i\dot{f} = -\Delta f/2m$.

## Problems (1.3.16)

1. Show that the representations of $\mathscr{A}_F$ and $\mathscr{A}_B$ on $\mathscr{H}_F$ are irreducible.
2. Construct the generators of free time-evolution and of translation.
3. Find dense domains of definition for the quadratic forms $\rho(\mathbf{x})$ and $\mathbf{j}(\mathbf{x})$.
4. Define the number of particles in the volume $V$, $N_V = \int_V d^3x\rho(\mathbf{x})$, as an unbounded, self-adjoint operator.
5. For $\mathscr{A} \subset \mathscr{B}(\mathscr{H})$ and $\Omega \in \mathscr{H}$, show that $\Omega$ is cyclic for $\mathscr{A}$ iff $\Omega$ separates $\mathscr{A}'$.
6. The mapping $a \to b: b(f) = a(f) + L(f)$ is an automorphism $\alpha_L$ of the Bose algebra whenever $L$ is a linear, but not necessarily continuous, functional. Show that $\alpha_L$ is unitarily implementable on $\mathscr{H}_F$, i.e., there exists a $U_L \in \mathscr{B}(\mathscr{H}_F)$ such that $\mathbf{1} = U_L^* U_L = U_L U_L^*$ and $U_L a(f) U_L^{-1} = b(f)$, iff $L$ is continuous, which means that it can be written as $L(f) = (\rho | f)$ for some $\rho \in \mathscr{H}_1$.
7. Let $b(f) = a(\Phi f) + a^*(\overline{\Psi} f)$, $\Phi, \Psi \in \mathscr{B}(\mathscr{H}_1)$, $\Phi$ invertible. Show

   (i) that $a \to b$ is an automorphism of the Bose (resp. Fermi) field algebra if

$$\Phi\Phi^* \mp \Psi\Psi^* = 1 = \Phi^*\Phi \mp (\Psi^*\Psi)^t,$$

$$\Phi\Psi^t \mp \Psi\Phi^t = 0 = (\Psi^*\Phi)^t \mp \Psi^*\Phi,$$

   where $\overline{\Psi} = \Psi^{*t}$; and

   (ii) that it can be represented as a unitary operator on $\mathscr{H}_F$ iff $\Phi^{-1}\Psi \in \mathscr{C}_2(\mathscr{H}_1)$.
8. Show that although the $a(\mathbf{x})$ of (1.3.3; 7) is densely defined, it is not closeable, and the domain of definition of its adjoint $a^*(\mathbf{x})$ contains only the zero vector.

## Solutions (1.3.17)

1. Let $b$ be an operator such that $[b, a(f)] = [b, a^*(f)] = 0$ for all $f \in \mathscr{H}_F$. From the commutation relations of (1.3.3; 2) and $a(f)|0\rangle = 0$, it follows that $\langle 0|a(f_1)\ldots a(f_m)ba^*(g_1)\ldots a^*(g_n)|0\rangle = \langle 0|b|0\rangle \cdot \langle 0|a(f_1)\ldots a^*(g_n)|0\rangle$, which implies that $\langle x|bx\rangle = \langle 0|b|0\rangle \|x\|^2$ on a dense set, and therefore $b = \langle 0|b|0\rangle \cdot \mathbf{1}$.
2. With Theorem (1.3.5) and the fact that the $\mathscr{H}_n$ are invariant, by reasoning as in (1.3.13) we find that the two generators are

$$\underset{M\to\infty}{\text{s-lim}} \sum_{i,j}^{M} \int \nabla f_j^*(\mathbf{x}) \cdot \nabla f_i^*(\mathbf{x}) a^*(f_j) a(f_i)\, d^3x$$

   and

$$\underset{M\to\infty}{\text{s-lim}}\, i \sum_{k,j}^{M} \int \nabla f_j^*(\mathbf{x}) f_k(\mathbf{x}) a(f_k)\, d^3x,$$

   where the strong limit is defined as in (III: 2.5.8; 3). Formally, these can be written as $\int d^3x \nabla a^*(\mathbf{x}) \cdot \nabla a(\mathbf{x})$ and $i \int d^3x a^*(\mathbf{x}) \overleftrightarrow{\nabla} a(\mathbf{x})$.
3. For $\rho(\mathbf{x})$, linear combinations of $\prod_j a^*(f_j)|0\rangle$ with continuous $f_j$. For $\mathbf{j}(\mathbf{x})$, the $f_k$ have to be continuously differentiable.

4. $N_V = \sum_{j,k} a^*(f_j)a(f_k)\int_V d^3x f_j^*(\mathbf{x})f_k(\mathbf{x})$, $0 \leq N_V \leq N$, is a Hermitian operator on $D_N$ (1.3.13), and hence the domain of its Friedrichs extension contains $D_N$.

5. "If": Let $P$ be the projection onto the orthogonal complement of $\{a|\Omega\rangle\}$ for $a \in \mathscr{A}$. Then $P \in \mathscr{A}'$ and $P|\Omega\rangle = 0$, so $P = 0$.
"Only if": Let $a' \in \mathscr{A}'$, $a'|\Omega\rangle = 0$. Then $a'a|\Omega\rangle = 0$ for all $a \in \mathscr{A}$, which implies that $a' = 0$ on a dense set, so $a' = 0$.

6. The mapping $a \to b$ is unitarily implementable on $\mathscr{H}_F$ iff there exists a vector $|0_b\rangle \in \mathscr{H}_F$ such that $b(f)|0_b\rangle = 0$ for all $f \in \mathscr{H}_1$. It is clear that the existence of $U$ implies that of $|0_b\rangle = U|0\rangle$. On the other hand, the mapping

$$\prod_{i=1}^{n} a_i^*|0\rangle \to \prod_{i=1}^{n} b_i^*|0_b\rangle,$$

where $a_i = a(f_i)$, $b_i = b(f_i)$, and $\{f_i\}$ is an orthonormal basis, defines a unitary operator $U$, since this set of vectors is total. (Every vector is cyclic for an irreducible representation.) If $L$ is not continuous, then ker $L$ is dense in $\mathscr{H}_1$, and therefore $a(f)|0_b\rangle = 0$ for a dense set of $f$'s. This implies that $|0_b\rangle = |0\rangle$ and thus that $L \equiv 0$, which is continuous. Therefore $|0_b\rangle \notin \mathscr{H}_F$. If, however, $L(f) = (g|f)$, $g \in \mathscr{H}_1$, it is possible to choose $f_1 = g/\|g\|$. Because $a\exp[-a^*\|g\|] = \exp[-a^*\|g\|](a - \|g\|)$, the vector $|0_b\rangle = \exp[-a_1^*\|g\|]|0\rangle$ formally satisfies $b_k|0_b\rangle = (a_k + \delta_{k1}\|g\|)|0_b\rangle = 0$. It is also normalizable provided that

$$\infty > \langle 0|\exp[-\|g\|a_1]\exp[-\|g\|a_1^*]|0\rangle = \sum_{n=0}^{\infty} \frac{1}{(n!)^2}\|g\|^{2n}n! = \exp\|g\|^2,$$

so $\langle 0_b|0_b\rangle < \infty$ if $\|g\|^2 < \infty$.

7. (i) In matrix notation, for $b = \Phi a + \Psi a^*$, (i) must hold: $1 = [b, b^*]_\mp = \Phi\Phi^* \mp \Psi\Psi^*$, and $0 = [b, b]_\mp = \Phi\Psi^t \mp \Psi\Phi^t$. Written as block matrices, this becomes

$$\begin{pmatrix} \Phi & \Psi \\ \Psi^{*t} & \Phi^{*t} \end{pmatrix}\begin{pmatrix} \Phi^* & \mp\Psi^t \\ \mp\Psi^* & \Phi^t \end{pmatrix} = \mathbf{1}.$$

For invertibility it is necessary that

$$\begin{pmatrix} \Phi^* & \mp\Psi^t \\ \mp\Psi^* & \Phi^t \end{pmatrix}\begin{pmatrix} \Phi & \Psi \\ \Psi^{*t} & \Phi^{*t} \end{pmatrix} = \mathbf{1},$$

which produces the second line of the conditions.

(ii) The Fock vacuum $|0_b\rangle$ satisfies $0 = (\Phi^{-1}b)_k|0_b\rangle = (a_k + M_{kl}a_l^*)|0_b\rangle$, where $M = \Phi^{-1}\Psi$. Because $[a, a^*Ma^*] = 2Ma^*$, it can be written formally as $|0_b\rangle = c\exp[-a^*Ma^*/2]|0\rangle$ (Observe that by (i), $M = M^t$ (resp. $M = -M^t$).) To determine the normalization constant $c$, we shall calculate

$$\langle 0|\exp[-\tfrac{1}{2}aNa]\exp[-\tfrac{1}{2}a^*Ma^*]|0\rangle$$

when $M = \pm M^t$, $N = \pm N^t$, $[M, N^*] = 0$ and $M$ and $N$ are for the moment real. They can then simultaneously be put into the normal forms

$$\begin{pmatrix} n_1 & & & \\ & n_2 & & \\ & & n_3 & \\ & & & \ddots \end{pmatrix}, \quad \begin{pmatrix} m_1 & & & \\ & m_2 & & \\ & & m_3 & \\ & & & \ddots \end{pmatrix}$$

and respectively

$$\begin{pmatrix} & n_1 & & & \\ -n_1 & & & & \\ & & & n_2 & \\ & & -n_2 & & \\ & & & & \ddots \end{pmatrix}, \quad \begin{pmatrix} & m_1 & & & \\ -m_1 & & & & \\ & & & m_2 & \\ & & -m_2 & & \\ & & & & \ddots \end{pmatrix}$$

with real, orthogonal transformations. The transformations preserve the commutation relations of the field operators, so we may use this basis to calculate

$$\langle 0|\exp\left[-\frac{n_1}{2}a_1^2\right]\exp\left[-\frac{m_1}{2}a_1^{*2}\right]|0\rangle = \sum_{n=1}^{\infty}\frac{(n_1 m_1)^n}{4^n(n!)^2}(2n)! = (1 - n_1 m_1)^{-1/2}$$

and, respectively, for fermions,

$$\langle 0|\exp[-n_1 a_2 a_1]\exp[-m_1 a_1^* a_2^*]|0\rangle = 1 + n_1 m_1.$$

Therefore,

$$\langle 0|\exp[-\tfrac{1}{2}aNa]\exp[-\tfrac{1}{2}a^*Ma^*]|0\rangle = \prod_i (1 - n_i m_i)^{-1/2} = \left(\mathrm{Det}\begin{pmatrix} 1 & M \\ N & 1 \end{pmatrix}\right)^{-1/2}$$

and, respectively,

$$\prod_i (1 + n_i m_i) = \left(\mathrm{Det}\begin{pmatrix} 1 & M \\ N & 1 \end{pmatrix}\right)^{1/2}.$$

This can be continued analytically to complex matrix elements, and, in particular, in our case,

$$|c|^2\left(\mathrm{Det}\begin{pmatrix} 1 & M \\ M^* & 1 \end{pmatrix}\right)^{\mp 1/2} = 1.$$

The determinant is finite for $M \in \mathscr{C}_2$. Observe that in the case of bosons, $\Phi^*\Phi \geq 1$, and so $\Phi = V(\Phi^*\Phi)^{1/2}$ is always invertible. The result for fermions is valid for $M$ acting on either even or odd dimensional spaces.

8. The dense domain of definition of $a(\mathbf{x})$ consists of vectors with continuous, bounded $f$'s. For example, for fermions,

$$a(\mathbf{x})|f_{j_1} \wedge \cdots \wedge f_{j_n}\rangle = f_{j_1}(\mathbf{x})|f_{j_2} \wedge \cdots \wedge f_{j_n}\rangle - f_{j_2}(\mathbf{x})|f_{j_1} \wedge f_{j_3} \wedge \cdots \wedge f_{j_n}\rangle + \cdots + (-1)^{n+1} f_{j_n}(\mathbf{x})|f_{j_1} \wedge \cdots \wedge f_{j_n}\rangle.$$

The operator $a(\mathbf{x})$ is not closeable. Suppose that $f_\lambda(\mathbf{x}') = \exp[-|\mathbf{x} - \mathbf{x}'|^2\lambda]$; then $|f_\lambda\rangle \to 0$ as $\lambda \to \infty$, but $a(\mathbf{x})|f_\lambda\rangle = |0\rangle \nrightarrow 0$. Formally, $a^*(\mathbf{x})$ creates a particle with wave-function $f(\mathbf{x}') = \delta^3(\mathbf{x} - \mathbf{x}')$. Since this is not normalizable, $a^*(\mathbf{x})$ makes every vector $|f_{j_k} \wedge \cdots \wedge f_{j_n}\rangle$ infinitely long.

## 1.4 Representations with $N = \infty$

*Systems of* N *particles are represented on a Hilbert space that is the tensor product of* N *Hilbert spaces for single particles. The infinite tensor product opens the door to the new mathematical features of field theory.*

The scalar product on an $N$-fold tensor product of spaces $\mathscr{H}_1$ was defined multiplicatively by

$$\langle x|x\rangle = \prod_{i=1}^{N}(x_i|x_i), \qquad |x\rangle = |x_1)\otimes|x_2)\ldots|x_n), \qquad x_i\in\mathscr{H}_1. \quad (1.4.1)$$

If $N = \infty$, the vectors $|x\rangle$ that can be used in this formula are initially only those for which the infinite product converges. The product might well converge to 0 even though $(x_i|x_i) > 0$ for all $i$. In order to form the quotient space with respect to the zero vectors, it will first be necessary to form the equivalence class not only of vectors with some factor zero but also containing the vectors for which the product

$$\prod_{i=1}^{\infty}(x_i|x_i)$$

converges to zero. On the quotient space, (1.4.1) defines a separating norm, so the space can be completed to a Hilbert space $\mathscr{H}$, with the linear structure defined in the usual way

This does not yet, however, suffice to define the scalar product of different vectors $|x\rangle$ and $|y\rangle$. Though only vectors such that $(x_i|x_i) = (y_i|y_i) = 1$ for all $i$ need to be considered, there are still two possibilities, namely

$$\text{(I)} \qquad \prod_{i=1}^{\infty}|(x_i|y_i)| \to c > 0,$$

and

$$\text{(II)} \qquad \prod_{i=1}^{\infty}|(x_i|y_i)| \to 0,$$

where $\to$ means unconditional convergence. In case (II), $\prod_{i=1}^{\infty}(x_i|y_i) \to 0$ as well, and the vectors may be considered orthogonal. Possibility (I), on the other hand, does not guarantee that $\prod_i(x_i|y_i)$ converges. If $(x_j|y_j) = \exp(i\varphi_j)|(x_j|y_j)|$, then their product is said to converge if not only $\prod_i|(x_i|y_i)|$ but also $\sum_i|\varphi_i|$ converges. One now encounters the convention that vectors may be deemed orthogonal whenever $\sum_i|\varphi_i| \to \infty$ (case $(I_b)$). Let us thus agree on a

**Definition of the Scalar Product** (1.4.2)

$$\langle x|y\rangle = c \quad \text{provided that} \prod_i(x_i|y_i) \to c \neq 0, \quad \text{(case (Ia));}$$

$$\langle x|y\rangle = 0 \quad \text{provided that} \prod_i(x_i|y_i) \to 0 \quad \text{(case (II), or in the divergent sense (I}_b\text{)).}$$

**Remarks** (1.4.3)

1. It is easy to see that the scalar product this defines on $\mathscr{H}$ obeys all the rules of the game.
2. The space $\mathscr{H}_1$ has been assumed separable, yet even if $\mathscr{H}_1 = \mathbb{C}^2$, the larger space $\mathscr{H}$ is nonseparable. Let $|\mathbf{n}) \in \mathbb{C}^2$ be defined such that $(\mathbf{n}|\mathbf{n}) = 1$, $(\mathbf{n}|\boldsymbol{\sigma}|\mathbf{n}) = \mathbf{n} \in \mathbb{R}^3$, $|\mathbf{n}|^2 = 1$, and $|\mathbf{n}\rangle = |\mathbf{n}) \otimes |\mathbf{n}) \otimes \ldots$. Then $\langle \mathbf{n}|\mathbf{n}'\rangle = 1$ if $\mathbf{n} = \mathbf{n}'$ and is otherwise 0, showing that there is an uncountable orthonormal system of vectors.
3. Possibilities (Ia) and (I) create equivalence relations between vectors, because the convergence of $\prod_i (x_i|y_i)$ and $\prod_i (y_i|z_i)$ implies that of $\prod_i (x_i|z_i)$, and, likewise, that of $\prod_i |(x_i|y_i)|$ and $\prod_i |(y_i|z_i)|$ implies that of $\prod_i |(x_i|z_i)|$ (Problem 2). It is accordingly necessary to distinguish between strong (Ia) and weak (I) equivalence classes:

$$\text{(Ia)}: \prod_i{}' (x_i|y_i) \to c \neq 0, \qquad \text{(I)}: \prod_i{}' |(x_i|y_i)| \to c > 0.$$

The symbol $\prod'$ means that any finite number of factors 0 are to be left out. The equivalence classes span linear subspaces, so $\mathscr{H}$ can be decomposed into (uncountably) many weak equivalent classes, for which vectors of different classes are orthogonal. Each weak equivalence class can be further decomposed into mutually orthogonal strong equivalence classes. Since the latter differ only by phase factors within a given weak equivalence class, they contain the same physical information.

**Representations of $\mathscr{A}$ on Infinite Tensor Products** (1.4.4)

For the reasons stated in §1.1 and §1.3 we shall be interested in the algebra generated by the operators $\mathscr{B}(\mathscr{H}_i)$. More precisely, let $\mathscr{A}$ be the algebra generated by $\mathscr{B}(\mathscr{H}_1) \otimes \mathbf{1} \otimes \mathbf{1} \ldots, \mathbf{1} \otimes \mathscr{B}(\mathscr{H}_2) \otimes \mathbf{1} \ldots$, etc., and let $\mathscr{A}''$ be its strong (= weak) closure. The first thing to notice is that an element $a$ of $\mathscr{A}$ sends no vector of $\mathscr{H}$ out of its strong equivalence class; since other than a finite number of entries there is always an infinite $\mathbf{1} \otimes \mathbf{1} \otimes \mathbf{1} \ldots$, nothing alters the convergence of $\prod_{i=1}^{\infty} (x_i|y_i)$. The representation of $\mathscr{A}$ on $\mathscr{H}$ is consequently reducible to a high degree; every strong equivalence class is an invariant subspace. The formation of the weak closure changes nothing, since $\langle x|a_n y\rangle = 0$ for $|x\rangle$ and $|y\rangle$ in different equivalence classes, and if $a_n \rightharpoonup a$, then clearly $\langle x|ay\rangle = 0$. Thus every strong equivalence class provides a representation of $\mathscr{A}$ and of $\mathscr{A}''$, and it is a peculiarity of the infinite tensor product that these representations are inequivalent so long as they arise from different weak equivalence classes.

**Example** (1.4.5)

Return to the simple case of (1.4.3; 2), and define $\boldsymbol{\sigma}_j \cdot \mathbf{n} = \sigma_j$, and $\sigma_j^{\pm}$ in analogy with (1.1.2) such that $\sigma_j^{-}|\mathbf{n}) = |-\mathbf{n})$, $\sigma_j^{+}|-\mathbf{n}) = |\mathbf{n})$, $\sigma_j^{+}|\mathbf{n}) = \sigma_j^{-}|-\mathbf{n}) = 0$. Let $\mathscr{A}$ be the algebra generated by $\sigma_j$ and $\sigma_j^{\pm}$, $j = 1, 2, \ldots$, let

$\pi_{\mathbf{n}}$ be its representation on the strong equivalence class of $|\mathbf{n}\rangle$, and define $\mathscr{A}_{\mathbf{n}} \equiv \pi_{\mathbf{n}}(\mathscr{A})$. The representation is constructed like the Fock representation, the operators $\pi_{\mathbf{n}}(\sigma_j^{\pm})$ corresponding to creation and annihilation operators and $|\mathbf{n}\rangle$ to the vacuum: $\pi_{\mathbf{n}}(\sigma_j^{+})|\mathbf{n}\rangle = 0$ for all $j$. The vectors $\pi_{\mathbf{n}}(\sigma_{j_1}^{-} \ldots \sigma_{j_k}^{-})|\mathbf{n}\rangle$ are total for the (strong) equivalence class, and the representation $\mathscr{A}_n$ is irreducible (likewise for $\mathscr{A}_n''$ *a fortiori*).

**Remarks** (1.4.6)

1. These representations of the $\sigma$'s are always equivalent on finite tensor products; the Hilbert space constructed with the GNS procedure contains every vector $|\mathbf{n}'\rangle$, in contrast to the infinite case, where the $\sigma$'s never send vectors out of equivalence classes, which, however, contain no vectors $|\mathbf{n}'\rangle$ with $\mathbf{n}' \neq \mathbf{n}$.
2. The mean magnetization

$$\mathbf{s} = \lim_{N \to \infty} \sum_{j=1}^{N} \frac{1}{N} \pi_{\mathbf{n}}(\boldsymbol{\sigma}_j)$$

exists as a strong limit, so $\mathbf{s} \in \mathscr{A}_n''$. As $N \to \infty$ the commutator of this observable with any element of the algebra goes to zero in the norm topology, so $\mathbf{s}$ is in the center of $\mathscr{A}_n''$. In any irreducible representation, $\mathbf{s}$ must be a multiple of the identity, and is thus the same as $\mathbf{n}$, its expectation value in the state $|\mathbf{n}\rangle$. If $\mathbf{n} \neq \mathbf{n}'$, then $\pi_{\mathbf{n}}$ and $\pi_{\mathbf{n}'}$ are inequivalent: If there existed a unitary transformation $U$ mapping the equivalence classes of $\mathbf{n}$ and $\mathbf{n}'$ onto each other and such that $U\pi_{\mathbf{n}}(\boldsymbol{\sigma}_j)U^{-1} = \pi_{\mathbf{n}'}(\boldsymbol{\sigma}_j)$, then this could be extended to a transformation of the strong closures $\mathscr{A}_n''$ and $\mathscr{A}_{n'}''$, and when applied to $\mathbf{s}$ it would imply that $U\mathbf{n}U^{-1} = \mathbf{n}'$. This is impossible, since two different multiples of the identity can not be unitarily related.
3. On the space $\mathscr{H}$ there exists a unitary transformation sending $|\mathbf{n}\rangle$ to $|\mathbf{n}'\rangle$. Let $n_j' = M_{jk}n_k$, $MM^t = 1$; then the transformation $|\mathbf{n}) \to |M\mathbf{n})$ (on every factor of $|\mathbf{n}\rangle$) is clearly the unitary transformation that brings this about. Upon restriction to an equivalence class, its action is

$$U\pi_{\mathbf{n}}(\sigma_j)U^{-1} = \pi_{\mathbf{n}}(\sigma_k)M_{kj},$$

in contrast to the previous $U$, and so it creates an isomorphism between $\pi_{\mathbf{n}}(\mathscr{A})$ and $\pi_{\mathbf{n}'}(\mathscr{A})$.
4. Within a given representation the rotation

$$\pi_{\mathbf{n}}(\sigma_j) \to \pi_{\mathbf{n}}(\sigma_k)M_{kj}$$

represents an automorphism of the $C^*$ algebra generated by the $\sigma$'s, and as such it preserves norms. Yet it can not be extended continuously to the weak closure. If there were such an extension, then $n_j \cdot \mathbf{1} \to n_k M_{kj} \cdot \mathbf{1}$, but $\lambda \cdot \mathbf{1}$ is invariant under every automorphism. Consequently, in the representation space of $\pi_{\mathbf{n}}$ there exists no unitary transformation $U^{-1}\pi_{\mathbf{n}}(\sigma_j)U = M_{jk}\pi_{\mathbf{n}}(\sigma_k)$, as it would extend to $\pi_{\mathbf{n}}(\mathscr{A})''$. Formally, it would turn $|\mathbf{n}\rangle$ into $|\mathbf{n}'\rangle$, but there is no vector $|\mathbf{n}'\rangle$ in the representation space of $\pi_{\mathbf{n}}$ (cf. Problems (1.3.16; 6) and (1.3.16; 7)).

5. Let $M(t)$ be a one-parameter group of rotations on $\mathbb{R}^3$—for definiteness about the 3-axis—and let $U(t)$ be its representation on $\mathscr{H}$ as discussed in Remark 3. On a formal level, $\sum_{j=1} \sigma_j^3$ could be regarded as the generator of the group. The unitary operators $U(t)$ map the equivalence class of $|\mathbf{n}\rangle$ into itself only if $\mathbf{n}$ points in the 3-direction, and in that case the restriction of $U(t)$ to this equivalence class belongs to $\mathscr{A}_{\mathbf{n}}''$. Although it is not possible to define $\sum_{j=1}^{\infty} \sigma_j^3$ densely, $\sum_{j=1}^{\infty} (\sigma_j^3 - 1)$ is essentially self-adjoint in the representation $\pi_{\mathbf{n}}$ on the dense set specified in (1.4.5) and is the generator of the rotations about the 3-axis. In other representations there is no workable definition of this operator, as all its matrix elements are infinite. It is natural to ask at this point what the generator of $U(t)$ looks like. It turns out, though, that $U(t)$ has no generator: By Stone's theorem (III: 2.4.24) the existence of a generator is equivalent to strong continuity of $U(t)$, but $U(t)$ is not even weakly continuous, for if $\mathbf{n}$ does not point in the 3-direction, then $\langle\mathbf{n}|U(t)|\mathbf{n}\rangle = 1$ if $t = 0$ and is otherwise 0. It is true that the mapping $t \to U(t)$ is weakly measurable, but the generalization of Stone's theorem for weakly measurable groups works only on separable Hilbert spaces.
6. "Local" rotations of $m$ spins are generated by $\sum_{j=1}^{m} \sigma_j^3$ and always exist.

The representations of the $\sigma$'s on the individual strong equivalence classes studied until now have all been irreducible, and correspond to GNS constructions using a pure state (cf. (III: 2.3.10; 5)). We shall also see in (2.1.6; 5) that mixed states likewise correspond to vectors in a larger Hilbert space on which the algebra is represented reducibly. That space is the tensor product of the irreducible representation space with another Hilbert space. The key fact to bear in mind when constructing such representations of the $\sigma$'s is that the infinite tensor product is no longer associative; for instance $\mathbb{C}^4 \otimes \mathbb{C}^4 \otimes \mathbb{C}^4 \otimes \cdots = (\mathbb{C}^2 \otimes \mathbb{C}^2) \otimes (\mathbb{C}^2 \otimes \mathbb{C}^2) \otimes (\mathbb{C}^2 \otimes \mathbb{C}^2) \otimes \cdots \neq \mathbb{C}^2 \otimes \mathbb{C}^2 \otimes \mathbb{C}^2 \otimes \ldots$: The vector

$$\frac{1}{\sqrt{2}}\left[\begin{pmatrix}1\\0\end{pmatrix} \otimes \begin{pmatrix}0\\1\end{pmatrix} + \begin{pmatrix}0\\1\end{pmatrix} \otimes \begin{pmatrix}1\\0\end{pmatrix}\right] \otimes \frac{1}{\sqrt{2}}\left[\begin{pmatrix}1\\0\end{pmatrix} \otimes \begin{pmatrix}0\\1\end{pmatrix} + \begin{pmatrix}0\\1\end{pmatrix} \otimes \begin{pmatrix}1\\0\end{pmatrix}\right] \otimes \ldots$$

on the left has no counterpart on the right. For this reason we shall not simply take the tensor product of the space examined in Example (1.4.5) with another Hilbert space, but shall instead proceed as follows.

**Thermal Representations** (1.4.7)

If there is only one spin, i.e., $\mathscr{A}$ is generated by $\mathbf{1}$ and $\boldsymbol{\sigma}$, then the GNS representation using the state given in (1.1.11) becomes a reducible representation on

$\mathbb{C}^4$: $\pi(\mathscr{A}) = \mathscr{B}(\mathbb{C}^2) \otimes \mathbf{1}$, $\pi(\boldsymbol{\sigma}) = \boldsymbol{\sigma} \otimes \mathbf{1}$, $\pi(\mathscr{A})' = \mathbf{1} \otimes \mathscr{B}(\mathbb{C}^2)$, $Z = \pi(\mathscr{A}) \cap \pi(\mathscr{A})' = \{\alpha \cdot \mathbf{1}\}$,

$$\Omega = \begin{pmatrix}1\\0\end{pmatrix} \otimes \begin{pmatrix}1\\0\end{pmatrix} \sqrt{\frac{1+s}{2}} + \begin{pmatrix}0\\1\end{pmatrix} \otimes \begin{pmatrix}0\\1\end{pmatrix} \sqrt{\frac{1-s}{2}}, \qquad 0 < s < 1,$$

$$\langle \boldsymbol{\sigma} \rangle = \langle \Omega | \boldsymbol{\sigma} \Omega \rangle = (0, 0, s).$$

Despite being reducible ($\mathscr{A}' \neq \{\alpha \cdot \mathbf{1}\}$), this representation is a factor (its center is $Z = \{\alpha \cdot \mathbf{1}\}$). Accordingly, when passing to infinitely many spins we consider the representation on $\mathbb{C}^4 \otimes \mathbb{C}^4 \otimes \mathbb{C}^4 \otimes \cdots$ constructed with $\Omega \otimes \Omega \otimes \Omega \otimes \ldots$. We find, analogously, that

$$\pi(\mathscr{A}) = (\mathscr{B}(\mathbb{C}^2) \otimes \mathbf{1}) \otimes (\mathscr{B}(\mathbb{C}^2) \otimes \mathbf{1}) \otimes \ldots,$$
$$\pi(\mathscr{A})' = (\mathbf{1} \otimes \mathscr{B}(\mathbb{C}^2)) \otimes (\mathbf{1} \otimes \mathscr{B}(\mathbb{C}^2)) \otimes \cdots + \text{weak limits}$$

$\pi(\mathscr{A})''$ is the weak closure of $\mathscr{A}$, and $Z = \{\alpha \cdot \mathbf{1}\}$,

which is a reducible factor representation.

**Remarks** (1.4.8)

1. This representation is not equivalent to any of those found in (1.4.5); as mentioned above, the vector $\Omega \otimes \Omega \otimes \Omega \otimes \ldots$ has no counterpart in the earlier representations $\pi_n$, since the corresponding functional in $\pi_n$ would then be strongly continuous. The state defined by $\Omega \otimes \Omega \otimes \Omega \otimes \ldots$ on $\mathscr{A}$.

$$\langle (\boldsymbol{\sigma}_{j_1} \cdot \mathbf{n}_1)(\boldsymbol{\sigma}_{j_2} \cdot \mathbf{n}_2) \ldots (\boldsymbol{\sigma}_{j_k} \cdot \mathbf{n}_k) \rangle = s^k n_1^z n_2^z \ldots n_k^z$$

is a (norm) continuous linear functional, and therefore extensible to the whole $C^*$ algebra generated by $\mathscr{A}$, but it still need not be strongly continuous in a representation: For instance, in the representation using $\pi_n$,

$$P_N = \prod_{i=N}^{2N} \frac{1 + \boldsymbol{\sigma}_i \cdot \mathbf{n}}{2}$$

converges strongly to $\mathbf{1}$, but $\langle P_N \rangle = ((1 + sn^z)/2)^N \to 0 \neq 1$. Recall that a refinement of the topology on the range space or a coarsening of the topology on the domain space may destroy the continuity of a mapping.

2. The fact that with only one spin, $\langle \boldsymbol{\sigma} \rangle = \operatorname{Tr} \boldsymbol{\sigma} \exp(-\eta\sigma_3)/\operatorname{Tr} \exp(-\eta\sigma_3)$, might mislead one into thinking that for infinitely many spins, in the notation of (1.1.1),

$$\langle \cdot \rangle = \operatorname{Tr} \cdot \rho, \qquad \rho = \frac{\exp(-\eta \sum_j \sigma_j)}{\operatorname{Tr} \exp(-\eta \sum_k \sigma_k)}.$$

What goes wrong is that

$$\frac{\exp(-\eta \sum_{j=1}^N \sigma_j)}{\operatorname{Tr} \exp(-\eta \sum_{j=1}^N \sigma_j)} \Rightarrow 0 \quad \text{as } N \to \infty.$$

3. In the thermal representation (1.4.7) it is of course possible to write $\langle \cdot \rangle = \mathrm{Tr} \cdot P_\Omega$, where $P_\Omega$ is the projection onto the cyclic vector, but $P_\Omega \notin \mathscr{A}''$.

**Decomposition of the Representations** (1.4.9)

Because of the analogy between $\sigma^\pm$ and the operators $a$ and $a^*$ for fermions, the phenomena we have discussed are also characteristic of systems of infinitely many fermions. It is not so important that the $\sigma$'s commute whereas the $a$'s anticommute; the distinction can be gotten around with the right transformation. For a system of bosons the individual factors of the tensor product are already infinite-dimensional, which causes additional complications. In either case there are a great number of inequivalent representations; the uniqueness theorem (III: 3.1.5) for finite systems does not hold any more. Thus it would be desirable to find a point of view that organizes them somehow. The concept of a factor was introduced in (III: 2.3.4), as an algebra with a trivial center, $Z = \{\alpha \cdot \mathbf{1}\}$. On a finite-dimensional space it amounts to a direct sum of equivalent irreducible representations. The first step in any decomposition is to collect the equivalent irreducible representations together in factors and then write the whole representation as a sum of various factors. In the finite-dimensional case this appears as shown in Figure 2.

It will be observed that the projections onto the space $\mathscr{H}_{ik}$ of the irreducible representations belong to $\pi(\mathscr{A})'$ and the projections onto the spaces $\mathscr{H}_i$ of the factors belong to the center. Both $\pi(\mathscr{A})$ and $\pi(\mathscr{A})'$ map

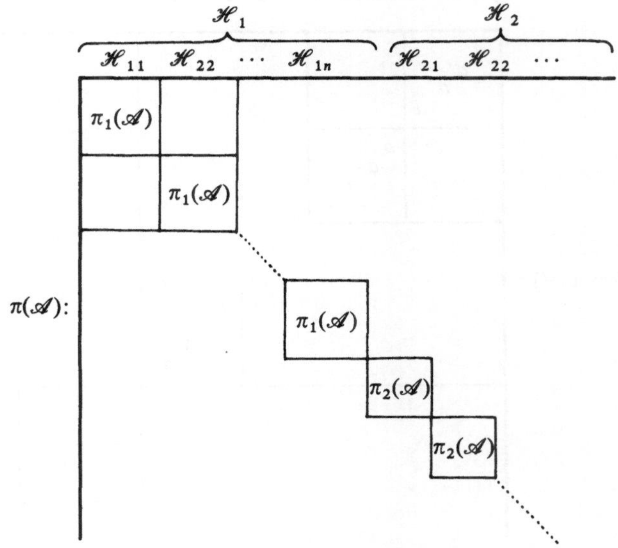


Figure 2a The representation of $\mathscr{A}$ in matrix form.

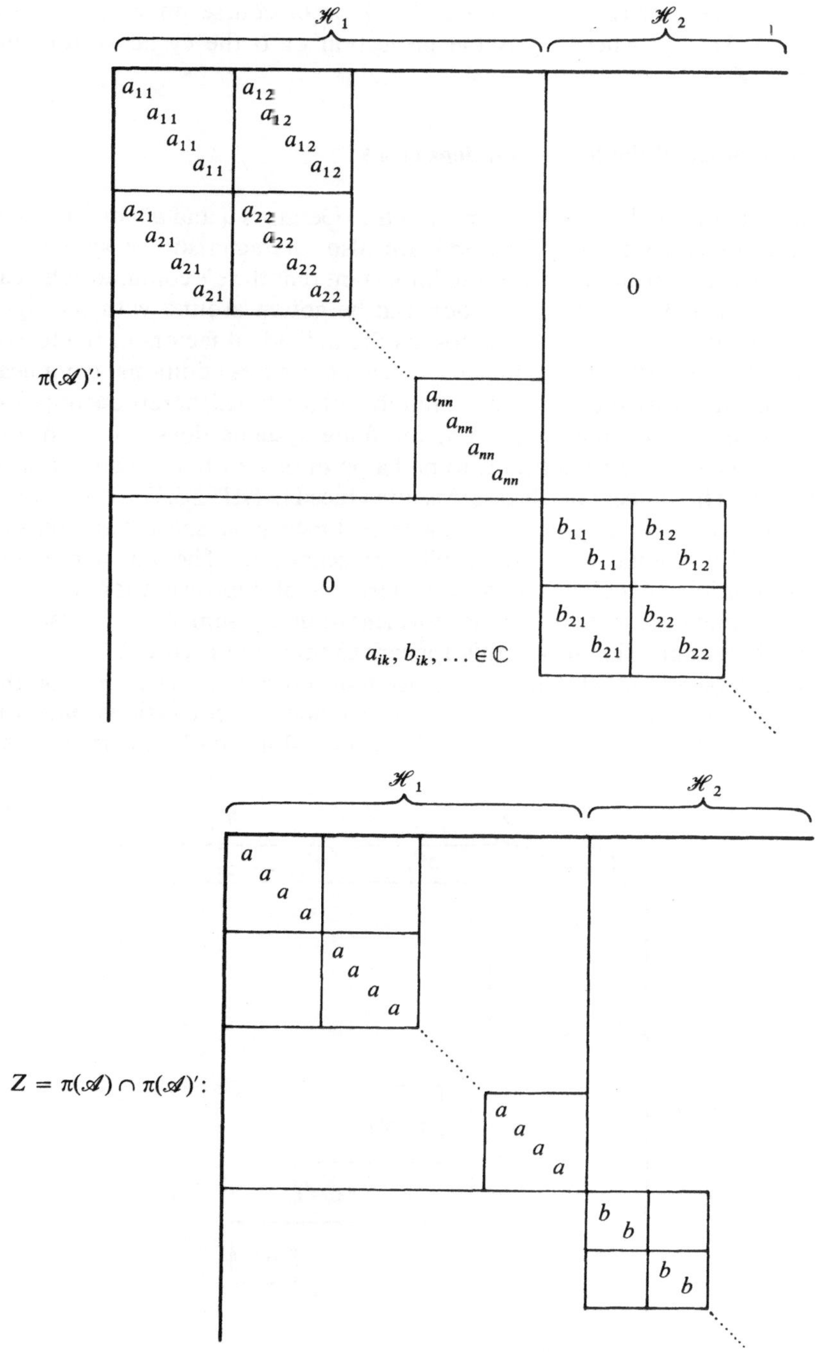


Figures 2b, c The representation of $\mathscr{A}'$ and the center $Z$ in matrix form.

$\mathcal{H}_i$ into itself. The elements of the center become multiples of the identity when projected onto $\mathcal{H}_i$; they can assume different values only on different $\mathcal{H}_i$. The decomposition into factors is thus uniquely fixed by $Z$ and consequently by $\pi(\mathcal{A})$. The further decomposition into irreducible representations is not likewise fixed; some arbitrariness is connected with the spaces $\mathcal{H}_{ik}$. If, for example, $\mathcal{H}_1 = \mathcal{H}_{11} \otimes \mathbb{C}^n = \mathcal{H}_{11} \otimes e_1 \oplus \mathcal{H}_{11} \otimes e_2 \oplus \cdots \oplus \mathcal{H}_{11} \otimes e_n$, then the choice of the basis $\{e_i\}$ for $\mathbb{C}^n$ remains free, since the space is the same for every choice of orthogonal basis. Different bases correspond to the different maximally Abelian subalgebras of $\pi(\mathcal{A})'$ that they diagonalize.

The passage to an infinite dimension requires the generalization of sums to integrals. The spectral theorem (III: 2.3.11) states that a Hermitian operator $a \in \mathcal{B}(\mathcal{H})$ may be represented as a multiplication operator on some space $L^2(d\mu, \text{Sp}(a))$. If there is degeneracy, then a spectral value $\alpha \in \text{Sp}(a)$ is associated not with a single complex number but with a many-dimensional Hilbert space $\mathcal{H}_\alpha$. If $v(\alpha)$ denotes the component of $v \in \mathcal{H}$ in $\mathcal{H}_\alpha$, then the scalar product on $\mathcal{H}$ can be written as

$$\langle v | w \rangle = \int d\mu(\alpha) \langle v(\alpha) | w(a) \rangle.$$

The action of $a$ on $v$ is $(av)(\alpha) = \alpha v(\alpha)$. The center $Z = \pi(\mathcal{A}) \cap \pi(\mathcal{A})'$ is a commutative algebra, and its elements may be simultaneously diagonalized, and so any $z \in Z$ may be written as $(zv)(\alpha) = f(\alpha)v(\alpha)$, where $f$ assigns a complex number to $\alpha$. Any element $a$ of $\mathcal{A}$ can then be represented by $[\pi(a)v](\alpha) = \pi_\alpha(a)v(\alpha)$, $\pi_\alpha(a) \in \mathcal{B}(\mathcal{H}_\alpha)$, and $b \in \pi(\mathcal{A})' \Rightarrow (bv)(\alpha) = b(\alpha)v(\alpha)$, $b(\alpha) \in \mathcal{B}(\mathcal{H}_\alpha)$, $[b(\alpha), \pi_\alpha(a)] = 0$ for all $a \in \mathcal{A}$. In a finite number of dimensions every $\mathcal{H}_\alpha$ can be written $\mathcal{H}_\alpha = \mathcal{H}_\alpha^{(1)} \otimes \mathcal{H}_\alpha^{(2)}$, $\pi_\alpha(\mathcal{A}) = \mathcal{B}(\mathcal{H}_\alpha^{(1)}) \otimes \mathbf{1}_{\mathcal{H}_\alpha^{(2)}}$, and $b(\alpha)$ is of the form $\mathbf{1}_{\mathcal{H}_\alpha^{(1)}} \otimes b$, $b \in \mathcal{B}(\mathcal{H}_\alpha^{(2)})$. This is as far as the finite-dimensional analogy goes; it will not be possible to write every factor $\pi_\alpha$ in the form $\mathcal{B}(\mathcal{H}) \otimes \mathbf{1}$.

**Classification of Factors** (1.4.10)

We pause now to take stock of the factors, which will function as basic building blocks. The possibility that comes to mind first for a preliminary, rough classification is to define a trace. In (III: 2.3.19) the trace was defined as a mapping from $\mathcal{A}_+$, the positive operators, to $\overline{\mathbb{R}}^+$, and it was extended to a linear mapping from the trace class $\mathcal{C}_1(\mathcal{H})$ to $\mathbb{C}$. The trace is discontinuous in all topologies weaker than the trace topology given by $\|\cdot\|_1$. It may even occur that the only element of an algebra $\mathcal{A}$ in the trace class is the zero operator, as for example with the factor $\mathcal{B}(\mathcal{H}) \otimes \mathbf{1}$, where $\mathbf{1}$ is the identity on an infinite-dimensional space. In this case there is plainly the possibility of defining a trace by $\Phi(a \otimes \mathbf{1}) = \text{Tr}_1\, a$, which has all the necessary properties. This observation suggests an abstract

**Definition of the Trace** (1.4.11)

Let $\mathscr{A}_+$ be the positive cone of a strongly closed algebra $\mathscr{A}$, i.e., a von Neumann algebra. A **trace** is a mapping $\Phi: \mathscr{A}_+ \to \overline{\mathbb{R}}^+$ with the following properties.

(i) $\Phi(\lambda_1 a_1 + \lambda_2 a_2) = \lambda_1 \Phi(a_1) + \lambda_2 \Phi(a_2)$ for $a_i \in \mathscr{A}_+$ and $\lambda_i \in \mathbb{R}^+$;

(ii) $\Phi(a) = \Phi(uau^{-1})$ for all $a \in \mathscr{A}_+$ and all unitary $u \in \mathscr{A}$.

The trace $\Phi$ is said to be

**faithful**, if $\Phi(a) = 0$ and $a \in \mathscr{A}_+ \Leftrightarrow a = 0$;
**finite**, if $\Phi(a) < \infty$ for all $a \in \mathscr{A}_+$;
**semifinite**, if for all $a \in \mathscr{A}_+$ there exists a nonzero $b < a$ such that $\Phi(b) < \infty$;
**normal**, if for every increasing filter (see (III: 2.2.21)) $F \subset \mathscr{A}_+$ with supremum $s$, $\Phi(s) = \sup_{a \in F} \Phi(a)$.

**Examples** (1.4.12)

1. $\Phi(a) = 0$ for all $a \in \mathscr{A}_+$. The trace is unfaithful, finite, and normal.
2. $\Phi(0) = 0$, $\Phi(a) = \infty$ for all $a \neq 0$. The trace is faithful, not semifinite, and normal (purely infinite).
3. Let $\mathscr{A}$ be the $n \times n$ matrices and $\Phi(a) = \operatorname{Tr} a$. The trace is faithful, finite, and normal.
4. $\mathscr{A} = \mathscr{B}(\mathscr{H})$, $\mathscr{H}$ infinite-dimensional, and $\Phi(a) = \operatorname{Tr}(a)$. The trace is faithful, semifinite, and normal.
5. $\mathscr{A} = \mathscr{B}(\mathscr{H}_1) \oplus \mathscr{B}(\mathscr{H}_2)$, $\Phi(a \oplus b) = \alpha \operatorname{Tr} a + \beta \operatorname{Tr} b$, $\alpha$ and $\beta \in \mathbb{R}^+$. The trace is faithful only if $\alpha$ and $\beta$ are nonzero and finite only if the $\mathscr{H}_i$ are finite-dimensional. In all cases it is semifinite and normal. (Note that although $\Phi$ is invariant under unitary transformations belonging to $\mathscr{A}$ for $\alpha \neq \beta$, it is not invariant under all unitary transformations in $\mathscr{B}(\mathscr{H}_1 \oplus \mathscr{H}_2)$.)
6. Let $\mathscr{A}$ be the algebra of multiplication operators $L^\infty(\mathbb{R}, d\mu)$ on $L^2(\mathbb{R}, d\mu)$, and $\Phi(a) = \int d\mu(x) a(x) \rho(x)$ for some non-negative, measurable $\rho$. If $\rho > 0$ a.e., then $\Phi$ is faithful; if $\rho \in L^1(\mathbb{R}, d\mu)$, then $\Phi$ is finite; and if $\rho < \infty$ a.e., then $\Phi$ is semifinite. In all cases the trace is normal.
7. Let $\mathscr{A}$ be the algebra of multiplication operators $l^\infty$ on $l^2$, and $\Phi(a) = \lim_{i \to \infty} a_i$ when the limit exists, and otherwise let the trace be defined by linear extension with the Hahn–Banach theorem. The trace is finite and neither faithful nor normal: If $F = \{(a_i)$, where $a_i = 1$ for finitely many $i$ and otherwise $= 0\}$, then $s = (a_i = 1)$, and $\Phi(s) = 1$, but $\Phi(a) = 0$ for all $a \in F$.

**Remarks** (1.4.13)

1. Property (ii) may be replaced with (ii)′: $\Phi(aa^*) = \Phi(a^*a)$ for all $a \in \mathscr{A}$ (Problem 3).
2. It can be shown in general that $\{a \in \mathscr{A}_+ : \Phi(a) < \infty\}$ consists of the positive elements of a two-sided self-adjoint ideal $\mathscr{M}_\Phi$, onto which $\Phi$ can be extended as a linear form (also denoted $\Phi$). It is discontinuous in every topology that is strictly coarser than the one defined by the norm $\|a\|_\Phi = \Phi((a^*a)^{1/2})$. All continuous linear functionals on $\mathscr{M}_\Phi$ with this topology are of the form $a \to \Phi(ab)$, $a \in \mathscr{M}_\Phi$, $b \in \mathscr{A}$ (Problem 4), and nonzero for $b \neq 0$.
3. Property (ii) implies for $a \in \mathscr{M}_\Phi$ and any unitary $u \in \mathscr{A}$ that $\Phi(ua) = \Phi(au)$. Moreover, since every element of $\mathscr{A}$ is a linear combination of unitary operators, $\Phi(ab) = \Phi(ba)$, $a \in \mathscr{M}_\Phi$, $b \in \mathscr{A}$.
4. The requirement of normality originates in the theory of integration, where monotonic convergence can be permuted with integration. The trace can consequently be regarded as a generalization of the integral to noncommutative integrands.
5. If $\Phi$ is normal, then $\mathscr{A}$ may be written as $\mathscr{A} = \mathscr{A}_1 \oplus \mathscr{A}_2 \oplus \mathscr{A}_3$, where $\Phi_{|\mathscr{A}_3}$ is faithful and semifinite, $\Phi_{|\mathscr{A}_1} = 0$, and $\Phi_{|\mathscr{A}_2}$ is purely infinite (Problem 5). As we shall be interested solely in normal traces and shall ignore the trivial cases of Examples 1 and 2, we may confine our attention to faithful, semifinite traces.

The ordering of operators induces an ordering of traces, whereby $\Phi \leq \Psi$ shall mean $\Phi(a) \leq \Psi(a)$ for all $a \in \mathscr{A}_+$. For the ordering of the trace there is a theorem on

**The Form of a Dominating Trace** (1.4.14)

*Let $\Phi$ and $\Psi$ be normal, semifinite traces on a von Neumann algebra $\mathscr{A}$. Then $\Phi \leq \Psi$ iff there exists $b \in \mathscr{A} \cap \mathscr{A}'$, $0 < b \leq 1$, such that $\Phi(a) = \Psi(ab)$ for all $a$.*

**Proof**

Let $\mathscr{M}_\Psi$ be the ideal on which $\Psi < \infty$, given the norm $\|a\| = \Psi((aa^*)^{1/2})$. The mapping $a \to \Phi(a)$ is then a continuous linear form on $\mathscr{M}_\Psi$, and by Remark (1.4.13; 2) it is $\Psi(ab)$ for some $b \in \mathscr{A}$. To prove that $b \in \mathscr{A}'$, observe that for all $a \in \mathscr{M}_\Phi$ and $c \in \mathscr{A}$, $0 = \Phi(ac - ca) = \Psi(acb - cab) = \Psi(a[c, b])$, so, according to (1.4.13; 2), $[c, b] = 0$. □

**Corollary** (1.4.15)

*Any two faithful, normal, semifinite traces on the same factor are proportional.* More specifically, if $\Phi_1$ and $\Phi_2$ are two such traces, then $\Phi_1 < \Phi_1 + \Phi_2$ and

$\Phi_2 < \Phi_1 + \Phi_2$. Since the center of the factor consists of multiples of the identity, $\Phi_i = \lambda_i(\Phi_1 + \Phi_2)$, $0 < \lambda_i < 1$, so $\Phi_1 = \lambda_1 \lambda_2^{-1} \Phi_2$.

Because the trace is essentially unique on any factor, it may be asked whether the trace of a projection is an integer $c$, which would allow a reasonable definition of the dimension of the subspaces onto which they project.

**The Types of Factors** (1.4.16)

**Factors of Type I**

The range of the trace of the projections of factors of type I is $c \in \mathbb{Z}^+$, and they are of the form $\mathscr{B}(\mathscr{H}) \otimes \mathbf{1}$, with $\mathscr{H}$ separable, i.e., a sum of identical copies of an irreducible algebra of operators. The trace is given by $\Phi(a \otimes \mathbf{1}) = c \operatorname{Tr} a$, and if the dimension of $\mathscr{H}$ is $n$, then it is finite for $n < \infty$ and not finite but only semifinite for $n = \infty$. This creates a distinction between subtypes $\mathrm{I}_n$ and $\mathrm{I}_\infty$.

**Factors of Type II**

On Factors of Type II there is a semifinite, normal, faithful trace the range of which when applied to the projections is either $[0, 1]$ or $\mathbb{R}^+$. Depending on whether the trace is finite or only semifinite, one distinguishes between subtypes $\mathrm{II}_1$ and $\mathrm{II}_\infty$. An example of type $\mathrm{II}_1$ is the algebra of infinitely many spins (1.1.2) represented with the GNS construction using the state $\Phi$: $\Phi(\mathbf{1}) = 1$, $\Phi(\prod \boldsymbol{\sigma}_j) = 0$ ((1.4.8) with $s = 0$). This state has the properties of a trace; commutativity (1.4.11(ii)) holds trivially, and this representation is a factor. Since the factor is obviously not isomorphic to anything of the form $\mathscr{B}(\mathscr{H}_n) \otimes \mathbf{1}$, $n < \infty$, and the trace is finite, it must be of type $\mathrm{II}_1$. It is reducible but not of type I, since it can not be written as a direct sum of identical irreducible algebras. Type $\mathrm{II}_\infty$ factors are of the form type $\mathrm{I}_\infty \otimes$ type $\mathrm{II}_1$, where the trace is defined multiplicatively on the tensor product.

**Factors of Type III**

They have no normal, faithful, semifinite trace. The infinite spin algebra (1.1.2) again provides an example, this time with the GNS representation using the state (1.1.11) with $s \neq 0$, in other words (1.4.8).

**Remarks** (1.4.17)

1. The type with the properties familiar from finite matrices is I, while types II and III are less intuitive. All three types occur in the GNS representation of the spin algebra with a state of the form (1.1.11), $\mathrm{I}_\infty$ with $s = 1$, $\mathrm{II}_1$ with

$s = 0$, and III with $0 < s < 1$. To the malicious delight of many mathematicians the initial impression that type III is the rule for infinite systems has panned out with the passage of time. Types I and II turn out to be peripheral possibilities.
2. It was ascertained in (III: 2.3.6; 5) that factor representations with maximally Abelian subalgebras are irreducible. As a result, representations of types II and III have no maximally Abelian subalgebras.
3. If a factor includes an irreducible subrepresentation, then a semifinite, normal trace can be defined on it, mapping the projections to a discrete set of values, and it must therefore be of type I. It was remarked in (III: 2.3.10; 5) that the GNS construction yields an irreducible representation iff the state it builds on is pure. This means that no vector in the Hilbert space of a representation of type II or III corresponds to a pure state on the algebra.
4. Any operator $a$ of an algebra of type III is of course bounded, so Tr $\rho a$ is well defined for any $\rho \in \mathscr{C}_1(\mathscr{H})$, only $\rho$ can not come from the algebra, which contains no element of a trace class (other than 0).

Let us end the section by recapitulating the physical significance of the new mathematical phenomena that make an appearance in infinite systems.

1. **Inequivalent Representations**
Since vectors that differ globally are always orthogonal, globally different situations lead to inequivalent representations. Within a given representation different elements of the algebra produce vectors that differ only locally.

2. **Non-normal States**
Expectation values with a vector of a different, inequivalent representation constitute a state on the algebra, but one that fails to be strongly continuous with respect to the original representation, and hence it is not normal. They are representations of different global circumstances, and thus assign different values to global observables like densities, which are only defined with strong limits.

3. **Factors**
Whereas $\pi(\mathscr{A})$ describes microscopic observables, $\pi(\mathscr{A})''$ covers macroscopic observables as well. Factors associate certain numerical values to the global observables lying in the center $\pi(\mathscr{A})'' \cap \pi(\mathscr{A})'$—factors are the macroscopically pure states. In factors, Khinchin's ergodic theorem applies to them, stating that these global quantities exhibit no fluctuation. Even if vectors of a factor are pure with respect to this subalgebra, they may produce mixed states. The ground state is associated with type I, finite temperature with type III, and infinite temperature with type II.

4. **Unitary Representation of the Time-Evolution**
If the algebra changes globally as time passes, then a representation may change at any moment into an inequivalent representation, and it is not

possible to represent the time-evolution with a group of unitary transformations within the representation. Yet if the representation is based on a time-invariant state, then the other vectors of the representation differ only locally, and thus do not change in time, from the global point of view. This establishes the possibility of a unitary time-evolution.

## Problems (1.4.18)

1. Show that with vectors $|x^{(1)}\rangle, \ldots, |x^{(n)}\rangle$ and $\lambda_1, \ldots, \lambda_n \in \mathbb{C}$, Definition (1.4.2) implies that $\sum_{i,k} \lambda_i^* \lambda_k \langle x^{(i)}|x^{(k)}\rangle \geq 0$. (Hint: it suffices to show this for the case where the $|x^{(i)}\rangle$ are strongly equivalent. Prove that $\sum_{i,k} \lambda_i^* \lambda_k \prod_{j=1}^N (x_j^{(i)}|x_j^{(k)}) \geq 0$ for any $N$ and take the limit $N \to \infty$.)

2. (i) Show that $|x\rangle$ and $|y\rangle$ are equivalent iff $\sum_i |1 - (x_i|y_i)| < \infty$ and weakly equivalent iff $\sum_i |1 - |(x_i|y_i)|| < \infty$.
   (ii) Conclude from (i) that the $\underset{\text{strong}}{\sim}$ of $|x\rangle \underset{\text{strong}}{\sim} |y\rangle$ has all the properties of an equivalence relation, namely reflexivity, symmetry, and transitivity. (Hint: use the inequality $|1 - (x|z)| \leq 4[|1 - (x|y)| + |1 - (y|z)|]$, which holds for unit vectors. This 4 is a generous constant.)
   (iii) Show that $|x\rangle \sim |y\rangle$ iff there exists a sequence $\{\varphi_j\}$ such that $|x\rangle \underset{\text{strong}}{\sim} |y'\rangle$, $|y'\rangle \equiv \exp(i\varphi_1)|y_1) \otimes \exp(i\varphi_2)(|y_2) \otimes \ldots$.
   (iv) Show that $\underset{\text{weak}}{\sim}$ is also an equivalence relation.

3. Show that condition (ii) of the definition of the trace (1.4.11), i.e., $\Phi(a) = \Phi(UaU^{-1})$, may be replaced with: $\Phi(a^*a) = \Phi(aa^*)$ for all $a$ in a von Neumann algebra $\mathscr{A}$.

4. Show that for a faithful, normal, semifinite trace $\Phi$, all continuous linear forms on $a \in \mathscr{M}_\Phi$ may be written as $a \to \Phi(ab)$ for some $b \in \mathscr{A}$. (Hint: use the inequality $|\Phi(ab)| \leq \Phi(|ab|) \leq \|b\|\Phi(|a|)$.)

5. Show that with any normal trace $\Phi$, $\mathscr{A}$ can be written $\mathscr{A} = \mathscr{A}_1 \oplus \mathscr{A}_2 \oplus \mathscr{A}_3$, where $\Phi_{|\mathscr{A}_1} \equiv 0$, $\Phi_{|\mathscr{A}_2}$ is faithful and semifinite, and $\Phi_{|\mathscr{A}_3}$ is purely infinite. (Use the following corollaries of von Neumann's density theorem (III: 2.3.24; 4):
   (I) Let $\mathscr{M} \subset \mathscr{A}$ be a strongly closed, two-sided ideal. Then $\mathscr{M}$ contains a projection operator $P$ such that $P \in \mathscr{A} \cap \mathscr{A}'$ and $P \geq Q$ for all projection operators $Q \in \mathscr{M}$.
   (II) Let $\mathscr{N}$ be a two-sided ideal and suppose $a$ is in the positive part of the weak closure of $\mathscr{N}$. Then there exists an increasing filter $\subset \mathscr{N}^+$ having $a$ for its supremum.)

## Solutions (1.4.19)

1. The $n \times n$ matrix

$$\begin{pmatrix} (x_j^{(1)}|x_j^{(1)}) & \ldots & (x_j^{(1)}|x_j^{(n)}) \\ \vdots & & \\ (x_j^{(n)}|x_j^{(1)}) & \ldots & (x_j^{(n)}|x_j^{(n)}) \end{pmatrix}$$

is Hermitian and non-negative, and is thus a sum of projections, i.e., matrices of the form

$$\begin{pmatrix} h_1^* h_1 & \dots & h_1^* h_n \\ \vdots & & \\ h_n^* h_1 & \dots & h_n^* h_n \end{pmatrix}.$$

This implies

$$(x_j^{(i)} | x_j^{(k)}) = \sum_{l_j = 1}^{n} h_i^{l_j *} h_k^{l_j},$$

and

$$\sum_{i,k} \lambda_i^* \lambda_k \prod_{j=1}^{N} (x_j^{(i)} | x_j^{(k)}) = \sum_{i,k} \lambda_i^* \lambda_k \sum_{l_1, \dots, l_N} h_i^{l_1 *} h_k^{l_1} \dots h_i^{l_N *} h_k^{l_N} \geq 0,$$

since

$$\sum_{i,k} \lambda_i^* \lambda_k h_i^{l_1 *} \dots h_k^{l_N} = \Big| \sum_k \lambda_k h_k^{l_1} \dots h_k^{l_N} \Big|^2 \geq 0.$$

2. (i) follows from the theory of infinite products [12].
 (ii) To prove the inequality, choose a basis for the subspace spanned by $|x)$, $|y)$, and $|z)$ such that they correspond to the vectors $(\alpha, \beta, 0)$, $(1, 0, 0)$ and $(\gamma, \delta, \varepsilon)$, where $|\alpha|^2 + |\beta|^2 = |\gamma|^2 + |\delta|^2 + |\varepsilon|^2 = 1$. Then $(x|y) = \alpha^*$, $(y|z) = \gamma$, $(x|z) = \alpha^*\gamma + \beta^*\delta$. $|1 - \alpha^*\gamma - \beta^*\delta| \leq |1 - \alpha^*\gamma| + |\beta||\delta| \leq 2|1 - \alpha| + 2|1 - \gamma| + (1 - |\alpha|^2)^{1/2}(1 - |\gamma|^2)^{1/2} \leq 2(|1 - \alpha|^{1/2} + |1 - \gamma|^{1/2})^2 \leq 4[|1 - \alpha| + |1 - \gamma|]$. The reflexivity and symmetry of the equivalence relation are trivial, and transitivity follows from (i) together with the inequality.
 (iii) $\Rightarrow$: Choose $\varphi_j = -\arg(x_j | y_j)$.
 $\Leftarrow$: This is trivial.
 (iv) follows from (ii) and (iii).
3. (ii) $\Rightarrow$ (ii′): With a polar decomposition, $a = V|a|$, where $a^*a = |a|^2 = V^*V|a|^2$, $aa^* = V|a|^2V^*$. Let $\mathscr{M}_\Phi$ be the trace-class ideal: $a \in \mathscr{M}_\Phi \Rightarrow a^*a \in \mathscr{M}_\Phi$ and $aa^* \in \mathscr{M}_\Phi$ $\Rightarrow Va^*a \in \mathscr{M}_\Phi$, since $V = w\text{-}\lim_{\varepsilon \downarrow 0} a(|a|^2 + \varepsilon)^{-1/2} \in \mathscr{A}$, which, with Remark (1.4.13; 3) implies $\Phi(V^*Va^*a) = \Phi(Va^*aV^*)$.
 (ii′) $\Rightarrow$ (ii): Let $a \geq 0$. $\Phi(UaU^{-1}) = \Phi(Ua^{1/2}a^{1/2}U^*) = \Phi(a^{1/2}U^*Ua^{1/2}) = \Phi(a)$, and every operator is a linear combination of positive operators.

4. To prove the inequality, let $a$ and $b$ be non-negative. $\Phi(ab) = \Phi(a^{1/2}ba^{1/2}) \leq \|b\|\Phi(a)$, since for any $a$ and $b$, $a^{1/2}ba^{1/2} \leq a^{1/2}\|b\|a^{1/2}$. Thus $|\Phi(ab)|^2 \leq \Phi(|a^*||b|)\Phi(|a||b^*|)$ and is consequently $\leq \||b|\|\Phi(|a^*|)\||b^*|\|\Phi(|a|) = \|b\|^2\Phi(|a|)^2$, in which the Cauchy–Schwarz inequality $|\Phi(ab)|^2 \leq \Phi(aa^*)\Phi(bb^*)$ (see (III: 2.2.20; 1)) was used in the form $|\Phi(ab)|^2 = |\Phi(U|a|V|b|)|^2$ (with the polar decompositions $a = U|a|$ and $b = V|b|$). This $= |\Phi(|b|^{1/2}U|a|^{1/2}|a|^{1/2}V|b|^{1/2})|^2 \leq \Phi(|b|^{1/2}U|a|^{1/2} \times |a|^{1/2}U^*|b|^{1/2}) \cdot \Phi(|b|^{1/2}V^*|a|^{1/2}|a|^{1/2}V|b|^{1/2}) = \Phi(|b|U|a|U^*)\Phi(V|b|V^*|a|) = \Phi(|b||a^*|)\Phi(|b^*||a|)$. Now let $ab = W|ab|$; then $\Phi(|ab|) = \Phi(W^*ab) \leq \|bW^*\| \times \Phi(|a|) \leq \|b\|\Phi(|a|)$. The first part of the inequality follows from $|\Phi(ab)| = |\Phi(ab \cdot \mathbf{1})| \leq \|\mathbf{1}\|\Phi(|ab|) = \Phi(|ab|)$.

 It is a corollary of the inequality that the norm of the mapping $a \to \Phi(ab)$ is $\|b\|$. This allows $\mathscr{A}$ to be identified with a closed subspace of $\mathscr{M}_\Phi^*$. To see that $\mathscr{A} = \mathscr{M}_\Phi^*$, first suppose $a \in \mathscr{M}_\Phi^+$. Then the mapping $\mathscr{A} \to \mathbb{C}$: $b \to \Phi(ab)$ is normal, entailing ultraweakly continuous (see (2.1.4)), which implies that for any $a \in \mathscr{M}_\Phi$, $b \to \Phi(ab)$ is ultraweakly continuous. Because of the inequality again, the norm of this mapping is $\Phi(|a|)$, which implies that $\mathscr{M}_\Phi$ can be imbedded isometrically and isomorphically in the predual $\mathscr{A}_*$, i.e., the space of ultraweakly continuous linear functions. Thus $\mathscr{M}_\Phi \subset \mathscr{A}_*$. We shall see in (2.1.3) and (2.1.4) that $\mathscr{C}_1 = \mathscr{B}(\mathscr{H})_*$ and $\mathscr{C}_1^* = \mathscr{B}(\mathscr{H})$. Since $\mathscr{A}$ is ultraweakly closed, $\mathscr{A}_\perp = \mathscr{C}_1/\mathscr{C}_1^\perp$ with $\mathscr{C}_1^\perp = \{\rho \in \mathscr{C}_1 : \mathrm{Tr}\, \rho a = 0 \text{ for all } a \in \mathscr{A}\}$, so $\mathscr{A} = (\mathscr{C}_1/\mathscr{C}_1^\perp)^*$. Therefore $\mathscr{M}_\Phi^* \subset (\mathscr{A}_*)^* = \mathscr{A}$, which implies $\mathscr{M}_\Phi^* = \mathscr{A}$. Remark: $\mathscr{M}_\Phi$ is dense in $\mathscr{A}_*$ but not in general closed.

5. For more about types I and II, see Chapter I, §3 of [4]. The set $\{a \in \mathscr{A}^+ : \Phi(a) = 0\}$ is the positive part of a two-sided ideal $\mathscr{N}$. Let $\mathscr{M}$ be the trace class, let $\bar{\mathscr{N}}$ and $\bar{\mathscr{M}}$ be the strong closures of $\mathscr{N}$ and $\mathscr{M}$, and $P_1$ and $P_2$ be respectively the largest projections they contain (see Corollary I). The Hilbert space $\mathscr{H}$ can be decomposed as $\mathscr{H}_1 \oplus \mathscr{H}_2 \oplus \mathscr{H}_3$, where $\mathscr{H}_1 \equiv P_1\mathscr{H}$, $\mathscr{H}_2 \equiv (P_2 - P_1)\mathscr{H}$, $\mathscr{H}_3 \equiv (1 - P_2)\mathscr{H}$ in which case $\mathscr{A} = \mathscr{A}_1 \oplus \mathscr{A}_2 \oplus \mathscr{A}_3$, where $\mathscr{A}_i \equiv \mathscr{A}_{|\mathscr{H}_i}$, since $P_1$ and $P_2$ belong to $\mathscr{A} \cap \mathscr{A}'$.
   It is obvious that $\Phi_{|\mathscr{A}_1} = 0$. To see that $\Phi_{|\mathscr{A}_2}$ is semifinite, apply Corollary II: Let $a \in \bar{\mathscr{M}}^+ \setminus \mathscr{M}^+$; then there exists an operator $b \in \mathscr{M}^+$, $b \leq a$, such that $\Phi(b) > 0$. The remaining claims are trivial.

# Thermostatics 2

## 2.1 The Ordering of the States

*The heuristic concepts of purer and more chaotic states can be made mathematically precise with reference to a lattice structure of the classes of equivalent density matrices.*

States are by definition (III: 2.2.18) normed, positive linear functionals on an algebra $\mathscr{A}$ of observables. If the dimension of the underlying space is finite, $\mathscr{A} = \mathscr{B}(\mathbb{C}^n)$, then all linear functionals are of the form $\mathscr{A} \ni a \to \operatorname{Tr} \rho a \equiv (\rho|a)$, $\rho \in \mathscr{B}(\mathbb{C}^n)$, and $\mathscr{B}(\mathbb{C}^n)$ is its own dual space. The inequality of (1.4.18; 4),

$$|(\rho|a)| \leq \|a\| \, \|\rho\|_1, \qquad \|\rho\|_1 = \operatorname{Tr}(\rho^*\rho)^{1/2} \tag{2.1.1}$$

then holds, and is optimal in the sense that

$$\sup_{\|\rho\|_1 = 1} |(\rho|a)| = \|a\|, \qquad \sup_{\|a\| = 1} |(\rho|a)| = \|\rho\|_1. \tag{2.1.2}$$

If the dimension of $\mathscr{H}$ is infinite, the inequality applies initially to the operators of finite rank (cf. (III: 2.3.21)), denoted $\mathscr{E}$ or $\mathscr{E}_1$, depending on whether the norm $\| \ \|$ or $\| \ \|_1$ is used. In these topologies continuous, linear functionals are of the form

$$\mathscr{E} \ni a \to \operatorname{Tr} \rho a \quad \text{with } \|\rho\|_1 < \infty$$

or

$$\mathscr{E}_1 \ni a \to \operatorname{Tr} \rho a \quad \text{with } \|\rho\| < \infty.$$

The linearity and continuity of the functionals thus defined are obvious, and it can be seen as follows that all functionals with these properties are of that form. By what was said earlier, a linear functional on $\mathscr{E}$ determines the restriction of an operator $\rho$ to any finite-dimensional subspace. To guarantee that $|(\rho|a)| \leq c\|a\|$ for all $a \in \mathscr{E}$ or $|(\rho|a)| \leq c\|a\|_1$ for all $a \in \mathscr{E}_1$, by (2.1.2) it is necessary to ensure that $\|\rho\|_1$ or respectively $\|\rho\|$ is bounded. If the spaces $\mathscr{E}$ and $\mathscr{E}_1$ are now completed, becoming the Banach spaces $\mathscr{C}$ and $\mathscr{C}_1$, of (III: 2.3.21), then their dual spaces are unaffected—the dual spaces of a space and of a dense subspace are the same. The state of affairs is analogous to that of $l^0$, $l^1$, and $l^\infty$, the spaces of sequences $(x_i)$ satisfying respectively $\lim_{i\to\infty} x_i = 0$, $\sum_i |x_i| < \infty$, and $\sup_i |x_i| < \infty$:

**Duality for the Subspaces of $\mathscr{B}(\mathscr{H})$** (2.1.3)

$\mathscr{C}^* = \mathscr{C}_1$, $\mathscr{C}_1^* = \mathscr{B}(\mathscr{H})$, where $\mathscr{C}$ and $\mathscr{B}(\mathscr{H})$ are given the norm $\| \ \|$, and $\mathscr{C}_1$ the norm $\| \ \|_1$. These norms on $\mathscr{C}_1$ and $\mathscr{B}(\mathscr{H})$ produce the strong topology on the dual spaces, as can be seen from a comparison of (2.1.2) with (III: 2.1.21).

The Banach space $\mathscr{C}$ is thus not reflexive, so $\mathscr{B}(\mathscr{H})^*$ is strictly larger than $\mathscr{C}_1$. If a Banach space $\mathscr{E}$ is nonreflexive, then the same is true of $\mathscr{E}^*$, $\mathscr{E}^{**}$, etc.: Let $a \in \mathscr{E}^{**}$ but $a \notin \mathscr{E}$. The functional $w: e + \lambda a \to \lambda$ defined on $\{E + \lambda a\}$ can be extended continuously to $\mathscr{E}^{**}$ by the Hahn–Banach theorem. Therefore, $w \in \mathscr{E}^{***}$, but $w_{|\mathscr{E}} = 0$. Hence $\mathscr{C}_1$ and $\mathscr{B}(\mathscr{H})$ are also not reflexive; $\mathscr{B}(\mathscr{H})^*$ is strictly larger than $\mathscr{C}_1$. All trace-class operators provide linear functionals on the bounded operators by $a \to \operatorname{Tr} \rho a$, and these linear functionals are even continuous if $\mathscr{B}(\mathscr{H})$ is equipped with a weaker topology than the one from $\| \ \|$: If the neighborhood basis is defined by

$$U_{\rho,\varepsilon}(a) = \{a' \in \mathscr{B}(\mathscr{H}): |\operatorname{Tr} \rho(a - a')| < \varepsilon\}, \qquad (2.1.4)$$

and $\rho$ ranges only over $\mathscr{E}$, then this is the weak topology. If $\rho$ is allowed to range over $\mathscr{C}_1$, then it is known as the ultraweak topology, and is genuinely finer than the weak topology but coarser than the $\| \ \|$-topology. The linear functionals $a \to \operatorname{Tr} \rho a$ for $\rho \in \mathscr{C}_1$ are, however, obviously continuous if $\mathscr{B}(\mathscr{H})$ has the ultraweak topology. These functionals have in addition the property of normality (III: 2.2.21): the order of taking weakly continuous linear functionals and suprema over bounded sets can be interchanged, since by Vigier's theorem (III: 2.3.24; 11) the supremum is the limit of a strongly, and therefore also weakly, convergent sequence. Since the weak and ultraweak topologies are equivalent on bounded sets, normality carries over to ultraweakly continuous, linear functionals. A somewhat deeper theorem ([4], I, §4, Theorem I) states that these include all normal linear functionals on $\mathscr{B}(\mathscr{H})$. We summarize by stating the

**Characterization of Normal States** (2.1.5)

The following properties are equivalent for a state $w$ on $\mathscr{B}(\mathscr{H})$:

(i) $w$ is normal (III: 2.2.21);
(ii) $w$ is given by a **density matrix** $\rho$ such that $w(a) = \operatorname{Tr} \rho a, \rho \geq 0, \operatorname{Tr} \rho = 1$;
(iii) $w$ is ultraweakly continuous.

**Remarks** (2.1.6)

1. The density matrices form a norm-closed, convex subset of the unit sphere of $\mathscr{C}_1$, the trace-class operators with the trace norm $\| \ \|_1$.
2. If the system is classical, then instead of $\mathscr{B}(\mathscr{H})$ there is an Abelian von Neumann algebra, and we are familiar with the normal traces in the guise of probability measures. Specifically, on the $L^\infty$ functions on phase space they are of the form $\rho(p, q)\, d\Omega$, $d\Omega$ being Liouville measure (I: 3.1.2; 3), $\rho \in L^1$, $\rho \geq 0$, $\int d\Omega \rho = 1$. Yet it may be that $|\rho| = \sup_{p,q} |\rho(p, q)| \not< 1$: Suppose that $\chi_A$ is the characteristic function of a set $A$ such that $\Omega(A) \equiv \int d\Omega \chi_A < 1$; then an example is furnished by $\rho = \chi_A/\Omega(A)$.
3. All states constructed with a vector of $\mathscr{H}$ are pure, normal, and even weakly continuous—the density matrix for them is a one-dimensional projection. Conversely, any one-dimensional projection yields a pure state on $\mathscr{B}(\mathscr{H})$.
4. The spectrum of a density matrix is discrete, as it is in the trace class (and hence compact). The sum of the eigenvalues $\rho_i$ is 1.
5. The density matrix can be thought of as a combination of the vectors that diagonalize it, or as a pure state on a larger Hilbert space $\mathscr{H}_g \equiv \mathscr{H} \otimes \mathscr{H}$, in which $\mathscr{B}(\mathscr{H})$ is imbedded as $\mathscr{B}(\mathscr{H}) \otimes \mathbf{1}$. The vector of $\mathscr{H}_g$ corresponding to $\rho = \sum_j |j\rangle\langle j| \rho_j$ is $\sum_j |j\rangle \otimes |j\rangle \sqrt{\rho_j}$ (cf. (1.4.7)). If $\mathscr{H}$ is separable, then the weak topology on $\mathscr{H}_g$ induces the ultraweak topology of $\mathscr{B}(\mathscr{H})$ on $\mathscr{B}(\mathscr{H}) \otimes \mathbf{1}$.
6. The normal states are weak-* dense in the positive unit sphere of $\mathscr{B}(\mathscr{H})^*$ (see (III: 2.1.19)), but are a proper subset rather than the whole of it. Hence they are not also weak-* compact.

Traces offer many advantages for doing calculations, owing to the commutativity property (1.4.13; 3). Inequalities for ordinary numbers often extend to traces, even when noncommutativity prevents them from extending directly to operators. Some of these inequalities will be used frequently later, and so are listed below. It will always be assumed that whatever the trace is taken of belongs to the trace class, though many of them have the generalization that if the lesser side of an inequality becomes infinite, then so does the greater side. For greater flexibility general forms are presented, while the name attached refers to the original version. The symbol Tr will always mean the trace on $\mathscr{B}(\mathscr{H})$. These inequalities apply trivially to factors of type I, and many also apply to type II.

**Basic Inequalities** (2.1.7)

1. Peierls's Inequality. Let $k$ be a convex function from $\mathbb{R}$ to $\mathbb{R}^+$ and $\{|i\rangle\}$ be a not necessarily complete, orthonormal set. Then
$$\operatorname{Tr} k(a) = \sup_{\{|i\rangle\}} \sum_i k(\langle i|a|i\rangle).$$
2. Convexity. Let $k$ be a convex function from $\mathbb{R}$ to $\mathbb{R}$ and $0 \le \alpha \le 1$. Then
$$\operatorname{Tr} k(\alpha a + (1-\alpha)b) \le \alpha \operatorname{Tr} k(a) + (1-\alpha) \operatorname{Tr} k(b).$$
3. The Peierls–Bogoliubov Inequality. Let $k$ be a strictly monotonically increasing, convex, differentiable function $\mathbb{R} \to \mathbb{R}$ (and thus the inverse function $k^{-1}$ exists), and suppose $k/k'$ is convex. Then
$$k^{-1}(\operatorname{Tr} k(\alpha a + (1-\alpha)b)) \le \alpha k^{-1}(\operatorname{Tr} k(a)) + (1-\alpha)k^{-1}(\operatorname{Tr} k(b)).$$
4. Monotony. If $m$ is a monotonically increasing function $\mathbb{R} \to \mathbb{R}$,
$$a \ge b \Rightarrow \operatorname{Tr} m(a) \ge \operatorname{Tr} m(b).$$
5. Klein's Inequality. Let $f$, $g$, and $h$ be functions $\mathbb{R} \to \mathbb{R}$ such that for all $\alpha \in \operatorname{Sp} a$, $\beta \in \operatorname{Sp} b$, and $c_k \in \mathbb{R}$,
$$\sum_k c_k f_k(\alpha) g_k(\beta) h_k(\alpha) \ge 0.$$
Then
$$\operatorname{Tr} \sum_k c_k f_k(a) g_k(b) h_k(a) \ge 0.$$
6. Hölder's Inequality. Suppose that $k_1$ and $k_2$ are convex, strictly monotonic functions $\mathbb{R} \to \mathbb{R}$, the mapping $(\alpha, \beta) \to k_1^{-1}(\alpha) k_2^{-1}(\beta)$ is concave, and $\mathscr{H}$ has dimension $N < \infty$. Then
$$\left|\frac{1}{N} \operatorname{Tr} ab\right| \le k_1^{-1}\left(\operatorname{Tr} \frac{1}{N} k_1(|a|)\right) k_2^{-1}\left(\operatorname{Tr} \frac{1}{N} k_2(|b|)\right).$$
7. The Cauchy–Schwarz Inequality. $|\operatorname{Tr}(ab)^2| \le \operatorname{Tr} a^* abb^*$.
8. Lieb's Theorem. Let $a$ and $b$ be non-negative, $a, b, c \in \mathscr{B}(\mathscr{H})$, and $0 \le \alpha \le 1$. Then the functions $a \to \operatorname{Tr} \exp(c + \ln a)$ and $(a, b) \to \operatorname{Tr} a^\alpha c b^{1-\alpha} c^*$ are concave.

**Proof**

1. By the spectral theorem and Jensen's inequality, for any unit vector $|i\rangle$, $\langle i|k(a)|i\rangle \ge k(\langle i|a|i\rangle)$, and therefore $\sum_i \langle i|k(a)|i\rangle \ge \sum_i k(\langle i|a|i\rangle)$.

Equality holds if the $|i\rangle$ are eigenvectors of $a$. It suffices to take the supremum over finite sets $\{|i\rangle\}$.

2. Let $|i\rangle$ be the eigenvectors of $\alpha a + (1-\alpha)b$. By Peierls's inequality,

$$\begin{aligned}\operatorname{Tr} k(\alpha a + (1-\alpha)b) &= \sum_i k(\alpha\langle i|a|i\rangle + (1-\alpha)\langle i|b|i\rangle)\\ &\le \alpha \sum_i k(\langle i|a|i\rangle) + (1-\alpha)\sum_i k(\langle i|b|i\rangle)\\ &\le \alpha \operatorname{Tr} k(a) + (1-\alpha)\operatorname{Tr} k(b).\end{aligned}$$

Note that the inequality $k(\alpha a + (1-\alpha)b) \le \alpha k(a) + (1-\alpha)k(b)$ can be false in the sense of operator ordering.

3. If $k/k'$ is convex, then for sequences of numbers $\{\beta_i\}$ and $\{\gamma_i\}$,

$$k^{-1}\left(\sum_i k(\beta_i\alpha + \gamma_i(1-\alpha))\right) \le \alpha k^{-1}\left(\sum_i k(\beta_i)\right) + (1-\alpha)k^{-1}\left(\sum_i k(\gamma_i)\right)$$

by Problem 2. Hence, as with Inequality 2,

$$\begin{aligned}k^{-1}(\operatorname{Tr} k(\alpha a + (1-\alpha)b)) &= k^{-1}\left(\sum_i k(\alpha\langle i|a|i\rangle + (1-\alpha)\langle i|b|i\rangle)\right)\\ &\le \alpha k^{-1}\left(\sum_i k(\langle i|a|i\rangle)\right)\\ &\quad + (1-\alpha)k^{-1}\left(\sum_i k(\langle i|b|i\rangle)\right)\\ &\le \alpha k^{-1}(\operatorname{Tr} k(a)) + (1-\alpha)k^{-1}(\operatorname{Tr} k(b)),\end{aligned}$$

using Inequality 1 again.

4. If $a \ge b$, then the min–max principle implies for their ordered eigenvalues that $a_i \ge b_i$, so $\sum_i m(a_i) \ge \sum_i m(b_i)$. Once again, the inequality $m(a) \ge m(b)$ may fail for operators.

5. Let $a_i$ and $b_i$ be the eigenvalues of $a$ and $b$, and $c_{ij}$ be the scalar product of the eigenvectors of $a$ with those of $b$. Then

$$\operatorname{Tr} \sum_k c_k f_k(a) g_k(b) h_k(a) = \sum_{i,j} |c_{ij}|^2 \sum_k c_k f_k(a_i) g_k(b_j) h_k(a_i) \ge 0.$$

6. Let $a_i$ and $b_i$ be the ordered eigenvalues of $|a|$ and $|b|$, and let $|i\rangle$ denote the eigenvectors of $a$. By the min–max principle (III: 3.5.21),

$$\begin{aligned}\operatorname{Tr} ab &= \sum_{i,j} \langle i|a|j\rangle\langle j|b|i\rangle \le \sum_i (a_i - a_{i+1}) \sum_{k=1}^{i} \langle k\|b\|k\rangle\\ &\le \sum_i (a_i - a_{i+1}) \sum_{k=1}^{i} b_k = \sum_i a_i b_i.\end{aligned}$$

The inequality

$$\frac{1}{N}\sum_{i=1}^{N} k_1^{-1}(\alpha_i)k_2^{-1}(\beta_i) \le k_1^{-1}\left(\frac{1}{N}\sum_{i=1}^{N}\alpha_i\right)k_2^{-1}\left(\frac{1}{N}\sum_{i=1}^{N}\beta_i\right),$$

$$\text{for } \alpha_i \equiv k_1(a_i) \text{ and } \beta_i \equiv k_2(b_i)$$

is just the assumption of concavity.

7. By the Cauchy–Schwarz inequality (III: 2.2.20; 1) for states,

$$|\mathrm{Tr}\, abab|^2 \le \mathrm{Tr}\, abb^*a^* \,\mathrm{Tr}\, b^*a^*ab = (\mathrm{Tr}\, a^*abb^*)^2.$$

The order of the operations is important; it is not true in general that $\mathrm{Tr}(ab)^2 \le \mathrm{Tr}\, a^*ab^*b$.

8. The proof of this rather deep proposition in the noncommutative case is too laborous to be repeated here—see [5]. □

**Corollaries** (2.1.8)

1. For any orthonormal system $\{|i\rangle\}$, $\beta F(H) \equiv -\ln \mathrm{Tr}\exp(-\beta H) \le -\ln \sum_i \exp(-\beta\langle i|H|i\rangle)$.
2. The function $H \to \mathrm{Tr}\exp(-\beta H)$ is convex.
3. In fact, even $H \to \ln \mathrm{Tr}\exp(-\beta H)$ is convex, so $F(H)$ is concave. By recourse to $(\partial/\partial\alpha)\,\mathrm{Tr}\, f(H+\alpha V)_{|\alpha=0} = \mathrm{Tr}\, Vf'(H)$, and the fact that $F$ is majorized by any tangent, one finds that

$$F(H_0) + \langle V\rangle_H \le F(H_0 + V) \le F(H_0) + \langle V\rangle_{H_0},$$

where $\langle a\rangle_H = \mathrm{Tr}\, a\exp(-\beta H)/\mathrm{Tr}\exp(-\beta H)$.

4. $H_1 \ge H_2 \Rightarrow F(H_1) \ge F(H_2)$.
5. If $k$ is convex, then $\mathrm{Tr}(k(a) - k(b) - (a-b)k'(b)) \ge 0$, so $\mathrm{Tr}(a\ln a - a\ln b - (a-b)) \ge 0$, too. If $f_1(\alpha) = \int_0^\alpha d\alpha' g(\alpha')$ and $f_2(\beta) = \int_0^\beta d\beta' g^{-1}(\beta')$, then by Young's inequality, $\alpha\beta \le f_1(\alpha) + f_2(\beta)$, and therefore $\mathrm{Tr}\, ab \le \mathrm{Tr}\, f_1(a) + \mathrm{Tr}\, f_2(b)$. In particular, if $p$ and $q$ are $\ge 1$ and related by $1/p + 1/q = 1$, and $a$ and $b$ are nonnegative, then $\mathrm{Tr}\, ab \le (1/p)\,\mathrm{Tr}\, a^p + (1/q)\,\mathrm{Tr}\, b^q$.
6. With $k_1(\alpha) = \alpha^p$, $k_2(\beta) = \beta^q$, Corollary 5 can be improved to $\mathrm{Tr}\, ab \le (\mathrm{Tr}\,|a|^p)^{1/p}(\mathrm{Tr}\,|b|^q)^{1/q}$; since this no longer involves $N$, it also holds when $N = \infty$. By iteration,

$$\left\|\prod_{i=1}^{n} a_i\right\|_p \le \prod_{i=1}^{n} \|a_i\|_{p_i},$$

$$\|a\|_p = (\mathrm{Tr}\,|a|^p)^{1/p}, \quad \text{where } \sum_i \frac{1}{p_i} = \frac{1}{p}, \quad p, p_i \ge 1.$$

As $p \to \infty$, $\|a\|_p \to \|a\|$, so $|\mathrm{Tr}\, ab| \le \|a\|\mathrm{Tr}|b|$; the trace class is a two-sided ideal of $\mathscr{B}(\mathscr{H})$ (cf. (III: 2.3.20; 3)).

7. If $a$ and $b$ are Hermitian, then $\mathrm{Tr}(ab)^2 \leq \mathrm{Tr}\, a^2b^2$, $a = a^*$, $b^{-1} = b^*$: $|\mathrm{Tr}(ab)^2| \leq \mathrm{Tr}\, a^2$. By iterating this, $|\mathrm{Tr}(ab)^{2^p}| \leq \mathrm{Tr}(abb^*a^*)^{2^{p-1}} = \mathrm{Tr}(|a|^2|b^2|)^{2^{p-1}} \leq \cdots \leq \mathrm{Tr}\,|a|^{2^p}|b|^{2^p}$. Because of the Trotter product formula $\exp(a + b) = s\text{-}\lim_{n\to\infty} (\exp(a/n)\exp(b/n))^n$ (see (III; 2.4.9)), $|\mathrm{Tr}\exp(\alpha a + \beta b)| \leq \mathrm{Tr}\,|\exp(\alpha a)|\,|\exp(\beta b)|$, for $\alpha, \beta \in \mathbb{C}$, and initially for Hermitian operators of finite rank. It then extends to $\exp(\alpha a + \beta b) \in \mathscr{B}_1(\mathscr{H})$, $\exp(\alpha a) \in \mathscr{B}_1(\mathscr{H})$, $\exp(\beta b) \in \mathscr{B}(\mathscr{H})$ and thereby yields a generalization of Corollary 3 known as the Golden–Thompson–Symanzik inequality [6], $\exp(-\beta\langle V\rangle_{H_0}) \leq \mathrm{Tr}\exp[-\beta(H_0 + V - F(H_0))] \leq \langle\exp(-\beta V)\rangle_{H_0}$.
8. The function $(a, b) \to \lim_{\alpha\downarrow 0} \mathrm{Tr}(1/\alpha)(a - a^{1-\alpha}b^\alpha) = \mathrm{Tr}\, a(\ln a - \ln b)$ is convex.

Our next task is to give the density matrices an ordering that indicates which of two $\rho$'s corresponds to the more chaotic state. The ordering must of course be independent of the basis, and so it can depend only on the eigenvalues $\rho_i$. If the eigenvalues are thought of as ordered by their magnitudes, then pure states are associated with sequences (1, 0, 0, ...), i.e., with the greatest possible first eigenvalue. Because $\sum_{i=1}^\infty \rho_i = 1$, two density matrices might not be strictly ordered by the natural ordering of Hermitian operators. However, by the min–max principle (III: 3.5.21),

$$\rho(n) \equiv \sum_{i=1}^{n} \rho_i = \sup_{\mathscr{H}_n} \mathrm{Tr}_{\mathscr{H}_n}\rho,$$

which permits the following

**Definition of the Ordering of the Density Matrices** (2.1.9)

A density matrix $\tilde{\rho}$ is said to be **more mixed**, or **more chaotic**, than $\rho$ if $\tilde{\rho}(n) \leq \rho(n)$ for all $n$. In symbols, $\tilde{\rho} \succsim \rho$ (or $\rho \precsim \tilde{\rho}$).

**Remarks** (2.1.10)

1. This clearly defines a preordering of the density matrices, i.e., $\rho \precsim \rho$; and if $\rho \precsim \tilde{\rho}$ and $\tilde{\rho} \precsim \tilde{\tilde{\rho}}$, then $\rho \precsim \tilde{\tilde{\rho}}$. If two density matrices are equivalent, that is, $\rho \precsim \tilde{\rho}$ and $\tilde{\rho} \precsim \rho$, then $\rho_i = \tilde{\rho}_i$, and so they are related by $\tilde{\rho} = V\rho V^*$. If the space is finite-dimensional, then $V$ can be chosen unitary, and otherwise it is only an isometric mapping $(\mathrm{Ker}\,\rho)^\perp \to (\mathrm{Ker}\,\tilde{\rho})^\perp$; if $\mathrm{Dim\,Ker}\,\rho \neq \mathrm{Dim\,Ker}\,\tilde{\rho}$, then it has no unitary extension.
2. If the equivalent density matrices are classed together, then (2.1.9) gives the classes a lattice structure, characterized by the sequences of numbers $\{\rho(n)\}$. The sequence $\{\min(\rho(n), \tilde{\rho}(n))\}$ yields the equivalence class of the purest states more mixed than either $\rho$ or $\tilde{\rho}$. The concave hull of $\max(\rho(n), \tilde{\rho}(n))$ with respect to $n$ characterizes the most mixed states

purer than either $\rho$ or $\tilde{\rho}$. The sequences thus defined are positive, increasing, and concave in $n$, and tend to 1 as $n \to \infty$ (or equal 1 when $n = \text{Dim}\, \mathscr{H}$). Their successive differences are therefore decreasing sequences of positive numbers summing to 1, which correspond to an equivalence class of density matrices. The lattice contains a class of purest elements, namely the extremal states. If the dimension of $\mathscr{H}$ is finite, then there is also a most mixed state with $\rho = \mathbf{1}/\text{Dim}\, \mathscr{H}$, but if it is infinite, there is none.

3. The ordering and convexity are compatible on the space of states in the sense that if $\rho \precsim \mu$ and $\rho \precsim \nu$ then $\rho \precsim \alpha\mu + (1-\alpha)\nu$ for $0 \le \alpha \le 1$:

$$\sup_{\mathscr{H}_n} \text{Tr}_{\mathscr{H}_n}(\alpha\mu + (1-\alpha)\nu) \le \alpha \sup_{\mathscr{H}_n} \text{Tr}_{\mathscr{H}_n} \mu + (1-\alpha) \sup_{\mathscr{H}_n} \text{Tr}_{\mathscr{H}_n} \nu \le \rho(n).$$

4. Since the operators $\rho(n)$ are suprema of the weakly continuous functions $\text{Tr}_{\mathscr{H}_n}\rho$, they are weakly lower semicontinuous. Moreover, it will be shown later (2.4.19; 1) that sequences of density matrices converging weakly to a density matrix are convergent even in the trace norm. Hence the maps $\rho \to \rho(n)$ are actually weakly continuous, and the limit belongs to the same mixing class.

5. The ordering of the density matrices is not total—for instance

$$\begin{pmatrix} \frac{1}{2} & & \\ & \frac{1}{2} & \\ & & 0 \end{pmatrix} \quad \text{and} \quad \begin{pmatrix} \frac{3}{4} & & \\ & \frac{1}{8} & \\ & & \frac{1}{8} \end{pmatrix}$$

are not related by it.

**Examples** (2.1.11)

1. In the Schrödinger picture the time-evolution of a system is given by $\rho \to \rho_t \equiv U(t)\rho U^{-1}(t)$, which shows that density matrices remain in their equivalence classes.
2. The time-average $(1/T)\int_0^T dt\rho_t$ is more mixed than the original density matrices. This operation involves combinations and weak limits, which can only make density matrices more chaotic.
3. If the time-evolution of a density matrix is a linear transformation of the eigenvalues, $\rho_i(t) = M_{ik}(t)\rho_k(0)$, then for $\text{Tr}\,\rho = 1$ and $\rho \ge 0$ it must be true that $\sum_i M_{ik} = 1$ for all $k$, and $M_{ik} \ge 0$ for all $i$ and $k$. If, for finite dimension $N$, it is also required that the chaotic state $\rho_i = 1/N$ be stationary for all $i$, then, moreover, $\sum_k M_{ik} = 1$ for all $i$. The matrix $M$ is then said to be **doubly stochastic**. Such matrices clearly form a convex set, and are consequently convex combinations of the extremal elements by the Krein–Milman theorem. The extremal elements have entries $M_{ik} = 0$ or 1, and so $1 = \sum_i M_{ik} = \sum_k M_{ik}$ implies that each row and each column has exactly one 1; this makes them permutation matrices, mapping any $\rho$ to an equivalent $\rho$. Therefore, $\rho(t) \succsim \rho(0)$, as $\rho(t)$ is a convex combina-

tion of $\rho$'s equivalent to $\rho(0)$. This kind of time-evolution thus increases the mixing. Its differential version $\dot{\rho}(t) = \dot{M}\rho(0)$ is a **master equation** $\dot{\rho}_i = \sum_k W_{ik}(\rho_k - \rho_i)$, where $W$ satisfies $\sum_i W_{ik} = \sum_i W_{ki}$.

4. If an observable has one-dimensional projections $P_i$, then the state is immediately converted to $\tilde{\rho} \equiv \sum_i P_i \rho P_i$ when the observable is measured. Once it is perceived that the $k$th eigenvalue has been measured, $\rho$ becomes $P_k$. The first stage of the measurement increases the mixing of the state, $\tilde{\rho} \succeq \rho$. This follows from the min–max principle: If $P_i|i\rangle = |i\rangle$, then

$$\tilde{\rho}(n) = \sum_{i=1}^{n} \langle i|\rho|i\rangle \leq \rho(n) = \sup_{\mathscr{H}_n} \mathrm{Tr}_{\mathscr{H}_n} \rho.$$

The second stage makes the state pure. This can be interpreted in that the interaction with the measuring apparatus extracts information, which unmixes the state upon transmission to the human mind.

5. The "coarse-grained" density matrix $\tilde{\rho} \equiv \sum_i P_i \lambda_i$, $\lambda_i = \mathrm{Tr}\, \rho P_i$, is more mixed than $\sum_i P_i \rho P_i$ by Problem 1, and *a fortiori* $\tilde{\rho} \succeq \rho$.
6. Suppose that the function $k$ is convex from $\mathbb{R}^+$ to $\mathbb{R}^+$ and $k(0) = 0$; then clearly the smaller eigenvalues are suppressed to a greater degree in $k(\rho)$. In fact, $\rho \succeq k(\rho)/\mathrm{Tr}\, k(\rho)$ by Problem 3, and the resulting states are purer. In particular, if $k(x) = x^{\beta'/\beta}$, $\beta' > \beta$, then $\exp(-\beta H)/\mathrm{Tr} \exp(-\beta H) \succeq \exp(-\beta' H)/\mathrm{Tr} \exp(-\beta' H)$. The physical significance is that the mixing of the canonical density matrices is greater at higher temperatures.

We have seen that convex combinations of $U\rho U^{-1}$ and weak limits increase the mixing of $\rho$. This exhausts the possibilities:

**Theorem** (2.1.12)

*$\tilde{\rho} \succeq \rho$ iff $\tilde{\rho}$ is in the weakly closed convex hull of $\{U\rho U^{-1}\}$.*

**Remark** (2.1.13)

The weak closure of $\{a \in \mathscr{B}^+(\mathscr{H}), \|a\| = 1\}$ is $\{a \in \mathscr{B}^+(\mathscr{H}), \|a\| \leq 1\}$, and density matrices may converge weakly to zero. This means that the set of density matrices is not closed, which causes technical difficulties in the proof, which is put off to Problem 4 for that reason.

**Corollary** (2.1.14)

*If $\tilde{\rho} \succeq \rho$, then for any convex function $k$,* $\mathrm{Tr}\, k(\tilde{\rho}) \leq \mathrm{Tr}\, k(\rho)$.

**Proof**

If $\tilde{\rho} = \sum_i c_i U_i \rho U_i^{-1}, 0 \leq c_i \leq 1, \sum_i c_i = 1$, and the sum is finite, then by the convexity inequality (2.1.7; 2), $\operatorname{Tr} k(\tilde{\rho}) \leq \sum_i c_i \operatorname{Tr} k(U_i \rho U_i^{-1}) = \operatorname{Tr} k(\rho)$. Moreover, $\rho \to \operatorname{Tr} k(\tilde{\rho})$ is weakly lower semicontinuous, so the limiting case of an infinite sum is likewise bounded by $\operatorname{Tr} k(\rho)$. □

Corollary (2.1.14) gives rise to the possibility of defining mappings of the density matrices to the real numbers, monotonic with respect to the ordering $\succeq$, and so enables the degree of disorder to be measured. For instance, if $k(\rho) = \rho^2$, then $\operatorname{Tr} k(\rho)$ can equal 1 only for pure states, and is otherwise smaller. The next section will discuss some other properties distinguished by the function $-k(\rho) = -\rho \ln \rho$ used to define the entropy. For now, note that the converse of (2.1.14) is also true:

**Theorem** (2.1.15)

*$\tilde{\rho} \succeq \rho$ iff for every convex function k,* $\operatorname{Tr} k(\tilde{\rho}) \leq \operatorname{Tr} k(\rho)$.

**Proof**

Because of (2.1.14), we need only show that if $\tilde{\rho} \not\succeq \rho$, then there exists a function $k$ such that $\operatorname{Tr} k(\tilde{\rho}) \geq \operatorname{Tr} k(\rho)$. Let $m$ be the first integer such that $\tilde{\rho}_1 + \tilde{\rho}_2 + \cdots + \tilde{\rho}_m > \rho_1 + \rho_2 + \cdots + \rho_m$, and let $k(x) = (x - \rho_m)$, when $x \geq \rho_m$, and otherwise 0. Then $k(\rho_1) = \rho_1 - \rho_m, \ldots, k(\rho_m) = \rho_m - \rho_m = 0 = k(\rho_{m+1}) = k(\rho_{m+2}) = \cdots$. By assumption, $\tilde{\rho}_1 + \tilde{\rho}_2 + \cdots + \tilde{\rho}_{m-1} \leq \rho_1 + \rho_2 + \cdots + \rho_{m-1}$, so $\tilde{\rho}_m > \rho_m$, which implies $k(\tilde{\rho}_i) = \tilde{\rho}_i - \rho_m > 0$ for all $i \leq m$. Therefore, $\operatorname{Tr} k(\rho) = \rho_1 + \rho_2 + \cdots + \rho_m - m\rho_m < \tilde{\rho}_1 + \tilde{\rho}_2 + \cdots + \tilde{\rho}_m - m\rho_m \leq \operatorname{Tr} k(\tilde{\rho})$. □

Since expectation values in mixed states are averages of different spectral values, they do not reach the extremes of the spectrum so easily. This observation creates a new way to define the ordering relationship.

**Theorem** (2.1.16)

(i) $\tilde{\rho} \succeq \rho \Leftrightarrow \sup\limits_{\substack{U \\ U^* = U^{-1}}} \operatorname{Tr} U\tilde{\rho}U^{-1}a \leq \sup\limits_{\substack{U \\ U^* = U^{-1}}} \operatorname{Tr} U\rho U^{-1}a$ for all $a \in \mathscr{B}^+(\mathscr{H})$,

(ii) $\tilde{\rho} \succeq \rho \Leftrightarrow \inf\limits_{\substack{U \\ U^* = U^{-1}}} \operatorname{Tr} U\tilde{\rho}U^{-1}a \geq \inf\limits_{\substack{U \\ U^* = U^{-1}}} \operatorname{Tr} U\rho U^{-1}a$ for all $a \in \mathscr{B}^+(\mathscr{H})$.

**Proof**

See Problem 5. □

**Corollary** (2.1.17)

*Let* $(\Delta_\rho a)^2 \equiv \mathrm{Tr}\,\rho a^2 - (\mathrm{Tr}\,\rho a)^2 = \inf_\lambda \mathrm{Tr}\,\rho(a-\lambda)^2$. *Then* $\tilde{\rho} \succeq \rho$ *implies that*

$$\inf_U \Delta_{U\tilde{\rho}U^{-1}} a \geq \inf_U \Delta_{U\rho U^{-1}} a \quad \textit{for all } a.$$

This means that if one is interested in the least deviation $\Delta a$ of $a$ within the equivalence classes of $\rho$ and $\tilde{\rho}$, then it is smaller for the state that is less mixed.

The various aspects of the relationship can be summarized as follows:

**Conditions for Density Matrices to be Compared** (2.1.18)

The ordering relationship $\tilde{\rho} \succeq \rho$ is equivalent to each of the following:

(i) $\tilde{\rho}(n) \leq \rho(n)$ for all $n$;
(ii) $\tilde{\rho} = w\text{-}\lim_\alpha \sum_i c_{i\alpha} U_{i\alpha} \rho U_{i\alpha}^{-1}$, $c_{i\alpha} > 0$, $\sum_i c_{i\alpha} = 1$, $U_{i\alpha}^{-1} = U_{i\alpha}^*$;
(iii) $\mathrm{Tr}\, k(\tilde{\rho}) \geq \mathrm{Tr}\, k(\rho)$ for every concave function $k$;
(iv) $\begin{matrix}\sup \\ \inf \\ U\end{matrix} \mathrm{Tr}\, U\tilde{\rho}U^{-1}a \begin{matrix}\leq \\ \leq\end{matrix} \begin{matrix}\sup \\ \inf \\ U\end{matrix} \mathrm{Tr}\, U\rho U^{-1}a$, $a \in \mathscr{B}^+(\mathscr{H})$, $U^{-1} = U^*$.

**Problems** (2.1.19)

1. Let $P_i$ be pairwise orthogonal projections of dimensions $n_i < \infty$ and $\sum_i P_i = \mathbf{1}$. Show that $\sum_i (1/n_i) P_i \,\mathrm{Tr}\, P_i \rho \succeq \sum_i P_i \rho P_i$.

2. Let $k(x) > 0$, $k' > 0$, $k'' > 0$, $k/k'$ convex. Show that the mapping $(\beta_1, \ldots, \beta_n) \to k^{-1}(\sum_{i=1}^n k(\beta_i))$ of $\mathbb{R}^n$ to $\mathbb{R}$ is convex. (Hint: note that: (i) A mapping $f(\beta_1, \ldots, \beta_n)$ is convex if $\chi''(0) \geq 0$, where $\chi$ is the function $\chi(t) = f(\beta_1 + u_1 t, \ldots, \beta_n + u_n t)$ and $(u_1, \ldots, u_n)$ and $(\beta_1, \ldots, \beta_n)$ are arbitrary. (ii) If the function $K(\delta)/\delta$ increases monotonically, then $K(\sum_i \delta_i) \geq \sum_i K(\delta_i)$, $\delta_i > 0$.)

3. Let $k$ be a convex, monotonically increasing function, $k(x) \geq 0$ for $x \geq 0$, and $k(0) = 0$. Show that $\rho \succeq k(\rho)/\mathrm{Tr}\, k(\rho)$.

4. Show that $\tilde{\rho} \succeq \rho \Leftrightarrow \tilde{\rho} \in \mathrm{Conv}\overline{\{U\rho U^{-1}\}}^{\text{weak}}$.

   (i) Let $\mathscr{K}(\rho) = \{a \geq 0\colon a$ is compact, and $\alpha_1 + \cdots + \alpha_n \leq \rho(n)$ for all $n$, where $\alpha_i$ are the eigenvalues in increasing order$\}$. Show that $\mathscr{K}(\rho)$ is convex and weakly compact.
   (ii) Let $\mathscr{E}(\rho) = \{a \in \mathscr{K}(\rho)\colon \alpha_1 = \rho_1, \ldots, \alpha_n = \rho_n, \alpha_{n+1} = \cdots = 0$ or $\alpha_i = \rho_i$ for all $i\}$. Show that $\mathscr{E}(\rho)$ contains the extremal points of $\mathscr{K}(\rho)$.
   (iii) Show that $\mathscr{E}(\rho) \subset \overline{\{U\rho U^{-1}\}}^{\text{weak}}$.
   (iv) Finish the proof by applying the Krein–Milman theorem: Every compact, convex set equals the closure of the convex hull of its extremal points.

5. Prove Theorem (2.1.16).

**Solutions** (2.1.20)

1. Let $d\mu(U)$ be the invariant measure on the compact group $U(n)$, normalized to 1. For all $a \in \mathscr{B}(\mathbb{C}^n)$, $\mathbf{1}_{|\mathbb{C}^n}(1/n) \operatorname{Tr} a = \int d\mu U a U^{-1}$, since the right side is invariant under all $U$ and hence proportional to $\mathbf{1}_{|\mathbb{C}^n}$, and $\operatorname{Tr} \int d\mu U a U^{-1} = \operatorname{Tr} a$. Similarly,

$$\frac{1}{n} P \operatorname{Tr} P\rho + (\mathbf{1} - P)\rho(\mathbf{1} - P) = \int d\mu_P U_P \rho U_P^{-1}$$

$$= \int d\mu_P U_P(P\rho P + (\mathbf{1} - P)\rho(\mathbf{1} - P))U_P^{-1},$$

if the operators $U_P$ vary over the unitary transformations of $\mathscr{H}$ equaling $\mathbf{1}$ on $(\mathbf{1} - P)\mathscr{H}$. Therefore, $(1/n)P \operatorname{Tr} P\rho + (\mathbf{1} - P)\rho(\mathbf{1} - P) \succeq P\rho P + (\mathbf{1} - P)\rho(\mathbf{1} - P)$, which proves the claim by iteration.

2. (i) is trivial, and (ii) follows from

$$\sum_i \delta_i \geq \delta_k \Rightarrow \delta_k K\left(\sum_i \delta_i\right) \geq \left(\sum_i \delta_i\right) K(\delta_k), \left(\sum_k \delta_k\right) K\left(\sum_i \delta_i\right) \geq \left(\sum_i \delta_i\right) \sum_k K(\delta_k).$$

Now let $\chi(t) \equiv k^{-1}(\sum k(\beta_i + u_i t))$. The function $\chi(t)$ is convex iff $\chi''(t) \geq 0$. $[k'(\chi)]^3 \chi'' = [k'(\chi)]^2 [\sum_i u_i^2 k''(\beta_i)] - k''(\chi)[\sum_i u_i k'(\beta_i)]^2$ (where $\chi \equiv \chi(0)$, $\chi'' \equiv \chi''(0)$), so it remains to show that $[k'(\chi)]^2 \sum_i u_i^2 k''(\beta_i) \geq k''(\chi)[\sum_i u_i k'(\beta_i)]^2$. By the Cauchy–Schwarz inequality, $[\sum_i u_i k'(\beta_i)]^2 = [\sum_i u_i \sqrt{k''(\beta_i)} \sqrt{k'(\beta_i)^2/k''(\beta_i)}]^2 \leq [\sum_i u_i^2 k''(\beta_i)] \times [\sum_i k'(\beta_i)^2/k''(\beta_i)]$, and the desired inequality is certainly satisfied if $\psi(\chi) \equiv k'(\chi)^2/k''(\chi) \geq \sum_i k'(\beta_i)^2/k''(\beta_i) = \sum_i \psi(\beta_i)$. By (ii), this is the case if $K(\delta)/\delta$ increases monotonically, where $K$ is defined by $\delta_i = k(\beta_i)$, $K(\delta_i) = \psi(\beta_i)$. Finally, $K(\delta)/\delta$ increases monotonically $\Leftrightarrow k'^2/kk''$ increases monotonically $\Leftrightarrow k/k'$ is convex.

3. If $0 \leq x \leq y$, then $x = (x/y)y + (1 - (x/y))0$, and hence $k(x) \leq (x/y)k(y)$, $yk(x) \leq xk(y)$. Consequently

$$\sum_{i=1}^{m} k(\rho_i)\left(\sum_{j=1}^{m} \rho_j + \sum_{j=m+1}^{\infty} \rho_j\right) \geq \sum_{j=1}^{m} \rho_j\left(\sum_{i=1}^{m} k(\rho_i) + \sum_{i=m+1}^{\infty} k(\rho_i)\right),$$

i.e.,

$$[k(\rho_1) + \cdots + k(\rho_m)]\left(\sum_{i=1}^{\infty} k(\rho_i)\right)^{-1} \geq [\rho_1 + \cdots + \rho_m]\left(\sum_{i=1}^{\infty} \rho_i\right)^{-1}.$$

Remark: If $k$ is concave, then $\rho \preceq k(\rho)/\operatorname{Tr} k(\rho)$.

4. (i) By (2.1.10; 3) the set $\mathscr{K}(\rho)$ is convex. Moreover, $\alpha_1 + \cdots + \alpha_n = \sup_{\mathscr{H}_n} \operatorname{Tr}_{\mathscr{H}_n} a$ is weakly lower semicontinuous in $a$, so $\mathscr{K}(\rho)$ is weakly closed and, since $\|a\| = \alpha_1 \leq \rho_1 = \|\rho\| \leq \operatorname{Tr} \rho = 1$, also weakly compact.
   (ii) By considering all the possibilities, one realizes that it is possible to write any $a \in \mathscr{K}(\rho)$ as $\alpha\rho_1 + (1 - \alpha)\rho_2$, $0 < \alpha < 1$, with $\rho_i \in \mathscr{K}(\rho)$, unless $a \in \mathscr{E}(\rho)$.
   (iii) Let $a = \sum_{i=1}^{n} \rho_i |1, i\rangle\langle 1, i|$, $\rho = \sum \rho_i |2, i\rangle\langle 2, i|$, where $\{|1, i\rangle\}$ and $\{|2, i\rangle\}$ are two orthonormal systems. Let $U|2, i\rangle = |1, i\rangle$, $U_l|1, n + i\rangle = |1, n + l - i\rangle$ for $1 \leq i \leq l - 1$, $U_l|1, i\rangle = |1, i\rangle$ otherwise. $a = s\text{-}\lim_{l \to \infty} U_l U \rho U^{-1} U_l^{-1}$.

(iv) By the Krein–Milman theorem, $\mathscr{K}(\rho) = \overline{\text{Conv } \mathscr{E}(\sigma)}^{\text{weak}} = \overline{\text{Conv}\{U\rho U^{-1}\}}^{\text{weak}}$ (by (iii)), and $\tilde{\rho} \in \mathscr{K}(\rho)$, if $\tilde{\rho} \gtrsim \rho$.

5. By a replacement of $a$ with $a + \|a\|$ if necessary, $a$ may be assumed positive. Then $\text{Tr } \rho a = \sup_n \text{Tr } \rho^{(n)} a$, $\rho^{(n)}$, where the $\rho^{(n)}$ have the eigenvalues $\rho_1, \rho_2, \ldots, \rho_n, 0, 0, \ldots$. The changes of the orders of operation in what follows are justified for the $\rho^{(n)}$, and the suprema can also be interchanged:

(i) $\Rightarrow$: Let $\alpha_1 \geq \alpha_2 \geq \cdots$ be the decreasing sequence of eigenvalues of $a$ and $\alpha_\infty$ be the upper boundary of $\sigma_{\text{ess}}(a)$ (to be understood in the sense analogous to (III: 3.5.21)). If $\rho = \sum \rho_i |i\rangle\langle i|$, then

$$\begin{aligned}\text{Tr } \rho a &= \sum \rho_i \langle i|a|i\rangle = (\rho_1 - \rho_2)\langle 1|a|1\rangle + (\rho_2 - \rho_3)[\langle 1|a|1\rangle + \langle 2|a|2\rangle] + \cdots \\ &\leq (\rho_1 - \rho_2)\alpha_1 + (\rho_2 - \rho_3)(\alpha_1 + \alpha_2) + \cdots = \sum \rho_i \alpha_i,\end{aligned}$$

and $\sup \text{Tr } U\rho U^{-1} a = \sum \rho_i \alpha_i$.

$$\begin{aligned}\sum_i \tilde{\rho}_i \alpha_i &= \tilde{\rho}_1(\alpha_1 - \alpha_2) + (\tilde{\rho}_1 + \tilde{\rho}_2)(\alpha_2 - \alpha_3) + \cdots + \alpha_\infty \\ &\leq \rho_1(\alpha_1 - \alpha_2) + (\rho_1 + \rho_2)(\alpha_2 - \alpha_3) + \cdots + \alpha_\infty = \sum \rho_i \alpha_i.\end{aligned}$$

$\Leftarrow$: Choose an $n$-dimensional projection for $a$ and use the min–max principal.

The proof of (ii) is similar.

## 2.2 The Properties of Entropy

*The information about a system in a mixed state is incomplete. The entropy is a measure of how far from maximal the information is.*

In statistical physics, entropy is not an observable in the sense of an operator on Hilbert space, but rather a property of the state of the system, measuring the lack of our knowledge as expressed in the specification of the state. This section will consider what sorts of conditions single out a particular measure of this lack of knowledge and will see what conclusions can be drawn from it.

A primary requirement would be monotony with respect to the ordering introduced in the preceding section (we consider only normal states). In other words, a density matrix that is more mixed should have more entropy, which we denote $S$: $\tilde{\rho} \gtrsim \rho \Rightarrow S(\tilde{\rho}) \geq S(\rho)$. This leaves many possibilities open for the definition; every monotonic function of the trace of a concave function of $\rho$ would satisfy this requirement (cf. (2.1.14)). A further reasonable requirement is the additivity of the entropies of independent systems. If their combination is represented on the tensor product of their Hilbert spaces, this means

$$S(\rho' \otimes \rho'') = S(\rho') + S(\rho''). \tag{2.2.1}$$

The two requirements together do not yet quite determine $S$ uniquely. The whole one-parameter family of

**α-Entropies** (2.2.2)

$$S_\alpha(\rho) = \frac{1}{1-\alpha} \ln \operatorname{Tr} \rho^\alpha, \qquad \alpha \in \mathbb{R}^+ \setminus \{1\},$$

satisfy the general

**Properties of Entropy** (2.2.3)

(i) $0 \leq S_\alpha(\rho) \leq \ln \dim \mathscr{H}$;
(ii) $\tilde{\rho} \succeq \rho \Rightarrow S_\alpha(\tilde{\rho}) \geq S_\alpha(\rho)$;
(iii) $S_\alpha(\rho' \otimes \rho'') = S_\alpha(\rho') + S_\alpha(\rho'')$;
(iv) If $\rho = P/\dim P$, $P = P^2 = P^*$, then $S_\alpha(\rho) = \ln \dim P$.

(In particular, $S_\alpha(\rho) = 0$ iff $\rho$ is a pure state, and $S_\alpha(\rho) = \ln \dim \mathscr{H}$ iff $\rho$ is the chaotic state $1/\dim \mathscr{H}$.)

**Proof**

(i) If $\alpha > 1$, then $\sum_i \rho_i^\alpha \leq (\sum_i \rho_i)^\alpha = 1$, and if $\alpha < 1$, then $\sum_i \rho_i = 1 \leq (\sum_i \rho_i^\alpha)^{1/\alpha}$. This shows the left side of the inequality, and the right follows from (iv) and (ii).
(ii) The function $\rho^\alpha$ is concave for $\alpha < 1$ and convex for $\alpha > 1$. The logarithm is monotonic, and the $1 - \alpha$ accounts for the sign (see (2.1.18(iii)).
(iii) $\operatorname{Tr}(\rho' \otimes \rho'')^\alpha = \operatorname{Tr}[(\rho')^\alpha \otimes (\rho'')^\alpha] = \operatorname{Tr}(\rho')^\alpha \cdot \operatorname{Tr}(\rho'')^\alpha$.
(iv) If $n = \dim P$, then $S_\alpha(\rho) = (1/(1-\alpha)) \ln(nn^{-\alpha})$. □

The entropy can be fixed uniquely by a more stringent assumption of additivity (2.2.1), with which monotony emerges as a consequence rather than a separate axiom:

**Characterization of the von Neumann Entropy** (2.2.4)

*The only entropy satisfying the following conditions is* $S(\rho) = -\operatorname{Tr} \rho \ln \rho$

(i) $S(\rho)$ *is a continuous function of the eigenvalues of* $\rho$;

(ii) $S\begin{pmatrix} \frac{1}{2} & 0 \\ 0 & \frac{1}{2} \end{pmatrix} = \ln 2$;

(iii) *If*

$$\mathscr{H} = \bigoplus_{n=1}^{N} \mathscr{H}_n, \qquad \rho = \bigoplus_{n=1}^{N} p_n \rho_n, \qquad \sum_n p_n = 1,$$

$$0 \leq p_i \leq 1, \ \operatorname{Tr} \rho_n = 1,$$

*then, regardless of the dimension of* $\mathscr{H}_n$, $S(\rho) = \sum_{n=1}^N p_n S(\rho_n) + S(p)$, *where* $p$ *is the diagonal matrix on* $\mathbb{C}^n$ *having eigenvalues* $\rho_n$.

**Remark** (2.2.5)

1. Since the representation should make no difference, $S$ can only depend on the eigenvalues. It certainly does not seem unreasonable to demand continuity.
2. Condition (ii) is a normalization.
3. If all the $\mathscr{H}_n$ in condition (iii) have the same dimension and all $\rho_n$ are equal, then $\mathscr{H} = \mathscr{H}_1 \otimes \mathbb{C}^n$, and (iii) reduces to (2.2.1). This generalization of (2.2.1), which makes possible an inductive proof, has the following interpretation: Suppose a system consists of two subsystems, one described by $\mathbb{C}^n$ and the other having several variants according to the position of the state vector of the first in $\mathbb{C}^n$. Then the entropy of the total system is just the sum of the entropy of the first subsystem and those of the second, averaged according to their probabilities.
4. The formula $S = -\mathrm{Tr}\,\rho \ln \rho$ can be justified in the spirit of Boltzmann as follows. Let the state corresponding to $\rho$ be realized as a vector of a reducible representation of the algebra $\mathscr{A}$ of observables consisting of $N$ identical representations. The ensemble described by $\rho$ can be thought of as having been subjected to a sequence of $N$ measurements, where $\rho_i$ is $N_i/N$, $N_i$ being the number of times the eigenvector $e_i$ has been measured. The Hilbert space is $\mathscr{H} = \bigoplus_{j=1}^N \mathscr{H}_j$, where the spaces $\mathscr{H}_j$ are all identical and are spanned by $\{e_i\}$. The observables are represented as a direct sum of $N$ identical representations. With the use of doubled indices, this can be written as $\mathscr{H}_j = \bigoplus_{i=1}^\infty e_{i,j}$. A $\rho$ of rank $r$ and with $\rho_i = N_i/N$, $i = 1, \ldots, r$, is represented by the vector

$$\frac{1}{\sqrt{N}}(e_{1,1} + e_{1,2} + \cdots + e_{1,N_1} + e_{2,N_1+1} + \cdots + e_{2,N_1+N_2} + \cdots + e_{r,N_1+N_2+\cdots+N_{r-1}+1} + \cdots + e_{r,N})$$

of $\mathscr{H}$. If the $e_i$ are chosen from other spaces $\mathscr{H}_j$, the same state results, and there are clearly $W \equiv N!/\prod_i N_i!$ different vectors for the same $\rho$. If the numbers $N_i$ are large enough, then $\ln W \cong N \ln N - \sum_i N_i \ln N_i = -N \sum_i \rho_i \ln \rho_i$, so $(1/N) \ln W \to -\mathrm{Tr}\,\rho \ln \rho$. Assuming that every vector of $\mathscr{H}$ is assigned the same probability, $S$ turns out to be roughly the logarithm of the probability of the configuration, and there is an identification: the most mixed state = the state of greatest entropy = the most probable state.
5. $S(\rho) = \lim_{\alpha\to 1} S_\alpha(\rho)$, yet if the dimension is infinite, then $S(\rho)$ may become $+\infty$. However, Properties (2.2.3) remain valid in this limit, and apply to $S$ as well.

6. A particular consequence of (2.2.3(ii)) is that $S(\alpha\rho + (1-\alpha)U\rho U^{-1}) \geq S(\rho)$. More generally, (2.1.7; 2) implies that the mapping $\rho \to S(\rho)$ is concave: $S(\alpha\rho_1 + (1-\alpha)\rho_2) \geq \alpha S(\rho_1) + (1-\alpha)S(\rho_2)$. This means that the entropy of a mixed state is greater than the constituent entropies weighted as in the mixing. If $\rho = \sum_n p_n\rho_n$, $0 \leq p_n \leq 1$, $\sum_n p_n = 1$, then the inequalities

$$\sum_n p_n S(\rho_n) \leq S(\rho) \leq \sum_n p_n S(\rho_n) + \sum_n p_n \ln \frac{1}{p_n}$$

necessarily follow (Problem 4). They are optimal in the sense that equality holds on the left if all $\rho_n$ are equal, and on the right if all $\rho_n$ have disjoint support, by (2.2.4(iii)).
7. Although by (2.2.3(iv)) all the $S_\alpha$ are the same with the chaotic state, with the canonical state $\rho = \exp(-\beta(H - F(\beta)))$, $\operatorname{Tr}\exp(-\beta H) = \exp(-\beta F(\beta))$, they are different (Problem 6).

**Proof of (2.2.4)**

We write $S(\rho_1, \rho_2, \ldots)$ for $S(\rho)$.

(a) Let $\mathscr{H} = \mathbb{C}^1$. Then $S(1) = 0$, because on $\mathbb{C}^2$, $S(\rho_1, \rho_2) = \rho_1 S(1) + \rho_2 S(1) + S(\rho_1, \rho_2)$.
(b) Let $\mathscr{H} = \mathbb{C}^n$, $f(n) \equiv S(1/n, 1/n, \ldots, 1/n)$, and let $n = m_1 m_2$. We write $\mathbb{C}^n = \mathbb{C}^{m_1} \oplus \mathbb{C}^{m_1} \oplus \cdots \oplus \mathbb{C}^{m_1}$ and use (iii) with

$$N = m_2, p_i = m_2^{-1}, \qquad \rho_i = \begin{pmatrix} 1/m_1 & & \\ & \ddots & \\ & & 1/m_1 \end{pmatrix},$$

$$f(m_1 m_2) = m_2 \frac{1}{m_2} f(m_1) + f(m_2) = f(m_1) + f(m_2).$$

The solution of this equation is $f(n) = C \ln n$, and the normalization (ii) makes $C = 1$. Other solutions are excluded by the continuity requirement (Problem 1).

(c)
$$f(m) = S\Bigl(\overbrace{\frac{1}{m}, \frac{1}{m}, \ldots, \frac{1}{m}}^{n}, \overbrace{\frac{1}{m}, \ldots, \frac{1}{m}}^{m-n}\Bigr)$$
$$= \frac{n}{m} f(n) + \frac{m-n}{m} f(m-n) + S\left(\frac{n}{m}, \frac{m-n}{m}\right),$$

so by step (b),

$$S\left(\frac{n}{m}, 1 - \frac{n}{m}\right) = -\frac{n}{m} \ln \frac{n}{m} - \left(1 - \frac{n}{m}\right) \ln\left(1 - \frac{n}{m}\right).$$

This holds initially only for integers $n$ and $m$, and then by continuity holds generally, $S(\rho_1, \rho_2) = -\sum_{i=1}^{2} \rho_i \ln \rho_i$.

(d) The rest of the proof proceeds inductively: with $\mathbb{C}^{n+1} = \mathbb{C}^n \oplus \mathbb{C}$, $p_1 = 1 - \rho_n, p_2 = \rho_n$,

$$S(\rho_1, \rho_2, \ldots, \rho_{n-1}, \rho_n) = (1 - \rho_n)S\left(\frac{\rho_1}{1 - \rho_n}, \ldots, \frac{\rho_{n-1}}{1 - \rho_n}\right) + \rho_n S(1)$$
$$+ S(1 - \rho_n, \rho_n)$$
$$= -\sum_{j=1}^{n-1} \rho_i \ln \frac{\rho_i}{1 - \rho_n} - \rho_n \ln \rho_n$$
$$-(1 - \rho_n)\ln(1 - \rho_n) = -\sum_{j=1}^{n} \rho_i \ln \rho_i. \qquad \square$$

**The Classical Entropy** (2.2.6)

For a classical density $\rho_{\mathrm{cl}}(\mathbf{x}, \mathbf{p})$ on phase space the entropy would be defined as $-\int d\Omega \rho_{\mathrm{cl}} \ln \rho_{\mathrm{cl}}$. This is not *a priori* positive-definite; for instance $\rho_{\mathrm{cl}} = \chi(A)/\Omega(A)$ as in (2.1.6; 2) leads to $-\int d\Omega \chi \ln \chi = \ln \Omega(A)$, which is negative if $\Omega(A) < 1$. It is easy to see that this entropy also depends on the measure of volume in phase space. There are many ways to associate a density $\rho_{\mathrm{cl}}$ with a density matrix $\rho$ or vice versa.

The most useful such expressions are obtained with a method of A. Wehrl, in which for a given density matrix $\rho$ one calculates expectation values in coherent states, and, conversely, a classical density is used to mix coherent states. The coherent states $W(\mathbf{z})|u\rangle \equiv |\mathbf{z}\rangle$ of (III: 3.1.13) can be generalized for functions $u$ that are even and normalized, but not necessarily Gaussian. The state $|\mathbf{z}\rangle$ has the wavefunction $\exp(i\mathbf{k} \cdot \mathbf{x})u(\mathbf{x} - \mathbf{q})$ if $\mathbf{z} = \mathbf{q} + i\mathbf{k} \in \mathbb{C}^{dN}$, which is the phase space for $N$ particles in a physical space of dimension $d$. It is easy to check that $\mathbf{z} = \langle \mathbf{z}|\mathbf{x}|\mathbf{z}\rangle + i\langle \mathbf{z}|\mathbf{p}|\mathbf{z}\rangle$ still holds and that the states are complete, $\int d^{2Nd}z(2\pi)^{-Nd}|\mathbf{z}\rangle\langle\mathbf{z}| = \mathbf{1}$.

**The Density Matrix and the Phase-Space Density** (2.2.7)

*If to an $N$-particle density matrix $\rho$ we associate the phase-space density $\rho_{\mathrm{cl}}(\mathbf{z}) = \langle \mathbf{z}|\rho|\mathbf{z}\rangle$, and to a classical density $f(\mathbf{z})$ on phase space we associate the density matrix $\rho_{\mathrm{qu}} = \int d\Omega_z f(\mathbf{z})|\mathbf{z}\rangle\langle\mathbf{z}|$, $d\Omega_z^N = (2\pi)^{-Nd}d^{2N}z$, then*

$$\rho \geq 0, \qquad \mathrm{Tr}\, \rho = 1 \Rightarrow 0 \leq \rho_{\mathrm{cl}}(\mathbf{z}) \leq 1, \qquad \int d\Omega_z^N \rho_{\mathrm{cl}}(\mathbf{z}) = 1,$$

$$f \geq 0, \qquad \int d\Omega_z f(z) = 1 \Rightarrow 0 \leq \rho_{\mathrm{qu}} \leq 1, \qquad \mathrm{Tr}\, \rho_{\mathrm{qu}} = 1. \qquad (2.2.8)$$

**Proof**

Positivity is trivial, and the connection between the trace and the phase-space integral follows from the $n$-dimensional version of a formula of (III: 3.1.14; 1):

$$\mathbf{1} = \int d\Omega_{\mathbf{z}}^N |\mathbf{z}\rangle\langle\mathbf{z}| \Rightarrow \operatorname{Tr} a = \sum_i \langle i|a|i\rangle = \sum_i \int d\Omega_{\mathbf{z}}^N \langle i|\mathbf{z}\rangle\langle\mathbf{z}|a|i\rangle$$
$$= \int d\Omega_{\mathbf{z}}^N \langle\mathbf{z}|a|\mathbf{z}\rangle.$$

Conversely, $\operatorname{Tr} \int d\Omega_{\mathbf{z}} f(\mathbf{z})|\mathbf{z}\rangle\langle\mathbf{z}| = \sum_i \int d\Omega_{\mathbf{z}}^N f(\mathbf{z})|\langle\mathbf{z}|i\rangle|^2 = \int d\Omega_{\mathbf{z}} f(\mathbf{z})$, since $\langle\mathbf{z}|\mathbf{z}\rangle = 1$. The denominator $(2\pi)^{dN}$ in $d\Omega_{\mathbf{z}}^N$ reveals that the phase-space volume is measured in units of $h$ rather than $\hbar = h/2\pi = 1$. □

**Inequalities for the Classical and Quantum-Mechanical Entropies** (2.2.9)

(i) $S(\rho) \leq -\int d\Omega_{\mathbf{z}}^N \rho_{\mathrm{cl}}(\mathbf{z}) \ln \rho_{\mathrm{cl}}(\mathbf{z}) \equiv S_{\mathrm{cl}}(\rho)$;
(ii) $-\int d\Omega_{\mathbf{z}}^N f(\mathbf{z}) \ln f(\mathbf{z}) \leq S(\rho_{\mathrm{qu}})$.

**Remarks** (2.2.10)

1. Inequality (i) implies that the $\rho_{\mathrm{cl}}$ of (2.2.7) always has more entropy than $S(\rho)$. This classical entropy is therefore always positive; the density $\rho_{\mathrm{cl}}$ defined in (2.2.7) can never be so concentrated as to make the classical entropy negative, and indeed $\rho_{\mathrm{cl}} \leq 1$.
2. It can also be shown that this classical entropy equals 1 if $\rho$ is extremal, and otherwise it is greater than 1 [32].
3. If a quantum-mechanical density is associated with a classical density $f$ by mixing the coherent states with $f$, then Inequality (ii) states that the quantum-mechanical entropy is greater than the classical entropy. The latter may even tend to $-\infty$, for instance if $f$ tends to a delta function.
4. Inequality (ii) shows that the continuous analogue of the last inequality of (2.2.3; 6) is false: $S(|\mathbf{z}\rangle\langle\mathbf{z}|) = 0$, and in this case the inequality goes in the other direction, with the replacements $p_n \to f(\mathbf{z})$, $\sum_n \to \int d\Omega_{\mathbf{z}}^N$:
$$-\int d\Omega_{\mathbf{z}}^N f(\mathbf{z}) \ln f(\mathbf{z}) + \int d\Omega_{\mathbf{z}}^N f(\mathbf{z}) S(|\mathbf{z}\rangle\langle\mathbf{z}|) \leq S\left(\int d\Omega_{\mathbf{z}}^N f(\mathbf{z})|\mathbf{z}\rangle\langle\mathbf{z}|\right).$$
5. If the particles are identical, states must be either symmetrized or antisymmetrized according to the statistics. For bosons this is accomplished most easily with the aid of the creation operator
$$a_{\mathbf{z}}^* \equiv a^*(\exp[i\mathbf{k}\cdot\mathbf{x}]u(\mathbf{q}-\mathbf{x})), \qquad |\mathbf{z}_1, \ldots, \mathbf{z}_N\rangle = \mathbf{a}_{\mathbf{z}_1}^* \cdots \mathbf{a}_{\mathbf{z}}^*|0\rangle,$$

with which

$$\mathbf{1} = \sum_{n=0}^{\infty} \frac{(2\pi)^{-nd}}{n!} \int d^{2d}\mathbf{z}_1 \cdots d^{2d}\mathbf{z}_n |\mathbf{z}_1, \ldots, \mathbf{z}_n\rangle\langle \mathbf{z}_1, \ldots, \mathbf{z}_n|.$$

So, with identical bosons, when the trace is taken the volume of the classical phase space has to be divided by $n!$. The states are not yet normalized to norm 1.

$$\langle \mathbf{z}'_1, \ldots, \mathbf{z}'_N | \mathbf{z}_1, \ldots, \mathbf{z}_N\rangle = \sum_P (\pm 1)^P \prod_{i=1}^{N} \langle \mathbf{z}'_i | \mathbf{z}_{P_i}\rangle \equiv \frac{\text{Per}}{\text{Det}} (\langle \mathbf{z}'_i | \mathbf{z}_k\rangle),$$

where $P_1, \ldots, P_n$ is a permutation of $1, \ldots, n$, because the coherent states are not orthogonal:

$$\langle \mathbf{z}' | \mathbf{z}\rangle = \int d^d x \exp[i\mathbf{x} \cdot (\mathbf{k} - \mathbf{k}')] u^*(\mathbf{x} - \mathbf{q}') u(\mathbf{x} - \mathbf{q}).$$

These determinants and permanents crop up along with $d\Omega_{\mathbf{z}}^N$ in the calculations of expectation values, making them more laborious.

6. Since these inequalities are valid for coherent states with a great degree of arbitrariness in $u$, they can be optimized by varying $u$.

Inequalities (2.2.9) will follow from a lemma of Berezin on the

**Relationship between the Trace and the Phase-Space Integral** (2.2.11)

*Let $K$ be a convex function and suppose $a^* = a$. Then*

(i) $\text{Tr}\, K(a) \geq \int d\Omega_{\mathbf{z}}^N K(\langle \mathbf{z}|a|\mathbf{z}\rangle)$;
(ii) $\int d\Omega_{\mathbf{z}}^N K(f(\mathbf{z})) \geq \text{Tr}\, K(a)$, *where* $a = \int d\Omega_{\mathbf{z}}^N f(\mathbf{z}) |\mathbf{z}\rangle\langle \mathbf{z}|$, $K(a) \in C^1$, *and $f$ is a measurable function* $\mathbb{C}^N \to \mathbb{R}$.

**Proof**

(i) As noted in the proof of Peierls's inequality, $\langle |K(a)|\rangle \geq K(\langle |a|\rangle)$ for expectation values in an arbitrary vector, so

$$\text{Tr}\, K(a) = \int d\Omega_{\mathbf{z}}^N \langle \mathbf{z}|K(a)|\mathbf{z}\rangle \geq \int d\Omega_{\mathbf{z}}^N K(\langle \mathbf{z}|a|\mathbf{z}\rangle).$$

(ii) If $|j\rangle$ denotes an eigenfunction of $a$, then

$$\begin{aligned}\text{Tr}\, K(a) &= \sum_j K(\langle j|a|j\rangle) = \sum_j K\left(\int d\Omega_{\mathbf{z}}^N f(\mathbf{z}) |\langle \mathbf{z}|j\rangle|^2\right) \\ &\leq \sum_j \int d\Omega_{\mathbf{z}}^N |\langle \mathbf{z}|j\rangle|^2 K(f(\mathbf{z})) \\ &= \int d\Omega_{\mathbf{z}}^N K(f(\mathbf{z})).\end{aligned}$$

□

**Proof of (2.2.9)**

The function $x \ln x$ is convex, and for the concave function $-x \ln x$ the inequalities for convex functions are reversed. □

The additivity of the entropy when $\rho = \rho_1 \otimes \rho_2$ generalizes to an inequality when $\rho$ is not in the form of a product. To cover general $\rho_1$ and $\rho_2$ requires the

**Definition of Partial Traces** (2.2.12)

Let $\mathscr{H} = \mathscr{H}_1 \otimes \mathscr{H}_2$. The **partial traces** $\mathrm{Tr}_1$ and $\mathrm{Tr}_2$ are defined by $\mathrm{Tr}_{1,2}\, a = \sum_j \langle j|a|j\rangle \in \mathscr{B}(\mathscr{H}_{2,1})$ for any $a \in \mathscr{C}_1(\mathscr{H})$, where $\{|j\rangle\}$ is any complete orthonormal set in $\mathscr{H}_{1,2}$.

A consequence of this is the

**Subadditivity of the Entropy** (2.2.13)

*Let* $\rho_{1,2} = \mathrm{Tr}_{2,1}\, \rho$. *Then* $S(\rho) \leq S(\rho_1) + S(\rho_2)$.

**Remarks** (2.2.14)

1. If $\rho = \rho_1 \otimes \rho_2$, then $\rho_{1,2} = \mathrm{Tr}_{2,1}\, \rho$ and by (2.2.3(iii)) equality holds in (2.2.13).
2. The partial traces reproduce the reduced density matrices used in §1.1. At that time we noticed that the reduction entailed a loss of information. Inequality (2.2.13) indicates that there is less information in $\rho_1$ and $\rho_2$ than in the original $\rho$.
3. If $\alpha \neq 1$, then the $\alpha$-entropies $S_\alpha$ (2.2.3) are not subadditive (Problem 2). It is consequently not necessarily true that $\rho_1 \otimes \rho_2 \succeq \rho$.
4. Subadditivity allows axiom (iii) of (2.2.4) to be replaced [7] with
   (iii (a)) $S(\rho) = S(V^*\rho V)$ for all isometries $V$; and
   (iii (b)) $S(\rho) \leq S(\rho_1) + S(\rho_2)$, equality holding iff $\rho = \rho_1 \otimes \rho_2$.

**Proof**

By Klein's inequality (2.1.7; 5), $\mathrm{Tr}\, a \ln a - \mathrm{Tr}\, a \ln b \geq \mathrm{Tr}\,(a - b)$. Put $a = \rho$ and $b = \rho_1 \otimes \rho_2$ and note that $\ln \rho_1 \otimes \rho_2 = \ln \rho_1 \otimes \mathbf{1} + \mathbf{1} \otimes \ln \rho_2$. □

**Corollary** (2.2.15)

Consider a sequence of ever larger systems on the tensor product $\mathscr{H}^n$, $n = 1, 2, 3, \ldots$. Suppose that the density matrices $\rho_n$ are compatible so that when reduced to a subsystem they always become the density matrix of the smaller system: $\rho_m = \mathrm{Tr}_{n-m}\rho_n$, $m \leq n$. If $\sigma_n = -(1/n)\mathrm{Tr}\,\rho_n \ln \rho_n$, then $n\sigma_n \leq m\sigma_m + (n-m)\sigma_{n-m}$. In particular, $\sigma_{2n} \leq \sigma_n$, and hence the limits $\lim_{n\to\infty} \sigma_n = \inf_n \sigma_n$ must exist and be $\geq 0$. Although the entropy itself does not tend to a limit as the size of the system gets arbitrarily large, the specific entropy does.

It will be asked by how far (2.2.13) misses equality. More precisely, it might be supposed that the entropy of a united system is always greater than that of any single one of its parts. Surprisingly, this is not necessarily so with quantum statistics; $\rho$ could be a pure state, thus having entropy zero, while the $\rho_i$ correspond to mixtures. This is the case that arose in the discussion of the time-evolution in §1.1; the additional information contained in $\rho$ has to do with the correlations between the subsystems. The correlations are precisely pinned down in

**Lemma** (2.2.16)

*Let $\rho$ be pure; then $\rho_1$ and $\rho_2$ have the same spectrum with the same multiplicities, except possibly for an eigenvalue at* 0.

**Proof**

See Problem 3. □

**Corollary** (2.2.17)

If $\rho$ is pure, then $S(\rho_1) = S(\rho_2)$. Our information about the subsystems is correlated, so they possess the same amount of disorder.

In this case, $S(\rho) = S(\rho_1) - S(\rho_2)$; more generally there is a

**Triangle Inequality** (2.2.18)

$$|S(\rho_1) - S(\rho_2)| \leq S(\rho) \leq S(\rho_1) + S(\rho_2).$$

(Lieb and Araki [8]).

**Remarks** (2.2.19)

1. This inequality has no classical analogy; a counterexample is provided by a $\rho$ with $S(\rho) < 0$ but $S(\rho_1) = S(\rho_2)$.
2. Even if the entropy of a subsystem can be greater than that of the whole system, the triangle inequality reveals that it can not exceed the sum of the total entropy and the entropy of the complementary subsystem.
3. Astonishingly, the classical entropy (2.2.9) of a quantum-mechanical density matrix is monotonic; it is always larger for the whole than for a part: $S_{\mathrm{cl}}(\rho) \geq S_{\mathrm{cl}}(\rho_1)$. (For the proof see Problem 5.)

**Proof**

According to Remark (2.1.6; 5), $\rho$ may be regarded as a pure state $\rho_{123}$ on a large Hilbert space $\mathscr{H}_1 \otimes \mathscr{H}_2 \otimes \mathscr{H}_3$, for which $\rho = \mathrm{Tr}_3\, \rho_{123}$. Let $\rho_3 = \mathrm{Tr}_{12}\, \rho_{123}$, $\rho_{23} = \mathrm{Tr}_1\, \rho_{123}$; then by Corollary (2.2.17), $S(\rho) = S(\rho_3)$, $S(\rho_1) = S(\rho_{23})$. Because of subadditivity, $S(\rho_1) = S(\rho_{23}) \leq S(\rho_3) + S(\rho_2) = S(\rho) + S(\rho_2)$, and along with the same thing with 1 and 2 interchanged, this yields the left inequality of (2.2.18). □

An ideal measurement leaves the system in a pure state, reducing the entropy to 0. For this reason, $S(\rho)$ may be regarded as a measure of the amount of information to be gained by an ideal measurement. The difference $S(\rho) - S(\rho_1)$ specifies how much more information a measurement of the total system can yield than a measurement of a subsystem. Inequality (2.2.18) bounds this relative information gain by $S(\rho_2)$:

$$|S(\rho) - S(\rho_1)| \leq S(\rho_2).$$

With quantum statistics the difference can be either positive or negative. If $\rho$ is pure, so that the greatest possible information about the total system is available, but $\rho_1$ is a mixture, then more information can be obtained by measuring the subsystem. On the other hand, there are some inequalities for this entropy difference that are analogous to those of the classical entropy:

**Inequalities for the Entropy Difference** (2.2.20)

*Let $\rho_{123}$ be given on $\mathscr{H}_1 \otimes \mathscr{H}_2 \otimes \mathscr{H}_3$, and $\rho_{12} = \mathrm{Tr}_3\, \rho_{123}$, $\rho_1 = \mathrm{Tr}_2\, \rho_{12}$, etc. Then*

(i) $S(\rho_{12}) - S(\rho_1)$ *is concave in* $\rho_{12}$;
(ii) $S(\rho_{13}) - S(\rho_1) + S(\rho_{23}) - S(\rho_2) \geq 0$ (Lieb and Ruskai [8]); *and*
(iii) $S(\rho_{123}) - S(\rho_2) \leq S(\rho_{12}) - S(\rho_2) + S(\rho_{32}) - S(\rho_2)$.

**Remarks** (2.2.21)

1. Proposition (i) implies that mixing increases the relative information gain. In particular, the relative information gain is a monotonic function in $\rho_{12}$ with the ordering introduced in (2.1.9).
2. If Roman numerals are used to denote the systems corresponding to the Hilbert spaces $\mathscr{H}_i$, then Inequality (ii) implies that more information can be obtained by measuring I ∪ III and II ∪ III than I and II. If $\mathscr{H}_2$ is one-dimensional, so $S(\rho_2) = 0$ and $S(\rho_{23}) = S(\rho_3)$, then this proposition reduces to (2.2.18).

3. Inequality (iii) is subadditivity for the entropy difference. The information content of I ∪ II and III ∪ II relative to II is greater than that of I ∪ II ∪ III relative to II.

**Proof**

(i) Let $\rho_{12} = \alpha\rho'_{12} + (1-\alpha)\rho''_{12}$, $\rho_1 = \alpha\rho'_1 + (1-\alpha)\rho''_1$. Then

$$\begin{aligned}-S(\rho_{12}) &+ \alpha S(\rho'_{12}) + (1-\alpha)S(\rho''_{12}) + S(\rho_1) - \alpha S(\rho'_1) - (1-\alpha)S(\rho''_1)\\ &= \alpha \operatorname{Tr}_{12} \rho'_{12}[\ln \rho_{12} - \ln \rho'_{12} - \ln \rho_1 + \ln \rho'_1]\\ &\quad + (1-\alpha) \operatorname{Tr}_{12} \rho''_{12}[\ln \rho_{12} - \ln \rho''_{12} - \ln \rho_1 + \ln \rho''_1]\\ &\equiv \alpha\Delta' + (1-\alpha)\Delta''.\end{aligned}$$

If $a = -\beta H_0 - \ln \operatorname{Tr} \exp(-\beta H_0)$ and $b = -\beta V$, then because of (2.1.8; 3) and $\operatorname{Tr} \exp(a) = 1$, $\exp(\operatorname{Tr} b \exp(a)) \leq \operatorname{Tr} \exp(a + b)$, so with $a = \ln \rho'_{12}, b = [\cdots]$, we find $\exp(\Delta) \leq \operatorname{Tr}_{12} \exp(\ln \rho_{12} - \ln \rho_1 + \ln \rho'_1)$. Therefore, with Lieb's theorem (2.1.7; 8),

$$\begin{aligned}\exp(\alpha\Delta' + (1-\alpha)\Delta'') &\leq \alpha \exp(\Delta') + (1-\alpha)\exp(\Delta'')\\ &\leq \alpha \operatorname{Tr}_{12} \exp(\ln \rho_{12} - \ln \rho_1 + \ln \rho'_1)\\ &\quad + (1-\alpha) \operatorname{Tr}_{12} \exp(\ln \rho_{12} - \ln \rho_1 + \ln \rho''_1)\\ &\leq \operatorname{Tr}_{12} \exp(\ln \rho_{12} - \ln \rho_1 + \ln(\alpha\rho'_1\\ &\quad + (1-\alpha) \ln \rho''_1)) = \operatorname{Tr}_{12} \exp(\ln \rho_{12}) = 1.\end{aligned}$$

(ii) Since $\rho_{ik}$ and $\rho_i$ can be expressed linearly in $\rho_{123}$, part (i) makes the left side concave in $\rho_{123}$. The minimum is consequently attained when $\rho_{123}$ is pure. But by Corollary (2.2.17), in this case $S(\rho_{13}) = S(\rho_2)$ and $S(\rho_{23}) = S(\rho_1)$, and the minimum is zero.

(iii) Choose a pure $\rho_{1234}$ on $\mathscr{H}_{123} \otimes \mathscr{H}_4$, such that $\operatorname{Tr}_4 \rho_{1234} = \rho_{123}$. Then by Corollary (2.2.17), $S(\rho_{123}) + S(\rho_2) - S(\rho_{12}) - S(\rho_{23}) = S(\rho_4) + S(\rho_2) - S(\rho_{12}) - S(\rho_{14})$ which is $\leq 0$ by (ii). □

These general inequalities for density matrices reflect mixing properties of the entropy like those used in phenomenological thermodynamics, and thereby provide a deeper foundation for those classical rules.

Nearly as important as the entropy differences is the

**Relative Entropy** (2.2.22)

$$S(\sigma | \rho) \equiv \operatorname{Tr} \rho(\ln \rho - \ln \sigma), \qquad \rho, \sigma \geq 0, \qquad \operatorname{Tr} \rho = \operatorname{Tr} \sigma = 1,$$

for which

(i) $S(\sigma | \rho) \geq 0$;
(ii) *the function* $(\sigma, \rho) \to S(\sigma | \rho)$ *is strictly convex and lower semicontinuous*;
(iii) $S(\sigma \otimes \tau | \rho \otimes \tau) = S(\sigma | \rho)$ *for any density matrix* $\tau$; *and*
(iv) $S(\sigma_1 | \rho_1) \leq S(\sigma | \rho)$ *for* $\mathscr{H} = \mathscr{H}_1 \otimes \mathscr{H}_2$, $(\sigma_1, \rho_1) = \operatorname{Tr}_2(\sigma, \rho)$.

**Proof of the Properties of the Relative Entropy**

(i) This was shown in the proof of subadditivity (2.2.13).

(ii) Convexity follows from (2.1.7; 8) when $\alpha \to 0$. The function is lower semicontinuous because $S(\sigma|\rho)$ can be written as the supremum of a set of continuous functions (Problem 7).

(iii)
$$\begin{aligned} S(\sigma \otimes \tau | \rho \otimes \tau) &= \mathrm{Tr}_{12}\, \rho \otimes \tau[(\ln \rho) \otimes \mathbf{1} + \mathbf{1} \otimes \ln \tau \\ &\qquad - (\ln \sigma) \otimes \mathbf{1} - \mathbf{1} \otimes \ln \tau] \\ &= \mathrm{Tr}_1\, \rho(\ln \rho - \ln \sigma)\mathrm{Tr}_2\, \tau = S(\sigma|\rho). \end{aligned}$$

(iv) As in Problem (2.1.19; 1), write $\rho_1 \otimes \mathbf{1}/d_2 = \int d\mu U_2 \rho U_2^{-1}$, $d_2 = \dim \mathscr{H}_2$, and similarly for $\sigma$. By (iii) and (ii),

$$\begin{aligned} S(\sigma_1|\rho_1) &= S\left(\sigma_1 \otimes \frac{\mathbf{1}}{d_2}\middle| \rho_1 \otimes \frac{\mathbf{1}}{d_2}\right) \\ &= \int S\left(\int d\mu U_2 \sigma U_2^{-1} \middle| \int d\mu U_2 \rho U_2^{-1}\right) \\ &\leq \int d\mu S(U_2 \sigma U_2^{-1} | U_2 \rho U_2^{-1}) = S(\sigma|\rho). \end{aligned}$$

Since $d_2$ drops out of the expression, this proof for $d_2 < \infty$ extends to the infinite-dimensional case. □

**Remarks** (2.2.23)

1. If $\sigma$ is the canonical density matrix $\sigma = \exp(-\beta H)/\exp(-\beta F)$, and the free energy is $F = -\beta^{-1} \ln \mathrm{Tr} \exp(-\beta H)$, then $S(\sigma|\rho) = \beta(\mathrm{Tr}\, \rho H - F) - S(\rho)$. If a free energy $F(\rho) \equiv \mathrm{Tr}\, \rho H - \beta^{-1} S(\rho)$ is ascribed to $\rho$, then $S(\sigma|\rho) = F(\rho) - F$. The relative entropy $S(\sigma|\rho)$ measures the difference from the canonical free energy $F(\sigma) = F$, which always lies lower because of (i).
2. By Property (ii), mixing and passing to limits bring the free energy closer to the canonical free energy.
3. Property (iii) states that the difference from the canonical free energy is the same for $\rho_1$ and $\rho$ if there are two independent subsystems 1 and 2, where $\rho = \rho_1 \otimes \rho_2$, and $\rho_2$ is the canonical density matrix of system 2.
4. If a subsystem is weakly coupled, $H_{12} \cong H_1 \otimes \mathbf{1}_2 + \mathbf{1}_1 \otimes H_2$, i.e., $\exp(-\beta(H_1 - F_1)) \cong \mathrm{Tr}_1 \exp(-\beta(H_{12} - F_{12}))$, then its difference from its canonical free energy is always less than that of the whole system. The analogous argument for the entropy only leads to $S(\rho_1) \leq S(\rho) + \ln d_2$, which already follows from (2.2.5; 3).

A final matter to investigate is how sensitive $S$ is to small changes in $\rho$.

**Theorem** (2.2.24)

*The mapping $\mathscr{C}_1^+ \to \mathbb{R}^+ : \rho \to S(\rho)$ is lower semicontinuous in the trace-norm topology of $\mathscr{C}_1$.*

**Remarks** (2.2.25)

1. The set $\mathscr{C}_1$ is topologized with the trace norm $\| \ \|_1$. If a sequence $\{\rho_N\}$ converges in this topology to $\rho$, then $S(\rho)$ is at most $\lim_{N\to\infty} S(\rho_N)$. However, we shall see in (2.4.19; 1) that for density matrices all topologies between the trace topology and the weak topology are equivalent.
2. Continuity does not occur, because in every $\| \ \|_1$-neighborhood of $\rho$ there are density matrices with arbitrarily much entropy. This follows directly from concavity,
$$S\left(\frac{1}{N}\rho_N + \left(1 - \frac{1}{N}\right)\rho\right) \geq \frac{1}{N} S(\rho_N) + \left(1 - \frac{1}{N}\right) S(\rho).$$
Let $S(\rho) = 0$, and $S(\rho_N) = N^2$; then $S((1/N)\rho_N + (1 - 1/N)\rho) \geq N$, although
$$\left\| \frac{1}{N}\rho_N + \left(1 - \frac{1}{N}\right)\rho - \rho \right\|_1 \leq \frac{2}{N},$$
so the density matrices converge to $\rho$. The two terms in the expression $(1/N)\rho_N + (1 - (1/N))\rho$, however, can not be comparable in the sense of (2.1.9); that would contradict (2.1.10; 4), by which the limit of a sequence of equivalent density matrices can not be purer than the elements of the sequence.
3. The mappings $\mathscr{C}_1^+ \to \mathbb{R}^+ : \rho \to S_\alpha(\rho)$, $\alpha > 1$ are continuous (see below).
4. By lower semicontinuity the sets $S_n \equiv \{\rho : S(\rho) \leq n\}$ are closed, and by Remark 2 they are nowhere dense. This means that the set $\bigcup_n S_n$ of $\rho$'s of finite entropy is of the first category, the topological analogue of a null set. In this sense the entropy is almost always $+\infty$.

**Proof**

Because $\operatorname{Tr} \rho^\alpha = \|\rho\|_\alpha^\alpha \leq \|\rho\|^{\alpha-1} \cdot \|\rho\|_1$, the mapping of $\mathscr{C}_1$ to $\mathbb{R}^+ : \rho \to S_\alpha(\rho)$ is continuous. As the supremum of a set of continuous functions, $S(\rho) = \sup_{\alpha > 1} S_\alpha(\rho)$ is lower semicontinuous. □

The failure of $S(\rho)$ to be continuous does not diminish its usefulness. The density matrices $\rho$ of very large $S$ have their eigenvalues $\rho_i$ spread so far apart that the average of the energy diverges.

**The Continuity of the Entropy at Finite Energy** (2.2.26)

*Suppose that* $H \geq 0$ *and* $\text{Tr}\exp(-\beta H) < \infty$ *for some* $\beta > 0$. *If the density matrices having* $\text{Tr}\,\rho H < \infty$ *are topologized with the norm* $\|\rho\|_H = \text{Tr}\,\rho(1 + H)$, *then* $S(\rho)$ *is a continuous mapping* $\mathscr{C}_H \to \mathbb{R}^+$, *where* $\mathscr{C}_H = \{\rho \in \mathscr{C}_1, \|\rho\|_H < \infty\}$.

**Proof**

According to Remark (2.2.23; 1), $S(\rho) = \beta(\text{Tr}(\rho H) - F) - S(\sigma|\rho)$, where $\sigma = \exp(-\beta H)/\exp(-\beta F)$. The function $\text{Tr}\,\rho H$ is continuous in the $\| \ \|_H$-topology, and $-S(\sigma|\rho)$ is upper semicontinuous, because the $\| \ \|_H$-topology is finer than the trace topology. Since $S(\rho)$ is lower semicontinuous in the trace-norm topology, it is also lower semicontinuous in $\| \ \|_H$, and hence continuous in $\| \ \|_H$. □

**Problems** (2.2.27)

1. (i) Show for the functions $f(n) \equiv S(1/n, \ldots, 1/n)$ that $\lim_{n\to\infty} [f(n) - f(n-1)] = 0$.

   (ii) Conclude from (i) that the only solution of the equation $f(mn) = f(m) + f(n)$ is of the form $f(n) = C \ln n$, supposing that $S$ is continuous according to (2.2.4(i)).

2. For $\alpha \neq 1$, show that the $\alpha$-entropies $S_\alpha$ of (2.2.2) are not subadditive.

3. Prove Lemma (2.2.16).

4. Show that $S(\sum_i \lambda_i \rho_i) \leq \sum_i \lambda_i S(\rho_i) - \sum_i \lambda_i \ln \lambda_i$, $\lambda_i > 0$, $\sum_i \lambda_i = 1$.

5. Show that $S_{\text{cl}}(\rho_1) \leq S_{\text{cl}}(\rho)$ if $\mathscr{H} = \mathscr{H}_1 \otimes \mathscr{H}_2$, where $\mathscr{H}_i$ are one-particle Hilbert spaces, particles 1 and 2 are distinguishable, $\rho_1 = \text{Tr}_2\,\rho$, and $S_{\text{cl}}(\rho)$ is defined as in (2.2.9).

6. Calculate $S_\alpha(\exp(-\beta[H - F(\beta)])$, where $\exp(-\beta F(\beta)) = \text{Tr}\exp(-\beta H)$.

7. Show that $S(\sigma|\rho)$ is lower semicontinuous. Hint: use

   (i) $S(\sigma|\rho) = \sup_{0<\lambda<1} S_\lambda(\sigma|\rho)$, $S_\lambda(\sigma|\rho) \equiv (1/\lambda)(S(\lambda\rho + (1-\lambda)\sigma) - \lambda S(\rho) - (1-\lambda)S(\sigma)) \geq 0$;

   (ii) if $a \geq 0$ then $\text{Tr}\,a = \sup_n P_n a$, $P_n \to \mathbf{1}$, is an increasing sequence of finite-dimensional projections; and

   (iii) the operator inequalities (III: 2.2.38; 11), to show that the function $s(x) \equiv -x \ln x$ is concave for operators, i.e., $s(\lambda a + (1-\lambda)b) \geq \lambda s(a) + (1-\lambda)s(b)$ for all $a, b \in \mathscr{B}(\mathscr{H})$.

8. Prove the formula for the identity operator in (2.2.10; 5).

**Solutions** (2.2.28)

1. (i) Let $d_n = f(n) - f(n-1)$ and $\delta_n = S(1/n, 1-(1/n))$. Because $S$ is continuous, $\delta_n \to 0$.

$$f(n-1) = d_{n-1} + \cdots + d_2 \Rightarrow d_n = \delta_n + \frac{d_{n-1} + \cdots + d_2}{n},$$

$$\sum_{n=2}^{N} (n\, d_n + d_{n-1} + \cdots + d_2) = \sum_{n=2}^{N} n\, \delta_n,$$

$$\sum_{n=2}^{N} d_n = \frac{1}{N} \sum_{n=2}^{N} n\, \delta_n \Rightarrow d_N = \delta_N - \frac{1}{N(N-1)} \sum_{n=2}^{N-1} n\, \delta_n,$$

$$|d_N - \delta_N| \leq \frac{1}{N-1} \sup_n \delta_n + \sup_{n \geq \sqrt{N}} \delta_n \quad \text{for all } N \geq 2, \text{ which} \Rightarrow \lim d_N = 0,$$

because $\sup_n \delta_n < \infty$ and $\lim_{n\to\infty} \delta_n = 0$.

(ii) It suffices to show that $\lim_{n\to\infty} f(n)/\ln n = f(n_0)/\ln n_0$ (for any fixed $n_0 \geq 2$); because $f(n^k) = kf(n)$ this implies that $f(n) = (f(n_0)/\ln n_0) \ln n$ for all $n \geq 2$. Define $g(n) = f(n) - (f(n_0)/\ln n_0) \ln n$; then it suffices to show that $g(n)/\ln n \to 0$. Let $n = n_1 n_0 + l_1$, with $0 \leq l_1 \leq n_0$. Because $g(n_0) = 0$, $g(n) = \sum_{n_1 n_0}^{n-1} \varepsilon_k + g(n_1 n_0)$, where $\varepsilon_k \equiv g(k+1) - g(k)) = \sum_k \varepsilon_k + g(n_1)$. Now let $n_1 = n_2 n_0 + l_2$; then $g(n) = g(n_2) + \sum_{j=1}^{2} \sum_{i=0}^{l_j - 1} \varepsilon_{n_j + i}$, etc. After $k_0 < \ln n/\ln n_0$ steps, $n_{k_0} < n_0$, and therefore $g(n) = \sum_k \varepsilon_k$. The sum has fewer than $n_0 k_0$ summands, and therefore $\lim g(n)/\ln n = 0$. since $\varepsilon_k \to 0$.

2. Let $\mathscr{H} = \mathbb{C}^2 \otimes \mathbb{C}^2$ and $\rho = (\rho_{ik,jl})$, where $\rho_{ik,jl} = \delta_{ij}\delta_{kl} r_{ik}$; $r_{11} = pq + \varepsilon$, $r_{12} = p(1-q) - \varepsilon$, $r_{21} = (1-p)q - \varepsilon$, $r_{22} = (1-p)(1-q) + \varepsilon$ with $0 < p,\ q < 1$, $p, q \neq \frac{1}{2}$. Since $\rho$ is diagonal, this allows $S_\alpha(\rho)$ to be read off with no further ado: If $\varepsilon = 0$, then $\rho = \rho_1 \otimes \rho_2$,

$$\rho_1 = \mathrm{Tr}_2\, \rho = \begin{pmatrix} p & 0 \\ 0 & 1-p \end{pmatrix}, \qquad \rho_2 = \mathrm{Tr}_1\, \rho = \begin{pmatrix} q & 0 \\ 0 & 1-q \end{pmatrix}.$$

If $S_\alpha(\rho)$ were $\leq S_\alpha(\rho_1) + S_\alpha(\rho_2)$, then the function $g(\varepsilon) \equiv (pq + \varepsilon)^\alpha + (p(1-q) - \varepsilon)^\alpha + ((1-p)q - \varepsilon)^\alpha + ((1-p)(1-q) + \varepsilon)^\alpha$ would have an extremum at $\varepsilon = 0$, but $g'(0) \neq 0$ if $\alpha \neq 1$.

3. Let $|x\rangle \in \mathscr{H}_1 \otimes \mathscr{H}_2$. $|x\rangle = \sum_{i,k} c_{ik} |i\rangle_1 \otimes |k\rangle_2$, where $\{|i\rangle_1\}$ and $\{|k\rangle_2\}$ are orthonormal sets in $\mathscr{H}_1$ and $\mathscr{H}_2$ respectively, and $\rho = |x\rangle\langle x|$.

$$\mathrm{Tr}_2 |x\rangle\langle x| = \mathrm{Tr}_2 \sum_{ijkl} c_{ik} c_{jl}^* |i\rangle_{11}\langle j| \otimes |k\rangle_{22}\langle l|$$

$$= \sum_{ijkl} c_{ik} c_{jl}^* |i\rangle_{11}\langle j| \delta_{kl} = \sum_{ijk} c_{ik} c_{jk}^* |i\rangle_{11}\langle j|,$$

which implies that the positive eigenvalues of $\mathrm{Tr}_2\, |x\rangle\langle x|$ are the same as those of the matrix $CC^*$, where $C = (c_{ij})$. A similar argument shows that the positive eigenvalues of $\mathrm{Tr}_1\, |x\rangle\langle x|$ are the same as those of $C^*C$ and thus of $CC^*$.

4. Let $\lambda_i \rho_i = a_i$; then the proposition is equivalent to $S(\sum_i a_i) \leq \sum_i S(a_i)$ for all $a_i \in \mathscr{C}_1^+$. Since $\ln x$ is monotonic as an operator function (III: 2.2.38; 11), if $a_k \geq 0$,

then $\ln a_i \le \ln(\sum_j a_j)$, which implies $a_i^{1/2}(\ln a_i)a_i^{1/2} \le a_i^{1/2}(\ln \sum_j a_j)a_i^{1/2}$, and therefore $\sum_i \mathrm{Tr}(a_i \ln a_i) \le \mathrm{Tr}[(\sum_i a_i)\ln(\sum_i a_i)]$.

5. $\rho_1(\mathbf{z}_1) \equiv \langle \mathbf{z}_1|\rho_1|\mathbf{z}_1\rangle = \sum_i \langle \mathbf{z}_1 \otimes e_i|\rho|\mathbf{z}_1 \otimes e_i\rangle \ge \langle \mathbf{z}_1 \otimes \mathbf{z}_2|\rho|\mathbf{z}_1 \otimes \mathbf{z}_2\rangle \equiv \rho(\mathbf{z}_1, \mathbf{z}_2)$, since $\{e_i\}$ may be chosen to be an arbitrary basis. Therefore

$$S_{\mathrm{cl}}(\rho) - S_{\mathrm{cl}}(\rho_1) = \int d\Omega_z^2 \rho(\mathbf{z}_1, \mathbf{z}_2) \ln\left(\frac{\rho(\mathbf{z}_1)}{\rho(\mathbf{z}_1, \mathbf{z}_2)}\right) \ge 0.$$

6. $$S_\alpha = \frac{1}{1-\alpha} \ln \mathrm{Tr} \exp(-\alpha\beta(H - F(\beta))) = \frac{1}{\alpha - 1}[F(\alpha\beta) - \alpha F(\beta)].$$

As $\alpha \to 1$, $S \to \partial F(\beta)/\partial\beta$

7. (i) The function $\lambda S_\lambda$ is concave in $\lambda$, because

$$S(\alpha(\lambda_1\sigma + (1-\lambda_1)\rho) + (1-\alpha)(\lambda_2\sigma + (1-\lambda_2)\rho)) \ge \alpha S(\lambda_1\sigma + (1-\lambda_1)\rho) + (1-\alpha)S(\lambda_2\sigma + (1-\lambda_2)\rho),$$

so

$$S(\sigma|\rho) = \frac{d}{d\lambda}\lambda S_\lambda|_{\lambda=0} = \sup_{0<\lambda<1} S_\lambda.$$

(ii) This is the normality of the trace.
(iii) The operator concavity of $-x \ln x = \int_0^\infty (1 - \alpha/(x+\alpha) - x/(1+\alpha))d\alpha$ is equivalent to the operator convexity of $1/(x+1)$, and it suffices to show convexity with $\alpha = \frac{1}{2}$:

$$\frac{1}{(A+B)/2+1} \le \frac{1}{2(A+1)} + \frac{1}{2(B+1)} \Leftrightarrow \frac{4}{A+1+B+1} \le \frac{1}{A+1} + \frac{1}{B+1}$$

$$\Leftrightarrow \frac{4}{(B+1)^{-1/2}(A+1)(B+1)^{-1/2}+1} \le (B+1)^{1/2}(A+1)^{-1}(B+1)^{1/2} + 1.$$

Since $4/(x+1) \le (1/x) + 1$ for all $x \in \mathbb{R}^+$, this is also valid for positive operators. Therefore, $(1/\lambda)[s(\lambda\sigma + (1-\lambda)\rho) - \lambda s(\sigma) - (1-\lambda)s(\rho)] \ge 0$, which implies $S(\sigma|\rho) = \sup_n \sup_{0<\lambda<1} (1/\lambda)\,\mathrm{Tr}\, P_n[s(\lambda\sigma + (1-\lambda)\rho) - \lambda s(\sigma) - (1-\lambda)s(\rho)]$, and $s(\rho)$ is continuous in finite dimensions. This also provides a new proof of the lower semicontinuity of $S(\rho)$.

8. The right side of the equation clearly leaves the number of particles invariant. Hence the formula is shown by

$$\langle f_1, \ldots, f_N| \int \frac{dz_1 \cdots dz_N}{N!(2\pi)^{dN}} |\mathbf{z}_1, \ldots, \mathbf{z}_N\rangle\langle \mathbf{z}_1, \ldots, \mathbf{z}_N|g_1, \ldots, g_N\rangle$$

$$= \sum_{P,Q} (\pm 1)^{P+Q} \int \frac{dz_1 \cdots dz_N}{N!(2\pi)^{dN}} \prod_{i=1}^N \langle f_{P_i}|\mathbf{z}_i\rangle\langle \mathbf{z}_i|g_{Q_i}\rangle$$

$$= \sum_{P,Q} (\pm 1)^{P+Q} \prod_i \langle f_{P_i}|g_{Q_i}\rangle \frac{1}{N!}$$

$$= \sum_{P'} (\pm 1)^{P'} \prod_i \langle f_i|g_{P'_i}\rangle = \langle f_1, \ldots, f_N|g_1, \ldots, g_N\rangle.$$

## 2.3 The Microcanonical Ensemble

*Insight into the fundamental thermodynamic laws is gained by investigating the chaotic state below the energy surface.*

Two trains of thought are usually followed to justify regarding the equilibrium state as predominating for macroscopic systems. Like Boltzmann, one can investigate the time-evolution of a system and show that most states tend to equilibrium. Alternatively, one can follow Gibbs and examine an ensemble of identical copies of the system and identify states of scanty information with equilibrium states. The set of problems connected with the first procedure is the subject of the next chapter, while in this section we shall study systems for which the only information concerns the energy. If it is known that the energy does not exceed some maximum value $E_m$, then, as remarked in (2.1.10; 2), the most mixed state containing no further information corresponds to the

**Microcanonical Density Matrix** (2.3.1)

$$\rho = \Theta(E_m - H)/\mathrm{Tr}\,\Theta(E_m - H), \qquad \Theta(x) = \begin{cases} 1 & \text{for } x \geq 0 \\ 0 & \text{for } x < 0 \end{cases},$$

where $E_m \geq \varepsilon_1 \equiv$ the lowest eigenvalue of $H$. Its

**Entropy and Average Energy** (2.3.2)

Are

$$S = \ln \mathrm{Tr}\,\Theta(E_m - H), \qquad E = \exp(-S)\mathrm{Tr}\,H\Theta(E_m - H).$$

**Remarks** (2.3.3)

1. The discontinuous function $\Theta$ of a self-adjoint operator is defined with the spectral representation of the operator.
2. It is assumed that $H$ is bounded below and that $\sigma_{\mathrm{ess}}(H)$ is empty, so the traces in (2.3.2) are finite.
3. The entropy $S$ is a discontinuous function of $E_m$, and has no well-defined inverse. On the other hand, $E$ may be construed as a function of $S$, as shown in Figure 3. The function $E(S)$ increases monotonically.
4. By the min–max principle, $E(S)$ is also given by $E(S) = \exp(-S)\inf_{\mathscr{H}_n} \mathrm{Tr}_{\mathscr{H}_n} H$, where $\mathscr{H}_n$ is an $n$-dimensional subspace of $D(H)$ and $n = \exp(S)$. It is consequently a concave function of all parameters on which the dependence of $H$ is concave.

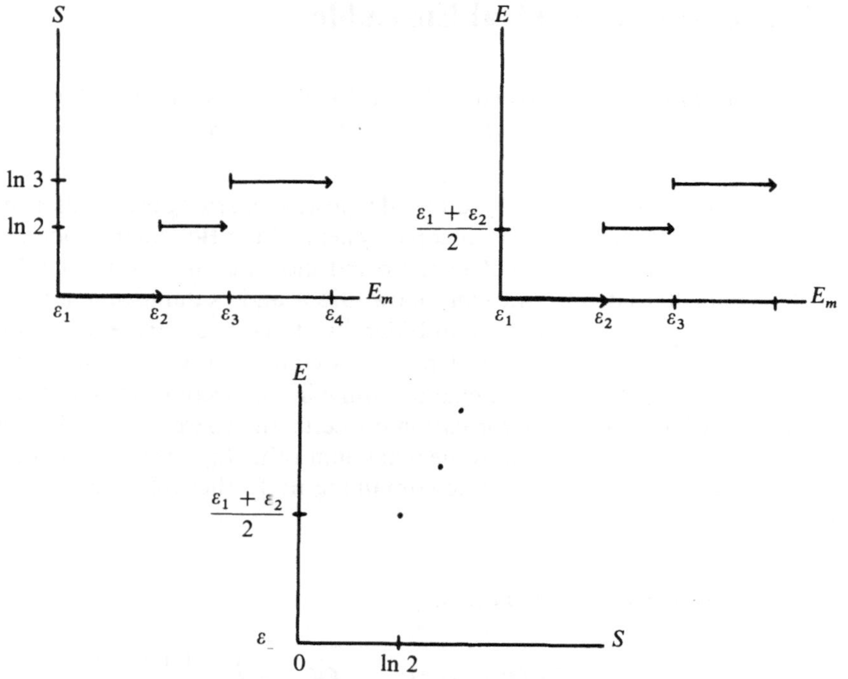


Figure 3 The thermodynamic functions for a finite system.

5. By Property (2.2.3(iv)), all $\alpha$-entropies $S_\alpha$ lead to the same $S$ (2.3.2), which can be identified as the entropy of phenonemological thermodynamics.
6. It will be seen shortly that in the systems under consideration here the density of states increases so rapidly with the energy that in the limit of an infinite system, any density matrix $\rho \sim \Theta(E - H) - \Theta(E(1 - \varepsilon) - H)$ yields the same entropy density for all $\varepsilon > 0$.

The further properties of $E(S)$ follow from the special form of the Hamiltonian,

$$H_N = \sum_{i=1}^{N} \frac{|\mathbf{p}_i|^2}{2m_i} + \sum_{i>j} v(\mathbf{x}_i - \mathbf{x}_j),$$

where $v$ is assumed bounded relative to the kinetic energy. It will be most convenient to deal with the quadratic form associated with $H_N$ (cf. (III: 2.5.18; 2)). The quadratic-form domain $Q(H_N)$ consists of functions $\psi$ such that $\sum_i (1/2m_i) \int |\nabla_i \psi|^2 < \infty$ and with some other restrictions from the boundary conditions. The formula of Remark (2.3.3; 4) then holds with $\mathscr{H}_n \subset Q(H_N)$. The boundary conditions we shall choose are Dirichlet conditions on the surface of a volume $V \subset \mathbb{R}^3$, which mean specifically that: $\mathscr{H} \subset L^2(V^N)$

and $\psi|_{\partial(V^N)} = 0$. The Hilbert space $\mathscr{H}$ is $L^2(V^N)$ if the particles are distinguishable, and if they are identical bosons or fermions, then $\mathscr{H}$ must be restricted to functions of the appropriate symmetry. The energy can be treated as a function of $S$, $V$, and $N$, and its dependence on $V$ is described by the following theorem.

**Monotony of the Energy** (2.3.4)

*If $V' \supset V$, then*

$$E(S, V', N) \leq E(S, V, N).$$

**Proof**

This follows from (2.3.3; 4) because $Q(H(V')) \supset Q(H(V))$, where $\supset$ is intended in the sense of the natural imbedding, i.e., functions $\psi$ such that $\psi|_{\partial V} = 0$ are set to 0 in $V' \setminus V$. □

Subadditivity generalizes this monotony when particles in separated volumes do not repel one another.

**Subadditivity of the Energy** (2.3.5)

*If $V_1 \cap V_2 = \emptyset$ and $v(\mathbf{x}_i - \mathbf{x}_j) \leq 0$ for all $\mathbf{x}_i \in V_1$, $\mathbf{x}_j \in V_2$, then*

$$E(S_1 + S_2, V_1 \cup V_2, N_1 + N_2) \leq E_1(S_1, V_1, N_1) + E(S_2, V_2, N_2).$$

**Proof**

This again follows from (2.3.3; 4), since the right side results from taking the infimum over a subspace of $Q(H)$, which consists of tensor products of $\exp(S_1)$ vectors, for which $N_1$ particles lie within the volume $V_1$, with $\exp(S_2)$ vectors having $N_2$ particles within $V_2$. The tensor products have to be symmetrized or antisymmetrized if there are Bose or Fermi statistics. However, since symmetrization does not affect the expectation values of (2.3.5) when the functions have disjoint supports, (2.3.5) is independent of the statistics. □

The existence of $\lim_{V \to \infty} E/V$ can be derived from the subadditivity, though it is rather difficult to go beyond the restriction $v \leq 0$. This problem will have to be investigated later for each of the systems discussed in §1.2, and for now convergence will simply be assumed. The condition is satisfied trivially for free particles ($v = 0$). To draw conclusions like those of (2.2.15), assume that $V$ is a cube, the volume of which will also be fearlessly denoted

$V \in \mathbb{R}^+$. If eight cubes are packed together as a single cube of double the side, then (2.3.5) implies

$$E(8S, 8V, 8N) \leq 8E(S, V, N). \tag{2.3.6}$$

Assuming in addition that there exists $A \in \mathbb{R}^+$ such that

$$H_N \geq -AN \quad \text{for all } N \in \mathbb{Z}^+, \tag{2.3.7}$$

the limit

$$\lim_{\mathbb{Z}^+ \ni \nu \to \infty} 8^{-\nu} E(8^\nu S, 8^\nu V, 8^\nu N) = \inf_\nu 8^{-\nu} E(8^\nu S, 8^\nu V, 8^\nu N)$$

exists. This allows the passage to an infinite system, for which the energy, entropy, and particle densities are defined by $E/V = \varepsilon$, $S/V = \sigma$, and $N/V = \rho$.

**The Thermodynamic Limit of the Energy Density** (2.3.8)

$$\varepsilon(\sigma, \rho) = \inf_{\mathbb{Z}^+ \ni \nu} 8^{-\nu} \rho E(8^\nu \sigma \rho^{-1}, 8^\nu \rho^{-1}, 8^\nu).$$

**Remarks** (2.3.9)

1. Equation (2.3.7) guarantees that $\varepsilon > -\infty$, so the infimum always exists; but (2.3.8) is only of interest when there is a well-defined limit, for only then is it certain that the thermodynamic properties do not depend on the exact number of particles. Even if the limit exists, as in the case of (2.3.6), it does not guarantee that the resulting $\varepsilon$ is nontrivial. If, say, the particles can be distinguished (which does not invalidate the general conclusions), then classically,

$$\exp(S) = \int_{V^N} d^{3N}x \int d^{3N}p \Theta\left(E_m - \sum_{i=1}^{N} |\mathbf{p}_i|^2\right) = \pi^{3N/2} \frac{E_m^{3N/2} V^N}{(3N/2)!},$$

and

$$E = \frac{E_m}{1 + 2/3N}.$$

Therefore, as $N \to \infty$,

$$\frac{E}{V} = \frac{3}{2\pi e} \frac{\rho^{5/3}}{N^{2/3}} \exp(\tfrac{2}{3}\sigma\rho^{-1}) \to 0.$$

The familiar result obtains only with the replacement $\exp(S) \to (1/N!)\exp(S)$ to account for the particles being identical. A later calculation of $\varepsilon(\sigma, \rho)$ will reveal that (2.3.8) is then not without content.
2. Though the result has been derived only for cubes, the limit clearly exists for other shapes if they are not too different from cubes.

3. The effect of dilatations on the kinetic energy (cf. (III: 3.3.21; 8) and (III: 4.1.4)) of free particles implies, moreover, that
$$E(S, V, N) = \exp(2\tau)E(S, \exp(-3\tau)V, N).$$
Hence the one-parameter family of limits
$$\lim_{\nu\to\infty} 8^{-\nu(1-2\tau)}E(8^\nu S, 8^{\nu(1-3\tau)}V, 8^\nu N)$$
exist (cf. (1.2.1)). Ordinarily, the limit is taken with $\tau = 0$, and quantities proportional to $N$, like $E$, $S$, and $V$, are described as extensive, while $N$-independent quantities like $\varepsilon$, $\rho$, and $\sigma$ are called intensive. The existence of some limit is important, for, whatever it may be like, it enables precise propositions to be formulated. In reality systems are large but still finite, but if a quantity converges as $N \to \infty$ the limit may be expected to be attained for practical purposes when, say, $N = 10^{24}$. Indeed, it will be shown in realistic situations that the limit is sometimes attained to $O(N^{-1/6})$, which is sufficient accuracy for macroscopic bodies. There are various ways to interpret the limit $N \to \infty$. As has been done here, the system may be thought of as becoming larger and larger, or, alternatively, the atoms may be imagined smaller and smaller with their number in the fixed volume of the container being increased at the same time.

Since monotony and convexity survive pointwise limits, there are the following

**Properties of the Energy Density** (2.3.10)

For the function $\mathbb{R}^+ \times \mathbb{R}^+ \to \mathbb{R}^+: \sigma, \rho \to \varepsilon(\sigma, \rho)$,

(i) *$\varepsilon$ increases monotonically in $\sigma$*:
(ii) *$\rho^{-1}\varepsilon(\alpha\rho, \rho)$ increases monotonically in $\rho$*;
(iii) *$\varepsilon$ is convex in $(\sigma, \rho)$*;
(iv) *moreover, for free particles, $\varepsilon(\sigma, \rho) = \rho^{5/3}f(\sigma/\rho)$.*

**Proof**

Property (i) holds as remarked in (2.3.3; 3), and Property (ii) follows from Theorem (2.3.4). From subadditivity (2.3.5),
$$\varepsilon(\tfrac{1}{2}(\sigma_1 + \sigma_2), \tfrac{1}{2}(\rho_1 + \rho_2)) \leq \tfrac{1}{2}(\varepsilon(\sigma_1, \rho_1) + \varepsilon(\sigma_2, \rho_2)),$$
which implies (iii), and (iv) follows from (2.3.9; 3). □

**Remarks** (2.3.11)

1. Since $N \in \mathbb{Z}^+$, $S \in \ln \mathbb{Z}^+$, $\varepsilon$ is at first defined only on the dense set for which $\sigma\rho^{-1}$ is a power of $(\ln z)/2$, $z \in \mathbb{Z}^+$. It extends continuously to $\mathbb{R}$, because monotony and concavity with the coefficient $\frac{1}{2}$ imply uniform

continuity. There are discontinuous functions that are concave with coefficient $\frac{1}{2}$, such as

$$f(x) = \begin{cases} x, & x \text{ rational}, \\ 0, & \text{otherwise}, \end{cases}$$

for which the equation $f(\alpha x) = \alpha f(x)$ holds for all rational $\alpha$. However, this can not occur if the function is monotonic. The extension then in addition satisfies the inequality

$$\varepsilon(\alpha\sigma_1 + (1-\alpha)\sigma_2, \alpha\rho_1 + (1-\alpha)\rho_2) \leq \alpha\varepsilon(\sigma_1, \rho_1) + (1-\alpha)\varepsilon(\sigma_2, \rho_2)$$
$$\text{for all } \alpha \in \mathbb{R}, 0 \leq \alpha \leq 1.$$

2. Subadditivity (2.3.5) is sufficient but not necessary for Property (iii); (2.3.5) may be violated if the interaction is partially repulsive, which is a necessary assumption or $H_N \geq -AN$ when the particles interact. However, if the potential goes to zero rapidly enough at infinity, the correction to (2.3.5) on any finite region is a surface effect, so the convexity of the energy density is still guaranteed in the thermodynamic limit. On the other hand, the special form (2.3.8) is crucial, and in §4.2 it will be seen that convexity (2.3.10(iii)) is violated in gravitating systems, although (2.3.5) is valid.
3. Since the limiting function is continuous, Dini's theorem ensures that the monotonic limit (2.3.8) is uniform on compact sets.
4. Let $H$ be defined so that $\inf \varepsilon = 0$. Since $\varepsilon$ is convex in $\sigma$, unless $\varepsilon \equiv 0$, there exists a $\sigma_0$ such that $\varepsilon$ is strictly monotonic in $\sigma$ for all $\sigma > \sigma_0$. There is consequently an inverse function $\sigma(\varepsilon, \rho)$ (see Figure 4), which is concave and monotonically increasing in $\varepsilon$.
5. As long as $\sigma$ is strictly monotonically increasing in $\varepsilon$, the density matrices

$$\rho = \Theta(E_m - H)\exp(-S)$$

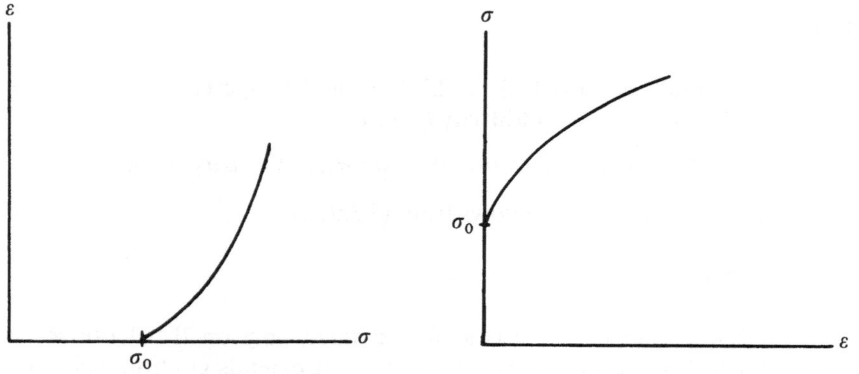


Figure 4 The thermodynamic functions for an infinite system.

and

$$\rho_\delta = (\Theta(E_m - H) - \Theta(E_m - V\delta - H))\exp(-S_\delta)$$

yield the same entropy densities in the limit $N \to \infty$:

$$\sigma_\delta = \lim_{V\to\infty} \frac{1}{V} \ln \operatorname{Tr}(\Theta(E_m - H) - \Theta(E_m - V\delta - H))$$

$$= \sigma(\varepsilon, \rho) + \lim_{V\to\infty} \frac{1}{V} \ln(1 - \exp[-V(\sigma(\varepsilon, \rho) - \sigma(\varepsilon - \delta, \rho))]) = \sigma.$$

This means that as $N \to \infty$ most of the states crowd just under the energy surface with arbitrarily high density.

6. For some systems $\sigma(\varepsilon)$ is constant for $\varepsilon$ greater than some $\varepsilon_1$, in which case $\rho$ and $\rho_\delta$ may have different entropies. Consider for example $N$ spins in an external field ((1.1.3) with $\varepsilon = 0$). The density of states $(\partial/\partial E)\exp[S(E)]$ is invariant under $\sigma \to -\sigma$ and thus an even function in $E$. This makes $\operatorname{Tr}\rho_\delta$ a decreasing function of $E_m$ when $E_m + \delta > 0$, which is impossible for $\operatorname{Tr}\rho$ (see Figure 5); Definition (2.3.1) rules negative temperatures out.
7. The number of energy levels below $E_m$ is $\exp(N\sigma/\rho)$, which is immense for macroscopic bodies, $N \sim 10^{24}$. It would never be possible to isolate the energy levels completely—their widths are on the order of (macroscopic time)$^{-1}$, which is much larger than their spacing. Systems will later be idealized as infinite, having continuous energy spectra, which comes closer to reality than does the fiction of a discrete spectrum.

After this first exposure to these ideas, let us consider two systems the interaction between which is so weak that it can be neglected in comparison with other energies. They are to be considered as parts of a larger system with $\mathscr{H} = \mathscr{H}_1 \otimes \mathscr{H}_2$, $H = H_1 + H_2$. The question is how the energy and entropy are shared by the two subsystems. Even though $H$ is a sum, the microcanonical density matrix (2.3.1) is not in the form of a product $\rho = \rho_1 \otimes \rho_2$, and we will have to see how the entropy of this state can nonetheless be additive for independent, macroscopic systems. Assume to

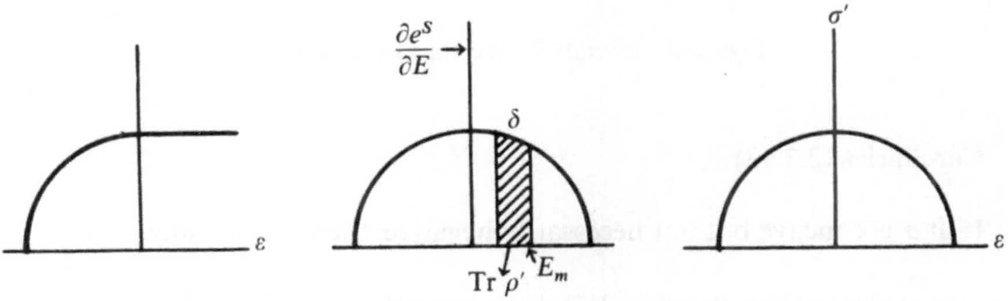


Figure 5 Inequivalence of the microcanonical ensembles for spins in a magnetic field.

this end that the systems are large and that the sequence (2.3.8) converges and has all the necessary kinds of continuity so that $\varepsilon = E/V$ can be regarded as a continuous variable for the purposes of integration and differentiation. For the problem at hand and other estimates we shall need

**Lemma** (2.3.12)

*Let* $\sigma(\varepsilon) \leq 0$ *and be concave on* $[0, 1]$, *and* $\sigma(1) = 0$, $-\infty < \sigma(0) < 0$; *this implies that* $\sigma$ *is nondecreasing and that there exists an* $\varepsilon_0$, $0 < \varepsilon_0 \leq 1$, *such that* $\sigma' \equiv \sigma'(\varepsilon_0) > 0$. *Then*

$$\frac{1 - \exp(-V|\sigma(0)|)}{V|\sigma(0)|} \leq \int_0^1 d\varepsilon \exp(V\sigma(\varepsilon)) \leq 1 - \varepsilon_0 + \frac{1 - \exp(-V\varepsilon_0\sigma')}{V\sigma'}.$$

**Proof**

By assumption (see Figure 6),

$$(1 - \varepsilon)\sigma(0) \leq \sigma(\varepsilon) \leq \begin{cases} 0 & \text{for } \varepsilon_0 \leq \varepsilon \leq 1 \\ -(\varepsilon_0 - \varepsilon)\sigma' & \text{for } 0 \leq \varepsilon \leq \varepsilon_0. \end{cases}$$ □

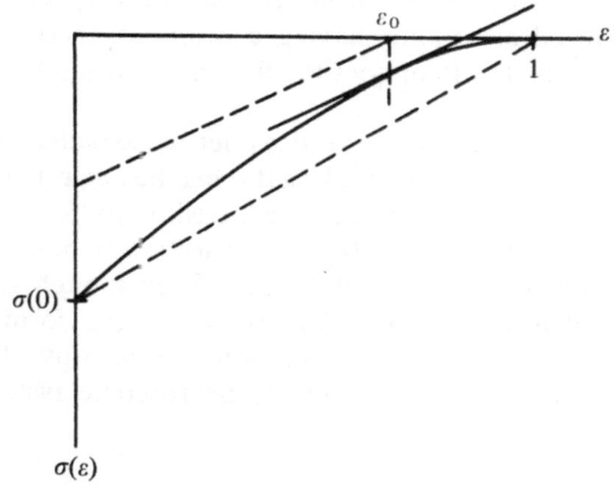


Figure 6 Bounds for the concave function $\sigma(\varepsilon)$.

**Corollaries** (2.3.13)

1. If $\sigma$ is concave but not necessarily negative, then the formula

$$\int_a^b V\,d\varepsilon \exp(V\sigma(\varepsilon)) = \exp(V\bar{\sigma}) \int_a^b d\varepsilon V \exp(V(\sigma(\varepsilon) - \bar{\sigma})) \text{ with } \bar{\sigma} = \max_{a \leq \varepsilon \leq b} \sigma(\varepsilon)$$

can be used instead, since $-\infty < \bar{\sigma} < \infty$ unless $\sigma \equiv \pm\infty$. By an application of the lemma, possibly after subdivision of the region of integration,

$$\lim_{V\to\infty} \frac{1}{V} \ln \int_a^b d\varepsilon V \exp(V(\sigma(\varepsilon) - \bar{\sigma})) = 0.$$

Thus only the maximum value of $\sigma$ contributes in the infinite limit:

$$\lim_{V\to\infty} \frac{1}{V} \ln \int_a^b V d\varepsilon \exp(V\sigma(\varepsilon)) = \sup_{a\le\varepsilon\le b} \sigma(\varepsilon) = \bar{\sigma}(\varepsilon_1) = \sup_{a\le\varepsilon\le b} \bar{\sigma}(\varepsilon).$$

2. Remark (2.3.11; 5) leads one to expect that $E_m$ and $E$ may become equal for large systems. More precisely, if $\sigma$ is concave in $\varepsilon$, $d\sigma/d\varepsilon > 0$, $\lim_{V\to\infty} (E - E_m)/V = 0$. This follows because $E$ may be written as

$$E = \exp(-S)\operatorname{Tr} H\Theta(E_m - H) = \int_0^{E_m} dE' E' \frac{\partial}{\partial E'} \operatorname{Tr} \Theta(E' - H)\exp(-S)$$

$$= E_m - \int_0^{E_m} dE' \operatorname{Tr} \Theta(E' - H)\exp(-S).$$

With $\varepsilon_0 = 1$ and $E' = \varepsilon V$ the lemma now implies that the last integral is $O(1)$, whereas $E_m \sim V$.

3. We next calculate $\exp(S(E)) = \operatorname{Tr} \Theta(E - H_1 - H_2)$, $H_i \ge 0$, as $V = V_1 + V_2 \to \infty$ with $V_i/V$ fixed. Because of the assumption of subadditivity,

$$\sigma_{1,V_1}(\varepsilon) \equiv \frac{1}{V_1} \ln \operatorname{Tr}_1 \Theta(V_1\varepsilon - H_1)$$

is concave in $\varepsilon$ and increases monotonically to $\sigma_1(\varepsilon)$. Let $E_2[n]$ denote the ordered sequence of eigenvalues of $H_2$. If the entropies are considered as functions of the maximum energy, which leads to the same function in the limit $V \to \infty$ because of Corollary 2, then $n$ may be identified with $\exp S$, and $E_2(S_2) \equiv E_2[\exp(S_2)]$ becomes the function introduced in (2.3.3; 3). With $E = \varepsilon V$,

$$\sigma(\varepsilon) = \lim_{V\to\infty} \frac{1}{V} \ln \operatorname{Tr} \Theta(E - H_1 - H_2)$$

$$= \lim_{V\to\infty} \frac{1}{V} \ln \sum_{n=1}^{\exp(S_2(E))} \exp(S_1(E - E_2[n])).$$

Now regard $n$ as a continuous variable, and interpolate $E_2[n]$ linearly. Since the integrand decreases monotonically, the sum $\sum_{n=1}^{\exp(S_2(E))} \cdots$ lies between $\int_0^{\exp(S_2(E))} dn \cdots$ and $\int_1^{\exp(S_2(E))+1} dn \cdots$, and the evaluation of the error is unnecessary, since $\exp(S_2(E)) \sim \exp(10^{23})$. With the variables $\sigma_2 = (1/V_2)\ln n$, $\sigma(\varepsilon)$ can be written as

$$\lim_{V\to\infty} \frac{1}{V} \ln \int_0^{\sigma_2(\varepsilon)} V_2\, d\sigma_2 \exp\left[V_1\sigma_{1,V_1}\left(\frac{V}{V_1}\varepsilon - \frac{V_2}{V_1}\varepsilon_{2,V_2}(\sigma_2)\right) + V_2\sigma_2\right].$$

Now note that $\sigma_2 \to a - b\varepsilon_2(\sigma_2)$ is concave if $b \geq 0$, $\sigma_{1,V_1}$ is concave and increasing, and that (concave, increasing) $\circ$ concave = concave. This allows the lemma to be applied, to show

$$\sigma(\varepsilon) = \lim_{V\to\infty} \sup_{0\leq\sigma_2\leq\sigma_2(\varepsilon)} \left[\frac{V_1}{V}\sigma_{1,V_1}\left(\frac{V}{V_1}\varepsilon - \frac{V_2}{V_1}\varepsilon_{2,V_2}(\sigma_2)\right) + \frac{V_2}{V}\sigma_2\right]$$
$$= \sup_{0\leq\sigma_2\leq\sigma_2(\varepsilon)} \left[\frac{V_1}{V}\sigma_1\left(\frac{V}{V_1}\varepsilon - \frac{V_2}{V_1}\varepsilon_2(\sigma_2)\right) + \frac{V_2}{V}\sigma_2\right].$$

The interchange of the limit $V \to \infty$ and the supremum is justified because $\varepsilon_{2,V_2}(\sigma_2)$ increases monotonically in $\sigma_2$ for all $V_2$, and since $\sigma_{1,V_1}(\varepsilon)$ likewise increases in $\varepsilon$, it decreases in $\sigma_2$, and consequently the first term in the brackets [ ] converges uniformly on compact sets to

$$\sigma_2 \to \sigma_1\left(\frac{V}{V_1}\varepsilon - \frac{V_2}{V_1}\varepsilon_2(\sigma_2)\right).$$

Although the concavity of $\sigma$ is preserved in the limit $V \to \infty$, strict concavity, which is needed to guarantee that the maximum is attained at only one point, may break down. A lack of strict concavity means that there is a phase transition, and will be examined in detail later. If, however, $\sigma_i(\varepsilon_i)$ are strictly concave and continuously differentiable, then the result of Corollary 3 can be improved upon and the additivity of the entropies demonstrated.

**Equilibrium Condition** (2.3.14)

*Let* $\sigma_i(\varepsilon_i) = \lim_{V_i\to\infty} (1/V_i) \ln \operatorname{Tr} \Theta(V_i\varepsilon_i - H_i)$ *be strictly concave and continuously differentiable,* $\lim_{\varepsilon\to 0} \sigma'(\varepsilon) = \infty$ *and* $\lim_{V\to\infty} V_i/V \equiv \alpha_i, \alpha_1 + \alpha_2 = 1$. *Then*

$$\lim_{V\to\infty} \frac{1}{V} \ln \operatorname{Tr} \Theta(V\varepsilon - H_1 - H_2) \equiv \sigma(\varepsilon) = \alpha_1\sigma_1(\varepsilon_1) + \alpha_2\sigma_2(\varepsilon_2),$$

*where* $\varepsilon_i$ *are determined uniquely by*

$$\alpha_1\varepsilon_1 + \alpha_2\varepsilon_2 = \varepsilon, \qquad \frac{\partial}{\partial\varepsilon_1}\sigma_1(\varepsilon_1) = \frac{\partial}{\partial\varepsilon_2}\sigma_2(\varepsilon_2).$$

**Remarks** (2.3.15)

1. The energy densities can equally well be regarded as functions of the entropy densities, which reformulates the equilibrium condition as

$$\frac{\partial}{\partial\sigma_1}\varepsilon_1(\sigma_1) = \frac{\partial}{\partial\sigma_2}\varepsilon_2(\sigma_2) \quad \text{and} \quad \alpha_1\sigma_1 + \alpha_2\sigma_2 = \sigma.$$

2. Convexity of $\varepsilon(\sigma)$ is equivalent to concavity of $\sigma(\varepsilon)$, which is equivalent to the number of states below $E_m$ not increasing faster than exponentially with the energy. This is not a general property of quantum-mechanical systems, and has to be checked in individual cases. A simple counter-example is the hydrogen atom, for which $E_n \sim -1/n^2$, $\exp(S(E_n)) \sim n^3$, where $n$ is the principal quantum number, and therefore

$$E \sim -\exp(-\tfrac{2}{3}S), \qquad \frac{\partial E}{\partial S} \sim \tfrac{2}{3}\exp(-\tfrac{2}{3}S) > 0, \qquad \frac{\partial^2 E}{\partial S^2} \sim -\tfrac{4}{9}\exp(-\tfrac{2}{3}S) < 0.$$

In such cases there may be many solutions of the equilibrium condition (see Figure 7).
3. Condition (2.3.14) implies that the energy is apportioned between the two systems so as to maximize the total entropy. From the point of view of $\varepsilon(\sigma)$ this means distributing entropy so as to minimize the total energy. As a consequence, the subadditivity inequality (2.3.5) becomes an equality in the limit $V \to \infty$.
4. If $\varepsilon_i(\sigma) \in \mathbb{C}^2$, then at the minimum, $\varepsilon_1''/\alpha_1 + \varepsilon_2''/\alpha_2 \geq 0$, where $\varepsilon'' = \partial^2\varepsilon/\partial\sigma^2$. Then by Problem 4, at the minimum, $1/\varepsilon'' = \alpha_1/\varepsilon_1'' + \alpha_2/\varepsilon_2''$.

If the total system consists of a system immersed in a thermal reservoir, then the system of interest is not affected by the fine details of the reservoir, but only by $\partial\sigma_2/\partial\varepsilon_2$, which not only determines $\partial\sigma_1/\partial\varepsilon_1$, but also equals $\partial\sigma/\partial\varepsilon$, because

$$\frac{d}{d\varepsilon}\left(\alpha_1\sigma_1(\varepsilon_1(\varepsilon)) + \alpha_2\sigma_2\left(\frac{\varepsilon}{\alpha_2} - \frac{\alpha_1}{\alpha_2}\varepsilon_1(\varepsilon)\right)\right)$$
$$= \sigma_2'(\varepsilon_2(\varepsilon)) + \alpha_1\frac{d\varepsilon_1}{d\varepsilon}(\sigma_1'(\varepsilon_1(\varepsilon)) - \sigma_2'(\varepsilon_2(\varepsilon))),$$

where

$$\varepsilon_2(\varepsilon) \equiv \frac{\varepsilon}{\alpha_2} - \frac{\alpha_1}{\alpha_2}\varepsilon_1(\varepsilon),$$

and the latter term vanishes because of (2.3.14). This is the justification for

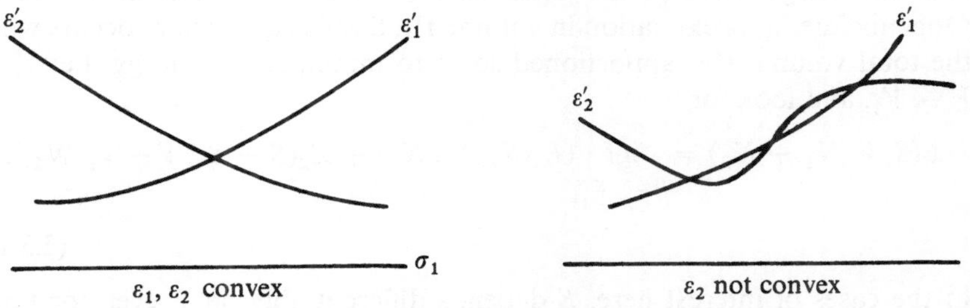


Figure 7 Uniqueness of the equilibrium temperature.

**Definition** (2.3.16)

The **temperature** is

$$T = \frac{\partial \varepsilon}{\partial \sigma}.$$

**Remarks** (2.3.17)

1. The temperature has the dimension of energy in units where Boltzmann's constant $k$ is set to 1.
2. The temperature is always positive with the microcanonical $\rho$ (2.3.1), but $\rho_\delta$ gives the spin system of (2.3.11; 6) a negative temperature at $E > 0$.
3. The concavity of $\sigma$ means that the specific heat at constant volume,

$$V^{-1}C_N \equiv c_V = \frac{d\varepsilon}{dT} = \frac{d\varepsilon}{d\sigma}\left(\frac{dT}{d\sigma}\right)^{-1} = \frac{T}{d^2\varepsilon/d\sigma^2}$$

is positive. In particular, by Remark (2.3.15; 4), the heat capacity (at constant volume) $C_V = V \cdot 1/\varepsilon''$ of the total system is the sum of the heat capacities $V_i \cdot 1/\varepsilon_i''$ of the subsystems. The condition of stability $\varepsilon_1''/\alpha_1 + \varepsilon_2''/\alpha_2 \geq 0$ implies that two systems of negative specific heat can not coexist in equilibrium. Heat transferred from the hotter system to the colder one would make the hot one hotter and the cold one colder. Large temperature fluctuations would arise, making the situation unstable. If only subsystem 1 has negative specific heat, while that of subsystem 2 is positive, then the heat capacities must satisfy $|C_1| > C_2$: The transfer of heat from 1 to 2 would warm subsystem 1 less than 2, so 2 would immediately cool off by transferring heat back to 1, making the temperature equilibrium between the subsystems stable. This means that the temperature of a system of negative specific heat should be taken with a small thermometer, and never with a large thermal reservoir.

Now allow the wall between the subsystems to be slowly moveable. The energy as a function of $V$ acts as a potential energy for the wall, just as the electron energy acted as the potential for the atomic nuclei in the Born–Oppenheimer approximation in volume III. Stable equilibrium occurs when the total volume $V$ is apportioned so as to minimize the energy. Let $V_2 = V - V_1$, and look for

$$E(S, V, N_1 + N_2) = \inf_{\substack{0 \leq S_1 \leq S \\ 0 \leq V_1 \leq V}} (E_1(S_1, V_1, N_1) + E_2(S - S_1, V - V_1, N_2)). \tag{2.3.18}$$

In the cases of interest here, $E$ depends differentiably on $V$ even for finite systems, and $E \to \infty$ if $V \to 0$. Hence the infimum is attained within the interval $0 < V_1 < V$, and is determined by the

**Equilibrium Condition** (2.3.19)

For $E$ of (2.3.18), the equilibrium volume $V_1$ satisfies

$$\frac{\partial E_1}{\partial V_1} = \left.\frac{\partial E_2}{\partial V_2}\right|_{V_2 = V - V_1}.$$

**Remarks** (2.3.20)

1. Because the energy is monotonic (2.3.4), with the boundary conditions $\psi_{|\partial V} = 0$, it follows that $\partial E/\partial V < 0$, and so (2.3.19) definitely has a solution $V_1$. At that minimum,
$$\frac{\partial E}{\partial V} = \frac{\partial E_1}{\partial V_1} = \left.\frac{\partial E_2}{\partial V_2}\right|_{V_2 = V - V_1},$$
and
$$\frac{\partial^2 E_1}{\partial V_1^2} + \frac{\partial^2 E_2}{\partial V_2^2} \geq 0, \qquad \left(\frac{\partial^2 E}{\partial V^2}\right)^{-1} = \left(\frac{\partial^2 E_1}{\partial V_1^2}\right)^{-1} + \left(\frac{\partial^2 E_2}{\partial V_2^2}\right)^{-1}.$$
2. With other boundary conditions it may not be true that $\partial E/\partial V < 0$. For example, if a hydrogen atom is confined to a sphere on the surface of which $d\psi_{|\partial V} = 0$, then $E = E_\infty - \alpha V^{-1/3}$, so $\partial E/\partial V > 0$. This kind of boundary condition can be approximately realized physically with a very strong $\delta'$ potential. The lesson of this is that it is necessary to verify the hope that in infinite systems the pressure (see (2.3.21)) satisfies $P \equiv -\partial E/\partial V \geq 0$. It is not guaranteed that $\partial^2 E/\partial V^2 \geq 0$ even with the boundary condition $\psi_{|\partial V} = 0$, which makes the proof of the convexity of $\varepsilon(\sigma, \rho)$ all the more important for real matter.
3. Since $\partial E/\partial V|_S = -\partial E/\partial S|_V \, \partial S/\partial V|_E$, another interpretation of (2.3.19) is that the condition $\partial(S_1(E_1, V_1) + S_2(E_2, V - V_1))/\partial V_1 = 0$ determines $V_1$; that is, the volumes arrange themselves to maximize the total entropy.

Analogous to (2.3.16) is

**Definition** (2.3.21)

The **pressure** is $P \equiv -\partial E/\partial V$. In the limit $V \to \infty$ it becomes

$$P = -\varepsilon + \rho\frac{\partial \varepsilon}{\partial \rho} + \sigma\frac{\partial \varepsilon}{\partial \sigma} = T\left(\sigma - \varepsilon\frac{\partial \sigma}{\partial \varepsilon} - \rho\frac{\partial \sigma}{\partial \rho}\right).$$

**Remarks** (2.3.22)

1. For realistic systems it can be shown how the pressure defined in (2.3.21) arises from the forces exerted by the system on the wall [9].
2. The equilibrium condition states that the pressures of the two subsystems are equal, with the same value as the total system has.

3. Remark (2.3.20; 1) implies for the compressibility
$$\kappa = -\left[V\frac{\partial P}{\partial V}\right]^{-1} \xrightarrow{V\to\infty} \left[\rho^2\frac{\partial^2\varepsilon}{\partial\rho^2} + 2\rho\sigma\frac{\partial^2\varepsilon}{\partial\rho\partial\sigma} + \sigma^2\frac{\partial^2\varepsilon}{\partial\sigma^2}\right]^{-1}$$
that
$$\kappa = \frac{V_1}{V}\kappa_1 + \frac{V_2}{V}\kappa_2.$$
4. For the systems to be stable against displacements of their interface, their volumes and compressibilities must be related by $(\kappa_1 V_1)^{-1} + (\kappa_2 V_2)^{-1} \geq 0$. For reasons like those of (2.3.17; 3) it is not possible for two systems of negative compressibility to coexist, because the pressure of one system would increase with its volume and force that of the other one down. If only subsystem 1 has negative compressibility, then a necessary condition for stable equilibrium is $V_1 \geq V_2 \cdot \kappa_2/|\kappa_1|$. The increase of pressure in subsystem 1 when it expands is then less than that of 2 when it contracts. If $V_1$ is large enough in comparison with $V_2$, then subsystem 2 undergoes a large relative compression and exerts more pressure back on 1 than 1 exerts on 2. The volumes adjust in the other direction and stable equilibrium is established.

Consider finally what happens to the particle configuration if the subsystems can exchange particles to maximize the entropy. Formally, this means that the Hilbert space is
$$\mathscr{H} = \bigoplus_{N_1=1}^{N} \mathscr{H}_{N_1,V_1} \otimes \mathscr{H}_{N_2,V_2},$$
and the quantity to be calculated is
$$\operatorname{Tr}\Theta(E - H) = \sum_{N_1=0}^{N} \exp(S(N_1))\exp(S(N - N_1)). \tag{2.3.23}$$
In the limit $V \to \infty$, $N \to \infty$, $V_i/V \to \alpha_i$, $N_i/V_i \to \rho_i$, if $S$ is concave in $N$, then arguments like those made earlier yield
$$\sigma(\rho) = \sup_{\alpha_1\rho_1 + \alpha_2\rho_2 = \rho} (\alpha_1\sigma_1(\rho_1) + \alpha_2\sigma_2(\rho_2)). \tag{2.3.24}$$
If the functions $\sigma_i(\rho_i)$ are nice, we obtain the

**Equilibrium Condition** (2.3.25)

Let $\sigma_i(\rho_i)$ be strictly concave and continuously differentiable. Then $\sigma(\rho) = \alpha_1\sigma_1(\rho_1) + \alpha_2\sigma_2(\rho_2)$, where $\rho_i$ are determined uniquely by the conditions
$$\alpha_1\rho_1 + \alpha_2\rho_2 = \rho \quad \text{and} \quad \frac{\partial\sigma_1}{\partial\rho_1} = \frac{\partial\sigma_2}{\partial\rho_2}.$$

**Remarks** (2.3.26)

1. For a given $\varepsilon$ and a given $\rho$, the six variables $\varepsilon_i$, $\rho_i$, $\alpha_i$ satisfy the three equations $\alpha_1\varepsilon_1 + \alpha_2\varepsilon_2 = \varepsilon$, $\alpha_1\rho_1 + \alpha_2\rho_2 = \rho$, $\alpha_1 + \alpha_2 = 1$. The three variations $\delta E$, $\delta V$, and $\delta N$ corresponding to the equilibrium conditions are not independent, because $S(E, V, N)$ is of the special form $V\sigma(E/V, N/V)$, and there is one equation too few to fix six variables. Suppose for simplicity that the two subsystems are identical, $\sigma_1 = \sigma_2 = \sigma$; then because of the concavity, the maximum of $\alpha_1\sigma(\varepsilon_1, \rho_1) + \alpha_2\sigma(\varepsilon_2, \rho_2)$ is assumed when $\varepsilon_1 = \varepsilon_2 = \varepsilon$, $\rho_1 = \rho_2 = \rho$, and $\alpha_1 = 1 - \alpha_2$ is not determined by (2.3.25) and can be specified arbitrarily. Equality of the temperatures and the chemical potentials (see (2.3.27)) suffices to guarantee that the pressures are equal. After the onset of equilibrium, the wall allowing the exchange of energy and particles no longer exerts any force, and can be placed anywhere.
2. It is still possible to minimize the energy instead of maximizing the entropy. But this does not furnish a new stability condition, since if $\partial\varepsilon/\partial\sigma > 0$ the concavity of $(\varepsilon, \rho) \to \sigma(\varepsilon, \rho)$ is equivalent to the convexity of $(\sigma, \rho) \to \varepsilon(\sigma, \rho)$ (Problem 2). Besides $c_V > 0$ and $\kappa > 0$, this requires that

$$\frac{\partial^2 E}{\partial S^2}\frac{\partial^2 E}{\partial V^2} > \left(\frac{\partial^2 E}{\partial S\,\partial V}\right)^2,$$

or, in terms of the adiabatic expansivity

$$\alpha = \frac{1}{V}\frac{\partial V}{\partial T}\bigg|_S, \qquad \alpha^2 > c_V\kappa/T.$$

This amounts physically to the requirement of stability under a simultaneous change in the entropy and volume, related by

$$\delta S \sim \frac{\partial^2 E}{\partial V^2}, \qquad \delta V \sim -\frac{\partial^2 E}{\partial S\,\partial V}.$$

The equilibrium condition (2.3.25) requires the chemical potentials of the subsystems to be equal, if they are defined as with (2.3.26; 2) by minimizing the energy:

**Definition** (2.3.27)

The **chemical potential** is

$$\mu = \frac{\partial \varepsilon}{\partial \rho} = -\frac{\partial \varepsilon}{\partial \sigma}\bigg|_\rho \frac{\partial \sigma}{\partial \rho}\bigg|_\varepsilon.$$

**Remarks** (2.3.28)

1. The intuitive meaning of the temperature is the amount of energy it would take to raise the system from the quantum number $n$ to $en$ ($e = 2.718 \cdots$). Analogously, the chemical potential is the energy increase when a particle is added to the system without changing $V$ or $S$.
2. Although $T$ and $P$ are always positive with the assumptions and boundary conditions that have been postulated, $\mu$ can in general have either sign. Because the density of states increases with $N$, the $e^S$th eigenvalue may decrease with $N$ even if $H \geq 0$.

In phenomenological thermodynamics entropy increases if the energy, volume, or particle number increases, according to the relationship $T\,dS = dE + P\,dV - \mu\,dN$. As we have seen, some of these differentials are well defined only in the thermodynamic limit, and are then considered as intensive properties. For future convenience, we collect the

**Interrelationships among the Thermodynamic Properties** (2.3.29)

$$T = \frac{\partial \varepsilon}{\partial \sigma}, \qquad \mu = \frac{\partial \varepsilon}{\partial \rho} = -T\,\frac{\partial \sigma}{\partial \rho},$$

$$P = -\varepsilon + \sigma\frac{\partial \varepsilon}{\partial \sigma} + \rho\frac{\partial \varepsilon}{\partial \rho} = T\left(\sigma - \varepsilon\frac{\partial \sigma}{\partial \varepsilon} - \rho\frac{\partial \sigma}{\partial \rho}\right),$$

$$c_V = T\left[\frac{\partial^2 \varepsilon}{\partial \sigma^2}\right]^{-1}, \qquad \kappa = \left[\sigma^2\frac{\partial^2 \varepsilon}{\partial \sigma^2} + 2\rho\sigma\,\frac{\partial^2 \varepsilon}{\partial \rho \partial \sigma} + \rho^2\frac{\partial^2 \varepsilon}{\partial \rho^2}\right]^{-1}.$$

**Gloss**

The sense of the partial derivatives is that, of the two variables on which a function has been regarded as depending, the one not written explicitly is to be held fixed. In any doubtful case the fixed argument will be indicated explicitly.

**Remark** (2.3.30)

Without knowledge of the Hamiltonian nothing can be said about the values the thermodynamic functions can assume. In (2.3.11; 6) there was an example in which $\varepsilon(\sigma)$ was even bounded above. If the function $\varepsilon(\sigma)$ is convex and asymptotically linear, then there is a maximum temperature. This is quite possibly the case realized in Nature, and $T_{max} = 140$ MeV. In a model to be investigated shortly (2.3.32; 2), the function $\varepsilon(\sigma)$ has a kink, so $T$ skips over certain values. It depends on the system whether the minimum entropy $\sigma_0$ defined in (2.3.11; 4) equals zero as postulated in the third law of thermodynamics. For instance, with a system consisting of $N$ spins without energy $\otimes$ a system with entropy $N\sigma$, the total entropy divided by $N$ equals $\sigma + \ln 2$, and when $\sigma \to 0$ the total entropy is the $\ln 2$ left over. It is true that the ground state of this system is degenerate, but it is also easy to find examples with nondegenerate ground states for which the third law fails, simply by taking the previous Hamiltonian $\oplus$ a one-dimensional system with a lower energy level. The resulting ground state is simple, but that has no effect on what happens as $N \to \infty$.

It has been seen that the concavity of the function $\sigma(\varepsilon, \rho)$ is at the root of thermodynamic stability. Concavity is jeopardized when $\sigma$ is maximized with respect to all of its parameters—the supremum of a set of concave functions is not necessarily concave, in contrast to the infimum. However, there is a useful

**Lemma on the Envelope of a Set of Concave Functions** (2.3.31)

*If $\sigma(\varepsilon, \alpha)$ is jointly concave in $\varepsilon$ and $\alpha$, then $\bar{\sigma}(\varepsilon) = \sup_\alpha \sigma(\varepsilon, \alpha)$ is concave in $\varepsilon$.*

**Picture of the Proof**

Think of the silhouette of a concave mountain slope and of a mountain with hollows.

**Formal Proof if $\sigma(\epsilon, \alpha) \in C^2(K)$**

With this assumption, the maximum is attained at a point $\alpha(\varepsilon)$, $\bar{\sigma}(\varepsilon) = \sigma(\varepsilon, \alpha(\varepsilon))$, and

$$\sigma_{,\alpha}(\varepsilon, \alpha(\varepsilon)) = 0 \Rightarrow \sigma_{,\alpha\varepsilon} + \frac{d\alpha(\varepsilon)}{d\varepsilon}\sigma_{,\alpha\alpha} = 0.$$

Then

$$\frac{d^2\bar{\sigma}}{d\varepsilon^2} = \sigma_{,\varepsilon\varepsilon} + \sigma_{,\varepsilon\alpha}\frac{d\alpha(\varepsilon)}{d\varepsilon} = \frac{\sigma_{,\varepsilon\varepsilon}\sigma_{,\alpha\alpha} - (\sigma_{,\varepsilon\alpha})^2}{\sigma_{,\alpha\alpha}}.$$

Since $\sigma_{,\alpha\alpha} \le 0$ and $\sigma_{,\varepsilon\varepsilon}\sigma_{,\alpha\alpha} - (\sigma_{,\varepsilon\alpha})^2 \ge 0$, $d^2\bar{\sigma}/d\varepsilon^2 \le 0$. If $\sigma_{,\alpha\alpha} = 0$, it follows that $\sigma_{,\alpha\varepsilon}(\varepsilon, \alpha(\varepsilon)) = 0$, and therefore $\bar{\sigma}_{,\varepsilon\varepsilon} = \sigma_{,\varepsilon\varepsilon} \le 0$. (For the proof without the assumption that $\sigma(\varepsilon, \alpha) \in C^2$, see Problem 3.) □

If the entropy is maximized with respect to parameters in the absence of joint concavity, then thermodynamic stability may be lost, and it will be necessary to reconsider the foregoing assumptions.

**Examples** (2.3.32)

1. Model of a star
   Consider $N$ classical particles in a container $V$ and attracting each other pairwise only within some $V_0 \subset V$. Suppose the potentials are constant in $V_0$ and $\sim N^{-1}$, to ensure that $E$ be extensive.

$$H_N = \sum_{i=1}^{N} |\mathbf{p}_i|^2 - \frac{1}{N} \sum_{i,j=1}^{N} \chi_{V_0}(\mathbf{x}_i)\chi_{V_0}(\mathbf{x}_j),$$

$$\chi_{V_0}(\mathbf{x}) = \begin{cases} 1 & \text{for } \mathbf{x} \in V_0 \\ 0 & \text{otherwise.} \end{cases}$$

With indistinguishable particles, the volume of phase space below the energy surface,

$$\begin{aligned} &\exp(S(E, V, N)) \\ &= \frac{1}{N!} \int d^{3N}p\, d^{3N}x \Theta\left(E - \sum_{i=1}^{N} |\mathbf{p}_i|^2 + \frac{1}{N} \sum_{i,j=1}^{N} \chi_{V_0}(\mathbf{x}_i)\chi_{V_0}(\mathbf{x}_j)\right) \\ &= \frac{\pi^{3N/2}}{N!(3N/2)!} \int_{(\cdots)>0} d^{3N}x \left(E + \frac{1}{N} \sum_{i,j=1}^{N} \chi_{V_0}(\mathbf{x}_i)\chi_{V_0}(\mathbf{x}_j)\right)^{3N/2}, \end{aligned}$$

can be calculated exactly, because the integrand is piecewise constant. Let $N_0$ be the number of the $\mathbf{x}_i$ in $V_0$. Then

$$\begin{aligned} \exp(S) &= \frac{V_0^N \pi^{3N/2}}{(3N/2)!} \sum_{-NE \le N_0^2 \le N^2} \left(\frac{V}{V_0} - 1\right)^{N-N_0} \frac{(E + N_0^2/N)^{3N/2}}{N_0!(N - N_0)!} \\ &\equiv \sum_{N_0=1}^{N} \exp(S(E, V, N; N_0)). \end{aligned}$$

Only the dependence on $E$ matters, so let $E = \varepsilon \cdot N$, $\rho = N/V = 1$, $N_0/N \equiv \alpha$, $(\max(0, -\varepsilon))^{1/2} \le \alpha \le 1$. Then it remains to evaluate

$$\sigma(\varepsilon) = \sup_{\alpha} \lim_{N\to\infty} \frac{1}{N} S(N\varepsilon, N, N; \alpha N) \equiv \sup_{\alpha} \sigma(\varepsilon, \alpha),$$

and with the help of Stirling's formula,

$$\begin{aligned} \sigma(\varepsilon, \alpha) &= \tfrac{3}{2} \ln(\varepsilon + \alpha^2) - \alpha \ln \alpha - (1 - \alpha)\ln(1 - \alpha) + F(1 - \alpha) + \text{constant}, \\ F &= \ln\left(\frac{V}{V_0} - 1\right). \end{aligned} \tag{2.3.33}$$

A calculation of the derivatives yields

$$\sigma_{,\varepsilon} = \frac{\frac{3}{2}}{\varepsilon + \alpha^2}, \qquad \sigma_{,\alpha} = \frac{3\alpha}{\varepsilon + \alpha^2} + \ln\left(\frac{1}{\alpha} - 1\right) - F,$$

$$\sigma_{,\varepsilon\varepsilon} = \frac{-\frac{3}{2}}{(\varepsilon + \alpha^2)^2}, \qquad \sigma_{,\varepsilon\alpha} = \frac{-3\alpha}{(\varepsilon + \alpha^2)^2}, \qquad \sigma_{,\alpha\alpha} = \frac{3\varepsilon - 3\alpha^2}{(\varepsilon + \alpha^2)^2} - \frac{1}{\alpha(1 - \alpha)}.$$

The maximum is achieved on the curve

$$\varepsilon(\alpha) = -\alpha^2 + \frac{3\alpha}{F - \ln(1/\alpha - 1)},$$

and the ranges of values of the variables are such that $\varepsilon + \alpha^2 \geq 0$, so only the branch of $F > \ln(1/\alpha - 1)$ comes into consideration. Because

$$\sigma_{,\alpha\alpha} = -\frac{(\varepsilon - \varepsilon_1(\alpha))(\varepsilon - \varepsilon_2(\alpha))}{(\varepsilon + \alpha^2)^2\alpha(1 - \alpha)},$$

$$\varepsilon_{1,2} = \frac{3\alpha}{2}\left(1 - \tfrac{5}{3}\alpha \pm \sqrt{1 - \alpha}\sqrt{1 - 11\alpha/3}\right),$$

$\sigma(\varepsilon, \alpha)$ is concave in $\alpha$ except when $\varepsilon_2 < \varepsilon < \varepsilon_1$. The sign of $d\varepsilon/d\alpha = -\sigma_{,\alpha\alpha}/\sigma_{,\alpha\varepsilon}$ changes in the interval $\varepsilon_2 < \varepsilon < \varepsilon_1$, so three values of $\alpha$ belong to a single $\varepsilon$, and the maximum needed is the greater of the two. Joint concavity requires that

$$\sigma_{,\varepsilon\varepsilon}\sigma_{,\alpha\alpha} - (\sigma_{,\alpha\varepsilon})^2 = \frac{3(\varepsilon - 3\alpha + 4\alpha^2)}{2(\varepsilon + \alpha^2)^3\alpha(1 - \alpha)} \geq 0$$

and implies $\varepsilon \geq 3\alpha - 4\alpha^2$. If $\varepsilon(\alpha)$ lies in this range of values, then the system has positive specific heat, and otherwise not (see Figure 8). Indeed,

$$\tfrac{3}{2}T = \varepsilon + \alpha^2 = \frac{3\alpha}{F - \ln(1/\alpha - 1)}$$

behaves as a function of $\varepsilon$ as shown in Figure 9. The physical significance is that if energy is removed, the temperature falls until a certain fraction of the particles reside in $V_0$, which causes the system to start heating back up. If most of the particles are eventually in $V_0$, then they behave normally again. The system can be thought of as a normal system with

$$\sigma(\varepsilon, \rho) = \rho(\tfrac{3}{2}\ln\varepsilon - \tfrac{5}{2}\ln\rho)$$

put into contact with a peculiar system with

$$\sigma(\varepsilon, \rho) = \rho(\tfrac{3}{2}\ln(\varepsilon + \rho^2) - \tfrac{5}{2}\ln\rho) - F\rho.$$

If the energy is apportioned between them according to

$$\sigma(\varepsilon, \alpha) = \sup_{\varepsilon_1}(\tfrac{3}{2}(\alpha\ln(\varepsilon_1 + \alpha^2) + (1 - \alpha)\ln(\varepsilon - \varepsilon_1)) - \alpha F$$
$$- \tfrac{5}{2}(\alpha\ln\alpha + (1 - \alpha)\ln(1 - \alpha))),$$

then the entropy becomes exactly that of (2.3.33).

2. Model of a Ferromagnet
This problem is quantum-mechanical, but its analysis soon begins to resemble that of Example 1, for which reason we shall boldly plunge on

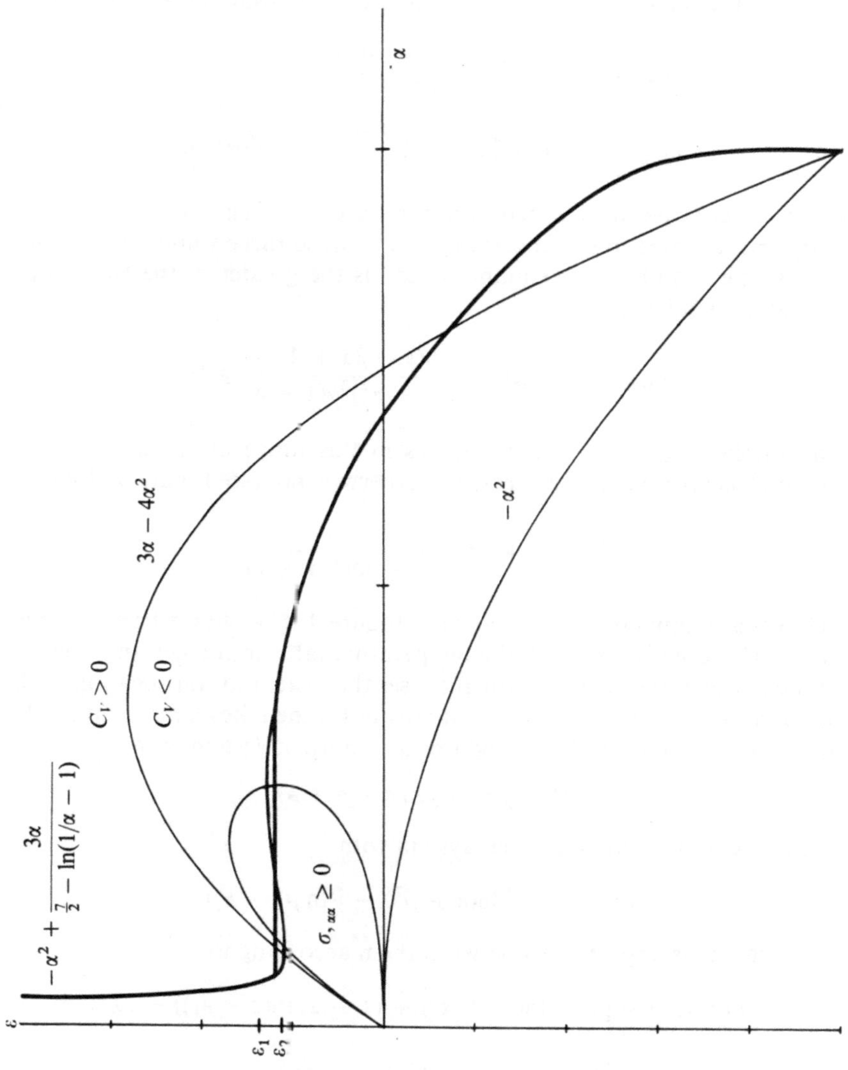


Figure 8 $\varepsilon(\alpha)$ in Example (2.3.32; 1).

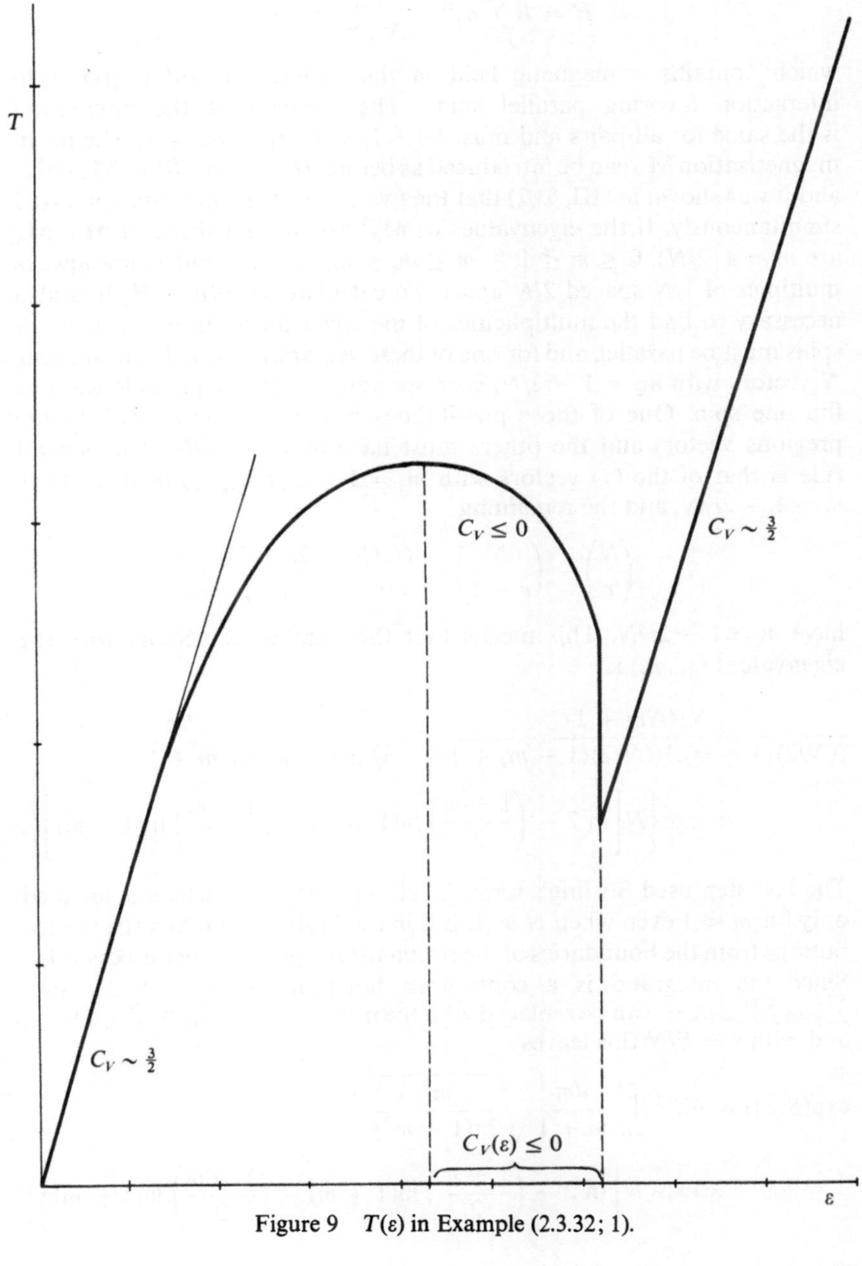


Figure 9 $T(\varepsilon)$ in Example (2.3.32; 1).

to the estimates without wasting time about epsilontic details. The Hamiltonian of (1.1.3) is modified to

$$H = B\sum_{j=1}^{N} \sigma_j^{(z)} - \frac{1}{N}\sum_{i,j=1}^{N} \sigma_i \cdot \sigma_j,$$

which contains a magnetic field in the $z$-direction and a spin–spin interaction favoring parallel spins. The strength of the interaction is the same for all pairs and must be $\sim 1/N$ for $H$ to be $\sim N$. The mean magnetization $\mathbf{M}_N$ can be introduced as before, $H/N = BM_N^{(z)} - \mathbf{M}_N \cdot \mathbf{M}_N$, and it was shown in (III, §3.2) that the two parts of $H$ can be diagonalized simultaneously. If the eigenvalues of $M_N^{(z)}$ are $m_z$ and those of $\mathbf{M}_N \cdot \mathbf{M}_N$ are $m(m + 2/N)$, $0 \leq m \leq 1$, $-m \leq m_z \leq m$, then $m_z$ and $m$ are always multiples of $1/N$ spaced $2/N$ apart. To calculate Tr $\Theta(E - H)$ it is also necessary to find the multiplicities of the eigenvalues: If $m = 1$, then all spins must be parallel, and for one of these vectors, $m_z = 1$. There are now $N$ vectors with $m_z = 1 - 2/N$, corresponding to the $N$ possible ways to flip one spin. One of those possibilities has $m = 1$ (apply $M^-$ to the previous vector) and the others must have $m = 1 - 2/N$. The general rule is that of the $\binom{N}{r}$ vectors with $m_z = 1 - 2r/N$, $\binom{N}{r-1}$ of them have $m > 1 - 2r/N$, and the remaining

$$\binom{N}{r} - \binom{N}{r-1} = \frac{N!(N - 2r + 1)}{r!(N - r + 1)!}$$

have $m = 1 - 2r/N$. This means that the number of vectors with the eigenvalues $(m, m_z)$ is

$$\frac{N!(Nm + 1)}{((N/2)(1 - m))!((N/2)(1 + m) + 1)!} \sim \sqrt{\frac{2}{\pi(1 - m^2)N}}\frac{2m}{m + 1}$$
$$\times \exp\left\{N\left[\ln 2 - \left(\frac{1 + m}{2}\right)\ln(1 + m) - \left(\frac{1 - m}{2}\right)\ln(1 - m)\right]\right\}.$$

The last step used Stirling's formula $x! \sim (x/e)^x\sqrt{2\pi x}$, which is justified only for $m < 1$ even when $N \gg 1$, but in the limit being taken the contributions from the boundaries of the summation region are inconsequential. Since the integrand is a continuous function, as $N \to \infty$ the sum $\sum_{m=0}^{1}\sum_{m_z=-m}^{m}\cdots$ can be replaced with the integral $(N/2)^2 \int_0^1 dm \int_{-m}^{m} dm_z \cdots$, and with $\varepsilon = E/N$ this leaves

$$\exp(S(\varepsilon)) = N^{3/2}\int_0^1 \frac{dm}{m + 1}\sqrt{\frac{m^2}{2\pi(1 - m^2)}}$$
$$\times \exp\left\{N\left[\ln 2 - \left(\frac{1 + m}{2}\right)\ln(1 + m) - \left(\frac{1 - m}{2}\right)\ln(1 - m)\right]\right\}$$
$$\times \int_{-m}^{m} dm_z \Theta(\varepsilon + m^2 - Bm_z). \tag{2.3.34}$$

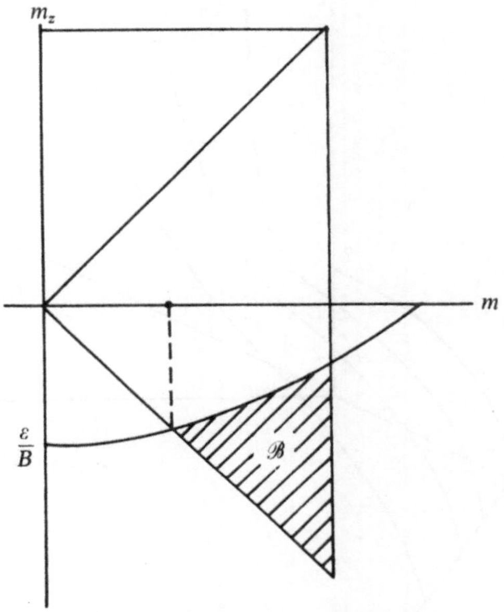


Figure 10 The region of integration in the $m - m_z$-plane.

Therefore the domain $\mathscr{B}$ of integration is $\{(m, m_z): 0 \leq m \leq 1, -m \leq m_z \leq m\} \cap \{(m, m_z): m_z \leq (\varepsilon + m^2)/B\}$. The entropy $S$ is obviously even in $B$, so we may restrict consideration to $B \geq 0$ (see Figure 10).

Since the exponential function decreases rapidly with $m$, the appropriate generalization of Lemma (2.3.12) makes $\sigma = \lim_{N\to\infty} S/N$ sensitive only to $m_0 \equiv \inf_{m, m_z \in \mathscr{B}} m$ (the exponent in (2.3.34) decreases monotonically in $m$):

$$m_0 = \Theta(-\varepsilon)\left(\sqrt{\frac{B^2}{4} - \varepsilon} - \frac{B}{2}\right),$$
$$\sigma = \ln 2 - \tfrac{1}{2}\Theta(-\varepsilon)[(1 + m_0)\ln(1 + m_0) + (1 - m_0)\ln(1 - m_0)], \tag{2.3.35}$$

if $\varepsilon \geq -1 - B$, and is otherwise 0. Since $\sigma$ is concave but decreasing in $m_0$, the concavity in $\varepsilon$ remains to be verified:

$$T^{-1} = \frac{d\sigma}{d\varepsilon} = \frac{\Theta(-\varepsilon)}{4(B^2/4 - \varepsilon)^{1/2}} \ln\frac{1 + m_0}{1 - m_0} \geq 0,$$
$$-\frac{1}{T^2 c} = \frac{d^2\sigma}{d\varepsilon^2} = -\frac{\Theta(-\varepsilon)}{8}$$
$$\times\left[-\left(\frac{B^2}{4} - \varepsilon\right)^{-3/2} \ln\frac{1 + m_0}{1 - m_0} + \frac{2}{(B^2/4 - \varepsilon)(1 - m_0^2)}\right]. \tag{2.3.36}$$

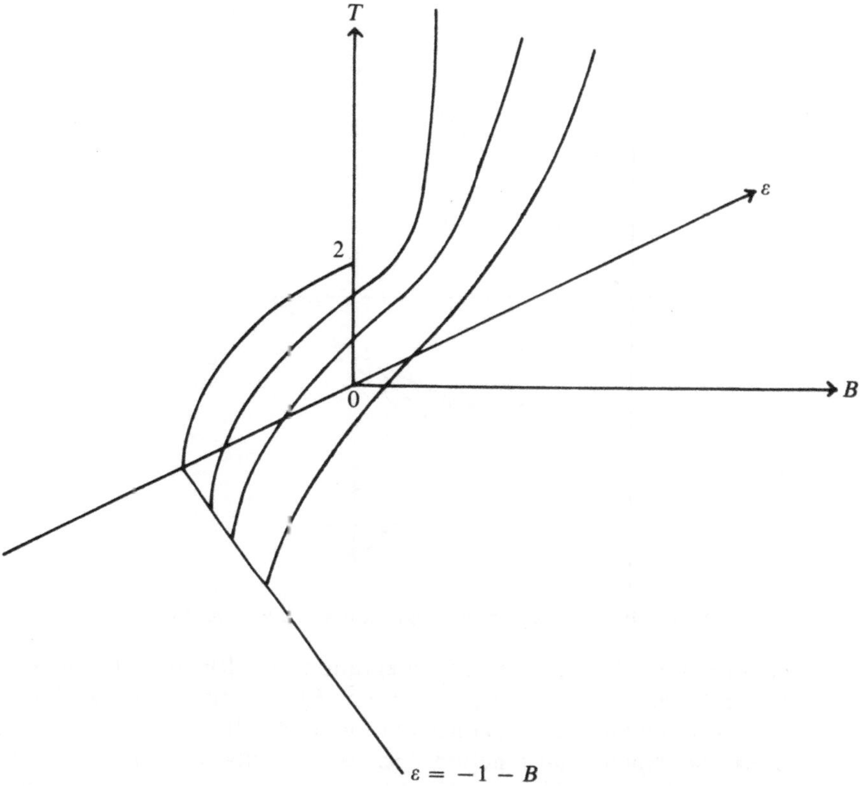


Figure 11 The surface of states in $T - \varepsilon - B$-space.

In a lucky break, the positive term in the brackets $[\cdots]$ is always greater than the negative one, and $c_V$ is always positive. If $-1 > -1 - B \leq \varepsilon \leq 0$, then $T$ increases continuously from 0 to $\infty$. The heat capacity $c_V$ increases from 0 to a maximum value and then falls back to 0. If $B = 0$, then $T$ reaches the value 2 for $\varepsilon = 0$, at which $c_V$ has risen to $\frac{3}{2}$. Afterwards, $T$ jumps up to $\infty$ and $c_V$ falls back to 0 (see Figure 11).

Thus if $B = 0$ and $T < 2$, the thermal motion is no longer strong enough to counter the ordering tendency of $H$, and a spontaneous magnetization $m_0$ appears. As no direction is preferred, the thermal expectation value $|\mathrm{Tr}\,\rho\mathbf{M}|$ remains 0. We shall learn later that as $N \to \infty$, the GNS representations of the $\sigma$'s constructed with $\rho$ become integrals over all directions of thermal representations (1.4.7). If $B > 0$, then $\mathrm{Tr}\,\rho\mathbf{M}$ points in the $z$-direction, and $m_0$ grows smoothly from 0 to 1 as $T$ decreases.

The interactions in these examples could have been replaced with average fields. This is typical of forces of long range like gravity. If the long-range forces neutralize each other—for instance if they are electric—then the system is basically the sum of its parts, i.e., it can be decomposed into parts in such a way that the entropy, energy, volume, and particle number are all additive. In that case the maximum entropy is concave.

**Thermodynamic Stability of Decomposable Systems** (2.3.37)

*For an arbitrary function* $\sigma$,

$$\bar{\sigma}(\varepsilon, \rho) \equiv \sup_n \sup_{K_n} \sum_{i=1}^{n} \alpha_i \sigma(\varepsilon_i, \rho_i),$$

*where*

$$K_n = \left\{ (\alpha_i), (\varepsilon_i), (\rho_i) \,\middle|\, \sum_{i=1}^{n} \alpha_i = 1, \sum_{i=1}^{n} \alpha_i \varepsilon_i = \varepsilon, \sum_{i=1}^{n} \alpha_i \rho_i = \rho \right\},$$

*is jointly concave in its two variables.*

**Proof**

Let $\varepsilon = \gamma\varepsilon' + (1 - \gamma)\varepsilon''$, $\rho = \gamma\rho' + (1 - \gamma)\rho''$. Divide $(\alpha_i)$ into $(\alpha_i')$ and $(\alpha_i'')$, and take the supremum over $K_{n'}$ and $K_{n''}$:

$$K_{n'} \equiv \left\{ (\alpha_i'), (\varepsilon_i'), (\rho_i') \,\middle|\, \sum_{i=1}^{n'} \alpha_i' = \gamma, \sum_{i=1}^{n'} \alpha_i' \varepsilon_i' = \varepsilon', \sum_{i=1}^{n'} \alpha_i' \rho_i' = \rho' \right\},$$

$$K_{n''} \equiv \left\{ (\alpha_i''), (\varepsilon_i''), (\rho_i'') \,\middle|\, \sum_{i=1}^{n''} \alpha_i'' = 1 - \gamma, \sum_{i=1}^{n''} \alpha_i'' \varepsilon_i'' = \varepsilon'', \sum_{i=1}^{n''} \alpha_i'' \rho_i'' = \rho'' \right\}.$$

Since this is only a particular division,

$$\begin{aligned} \bar{\sigma}(\varepsilon, \rho) &\geq \sup_{n', n''} \sup_{K_{n'}, K_{n''}} \left( \sum_i \alpha_i' \sigma(\varepsilon_i', \rho_i') + \sum_i \alpha_i'' \sigma(\varepsilon_i'', \rho_i'') \right) \\ &= \gamma \bar{\sigma}(\varepsilon', \rho') + (1 - \gamma)\bar{\sigma}(\varepsilon'', \rho''). \end{aligned}$$

□

**Remark** (2.3.38)

The construction (2.3.37) gives the concave envelope of $\sigma$, but nothing guarantees that $\bar{\sigma}$ is strictly concave. If $\sigma$ is linear, then $\bar{\sigma} = \sigma$, and $\sigma$ is of the form of Example (2.3.32; 1). The convex part of the curve gets bridged by a straight line, as shown in Figure 12.

The function $\bar{\sigma}$ is simply $\alpha\sigma(\varepsilon_1) + (1 - \alpha)\sigma(\varepsilon_2)$ in the intervening region where $\varepsilon = \alpha\varepsilon_1 + (1 - \alpha)\varepsilon_2$ for fixed $\varepsilon_1$ and $\varepsilon_2$. An interpretation is that the system consists of two phases in this region, having energies $\varepsilon_1$ and $\varepsilon_2$, and the temperature remains constant as the total energy varies, while the proportions of the phases present change. This suggests a

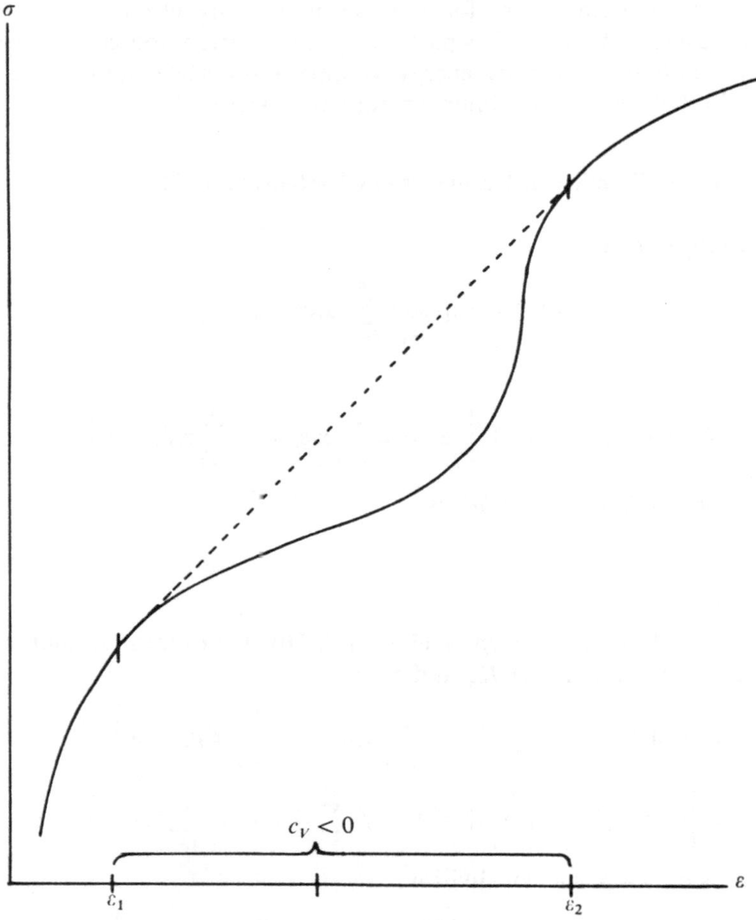


Figure 12 The region of negative specific heat.

**Rough Definition of the Thermodynamic Phases** (2.3.39)

The extreme points of the concave function $\sigma(\varepsilon, \rho)$ correspond to pure phases, and in the regions of coexistence of more than one phase the function $\sigma$ is not strictly concave.

**Examples** (2.3.40)

1. If the graph of $\sigma(\varepsilon, \rho)$ shows a belt-like region the curvature of which vanishes in only one direction, then two phases coexist in its interior. The sides of the belt correspond to pure phases and the end to a critical point (see Figure 13):

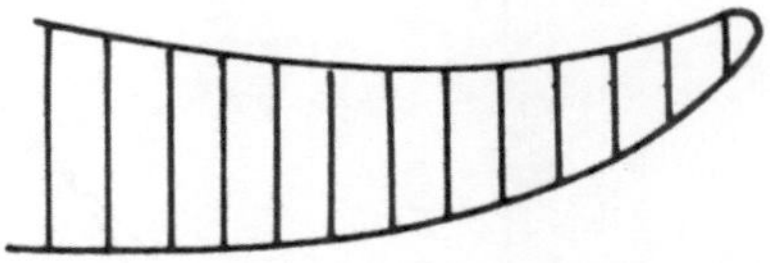

Figure 13 The region of coexistence of two phases.

2. In the usual solid–liquid–gas phase diagram, the triple point occurs in a region at which the curvature of $\sigma(\varepsilon, \rho)$ vanishes in both directions (Figure 14):

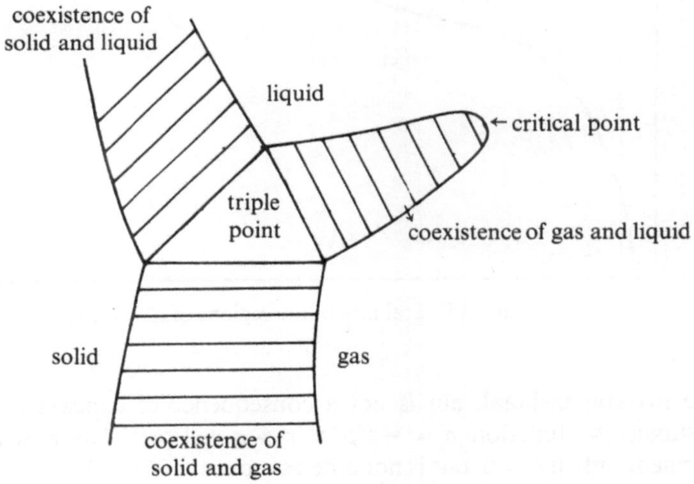


Figure 14 Regions of coexistence.

**Remarks** (2.3.41)

1. The sum, in the sense of (2.3.37), of many copies of Example (2.3.32; 1) produces a concave $\bar{\sigma}$, since the convex part lies below the phase-transition line. Some concave pieces of the curve are also bridged over, and are known as metastable phases, which arise in superheated stars and supercooled gases. They have positive specific heats and are locally stable (see Figure 15):
2. Gibbs's phenomenological phase rule states that whenever a material has two coexisting phases, there is always a one-parameter family of coexisting phases described by $T(\alpha)$ and $\mu(\alpha)$. Three coexisting phases can only exist at discrete values of $(T, \mu)$. This is exactly what went on in (2.3.40; 1) and (2.3.40; 2), where the parts that are flat in one direction

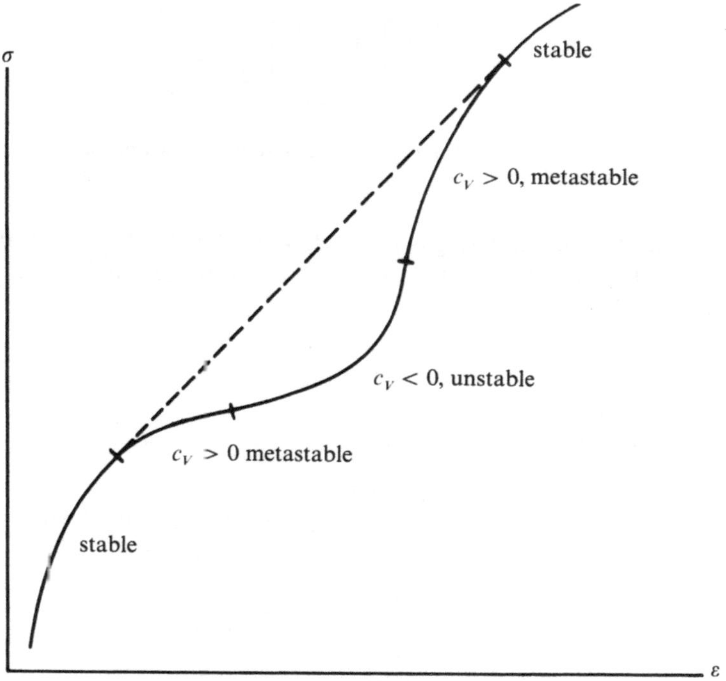


Figure 15 Stability of the regions of (2.3.32; 1).

are two-dimensional, but is not a consequence of concavity alone; for instance the function $\sigma = -\varepsilon^p \rho^{-q}$, $p > q + 1 > 1$, has a straight line segment only if $\varepsilon = 0$, but is nonetheless concave in $(\varepsilon, \rho)$.

3. A quadruple point of a substance would be a flat rectangle in the energy surface. The nonexistence of quadruple points does not follow from concavity, but amounts to the assumption that the flat pieces of the energy surface form a simplex. If they do not form a simplex, then the ratio of the phases in the mixture is not even necessarily determined by $\varepsilon$ and $\rho$:

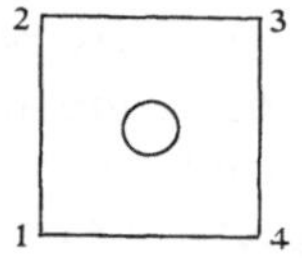


$$\bigcirc = \tfrac{1}{2}(1 + 3) = \tfrac{1}{2}(2 + 4).$$

At this point we have no arguments that would show that quadruple points do not occur, and in fact it is easy to construct models with quadruple points by taking the sum of two independent systems each of which has a phase transition. We shall have to take the issue up anew in (3.2.12; 2).

**Problems** (2.3.42)

1. Show that if $\sigma(\varepsilon, \rho)$ is concave, then $(E, N, V) \to S(E, N, V)$ is concave.
2. Show that for $\varepsilon_{,\sigma} > 0$, $\sigma(\varepsilon, \rho)$ is concave iff $\varepsilon(\sigma, \rho)$ is convex.
3. Without assuming differentiability, show that if $\sigma(\varepsilon, \alpha)$ is concave, then $\bar{\sigma}(\varepsilon) = \sup_\alpha \sigma(\varepsilon, \alpha)$ is concave.
4. Prove the relationship $V/\varepsilon'' = V_1/\varepsilon_1'' + V_2/\varepsilon_2''$ of (2.3.15; 4).

**Solutions** (2.3.43)

1. For simplicity assume that $\sigma$ is twice differentiable. Then

$$D^2S = \frac{1}{V}\begin{vmatrix} \sigma_{,\varepsilon\varepsilon} & \sigma_{,\varepsilon\rho} & -\varepsilon\sigma_{,\varepsilon\varepsilon} - \rho\sigma_{,\varepsilon\rho} \\ \sigma_{,\varepsilon\rho} & \sigma_{,\rho\rho} & -\varepsilon\sigma_{,\varepsilon\rho} - \rho\sigma_{,\rho\rho} \\ -\varepsilon\sigma_{,\varepsilon\varepsilon} - \rho\sigma_{,\varepsilon\rho} & -\varepsilon\sigma_{,\varepsilon\rho} - \rho\sigma_{,\rho\rho} & \varepsilon^2\rho_{,\varepsilon\varepsilon} + 2\varepsilon\rho\sigma_{,\varepsilon\rho} + \rho^2\sigma_{,\rho\rho} \end{vmatrix}.$$

Observe that the concavity of $S$ is equivalent to $D^2S \le 0$, which means that $D^2\sigma \le 0$ and $\det D^2S \le 0$. However, $D^2S = 0$ because the mapping $\lambda \to S(\lambda E, \lambda N, \lambda V)$ is affine.

2. The function $\sigma$ is concave iff the concave hull $\bar{\Gamma} = \{(x, y, z) = \sum \lambda_i(x_i, y_i, z_i), (x_i, y_i, z_i) \in \Gamma, 0 \le \lambda_i \le 1, \sum_i \lambda_i = 1\}$ of the graph $\Gamma = \{(x, \varepsilon, \rho): x = \sigma(\varepsilon, \rho)\}$ lies completely below $\Gamma$. However, looked at from the other side, $\Gamma$ is also the graph of the inverse function $\varepsilon(\sigma, \rho)$, except that "below" becomes "above" and vice versa.

3. Let $\varepsilon = \gamma\varepsilon_1 + (1 - \gamma)\varepsilon_2$, and choose $\alpha_{1,2}$ so that $\sup_\alpha \sigma(\varepsilon_i, \alpha) = \sigma(\varepsilon_i, \alpha_j)$, $i = 1, 2$, or at least comes arbitrarily close to equality.

$$\begin{aligned} \sup_\alpha \sigma(\varepsilon, \alpha) &\ge \sigma(\gamma\varepsilon_1 + (1 - \gamma)\varepsilon_2, \gamma\alpha_1 + (1 - \gamma)\alpha_2) \\ &\ge \gamma\sigma(\varepsilon_1, \alpha_1) + (1 - \gamma)\sigma(\varepsilon_2, \alpha_2) \\ &= \gamma\bar{\sigma}(\varepsilon_1) + (1 - \gamma)\bar{\sigma}(\varepsilon_2). \end{aligned}$$

4. $$\varepsilon_1'(\sigma_1) = \varepsilon_2'\left(\frac{V\sigma - V_1\sigma_1}{V_2}\right) \Rightarrow \sigma_1'\varepsilon_1'' = \left(\frac{V}{V_2} - \frac{V_1}{V_2}\sigma_1'\right)\varepsilon_2'' \Rightarrow \sigma_1' = \frac{V}{V_2}\frac{\varepsilon_2''}{\varepsilon_1'' + \varepsilon_2'' V_1/V_2},$$

$$\varepsilon'(\sigma) = \varepsilon_1'(\sigma_1) \Rightarrow \varepsilon'' = \sigma_1'\varepsilon_1'' = \frac{V}{V_2}\varepsilon_2''\frac{\varepsilon_1''}{\varepsilon_1'' + \varepsilon_2'' V_1/V_2} \Rightarrow \frac{V}{\varepsilon''} = \frac{V_1}{\varepsilon_1''} + \frac{V_2}{\varepsilon_2''}.$$

## 2.4 The Canonical Ensemble

*The Maxwell–Boltzmann distribution arises from the state of a system in contact with a thermal reservoir. If the system is large, this state is indistinguishable from that of the microcanonical ensemble.*

In the preceding section it was shown that the entropy of two large subsystems without interaction is additive. The entropy was always defined with the microcanonical density matrix (2.3.1), but when the density matrix is restricted to a subsystem,

$$\rho_1 = \frac{\operatorname{Tr}_2 \Theta(E - H_1 - H_2)}{\operatorname{Tr} \Theta(E - H_1 - H_2)} \equiv \exp(S(E - H_1))/\operatorname{Tr}_1 \exp(S(E - H_1)), \tag{2.4.1}$$

it appears quite different. It will now be shown that $\rho_1$ does not depend on the nature of the second system if it is infinitely large (a thermal reservoir). We shall also find out that this so-called canonical density matrix is equivalent to the microcanonical density matrix if the system is large. The convergence of $\rho_1$ as the second subsystem becomes infinitely large is described by

**Lemma** (2.4.2)

*Suppose that the concave, increasing functions* $(1/V)S(E) \equiv \sigma_V(E/V)$ *and their derivatives converge uniformly on some neighborhood of* $\varepsilon = E/V$ *to a function* $\sigma(\varepsilon) \in C^1$ *and to* $\sigma'(\varepsilon)$. *Then as* $V \to \infty$,

$$\rho_V \equiv \frac{\exp[V\sigma_V((E - H_1)/V)]}{\operatorname{Tr} \exp[V\sigma_V((E - H_1)/V)]} \to \frac{\exp(-H_1\sigma'(\varepsilon))}{\operatorname{Tr} \exp(-H_1\sigma'(\varepsilon))}$$

*in the trace norm, provided that* $\exp(-H_1\sigma'(\varepsilon))$ *is of the trace class* $\mathscr{C}_1$.

**Remarks** (2.4.3)

1. As in (2.3.13; 2), $E$ and $E_m$ can be identified.
2. *A priori*, $S(E)$ has been defined only for discrete values. We assume that it can be interpolated with a concave, strictly increasing, continuously differentiable function
3. The facts $\sigma_{\text{ess}}(H) = \varnothing$ and $H \geq 0$ do not suffice to make $\exp(-\beta H) \in \mathscr{C}_1$; $\operatorname{Sp}(H)$ could be $\mathbb{Z}^+$ and the eigenvalues $n \in \mathbb{Z}^+$ could have multiplicity $n^n$. More assumptions are needed than (2.3.3; 2).
4. The significance of the lemma is that temperature is the only property of a reservoir in the infinitely large limit that enters into the reduced density matrix. The reduced density matrix has the canonical form regardless of the structure of the reservoir, when the energy of interaction can be neglected.

**Proof of (2.4.2)**

With $\operatorname{Tr}_1 \Theta(E_1 - H_1) = \exp(S_1(E_1))$, $\operatorname{Tr} \Theta(E - H_1 - H_2) = \int dE_1 \exp(S(E - E_1) + S_1(E_1))S_1'(E_1)$, $\rho_V$ can be written as

$$\rho_V = \frac{\exp\{V[\sigma_V(\varepsilon - (H_1/V)) - \sigma_V(\varepsilon)]\}}{\int dE_1 \exp\{S_1(E_1) + \ln S_1'(E_1) + V[\sigma_V(\varepsilon - (E_1/V)) - \sigma_V(\varepsilon)]\}}.$$

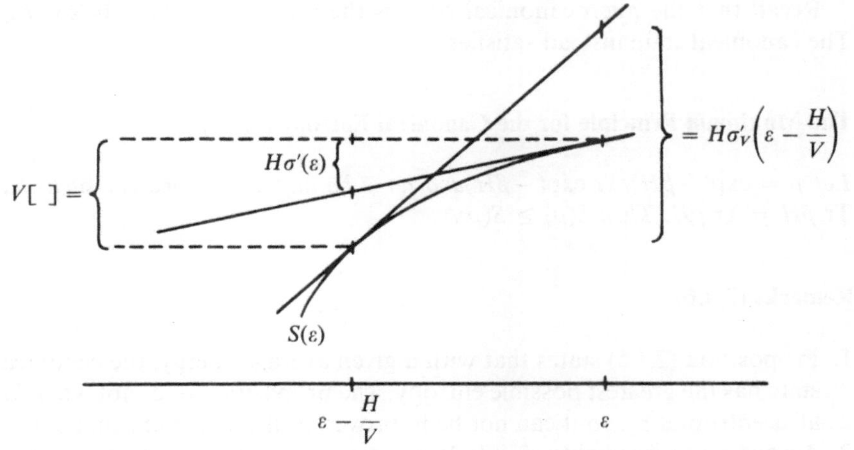


Figure 16 Estimating the slope of $S(\varepsilon)$.

Because of concavity, if $H_1 \geq 0$, then $H_1\sigma'_V(\varepsilon) \leq V[\sigma_V(\varepsilon) - \sigma_V(\varepsilon - (H_1/V)]$ $\leq H_1\sigma'_V(\varepsilon - (H_1/V))$ (see Figure 16).

The assumption that $\sigma'$ converges uniformly then makes $V[\sigma_V(\varepsilon - (H_1/V)) - \sigma_V(\varepsilon)]$ converge uniformly to $-H_1\sigma'(\varepsilon)$ on compact sets in $\mathrm{Sp}(H_1)$. Moreover, there exist $V'$ and $\beta$ such that for all $V > V'$, there is an operator inequality, $\exp[V(\sigma_V(\varepsilon - (H_1/V)) - \sigma_V(\varepsilon))]$ $\leq \exp(-\beta H_1)$. In the spectral representation of $H_1$, $\exp[V(\sigma_V(\varepsilon - (H_1/V)$ $- \sigma_V(\varepsilon))] \to \exp(-H_1\sigma'(\varepsilon))$ in the strong topology, by the Lebesgue dominated convergence theorem. If the operator on the right belongs to $\mathscr{C}_1$, then by the dominated convergence theorem again,

$$\begin{aligned}\mathrm{Tr}\exp[-H_1\sigma'(\varepsilon)] &= \int dE_1 \exp[S_1(E_1) + \ln S'_1(E_1) - E_1\sigma'(\varepsilon)] \\ &= \lim_{V\to\infty} \int dE_1 \exp\Big\{S_1(E_1) + \ln S'_1(E_1) \\ &\qquad + V\left[\sigma_V\left(\varepsilon - \frac{E_1}{V}\right) - \sigma_V(\varepsilon)\right]\Big\}.\end{aligned}$$

The proof is completed by appealing to the theorem (Problem 1) that strong convergence of density matrices implies convergence in the trace norm. □

**Corollaries** (2.4.4)

1. Since $\rho_V$ converges in the sense of the strong topology of $\mathscr{B}(\mathscr{H})^*$ (cf. (2.1.2)), $\mathrm{Tr}\,\rho_V a \to \mathrm{Tr}\, a \exp[-\beta(H_1 - F)]$ for all $a \in \mathscr{B}(\mathscr{H}_1)$, where $\beta \equiv \sigma'(\varepsilon)$, $\exp(-\beta F) = \mathrm{Tr}\exp(-\beta H_1)$.
2. Because of Theorem (2.2.24), $S(\exp[-\beta(H_1 - F)]) \leq \underline{\lim}_{V\to\infty} S(\rho_V)$.

Recall that the microcanonical state is the most mixed state below $E_m$. The canonical state instead satisfies

**The Maximum Principle for the Canonical Entropy** (2.4.5)

*Let* $\rho = \exp(-\beta H)/\mathrm{Tr}\,\exp(-\beta H)$ *and let* $\bar{\rho}$ *be any density matrix such that* $\mathrm{Tr}\,\bar{\rho}H = \mathrm{Tr}\,\rho H$. *Then* $S(\rho) \geq S(\bar{\rho})$.

**Remarks** (2.4.6)

1. Proposition (2.4.5) states that with a given average energy, the canonical state has the greatest possible entropy. The proposition does not work for all $\alpha$-entropies $S_\alpha$, so it can not be improved to the statement that $\rho \succeq \bar{\rho}$.
2. According to inequality (2.1.7; 2), since $x \to -x \ln x$ is strictly concave, $S$ is a strictly concave function on the convex set of density matrices $\rho$ such that $\mathrm{Tr}\,\rho H = E$. This means that the maximum is unique, and there can not even be local maxima elsewhere.
3. Not all $S_\alpha(\rho)$ are equal with the canonical $\rho$: $S_\alpha = \beta(F(\alpha\beta) - F(\beta))/(\alpha - 1)$.
4. This maximum principle is sometimes invoked as the motivation for the canonical density matrix, without appealing to the microcanonical state.
5. The free energy satisfies the inequality $F(\bar{\rho}) \geq F(\rho)$ without the assumption that $\mathrm{Tr}\,\bar{\rho}H = \mathrm{Tr}\,\rho H$.

**Proof**

Proposition (2.4.5) follows directly from Remark (2.2.23; 1). □

The canonical **partition function** $Z \equiv \mathrm{Tr}\,\exp(-\beta H)$ is easier to work with than the microcanonical partition function, because it does not involve discontinuous functions; if the dimension is finite, it is even an entire function of $\beta$. If the dimension is infinite, then $\exp(-\beta H)$ is required to belong to $\mathscr{C}_1$, so the spectrum of $H$ must be bounded below and extend to $+\infty$. This, however, means that $\exp(-\beta H) \notin \mathscr{C}_1$ for $\beta < 0$, so the most that can be hoped for is analyticity in $\mathbb{C}^+ \equiv \{x + iy: x > 0\}$. For the cases of interest, there is in fact a proposition on

**The Analyticity of the Partition Function of Finite Systems** (2.4.7)

*Let* $\exp(-\beta H_0) \in \mathscr{C}_1$ *for all* $\beta > 0$ *and suppose* $V$ *is* $\varepsilon$*-bounded with respect to* $H_0$ (cf. (III: 3.4.1)). *Then the mapping* $\mathbb{C} \times \mathbb{C}^+ \to \mathbb{C}: (\alpha, \beta) \to \mathrm{Tr}\,\exp[-\beta(H_0 + \alpha v)]$ *is analytic, and* $(\partial/\partial\alpha)\,\mathrm{Tr}\,\exp[-\beta(H_0 + \alpha v)]_{|\alpha=0} = -\mathrm{Tr}\,\beta v\,\exp[-\beta H_0]$.

**Remarks** (2.4.8)

1. Since the operator $H_0 + \alpha v$ is not normal when $\alpha$ is nonreal, the exponential function has to be defined. This can be done as in (2.1.8; 7) or by integrating the resolvent,

$$\exp[-\beta(H_0 + \alpha v)] = \int_C \frac{dz}{2\pi i} \frac{\exp(-\beta z)}{(H_0 + \alpha v - z)},$$

in which the integration contour runs through the region of analyticity (cf. (III: 3.5.13)) so that the integral converges in norm.
2. The next task is to make sense of Tr $\exp[-\beta(H_0 + av)]$ and show that it belongs to $\mathscr{C}_1$ for $(\alpha, \beta) \in \mathbb{C} \times \mathbb{C}^+$. If $\alpha, \beta \in \mathbb{R} \times \mathbb{R}^+$, then this follows from $H_0 + \alpha v \geq H_0/2 - C(\alpha)$, $\exp(-\beta H_0) \in \mathscr{C}_1$, and the observation that if $0 < a < b \in \mathscr{C}_1$ then $a \in \mathscr{C}_1$. If $\alpha$ and $\beta$ are complex, then Corollary (2.1.8; 7) can be appealed to for $|\mathrm{Tr}\exp(\alpha a + \beta b)| \leq \mathrm{Tr}\,|\exp(\alpha a)||\exp(\beta b)|$, with $\exp(a)$ and $\exp(b)$ Hermitian, and in particular $|\mathrm{Tr}\exp[-aH_0 - bv + i(cH_0 + dv)] \leq \mathrm{Tr}\exp(-aH_0 - bv)$ for all real $a$, $b$, $c$, and $d$.
3. The proposition implies that the free energy $F = -T \ln z$ can have singularities only at the zeros of $Z$. If $(\alpha, \beta) \in \mathbb{R} \times \mathbb{R}^+$ then $Z > 0$, so $F$ is analytic in a neighborhood of $\mathbb{R} \times \mathbb{R}^+$. In addition, Corollary (2.1.8; 3) states that $-\ln Z$ is concave in $(\beta, \alpha\beta) \in \mathbb{R} \times \mathbb{R}^+$, so $F$ is concave in $(T, \alpha/T)$ (cf. (III: 3.5.24)). The equation $\partial F/\partial\alpha = \langle v \rangle$ generalizes the Feynman–Hellmann formula (III: 3.5.19; 2).

**Proof**

See Problem 2. □

Since the exponential function is convex, the free energy can be bounded in terms of phase-space integrals by means of (2.2.11), and the upper bound of (2.2.11) can be improved upon with Corollary (2.1.8: 7).

**The Connection with the Classical Free Energy** (2.4.9)

Let

$$H = \sum_{i=1}^{N} |\mathbf{p}_i|^2 + v(\mathbf{x}), \qquad \exp(-\beta F) = \mathrm{Tr}\exp(-\beta H) < \infty,$$

and

$$\exp[-\beta F_{cl}(v)] = \int d^{3N}x \frac{d^{3N}p}{(2\pi)^{3N}} \exp\left[-\beta\left(\sum_{i=1}^{N} |\mathbf{p}_i|^2 + v(\mathbf{x})\right)\right].$$

Then

$$F_{cl}(v) \leq F \leq \inf_u F_{cl}(v_u),$$

where

$$v_u(\mathbf{x}) = \int d^{3N}x' v(\mathbf{x}')|u(\mathbf{x} - \mathbf{x}')|^2 + \int d^{3N}x |\nabla u(\mathbf{x})|^2.$$

**Remarks** (2.4.10)

1. The function $v(\mathbf{x})$ contains the interaction between the particles, as well as a possible external field. It must even account for the box confining the system, as the Hilbert space is $L^2(\mathbb{R}^{3N})$.
2. The proposition shows that quantum effects can only increase the free energy, either with a kinetic zero-point energy or a smeared-out effective potential.
3. The particles have been assumed distinguishable; the modifications needed for indistinguishable particles will be discussed below.
4. Countless attempts at expansions in $\hbar$ have been made in the literature, but the results are not conclusive because rigorous bounds on the higher-order contributions have not been obtained.
5. If $\hbar$ is not set to 1, the dimensionless volume in phase space becomes $d^{3N}x\, d^{3N}p h^{-3N}$, rather than $d^{3N}x\, d^{3N}p\hbar^{-3N}$.

**Proof**

The lower bound for $F$. By Corollary (2.1.8; 7),

$$\begin{aligned} \operatorname{Tr} \exp[-\beta(H_0 + v)] &\leq \operatorname{Tr} \exp(-\beta H_0)\exp(-\beta v) \\ &= \int d^{3N}x \langle \mathbf{x} | \exp(-\beta H_0) | \mathbf{x} \rangle \exp(-\beta v(\mathbf{x})), \end{aligned}$$

and it was observed in (III: 3.3.3) that $\exp(-\beta H_0)$ has the integral kernel

$$K(\mathbf{x}, \mathbf{x}) = \left(\frac{1}{4\pi\beta}\right)^{3N/2} = \int \frac{d^{3N}p}{(2\pi)^{3N}} \exp\left(-\beta \sum_{k=1}^{N} |\mathbf{p}_k|^2\right).$$

The upper bound for $F$ follows immediately from (2.2.11), for $\langle z \| \mathbf{p} |^2 | z \rangle = (\operatorname{Im} z^2) - \int dx |\nabla u|^2$. □

**Example** (2.4.11)

The one-dimensional harmonic oscillator; $u(x) = \exp(-bx^2/2)/\sqrt[4]{\pi}$, $H = p^2 + \omega^2 x^2$,

$$\operatorname{Tr} \exp(-\beta H) = \sum_{n=0}^{\infty} \exp[-\beta\omega(2n + 1)] = \frac{\exp(-\omega\beta)}{1 - \exp(-2\omega\beta)},$$

$$v_u = \omega^2\left(x^2 + \frac{1}{2b}\right) + \frac{b}{2},$$

which has the minimum $\omega^2 x^2 + \omega$ when $b = \omega$. Since

$$\int_{-\infty}^{\infty} \frac{dp\,dx}{2\pi} \exp[-\beta(p^2 + \omega^2 x^2)] = \frac{1}{2\omega\beta},$$

the bounds (2.4.9) yield the inequalities

$$\frac{\exp(-\alpha/2)}{\alpha} \le \frac{\exp(-\alpha/2)}{1 - \exp(-\alpha)} \le \frac{1}{\alpha}, \qquad \alpha \equiv 2\omega\beta \in \mathbb{R}^+.$$

The interest in the bounds (2.4.9) is mainly academic, since the particles in real physics are either fermions or bosons. In addition to multiplying the volume element of the phase-space integral by $1/N!$, the generalization for indistinguishable particles entails an effective interaction that vanishes as $mT \to \infty$, and is repulsive for fermions and attractive for bosons.

**Bounds on F for Indistinguishable Particles** (2.4.12)

*Suppose that*

$$H = \frac{1}{2m} \sum_{i=1}^{N} |\mathbf{p}_i|^2 + v(\mathbf{x}_1, \ldots, \mathbf{x}_N),$$

$$\exp[-\beta F_{cl}(H)] = \frac{1}{(2\pi)^{3N} N!} \int d^{3N}x\, d^{3N}p \exp[-\beta H(\mathbf{p}_1, \ldots, \mathbf{p}_N, \mathbf{x}_1, \ldots, \mathbf{x}_N)],$$

*and that $F_B(H)$ and $F_F(H)$ equal $-T \ln \operatorname{Tr} \exp(-\beta H)$, where the trace is taken over the symmetric (resp. antisymmetric) tensor product of the one-particle spaces. Then*

$$F_{cl}(H) \le F_F(H) \le F_{cl}(h + v_F),$$

$$F_{cl}(H + v_B) \le F_B(H) \le F_{cl}(h),$$

*where the function $h(\mathbf{p}_i, \mathbf{x}_i)$ is the expectation value of $H$ in the symmetrized (resp. antisymmetrized) states of* (2.2.10; 5):

$$h(\mathbf{z}_1, \ldots, \mathbf{z}_N) = \frac{\langle \mathbf{z}_1, \ldots, \mathbf{z}_N | H | \mathbf{z}_1, \ldots, \mathbf{z}_N \rangle}{\langle \mathbf{z}_1, \ldots, \mathbf{z}_N | \mathbf{z}_1, \ldots, \mathbf{z}_N \rangle}, \qquad \mathbf{z}_i = \mathbf{x}_i + i\mathbf{p}_i.$$

If the coherent states are chosen with $u(\mathbf{x}) = \exp(-mT|\mathbf{x}|^2/2)$, then the effective potentials are

$$v_F = \begin{cases} T \ln 2 \sum_{i \ne k} \exp(-mT|\mathbf{x}_i - \mathbf{x}_k|^2) & \text{if } \sup_j \sum_{i \ne j} \exp\left(\dfrac{-mT|\mathbf{x}_i - \mathbf{x}_j|^2}{2}\right) \le \dfrac{1}{2}, \\ \infty & \text{otherwise}; \end{cases}$$

and

$$v_B = -T \sum_{i,k} \exp\left(\frac{-mT|\mathbf{x}_i - \mathbf{x}_k|^2}{2}\right).$$

**Proof**

The lower bounds. For one particle in $x$-space (see (III: 3.3.3)),

$$\langle \mathbf{x}|\exp\left(\frac{-\beta|\mathbf{p}|^2}{2m}\right)|\mathbf{x}'\rangle = \left(\frac{mT}{2\pi}\right)^{3/2}\exp\left(\frac{-mT|\mathbf{x}-\mathbf{x}'|^2}{2}\right),$$

so in the properly symmetrized or antisymmetrized basis, if there are $N$ particles, then

$$\langle \mathbf{x}_1,\ldots,\mathbf{x}_N|\exp\left(\frac{-\beta\sum_i|\mathbf{p}_i|^2}{2m}\right)|\mathbf{x}_1,\ldots,\mathbf{x}_N\rangle$$
$$= \frac{1}{N!}\left(\frac{mT}{2\pi}\right)^{3N/2}\sum_P(\pm 1)^P\exp\left(\frac{-mT\sum_i|\mathbf{x}_i-\mathbf{x}_{P_i}|^2}{2}\right).$$

The sum over permutations amounts to just a permanent or determinant of the form $\langle \mathbf{z}_1,\ldots,\mathbf{z}_N|\mathbf{z}_1,\ldots,\mathbf{z}_N\rangle$, by (2.2.10; 5). It is therefore $\geq 1$ or, respectively, $\leq 1$, since the length of a vector is increased or, respectively, decreased when acted upon by $a_f^*$ with $\|f\| = 1$:

$$\|a_f^*|\ \rangle\|^2 = \langle|a_f a_f^*|\rangle = \langle\ |\ \rangle \pm \langle|a_f^* a_f|\rangle \gtreqless \langle\ |\ \rangle.$$

For fermions, $\mathrm{Det}(\langle \mathbf{z}_i|\mathbf{z}_k\rangle) \leq 1$, whereas for bosons the permanent has an upper bound from Problem 4, $\mathrm{Per}(\langle \mathbf{z}_i|\mathbf{z}_k\rangle) \leq \exp[\sum_{i,k}|\langle \mathbf{z}_i|\mathbf{z}_k\rangle|]$. The rest of the proof is similar to that of the lower bound of (2.4.10):

$$\mathrm{Tr}\exp[-\beta(H_0+v)] \leq \mathrm{Tr}\exp(-\beta H_0)\exp(-\beta v)$$
$$= \frac{1}{N!(2\pi)^{3N}}\int d^{3N}x\, d^{3N}p\exp[-\beta(H_0(\mathbf{p}_1,\ldots,\mathbf{p}_N) + v(\mathbf{x}_1,\ldots,\mathbf{x}_N))]$$
$$\times \begin{matrix}\mathrm{Per}\\ \mathrm{Det}\end{matrix}\left(\exp\left(-\frac{m}{2}|\mathbf{x}_i-\mathbf{x}_j|^2 T\right)\right)$$
$$\leq \frac{1}{N!(2\pi)^{3N}}\int d^{3N}x\, d^{3N}p\exp\left[-\beta\left(H_0(\mathbf{p}_1,\ldots,\mathbf{p}_N) + v(\mathbf{x}_1,\ldots,\mathbf{x}_N)\right.\right.$$
$$\left.\left.-T\begin{Bmatrix}\exp(-\sum_{i,j}(mT/2)|\mathbf{x}_i-\mathbf{x}_j|^2)\\ 0\end{Bmatrix}\right)\right].$$

The upper bounds. Since the symmetrized and antisymmetrized coherent states are not normalized,

$$\langle \mathbf{z}_1,\ldots,\mathbf{z}_N|\mathbf{z}_1,\ldots,\mathbf{z}_N\rangle \equiv n(z) = \begin{matrix}\mathrm{Per}\\ \mathrm{Det}\end{matrix}(\langle \mathbf{z}_i|\mathbf{z}_k\rangle) \gtreqless 1,$$

the normalization has to be accounted for in (2.2.11(i)):

$$\mathrm{Tr}\, k(a) \geq \int d\Omega_z n(z) k\left(\frac{\langle z|a|z\rangle}{n(z)}\right).$$

For bosons the inequality follows now from $n(z) \geq 1$. For fermions, with $u(\mathbf{x}) = \exp(-mT|\mathbf{x}|^2/4)$, it is necessary to estimate $\operatorname{Det}(1 + K)$, where

$$K_{ij} = \begin{cases} \exp(-(mT/2)|\mathbf{x}_i - \mathbf{x}_j|^2), & i \neq j \\ 0, & i = j. \end{cases}$$

Since

$$\|K\| \leq \sup_j \sum_{i \neq j} \exp\left(-\frac{mT}{2}|\mathbf{x}_i - \mathbf{x}_j|^2\right),$$

we find

$$\begin{aligned} \ln \operatorname{Det}(\langle \mathbf{z}_i | \mathbf{z}_j \rangle) &= \ln \operatorname{Det}(1 + K) = \operatorname{Tr} \ln(1 + K) \\ &= \sum_{n=2}^{\infty} \operatorname{Tr} K^n \frac{(-1)^{n+1}}{n} \leq \operatorname{Tr} K^2 \sum_{n=0}^{\infty} \frac{\|K\|^n}{n+2} \\ &\leq \ln \frac{1}{1 - \|K\|} \operatorname{Tr} K^2 \\ &\leq \begin{cases} \ln 2 \operatorname{Tr} K^2 & \text{for } \|K\| \leq \frac{1}{2} \\ \infty & \text{otherwise.} \end{cases} \end{aligned}$$

Finally,

$$\operatorname{Tr} K^2 = \sum_{i \neq j} \exp[-mT|\mathbf{x}_i - \mathbf{x}_j|^2]. \qquad \square$$

**Remarks** (2.4.13)

1. If $\min_{i,j}|\mathbf{x}_i - \mathbf{x}_j| \equiv b > 0$, then $\|K\| \cong b^{-3} \int_b^\infty dr r^2 \exp(-r^2 mT/2) \cong \exp(-mTb^2/2)$, so $v_F$ can be replaced with a hard-core potential with a radius depending on $T$ and energy $\sim N$.
2. The ranges of the potentials $v_B$ and $v_F$ are approximately the thermal wavelength, i.e., the wavelength of a particle with kinetic energy $3T/2$, so when the particles are about this close together, as in a degenerate quantum gas, the bounds spread wide apart.

In closing, let us study the limit $N \to \infty$ in the framework of the canonical ensemble. Not only the reservoir but also the subsystem will be made infinite at the same time, and we wish to know whether the free energy density $F/V$ tends to a limit $\varphi$. This should be the case whenever this limit exists microcanonically. Then the issue is how to recover the microcanonical quantities from knowledge of $\varphi$:

**Theorem** (2.4.14)

*Suppose that, with $H \geq 0$, $\sigma_V(\varepsilon, \rho) = (1/V) \ln \operatorname{Tr} \Theta(V\varepsilon - H)$ converges uniformly on compact sets to a concave function $\sigma(\varepsilon, \rho)$ and is bounded above* *by a function* $s(\varepsilon, \rho)$ *such that* $0 = s(\varepsilon_0, \rho) = \lim_{\varepsilon\to\infty} s(\varepsilon, \rho)/\varepsilon$, *when* $V$ *is big enough. Writing as usual* $\beta = 1/T$, *then*

$$\lim_{V\to\infty}\left(-\frac{T}{V}\ln \operatorname{Tr}\exp(-\beta H)\right) = \inf_{\varepsilon}(\varepsilon - T\sigma(\varepsilon, \rho)) \equiv \varphi(T, \rho).$$

**Remarks** (2.4.15)

1. Since $\sigma$ is concave, it has a right derivative,

$$\sigma \equiv \lim_{\delta\downarrow 0}(\sigma(\varepsilon + \delta, \rho) - \varepsilon(\varepsilon, \rho))\frac{1}{\delta}.$$

   The infimum is attained at the point $\varepsilon(T, \rho)$ for which $\sigma'(\varepsilon(T, \rho), \rho) = 1/T$ (see Figure 17). If $\sigma'$ has a discontinuity, jumping over the value $1/T$, then $\varepsilon(T, \rho)$ is the point at which the jump takes place. The usual thermodynamic relationship $\varphi(T, \rho) = \varepsilon(T, \rho) - T\sigma(\varepsilon(T, \rho), \rho)$ holds for the free energy.
2. The function $\beta\varphi$ is a Legendre transform $(\mathscr{L}(\sigma))(\beta) = \inf_\varepsilon(\beta\varepsilon - \sigma(\varepsilon))$. The transformation $\mathscr{L}$ has the following properties:
   (i) $\mathscr{L}\circ\mathscr{L}$ produces the concave envelope of any function so $\mathscr{L}\circ\mathscr{L} = 1$ on concave functions;
   (ii) $\mathscr{L}$ maps a linear piece of a concave function to the point of a corner and vice versa;
   (iii) $\mathscr{L}$ maps the set of strictly concave, continuously differentiable functions into itself. By Property (i),

$$\sigma(\varepsilon) = \inf_\beta(\beta\varepsilon - \mathscr{L}(\sigma)(\beta)) = \inf_T\frac{\varepsilon - \varphi(T)}{T}.$$

3. If $\sigma(\varepsilon)$ is strictly concave and continuously differentiable, then by Problem 3 the limit $V \to \infty$ and the derivative by $\beta$ can be taken in either order.

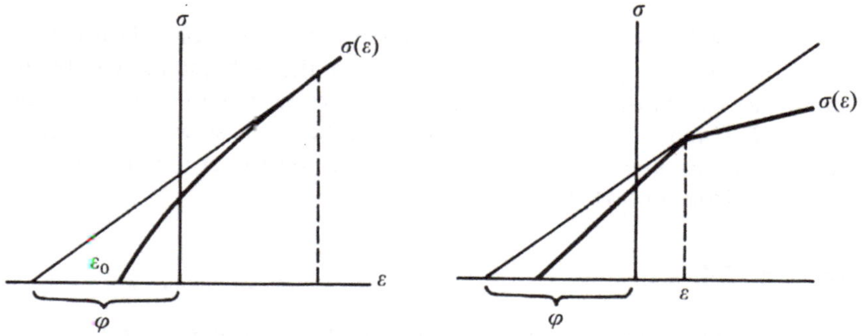


Figure 17 The geometric meaning of the free energy.

The energy and entropy densities calculated with the canonical density matrix are

$$\lim_{V\to\infty} \operatorname{Tr} \frac{H}{V} \frac{\exp(-\beta H)}{\operatorname{Tr}\exp(-\beta H)} = -\lim_{V\to\infty} \frac{\partial}{\partial\beta}\frac{1}{V} \ln \operatorname{Tr}\exp(-\beta H)$$
$$= -T^2 \frac{\partial}{\partial T}\frac{\varphi}{T} = \varphi + T\sigma$$

and

$$\lim_{V\to\infty} \frac{T}{V} S\left(\frac{\exp(-\beta H)}{\operatorname{Tr}\exp(-\beta H)}\right) = \varepsilon - \varphi,$$

which are obviously identical to the microcanonical energy and entropy densities. This fact is known as the **equivalence of the ensembles**.

4. The concavity of $\sigma$ in $\varepsilon$ is a necessary condition for the ensembles to be equivalent, since the specific heat in the canonical ensemble,

$$\frac{\partial\varepsilon}{\partial T} = \frac{\beta^2}{V}\frac{\partial^2}{\partial\beta^2} \ln \operatorname{Tr}\exp(-\beta H)$$

is automatically positive by Corollary (2.1.8; 3).

5. The bounding function $s$ is necessary to ensure that

$$\lim_{V\to\infty} \sup_{\varepsilon}(T\sigma_V(\varepsilon, \rho) - \varepsilon) = \sup_{\varepsilon}(T\sigma(\varepsilon, \rho) - \varepsilon);$$

without it, $T\sigma_V(\varepsilon) - \varepsilon = 1 - (1 - \varepsilon/V)^2$ is a counterexample.

(The assumption that $H \geq 0$ is a normalization.)

**Proof**

$$\operatorname{Tr}\exp(-\beta H) = \int_0^\infty dE \exp(-\beta E) \frac{\partial}{\partial E} \operatorname{Tr} \Theta(E - H)$$
$$= \beta \int_0^\infty dE \exp[-\beta E + S(E)]$$
$$= \beta V \exp[-\beta V \varphi_V(T, \rho)]$$
$$\times \int_0^\infty d\varepsilon \exp[-\beta V(\varepsilon - T\sigma_V(\varepsilon) - \varphi_V)],$$

where

$$\varphi_V(T, \rho) = \inf_{\varepsilon}(\varepsilon - T\sigma_V(\varepsilon, \rho)).$$

If $V$ is taken large enough, then the infimum lies between 0 and $\varepsilon_0$: $\varepsilon_0 - T\sigma(\varepsilon_0, \rho) = 0$. By assumption the functions $\sigma_V$ converge uniformly on this compact interval, so $\varphi_V(T, \rho) \to \varphi(T, \rho)$. A modification of Lemma (2.3.12) shows that the contribution of the integral to $\varphi$ is negligible in this limit. This step uses the assumption to ensure that for all $T > 0$ the exponent is dominated by $-\beta E$ for large $E$, so that the dominated convergence lemma applies. □

Several general properties of the Legendre transform of $\sigma$ can be deduced from those of the microcanonical energy density (2.3.10), and are listed below:

**Properties of the Free Energy Density** (2.4.16)

1. As the infimum of a set of linear functions, $\varphi(T, \rho)$ is concave in $T$. If $H \geq 0$, then $\varphi(T, \rho) \leq 0$, and $\varphi(0, \rho) = 0$.
2. The function $\varphi(T, \rho)$ is convex in $\rho$, because $f(x, y)$ being convex in $(x, y)$ implies that $\inf_x f(x, y)$ is convex in $y$ (see (2.3.31)).
3. $\rho^{-1}\varphi(T, \rho)$ is an increasing function of $\rho$, since $\operatorname{Tr} \exp(-\beta H)$ is an increasing function of $V$ when $N$ and $\beta$ are fixed.
4. $T^{-1}\varphi(T, \rho)$ is a decreasing function of $T$, since for $H \geq 0$, $\exp(-\beta H)$ is a decreasing function of $\beta$.

**Remark** (2.4.17)

Although convexity survives the thermodynamic limit, the analyticity (2.4.8; 3) of $F$ is less hardy. The zeros of $Z$ may approach the real axis as the system is made infinite, causing discontinuities in the derivatives of $\varphi$. Example (2.3.32; 2) can be modified to a degenerate BCS model, with

$$H = B\sum_{j=1}^{N} \sigma_j^{(z)} - \frac{1}{N}\sum_{i,j=1}^{N} (\boldsymbol{\sigma}_i \cdot \boldsymbol{\sigma}_j - \sigma_i^{(z)}\sigma_j^{(z)}).$$

This Hamiltonian has the eigenvalues $N(Bm_z - m(m + 2/N) + m_z^2)$, and, as in (2.3.34),

$$\varphi(T, B) = \inf_{0 \leq |m_z| \leq m \leq 1} \left(-m^2 + \left(m_z + \frac{B}{2}\right)^2 - \frac{B^2}{4} - T\sigma(m)\right),$$

$$\sigma(m) = \ln 2 - \frac{1+m}{2}\ln(1+m) - \frac{1-m}{2}\ln(1-m).$$

The infimum with respect to $m_z$ is attained at $\max\{-B/2, -m\}$, assuming $B \geq 0$. If $m_z = -B/2$, then setting the derivative by $m$ to zero leads to the equation

$$m(T) = \tanh\left(\frac{2m(T)}{T}\right).$$

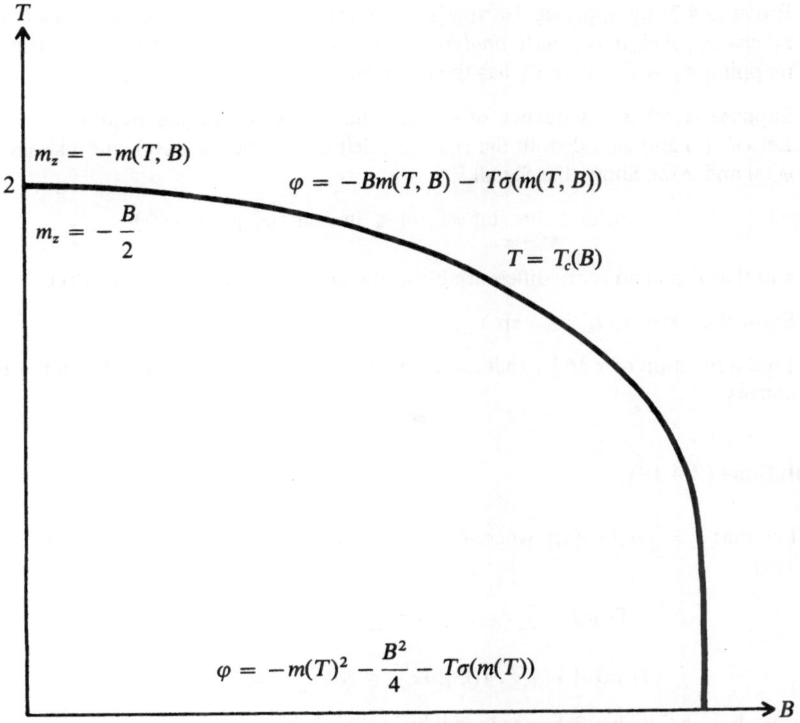


Figure 18 The free energy in Example (2.2.32; 2).

If $m_z = -m$, then instead of this, the minimizing value is $m(T, B) = \tanh(B/T)$. The two different possibilities give critical temperatures

$$T_c(B) \equiv \begin{cases} B/\operatorname{arctanh}(B/2) & \text{if } 0 < B < 2, \\ 0 & \text{if } 2 \leq B. \end{cases}$$

Figure 18 depicts $\varphi(T, B)$. The values of $m$ and $m_z$ are continuous at the transition point, but their derivatives are not. The function $\varphi$ remains continuous along with its first derivatives—the derivatives by $m$ and $m_z$ vanish—but the second derivatives of $\varphi(T, B)$ are discontinuous at $T = T_c(B)$. Such properties as the specific heat display the discontinuity characteristic of a phase transition.

**Problems** (2.4.18)

1. Let $\rho_n$ and $\rho$ be density matrices for which $\rho_n \to \rho$. Show that $\operatorname{Tr} |\rho_n - \rho| \to 0$. (Hint: use the following lemma: *If $\rho$ is a density matrix and $Q$ a projection such that $\operatorname{Tr} \rho Q < \varepsilon$, then for all $a \in \mathscr{B}(\mathscr{H})$, $|\operatorname{Tr} \rho Q a| < \|a\|\sqrt{\varepsilon}$.*)

2. Prove (2.4.7) by applying Hartogs's theorem: *If $f(z_1, z_2)$ is separately analytic in $z_1$ and $z_2$, then it is jointly analytic.* Also observe that the trace is a continuous mapping $\mathscr{C}_1 \to \mathbb{C}$, where $\mathscr{C}_1$ has the norm $\|\cdot\|_1$.

3. Suppose $\varphi_V(\varepsilon)$ is a sequence of concave functions converging pointwise to $\varphi(\varepsilon)$. Let $\varphi'_{V,r}(\varepsilon)$ and $\varphi'_{V,l}$ denote the right and left derivatives of $\varphi_V(\varepsilon)$, and likewise for $\varphi'_r(\varepsilon)$ and $\varphi'_l(\varepsilon)$. Show that for all $\varepsilon$,
$$\varphi'_r(\varepsilon) \leq \liminf_{V\to\infty} \varphi'_{V,r}(\varepsilon) \leq \limsup_{V\to\infty} \varphi'_{V,l}(\varepsilon) \leq \varphi'_l(\varepsilon),$$
and that if $\varphi_V$ and $\varphi$ are differentiable at the point $\varepsilon$, then $\lim \varphi'_V(\varepsilon) = \varphi'(\varepsilon)$.

4. Show that $|\mathrm{Per}\langle \mathbf{z}_i | \mathbf{z}_k \rangle| \leq \exp \sum_{i,k} |\langle \mathbf{z}_i | \mathbf{z}_k \rangle|$.

5. Find a function of $x$ and $y$ that is convex in each variable separately but not jointly convex.

## Solutions (2.4.19)

1. Lemma: $\rho = \sum c_i |x_i)(x_i|$, where $c_i \geq 0$, $\sum_i c_i = 1$, and $\{x_i\}$ is an orthonormal basis. Then
$$\mathrm{Tr}\, \rho Q = \sum_i c_i(x_i | Q x_i) = \sum_i c_i \|Q x_i\|^2 < \varepsilon,$$
$$|\mathrm{Tr}\, \rho Q a| = |\sum c_i(Q x_i | a x_i)| \leq \|a\| \cdot \sum_i c_i \|Q x_i\| < \|a\| \sqrt{\varepsilon},$$
since by the Cauchy–Schwarz inequality,
$$\sum_i c_i \|Q x_i\| = \sum_i \sqrt{c_i}\, \|Q x_i\| \sqrt{c_i} \leq \left(\sum_i c_i \|Q x_i\|^2\right)^{1/2} \cdot \left(\sum_i c_i\right)^{1/2} = \left(\sum_i c_i \|Q x_i\|^2\right)^{1/2}.$$
Proof of the proposition: For any finite-rank operator $a$, $\mathrm{Tr}\, \rho_n a \to \mathrm{Tr}\, \rho a$, and $\mathrm{Tr}\, \rho_n(\mathbf{1} - a) \to \mathrm{Tr}\, \rho(\mathbf{1} - a)$. Now let $P$ be the projection onto the first $N$ eigenvalues of $\rho$ and choose $N$ such that $\mathrm{Tr}\, \rho(\mathbf{1} - P) < \varepsilon$. Then
$$\mathrm{Tr}(\rho_n - \rho) a = \mathrm{Tr}\, \rho_n(\mathbf{1} - P) a + \mathrm{Tr}(\mathbf{1} - P)\rho_n P a + \mathrm{Tr}(\rho_n - \rho) P a P + \mathrm{Tr}(P \rho P - \rho) a.$$
$$\mathrm{Tr}(\rho_n - \rho) P a P < \varepsilon \|P a P\| < \varepsilon \|a\|$$
for sufficiently large $n$, since all topologies are equivalent on the finite-dimensional space $P\mathscr{B}(\mathscr{H})P$, and $\mathrm{Tr}(\rho_n - \rho) P a P \to 0$. $|\mathrm{Tr}(P \rho P - \rho) a| \leq \|a\| \mathrm{Tr}(\mathbf{1} - P)\rho < \|a\| \cdot \varepsilon$. $\mathrm{Tr}\, \rho_n(\mathbf{1} - P) \to \mathrm{Tr}\, \rho(\mathbf{1} - P) < \varepsilon$, which implies that for $n$ large enough, $\mathrm{Tr}\, \rho_n(\mathbf{1} - P) < 2\varepsilon$. Hence, by the lemma,
$$|\mathrm{Tr}\, \rho_n(\mathbf{1} - P) a| < \sqrt{2\varepsilon}\, \|a\|,$$
$$|\mathrm{Tr}(\mathbf{1} - P)\rho_n P a| = |\mathrm{Tr}\, \rho_n(\mathbf{1} - P) a^* P| \leq \sqrt{2\varepsilon}\, \|a^* P\| \leq \sqrt{2\varepsilon}\, \|a\|.$$
Consequently,
$$|\mathrm{Tr}(\rho_n - \rho) a| < (2\varepsilon + 2\sqrt{2\varepsilon})\, \|a\|,$$
$$\mathrm{Tr}\, |\rho_n - \rho| = \sup_{\|a\| \leq 1} |\mathrm{Tr}(\rho_n - \rho) a| < 2\varepsilon + 2\sqrt{2\varepsilon}.$$

2. $U(\alpha, \beta) \equiv \exp[-\beta(H_0 + \alpha v)] \in \mathscr{C}_1$

   (i) Analyticity (=complex differentiability) in $\beta$:

$$\left\| \frac{U(\alpha, \beta + \beta') - U(\alpha, \beta)}{\beta'} + (H_0 + \alpha v)U(\alpha, \beta) \right\|_1$$

$$\leq \left\| \left( \frac{U(\alpha, \beta') - 1}{\beta'} + (H_0 + \alpha v) \right) U\left(\alpha, \frac{\beta}{2}\right) \right\| \left\| U\left(\alpha, \frac{\beta}{2}\right) \right\|_1 \to 0$$

   as $\beta' \to 0$, since $U$ is a $\|\cdot\|$-convergent integral of $\|\cdot\|$-analytic functions and therefore a $\|\cdot\|$-analytic mapping, $\mathbb{C} \times \mathbb{C}^+ \to \mathscr{B}$.

   (ii) Analyticity in $\alpha$:

$$U(\alpha + \alpha', \beta) - U(\alpha, \beta) = -\beta\alpha' \int_0^1 d\tau U(\alpha + \alpha', \beta(1 - \tau))vU(\alpha, \tau\beta),$$

$$\|U(\alpha + \alpha', \beta(1 - \tau))vU(\alpha, \tau\beta)\|_1 \leq \|U(\alpha + \alpha', \beta(1 - \tau))\| \times \left\| vU\left(a, \frac{\tau\beta}{2}\right) \right\| \left\| U\left(\alpha, \frac{\tau\beta}{2}\right) \right\|_1 \leq \text{constant},$$

   when $\frac{1}{2} \leq \tau \leq 1$. If $0 \leq \tau \leq \frac{1}{2}$, then the first factor has to be divided up. This shows that the mapping $\mathbb{C} \times \mathbb{C}^+ \to \mathscr{B}_1: (\alpha, \beta) \to U(\alpha, \beta)$ is analytic, and therefore the mapping $\mathbb{C} \times \mathbb{C}^+ \to \mathbb{C}: (\alpha, \beta) \to \operatorname{Tr} U(\alpha, \beta)$ is analytic, because the trace is continuous and linear $\mathscr{B}_1 \to \mathbb{C}$, and thus also analytic.

3. Concavity yields $(1/\varepsilon')(\varphi_V(\varepsilon + \varepsilon') - \varphi_V(\varepsilon)) \leq \varphi'_{V,r}(\varepsilon) \leq \varphi'_{V,l}(\varepsilon) \leq (1/\varepsilon')(\varphi_V(\varepsilon - \varepsilon') - \varphi_V(\varepsilon))$ for all $\varepsilon' > 0$, and the statement follows from this with the limits $\lim_{\varepsilon' \to 0} \lim_{V \to \infty}$.

4. $$\operatorname{Per}\langle \mathbf{z}_i | \mathbf{z}_k \rangle \leq \operatorname{Per}|\langle \mathbf{z}_i | \mathbf{z}_k \rangle| = \sum_P \prod_{i=1}^N |\langle \mathbf{z}_i | \mathbf{z}_{P_i} \rangle| \leq \prod_{(i, j)} (1 + |\langle \mathbf{z}_i | \mathbf{z}_j \rangle|)$$

$$\leq \exp\left( \sum_{i, j} |\langle \mathbf{z}_i | \mathbf{z}_j \rangle| \right).$$

5. $f(x, y) = -xy$. The Hessian matrix $\begin{pmatrix} 0 & -1 \\ -1 & 0 \end{pmatrix}$ is not positive.

## 2.5 The Grand Canonical Ensemble

*The thermodynamic functions are easier to calculate explicitly if the constraint of a fixed number N of particles is dropped. It is physically realistic for a system coupled to a reservoir of particles.*

This section will investigate the situation of a system with a reservoir with which it can exchange particles as well as heat. As in (2.3.23), the underlying Hilbert space is taken as

$$\bigoplus_{N_1=0}^{N} \mathscr{H}_{N_1, V_1} \otimes \mathscr{H}_{N-N_1, V_2},$$

and the Hamiltonian is

$$H = \bigoplus_{N_1=0}^{N} (H_1(V_1, N_1) + H_2(V_2, N - N_1)).$$

We consider the limit as $N \to \infty$ and $V_2 \to \infty$, and begin by collecting the immediate generalizations of some of the results of §2.4. Proofs will not be given, as they entail only slight modifications of the earlier ones.

**Convergence of the Reduced Density Matrix** (2.5.1)

*Suppose that the concave, increasing functions*

$$\frac{1}{V_2}\operatorname{Tr}_2 \Theta(E_2 - H_2) = \frac{1}{V_2} S_2(E_2, V_2, N_2) \equiv \sigma_{V_2}\left(\frac{E_2}{V_2}, \frac{N_2}{V_2}\right)$$

*and their derivatives converge uniformly on a neighborhood of* $\varepsilon = E_2/V_2$ *and* $\rho = N_2/V_2$ *to* $\sigma(\varepsilon, \rho)$, $\partial\sigma/\partial\varepsilon$, *and* $\partial\sigma/\partial\rho$. *Then with* $V = V_1 + V_2$, $N = N_1 + N_2$,

$$\lim_{V_2 \to \infty} \frac{\operatorname{Tr}_2 \Theta(E - H)}{\operatorname{Tr} \Theta(E - H)} \to \frac{\exp\left[-H_1(V_1, N_1)\dfrac{\partial\sigma}{\partial\varepsilon} - N_1\dfrac{\partial\sigma}{\partial\rho}\right]}{\operatorname{Tr}_1 \exp\left[-H_1(V_1, N_1)\dfrac{\partial\sigma}{\partial\varepsilon} - N_1\dfrac{\partial\sigma}{\partial\rho}\right]} = \rho_{GC}$$

*in the trace norm.*

**Remarks** (2.5.2)

1. The symbol $\operatorname{Tr}_2$ denotes the trace in the second factor of

$$\bigoplus_{N_1=0}^{N} \mathscr{H}_{N_1, V_1} \otimes \mathscr{H}_{N-N_1, V_2},$$

so in the limit $N \to \infty$, $H_1(N_1, V_1)$ operates on $\sum_{N_1=0}^{\infty} \mathscr{H}_{N_1, V_1}$. This operator on the Hilbert space of an indefinite number of particles is most conveniently written in terms of the field operators (1.3.2).

2. The values of $\mu$ for which $\exp[-\beta(H - \mu N)] \in \mathscr{C}_1$ depend on the problem. If, for instance,

$$-\ln \operatorname{Tr}_{|\mathscr{H}_{N_1}} \exp[-\beta H_1(N_1)] > -cN_1,$$

then the trace exists whenever $\operatorname{Re} \beta\mu < -c$.

Many of the results of §2.4 may be reformulated for the grand canonical ensemble merely be replacing $H$ with $H - \mu N$. An example is

**The Principle of Maximum Entropy** (2.5.3)

*Let $\bar{\rho}$ be a density matrix such that* $\text{Tr}\, \bar{\rho}H = \text{Tr}\, \rho_{GC} H$, $\text{Tr}\, \bar{\rho}N = \text{Tr}\, \rho_{GC} N$. *Then* $S(\rho_{GC}) \geq S(\bar{\rho})$.

If system 1 is now taken infinitely large, presupposing the extensivity following from $H > -N_c$, then $T/V$ times the logarithm of the grand canonical partition function has a limit, which may be identified as the pressure, with reference to (2.3.29).

**The Thermodynamic Limit** (2.5.4)

*If the assumptions of* (2.4.14) *are satisfied. then*

$$\lim_{V\to\infty} \frac{T}{V} \ln \text{Tr} \exp[-\beta(H - \mu N)]$$
$$= \lim_{V\to\infty} \frac{T}{V} \ln \sum_{N=0}^{\infty} \exp\left[-\beta V\left(\varphi_V\left(T, \frac{N}{V}\right) - \mu N\right)\right]$$
$$= \sup(\mu\rho - \varphi(T, \rho)) = P(T, \mu).$$

**Remarks** (2.5.5)

1. The supremum is attained where the right derivative
$$\lim_{\delta\downarrow 0} (\varphi(T, \rho + \delta) - \varphi(T, \rho))\delta^{-1} = \mu,$$
unless $\mu$ is on an endpoint of the interval on which $P(T, \mu)$ is defined. This means that with (2.3.29), $\mu$ can be identified with
$$\left.\frac{\partial\varepsilon}{\partial\rho}\right|_\sigma = \left.\frac{\partial\varepsilon}{\partial\rho}\right|_T - T\left.\frac{\partial\sigma}{\partial\rho}\right|_T = \left.\frac{\partial\varphi}{\partial\rho}\right|_T.$$
Because
$$\mu\rho - \varphi = \rho\left.\frac{\partial\varepsilon}{\partial\rho}\right|_\sigma + \sigma\left.\frac{\partial\varepsilon}{\partial\sigma}\right|_\rho - \varepsilon = P,$$
the grand canonical partition function turns out to be $\exp(PV/T)$. We shall also speak of $P$ as the pressure when the system is finite, although it does not exactly agree with the definition as the force per area on the wall.
2. As before, the ensembles are equivalent, on account of the identities
$$\rho = \left.\frac{\partial P}{\partial\mu}\right|_T, \qquad \varepsilon = T\left.\frac{\partial P}{\partial T}\right|_{\mu/T} - P = \mu\rho - T\left.\frac{\partial\varphi}{\partial T}\right|_\rho - \mu\rho + \varphi = \varphi + T\sigma,$$
$$T\sigma = \varepsilon - \mu\rho + P.$$
Observe that the grand canonical averages of $N/V$ and $H_N/V$ approach $\rho$ and $\varepsilon$, and that the entropy density of $\rho_{GC}$ equals $\sigma$.

**Properties of the Pressure** (2.5.6)

1. The function $(T, \mu) \to P$ is convex, since it is the supremum of convex functions.
2. The pressure increases with $\mu$, since it is the supremum of increasing functions.
3. If $H - \mu N \geq 0$, then $T^{-1}P$ is an increasing function of $T$, since $\exp[-\beta(H - \mu N)]$ is a decreasing function of $\beta$.

The grand canonical ensemble is particularly useful for identical particles, and allows the thermodynamic functions of bosons or fermions interacting with an external field to be evaluated more explicitly. For this purpose, we write the Hamiltonian and the particle number in terms of the field operators (1.3.2) and our orthogonal basis $\{f_m\}$, as

$$\begin{aligned} H &= \sum_{m,n} a_m^* a_n \left[ \int d^3x \nabla f_m^*(\mathbf{x}) \cdot \nabla f_n(\mathbf{x}) + f_m^*(\mathbf{x}) f_n(\mathbf{x}) v(\mathbf{x}) \right] \\ &\equiv \sum_{m,n} a_m^* a_n \langle f_m | h | f_n \rangle, \\ N &= \sum_m a_m^* a_m, \end{aligned} \tag{2.5.7}$$

where $h = |\mathbf{p}|^2 + v(\mathbf{x})$ is the one-particle Hamiltonian, and $a_m$ stands for $a(f_m)$. If $h$ has pure-point spectrum with eigenvalues $\varepsilon_m$, and $f_m$ are taken as the eigenvectors associated with $\varepsilon_m$, then

$$\operatorname{Tr} \exp[-\beta(H - \mu N)] = \operatorname{Tr} \exp\left[ -\beta \sum_m a_m^* a_m (\varepsilon_m - \mu) \right]. \tag{2.5.8}$$

Taking the trace leads to easily computed sums, since $a^*a$ has the eigenvalues 0 and 1 for fermions and 0, 1, 2, ..., for bosons. In these cases, $P_F$ and $P_B$ become

$$P_F(z) = -P_B(-z) = \frac{T}{V} \sum_m \ln(1 + z \exp(-\beta \varepsilon_m)), \tag{2.5.9}$$

where $z \equiv \exp(\beta\mu)$ is known as the **fugacity**. When written in terms of the one-particle Hamiltonian $h = |\mathbf{p}|^2 + V(\mathbf{x})$ and the trace tr on the one-particle space $L^2(\mathbb{R}^3)$,

**The Pressure of Fermions or Bosons in an External Field** (2.5.10)

becomes

$$P_F(T, z) = \frac{T}{V} \operatorname{tr} \ln(1 + z \exp(-\beta h)) = -P_B(T, -z).$$

**Remarks** (2.5.11)

1. In the limit $z \to 0$, $P_F(T, z) = P_B(T, z) = z(T/V) \sum_m \exp(-\beta\varepsilon_m)$, which corresponds to very dilute matter, for which both Bose and Fermi statistics become the same (Boltzmann statistics).
2. If $h \geq 0$ and $\exp(-\beta h) \in \mathscr{C}_1$, then the singularities of $\exp(P)$ occur where $z = -\exp(\beta\varepsilon_m) < -1$, $m = 0, 1, 2, \ldots$. The function $\exp(P)$ is analytic in $z$ until the singularities are reached, i.e., the power series in $z$ converges. The analytic function $P_F(T, z)$ describes all three kinds of statistics. Fermi statistics correspond to $z = \exp(\mu/T) > 0$, Boltzmann statistics to $z \to 0$, and Bose statistics to $-\exp(\varepsilon_0) < z < 0$ (see Figure 19).

It is easy to calculate expectation values as well as the partition function:

$$\langle a_m^* a_{m'} \rangle \equiv \operatorname{Tr} a_m^* a_{m'} \exp[-\beta(H - \mu N + PV)] = \frac{\delta_{mm'}}{\exp[\beta(\varepsilon_m - \mu)] \pm 1}. \tag{2.5.12}$$

Since every one-particle vector $|f\rangle \in L^2(\mathbb{R}^3)$ can be expanded in eigenvectors of $h$, and when restricted to $L^2(\mathbb{R}^3)$, $a_f^* a_f$ equals $P_f = |f\rangle\langle f|$, the information about the one-particle observables is contained in the

**Effective One-Particle Density Matrix** (2.5.13)

One-particle expectation values are given by $\rho_1 = (\exp[\beta(h - \mu)] \pm 1)^{-1}$ with the formula $\langle a_f^* a_f \rangle = \operatorname{Tr} \rho_1 P_f = \langle f | \rho_1 | f \rangle$. The density matrix $\rho_1$ has the properties

$$\operatorname{Tr} \rho_1 = \bar{N},$$

and

$$0 \leq \rho_1 \leq \begin{cases} 1 & \text{for fermions} \\ \bar{N} & \text{for bosons.} \end{cases}$$

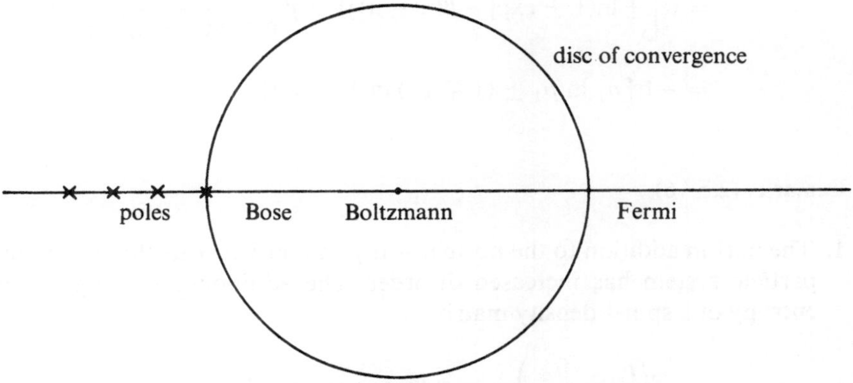


Figure 19 Singularities of $P$ in the complex $z$-plane.

**Remarks** (2.5.14)

1. The number $\bar{N}$ is defined by $\langle \sum_m a_m^* a_m \rangle = \text{tr}[\exp(\beta(h-\mu)) \pm 1]^{-1}$. If it is preferred to deal with these more understandable variables of the canonical ensemble, then this can be taken as the equation determining $\mu$.
2. Similarly, $\langle H \rangle = \text{tr}\, \rho_1 h$, etc.
3. If a reduced density matrix on the one-particle phase space is defined with coherent states (cf. (2.2.7) with (2.2.10; 5)), $\rho(\mathbf{x}, \mathbf{p}) = \langle a_z^* a_z \rangle = \langle z | \rho_1 | z \rangle$, then the properties of $\rho_1$ generalize as

$$\int \frac{d^3x\, d^3p}{(2\pi)^3} \rho(\mathbf{x}, \mathbf{p}) = \bar{N},$$

and

$$0 \leq \rho(\mathbf{x}, \mathbf{p}) \leq \begin{cases} 1 & \text{for fermions} \\ \bar{N} & \text{for bosons.} \end{cases}$$

This shows that the exclusion principle of fermions has the effect of reducing the maximum value $\bar{N}$ of $\rho(z)$ allowed in quantum mechanics to 1.

As well as the one-particle observables, global properties like $\langle H \rangle$ and $P$ can be calculated with $\rho_1$, and even the many-particle entropy can be expressed in terms of $\rho_1$:

**The Effective One-Particle Entropy** (2.5.15)

$$S(\rho_{GC}) = -\text{Tr}\, \rho_{GC} \ln \rho_{GC} = \frac{VP}{T} + \beta \langle H - \mu N \rangle$$

$$= \text{tr}\left[ \pm \ln(1 \pm \exp[-\beta(h-\mu)]) + \beta \frac{h-\mu}{\exp[\beta(h-\mu)] \pm 1} \right]$$

$$= -\text{tr}[\rho_1 \ln \rho_1 \pm (1 \mp \rho_1) \ln(1 \mp \rho_1)].$$

**Remarks** (2.5.16)

1. The part in addition to the normal $-\text{tr}\, \rho \ln \rho$ in $S$ reveals that the many-particle system has increased disorder. The addition shows up in the entropy of a spin-$\frac{1}{2}$ density matrix,

$$S\begin{pmatrix} \rho & 0 \\ 0 & 1-\rho \end{pmatrix} = -\rho \ln \rho - (1-\rho) \ln(1-\rho),$$

where $\rho$ is the probability for spin-up, and in the entropy of an oscillator,

$$S\left((1-x)\begin{pmatrix}1 & & & \\ & x & & \\ & & x^2 & \\ & & & \ddots\end{pmatrix}\right) = -\rho \ln \rho + (1+\rho)\ln(1+\rho),$$

where

$$\rho = \sum_{n=0}^{\infty} n\rho_{nn} = \frac{x}{1-x}$$

is the expectation value of the number of phonons.

2. In accordance with the maximum-entropy principle (2.5.3), the one-particle $\rho_1$ (2.5.13) is the $\rho \in \mathscr{C}_1(L^2(\mathbb{R}^3))$ that maximizes

$$\frac{PV}{T} = -\operatorname{tr}[\rho \ln \rho \pm (1 \mp \rho)\ln(1 \mp \rho) + \rho\beta(h-\mu)]$$

(Problem 4). Also, on a formal level,

$$0 = -\frac{V}{T}\frac{\delta P}{\delta \rho}\bigg|_{\rho=\rho_1} = \beta(h-\mu) + \ln \rho_1 - \ln(1 \mp \rho_1)$$

$$\Rightarrow \rho_1 = [\exp[\beta(h-\mu)] \pm 1]^{-1}.$$

The density matrix $\rho_1$ describes the distribution of bosons or fermions. Its significance is brought out most clearly in the classical limit.

**Classical Bounds for the Pressure of Particles in External Fields** (2.5.17)

*With notation like that of* (2.2.7), *let*

$$h = |\mathbf{p}|^2 + v(\mathbf{x}) = \int d\Omega_z f(\mathbf{z})|\mathbf{z}\rangle\langle\mathbf{z}|, \qquad h(\mathbf{z}) = \langle\mathbf{z}|h|\mathbf{z}\rangle,$$

$$\rho(\mathbf{z}) = \operatorname{Tr} a_z^* a_z \rho_{GC} = \langle\mathbf{z}|[\exp[\beta(h-\mu)] \mp 1]^{-1}|\mathbf{z}\rangle = \langle\mathbf{z}|\rho_1|\mathbf{z}\rangle,$$

*where $v$ is such that all expressions appearing are well defined. Then, with* $\mathbf{z} = \mathbf{q} + i\mathbf{p}$, *for bosons,*

$$-\int d\Omega_z \ln(1 - \exp[-\beta(h(\mathbf{z}) - \mu)]) \le \beta P(\beta, \mu)V$$

$$\le -\int d\Omega_z \ln(1 - \exp[-\beta(|\mathbf{p}|^2 + v(\mathbf{q}) - \mu)]),$$

$$\beta P(\beta, \mu)V \le \int d\Omega_z \ln(1 + \rho(\mathbf{z})),$$

*and for fermions,*

$$\int d\Omega_z \ln(1 + \exp[-\beta(h(\mathbf{z}) - \mu)]) \leq \beta P(\beta, \mu)V$$
$$\leq \int d\Omega_z \ln(1 + \exp[-\beta(f(\mathbf{z}) - \mu)]),$$
$$-\int d\Omega_z \ln(1 - \rho(\mathbf{z})) \leq \beta P(\beta, \mu)V.$$

*In analogy with (2.4.9), one gathers that*

$$h(\mathbf{q} + i\mathbf{p}) = |\mathbf{p}|^2 + v_u(\mathbf{q}) + \int |\nabla u(\mathbf{x})|^2 \, d^3x$$

*and*

$$f(\mathbf{q} + i\mathbf{p}) = |\mathbf{p}|^2 + v^u(\mathbf{q}) - \int |\nabla u(\mathbf{x})|^2 \, d^3x,$$

*where*

$$v_u(\mathbf{q}) = \int v(\mathbf{x})|u(\mathbf{x} - \mathbf{q})|^2 \, d^3x$$

*and*

$$v(\mathbf{x}) = \int v^u(\mathbf{q})|u(\mathbf{x} - \mathbf{q})|^2 \, d^3q,$$

*and u is an arbitrary vector of $L^2(\mathbb{R}^3)$ such that $\|u\|_2 = 1$ and $\|\nabla u\|_2 < \infty$.*

**Proof**

**Bosons.** The first two inequalities are the analogues of (2.4.12), where the lower bound relies on (2.2.11) with the convex function $x \to -\ln(1 - \exp(-x))$. The upper follows from Corollary (2.1.8: 7) if it is borne in mind that $h - \mu$ must be positive, so $\|\exp[-(h - \mu)]\| < 1$, and the series

$$-\ln(1 - \exp[-\beta(h - \mu)]) = \sum_{n=1}^{\infty} \frac{\exp[-n\beta(h - \mu)]}{n}$$

converges in the norm $\|\cdot\|$. It must even converge in the norm $\|\cdot\|_1$, since it was assumed that $-\ln(1 - \exp[-\beta(h - \mu)]) \in \mathscr{C}_1$, and the series is monotonic. With recourse again to (2.4.9), each term is bounded by

$$-\int d\Omega_z (1/n)\exp[-n\beta(|\mathbf{p}|^2 + V(\mathbf{q}) - \mu)],$$

which also converges by assumption. Since all terms are positive, $\sum_n$ and

$\int d\Omega_z$ can be interchanged. The final inequality follows from the concavity of the function $x \to \ln(1 + x)$:

$$-\langle \mathbf{z}|\ln(1 - \exp[-\beta(h - \mu)])|\mathbf{z}\rangle = \langle \mathbf{z}|\ln(1 + \rho_1)|\mathbf{z}\rangle \leq \ln(1 + \langle \mathbf{z}|\rho_1|\mathbf{z}\rangle)$$

implies that

$$-\operatorname{tr}\ln(1 - \exp[-\beta(h - \mu)]) \leq \int d\Omega_z \ln(1 + \rho(\mathbf{z})).$$

**Fermions.** The first two inequalities again come from (2.2.11) with the convex function $x \to \ln(1 + \exp(-x))$, and the last one is a consequence of the convexity of $x \to -\ln(1 - x)$. □

**Remarks** (2.5.18)

1. If $x > 0$, then $(\exp(x) \pm 1)^{-1}$ is convex, and if $x < 0$, then it is concave. For bosons, $x > 0$, and so
$$\rho(\mathbf{z}) = \langle \mathbf{z}|(\exp[\beta(h - \mu)] - 1)^{-1}|\mathbf{z}\rangle \geq (\exp[\beta(h(\mathbf{z}) - \mu)] - 1)^{-1}.$$
The analogous inequality for fermions is true only if $h - \mu > 0$.
2. In Problem 3 it is shown that $\langle \mathbf{z}|(-\Delta)|\mathbf{z}\rangle = |\mathbf{p}|^2 + K, K = \int d^3x|\nabla u(\mathbf{x})|^2$, where $\mathbf{z} = \mathbf{q} + i\mathbf{p}$, and on the other hand, $-\Delta = \int d\Omega_z(|\mathbf{p}|^2 - K)|\mathbf{z}\rangle\langle \mathbf{z}|$. Similarly, $\langle \mathbf{z}|v|\mathbf{z}\rangle = \int d^3x|u(\mathbf{x} - \mathbf{q})|^2 v(\mathbf{x}) = v_u(\mathbf{q})$, and $v = \int d\Omega_z v^u(\mathbf{q}) \times |\mathbf{z}\rangle\langle \mathbf{z}|$, if $v(\mathbf{x}) = \int d^3q|u(\mathbf{x} - \mathbf{q})|^2 v^u(\mathbf{q})$. What goes on with the lower bound is thus that the classical Hamiltonian $h$ is increased by the kinetic energy $K$ of $u$, and the potential is smeared out by convolution with $|u|^2$. With the upper bound the classical Hamiltonian is reduced by $K$ and the potential is unsmeared. If $v$ is of slow enough variation that even for $u$ with small $K$, $v^u(q)$ is approximately equal to $v_u(\mathbf{q}) = \langle \mathbf{z}|v|\mathbf{z}\rangle$, then the bounds draw close together.
3. In the very dilute limit of (2.5.11; 1) the bounds produce the classical result, if the indistinguishablity of the particles is accounted for by a $1/N!$ in the phase space:
$$\exp\left(\frac{PV}{T}\right) = \sum_{N=0}^{\infty} \frac{1}{N!} \int d^3x_1 \cdots d^3p_N$$
$$\times \exp[-\beta(|\mathbf{p}_1|^2 + \cdots |\mathbf{p}_N|^2 + v(\mathbf{x}_1) + \cdots + v(\mathbf{x}_N) - N\mu)]$$
$$= \exp[\exp(-\beta(F_{cl} - \mu))],$$
so by (2.5.4),
$$\frac{PV}{T} = \int d^3x\, d^3p \exp[-\beta(|\mathbf{p}|^2 + v(\mathbf{x}) - \mu)] = N,$$
which is the ideal gas law. Unless $\exp(\beta(h - \mu)) \gg 1$, the statistics matter. They are built into the bounds, but the indeterminacy relation forces the bounds apart.

4. In the classical limit, in which Inequalities (2.5.17) become equalities, $\rho_1(\mathbf{x}, \mathbf{p}) = (\exp[-\beta(|\mathbf{p}|^2 + V(\mathbf{x}) - \mu)] \pm 1)^{-1}$ is the density on phase space that optimizes

$$\frac{PV}{T} = S(\rho_1) - \beta\langle h - \mu\rangle \equiv -\int d\Omega_z[\rho_1(\mathbf{z}) \ln \rho_1(\mathbf{z})$$
$$\pm (1 \mp \rho_1(\mathbf{z})) \ln(1 \mp \rho_1(\mathbf{z})) + \rho_1(\mathbf{z})\beta(h(\mathbf{z}) - \mu)]$$

(Problem 4).

5. If, more generally, $\rho$ is a density matrix of the many particle system on Fock space, and

$$\rho_1 = \int d\Omega_z\, d\Omega_{z'}\, |\mathbf{z}\rangle\langle\mathbf{z}'|\, \mathrm{Tr}(\rho a_{z'}^* a_z)$$

and

$$\rho(\mathbf{z}) = \mathrm{Tr}(\rho a_z^* a_z) = \langle \mathbf{z}|\rho_1|\mathbf{z}\rangle$$

are the associated one-particle density matrix and density, then it follows from (2.5.3) and (2.5.15) that

$$S(\rho) = -\mathrm{Tr}\, \rho \ln \rho \leq -\mathrm{tr}[\rho_1 \ln \rho_1 \pm (1 \mp \rho_1) \ln(1 \mp \rho_1)]$$
$$\leq -\int d\Omega_z[\rho(z) \ln \rho(z) \pm (1 \mp \rho(z)) \ln(1 \mp \rho(z))],$$

where the $H$ in (2.5.3) is taken as the second quantization of $(1/\beta) \times [\ln(1 \mp \rho_1) - \ln \rho_1]$, and $\mu$ is set to 0. The first inequality becomes an equality with $\rho_{GC}$, which is the density matrix of greatest entropy for a given one-particle density matrix $\rho_1$. The second inequality follows from (2.2.11), since

$$x \to -[x \ln x \pm (1 \mp x) \ln(1 \mp x)]$$

is concave with the upper signs for $0 < x < 1$ and with the lower signs for $x < 0$.

The extent of the validity of the classical picture will be delineated through a series of examples.

**Free Bosons and Fermions in a Box with Soft Walls** (2.5.19)

With a harmonic potential $v(\mathbf{x}) = \omega^2|\mathbf{x}|^2$, the $N$-particle Hamiltonian is

$$H = \sum_{i=1}^{N} (|\mathbf{p}_i|^2 + \omega^2|\mathbf{x}_i|^2) = \sum_{i=1}^{N} |\mathbf{p}_i|^2 + \frac{\omega^2}{2N} \sum_{i,j=1}^{N} |\mathbf{x}_i - \mathbf{x}_j|^2 + \frac{\omega^2}{N}\left|\sum_{i=1}^{N} \mathbf{x}_i\right|^2,$$

containing harmonic forces between the particles and a harmonic force acting on the center of mass. As before (cf. (2.4.11) and (2.5.18; 1)), let $\mathbf{z} = \mathbf{q} + i\mathbf{k}$ and $u(\mathbf{x}) = \exp(-\omega|\mathbf{x}|^2/2)$: $h(\mathbf{z}) = |\mathbf{k}|^2 + \omega^2|\mathbf{q}|^2 + 3\omega$, $f(|\mathbf{z}|) = |\mathbf{k}|^2 + \omega^2|\mathbf{q}|^2 - 3\omega$. Because

$$\mp \int \frac{d^3k\, d^3q}{(2\pi)^3} \ln(1 \mp \exp[-\beta(|\mathbf{k}|^2 + \omega^2|\mathbf{q}|^2 - \mu)])$$
$$= \pm \frac{T^3}{(2\omega)^3} \sum_{\nu=1}^{\infty} (\pm 1)^\nu \frac{\exp(\nu\beta\mu)}{\nu^4},$$

(2.5.17) implies

$$\pm \frac{T^3}{(2\omega)^3} F_4(\pm(\exp[\beta(\mu - 3\omega)]) \le \ln \operatorname{Tr}_{\substack{B\\F}} \exp[-\beta(H - \mu N)]$$
$$\le \pm \frac{T^3}{(2\omega)^3} F_4(\pm\exp[\beta(\mu + 3\omega)]), \tag{2.5.20}$$

where

$$F_\sigma(x) \equiv \sum_{\nu=1}^{\infty} \frac{x^\nu}{\nu^\sigma}.$$

The result can be calculated exactly in this case, since the eigenvalues are $\varepsilon_\mathbf{m} = 3\omega + 2\omega(m_1 + m_2 + m_3)$, $\mathbf{m} \in (\mathbb{Z}^+)^3$, and so

$$\mp \sum_\mathbf{m} \ln(1 \mp \exp[-\beta(\varepsilon_\mathbf{m} - \mu)])$$
$$= \pm \sum_{\nu=1}^{\infty} (\pm 1)^\nu \frac{\exp[\nu\beta(\mu - 3\omega)]}{\nu} [(1 - \exp(-2\beta\omega\nu))^{-3} - 1].$$

The bounds draw together to this value in the limit $\omega \to 0$. This limit is related to the limit $V \to \infty$, since the average of, for instance, $|\mathbf{x}|^2$ is $\sim T/\omega^2$. Accordingly, we eliminate $\omega$ in favor of the effective volume $V = (\pi T)^{3/2}/\omega^3$ and take the limit $V \to \infty$. Then with $z = \exp(\beta\mu)$, (2.5.20) yields

$$P_{\substack{B\\F}}(T, z) = \pm \frac{T^{5/2}}{8\pi^{3/2}} F_4(\pm z). \tag{2.5.21}$$

**Remarks** (2.5.22)

1. As $\omega \to 0$, the potential $v$ goes to zero pointwise, and the density (2.5.14; 3) on phase space turns into the well-known Bose or Fermi distribution,
$$\rho(\mathbf{x}, \mathbf{p}) = [\exp(\beta(|\mathbf{p}|^2 - \mu)) \mp 1]^{-1}.$$
2. The energy spectrum of this example resembles that of a massless particle in a box $\{\mathbf{x}: |x_i| < L/2\}$, $E = (p_1^2 + p_2^2 + p_3^2)^{1/2}$, $p_i = m_i\pi/L$, $\mathbf{m} \in (\mathbb{Z}^+)^3$.

In the limit $L \to \infty$, this $E$ produces the same pressure up to a constant as $(m_1 + m_2 + m_3)\omega$, when $\omega$ is identified with $\pi/L$. Then

$$P_{\underset{F}{B}}(T, z) = \pm T^4 F_4(\pm z)\pi^{-3}.$$

**A Box with Hard Walls** (2.5.23)

Now suppose that the potential $v_L(\mathbf{x}) \geq 0$ is significantly smaller than $1/L^2$ for $|x_i| < L/2$ but increases exponentially as soon as $|x_i| > L/2$. Since what happens should not depend on the precise form of $v_L$, only certain bounds will be imposed on $v_L$. Because of the monotonic property, all the steps up to (2.5.17) and (2.5.18; 1) proceed as before.†

$$\gamma_-^3 \varphi(x)\varphi(y)\varphi(z) \leq v_L(\mathbf{x}) \leq \gamma_+^3 \varphi(x)\varphi(y)\varphi(z), \qquad 0 < \gamma_- < \gamma_+,$$

$$\varphi(x) = \exp\left(\frac{-cL}{2}\right) \cosh(cx),$$

$$\mathcal{N}^2 \int_{-\infty}^{\infty} dx' \exp(-bx'^2)\varphi(x + x') = \exp\left(\frac{c^2}{4b}\right)\varphi(x),$$

so for the other bound,

$$\varphi(x) = \int_{-\infty}^{\infty} dx \exp(-bx'^2) \exp\left(-\frac{c^2}{4b}\right)\varphi(x' + x)\mathcal{N}^2.$$

The $x$-space portion of the calculation of $\sum_{\nu=1}^{\infty} (-1)^{\nu+1}[\exp(\nu\beta\mu)/\nu] \times \int d\Omega_z \exp(-\beta\nu g(z))$, where $g(z) = f(z)$ or respectively $h(z)$ (cf. (2.5.17) and (2.2.11)), leads to

$$\int_{-\infty}^{\infty} dx \exp(-B_{\pm} \cosh cx) = \frac{2}{c}\int_1^{\infty} \frac{dv}{\sqrt{v^2 - 1}} \exp(-B_{\pm} v)$$

$$\overset{B_{\pm} \to 0}{=} \frac{2}{c}\left(\ln \frac{1}{B_{\pm}} + O(1)\right)$$

with $B_{\pm} = \gamma_{\pm} \beta \exp[\pm c^2/4b - cL/2])$, since it is being evaluated in the limit $V = L^3 \to \infty$. If a sequence $(v_L(\mathbf{x}))_{L\to\infty}$ of wall potentials has bounds of the above-mentioned form with $c(L) = o(L)$ and $\ln(\beta\gamma_{\pm}(L)) = o(c(L) \cdot L)$, then

$$\frac{2}{cL} \ln \frac{1}{B_{\pm}} = 1 \mp \frac{c}{2bL} - \frac{2}{cL} \ln \beta\gamma_{\pm}$$

converges to 1 for both bounds. The $p$-integral is the same as in (2.5.19), and so, finally,

$$P_{\underset{F}{B}}(T, z) = \pm \frac{T^{5/2}}{8\pi^{3/2}} F_{5/2}(\pm z). \tag{2.5.24}$$

† From this point until right before (2.5.24), + and − will indicate upper and lower bounds for the potential due to the wall rather than Bose and Fermi statistics.

**Remarks** (2.5.25)

1. This is the same result as that of summing over all the eigenvalues of a free particle in a box with Dirichlet boundary conditions on the wall (Problem 5). The bounds (2.5.17) show that in very large part it is only the total volume of $V$ that matters, rather than its detailed form.
2. The nature of the wall is expressed by $F_{5/2}$ in (2.5.23) and $F_4$ in (2.5.21). For lower densities, $z \ll 1$, they coincide, as $F_\sigma = z + O(z^2)$.

**The Thermodynamic Functions of Free Particles** (2.5.26)

All the thermodynamic functions can be obtained from $P(T, z)$, so (2.5.24) will allow the gaps left by (2.3.10) to be filled in, and the functions can be written down explicitly. We shall investigate the limiting cases where $z \to \infty$, $z \to 0$, and $z \to -1$, corresponding to the extremes of Fermi, Boltzmann, and Bose statistics. The limits $z \to \infty, -1$ are what is referred to as a **degenerate gas**. By Problem 1, $F$ has the asymptotic forms

$$-F_{5/2}(-z) \begin{cases} \xrightarrow{z \to -1} -\zeta(\tfrac{5}{2}) + (z+1)\zeta(\tfrac{3}{2}) \\ \xrightarrow{z \to 0} z - z^2 \cdot 2^{-5/2} \\ \xrightarrow{z \to \infty} \dfrac{4}{3\sqrt{\pi}}\left[\dfrac{2}{5}(\ln z)^{5/2} + \dfrac{\pi^2}{4}(\ln z)^{1/2}\right], \end{cases} \tag{2.5.27}$$

where $\zeta(\sigma)$ is the Riemann zeta function,

$$\zeta(\sigma) = \sum_{\nu=1}^{\infty} \frac{1}{\nu^\sigma} = F_\sigma(1), \qquad \sigma \in \mathbb{C}, \quad \operatorname{Re}\sigma > 1.$$

The zeta function has an analytic continuation to the punctured complex plane $\{\sigma \in \mathbb{C} \mid \sigma \neq 1\}$. In the three limits,

$$\frac{P}{T} = \frac{2\varepsilon}{3T} = \begin{cases} \xrightarrow{\text{Bose}} \dfrac{T^{3/2}}{8\pi^{3/2}}\left[\zeta(\tfrac{5}{2}) + (z-1)\zeta(\tfrac{3}{2})\right] \\ \xrightarrow{\text{Boltzmann}} \dfrac{T^{3/2}}{8\pi^{3/2}} z(1 \pm z \cdot 2^{-5/2}) \\ \xrightarrow{\text{Fermi}} \dfrac{T^{3/2}}{6\pi^2}\left[\dfrac{2}{5}(\ln z)^{5/2} + \dfrac{\pi^2}{4}(\ln z)^{1/2}\right], \end{cases} \tag{2.5.28}$$

so, writing

$$\varepsilon_{\substack{B\\F}} = T^2 \frac{\partial}{\partial T}\frac{1}{T} P_{\substack{B\\F}}(T, z) = \tfrac{3}{2} P_{\substack{B\\F}}(T, z) = \pm\tfrac{3}{2} T^{5/2} F_{5/2}(\pm z)\frac{1}{8\pi^{3/2}},$$

$$\rho_{\substack{B\\F}} = z\frac{\partial}{\partial z}\frac{1}{T} P_{\substack{B\\F}}(T, z) = \pm T^{3/2} F_{3/2}(\pm z)\frac{1}{8\pi^{3/2}},$$

$$\sigma_{\substack{B\\F}} = \pm\frac{T^{3/2}}{8\pi^{3/2}}\left[\tfrac{5}{2}F_{5/2}(\pm z) - \ln z F_{3/2}(\pm z)\right], \tag{2.5.29}$$

to the lowest nonvanishing order,

$$\rho = \begin{cases} \xrightarrow{\text{Bose}} z\zeta(\tfrac{3}{2})\dfrac{T^{3/2}}{8\pi^{3/2}} \\[2ex] \xrightarrow{\text{Boltzmann}} \dfrac{T^{3/2}}{8\pi^{3/2}} z(1 \pm z\cdot 2^{-3/2}) \\[2ex] \xrightarrow{\text{Fermi}} \dfrac{T^{3/2}}{6\pi^2}(\ln z)^{3/2} + \dfrac{T^{3/2}}{48}(\ln z)^{-1/2}, \end{cases} \tag{2.5.30}$$

$$\sigma = \begin{cases} \xrightarrow{\text{Bose}} \dfrac{T^{3/2}}{8\pi^{3/2}}\tfrac{5}{2}\zeta(\tfrac{5}{2}) \\[2ex] \xrightarrow{\text{Boltzmann}} \dfrac{T^{3/2}}{8\pi^{3/2}}(\tfrac{5}{2}z - z\ln z) \\[2ex] \xrightarrow{\text{Ferm}} \dfrac{T^{3/2}}{12}(\ln z)^{1/2}. \end{cases} \tag{2.5.31}$$

When expressed in terms of the more intuitively appealing variables $\rho$ and $T$,

$$P = \frac{2}{3}\varepsilon = \begin{cases} \xrightarrow{\text{Bose}} \rho T + T^{5/2}(\zeta(\tfrac{5}{2}) - \zeta(\tfrac{3}{2}))/8\pi^{3/2} \\[2ex] \xrightarrow{\text{Boltzmann}} \rho T \quad (\text{“ideal gas”}) \\[2ex] \xrightarrow{\text{Fermi}} \dfrac{(6\pi^2\rho)^{5/3}}{15\pi^2} + \dfrac{(6\pi^2\rho)^{1/3}}{24}T^2, \end{cases}$$

$$\sigma = \begin{cases} \xrightarrow{\text{Bose}} \tfrac{5}{2}\zeta(\tfrac{5}{2})\dfrac{T^{3/2}}{8\pi^{3/2}} \\[2ex] \xrightarrow{\text{Boltzmann}} \tfrac{5}{2}\rho - \rho\ln\dfrac{\rho 8\pi^{3/2}}{T^{3/2}} = \rho\ln\dfrac{T^{3/2}\exp(5/2)}{\rho 8\pi^{3/2}} \\[2ex] \xrightarrow{\text{Fermi}} \dfrac{T}{12}(6\pi^2\rho)^{1/3}. \end{cases} \tag{2.5.32}$$

**Remarks** (2.5.33)

1. As $z \to 0$, (2.5.32) gives the classical result (2.3.9; 1) with an additional factor $1/N!$ in the volume of phase space. If $V_P$ denotes the volume available in the one-particle phase space, and the $1/N!$ is incorporated into the general definition, then

$$S \sim \ln \frac{1}{N!}\left(\frac{V_P}{h^3}\right)^N$$

leads to

$$\frac{S}{N} \sim \ln \frac{V_P}{Nh^3}.$$

On the other hand, in configuration space and with units for which $\hbar = m = 1$, (2.5.32) informs us that $S \sim \ln VT^{3/2}/N$. Since $T^{-1/2}$ equals the thermal de Broglie wavelength $\lambda$ with these units, the following rule of thumb applies to the entropy: Entropy per particle = ln{volume of phase space per particle, as measured in $h^3$} = ln{volume of configuration space per particle, as measured in $\lambda^3$}.
2. Fermions have a zero-point energy $E_0 = V\varepsilon_0$ left over when $T \to 0$, where $\varepsilon_0 \equiv (6\pi^2\rho)^{5/3}/10\pi^2$, and a zero-point pressure $2\varepsilon_0/3$. Because

$$T = \frac{4(\varepsilon - \varepsilon_0)^{1/2}}{6(\pi^2\rho)^{1/6}},$$

it is also possible to write

$$\sigma = \left(\frac{\varepsilon}{\varepsilon_0} - 1\right)^{1/2} \frac{\rho 2\pi}{\sqrt{10}},$$

showing that the number $M$ of states in the interval $[E_0, E]$ is

$$M \cong \exp\left\{N\left(\frac{E}{E_0} - 1\right)^{1/2} \frac{2\pi}{\sqrt{10}}\right\}.$$

For example, in an atomic nucleus the kinetic energy is $E_0 \cong N \cdot 20$ MeV, so with a fixed kinetic excitation energy $\delta E = E - E_0$ the number of states in the interval is $\sim \exp 2\sqrt{N}\sqrt{\delta E/20 \text{ MeV}}$. If $\delta E \sim 1$ MeV, then for $N = 20$ there are about $e^2$, i.e., 7 or 8, states; whereas if $N \sim 200$, then the number increases to about $e^{6.5} \sim 0.5 \times 10^3$. This is in agreement with the experimental observation that the density of the energy states of heavy nuclei is on the order of $(\text{eV})^{-1}$.
3. If the energy of the ground state is redefined to zero, then $z$ must be less than 1 for bosons—otherwise by (2.5.12) $n_0 \equiv \langle a_0^* a_0 \rangle = z/(1 - z)$ is either infinite or negative. Because $F_{3/2}(z) < \zeta(\frac{3}{2})$ when $0 < z < 1$, it follows from (2.5.29) that $T > T_c \equiv (8\pi^{3/2}\rho/\zeta(\frac{3}{2}))^{2/3}$. On the other hand, $n_0$ can be made arbitrarily big by taking $z$ close enough to 1. The difficulty with this is that the two limits $z \to 1$ and $V \to \infty$ have to be taken jointly

if the density has been fixed. If $z(V) = 1 - 1/\rho_0 V$ and $T < T_c(\rho)$, then

$$\rho = \rho_0 + \zeta(\tfrac{3}{2})\frac{T^{3/2}}{8\pi^{3/2}},$$

$$P = \tfrac{2}{3}\varepsilon = \zeta(\tfrac{5}{2})\frac{T^{5/2}}{8\pi^{3/2}} = \lim_{V\to\infty}\frac{T}{V}\ln \operatorname{Tr}\exp\left[-\frac{1}{T}(H_V - \mu_V(T, \rho)N)\right],$$

with

$$\lim_{V\to\infty}\mu_V(T, \rho) = 0 \quad \text{for all } T \le T_c(\rho),$$

$$\sigma = \tfrac{5}{2}\zeta(\tfrac{5}{2})\frac{T^{3/2}}{8\pi^{3/2}}.$$

This shows that a nonzero fraction $\rho_0/\rho = 1 - (T/T_c)^{3/2}$ of the particles reside in the ground state and contribute nothing to the energy, pressure, or entropy (provided $H$ is replaced with $H - E_0$). The number of particles in the first excited state, $n_1 = 1/(z^{-1}\exp(\beta/L^2) - 1) \sim L^2$, is rather large, but $n_1/V \to 0$. For similar reasons, the relative mean-square deviation $(\Delta n)^2/\langle n_i\rangle^2$ remains positive for $n_0$ as $V \to \infty$, but goes to zero for the higher states. The specific heat

$$c_V = \left.\frac{\partial E}{\partial T}\right|_{V,N}$$

is continuous at $T_c$ and $\partial c_V/\partial T$ is discontinuous (Problem 2). If $T = T_c$, then the choice of $\rho_0$ has to depend on $V$.

4. The values $\mu = 0$ and $z = 1$ apply to a situation where $N$ is not conserved, such as a gas of photons or phonons (cf. (2.5.22; 2)). It is easy to calculate $\operatorname{Tr}\exp(-\beta H)$ with the $H$ of (2.5.7). The pressure $P = -\varphi$, and

$$\left.\frac{\partial P}{\partial \rho}\right|_T = -\left.\frac{\partial \varphi}{\partial \rho}\right|_T = \mu = 0,$$

so the compressibility is infinite. The system behaves much like a gas at the condensation point, the vacuum state, i.e., no particles, being analogous to the condensed state. It therefore has $\varepsilon = \sigma = P = V = 0$, and the system can be compressed into the vacuum. The entropy density $\sigma$ is then simply the quantity $\Delta s/\Delta v$ of the Clausius–Clapeyron equation which simply assumes the form

$$\left.\frac{\partial P}{\partial T}\right|_\rho = \sigma.$$

Since $P = -\varphi$, Theorem (2.4.14) implies that this equation holds identically. The quantities $\varepsilon/T \approx \rho \approx \sigma$ depend only on $T$ and correspond to a particle of energy $T$ in each wavelength cube. Consequently, entropy $\cong$ particle number $\cong$ energy/$T$.

**Particles in a Magnetic Field** (2.5.34)

The Hamiltonian was given in (III: 3.3.5; 3):

$$H = |\mathbf{p} - e\mathbf{A}|^2 = p_3^2 + 2eB(a^*a + \tfrac{1}{2}).$$

The boundary conditions are that the wave-function must vanish at $x_3 = 0$ and $x_3 = L$, the 3-axis pointing along $B$, so the eigenvalues of $p_3$ are $\pi m/L$, $m = 1, 2, 3, \ldots$. The center $\bar{\mathbf{x}}$ of the orbit is confined to $|\bar{\mathbf{x}}|^2 = (2/eB)(g + \frac{1}{2}) < R^2$ in the plane perpendicular to $B$, so the geometry is cylindrical. The "wall potential" $\infty \cdot \Theta(|\bar{\mathbf{x}}|^2 - R^2)$ confining the particle is not a multiplication operator by a real-valued function $V(x_1, x_2)$, but rather a function of the operator

$$|\bar{\mathbf{x}}|^2 = \tfrac{1}{4}(x_1^2 + x_2^2) + \frac{1}{e^2B^2}(p_1^2 + p_2^2) + \frac{1}{eB}(x_1p_2 - x_2p_1),$$

representing the sum of a two-dimensional harmonic oscillator in the $x_1 - x_2$-plane and the $x_3$-component of the angular momentum. The construction of such a momentum-dependent wall potential will be left to the ingenuity of the experimentalists. By (III: 3.5.3; 3), $|\bar{\mathbf{x}}|^2$ is quantized so that $g$ is a whole number, and $a^*a$ has the eigenvalues $n = 0, 1, 2, \ldots$. As $L \to \infty$, the sum $\sum_{g=0}^{R^2eB/2} \sum_{m=1}^{\infty} \sum_{n=0}^{\infty}$ turns into

$$\int_0^\infty dp_3 \frac{L}{\pi} \sum_{n=0}^{\infty} \frac{R^2eB}{2} = \frac{VeB}{2\pi^2} \int_0^\infty dp_3 \sum_{n=0}^{\infty},$$

where $V$ denotes the volume of the cylinder. The classical bounds amount to the replacement

$$\sum_{n=0}^{\infty} \to \int_0^\infty dn,$$

in which all magnetic effects are swept away. We have to resort to the exact expression (2.5.9), with which the grand canonical partition functions becomes

$$\beta P_{\substack{B\\F}}(z) = \mp \frac{eB}{2\pi^2} \int_0^\infty dp_3 \sum_{n=0}^{\infty} \ln(1 \mp z \exp[-\beta(p_3^2 + eB(2n + 1))])$$

$$= \pm \frac{T^{3/2}}{8\pi^{3/2}} \sum_{\nu=1}^{\infty} \frac{(\pm z)^\nu}{\nu^{5/2}} \frac{\nu eB\beta}{\sinh \nu eB\beta}, \tag{2.5.35}$$

where the $B$ in $P_{\substack{B\\F}}$ denotes Bose statistics as usual and has nothing to do with the magnetic field $B$. This reveals right away that, as in (2.3.33; 2), an arbitrarily weak magnetic field ruins the phase transition of the Bose gas, since for any $T$.

$$\rho_{\substack{B\\F}} = z\frac{\partial}{\partial z}\beta P_{\substack{B\\F}} = \pm\frac{T^{3/2}}{8\pi^{3/2}}\sum_{\nu=1}^{\infty}\frac{(\pm z)^{\nu}}{\nu^{1/2}}\frac{eB\beta}{\sinh \nu eB\beta}$$

can get arbitrarily big as $z \to \exp(\beta E_0) = \exp(\beta eB)$. This happens because the particles are free to move only parallel to $\mathbf{B}$ and are trapped in orbits in the direction perpendicular to $\mathbf{B}$ even though the radius of the cylinder goes to infinity. The system acts as though confined to a cylinder only the length of which tends to $\infty$, and in one dimension there is no Bose condensation. If the magnetic energy $eB$ is much less than the thermal energy $T$, then the next correction to the foregoing result is $\sim B^2$:

$$\beta P_{\substack{B\\F}} \to \pm\frac{T^{3/2}}{8\pi^{3/2}}\left[F_{5/2}(\pm z) - \frac{1}{6}\left(\frac{eB}{T}\right)^2 F_{1/2}(\pm z)\right]. \tag{2.5.36}$$

If this is used to calculate the magnetization per volume in the limit $B \to 0$ with $T$ fixed,

$$m = \frac{\partial P_{\substack{B\\F}}}{\partial eB} = \frac{1}{V}\left\langle \sum_{\substack{\text{all}\\\text{particles}}} (x_1 p_2 - x_2 p_1 - eB(x_1^2 + x_2^2))\right\rangle$$

$$= \mp\frac{T^{3/2}}{8\pi^{3/2}}\frac{eB}{3T}F_{1/2}(\pm z), \tag{2.5.37}$$

then with (2.5.26) and the formula $F_{\sigma-n}(z) = (z(d/dz))^n F_\sigma(z)$ (see (2.5.20)), its limits in the three extreme cases of the different statistics are

$$m \begin{cases} \xrightarrow{\text{Bose}} & -eB\cdot\infty \\ \xrightarrow{\text{Boltzmann}} & -\dfrac{eB}{3T}\rho \\ \xrightarrow{\text{Fermi}} & -\dfrac{eB}{12\pi^2}(6\pi^2\rho)^{1/3}. \end{cases} \tag{2.5.38}$$

**Remarks** (2.5 39)

1. The negative sign indicates diamagnetism, which is to be expected quantum-mechanically: By Lenz's law the classical orbits rotate in the direction with negative $L_z$. However, a current appears in the other direction when particles bounce off the wall of the box (see Figure 20).

   With classical statistics the circulating currents cancel out at every point of the interior, leaving only a current circulating along the surface,

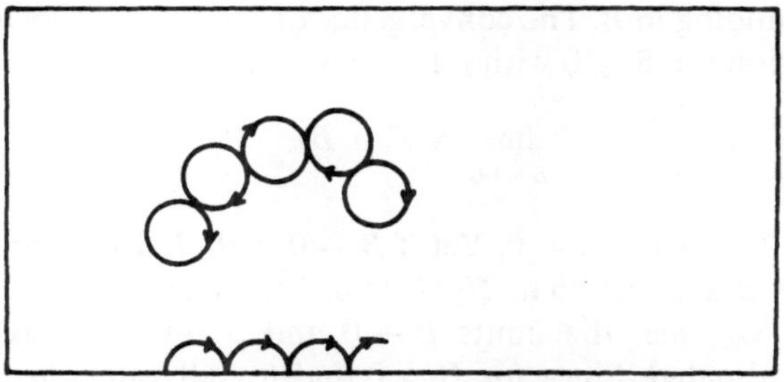

Figure 20 Classical trajectories of particles in a box with a magnetic field.

which is exactly compensated for by the "reflected" current, since the partition function

$$\int d^3x\, d^3p \exp[-\beta|\mathbf{p} - \mathbf{A}(\mathbf{x})|^2] = \int d^3x\, d^3p \exp(-\beta|\mathbf{p}|^2)$$

is completely independent of $B$. This means that if either $\rho$ is fixed and $T \to \infty$ or $T$ is fixed and $\rho \to 0$, then $m$ tends to 0. Diamagnetism is therefore a characteristically quantum-mechanical effect; if the sum $\sum_{n=0}^{\infty}$ is replaced with an integral $\int_0^\infty dn$, and $2n + 1$ becomes $2n$, which is in essence the limit $\hbar \to 0$, then $P$ becomes independent of $B$ (a theorem of Bohr and van Leeuwen).

2. In quantum theory, states with negative $L_z$ are energetically favored (III: 3.3.21; 4), so a quantum gas is diamagnetic. The reason that the magnetization $m$ of a completely degenerate Bose gas tends to $\infty$ is that $P$ fails to be analytic at $z = 1$, $B = 0$. This topic will shortly be discussed in more detail.
3. Since $P$ depends only on $R^2L$,

$$R\frac{\partial}{\partial R}P = 2L\frac{\partial}{\partial L}P,$$

i.e., the pressure remains isotropic.

In order to make sense of the limit of degenerate Bose gas, let $\beta\mu = \ln z$, and write

$$\frac{\beta P_B}{V} = \frac{T^{3/2}}{8\pi^{3/2}}\sum_{\nu=1}^{\infty}\frac{\exp[-\beta\nu(eB - \mu)]}{\nu^{5/2}}\frac{2eB\beta\nu}{1 - \exp[-2eB\beta\nu]},$$

$$\rho = \frac{T^{3/2}}{8\pi^{3/2}}\sum_{\nu=1}^{\infty}\frac{\exp[-\beta\nu(eB - \mu)]}{\nu^{3/2}}\frac{2eB\beta\nu}{1 - \exp[-2eB\beta\nu]},$$

$$m = -\rho + \frac{T^{3/2}}{4\pi^{3/2}}\sum_{\nu=1}^{\infty}\frac{\exp[-\beta\nu(eB - \mu)]}{\nu^{3/2}(1 - \exp[-2eB\beta\nu])^2} \times [1 - \exp[-2eB\beta\nu](1 + 2eB\beta\nu)], \qquad (2.5.40)$$

without expanding in $B$. The convergence of the series for $m$ and $\rho$ in (2.5.40) (by domination for $B \geq 0$ with $\mu$ fixed) implies that

$$\lim_{B \to +0} m(T, \mu, B) = 0$$

for all fixed $T > 0$ and $\mu < 0$. Yet if $B \to 0$ with $T$ fixed and $\mu < 0$ then all the densities $\rho$ are less than $\zeta(\frac{3}{2})T^{3/2}/8\pi^{3/2}$, as in (2.5.33; 3). If $T \leq T_c(\rho)$ (see (2.5.33; 3)), then the limits $B \to 0$ and $\mu \to 0$ must again be appropriately coordinated. Since for $B > 0$ and for all values $\rho > 0$ and $T > 0$ there exists a unique $\mu(T, \rho, B) < eB$ such that $\lim_{B \to 0} \mu(T, \rho, B) = 0$ for $T \leq T_c(\rho)$), and since the series for $m + \rho$ from (2.5.40) also converges uniformly in $B$ on an interval containing $\mu = eB$, the limit $B \to 0$ can be taken term by term. This yields

$$\lim_{B \to 0} m(T, \rho, B) = -\rho_0 = -\rho\left[1 - \left(\frac{T}{T_c(\rho)}\right)^{3/2}\right],$$

provided that $T \leq T_c(\rho)$ (cf. (2.5.33; 3)). If $T \geq T_c(\rho)$ then the limit is zero as observed earlier.

**Remarks** (2.5.41)

1. The physical interpretation of this result is that in the limit $B \to 0$ only the particles in the ground state contribute to the magnetization. The ground state has $L_z = -1$, so for $B = 0$ the contribution to $m$ is simply the sum of $L_z$ over the particles in a unit volume in the ground state.
2. The notation $B$ is perhaps misleading, since it stands only for the external field and not for that due to the system itself. Actually, the field due to the system has to be taken into account, as it screens $B$ throughout the interior of the system.

**Black-Body Radiation in Partial (i.e., Anisotropic) Equilibrium** (2.5.42)

If the particles are massless, as in (2.5.22; 2) and (2.5.33; 4), and they have a density matrix like $\rho_{GC}$ but containing only states in a certain dilatation-invariant part $D$ of $p$-space, then we can still write

$$\varphi = T \int_D \frac{d^3p}{(2\pi)^3} \ln(1 - \exp[-\beta|\mathbf{p}|]) = -cT^4,$$

where the constant $c$ depends on $D$ (but not on $T$). It is then still true that

$$\varepsilon = 3P = -3\varphi = \tfrac{3}{4}T\sigma = 3cT^4.$$

A realistic example of this situation is sunlight falling on the earth, for which essentially all the $p$-vectors come from the direction of the sun. The constant $c$ is reduced by a factor $\sim 10^{-5}$, the solid angle subtended by the sun, in comparison with the isotropic equilibrium value with $D = \mathbb{R}^3$. Once the radiation is made isotropic without changing $\varepsilon$ significantly by the time it reaches the earth, $T$ is lowered by a factor of about $10^{-5/4}$, from $\sim 6000°$ K to $\sim 300°$ K. At the same time, $\sigma = 4\varepsilon/3T$ is increased by this factor of 20. It is consistent with an increase in the total entropy that this physical process creates highly ordered structures with little entropy; their decrease of entropy is nothing compared with the gigantic increase of the radiation entropy. About $10^{20}$ photons per cm$^2$ arrive from the sun each minute, and this times 20 is the entropy increase/cm$^2$-min. In an hour this comes to roughly the total entropy of a cubic centimeter of matter for each square centimeter of ground, so, for example, a newly planted forest could grow to a height of 10 meters over a summer without violating the second law of thermodynamics. The sun thus expends entropy as well as energy. Although isotropic black-body radiation at 300° K would be just as energetic, the energy would be unusable for the creation of life (as would be the case as the universe subsided into heat death).

The grand canonical ensemble determines the expectation values of field operators as well as the thermodynamic functions. Equation (2.5.12) showed how to calculate quadratic expressions involving the field operators, and quartic expressions for particles in an external field can easily be calculated in the same way,

$$\begin{aligned}\langle a_m^* a_j^* a_{j'} a_{m'}\rangle &= (\delta_{mm'}\delta_{jj'} \pm \delta_{mj'}\delta_{jm'})(\exp[\beta(\varepsilon_m - \mu)] \mp 1)^{-1} \\ &\quad \times (\exp[\beta(\varepsilon_j - \mu)] \mp 1)^{-1} \\ &= \langle a_m^* a_{m'}\rangle\langle a_j^* a_{j'}\rangle \pm \langle a_m^* a_{j'}\rangle\langle a_j^* a_{m'}\rangle. \end{aligned} \tag{2.5.43}$$

**Remark** (2.5.44)

If the mean-square deviations of the occupation numbers are calculated in this way, then

$$\langle (a_m^* a_m)^2\rangle - \langle a_m^* a_m\rangle^2 = \langle a_m^* a_m\rangle(1 \pm \langle a_m^* a_m\rangle).$$

Independent particles would follow a Poisson distribution law $w(n) = \exp(-\bar{n})\bar{n}^n/n!$ for which the mean-square deviation would equal the expectation value of the occupation number. The deviation is greater with Bose statistics and less with Fermi statistics, which can be interpreted as meaning that bosons have a tendency to bunch up and fermions to keep at a distance.

In elementary quantum mechanics a state was characterized by the expectation values of the Weyl operators (cf. (III: 3.1.2; 1)), and likewise now the complete determination of the state requires the expectation value of, say, $\exp[i \int d^3x(a(\mathbf{x})f^*(\mathbf{x}) + a^*(\mathbf{x})f(\mathbf{x}))]$ for all $f \in C_0^\infty(\mathbb{R}^3)$. The best way for this to be calculated in the grand canonical ensemble for particles in an external field makes use of coherent states. In Problem 6 it is shown that

$$\frac{\operatorname{Tr}\exp[-\beta\omega a^*a]\exp[i(a^*\alpha + a\alpha^*)]}{\operatorname{Tr}\exp[-\beta\omega a^*a]} = \exp\left[-|\alpha|^2\left(\frac{1}{2} + \frac{1}{\exp(\beta\omega) - 1}\right)\right] \quad \text{if } [a, a^*] = 1.$$

Therefore:

**The Grand Canonical State for Bosons in an External Field** (2.5.45)

is

$$\left\langle \exp\left[i\sum_m (a_m^*\alpha_m + a_m\alpha_m^*)\right]\right\rangle = \exp\left[-\sum_m |\alpha_m|^2\left(\frac{1}{2} + \frac{z}{\exp(\beta\varepsilon_m) - z}\right)\right].$$

**Example** (2.5.46)

Free bosons in a cube of volume $V = L^3$, with periodic boundary conditions. Let

$$a_V(\mathbf{k}) = \int_V \frac{d^3x}{L^{3/2}} \exp(-i\mathbf{k}\cdot\mathbf{x})a(\mathbf{x}),$$

and

$$a_f = \sum_{k\in((2\pi/L)\mathbb{Z})^3} L^{-3/2}\tilde{f}(\mathbf{k})a_V(\mathbf{k}), \qquad \tilde{f}(\mathbf{k}) = \int d^3x \exp(i\mathbf{k}\cdot\mathbf{x})f(\mathbf{x}),$$

for $f \in L^2(V)$. Then because $\omega = |\mathbf{k}|^2$,

$$\langle\exp[i(a_f^* + a_{f^*})]\rangle = \exp\left[-\sum_{k\in((2\pi/L)\mathbb{Z})^3} L^{-3}|\tilde{f}(\mathbf{k})|^2\left(\frac{1}{2} + \frac{z}{\exp(\beta|\mathbf{k}|^2) - z}\right)\right].$$

A more convenient expression in the calculation of ordered products is $\exp[i\sum_m a_m^*\alpha_m]\exp[i\sum_m a_m\alpha_m^*]$. Its expectation values can be read off from the formula $\exp(A + B) = \exp A \exp B \exp(\frac{1}{2}[B, A])$, which holds provided that $[A, [A, B]] = [B, [A, B]] = 0$, which in this case is in accordance with the Weyl relations (III: 3.1.2; 1):

**The Generating Function for Ordered Products** (2.5.47)

$$\left\langle \exp\left[i\sum_m a_m^*\alpha_m\right]\exp\left[i\sum_m a_m\alpha_m^*\right]\right\rangle = \exp\left[-\sum_m |\alpha_m|^2 \frac{z}{\exp(\beta\varepsilon_m) - z}\right] \equiv E(\alpha_i, \alpha_k^*),$$

which can be written

$$\langle \exp(ia_f^*)\exp(ia_f)\rangle = \exp(-\langle f|\rho_1 f\rangle)$$

with the use of $\rho_1$ from (2.5.13).

The expectation values of polynomials in the field operators can be obtained by differentiating the generating function by $\alpha$ or $\alpha^*$. Note that all the factors within a given exponent of (2.5.47) commute, so nothing prevents the exponential functions from being differentiated:

$$\langle a_{m_1}^* \cdots a_{m_n}^* a_{j_1} \cdots a_{j_n}\rangle = (-i)^{n+n'} \frac{\partial}{\partial \alpha_{m_1}} \cdots \frac{\partial}{\partial \alpha_{m_n}} \frac{\partial}{\partial \alpha_{j_1}^*} \cdots \frac{\partial}{\partial \alpha_{j_n}^*} E \bigg|_{\substack{\alpha_i = 0 \\ \alpha_k^* = 0}}$$

$$= \delta_{nn'} \sum_P \prod_{i=1}^n \frac{\delta_{m_i j_{P_i}} z}{\exp(\beta \varepsilon_{m_i}) - z},$$

where $P$ stands for any permutation of $(1, 2, \ldots, n)$.

We have been confronted again with a permanent, and it is easy to understand that the analogous expression for fermions contains $(-1)^P$ and thus involves a determinant. The $-z$ in the denominator is also turned into $+z$, but there are no other changes. Linear extension covers the cases of expectation values of products of arbitrary $a_f$, which are most conveniently written in terms of the one-particle density matrix $\rho_1$, as before:

**The Grand Canonical Expectation Value of an Ordered Product** (2.5.48)

$$\langle a_{f_1}^* \cdots a_{f_n}^* a_{g_1} \cdots a_{g_{n'}}\rangle = \delta_{nn'} \operatorname*{Per}_{\mathrm{Det}}(\langle f_i|\rho_1 g_j\rangle).$$

This section will conclude with a further investigation into the thermodynamic limit of the grand canonical state of a system of particles in an external field. Such a state will exist under the circumstances in which $\rho_{1,V}$ converges weakly, as for example with free particles, for which:

**The Grand Canonical State of an Infinite System** (2.5.49)

$$\langle a_{f_1}^* \cdots a_{f_n}^* a_{g_1} \cdots a_{g_{n'}}\rangle = \delta_{nn'} \operatorname*{Per}_{\mathrm{Det}}(\langle f_i|\rho_1 g_j\rangle),$$

$$\langle f|\rho_1 g\rangle = \int \frac{d^3k}{(2\pi)^3} \frac{\tilde{f}^*(\mathbf{k})\tilde{g}(\mathbf{k})z}{\exp(\beta|\mathbf{k}|^2) \mp z},$$

where $\beta > 0$, and for bosons, $0 \le z < 1$, or for fermions, $z > 0$.

It was noticed in (2.5.33; 3) that with bosons at $T < T_c = (8\pi^{3/2}\rho/\zeta(\frac{3}{2}))^{2/3}$, the limits $V \to \infty$ and $z \to 1$ have to be taken jointly in order to have a given density $\rho$. This does not make the sum in (2.5.46) converge to the integral in

(2.5.49); rather, if $z = 1 - 1/\rho_0 V$, then the term with $k = 0$ survives separately:

$$\lim_{V\to\infty} \frac{1}{V} \sum_{k\in(2\pi/L)\mathbb{Z})^3} \frac{|\tilde{f}(\mathbf{k})|^2(1-(1/\rho_0 V))}{\exp(\beta|\mathbf{k}|^2) - 1 + (1/\rho_0 V)}$$

$$\to \rho_0|\tilde{f}(\mathbf{0})|^2 + \int \frac{d^3k}{(2\pi)^3} \frac{|\tilde{f}(\mathbf{k})|^2}{\exp(\beta|\mathbf{k}|^2) - 1}.$$

This formula is justified if $f \in L^2(\mathbb{R}^3)$ with compact support, which makes $\tilde{f} \in L^2(\mathbb{R}^3) \cap C_0^\infty(\mathbb{R}^3)$, so the integrand remains integrable even at $\mathbf{k} = \mathbf{0}$. Therefore we have:

**The Grand Canonical State in Bose Condensation** (2.5.50)

$$\lim_{V\to\infty} \langle \exp(ia_f^*)\exp(ia_f)\rangle_{\beta, z=1-(1/\rho_0 V)}$$

$$= \exp\left[-\rho_0|\tilde{f}(\mathbf{0})|^2 - \int \frac{d^3k}{(2\pi)^3} \frac{|\tilde{f}(\mathbf{k})|^2}{\exp(\beta|\mathbf{k}|^2) - 1}\right].$$

**Remarks** (2.5.51)

1. If $T < T_c$, then the grand canonical state of the Bose field algebra differs from the canonical state, which can be calculated as

$$\langle \exp(ia_f^*)\exp(ia_f)\rangle = \exp\left[-\int \frac{d^3k}{(2\pi)^3} \frac{|\tilde{f}(\mathbf{k})|^2}{\exp(\beta|\mathbf{k}|^2) - 1}\right]$$

$$\times \int_0^{2\pi} \frac{d\varphi}{2\pi} \exp[2i\sqrt{\rho_0}\,\mathrm{Re}(\tilde{f}(\mathbf{0})\exp(i\varphi))]$$

   for $T < T_c$ [13].
2. Other than for bosons at $T < T_c$, the representations in the individual factors are thermal (1.4.7). According to Remark (1.4.17; 1) the factors are of type III in the infinite system. They form a reducible representation $\pi$, the tensor product $\pi_1 \otimes \pi_2$ of two Fock-like representations of the field algebra (cf. (1.4.7)):

$$\pi(a_f) = \pi_1\left(a\frac{\tilde{f}(\mathbf{p})}{\sqrt{\mp\exp[-\beta(|\mathbf{p}|^2 - \mu)] + 1}}\right) \otimes \mathbf{1}$$

$$+ (-\mathbf{1})^N \otimes \pi_2\left(a^* \frac{\tilde{f}^*(\mathbf{p})}{\sqrt{\exp[\beta(|\mathbf{p}|^2 - \mu)] \mp 1}}\right),$$

   where $a_f N = (N+1)a_f$. It is straightforward to verify that

$$\langle a_{f_1}^* \cdots a_{f_n}^* a_{g_1} \cdots a_{g_{n'}}\rangle = \langle \Omega_1 \otimes \Omega_2 | \pi(a_{f_1}^*) \cdots \pi(a_{g_{n'}}) | \Omega_1 \otimes \Omega_2\rangle.$$

3. For bosons at $T < T_c$ there is no factor representation; the analogue of the mean magnetization $\mathbf{s}$ (1.4.6: 2) is

$$a_0 \equiv \underset{V\to\infty}{w\text{-lim}}\, a_0^V, \quad \text{where } a_0^V \equiv \frac{1}{V}\int_V d^3x a(\mathbf{x}).$$

All bounded functions of $a_0$ lie in the center of the von Neumann algebra $\pi(\mathscr{A})''$. Now

$$\langle a_0^{*n} a_0^m\rangle = \left(\frac{\partial}{\partial \tilde{f}(\mathbf{0})}\right)^n \left(\frac{\partial}{\partial \tilde{f}^*(\mathbf{0})}\right)^m E|_{f\equiv 0},$$

so for instance $\langle a_0\rangle = 0$, $\langle a_0^* a_0\rangle = \rho_0$. Thus $a_0$ is not represented as a multiple of the identity.

4. The canonical state (2.5.51; 1) is an integral over states $\omega_\varphi$ for which the exponent in the generating function

$$\omega_\varphi(\exp(i\lambda a_0^*)\exp(i\lambda a_0)) = \exp(2i\lambda\sqrt{\rho_0}\cos\varphi)$$

is linear in $\lambda \in \mathbb{R}$. These states produce factor representations:

$$\pi_\varphi(a_0) = \sqrt{\rho_0}\exp(-i\varphi)\cdot \mathbf{1}.$$

5. If a term $V^\alpha(a_0^V - \sqrt{\rho_0}\exp(-i\varphi))^*(a_0^V - \sqrt{\rho_0}\exp(-i\varphi))$ with $0 < \alpha < 1$ is added to the local Hamiltonian $H_V$, then the $\mathbf{k} = \mathbf{0}$ component of $\beta H_V$ becomes $\beta V^\alpha(a_0^V - \sqrt{\rho_0}\exp(-i\varphi))^*(a_0^V - \sqrt{\rho_0}\exp(-i\varphi))$. As will become more apparent below, the thermodynamic functions are unchanged for all $0 < T \le T_c(\rho)$ in the limit $V \to \infty$ if we set $z(V) \equiv 1$ and $\rho_0 = \rho(1 - (T/T_c(\rho))^{3/2})$ (cf. (2.5.33; 3)). Because

$$\begin{aligned}
&\mathrm{Tr}\{\exp[-\beta V^\alpha(a_0^V - \sqrt{\rho_0}\exp(-i\varphi))^*(a_0^V - \sqrt{\rho_0}\exp(-i\varphi))]\\
&\quad\times \exp(i\tilde{f}(\mathbf{0})a_0^{V*})\cdot\exp(i\tilde{f}^*(\mathbf{0})a_0^V)\}\\
&= \mathrm{Tr}[\exp(-\beta V_\alpha a_0^{V*} a_0^V)\exp(i\tilde{f}(\mathbf{0})a_0^{V*})\exp(i\tilde{f}^*(0)a_0^V)]\\
&\quad\times\exp(2i\sqrt{\rho_0}\,\mathrm{Re}(\tilde{f}(\mathbf{0})\exp(i\varphi))
\end{aligned}$$

and

$$\begin{aligned}
&\frac{\mathrm{Tr}[\exp(-\beta V^\alpha a_0^{V*}a_0^V)\cdot\exp(i\tilde{f}(\mathbf{0})a_0^{V*})\cdot\exp(i\tilde{f}^*(\mathbf{0})a_0^V)]}{\mathrm{Tr}\exp(-\beta V^\alpha a_0^{V*}a_0^V)}\\
&\quad= \exp\left[-|\tilde{f}(\mathbf{0})|^2/\beta V^\alpha + o\left(\frac{1}{V^\alpha}\right)\right]
\end{aligned}$$

(see Problem 6), in the limit $V \to \infty$ the perturbed grand canonical state reduces to $\omega_\varphi$, the integrand of the canonical state in the decomposition (2.5.51; 1), since the contribution to the generating function from the components of $H_V$ with $\mathbf{k} \ne \mathbf{0}$ is not affected by the extra term. Since the exponent in this generating function is linear in $\tilde{f}(\mathbf{0})$ and $\tilde{f}^*(\mathbf{0})$,

$$\pi_{\omega_\varphi}(a_0) = \sqrt{\rho_0}\exp(-i\varphi)\cdot\mathbf{1}.$$

This shows that $\omega_\varphi$ is a factor state, and the density of the particles in the ground state is represented by the (dispersionless) multiplication operator $\rho_0 \cdot \mathbf{1}$. Although the assumption that $\alpha > 0$ is essential (the limit state is not changed by perturbations bounded uniformly in $V$), the bound $\alpha < 1$ only serves to illustrate that a surface effect is enough to single out any given pure phase from a mixture as the limit $V \to \infty$ is taken.

This example appears at first only academic from the physical point of view. Since constant phases of the wave-functions are not observable properties, at least for free particles, the Bose algebra should be replaced with the gauge-invariant subalgebra $\mathscr{E}$, i.e., the subalgebra invariant under the automorphism induced by $f \to \exp(i\varphi)f$. All the states $\omega_\varphi$ are the same on the subalgebra, and the phase mixture of the ground state is not observable. However, these phases do have experimental consequences in superconductors, in the Josephson effect.

**Problems** (2.5.52)

1. Calculate the asymptotic forms of $F_{5/2}(z)$ (for $z \to 1$ use $zF'_\sigma(z) = F_{\sigma-1}(z)$, $F_\sigma(1) = \zeta(\sigma)$).
2. Calculate the heat capacity per particle of an ideal Bose gas at constant density, as well as its derivative by the temperature.
3. Verify (2.5.18; 2).
4. Show the maximum properties of (2.5.16; 2) and (2.5.18; 4).
5. Calculate $P_B$ and $P_F$ for particles in a box. Show that the result agrees with (2.5.24) in the limit $V \to \infty$.
6. Calculate $\operatorname{Tr} \exp[i(a^*\alpha - a\alpha^*)] \exp[-\beta a^* a]/\operatorname{Tr} \exp[-\beta a^* a]$, assuming that $[a, a^*] = 1$.

**Solutions** (2.5.53)

1. $z \to 0$: $F_{5/2}(z) = \sum_{\nu=1}^{\infty} \frac{z^\nu}{\nu^{5/2}} \sim z + \frac{z^2}{2^{5/2}} + \cdots$

$z \to 1$: $F_{5/2}(z) \sim F_{5/2}(1) + (z-1)F'_{5/2}(1) + \cdots = \zeta(\frac{5}{2}) + (z-1)\zeta(\frac{3}{2}) + \cdots$

$z \to \infty$: Let $\alpha = \ln z > 0$

$$\int_0^\infty dt\sqrt{t}\ln(1 + \exp(-t + \alpha)) = \int_{-\alpha}^\infty dt\sqrt{t + \alpha}\ln(1 + \exp(-t))$$

$$= \frac{2}{3}\int_{-\alpha}^\infty dt(t + \alpha)^{3/2}(1 + \exp(t))^{-1}$$

$$= \frac{2}{3}\left[\int_0^\alpha dt(\alpha - t)^{3/2} - \int_0^\alpha \frac{dt(\alpha - t)^{3/2}}{1 + e^t} + \int_0^\infty \frac{dt(t + \alpha)^{3/2}}{1 + e^t}\right]$$

$$= \frac{2}{3}\left[\int_0^\alpha dt(\alpha - t)^{3/2} + \int_0^\infty \frac{dt((t + \alpha)^{3/2} - |t - \alpha|^{3/2})}{1 + \exp(t)}\right] + O(\exp(-\alpha));$$

because

$$|(\alpha + t)^{3/2} - |\alpha - t|^{3/2} - 3t\alpha^{1/2}| \leq 2t^2\alpha^{-1/2}$$

and

$$\int_0^\infty \frac{dt\, t^{\sigma-1}}{1 + \exp(t)} = (1 - 2^{1-\sigma})\Gamma(\sigma)\zeta(\sigma),$$

with $\zeta(2) = \pi^2/6$, $\Gamma(2) = 1$, it follows that

$$\int_0^\infty dt\sqrt{t}\ln(1 + \exp(-t + \alpha)) = \tfrac{2}{3}\left[\frac{2}{5}\alpha^{5/2} + \alpha^{1/2}\frac{\pi^2}{4}\right] + O(\alpha^{-1/2}).$$

2.
$$\varepsilon = \begin{cases} \tfrac{3}{2}T^{5/2}\dfrac{1}{8\pi^{3/2}}F_{5/2}(z), & T > T_c, \text{ i.e., } 0 < z < 1, \\ \tfrac{3}{2}T^{5/2}\dfrac{1}{8\pi^{3/2}}\zeta(\tfrac{5}{2}), & T \leq T_c, \text{ i.e., } z = 1, \end{cases}$$

which implies

$$\gamma \equiv \lim_{N\to\infty}\frac{C_V}{N} = \begin{cases} \dfrac{15}{4}\dfrac{1}{8\pi^{3/2}\rho}T^{3/2}F_{5/2}(z) - \dfrac{9}{4}\dfrac{F_{3/2}(z)}{F_{1/2}(z)}, & T > T_c, \text{ i.e., } 0 < z < 1, \\ \dfrac{15}{4}\dfrac{T^{3/2}}{8\pi^{3/2}\rho}\zeta(\tfrac{5}{2}), & T \leq T_c, \text{ i.e., } z = 1, \end{cases}$$

because of the formula $F_{3/2}(z) = 8\pi^{3/2}\rho T^{-3/2}$ for $T > T_c$. The function $\gamma$ is continuous at $T = T_c$ and equals $(15/4)\zeta(\frac{5}{2})/\zeta(\frac{3}{2}) = 1.93$, and as $T \to \infty$, $F_\sigma(z) \sim z \sim 8\pi^{3/2}\rho T^{-3/2}$, and

$$\gamma \sim \frac{15}{4}\frac{1}{8\pi^{3/2}\rho}T^{3/2}z - \tfrac{9}{4} \sim \tfrac{15}{4} - \tfrac{9}{4} = \tfrac{3}{2}.$$

With the expansion $F_{5/2}(z) = 2.363t^{3/2} + 1.342 - 2.612t - 0.730t^2\ldots$, where $t \equiv -\ln z$, valid for $z \lesssim 1$, and the recursion formula

$$F_{\sigma-1}(\exp(-t)) = -(d/dt)F_\sigma(\exp(-t)),$$

there results

$$\left(\frac{\partial\gamma}{\partial T}\right)_{T=T_c-0} - \left(\frac{\partial\gamma}{\partial T}\right)_{T=T_c+0} = \frac{3.66}{T_c}.$$

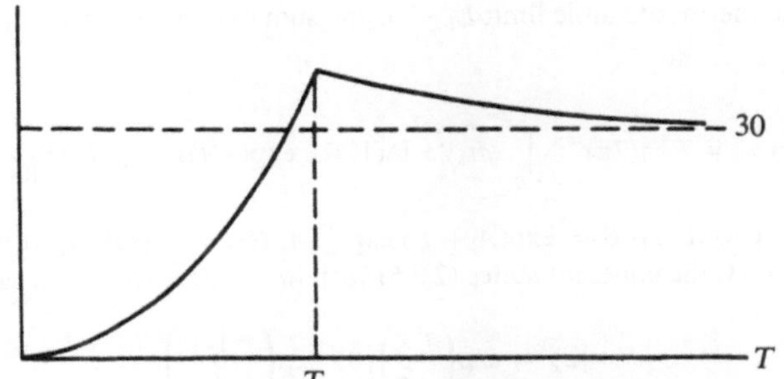


Figure 21 Specific heat of an ideal Bose gas.

3. If the wave-function of $|\mathbf{z}\rangle$ is $\exp(i\mathbf{k}\cdot\mathbf{x})u(\mathbf{x}-\mathbf{q})$ with $u$ real-valued, then $\langle\mathbf{z}||\mathbf{p}|^2|\mathbf{z}\rangle = \int d^3x|i\mathbf{k}u(\mathbf{x}-\mathbf{q}) - \nabla u(\mathbf{x}-\mathbf{q})|^2 = |\mathbf{k}|^2 + \int d^3x|\nabla u|^2$. At the same time, the expectation value of $\int d\Omega_z|\mathbf{z}\rangle\langle\mathbf{z}||\mathbf{k}|^2$ in a normalized $\psi$ equals

$$\int \frac{d^3q\,d^3k|\mathbf{k}|^2}{(2\pi)^3}\int d^3x\,d^3x'\psi^*(\mathbf{x})\exp(i\mathbf{k}\cdot\mathbf{x})u(\mathbf{x}-\mathbf{q})\exp(-i\mathbf{k}\cdot\mathbf{x}')u(\mathbf{x}'-\mathbf{q})\psi(\mathbf{x}')$$

$$= \int d^3q\,d^3x\nabla(\psi^*(\mathbf{x})u(\mathbf{x}-\mathbf{q}))\cdot\nabla(u(\mathbf{x}-\mathbf{q})\psi(\mathbf{x}))$$

$$= \int d^3x|\nabla\psi(\mathbf{x})|^2 + \int d^3q|\nabla u(\mathbf{q})|^2,$$

because the mixed terms drop out in the $q$ integration. Therefore,

$$\int d\Omega_z|\mathbf{z}\rangle\langle\mathbf{z}||\mathbf{k}|^2 = |\mathbf{p}|^2 + \int d\Omega_z|\mathbf{z}\rangle\langle\mathbf{z}|\int d^3q|\nabla u|^2.$$

4. Klein's inequality (2.1.8; 5) with $K(\rho) = \rho\ln\rho \pm (1\mp\rho)\ln(1\mp\rho)$, $K'(\rho) = -\ln(1/\rho \mp 1)$ and $\tilde\rho = [\exp(\beta(h-\mu)) \pm 1]^{-1}$ leads to

$$\mathrm{Tr}[K(\rho) - K(\bar\rho) + (\rho-\bar\rho)\beta(h-\mu)] \geq 0,$$

proving (2.5.16; 2). In the classical case, i.e., $\rho = \rho(z)$, $h = h(z)$, $\bar\rho = \bar\rho(z) =$ all being real,

$$K(\rho(z)) - K(\bar\rho(z)) + (\rho(z) - \bar\rho(z))\beta(h(z)-\mu) \geq 0$$

for all $z$, and consequently (2.5.18; 4).

5. Particles in a box. If the shape of the box is a parallelepiped with sides $L_1, L_2$, and $L_3$, and the wave-functions satisfy Dirichlet boundary conditions, then the eigenvalues are

$$\varepsilon_m = \pi^2\left(\frac{m_1^2}{L_1^2} + \frac{m_2^2}{L_2^2} + \frac{m_3^2}{L_3^2}\right), \qquad m_i \in \mathbb{Z}^+.$$

Consequently

$$\beta V P_{\substack{B\\F}}(z) = \mp \sum_{m_i=1}^{\infty} \ln(1 \mp z\exp(-\beta\varepsilon_m)),$$

and in the thermodynamic limit $L_i \to \infty$ the sum over $m_i$ becomes $L_1\cdot L_2\cdot L_3(2\pi)^{-2} \times \int_0^\infty d\varepsilon\sqrt{\varepsilon}\ldots$, so

$$P_{\substack{B\\F}}(T,z) = \mp T^{5/2}(2\pi)^{-2}\int_0^\infty dt\sqrt{t}\,\ln(1\mp z\exp(-t)) = \pm T^{5/2}\frac{1}{8\pi^{3/2}}F_{5/2}(\pm z).$$

6. Because $\exp A \exp B = \exp(A+B)\exp(\frac{1}{2}[A,B]) = \exp B\exp A\exp[A,B]$ for $[A,B] = c\,\mathbf{1}$, the coherent states (2.2.6) with $|u\rangle = |0\rangle$, $a|0\rangle = 0$, can be written

$$|z\rangle = \exp\left(\frac{a^*z}{\sqrt{2}}\right)|0\rangle\exp\left(\frac{-|z|^2}{4}\right).$$

As in Remark (III: 3.1.14; 1), with $\exp(-\beta a^*a)f(a^*)|0\rangle = f(a^*\exp(-\beta))|0\rangle$ it follows that

$$\operatorname{Tr}\exp(\alpha a^*)\exp(-\alpha^* a)\exp(-\beta a^* a)$$

$$= \int \frac{dz}{2\pi}\langle 0|\exp\left(\frac{az^*}{\sqrt{2}}\right)\exp(-\alpha^* a)\exp(-\beta a^* a)\exp(\alpha a^*)\exp\left(\frac{a^* z}{\sqrt{2}}\right)|0\rangle\exp\left(\frac{-|z|^2}{2}\right)$$

$$= \int \frac{dz}{2\pi}\langle 0|\exp\left[a\left(\frac{z^*}{\sqrt{2}} - \alpha^*\right)\right]\exp\left[\exp(-\beta)a^*\left(\frac{z}{\sqrt{2}} + \alpha\right)\right]|0\rangle\exp\left(\frac{-|z|^2}{2}\right)$$

$$= \int \frac{dz}{2\pi}\exp\left[-\frac{|z|^2}{2}(1 - \exp(-\beta)) + \exp(-\beta)\left(\frac{1}{\sqrt{2}}(z^*\alpha - z\alpha^*) - |\alpha|^2\right)\right]$$

$$= \exp\left[-|\alpha|^2\frac{1}{\exp(\beta) - 1}\right]\Big/(1 - \exp(-\beta)),$$

so by changing $\alpha$ to $i\alpha$,

$$\langle\exp[i(a^*\alpha + a\alpha^*)]\rangle = \langle\exp[\alpha a^* - \alpha^* a]\rangle = \langle\exp(\alpha a^*)\exp(-\alpha^* a)\rangle\exp(-\tfrac{1}{2}|\alpha|^2)$$

$$= \exp\left[-|\alpha|^2\left(\frac{1}{2} + \frac{1}{\exp(\beta) - 1}\right)\right].$$

# 3 Thermodynamics

## 3.1 Time-Evolution

*Whereas small systems evolve almost periodically in time, large systems appear chaotic and their time-evolution mixes the observables thoroughly.*

The framework for this discussion will be an algebra $\mathscr{A}$ of observables with a strongly continuous time-automorphism and a time-invariant state $\rho$. In the GNS representation the invariant state is made into a vector $|\Omega\rangle$, and the time-automorphism is represented as a unitary group of operators $U = \{\exp(iHt)\}$, $U|\Omega\rangle = |\Omega\rangle$. The time-evolution then extends to the weak closure $\mathscr{A}''$. If the representation is reducible, then it may occur that $U \not\subset \mathscr{A}''$, even if $U_t^{-1}\mathscr{A}U_t \subset \mathscr{A}$. The von Neumann algebra

$$\mathscr{R} \equiv \{\mathscr{A} \cup U\}'', \qquad \mathscr{R}' = \mathscr{A}' \cap U',$$

generated by $\mathscr{A}$ and $U$ is known as the **covariance algebra** and will figure prominently in what follows. If the only invariant elements of $\mathscr{A}'$ are of the form $\alpha \cdot \mathbf{1}$, then it is all of $\mathscr{B}(\mathscr{H})$, as $\mathscr{R}' = \alpha \cdot \mathbf{1} \Rightarrow \mathscr{R}'' = \mathscr{R} = \mathscr{B}(\mathscr{H})$.

An initial orientation to the various possibilities can be obtained by looking at some

**Examples** (3.1.1)

1. Classical dynamical systems. The Abelian algebra $\mathscr{A}$ of $C^\infty$ functions $a(p, q)$ on the phase space $T^*(M)$ is a special case of the general schema. If $d\mu$ is a probability measure on $T^*(M)$, then the elements $a \in \mathscr{A}$

are represented as multiplication operators on the Hilbert-space. $L^2(T^*(M), d\mu)$. The advantage of the Hilbert-space approach to classical mechanics is that it ignores exceptional trajectories making up null sets. If a time-invariant measure $d\mu$, such as the Liouville measure $dq_1 \cdots dp_{3N}$ is restricted to a time-invariant region $\Omega$ of finite volume and normalized, then the time-evolution $a(p, q) \to a(p(t), q(t))$ is represented unitarily on $L^2(\Omega, d\mu)$. It can be written formally as $U_t = \exp(-iht)$, where $h = iL_{X_H}$ is the Liouville operator (I:2.2.25; 1), and this unitary group of transformations extends to the von Neumann algebra $\mathscr{A}'' = L^\infty(\Omega, d\mu)$. Of course $U_t$ does not belong to $\mathscr{A}''$, which is maximally Abelian, $\mathscr{A}'' = \mathscr{A}' = \mathscr{Z}$. The algebra $\mathscr{R}$ is all of $\mathscr{B}(\mathscr{H})$ if and only if the system is ergodic, for then the only time-invariant functions are constant almost everywhere, and are thus the constant functions of $L^\infty(\Omega, d\mu)$.

2. A single spin in a magnetic field, cf. (1.1.1):

$$\mathscr{A} = \mathscr{B}(\mathbb{C}^2) = \{\mathbf{1}, \sigma, \sigma^\pm\}'', \qquad \rho(\cdot) = \left\langle \begin{pmatrix}1\\0\end{pmatrix} \middle| \cdot \middle| \begin{pmatrix}1\\0\end{pmatrix} \right\rangle,$$

$$U_t = \exp(iB(\mathbf{1} - \sigma)t) \qquad \mathscr{A}' = \mathscr{Z} = \mathscr{R}' = \{\alpha \cdot \mathbf{1}\}, \qquad \mathscr{A}'' = \mathscr{A} = \mathscr{R}.$$

Observe that while there is only one invariant vector, there is a second pure invariant state, $\langle (\begin{smallmatrix}0\\1\end{smallmatrix}) | \cdot | (\begin{smallmatrix}0\\1\end{smallmatrix}) \rangle$.

3. A single spin in a magnetic field, in a thermal representation (1.4.7):

$$\mathscr{A} = \{\mathbf{1}, \sigma, \sigma^\pm\}'' \otimes \mathbf{1}, \qquad \rho(\cdot) = \langle \Omega | \cdot | \Omega \rangle,$$

$$\Omega = \sqrt{\frac{1+s}{2}} \begin{pmatrix}1\\0\end{pmatrix} \otimes \begin{pmatrix}1\\0\end{pmatrix} + \sqrt{\frac{1-s}{2}} \begin{pmatrix}0\\1\end{pmatrix} \otimes \begin{pmatrix}0\\1\end{pmatrix},$$

$$\mathscr{A}' = \mathbf{1} \otimes \{\mathbf{1}, \tau, \tau^\pm\}'', \qquad U_t = \exp(iB(\tau - \sigma)t),$$

$$\mathscr{A}'' = \mathscr{A}, \ \mathscr{Z} = \{\alpha \cdot \mathbf{1}\}, \ \mathscr{R}' = \mathbf{1} \otimes \{\mathbf{1}, \tau\}'', \ \mathscr{R} = \{\mathbf{1}, \sigma, \sigma^\pm\}'' \otimes \{\mathbf{1}, \tau\}''.$$

This factor representation on $\mathbb{C}^4$ has a two-dimensional invariant subspace and a five-dimensional manifold of invariant states. Two of these are pure states corresponding to noninvariant vectors. Notice that the formal equation $h = -B\sigma$ has to be normalized not only with a constant but also by $B\tau \in \mathscr{A}'$, to ensure that $U|\Omega\rangle = |\Omega\rangle$. With a different choice of the basis for $\mathbb{C}^4$, $\Omega$ can also be written as $(\begin{smallmatrix}1\\0\end{smallmatrix}) \otimes (\begin{smallmatrix}1\\0\end{smallmatrix})$, which makes the representation $\pi$ of $\mathscr{A}$ somewhat more complicated (cf. (2.5.51; 2)):

$$\pi(\sigma^\pm) = \sqrt{\frac{1+s}{2}} \sigma^\pm \otimes \mathbf{1} - (-1)^\sigma \otimes \sqrt{\frac{1-s}{2}} \tau^\mp,$$

$$\pi(\sigma) = \frac{1+s}{2} \sigma \otimes \mathbf{1} - \mathbf{1} \otimes \tau \frac{1-s}{2} + \sqrt{1-s^2}\{\sigma^- \otimes \tau^- + \sigma^+ \otimes \tau^+\}.$$

It is easy to verify the algebraic relationships

$$\pi(\sigma^+)\pi(\sigma^-) \pm \pi(\sigma^-)\pi(\sigma^+) = \begin{cases} \mathbf{1}, \\ \pi(\sigma), \end{cases} \qquad \pi(\sigma^+)^2 = 0,$$

4. An infinite, interacting spin system. Consider the model of a ferromagnet (2.3.32; 2) in the limit $N \to \infty$. It is not hard to discover that the thermal expectation values converge to those with the vector

$$\otimes\left(\begin{pmatrix}1\\0\end{pmatrix} \otimes \begin{pmatrix}1\\0\end{pmatrix}\sqrt{\frac{1+s}{2}} + \begin{pmatrix}0\\1\end{pmatrix} \otimes \begin{pmatrix}0\\1\end{pmatrix}\sqrt{\frac{1-s}{2}}\right),$$

as with a type-III representation (1.4.7). The quantities

$$s = \langle\sigma\rangle = -\tanh B_{\text{eff}}\beta, \qquad B_{\text{eff}} = B - 2s,$$

are to be determined self-consistently, for the interaction can be written as

$$\frac{1}{N}\sum_{i,j} \boldsymbol{\sigma}_i \cdot \boldsymbol{\sigma}_j = \frac{1}{N}\sum_i (\boldsymbol{\sigma}_i - \langle\boldsymbol{\sigma}_i\rangle) \cdot \sum_j (\boldsymbol{\sigma}_j - \langle\boldsymbol{\sigma}_j\rangle) - 2\langle\sigma\rangle \sum_i \sigma_i + \text{const.}$$

If now $N \to \infty$, the first term on the right describes the fluctuations and becomes negligible compared with $-2\langle\sigma\rangle\sum_i \sigma_i$, and the commutators of $H$ approach those of $B_{\text{eff}} \sum_i \sigma_i$, $B_{\text{eff}} = -Z\langle\sigma\rangle$ (cf. (1.1.11)). The time-evolution is accordingly given by

$$U_t = \bigotimes_j \exp(iB_{\text{eff}}(\tau_j - \sigma_j)t).$$

The Hilbert space $\mathscr{H}$ contains infinitely many invariant vectors, viz., all the ones that differ from $\Omega$ in the replacement of finitely many factors with an invariant vector from Example 3. Since $B_{\text{eff}}$ depends on $\beta$, the time-automorphisms on representations with different $\beta$ are different. Therefore there is not any automorphism of the algebra $\mathscr{A}$ generated by the $\sigma$'s on the sum of two representations with different $\beta$. Although an isomorphism of $\pi(\mathscr{A})$, as a subalgebra of $\mathscr{B}(\mathscr{H}_\pi)$, is given by

$$\alpha_t(\pi(\mathscr{A})) = U_t^{\beta_1,\beta_2}\pi(\mathscr{A})(U_t^{\beta_1,\beta_2})^{-1}$$

with

$$U_t^{\beta_1,\beta_2} = U_t^{\beta_1} \oplus U_t^{\beta_2}, \qquad \pi = \pi_{\beta_1} \oplus \pi_{\beta_2},$$

it is not an automorphism, since there are times $t$ at which $\alpha_t(\pi(\mathscr{A})) \neq \pi(\mathscr{A})$. The smallest subalgebra of $\mathscr{B}(\mathscr{H}_\pi)$ for which $(\alpha_t)_{t\in\mathbb{R}}$ becomes a group of automorphisms is clearly $\bigcup_t \alpha_t(\pi(\mathscr{A}))$. If $B = 0$ and $T < 2$, then there is such a sum, or even an integral. There are nonzero solutions to the equation $B_{\text{eff}} = 2\tanh \beta B_{\text{eff}}$, but nothing favors any direction. Expectation values are averages over the unit sphere of expectation values with $\mathbf{B}_{\text{eff}} = \mathbf{n}B_{\text{eff}}$, by means of which the representation takes on the form

$$\pi(\mathscr{A}) = \int_{S_2} d\mathbf{n}\,\pi_{\mathbf{n}}(\mathscr{A}),$$

where $\pi_{\mathbf{n}}$ is specified by (1.4.7) with $\sigma \equiv (\boldsymbol{\sigma} \cdot \mathbf{n})$. The time-evolution on $\pi_{\mathbf{n}}(\mathscr{A})$ is the rotation $\sigma_j^\alpha(t) = (\exp(tR))^{\alpha\beta}\sigma_j^\beta$ having the matrix

$$R = B_{\text{eff}} \begin{pmatrix} 0 & n_3 & -n_2 \\ -n_3 & 0 & n_1 \\ n_2 & -n_1 & 0 \end{pmatrix}.$$

However, as the strong limit of $(1/N)\sum_{j=1}^{N} \boldsymbol{\sigma}_j$ as $N \to \infty$, $\mathbf{n}$ is contained in $\pi(\mathscr{A})''$ and lies in the center of this algebra but is not a multiple of $\mathbf{1}$. It is constant in time, and the $\mathbf{n}$-dependent time-evolution of the $\sigma$'s can be viewed as an automorphism of $\pi(\mathscr{A})''$.

5. Free fermions. The algebra $\mathscr{A}$ is generated by the field operators $a_f$ (1.3.2), and as in (1.3.3; 5) the free time-evolution

$$f(\mathbf{p}) \to \exp(-i|\mathbf{p}|^2 t) f(\mathbf{p}) \equiv f_t(\mathbf{p}),$$

provides a group of automorphisms on $\mathscr{A}: a_f \to a_{f_t}$. The thermal state (2.5.49) is clearly invariant in time and leads to a unitary time-evolution $U_t = \exp(-iHt)$. In order to tell the type of the representation, we can write it in a form like the one in Example 3. Let $|\Omega_{1,2}\rangle$ be two Fock vacua and $\pi_{1,2}(a_f)$ be the representations formed with $|\Omega_{1,2}\rangle$. Then with the tensor product

$$|\Omega\rangle = |\Omega_1\rangle \otimes |\Omega_2\rangle$$

we get

$$\pi(a(f)) = \pi_1\left(a\left(\frac{\tilde{f}(\mathbf{p})}{\sqrt{1 + \exp(-\beta(|\mathbf{p}|^2 - \mu))}}\right)\right) \otimes \mathbf{1}$$
$$+ (-\mathbf{1})^N \otimes \pi_2\left(a^*\left(\frac{\tilde{f}^*(\mathbf{p})}{\sqrt{1 + \exp(\beta(|\mathbf{p}|^2 - \mu))}}\right)\right),$$

where $aN = (N + 1)a$ (cf. (1.3.13)). It can be verified that

$$\langle a_{f_1}^* \cdots a_{f_n}^* a_{g_1} \cdots a_{g_n} \rangle = \langle \Omega | \pi(a_{f_1}^*) \cdots \pi(a_{f_n}^*) \pi(a_{g_1}) \cdots \pi(a_{g_n}) | \Omega \rangle,$$

so this representation is equivalent to the thermal representation with infinitely many spins. Consequently, if $T > 0$, then it is a factor of type III. The local field operators in momentum space can be used to write $H_\pi$ as

$$H_\pi = \int \frac{d^3p}{(2\pi)^3} |\mathbf{p}|^2 \{\pi_1(a^*(\mathbf{p})a(\mathbf{p})) \otimes \mathbf{1} - \mathbf{1} \otimes \pi_2(a^*(\mathbf{p})a(\mathbf{p}))\}.$$

The operator $a^*a$ differs from the usual one not only in that the infinite zero-point energy of field theory has been subtracted off, but also in the removal of an operator of $\mathscr{A}'$.

### **The Time-Evolution of Open Systems** (3.1.2)

It seems illusory to consider every single local property of a large system as belonging to the algebra of observables. It is certainly true that practically anything can be measured, but not all at once, and putting the system into a state that is dispersionless with respect to a maximally Abelian subalgebra is actually impossible. In reality only fairly small subsystems get measured, so it is of practical interest to divide the total system into the subsystem that is observed, called an "open" system, and all the rest, acting as a reservoir. Accordingly, let $\mathscr{H} = \mathscr{H}_S \otimes \mathscr{H}_R$ and let $\mathrm{Tr}^{S+R}$, $\mathrm{Tr}^S$, and $\mathrm{Tr}^R$ be the traces on $\mathscr{H}$, $\mathscr{H}_S$, and $\mathscr{H}_R$. The time-evolution $U_t$ will mix $\mathscr{H}_S$ and $\mathscr{H}_R$, so it does not create an automorphism of $\mathscr{B}(\mathscr{H}_S)$. However, if the initial state postulated can be factorized and written in terms of a density matrix $\rho \otimes \omega$, then a time-evolution $\tau_t$: $\mathscr{B}(\mathscr{H}_S) \to \mathscr{B}(\mathscr{H}_S)$ can be defined for the open system in the Heisenberg picture, or the dual time-evolution for the density matrices $\tau_t^*$: $\mathscr{C}_1(\mathscr{H}_S) \to \mathscr{C}_1(\mathscr{H}_S)$ can be defined in the Schrödinger picture. If $a \in \mathscr{B}(\mathscr{H}_S) \otimes \mathbf{1}$, then the time-dependence of the expectation values can be written as

$$\langle a(t)\rangle \equiv \mathrm{Tr}^{S+R}(\rho \otimes \omega)U_{-t}(a \otimes \mathbf{1})U_t = \mathrm{Tr}^S \rho\tau_t(a) = \mathrm{Tr}^S \tau_t^*(\rho)a,$$

where by definition

$$\tau_t(a) \equiv \mathrm{Tr}^R(\mathbf{1} \otimes \omega)U_{-t}(a \otimes \mathbf{1})U_t,$$
$$\tau_t^*(\rho) = \mathrm{Tr}^R U_t(\rho \otimes \omega)U_{-t}. \tag{3.1.3}$$

Note that the states transform with $U_t^* = U_{-t}$ rather than $U_t$.

**Properties of the Time-Evolution of the Subsystem** (3.1.4)

The operators $\tau_t$ and $\tau_t^*$ are

(i) one-parameter, strongly continuous families of completely positive linear mappings;
(ii) *not* groups: $\tau_{t_1} \circ \tau_{t_2} \neq \tau_{t_1+t_2}$;
(iii) *not* isomorphisms of the algebra: $\tau_t(a \cdot b) \neq \tau_t(a) \cdot \tau_t(b)$.

Equality holds in (ii) and (iii) only if $U_t$ factorizes.

**Gloss** (3.1.5)

A linear mapping $\Phi$: $\mathscr{B}(\mathscr{H}) \to \mathscr{B}(\mathscr{H})$ is said to be ***n*-positive** iff $\Phi \otimes \mathbf{1}$ acting on $\mathscr{B}(\mathscr{H}) \otimes \mathscr{B}(\mathbb{C}^n)$: $a \otimes M \to \Phi(a) \otimes M$ is positive for all $M \in \mathscr{B}(\mathbb{C}^n)$, i.e., it maps the cone of positive elements of $\mathscr{B}(\mathscr{H}) \otimes \mathscr{B}(\mathbb{C}^n)$ into itself. The mapping $\Phi$ is **completely positive** iff it is positive for all $n = 1, 2, \ldots$. It can be shown [14] that all completely positive mappings are obtained by taking tensor products of positive operators, composing with unitary operators, and then taking partial traces, just as in the construction of $\tau_t$ and $\tau_t^*$. The completely positive mappings form a semigroup with respect to composition.

**Examples** (3.1.6)

1. The classical harmonic oscillator
The observables are chosen as the position coordinates $q$, so

$$\mathrm{Tr}^{S+R} \to \int dp\, dq, \qquad \mathrm{Tr}^R \to \int dp, \qquad \mathrm{Tr}^S \to \int dq.$$

Let $\rho(q) = \pi^{-1/2}\exp(-(q - q_0)^2)$ be the probability distribution function of the coordinates and $\omega(p) = \pi^{-1/2}\exp(-(p - p_0)^2)$ be that of the momenta. The time-evolution of the total system, $q(t) = q\cos t + p\sin t$, $p(t) = p\cos t - q\sin t$, induces

$$\tau_t(q) = q\cos t + p_0 \sin t,$$

$$\tau_t^*(\rho) = \pi^{-1/2}\exp[-(q - q_0\cos t - p_0 \sin t)^2]$$

on the subsystem. However, $\tau_t$ is not an isomorphism,

$$\tau_t(q^2) = (q\cos t + p_0\sin t)^2 + \tfrac{1}{2}\sin^2 t \neq \tau_t(q)^2,$$

since $\omega$ is not free of fluctuations. The choice of equal widths for $\rho$ and $\omega$, as with quantum-mechanical coherent states, causes a rigid oscillation of $\rho$. If, instead, $\omega(p) = \delta(p - p_0)$, then there would be a periodic focusing and defocusing of $\rho$,

$$\tau_t^*(\rho) = \frac{\exp[-(q - q_0\cos t - p_0\sin t)^2\cos^{-2} t]}{\sqrt{\pi}\cos t}.$$

2. Quantum-mechanical coupled oscillators.
Let us return to the chain of oscillators (1.1.13) and take $\xi_0$ and $\xi_1$ as the open system. Instead of the pure state (1.1.21), suppose the system is in a thermal state

$$\left\langle \exp\left[i\sum_{n=-\infty}^{\infty}(\xi_{2n}r_n + \xi_{2n+1}s_n)\right]\right\rangle$$

$$= \exp\left[-\frac{1}{4}\tanh\frac{\eta}{2}\sum_{n=-\infty}^{\infty}(r_n^2 + s_n^2) + i\sum_{n=\infty}^{\infty}(r_n s_n' - r_n' s_n)\right].$$

As in (2.5.53.6),

$$\frac{\mathrm{Tr}\exp[-\eta((p-\bar{p})^2 + (q-\bar{q})^2)]\exp[i(pr + qs)]}{\mathrm{Tr}\exp[-\eta((p-\bar{p})^2 + (q-\bar{q})^2)]}$$

$$= \exp\left[-\frac{r^2+s^2}{4}\tanh\frac{\eta}{2} + i(\bar{p}r + \bar{q}s)\right], \qquad (3.1.7)$$

so this state is a Gibbs state with harmonic forces centered at $s'$, $-r'$. Under the time-evolution (1.1.18), the expectation values of the Weyl operators of the open system are

$$\langle \exp i(r\xi_0(t) + s\xi_1(t))\rangle = \exp\left\{\sum_n \left\{-\tfrac{1}{4}\tanh\frac{\eta}{2}[(rJ_{2n} + sJ_{2n+1})^2 + (rJ_{2n+1} + sJ_{2n})^2] + is'_n(rJ_{2n} + sJ_{2n-1}) - ir'_n(rJ_{2n+1} + sJ_{2n})\right\}\right\}.$$

At time $t$ the subsystem is in a state of the form (3.1.7) with

$$s'_0(t) = \sum_n (s'_n(0)J_{2n}(t) - r'_n(0)J_{2n+1}(t)),$$

$$r'_0(t) = \sum_n (r'_n(0)J_{2n}(t) - s'_n(0)J_{2n-1}(t)).$$

The average values $s'_0(t)$, $r'_0(t)$ move classically as in Example 1. They converge to zero, but not monotonically.

3. Coupled spins
Consider spin 1 of the chain (1.1.1) as the open system and the infinitely many others as the thermal reservoir. The coupling constants $\varepsilon(n)$ are chosen as in (1.1.9). The initial state

$$\rho_1 = \tfrac{1}{2}(1 + \sigma_1^+ \exp(-i\alpha) + \sigma_1^- \exp(i\alpha)),$$

$$\omega = \prod_{k \neq 1} \tfrac{1}{2}(1 + \sigma_k^+ \exp(-i\alpha) + \sigma_k^- \exp(i\alpha)),$$

((1.17) with $s = 0$) evolves as

$$\tau_t^*(\rho) = \tfrac{1}{2}\left(1 + \frac{\sin^2 t}{t^2}[\sigma^+ \exp(-i(\alpha + 2Bt) + \sigma^- \exp(i(\alpha + 2Bt)]\right)$$

if $N \to \infty$. The state $\rho$ oscillates as it approaches the equilibrium state $\frac{1}{2}\cdot\mathbf{1}$ as $T \to \infty$.

**Remarks** (3.1.8)

1. The failure of the time-evolution $\tau$ or $\tau^*$ to be a group is due to the effect of the system on the reservoir and the reaction of the reservoir on the system. The reaction influences the system at later times, so $(\partial/\partial t)\tau_t^*(\rho)$ depends on $\tau_s^*(\rho)$ not only for $s = t$ but for all $s \leq t$, i.e., on its whole history. The time-evolution of the density matrix of the reservoir can be written down formally and substituted into the equation for $(\partial/\partial t)\tau_t^*(\rho)$. The resulting **master equation** is an integrodifferential equation for $\rho$ including the memory effects just mentioned.
2. The requirement of complete positivity of the time-evolution is not a mere technicality but a genuine restriction, and it even has some experi-

mentally verifiable consequences. For instance, its implications for the motion of a spin in a thermal reservoir have been confirmed experimentally [15].

The retrospective effects of (3.1.8; 1) disappear in certain limiting cases, so the time-evolution $\tau$ becomes a semigroup. The limits involve the time-scale or the coupling constants. The most understandable case is that of a simplified version of electrodynamic radiative reaction of volume II, §2.4.

**Example** (3.1.9)

Model of Brownian motion

We modify Example (1.1.13) to take a single harmonic oscillator in three dimensions as the system and represent the rest of the system, functioning as a reservoir, as a continuous scalar field $\Phi(\mathbf{x})$. Suppose initially that the oscillator is coupled to an averaged field $\int d^3x \Phi(\mathbf{x})c(\mathbf{x})$, $c \in C_0^\infty(\mathbb{R}^3)$, and later take the limit $c(\mathbf{x}) \to \gamma\delta(\mathbf{x})$, $\gamma \in \mathbb{R}$. We shall study the quantum-theoretical time-evolution from the outset; since the equations of motion are linear it agrees with the classical time-evolution. If $Q$, $P$ and $\Phi(\mathbf{x})$, $\Pi(\mathbf{x})$ are the canonically conjugate coordinate and field variables, then the Hamiltonian is

$$H_S = \tfrac{1}{2}(P^2 + \omega_0^2 Q^2),$$

$$H_R = \tfrac{1}{2}\int d^3x \{\Pi(\mathbf{x})^2 + |\nabla\Phi(\mathbf{x})|^2\},$$

$$H' = \int d^3x c(\mathbf{x})\Phi(\mathbf{x})Q.$$

The resulting equations of motion,

$$\left(\frac{\partial^2}{\partial t^2} - \Delta\right)\Phi(\mathbf{x}, t) = c(\mathbf{x})Q(t),$$

$$\left(\frac{\partial^2}{\partial t^2} + \omega_0^2\right)Q(t) = \int d^3x \Phi(\mathbf{x}, t)c(\mathbf{x}),$$

can be integrated immediately with Green's formula (II: 1.2.36). This is the trivial case of a scalar field on $\mathbb{R}^4$, so with the Green function

$$D(\mathbf{x}, t) = \frac{\delta(r - t)}{4\pi r}$$

(II: 2.2.7), the solution of the initial-value problem is

$$\Phi(\mathbf{x}, t) = \int d^3x'(\Phi(\mathbf{x}', 0)\dot{D}(\mathbf{x} - \mathbf{x}', t) + \dot{\Phi}(\mathbf{x}', 0)D(\mathbf{x} - \mathbf{x}', t))$$

$$+ \int d^3x' \int_0^t dt' D(\mathbf{x} - \mathbf{x}', t - t')c(\mathbf{x}')Q(t')$$

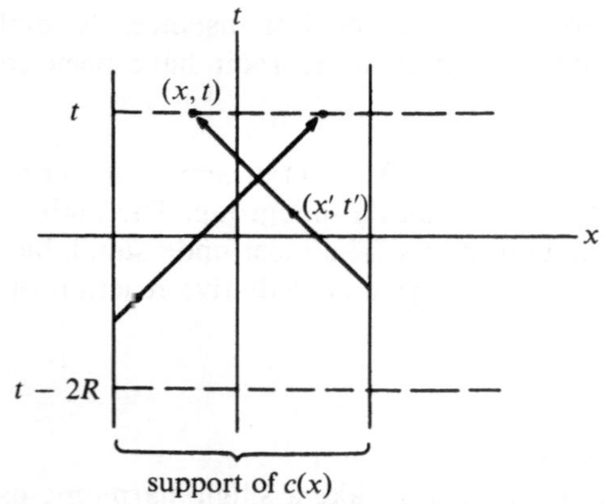


Figure 22 The domain of influence of $F_{\text{reaction}}$.

for all $t > 0$, where $\dot{\Phi} = \partial\Phi/\partial t$, etc. Hence the force exerted by the field on the oscillator is

$$\int d^3x \Phi(\mathbf{x}, t) c(\mathbf{x}) = F_{\text{field}}(t) + F_{\text{reaction}}(t),$$

$$F_{\text{field}}(t) = \int d^3x\, d^3x' c(\mathbf{x})(\Phi(\mathbf{x}', 0)\dot{D}(\mathbf{x} - \mathbf{x}', t) + \dot{\Phi}(\mathbf{x}', 0)D(\mathbf{x} - \mathbf{x}', t))$$

$$F_{\text{reaction}}(t) = \int \frac{d^3x d^3x'}{4\pi|\mathbf{x} - \mathbf{x}'|} c(\mathbf{x})c(\mathbf{x}')Q(t - |\mathbf{x} - \mathbf{x}'|)\Theta(t - |\mathbf{x} - \mathbf{x}'|).$$

In the reaction force $F_{\text{reaction}}(t)$, $Q(t')$ contributes only for $t - 2R \leq t' \leq t$ if $c(\mathbf{x}) = 0$ for all $\mathbf{x}$ such that $|\mathbf{x}| > R$ (see Figure 22).

Now if $c(\mathbf{x}) \to 2\sqrt{\pi}\gamma\delta(\mathbf{x})$, so $R \to 0$, then the retrospective effects disappear, and when the expansion

$$Q(t - |\mathbf{x} - \mathbf{x}'|) = Q(t) - |\mathbf{x} - \mathbf{x}'|\dot{Q}(t) + \tfrac{1}{2}|\mathbf{x} - \mathbf{x}'|^2\ddot{Q}(t) - \cdots$$

is substituted into $F_{\text{reaction}}$,

$$F_{\text{reaction}}(t) \to \delta\omega^2 Q(t) - \gamma^2\dot{Q}(t).$$

The quantity $\delta\omega^2$ is the formally infinite integral $\gamma^2 \int (d^3x\, d^3x'/|\mathbf{x} - \mathbf{x}'|) \times \delta(\mathbf{x})\delta(\mathbf{x}')$, so the limit $c(\mathbf{x}) \to \gamma\delta(\mathbf{x})$ must be taken jointly with a change in $\omega_0^2$. If $\varpi^2 \equiv \omega_0^2 - \delta\omega^2$, then the equation of motion becomes

$$\left(\frac{\partial^2}{\partial t^2} + \varpi^2 + 2\Gamma\frac{\partial}{\partial t}\right)Q(t) = F_{\text{field}}(t), \qquad \Gamma = \frac{\gamma^2}{2}, \quad t \geq 0.$$

For a thermal state with $\langle\Phi(\mathbf{x}, 0)\rangle = \langle\dot{\Phi}(\mathbf{x}, 0)\rangle = 0$, $\langle F_{\text{field}}(t)\rangle = 0$, and the time-evolution of the expectation value of $Q$ for $t \geq 0$ is

$$\langle Q(t)\rangle = \exp(-\Gamma t)\left(\langle Q(0)\rangle\left(\cos\omega t + \frac{\Gamma}{\omega}\sin\omega t\right) + \langle\dot{Q}(0)\rangle\frac{\sin\omega t}{\omega}\right),$$

provided that $\omega^2 \equiv \bar{\omega}^2 - \Gamma^2 > 0$. The expectation values of the canonical variables $\langle Q(t)\rangle$ and $\langle \dot{Q}(t)\rangle$ then evolve according to a symplectic semigroup,

$$\begin{pmatrix} \langle Q(t)\rangle \\ \langle \dot{Q}(t)\rangle \end{pmatrix} = \exp(-\Gamma t) \begin{pmatrix} \cos\omega t + \frac{\Gamma}{\omega}\sin\omega t & \frac{\sin\omega t}{\omega} \\ -\left(\omega + \frac{\Gamma^2}{\omega}\right)\sin\omega t & \cos\omega t - \frac{\Gamma}{\omega}\sin\omega t \end{pmatrix} \begin{pmatrix} \langle Q(0)\rangle \\ \langle \dot{Q}(0)\rangle \end{pmatrix}.$$

The time-evolution of an open system is not generally a unitary transformation of the density matrix, and so the entropy of a subsystem is not necessarily constant. Nothing can be said *a priori* about the sign of the change in entropy; the system might start off hotter than the reservoir and lose entropy as the temperature equalizes. However, the relative entropy introduced in (2.2.22) turns out to be a Liapunov function [16] for the time-evolution (3.1.3).

**The Decrease of the Relative Entropy** (3.1.10)

*For the time-evolution* $\tau^*$ *of* (3.13),

$$S(\tau_t^*(\sigma)\,|\,\tau_t^*(\rho)) \le S(\sigma\,|\,\rho).$$

**Proof**

With Definition (2.2.22) and the unitary invariance,

$$S(\mathrm{Tr}^R U_{-t}\sigma\otimes\omega U_t\,|\,\mathrm{Tr}^R U_{-t}\rho\otimes\omega U_t) \overset{\text{(iv)}}{\le} S(U_{-t}\sigma\otimes\omega U_t\,|\,U_{-t}\rho\otimes\omega U_t)$$
$$= S(\sigma\otimes\omega\,|\,\rho\otimes\omega) \overset{\text{(iii)}}{=} S(\sigma\,|\,\rho). \qquad \square$$

**Remarks** (3.1.11)

1. The relative entropy is always positive, and in the special case of (2.2.23; 1), it is $\beta$ times the difference between the free energy of the state $\rho$ and the free energy at equilibrium. Its decrease reflects the tendency of the system to equilibrium.
2. Monotony in time cannot be claimed if $\tau_{t_1+t_2} \neq \tau_{t_2}\circ\tau_{t_1}$. In Example (3.1.9) friction returned the oscillator monotonically to its rest-point, owing to the semigroup property, which was in turn a consequence of the absence of retrospective effects. The general fact is

**Monotony of the Relative Entropy with a Dynamic Semigroup** (3.1.12)

*If* $\tau_{t_1+t_2} = \tau_{t_2}\circ\tau_{t_1}$ *for all* $t_1$ *and* $t_2 \ge 0$, *then* $\tau_t$ *is said to be a dynamical semigroup. The function* $S(\tau_t^*(\sigma)\,|\,\tau_t^*(\rho))$ *is then a monotonically decreasing function of* $t$.

**Proof**

This is a direct consequence of (3.1.10). □

**Remarks** (3.1.13)

1. Because $S(\sigma|\rho) \geq 0$, the limit of $S(\tau_t^*(\sigma)|\tau_t^*(\rho))$ as $t \to \infty$ exists.
2. It cannot yet be claimed that the free energy approaches its equilibrium value as $t \to \infty$; $S(\sigma/\rho)$ might stop at some positive value and never fall to zero.
3. The apparent asymmetry in the direction of time comes from the requirement of (3.1.3) that the initial state factorizes. Starting at $t < 0$, the later state at $t = 0$ is factorized, so the relative entropy increases.
4. If the dynamical semigroup is governed by a master equation of the type of (2.1.11; 3), then $S(\rho)$ increases monotonically.

That finishes the orientation toward various phenomena connected with the time evolution. Let us now return to more global questions of time-dependence. The problem, put concisely, is that a finite system the Hamiltonian of which has pure point spectrum $\{\varepsilon_i\}$ has observables whose expectation values $\langle a(t)\rangle = \sum_{j,k} a_{jk} \exp(i(\varepsilon_j - \varepsilon_k)t)$ are almost-periodic functions, as superpositions of periodic functions. Only the average over time makes sense; the time-limit exists only for infinite systems the Hamiltonians of which have absolutely continuous spectra. Although in actuality only finite systems come under observation, the recurrence times are so long that they are indistinguishable from infinite systems within the times of relevance to human beings. In any event, the first issue to settle is how to define the time-average of a function $f(t) \in C(\mathbb{R})$, the set of bounded, continuous functions on $\mathbb{R}$. The obvious guesses would be

$$\lim_{T\to\infty} \frac{1}{2T}\int_{-T}^{T} dt f(t) \quad \text{or} \quad \lim_{\varepsilon\to 0}\frac{\varepsilon}{2}\int_{-\infty}^{\infty} dt \exp(-\varepsilon|t|) f(t),$$

but these limits do not converge for such functions as $\sin(\ln(|t| + 1)) \in C(\mathbb{R})$. Any suitable average would have to be linear, positive, and invariant under displacements in time. Every invariant state on the $C^*$ algebra $C(\mathbb{R})$ has the right qualifications, and the existence of many invariant states on $C(\mathbb{R})$ means that there are many possible time-averages. There is thus no question whether a time-average exists, but it is not unique.

**The Time-Average of an Observable** (3.1.14)

Let $\eta$ be an average over $C(\mathbb{R})$ and $t \to a_t$ be a weakly continuous mapping $\mathbb{R} \to \mathscr{B}(\mathscr{H})$ such that $\|a_t\| \leq \|a_0\|$ for all $t$. Then the average $\eta(a)$ is defined by

$$\langle x|\eta(a)|y\rangle = \eta(\langle x|a_t|y\rangle) \quad \text{for all } x, y \in \mathscr{H}.$$

**Remarks** (3.1.15)

1. Since $|\eta(\langle x|a_t|y\rangle)| \leq \|x\| \cdot \|y\| \cdot \|a_0\|$, this sesquilinear form defines a bounded operator $\eta(a)$.
2. In the Schrödinger picture, the average $\eta(\sigma)$ of a state $\sigma$ on the algebra generated by the operators $a_t$ is defined by $\eta(\sigma)(a) = \eta(\sigma(a_t))$.

**Examples** (3.1.16)

1. If $a_t = \exp(-iHt) \equiv U(t)$, then $\eta(U) = E_0 \equiv$ the projection onto the eigenvectors of $H$ with eigenvector 0.

**Proof**

(i) $\langle x|E_0\eta(U)y\rangle = \eta\langle E_0 x|U_t y\rangle = \langle x|E_0 y\rangle \Rightarrow E_0\eta(U) = E_0$.
(ii) $\langle x|U(t_0)\eta(U)y\rangle = \eta\langle x|U(t+t_0)y\rangle = \langle x|\eta(U)y\rangle \Rightarrow$ $U(t_0)\eta(U) = \eta(U) \Rightarrow E_0\eta(U) = \eta(U) = E_0$ by part (i). □

2. $a_t = U(t)aU^{-1}(t)$, where $U(t)$ has pure point spectrum. If the projections onto the eigenspaces are $E_i$, then $\eta(a) = \sum_i E_i a E_i$.

**Proof**

Take matrix elements with the eigenvectors of $H$ and note that $\eta(\exp(i\alpha t)) = 0$ for all $\eta$ and all $\alpha \in \mathbb{R}$ different from 0. □

3. $\eta(a_t E_0) = E_0 a E_0$, since $\eta(a_t E_0) = \eta(U(t)aE_0) = E_0 a E_0$, as in Example 1.

**Remarks** (3.1.17)

1. In these examples the concrete averages $(1/2T)\int_{-T}^{T} dt \exp(iHt)$ and $(\varepsilon/2)\int_{-\infty}^{\infty} dt \exp(-\varepsilon|t|)\exp(iHt)$ converge strongly (Problem 1). Hence $E_0$ belongs to $U''$ as well as $U'$.
2. In the Schrödinger picture the time-average of a vector $|x\rangle$ is defined by $|\eta(x)\rangle \equiv \eta(U(t)|x\rangle) = E_0|x\rangle$. It can be characterized as the vector with the least norm in the convex hull of its trajectory $\{U(t)|x\rangle, t \in \mathbb{R}\}$ (Problem 2). It is not, however, true in general for the state $\sigma(a) = \langle x|a|x\rangle$ formed with $|x\rangle$ that $\eta(\sigma)(a) = \langle \eta(x)|a|\eta(x)\rangle$.
3. There is no definition of $\eta(a)$ independent of the representation; since $\lim_{T\to\infty}(1/T)\int_0^T dt\, a_t$ belongs only to the weak closure of the algebra, $\eta$ may send operators out of their $C^*$ algebra. Our representations will usually be such that the time-automorphism $\alpha_t$ can be implemented

unitarily, and the image of $E_0$ will contain a cyclic vector for $\mathscr{A}$. If the averages $\eta(a)$ belong to $\mathscr{A}'$, then they are determined uniquely by

$$\eta(a)E_0 = \lim_{T\to\infty} \frac{1}{T}\int_0^T dta_t E_0 = \lim_{T\to\infty} \frac{1}{T}\int_0^T U_t a E_0 = E_0 a E_0,$$

since $E_0\mathscr{H}$ separates $\mathscr{A}'$ (Problem 5). However, as will be seen in (3.1.22; 4), $\eta(a)$ in general depends on the representation.

4. The time-average may be nonunique if $f(t)$ converges, as $t \to +\infty$ and $t \to -\infty$, but to different values. This situation is familiar to us from scattering theory. Whenever the time-average of a function $f$ is unique, it agrees with the "concrete average"

$$\lim_{T\to\infty} \frac{1}{2T}\int_{-T}^T dtf(t), \qquad \lim_{\varepsilon\to 0} \frac{\varepsilon}{2}\int_{-\infty}^\infty dt \exp(-\varepsilon|t|)f(t),$$

or even

$$\lim_{T\to\infty} \frac{1}{T}\int_0^T dtf(t).$$

These averages exist in classical ergodic theory, in which the Liouville measure on phase space provides the invariant cyclic vector. Some ergodic systems will be defined later, and for them $E_0$ is one-dimensional, projecting onto the cyclic vector. This projection is then constant on the energy shell, so the time-average $E_0 a E_0$ equals the average over the energy shell.

5. The point spectrum of $H$ can be turned into a continuum by an arbitrarily small perturbation, so averaging over time focuses unduly on the exact form of $H$, since $\eta$ is quite different depending on whether the spectrum is pointlike or continuous: If in the spectral representation of $H$ the operator $a$ on the subspace belonging to $\sigma_{ac}$ has a continuous integral kernel, then $\eta$ projects this part of $a$ to 0, and by Remark 2 only its point-spectrum part remains (cf. (I: 3.3.4; 6)).
6. Pure states of classical systems are points in phase space, and averages over pure states are averages over classical trajectories.
7. If the spectrum of $H$ is pure point and nondegenerate, then every normal, invariant state can be written as the time-average of a pure state. Normal, invariant states are of the form

$$\sigma(a) = \sum_i c_i\langle x_i|a\, x_i\rangle, \quad 0 \le c_i \le 1, \qquad \sum_i c_i = 1, \quad H|x_i\rangle = \varepsilon_i|x_i\rangle;$$

so

$$\sigma(a) = \langle x|\eta(a)|x\rangle, \qquad x = \sum_i \sqrt{c_i}|x_i\rangle.$$

Although the canonical state $\rho = \exp(-\beta(H - F))$ is an average over the trajectory of a pure state, it is certainly not true that every averaged pure state is the canonical state.

Our reasoning until this point has applied indifferently to all sorts of quantum systems, but not all quantum systems exhibit thermodynamic behavior. An isolated atom is rather like a frictionless perpetual-motion machine; only large systems are dissipative. The concept introduced in (1.3.10) of asymptotic commutativity turns out to be a useful characteristic of dissipative systems. If the local observables are asymptotically Abelian with respect to the time-automorphism $\alpha_t$, that means that local perturbations dissipate through the system as time passes. Of course, this is possible only if $H$ has continuous spectrum, and hence only if the system is infinite. We shall remain with Definition (1.3.10), although many of its consequences can be derived with weaker assumptions. Definition (1.3.10) applies to a system of free fermions, but it has not been possible to prove that even weakened versions of it apply to more realistic, interacting systems. It is trivial that classical systems are asymptotically Abelian, and (1.3.10) means roughly that asymptotically Abelian systems behave classically on a macroscopic time scale.

**Properties of Asymptotically Abelian Systems** (3.1.18)

*Let $\mathscr{A}$ be an asymptotically Abelian $C^*$ algebra with respect to a group of automorphisms $a \to a_t$, and let $\omega$ be an invariant state having a representation on a Hilbert space $\mathscr{H}$ with a cyclic vector $|\Omega\rangle$. Then, abbreviating $\mathscr{A}' = \pi_w(\mathscr{A})'$, etc.,*

1. *the invariant elements of $\mathscr{A}$ belong to $\mathscr{A}'$;*
2. *the invariant elements of $\mathscr{A}'$ lie in the center (i.e., $\mathscr{R}' = \mathscr{A}' \cap U' = \eta(\mathscr{A}')$ is a subalgebra of the center $\mathscr{A}' \cap \mathscr{A}''$), and so $\mathscr{R}' = \eta(\mathscr{A}'')$;*
3. *$E_0 \mathscr{A}'' E_0$ is maximally Abelian in $E_0 \mathscr{H}$, where $E_0$ is the projection onto the invariant vectors of $\mathscr{H}$; and*
4. *if $\sigma$ produces a factor (i.e., the GNS representation $\pi_\sigma(\mathscr{A})$ and $\pi_\sigma(\mathscr{A})'$ constructed with the cyclic vector $\Omega_\sigma$ generate all of $\mathscr{B}(\mathscr{H})$), then*

$$\lim_{t \to \pm\infty} (\sigma(a_t b) - \sigma(a_t)\sigma(b)) \to 0,$$

*even if $\sigma(a_t) \neq \sigma(a)$.*

**Remarks** (3.1.19)

1. Neither $E_0$ nor $E_0 \mathscr{A}'' E_0$ necessarily belongs to $\mathscr{A}''$. Moreover, $E_0 \mathscr{A}'' E_0$ may fail to be an algebra, and the somewhat loose phrasing of Property 3 is intended to mean that the algebra generated by $E_0 \mathscr{A}'' E_0$ is the same as its commutant.
2. The point of (3.1.18) is that invariant elements such as time-averages and time-limits form an Abelian algebra, and thus equal its center. Factor states are pure when restricted to the center, and are therefore

characters (see Definition (III: 2.2.25)), which explains why they factorize in time-limits and time-averages.

**Proof**

1. $[a, b] = \lim_{t\to\infty} [a_t, b] = 0$ for all invariant $a \in \mathscr{A}$ and all $b \in \mathscr{A}$.
2. By Property 3, $E_0 \mathscr{R} E_0 = E_0 \mathscr{A}'' E_0$ is maximally Abelian and so equal to $(E_0 \mathscr{R} E_0)' E_0$. Since $E_0 \in \mathscr{R}$, $(E_0 \mathscr{R} E_0)' E_0 = E_0 \mathscr{R}' E_0$ [17], and therefore $E_0 \mathscr{R}' E_0 = E_0(\mathscr{R}' \cap \mathscr{R}) E_0$. Since $|\Omega\rangle$ separates $\mathscr{A}'$, the equation $E_0 a' E_0 = a' E_0$ determines every $a' \in \mathscr{R}'$ uniquely, so $a' \in \mathscr{R}$. However, $\mathscr{R} \cap \mathscr{R}'$ is $\mathscr{A}'' \cap \mathscr{A}' \cap U'$, because $U \cap \mathscr{A}' = \{\mathbf{1}\}$.
3. The set $E_0 \mathscr{A} E_0$ must be Abelian, as otherwise some commutator would fail to vanish as $t \to \pm\infty$:

$$\eta_t[a_t, b] = 0 \Rightarrow \eta_t E_0(aU_t b - bU_{-t} a)E_0 = 0$$
$$\Rightarrow [E_0 a E_0, E_0 b E_0] = 0 \qquad \text{for all } a, b \in \mathscr{A}.$$

Hence $E_0 \mathscr{A}'' E_0 = (E_0 \mathscr{A} E_0)''$ is also Abelian, and in fact maximally Abelian, as otherwise $E_0 a E_0$ would be $\sim \mathbf{1}$ on a subspace of dimension greater than one for all $a \in \mathscr{A}$, and $|\Omega\rangle = E_0|\Omega\rangle$ would not be cyclic.
4. For every $b \in \pi_\sigma(\mathscr{A})$ there exist two operators $b_1$ and $b_2$ such that $b_2|\Omega_\sigma\rangle = b_1^*|\Omega_\sigma\rangle = 0$ and $b = \mathbf{1}\langle\Omega_\sigma|b|\Omega_\sigma\rangle + b_1 + b_2$. This is obvious for finite matrices:

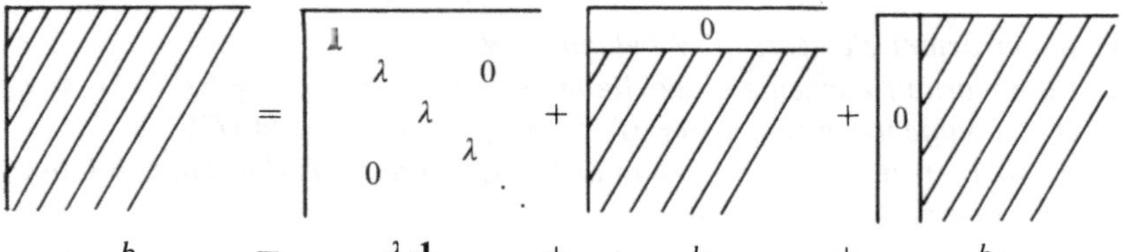


$b \quad = \quad \lambda \cdot \mathbf{1} \quad + \quad b_1 \quad + \quad b_2$

and it carries over to $\mathscr{B}(\mathscr{H})$. Then $\sigma(a_t b) - \sigma(a_t)\sigma(b) = \sigma([a_t, b_1])$. If $\sigma$ produces a factor, then $b_1$ can be approximated with a finite sum

$$\sum_{i=1}^{n} d_i d_i', \qquad d_i \in \pi_\sigma(\mathscr{A}), \qquad d_i' \in \pi_\sigma(\mathscr{A})',$$

and $\sum_i \sigma([a_t, d_i]d_i')$ tends to 0 as $t \to \pm\infty$ by Definition (1.3.10). Although the subalgebra of $\mathscr{B}(\mathscr{H})$ generated by $\pi_\sigma(\mathscr{A}) \cup \pi_\sigma(\mathscr{A})'$ is only strongly dense, operators with these properties can be approximated even in the norm sense ([18], V.1.4), which justifies these conclusions. □

The set of invariant states is convex, so any invariant state is a convex combination of the extremal points of the set or a limit of such combinations. As the purest among the time-invariant states, the extremal elements deserve a special term:

**Definition** (3.1.20)

An invariant state is **ergodic**, or **extremal invariant**, if it can not be written as a convex combination of other invariant states.

**Remarks** (3.1.21)

1. In classical dynamics an invariant submanifold $\mathcal{N}$ of phase space corresponds to an invariant state (= measure) $\mu_{\mathcal{N}} = \prod_i dq^i \wedge dp^i_{|\mathcal{N}}$, which is ergodic if $\mathcal{N}$ cannot be decomposed into invariant pieces with strictly positive measures $\mu_{\mathcal{N}}$.
2. A classical system is said to be ergodic if the surface of the energy shell $\rho(p, q) = \delta(E - H(p, q)) \exp(-S(E))$ corresponds to an ergodic state.
3. Every time-invariant state is a sum or integral of ergodic states, so it is tempting to interpret the ergodic states as the pure phases of the system. Mixtures would then be incoherent superpositions in the sense of quantum theory rather than coexisting, spatially separated phases. With any reasonable definition of pure phases, the decomposition into ergodic states should be unique, and the set of time-invariant states must be a simplex. This is indeed the case for asymptotically Abelian systems, which follows from the observation that $\mathcal{R}' = \mathcal{A}' \cap \{U_t\}'$ is Abelian: As was seen in (1.4.9) and (III: 2.3.24; 2), every Abelian subalgebra of $\mathcal{A}'$ corresponds to a unique decomposition of a state $\omega$; if $\{P_i\}$, $\sum_i P_i = \mathbf{1}$, are the orthogonal projections of this algebra, and
$$\omega_i(a) = \frac{\omega(P_i a)}{\omega(P_i)} \quad \text{for all } a \in \mathcal{A},$$
provided that $\omega(P_i) > 0$, and is otherwise arbitrary, then $\omega = \sum_i \lambda_i \omega_i$, $\lambda_i = \omega(P_i)$ and $\pi_\omega = \bigoplus_i \pi_{\omega_i}$, where $\pi_{\omega_i}$ acts on $P_i \mathcal{H}_\omega$. Now if $\omega$ is invariant and is to have a decomposition into other invariant states, then the projections $P_i$ must belong to $\mathcal{A}' \cap \{U_t\}'$, and in fact the extremal states correspond to the minimal projections. Since $\mathcal{A}' \cap \{U_t\}' \subset \mathcal{Z}$, the decomposition into ergodic states is never as fine as the factor decomposition. Hence if a factor representation is given by the invariant state $\omega$, it is necessarily ergodic.

Ergodicity in fact singles out the desired properties. This is shown by the

**Characterization of the Ergodic States** (3.1.22)

*Let $\mathcal{A}$ be an algebra that is asymptotically Abelian in time, $\rho$ an invariant state on $\mathcal{A}$, and $|\Omega\rangle$ the vector of the state $\rho$ in the* GNS *representation. Then the following conditions are equivalent:*

1. *$\rho$ is ergodic;*
2. $\mathcal{R}' = \{\alpha \cdot \mathbf{1}\}$;

3. *given any decomposition* $\rho = \int \sigma \, d\mu(\sigma)$ *and a* $\mu$*-measurable mean* $\eta$, $\eta(\sigma) = \rho$ *almost everywhere for* $\mu$;
4. $\eta(a) = \mathbf{1} \cdot \rho(a)$ *for all* $a \in \mathscr{A}$ *and all invariant means* $\eta$;
5. $(\mathscr{A} \cup \mathscr{A}') \cap U' = \{\alpha \cdot \mathbf{1}\}$;
6. $E_0 = |\Omega\rangle\langle\Omega|$;
7. $\rho$ *is a unique, invariant, normal state on* $\pi_\rho(\mathscr{A})''$;
8. $\eta(\rho(ab_t)) = \rho(a)\rho(b)$ *for all a and* $b \in \mathscr{A}$ *and all invariant means* $\eta$.

**Remarks** (3.1.23)

1. If the quantum system is finite, $H$ has pure point spectrum, with eigenvectors $\{|x_i\rangle\}$. As we have learned, the invariant states are of the form $a \to \sum_i c_i \langle x_i | a x_i \rangle$, so the extremal invariant states are of the form $a \to \langle x_i | a x_i \rangle$ and therefore pure. If the system is either infinite or classical, then ergodic does not imply pure. For example, the state of free fermions (2.5.49) produces a factor and is therefore ergodic, but $\mathscr{A}'$ is isomorphic to $\mathscr{A}$ and thus different from $\{\alpha \cdot \mathbf{1}\}$. It will be discovered later that this is the normal situation for equilibrium states.
2. According to (III: 2.3.10; 5), Condition 2 means that $\rho$ is a pure state on $\mathscr{R}$, and can also be written as $\mathscr{R} \cap \mathscr{R}' = \{\alpha \cdot \mathbf{1}\}$; in particular, every factor state is ergodic.
3. Condition 3 can be sharpened for classical systems with Birkhoff's ergodic theorem, according to which almost every trajectory fills the energy shell densely. In this case, with the decomposition into pure states, the Cesàro mean exists; $\eta(\sigma)$ is $\mu$-measurable, and the order of $\eta$ and $\int d\mu$ can be reversed.
4. By Condition 4 the time-average of operators in this situation is unique and a multiple of the identity. More particularly, the classical time-average of any set of positive $\rho$-measure is spread out over the whole support of $\rho$. Hence the time-average of states with a density function equals the equilibrium state. Since averaged observables are multiples of the identity, they exhibit no deviation.
5. The implication of Condition 5 for classical dynamics is that if the system is ergodic, then every measurable, time-independent function is constant on the energy shell. Note that $(\mathscr{A} \cup \mathscr{A}')''$ might contain additional time-invariant operators; for instance, for a factor this set is $\mathscr{B}(\mathscr{H})$ and therefore also contains $U$.
6. Condition 6 implies that 1 is a simple eigenvalue of $U$.
7. By Condition 7, all the other eigenvectors of $U$ lead to the same state as $\rho$. Classically, the eigenfunctions $\varphi(p, q)$ must always have $|\varphi|^2$ constant independently of $p$ and $q$. Thus ergodicity does not make it impossible for the spectrum to be purely pointlike, but only prevents 0 from being a degenerate eigenvalue of $H$. The extra word "normal" of Condition 7 is important. In Example (3.1.1; 5) of free fermions, equilibrium states at

different temperatures from that of the specified representation are invariant in time, but not normal. This means classically that different energy shells have disjoint support.

8. Condition 8 means that the autocorrelation function $\rho(ab_t) - \rho(a)\rho(b)$ has time-average 0. Also, according to Condition 4 the expectation values of operators in states of the form $a|\Omega\rangle$ have the same time-averages as those with the state $\rho$. Since the states $a|\Omega\rangle$ are dense, the time-average of every normal state is $\rho$. This is a sort of converse to Condition 3, in so far as $\eta(\sigma) = \rho$ for all $\sigma$'s that are pure and normal (as states on $\pi_\rho(\mathscr{A})''$). It may happen that the set of such $\sigma$'s is empty (cf. (1.4.17; 3)), and some non-normal, pure states converging to something other than the equilibrium state will make their appearance later.

**Proof**

$1 \Rightarrow 2$: Let $t \in \mathscr{R}'$, $0 < t < 1$; then the vector $|\Omega_\rho\rangle$ associated with $\rho$ in the GNS representation is cyclic for $\mathscr{R}$ and therefore separates $\mathscr{R}'$. With $|\Omega_\rho\rangle$,

$$0 < ||t^{1/2}\Omega_\rho\rangle|^2 = \langle\Omega_\rho|t\Omega_\rho\rangle \equiv \lambda < 1,$$

so if

$$\rho_1(a) = \frac{1}{\lambda}\langle\Omega_\rho|at\Omega_\rho\rangle,$$

and

$$\rho_2(a) = \frac{1}{1-\lambda}\langle\Omega_\rho|a(1-t)\Omega_\rho\rangle \quad \text{for all } a \in \mathscr{A},$$

then $\rho = \lambda\rho_1 + (1-\lambda)\rho_2$ has a genuine decomposition into invariant states.

$2 \Rightarrow 1$: Let $\rho = \lambda\rho_1 + (1-\lambda)\rho_2$, where $0 < \lambda < 1$. Then according to (III: 2.3.24; 2) there exists a $t \in \pi_\rho(\mathscr{A})'$ such that $0 \le t \le 1$ and $\rho_1(a) = \langle\Omega_\rho|t\Omega_\rho\rangle^{-1}\langle\Omega_\rho|at\Omega_\rho\rangle$ for all $a \in \mathscr{A}$. If $\rho_1$ is invariant, then $t$ is in $\mathscr{R}'$, and it follows from Condition 2 that $\rho = \rho_1 = \rho_2$.

$2 \Leftrightarrow 4$: $\mathscr{R}' \supset \{\eta(a): a \in \mathscr{A}\}$. (Cf. (3.1.18; 2).)

$1 \Rightarrow 3$: The state $\rho = \int \sigma\, d\mu(\sigma)$ is invariant in time, so $\rho(a) = \int d\mu(\sigma)\eta(\sigma(a))$. Therefore $\rho = \int d\mu(\sigma)\eta(\sigma)$, and, since $\rho$ is an extremal invariant, it equals the invariant state $\eta(\sigma)$ almost everywhere in $\mu$.

$3 \Rightarrow 1$: Suppose that $\rho$ is not ergodic. Then there exist invariant states $\rho_1 \neq \rho_2$ such that $\rho = \lambda\rho_1 + (1-\lambda)\rho_2$. This is a special case of a decomposition with $\rho_i = \eta(\rho_i) \neq \rho$, so Condition 3 would be violated.

$2 \Leftrightarrow 5$: The invariant elements of $\mathscr{A}$ and $\mathscr{A}'$ compose $\mathscr{R}'$.

$6 \Rightarrow 1$: Suppose that $\rho = \lambda\rho_1 + (1-\lambda)\rho_2$; then by (III: 2.3.24; 2), $\rho_1$ is of the form $\rho_1(a) = \langle\Omega_\rho|t\Omega_\rho\rangle^{-1}\langle t^{1/2}\Omega_\rho|at^{1/2}\Omega_\rho\rangle$ for $a \in \mathscr{A}$, and $t$ is in $\pi_\rho(\mathscr{A})' \cap U'_\rho$ if $\rho_1$ is invariant. Condition 6 implies that $|t^{1/2}\Omega_\rho\rangle \propto |\Omega_\rho\rangle$, because $|t^{1/2}\Omega_\rho\rangle \in E_0\mathscr{H}$, so $\rho = \rho_1 = \rho_2$.

$6 \Rightarrow 8$: $\eta(\rho(ab_t)) = \eta(\langle\Omega|aU_t b|\Omega\rangle) = \langle\Omega|aE_0 b|\Omega\rangle = \rho(a)\rho(b)$.

$7 \Rightarrow 6$: If there existed a second invariant vector $|\Omega'\rangle$, then all vectors $\sqrt{\alpha}|\Omega\rangle + \sqrt{1-\alpha}|\Omega'\rangle$ for $0 \le \alpha \le 1$ would give rise to the same state, but by Property (3.1.18; 3), since the algebra is maximally Abelian on the subspace, this would mean that $|\Omega\rangle = |\Omega'\rangle$.

$4 \Rightarrow 7$ and 8: $\omega$ invariant $\Rightarrow \omega = \eta(\omega) \Rightarrow \eta(\omega)(a) = \rho(a)$.

$8 \Rightarrow 4$: From $\eta([b_t, c]) = 0$ it follows that $\rho(ac)\rho(b) = \eta(\rho(acb_t)) = \eta(\rho(ab_t c))$, so the matrix elements of $\rho(b) \cdot \mathbf{1}$ and $\eta(b)$ are equal on a dense set.

□

**Examples** (3.1.24)

1. The only possible ergodic states on classical systems are those concentrated on $\delta(E - H(p, q))$; otherwise $\mathscr{A}$ would contain the additional invariant $F(H)$, contradicting Condition 4. Let us examine a chain of $N$ coupled oscillators (1.1.14). The Hamiltonian can be written in terms of action and angle variables $K_i$ (see (I: 3.3.3) and (I: 3.3.14)) and $\varphi_i \in T^1$ as

$$H = \sum_{i=1}^{N} \omega_i K_i,$$

and the time-evolution is $\varphi_i \to \varphi_i + \omega_i t$. If $N > 1$, the state $\sim \delta(E - H)$ is not ergodic, although the state $\sim \prod_i \delta(K_i - c_i)$ concentrated on $T^N$ is, provided that the angular velocities $\omega_i$ are rationally independent (cf. (I: 3.3.3)). To understand why, observe that the operator $h$ on $L^2(T^N)$ introduced in (3.1.1; 1) arises when $K_i$ is interpreted as the displacement operator, the eigenvalues of which are $2\pi n$, $n \in \mathbb{Z}$. The spectrum of $h$ is therefore purely pointlike, with eigenvalues $2\pi \sum_i \omega_i n_i$. If the $\omega_i$ are rationally independent, then the eigenvalue 0 (all $n_i = 0$) is nondegenerate and otherwise it is degenerate. According to (3.1.22; 6) this is a criterion for ergodicity. This example is also useful for illustrating the other criteria. For instance, Condition 4 states that every invariant $L^\infty$ function is constant almost everywhere on $T^N$. Roughly speaking, a function assuming one value on half the trajectories and a different value on the other half is not measurable.
2. Of the quantum-mechanical examples of (3.1.1), only the free fermions (3.1.1; 5) fall within the category covered by (3.1.22), as the others are not asymptotically Abelian. Since (3.1.1; 5) has a factor state, it is ergodic according to Condition 5. If we go through the other criteria, we notice

that Condition 8 holds in the sharpened form $\lim_{t \to \pm\infty} \rho(ab_t) = \rho(a)\rho(b)$ for all $a$ and $b \in \mathscr{A}$. This means that normal states approach $\rho$ not only in the mean, but also actually in the limit $t \to \pm\infty$. The situation is as described intuitively in §1.1, where the states converge to the equilibrium state.

Even though Example 1 is ergodic, it does not exhibit the sort of behavior appropriate for a thermodynamic system. The time-evolution is a rigid displacement in $T^N$, and this submanifold does not get thoroughly mixed. States like those given by pieces of $T^N$ do not converge as $t \to \infty$; only their means converge. Example 2 conforms better to the notion of a thermodynamic system, which suggests sharpening some of Criteria (3.1.22) as much as possible, by replacing the time-average with the time-limit.

**Definition** (3.1.25)

An invariant state on an asymptotically Abelian system is called **mixing** iff one of the following equivalent conditions is satisfied:

4′. $w\text{-}\lim_{t \to \pm\infty} \pi_\rho(a_t) = \mathbf{1} \cdot \rho(a)$ for all $a \in \mathscr{A}$ (The weak limit is that of the GNS representation);
6′. $U_t \xrightarrow{t \to \pm\infty} |\Omega\rangle\langle\Omega|$;
8′. $\lim_{t \to \pm\infty} \rho(ab_t) = \rho(a)\rho(b)$.

**Remarks** (3.1.26)

1. By Condition 4′, every operator converges to its equilibrium value and its deviation goes to zero. Hence, in the Schrödinger picture every normal state approaches the equilibrium state $\rho$. In classical dynamics probability distributions of normal states are described by functions—i.e., not by $\delta$ distributions—and so they spread out through all of $\rho$.
2. Criterion 6′ is satisfied if the spectrum of $U$ is absolutely continuous other than the eigenvalue associated with $|\Omega\rangle$. In any case, $|\Omega\rangle$ must be the only eigenvector.
3. Concerning Condition 8′, we have learned that for a factor the correlation functions vanish automatically as $t \to \pm\infty$. Therefore, for factors ergodic is equivalent to mixing. In general it is only true that mixing implies ergodic. It is also not true to say that mixing implies a factor, since there are classical mixing systems. However, it will be shown in the next section that in quantum theory equilibrium states are mixing iff the algebra is a factor. In the case of free particles with the spatial translations, as the group of automorphisms with respect to which their algebra of observables is asymptotically Abelian, this reasoning implies that the spatial correlation function goes to zero for factors.

4. If a state is a limit of pure states, then it is mixing: If $\sigma$ is pure and $\sigma_t \to \rho$ then $\rho(ab_t c) - \lim_{s\to\infty} \sigma(a_s b_{s+t} c_s) + \lim_{s\to\infty} \sigma(a_s c_s)\sigma(b_{t+s}) - \rho(ac)\rho(b) = 0$. A pure state is a factor state, so (3.1.18; 4) applies, showing that $\rho(ab_t c) \to \rho(ac)\rho(b)$. The converse is not true in general, since the pure states into which $\rho$ is decomposed need not converge as $t \to \pm\infty$. For example, the pure states for classical systems are points in phase space, which will keep moving forever.

**Proof of the Equivalence in** (3.1.25)

$8' \Leftrightarrow \rho(ab_t c) = \rho(a[b_t, c]) + \rho(acb_t) \to \rho(ac)\rho(b) \Leftrightarrow 4'$, and $\rho(a_t b) = \rho(aU_t b)$, hence $6' \Leftrightarrow 8'$. □

Classical systems that mix are of necessity complicated, and it requires a rather demanding example to show that the concept of (3.1.25) is not empty:

**Motion on a Surface of Constant, Negative Curvature** (3.1.27)

The ergodic system (3.1.24; 1) is not mixing; the spectrum of $U_t$ is purely discrete. This agrees with the perception that displacements in $T^2$ do not mix its parts together:

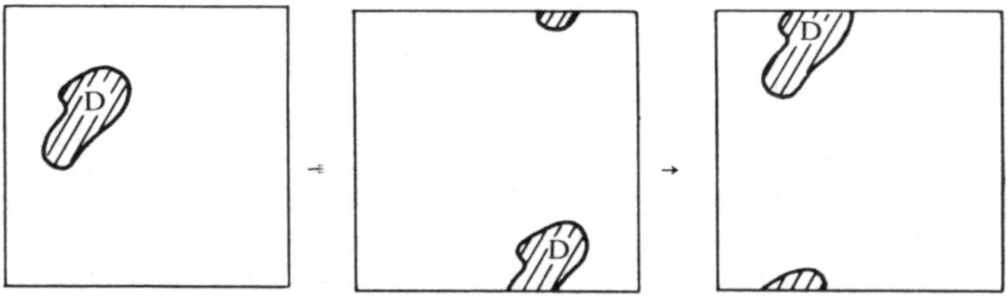


$t_0$ $\quad$ $t_1 > t_0$ $\quad$ $t_2 > t_1$

To produce mixing we need a somewhat geometrically irregular configuration; fortunately, as will now be demonstrated, it suffices to have a surface of constant negative curvature. The construction of the example makes use of the following more abstract reformation of (3.1.24; 1). Treat $\mathbb{R}^2$ as a two-dimensional group and the trajectory as a one-dimensional subgroup, and consider its image in the quotient space $T^2 = \mathbb{R}^2/\mathbb{Z}^2$. Conservation of angular momentum gets lost, and the trajectory can be dense in $T^2$. The present example will have an energy shell that is diffeomorphic to the Lorentz group SO(2, 1), and the trajectory will be a one-parameter subgroup. In order to destroy the other constants of the motion and have an energy shell of finite volume, map the space to SO(2, 1)/$\mathscr{Z}$, where $\mathscr{Z}$ is a discrete subgroup of SO(2, 1). The dynamics furnishes a unitary representation

$U_t = \exp(mt)$ of a one-parameter subgroup of SO(2, 1), but, unlike with $\mathbb{R}^2$, $U$ has only absolutely continuous spectrum other than the point 1, and so the system is mixing by (3.1.26; 2).

We realize these ideas in a classical system the Lagrangian of which is quadratic in the velocities. The motion thus proceeds in the absence of forces, but the invariance under SO(2, 1) brings about some unusual signs. The extended configuration space is the submanifold of $\mathbb{R}^3$ for which

$$(x|x) \equiv x_1^2 + x_2^2 - x_0^2 = -1. \tag{3.1.28}$$

If $\dot{x}$ denotes the derivative of $x$ by the proper time $t$, then the Lagrangian is

$$L = \tfrac{1}{2}(\dot{x}|\dot{x}).$$

The constraint (3.1.28) enters into the Euler–Lagrange equations through a Lagrange multiplier,

$$\ddot{x}_i = \lambda x_i, \tag{3.1.29}$$

and there are the following constants:

$$(x|x) = -1, \qquad (\dot{x}|x) = 0, \qquad (\dot{x}|\dot{x}) = 1 \tag{3.1.30}$$

(which normalize $t$). The three-dimensional manifold defined by the constants corresponds to the energy shell (recall that the configuration space is two-dimensional and the phase space is four-dimensional), and on it is the SO(2, 1)-invariant Liouville measure

$$d\Omega = d^3x\, d^3\dot{x}\,\delta((\dot{x}|x))\delta((x|x) + 1)\delta((\dot{x}|\dot{x}) - 1)\Theta(x_0). \tag{3.1.31}$$

There are also three constants associated with the angular momentum,

$$l_i = \varepsilon_{ikm} x_k \dot{x}_m, \tag{3.1.32}$$

which are connected by an algebraic relationship,

$$(l|l) = -(x|x)(\dot{x}|\dot{x}) = 1.$$

One dimension is left for the trajectory. Because $(l_i|x) = 0$, the projection of the trajectory onto configuration space is the intersection of the hyperboloid (3.1.28) with a plane passing through the origin and making an angle less than 45° with the $x_0$-axis (see Figure 23).

The energy is only apparently indefinite; $x_0$ can be eliminated, and then

$$L = \frac{\dot{x}_1^2 + \dot{x}_2^2 + (\dot{x}_1 x_2 - \dot{x}_2 x_1)^2}{x_1^2 + x_2^2 + 1}$$

describes motion in the $x_1 - x_2$-plane without forces, but with a positive effective mass that depends on the position.

The indefinite scalar product $(\cdot|\cdot)$ and consequently also the formalism that has been developed are invariant under SO(2, 1). The group SO(2, 1)

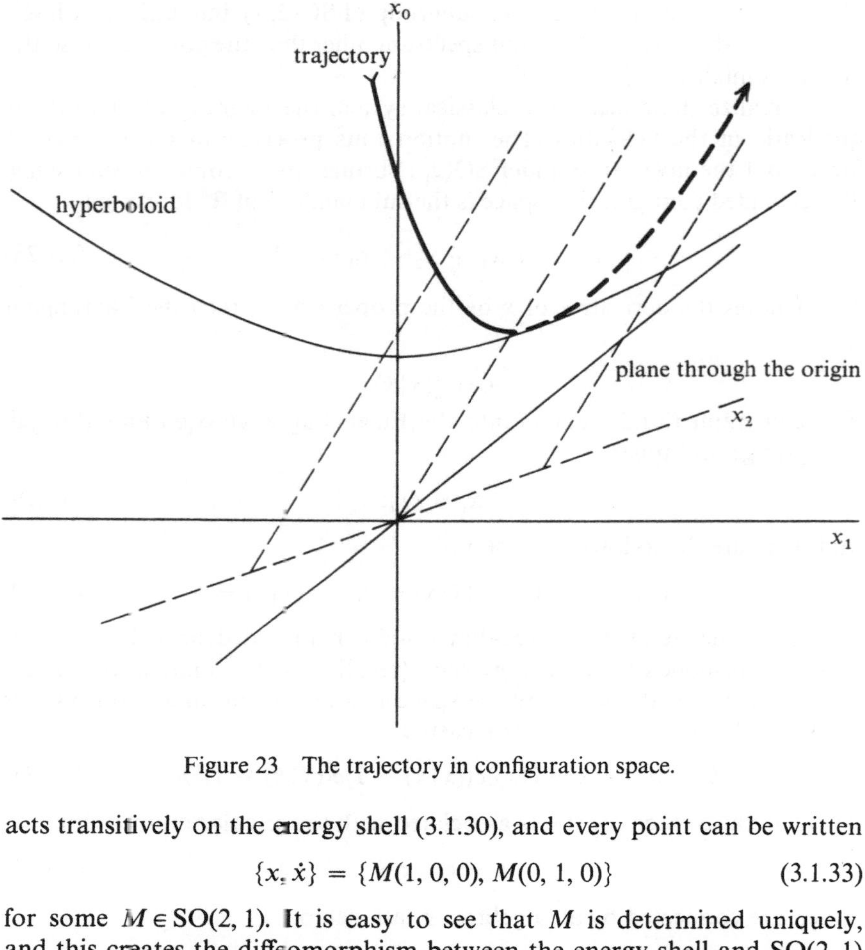


Figure 23 The trajectory in configuration space.

acts transitively on the energy shell (3.1.30), and every point can be written

$$\{x, \dot{x}\} = \{M(1, 0, 0), M(0, 1, 0)\} \tag{3.1.33}$$

for some $M \in SO(2, 1)$. It is easy to see that $M$ is determined uniquely, and this creates the diffeomorphism between the energy shell and SO(2, 1) that was mentioned above. Accordingly, every trajectory can be obtained by making Lorentz transformations of the group generated by

$$M(t) = \begin{bmatrix} \cosh t & \sinh t & 0 \\ \sinh t & \cosh t & 0 \\ 0 & 0 & 1 \end{bmatrix}.$$

The most convenient construction of the discrete subgroup makes use of the isomorphism between SO(2, 1) and $SL(2, \mathbb{R})/\{1, -1\}$, since $2 \times 2$ matrices are easier to handle than $3 \times 3$ matrices. The source of this isomorphism, like that of $SO(3) = SU(2, \mathbb{C})/\{1, -1\}$, lies in the observation that

$$\begin{pmatrix} \alpha & \beta \\ \gamma & \delta \end{pmatrix} \in SL(2, \mathbb{R}), \quad \text{i.e.,} (\alpha, \beta, \gamma, \delta) \in \mathbb{R}^4 : \alpha\delta - \beta\gamma = 1, \tag{3.1.34}$$

produces the Lorentz transformation $x \to x'$ by

$$\begin{pmatrix} x_0' + x_2' & x_1' \\ x_1' & x_0' - x_2' \end{pmatrix} = \begin{pmatrix} \alpha & \beta \\ \gamma & \delta \end{pmatrix} \begin{pmatrix} x_0 + x_2 & x_1 \\ x_1 & x_0 - x_2 \end{pmatrix} \begin{pmatrix} \alpha & \gamma \\ \beta & \delta \end{pmatrix}. \tag{3.1.35}$$

It is necessary to take the quotient by the center $\{1, -1\}$, since the Lorentz transformations corresponding to the matrix $m \in \mathrm{SL}(2, \mathbb{R})$ and $-m$ are the same. It is not hard to come up with discrete subgroups of SL(2, ℝ), such as

$$\mathscr{Z} = \left\{ \begin{pmatrix} \alpha & \beta \\ \gamma & \delta \end{pmatrix} \in \mathrm{SL}(2, \mathbb{R}) : \alpha, \beta, \gamma, \delta \text{ integers} \right\}.$$

Now let us investigate the motion on the quotient space $\Omega_0 = \mathrm{SO}(2, 1)/\mathscr{Z} \cong \mathrm{SL}(2, \mathbb{R})/\{1, -1\}/\mathscr{Z}$. Unlike the case of $T^2$, the quotient space is not a group, since $\mathscr{Z}$ is not a normal divisor, though for our purposes this does not matter. Thus $\Omega_0$ is the energy shell (3.1.30), on which points are identified if they are transformed into each other by $\mathscr{Z}$. For the trajectory this means that if it goes out one end of the domain of periodicity it reappears at the other. Conservation of angular momentum breaks down, leaving the possibility that the trajectory fills $\Omega_0$ densely.

To get a clearer picture of $\Omega_0$ we have to find out what corresponds to the square $0 \le \varphi_1, \varphi_2 \le 1$ of the earlier example, that is, a region containing no points equivalent under $\mathscr{Z}$, but for each boundary point of which there is a $z \neq 1$ of $\mathscr{Z}$ mapping it to another boundary point. The subgroup $\mathscr{Z}$ is generated by the matrices

$$\begin{pmatrix} 1 & 1 \\ 0 & 1 \end{pmatrix}, \quad \begin{pmatrix} 0 & 1 \\ -1 & 0 \end{pmatrix},$$

the latter of which is the reflection $(x_1, x_2) \to (-x_1, -x_2)$. It is therefore possible to restrict attention to the upper half plane $\{x_2 > 0\}$ in configuration space and choose a region symmetric about the $x_2$-axis. The boundary curves can be obtained by transforming the $x_2$-axis with the matrices

$$\begin{pmatrix} 1 & \pm\frac{1}{2} \\ 0 & 1 \end{pmatrix}$$

of SL(2, ℝ). They have the parametric representation

$$\begin{aligned} h_\pm &= \left\{ x_1' : \begin{pmatrix} x_0' + x_2' & 0 \\ x_1' & x_0' - x_2' \end{pmatrix} \right. \\ &= \left. \begin{pmatrix} 1 & \pm\frac{1}{2} \\ 0 & 1 \end{pmatrix} \begin{pmatrix} \sqrt{1 + x_2^2} + x_2 & 0 \\ 0 & \sqrt{1 + x_2^2} - x_2 \end{pmatrix} \begin{pmatrix} 1 & 0 \\ \pm\frac{1}{2} & 1 \end{pmatrix}, x_2 > 0 \right\}; \end{aligned} \tag{3.1.36}$$

note that

$$\begin{pmatrix} 1 & 1 \\ 0 & 1 \end{pmatrix} \cdot h_- = h_+,$$

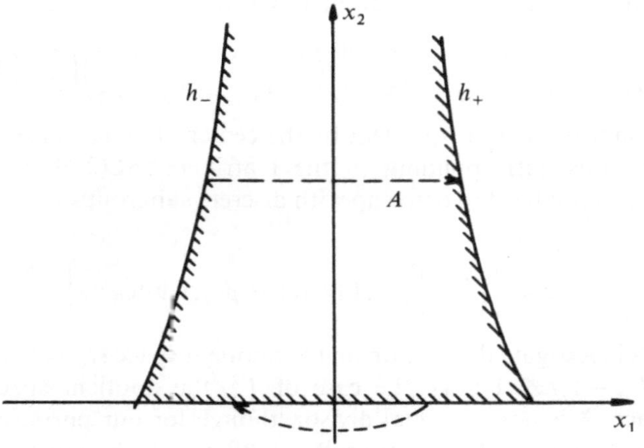


Figure 24

so $x_2' = \pm\frac{1}{4}(1/x_1' - 3x_1')$. The projection of $\Omega_0$ onto configuration space looks as depicted in Figure 24, where the lines $A$ indicate the identifications.

The identification of the boundary points by $\begin{pmatrix} 1 & 1 \\ 0 & 1 \end{pmatrix}$ means that if the trajectory leaves through one side, it reappears at the corresponding point of the other side (see Figure 25).

Now we are in a position to verify that the measure of $\Omega_0$ with $d\Omega$ (3.1.31) is actually finite. This follows from

$$\int d^3\dot{x}\delta[(\dot{x}|x)]\delta[(\dot{x}|\dot{x}) - 1] \equiv F(x, x) < \infty$$

and

$$F(-1)\int d^3x\delta[(x|x) + 1] < \infty,$$

where the integral runs over the region bounded by (3.1.36).

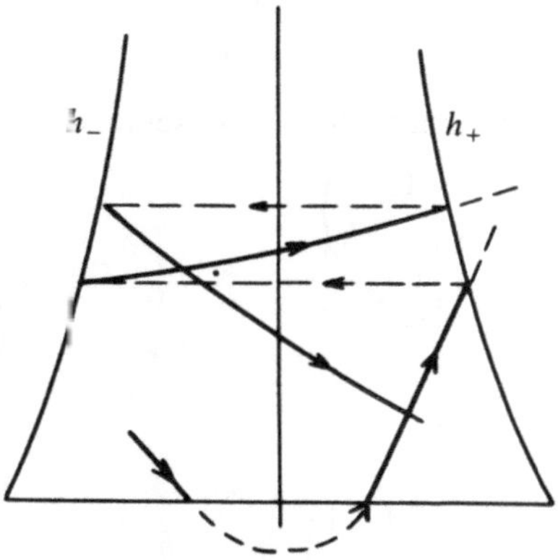


Figure 25

The time-evolution is controlled by the unitary group

$$U_t = \exp(mt), \qquad m = \left.\frac{\partial M}{\partial t}\right|_{t=0},$$

where the anti-Hermitian operator $m$ is one of the generators of SO(2, 1). If the other two generators are combined into $m_\pm \equiv m_1 \pm m_2$, then $m_\pm$ satisfy the commutation relations

$$[m, m_\pm] = \pm m_\pm \quad \text{and} \quad [m_+, m_-] = 2m.$$

Note that in contradistinction to SO(3), this time $(m_\pm)^* = -m_\pm$. This fact will be crucial, since the generators of SO(3) have purely discrete spectra. Instead of SO(2, 1), let us now examine the simpler two-parameter subgroups

$$U_\pm(a, t) = \exp(am_\pm)\exp(tm)$$

with the multiplication law

$$U_\pm(a, t)U_\pm(a', t') = U_\pm(a + \exp(\pm t)a', t + t').$$

Because $[m_+, m_-] = 2m$, the operators $U_+(a, 0)$ and $U_-(a, 0)$ generate the whole group, and $U(t) = U_+(0, t) = U_-(0, t)$.

Next consider the representation (3.1.1; 1) of classical dynamics on $\mathscr{H} = L^2(\Omega_0, d\Omega)$. Not just $U_t$, but in fact all of SO(2, 1) is represented unitarily on $\mathscr{H}$ by $f(x) \to f(Mx)$, and we shall now reduce this representation according to the irreducible representations of the subgroups $U_\pm$. We start by observing that $U_\pm(a, 0)$ is a normal divisor, and the factor groups $U_\pm(a, t)/U_\pm(a, 0)$ are isomorphic to $\mathbb{R}$. Hence there are irreducible, one-dimensional representations of the type

$$I: \quad U_\pm(a, t) = \exp(i\lambda t), \qquad \lambda \in \mathbb{R}.$$

In addition it is readily seen that $U_\pm$ can also be represented on $L^2(\mathbb{R}, dx)$ by

$$II: \quad [U_+(a, t)\psi](x) = \exp(iae^x)\psi(x + t), \qquad \psi \in L^2(\mathbb{R}, dx),$$

and similarly for $U_-$. It can be shown [19] that these possibilities exhaust the irreducible representations of SO(2, 1), so, decomposing into the irreducible representations of $U_\pm$,

$$L^2(\Omega_0, d\Omega) = \mathscr{H}_I^+ \oplus \mathscr{H}_{II}^+ = \mathscr{H}_I^- \oplus \mathscr{H}_{II}^-.$$

On the subspaces $\mathscr{H}_{II}^+$ and $\mathscr{H}_{II}^-$ the operator $U(t)$ acts as a translation on $L^2(\mathbb{R}, dx)$, and thus its spectrum is continuous. A discrete spectrum could only be found on $\mathscr{H}_I^+ \cap \mathscr{H}_I^-$, but every vector $\psi$ of $\mathscr{H}_I^+ \cap \mathscr{H}_I^-$ satisfies the equation

$$U_+(a, 0)\psi = \psi = U_-(a, 0)\psi.$$

Since $U_+(a, 0)$ and $U_-(a, 0)$ together suffice to generate all of SO(2, 1), $\psi$ is invariant under the action of every group element. Since the group acts transitively on $\Omega_0$, $\psi$ must be a constant. Because $\Omega_0$ has finite measure, any constant function belongs to $L^2(\Omega_0, d\Omega)$, so the situation is like that of (3.1.26; 2). Unless the quotient by $\mathscr{Z}$ is taken, $U$ has no point spectrum, as constant functions would not be integrable. In sum the argument is that the system is mixing because the spectrum of $U$ consists of a single nondegenerate eigenvalue 1 and an absolutely continuous portion. This is in contrast to the motion on the torus, for which the spectrum of $U_t$ was purely discrete, and the system was only ergodic, not mixing.

**Example** (3.1.37)

The quantum-mechanical example of an infinite system of free fermions was seen to be mixing. Despite the absence of interaction, a local perturbation spreads out to infinity through the diffusion of free wave-packets. From among the characterizations of ergodic states (3.1.22), let us look in particular at the third. It holds in the sharper form of (3.1.26; 4); the grand canonical state (2.5.49) is the time-limit of a pure state. The proof of this fact uses the transformations

$$a_\uparrow(f) = b_\uparrow(\beta f) + b_\downarrow^*(\sqrt{1 - |\beta|^2} f^*)$$

and

$$a_\downarrow(f) = b_\downarrow(\beta f) - b_\uparrow^*(\sqrt{1 - |\beta|^2} f^*). \tag{3.1.38}$$

We have directly taken up the realistic case of spin-$\frac{1}{2}$ fermions, where $\uparrow$ and $\downarrow$ indicate the direction of the spin that the field operator describes. In Fourier-transformed space $\beta$ is a function $\mathbf{k} \to \beta(\mathbf{k})\colon \mathbb{R}^3 \to \{z \in \mathbb{C}\colon |z|^2 \le 1\}$, and $\beta f$ is the function $\beta(\mathbf{k}) f(\mathbf{k})$. In $x$-space $\beta$ is a convolution. It is straightforward to verify that the $a$'s satisfy the usual commutation relations (1.3.3; 2),

$$[a_\uparrow(f), a_\uparrow^*(g)]_+ = [a_\downarrow(f), a_\downarrow^*(g)]_+ = (f|g),$$

$$[a_\uparrow(f), a_\uparrow(g)]_+ = [a_\uparrow(f), a_\downarrow(g)]_+ = [a_\uparrow(f), a_\downarrow^*(g)]_+ = [a_\downarrow(f), a_\downarrow(g)]_+ = 0, \tag{3.1.39}$$

supposing that the $b$'s satisfy the commutation relations. Clearly the $a$'s and the $b$'s generate the same $C^*$ algebra. The expectation values of the $a$'s in the Fock state $|0\rangle$ (1.3.2) for the $b$'s: $b_\uparrow(f)|0\rangle = b_\downarrow(f)|0\rangle = 0$, are

$$\langle 0|a_\uparrow(f)a_\uparrow^*(g)|0\rangle = \langle 0|a_\downarrow(f)a_\downarrow^*(g)|0\rangle = \int \frac{d^3\mathbf{k}}{(2\pi)^3} |\beta(\mathbf{k})|^2 f^*(\mathbf{k}) g(\mathbf{k}),$$

$$-\langle 0|a_\uparrow(f)a_\downarrow(g)|0\rangle = \langle 0|a_\downarrow(f)a_\uparrow(g)|0\rangle$$

$$= \int \frac{d^3\mathbf{k}}{(2\pi)^3} f^*(\mathbf{k}) g^*(\mathbf{k}) \beta(\mathbf{k}) \sqrt{1 - |\beta(\mathbf{k})|^2};$$

$$\langle 0|a_\uparrow(f)a_\uparrow(g)|0\rangle = \langle 0|a_\uparrow(f)a_\downarrow^*(g)|0\rangle = 0. \tag{3.1.40}$$

The state $|0\rangle$ was seen to be pure in (1.3.16; 1). Under the time-evolution $f(\mathbf{k}) \to \exp(-it|\mathbf{k}|^2) f(\mathbf{k})$, the quantity $-\langle 0|a_\uparrow a_\downarrow|0\rangle = \langle 0|a_\downarrow a_\uparrow|0\rangle$ goes to 0 as $t \to \pm\infty$ by the Riemann–Lebesque lemma. If

$$\beta(\mathbf{k}) = (1 + \exp(-\beta(|\mathbf{k}|^2 - \mu)))^{-1/2},$$

then in the limit $t \to \pm\infty$ the generalization of the state (2.5.49) for spin $\frac{1}{2}$ is all that is left over.

**Remarks** (3.1.41)

1. The limit of a pure state is clearly not always an equilibrium state; other functions could be chosen for $\beta(\mathbf{k})$.
2. Since the thermal representation of free fermions (3.1.1; 5) is a factor of type III, the pure state $|0\rangle$ associated with the thermal representation cannot be normal (cf. (1.4.17; 3)). Likewise, any other states of the latter formed with different $\beta(\mathbf{k})$ are not normal because of (3.1.22; 7), even though they are invariant.
3. The state given by $|0\rangle$ is not invariant in time, and in this representation the time-evolution is certainly not a unitary group (cf. (1.3.16; 7)). If it were, then the time displacement $\tau_t: a \to a_t$ would be weakly continuous and hence extensible to $\pi(\mathscr{A})''$, which would lead to a contradiction: $\mathscr{A}_e$ is asymptotically Abelian with respect to the spatial translation $T_x$, so in the representation with the translation-invariant state $|0\rangle$, $\lim_{|x|\to\infty} T_x a = \mathbf{1}\cdot\langle 0|a|0\rangle$ for all $a \in \mathscr{A}_e$. Since $T_x$ commutes with $\tau_t$, it would follow that $\lim_{x\to\infty} T_x \tau_t(a) = \mathbf{1}\cdot\langle 0|a_t|0\rangle = \lim_{x\to\infty} \tau_t T_x(a) = \mathbf{1}\cdot\langle 0|a|0\rangle$, which would then imply that the state $\langle 0|\cdot|0\rangle$ would be invariant in time.

**Problems** (3.1.42)

1. (i) Prove von Neumann's statistical ergodic theorem, $(1/2T)\int_{-T}^{T} \exp(iHt)\,dt \to E_0$. (Show that on all vectors of the form $x = \exp(iHs)y - y$, $y \in \mathscr{H}$, $s \in \mathbb{R}$, we have $(1/2T)\int_{-T}^{T} \exp(iHt)x\,dt \to 0$. Let $\mathscr{H}_1$ be the closed linear hull of these vectors, and note that the same fact applies to all $x \in \mathscr{H}_1$. Finally, show that $\mathscr{H}_1^{\perp} = \{x : \exp(iHs)x = x \text{ for all } s\} = E_0\mathscr{H}$.)
   (ii) Show similarly that $(\varepsilon/2)\int_{-\infty}^{\infty} \exp(-\varepsilon|t|)\exp(iHt)\,dt \to E_0$.
2. Show that in the Schrödinger picture the time-average of a vector $x$ has the following characterization: $\eta(x)$ is the vector of least norm of the norm-closed, convex hull of $\{U(t)x\}$, denoted $\mathscr{K}$. (Hint: see the example given earlier for $\eta(x) \in \mathscr{K}$. Show (i) that $\mathscr{K}$ contains a unique vector $\xi$ of least norm; (ii) that $\xi$ is invariant under all $U(t)$; and (iii) that $\mathscr{K}$ contains no other fixed point.)
3. Show that $\mathscr{Z} = \{\alpha\cdot\mathbf{1}\}$ iff $w(ab) = w(a)w(b)$ for all $w \in \mathscr{A}^*$, $a \in \mathscr{A}$, and $b \in \mathscr{Z}$.
4. Show that for a classical system, if there exists a constant $f(p, q)$ not of the form $\alpha\cdot\mathbf{1}$, then $\rho$ is not ergodic.
5. Show that a set $E \subset \mathscr{H}$ is a totalizer for $\mathscr{A}$ iff $E$ separates $\mathscr{A}'$. (Cf. (III: 2.3.4); a totalizer is a set $E$ such that $\mathscr{A}E$ is dense in $\mathscr{H}$, and separating means that $a'E = 0 \Rightarrow a' = 0$.)

6. Boson states of the form (2.5.49) with $\langle f | \rho g \rangle = \int d^3k \rho(\mathbf{k}) \tilde{f}^*(\mathbf{k}) \tilde{g}(\mathbf{k})$, $0 \leq \rho(\mathbf{k})$, are factor states and consequently mixing. Express such a state as a time-limit of a pure state (cf. (3.1.37)).

## Solutions (3.1.43)

1. (i) If $x = \exp(iHs)y - y$, then

$$\left\| \frac{1}{2T} \int_{-T}^{T} \exp(iHt)x \, dt \right\| = \left\| \frac{1}{2T} \left\{ \int_{T}^{T+s} \exp(iHt)y \, dt - \int_{-T}^{-T+s} \exp(iHt)y \, dt \right\} \right\|$$
$$\leq \frac{|s| \, \|y\|}{T} \to 0.$$

Because $\|(1/2T) \int_{-T}^{T} \exp(iHt) \, dt\| \leq 1$, this holds for all $x \in \mathscr{H}_1$.

$$x \in \mathscr{H}_1^{\perp} \Leftrightarrow (x | \exp(iHs)y - y) = (\exp(-iHs)x - x | y) = 0 \qquad \text{for all } y \in \mathscr{H}$$
$$\Leftrightarrow \exp(iHs)x = x \qquad \text{for all } s \Leftrightarrow E_0 x = x$$

by the spectral theorem.

(ii) It suffices to show that $\varepsilon \int_0^\infty \exp(-\varepsilon t) \exp(iHt) \, dt \to E_0$, which will follow if $\varepsilon \int_0^\infty \exp(-\varepsilon t) \exp(iHt)x \, dt \to 0$ for vectors $x = \exp(iHs)y - y$.
This integral equals

$$\varepsilon \exp(\varepsilon s) \int_s^\infty \exp(-\varepsilon t) \exp(iHt)y \, dt - \varepsilon \int_0^\infty \exp(-\varepsilon t) \exp(iHt)y \, dt$$

$$= (\exp(\varepsilon s) - 1)\varepsilon \int_s^\infty \exp(-\varepsilon t) \exp(iHt)y \, dt - \varepsilon \int_0^s \exp(-\varepsilon t) \exp(iHt)y \, dt \to 0,$$

since $\|\varepsilon \int_0^\infty \exp(-\varepsilon t) \exp(iHt)y \, dt\| \leq \|y\|$.

2. (i) Let $\lambda = \inf\{\|x\| : x \in \mathscr{K}\}$. There exists a sequence $\{x_n\}$ in $\mathscr{K}$ such that $\|x_n\| \to \lambda$. By the parallelogram law,

$$\left\| \frac{x_n - x_m}{2} \right\|^2 + \left\| \frac{x_n + x_m}{2} \right\|^2 = \tfrac{1}{2}(\|x_n\|^2 + \|x_m\|^2),$$

$x_n$ is a Cauchy sequence, so it has a limit $\xi$. If $\|x\| = \|\xi\|$, then

$$\left\| \frac{x - \xi}{2} \right\|^2 = \tfrac{1}{2}(\|x\|^2 + \|\xi\|^2) - \left\| \frac{x + \xi}{2} \right\|^2 \leq 0, \text{ which implies that } x = \xi.$$

(ii) $\|U(t)\xi\| = \|\xi\| \Rightarrow U(t)\xi = \xi$.

(iii) Suppose that $\eta$ is a second fixed point. For all $\varepsilon > 0$, there exist $\lambda_1, \ldots, \lambda_n$ and $\lambda'_1, \ldots, \lambda'_m$ such that $\sum_i \lambda_i = \sum_i \lambda'_i = 1$, with $\lambda_i, \lambda'_i \geq 0$, and there exist $t_1, \ldots, t_n$ and $t'_1, \ldots, t'_m$ such that if $V \equiv \lambda_1 U(t_1) + \cdots + \lambda_n U(t_n)$, and $W \equiv \lambda'_1 U(t'_1) + \cdots + \lambda'_m U(t'_m)$, then $\|Vx - \xi\| < \varepsilon$, and $\|Wx - \eta\| < \varepsilon$. However, then

$$\|\xi - \eta\| \leq \|\xi - VWx\| + \|VWx - \eta\| = \|W\xi - VWx\| + \|VWx - V\eta\|$$
$$\leq \|W\| \, \|Vx - \xi\| + \|V\| \, \|Wx - \eta\| < 2\varepsilon,$$

so $\xi = \eta$.
Remark: The strong and weak closures of a convex set are identical.

3. $\Rightarrow$: This part is trivial.
   $\Leftarrow$: Let $P_1$ and $P_2$ be projections in $\mathscr{Z}$, such that $P_1 \perp P_2$, and let $w_i(\cdot) = w(P_i\cdot)$, $a_i = P_i a$, $b_i = P_i b$ for $i = 1, 2$. If $w = \alpha w_1 + (1 - \alpha)w_2$, then

$$\begin{aligned} w(ab) &= \alpha w(a_1)w(b_1) + (1 - \alpha)w(a_2)w(b_2) \\ &\neq (\alpha w(\alpha_1) + (1 - \alpha)w(a_2))(\alpha w(b_1) + (1 - \alpha)w(b_2)). \end{aligned}$$

4. Let $\bar{f}(p, q) = \inf(1, |f(p, q)|)$ (if necessary multiply $f$ by a suitable constant to ensure that $\bar{f}$ is not identically 1). Then $d\rho$ is the sum of two invariant states,

$$d\rho = \tfrac{1}{2}(1 + \bar{f})\, d\rho + \tfrac{1}{2}(1 - \bar{f})\, d\rho.$$

5. $\Rightarrow$: Let $a' \in \mathscr{A}'$. $a'E = 0 \Rightarrow a'\mathscr{A}E = 0 \Rightarrow a' = 0$ on a dense set, which implies that $a' = 0$.
   $\Leftarrow$: Let $E_\perp$ be the orthogonal complement of $\mathscr{A}E$. Then $\mathscr{A}E_\perp = E_\perp$, so the projection $P_\perp$ onto $E_\perp$ belongs to $\mathscr{A}'$, but $P_\perp E = 0$, so $E$ does not separate $\mathscr{A}'$.

6. In a Fock representation of the free fields $b$, $b(k)|0\rangle = 0$, write

$$a(\mathbf{k}) = \sqrt{\rho(\mathbf{k})}b^*(\mathbf{k}) + \sqrt{1 + \rho(\mathbf{k})}b(\mathbf{k}),$$

and

$$a^*(\mathbf{k}) = \sqrt{\rho(\mathbf{k})}b^*(\mathbf{k}) + \sqrt{1 + \rho(\mathbf{k})}b(\mathbf{k}).$$

These operators $a$ likewise satisfy the commutation relations

$$a(\mathbf{k})a^*(\mathbf{k}') - a^*(\mathbf{k}')a(\mathbf{k}) = \delta(\mathbf{k} - \mathbf{k}'),$$

and

$$\begin{aligned} \langle 0|a(\mathbf{k})a^*(\mathbf{k}')|0\rangle &= \delta(\mathbf{k} - \mathbf{k}')\rho(\mathbf{k}), \\ \langle 0|a(\mathbf{k})a(\mathbf{k}')|0\rangle &= \delta(\mathbf{k} - \mathbf{k}')\sqrt{\rho(\mathbf{k})}\sqrt{1 + \rho(\mathbf{k})}. \end{aligned}$$

Hence

$$\langle |a_{f_t} a^*_{g_t}|0\rangle = \int dk \rho(\mathbf{k})\tilde{f}^*(\mathbf{k})\tilde{g}(\mathbf{k}),$$

$$\langle 0|a_{f_t} a_{g_t}|0\rangle = \int dk \exp(2i|\mathbf{k}|^2 t)\sqrt{\rho(\mathbf{k})}\sqrt{1 + \rho(\mathbf{k})}\,\tilde{f}^*(\mathbf{k})\tilde{g}^*(\mathbf{k});$$

this last integral goes to zero as $t \to \pm\infty$ by the Riemann–Lebesgue lemma, and therefore its time-average is zero. The analogous fact holds for the higher correlation functions, so the time-average of the pure Fock state $|0\rangle$ is of the form (2.5.49).

## 3.2 The Equilibrium State

*In the course of time the Maxwell–Boltzmann distribution has proved more and more fundamental, and has become deeply rooted in the mathematical description of infinite quantum systems.*

With a certain normalization of $H$ the canonical state has the form $w(a) = \mathrm{Tr}\exp(-\beta H)a$, as we have seen. The appearance of the Hamiltonian $H$ in both the time-evolution and the state creates all sorts of important connections between them. To avoid technical complications at first we shall concentrate only on the finite-dimensional case. The commutativity of the trace gives rise to a symmetry between the representation of the algebra and its commutant.

**The GNS Representation of $\mathscr{B}(\mathbb{C}^n)$ with a Faithful State** (3.2.1)

*Let $\mathscr{A} = \mathscr{B}(\mathbb{C}^n)$ be given the inner product $\langle a|b\rangle = \mathrm{Tr}\, a^*b$ so that it becomes a Hilbert space isomorphic to $\mathbb{C}^{n^2}$, and define*

$$\pi: \mathscr{A} \to \mathscr{B}(\mathbb{C}^{n^2}): \pi(a)|b\rangle = |ab\rangle,$$
$$\pi': \mathscr{A} \to \mathscr{B}(\mathbb{C}^{n^2}): \pi'(a)|b\rangle = |ba^*\rangle,$$
$$J: \mathbb{C}^{n^2} \to \mathbb{C}^{n^2}: J|b\rangle = |b^*\rangle.$$

*Then*

(i) *$\pi$ is a factor representation (*-isomorphism);*
(ii) *$\pi'$ is a *-antiisomorphism, i.e.,*

$$\pi'(ab) = \pi'(a)\pi'(b), \qquad \pi'(\lambda a) = \bar{\lambda}\pi'(a), \qquad \pi'(a^*) = (\pi'(a))^*,$$
$$\pi'(a + b) = \pi'(a) + \pi'(b)) \quad \textit{with } \pi'(\mathscr{A}) = \pi(\mathscr{A})';$$

(iii) *the conjugate-linear operator $J$ preserves norms and $J^2 = \mathbf{1}$;*
(iv) *$J\pi(\mathscr{A})J = \pi'(\mathscr{A})$, $J\pi'(\mathscr{A})J = \pi(\mathscr{A})$;*
(v) *let $w$ be a faithful state, that is, if $a > 0$, then $w(a) > 0$, so by (2.1.5(ii)), $w(a) = \mathrm{Tr}\,\rho a = \langle\sqrt{\rho}|a|\sqrt{\rho}\rangle$, $\rho > 0$, $\mathrm{Tr}\ \rho = 1$. The vector $|\sqrt{\rho}\rangle$ is cyclic and separating for $\pi$ and $\pi'$, i.e., $\pi(a)|\sqrt{\rho}\rangle = 0 \Rightarrow a = 0$. Hence the* GNS *representation using $w$ is unitarily equivalent to $\pi$.*

**Proof**

The isomorphism and antiisomorphism properties are obvious.

(ii) $\pi'(a)\pi(b)|c\rangle = \pi'(a)|bc\rangle = |bca^*\rangle = \pi(b)\pi'(a)|c\rangle$, and therefore $\pi'(\mathscr{A}) \subset \pi(\mathscr{A})'$. On the other hand, if $B \in \pi(\mathscr{A})'$ then $B|1\rangle$ is $|b^*\rangle$ for some $b \in \mathscr{A}$. Hence

$$B|a\rangle = B\pi(a)|1\rangle = \pi(a)B|1\rangle = \pi(a)|b^*\rangle = \pi(a)\pi'(b)|1\rangle = \pi'(b)|a\rangle$$

for all $a \in \mathscr{A}$, so $B = \pi'(b)$ and $\pi'(\mathscr{A}) = \pi(\mathscr{A})'$.

(i) Let $\pi(a) \in \pi(\mathscr{A})'$. Then by part (ii) it equals $\pi'(b^*)$ for some $b$. Hence $\pi(a)|c\rangle = |ac\rangle = \pi'(b^*)|c\rangle = |cb\rangle$, so $ac = cb$ for all $c \in \mathscr{A}$, and therefore $a = b = \alpha \cdot \mathbf{1}$. Thus $\pi(\mathscr{A})$ is a factor.

(iii) $\|J|a\rangle\|^2 = \operatorname{Tr} aa^* = \operatorname{Tr} a^*a = \||a\rangle\|^2$, and $J^2 = \mathbf{1}$ since $b^{**} = b$.
(iv) $J\pi(a)J|b\rangle = J\pi(a)|b^*\rangle = J|ab^*\rangle = |ba^*\rangle = \pi'(a)|b\rangle \Rightarrow J\pi(a)J = \pi'(a) \Rightarrow \pi(a) = J\pi'(a)J$, because $J^2 = \mathbf{1}$.
(v) Since $\rho^{-1}$ exists, $|a\rangle$ may be written as $|b\sqrt{\rho}\rangle = \pi(b)|\sqrt{\rho}\rangle, b = a\rho^{-1/2}$, which shows that $\sqrt{\rho}$ is cyclic for $\pi$. If $\rho_i > 0$ are the eigenvalues of $\rho$, then in the diagonal representation of $\rho$,

$$\|\pi(a)|\sqrt{\rho}\rangle\|^2 = \operatorname{Tr} \rho a^* a = \sum_{i,k} \rho_i |a_{ik}|^2 = 0,$$

which implies that $a_{ik} = 0$, and similarly for $\pi'$. By (III: 2.3.10; 6) $\pi_\rho$ is equivalent to $\pi$. □

**Remarks** (3.2.2)

1. An anti-isomorphism came up once before, in the reversal of the motion (III: 3.3.18), and $J$ is like the conjugate-linear operator $\Theta'$ (3.3.19; 2).
2. The representation $\pi$, being a finite-dimensional factor of type I, is of the form $\pi(a) = a \otimes \mathbf{1}_{|\mathbb{C}^n}$, so $\pi'(a)$ is $\mathbf{1}_{|\mathbb{C}^n} \otimes a^*$.

Consider next how to represent the time-evolution $a \to a_t = \exp(iht)\, a \exp(-iht)$. At first thought it might be represented by $\exp(i\pi(h)t)$, but this would not leave the cyclic vector $|\sqrt{\rho}\rangle$ invariant. The correct way to proceed is as in Example (3.1.1; 3).

**The Time-Evolution on $\mathscr{B}(\mathbb{C}^n)$** (3.2.3)

*The unitary representation* (1.3.5) *of the time-evolution $a \to a_t$ on the invariant state $a \to \operatorname{Tr} \rho a$, $\rho = \exp(-\beta h)$, is given by $U_t = \exp(-iHt)$, $H = \pi(h) - \pi'(h)$. It satisfies the following*:

(i) $JHJ = -H, JU_tJ = U_t$;
(ii) $U_{-i\beta/2}\pi(a)|\sqrt{\rho}\rangle = J\pi(a^*)|\sqrt{\rho}\rangle$;
(iii) $\langle\sqrt{\rho}|\pi(a)\pi(b)|\sqrt{\rho}\rangle = \langle\sqrt{\rho}|\pi(b)\pi(a_{i\beta})|\sqrt{\rho}\rangle$.

**Proof**

It is immediately clear that $\exp(iHt)\pi(a)\exp(-iHt) = \pi(a_t)$. Moreover, $\exp(iHt)|\sqrt{\rho}\rangle = |\exp(iht)\exp(-\beta h)\exp(-iht)\rangle = |\sqrt{\rho}\rangle$.

(i) This follows from (3.2.1(iv)).
(ii) $U_{-i\beta/2}\pi(a)|\sqrt{\rho}\rangle = U_{-i\beta/2}|a \exp(-\beta h/2)\rangle = |\exp(-\beta h/2)a\rangle = J|a^* \exp(-\beta h/2)\rangle = J\pi(a^*)|\sqrt{\rho}\rangle$.
(iii) $\operatorname{Tr} \exp(-\beta h)ab = \operatorname{Tr} \exp(-\beta h) a \exp(\beta h) \exp(-\beta h) b = \operatorname{Tr} \exp(-\beta h) b a_{i\beta}$. □

**Remarks** (3.2.4)

1. The density matrix $\rho$ was written simply as $\exp(-\beta H)$ under the assumption that $h$ had been redefined by the addition of a multiple of the identity so that $\operatorname{Tr}\exp(-\beta h) = 1$. This affects neither the time-evolution nor $H$.
2. Note that $J$ does not reverse the direction of time.
3. The operator $\rho = \exp(-h)$ is always positive. Conversely, if $\rho > 0$ (i.e., all eigenvalues $\rho_i > 0$), then $\ln \rho = -h$ is well defined. This shows that groups of automorphisms and faithful states are bijectively related. There is a special term for their relationship.

**The Modular Automorphism** (3.2.5)

For each faithful state $w$ on $\mathscr{B}(\mathbb{C}^n)$ there is a unique one-parameter group of automorphisms $\tau_t: a \to a_t$ such that

(i) $w$ is invariant in the sense that $w(a_t) = w(a)$.
(ii) $w$ satisfies the **Kubo–Martin–Schwinger** (KMS) condition, $w(ab) = w(ba_i)$.
(iii) there exists an anti-isomorphism $\pi_w(\mathscr{A}) \to J\pi_w(\mathscr{A})J$ onto $\pi_w(\mathscr{A})'$ such that

$$U_{-i/2}\,\pi(a)|\Omega\rangle = J\pi(a^*)|\Omega\rangle,$$

where $|\Omega\rangle$ is the cyclic vector and $U_t$ is the unitary operator representing $\tau_t$ in the GNS representation with $w$.

If the dimension of the Hilbert space is now infinite, but the state is still given by a density matrix $\rho = \exp(-\beta h)$, then there are a few technical difficulties to clear up.

**The Temporal Correlation Functions of Finite Quantum Systems** (3.2.6)

If the time is made complex, then in general

$$a_{x+iy} \equiv \exp((ix - y)h)a\exp(-(ix - y)h)$$

is unbounded, and hence does not belong to the algebra. However, we shall continue to use this notation, as this operator will never act on anything outside its domain of definition.

(i) Continuity in the strip $-\beta \le \operatorname{Im} t \le 0$. $w(a_t b) = \langle\Omega|a\exp(-iHt)b|\Omega\rangle$, and if $t$ is complex, then by (3.2.3(ii)), $b|\Omega\rangle$ is in the form domain of $\exp(yH)$ for $y \ge -\beta$. In a spectral representation it is apparent that the vector $\exp(yH/2)b|\Omega\rangle$ is norm-continuous in $y$, so $\rho(a_t b)$ is norm-continuous in $t$.

(ii) Boundedness in the strip $-\beta \leq \operatorname{Im} t \leq 0$. Let $H = \pi(h) - \pi'(h)$ as in (3.2.3), so $H|\Omega\rangle = 0$. Because

$$\begin{aligned} a_{x+iy} &= \exp((ix - y)H)a\exp(-(ix - y)H), \\ |w(a_{x+iy}b)|^2 &= |\langle\Omega|a_x\exp(yH)b|\Omega\rangle|^2 \\ &\leq \langle\Omega|a_x\exp(yH)a_x^*|\Omega\rangle\langle\Omega|b^*\exp(yH)b|\Omega\rangle. \end{aligned}$$

The function $\langle\Omega|a\exp(yH)a^*|\Omega\rangle$ is positive and, because

$$\frac{\partial^2}{\partial y^2}\langle\Omega|a\exp(yH)a^*|\Omega\rangle = \|H\exp(yH/2)a^*|\Omega\rangle\|^2 \geq 0,$$

convex, achieving its maximum at $y = 0$ or $y = -\beta$. It is clear that $w(aa^*) \leq \|a\|^2$, but even at the lower edge it is bounded, as shown by

$$\begin{aligned} w(a_{i\beta/2}a^*_{-i\beta/2}) &= \operatorname{Tr}\exp(-\beta h)\exp(\beta h/2)a^*\exp(-\beta h)a\exp(\beta h/2) \\ &= \operatorname{Tr}\exp(-\beta h)aa^* \leq \|a\|^2, \end{aligned}$$

since $\operatorname{Tr}\exp(-\beta h) = 1$. Therefore

$$|w(a_t b)| \leq \|a\|\,\|b\| \quad \text{for } -\beta \leq \operatorname{Im} t \leq 0.$$

(iii) Analyticity in the strip $-\beta < \operatorname{Im} t < 0$. The function $w(a_t b)$ is not differentiable on the real axis for generic $a$'s, but only for complex times within the strip. The proof is similar to that of (2.4.7) and will not be repeated here. The relationship $w(ab) = w(ba_{i\beta})$, named for Kubo, Martin, and Schwinger, which follows from the invariance of the trace, can be continued analytically to the strip: The functions $w(a_t b)$ and $w(ba_t)$ are analytic respectively in $-\beta < \operatorname{Im} t < 0$ and $0 < \operatorname{Im} t < \beta$, where they satisfy the KMS condition $w(a_t b) = w(ba_{t+i\beta})$, which determines the value of $w(a_t b)$ at $y = -\beta$ as $w(ba)$ (see Figure 26).

(iv) The physical significance of the KMS condition. For a finite system the canonical state with $\rho = \exp(-\beta H)$ is not an eigenstate of the energy. The modular Hamiltonian (also denoted $H$) has $|\Omega\rangle$ as an eigenvector, $H|\Omega\rangle = 0$. This operator $H$ is not generally bounded below; however, the KMS condition distinguishes positive energies because of the positive sign of $\beta$. The energy spectrum of $\pi(a)|\Omega\rangle$ for $a = a^* \in \mathscr{A}$ consists predominately of positive energies,

$$\begin{aligned} f(E) \equiv \langle\Omega|\pi(a)\delta(H - E)\pi(a)|\Omega\rangle &= \int_{-\infty}^{\infty}\frac{dt}{2\pi}\exp(iEt)\rho(a_t a) \\ &= \int_{-\infty}^{\infty}\frac{dt}{2\pi}\exp(iEt)\rho(aa_{t+i\beta}) \\ &= \exp(\beta E)\langle\Omega|\pi(a)\delta(H + E)\pi(a)|\Omega\rangle, \end{aligned}$$

and therefore

$$\frac{f(E)}{f(-E)} = \exp(\beta E).$$

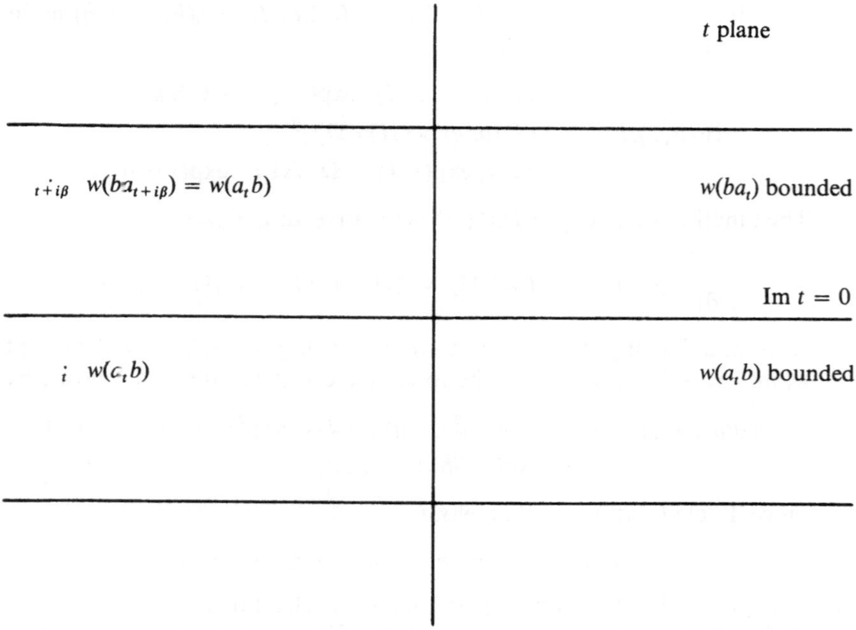


Figure 26 The connection between $w(ba_t)$ and $w(a_t b)$ on their domain of analyticity.

It is thus not possible to remove arbitrary amounts of energy from a system in equilibrium, even though $|\Omega\rangle$ is not its ground state.

(v) Analytic operators. If the dimension of the space is finite, the mapping $t \to a_t$ is analytic, and thus so is $t \to w(a_t b)$. If it is only known that $h$ is semibounded, this is not necessarily the case, and the question arises of which $a$'s are analytic in $t$. One way to construct such elements of $\mathscr{A}$ is to average over time,

$$a(f) \equiv \int_{-\infty}^{\infty} dt' a(t') f(t').$$

If the Fourier transform $\tilde{f} \in \mathscr{C}^2$, and $\operatorname{supp} \tilde{f} \subset [-\alpha, \alpha]$, then $f(t)$ is analytic and satisfies the estimate

$$|f(x + iy)| \leq \frac{\exp(\alpha|y|)}{(1 + x^2)} \gamma, \quad \text{where } \gamma = (2\pi)^{-1/2}(\|\tilde{f}\|_1 + \|\tilde{f}''\|_1).$$

The time-translate of $a(f)$,

$$\tau_t(a(f)) = \int_{-\infty}^{\infty} dt' a(t') f(t' - t),$$

is then an entire function in $t$ such that $\|\tau_{x+iy}(a(f))\| \leq \pi\gamma\|a\| \exp(\alpha|y|)$. It is easy to see from the continuity of $\tau_t$ that the set $\tilde{\mathscr{A}}$ of such regularized

$a$'s (for variable $f$ and $\alpha$) is dense in $\mathscr{A}$ in norm. Within the set $\tilde{\mathscr{A}}$ it is always possible to continue analytically with controlled growth.

If we now think about an infinite system, the density matrix

$$\exp(-\beta H)/\operatorname{Tr}\exp(-\beta H)$$

no longer makes sense. However, the characterization of certain states made in (3.2.5(ii)) may continue to work in the infinite limit.

**Definition** (3.2.7)

Given a $C^*$ algebra $\mathscr{A}$ with a continuous time-automorphism $a \to a_t$, a state $w$ on the algebra is called a **KMS state** with respect to temperature $1/\beta$ whenever the functions $t \to w(a_t b)$ and $t \to w(ba_t)$ can be continued analytically to the strips $-\beta < \operatorname{Im} t < 0$ and, respectively, $0 < \operatorname{Im} t < \beta$, and are continuous on the closures of the strips, where they satisfy the condition

$$w(a_t b) = w(ba_{t+i\beta}).$$

**Examples** (3.2.8)

1. Free fermions. The grand canonical state (2.5.49) is KMS with respect to the combination of free time-evolution and gauge transformations,

$$\tau_t\colon a_f \to a_{f_t}, \qquad \tilde{f}_t(\mathbf{k}) = \exp[it(|\mathbf{k}|^2 - \mu)]\tilde{f}(\mathbf{k}).$$

First, note that clearly

$$\begin{aligned}\rho(a_f a^*_{g_{i\beta}}) &= \int \frac{d^3k}{(2\pi)^3}\tilde{f}^*(\mathbf{k})\tilde{g}(\mathbf{k})\exp[-\beta(|\mathbf{k}|^2 - \mu)] \\ &\quad\times\left(1 - \frac{1}{\exp[\beta(|\mathbf{k}|^2 - \mu)] + 1}\right) \\ &= \rho(a^*_g a_f),\end{aligned}$$

and likewise

$$\begin{aligned}\rho(a^*_g a_{f_{i\beta}}) &= \int \frac{d^3k}{(2\pi)^3}\tilde{f}^*(\mathbf{k})\tilde{g}(\mathbf{k})\exp[\beta(|\mathbf{k}|^2 - \mu)]\left(\frac{1}{\exp[\beta(|\mathbf{k}|^2 - \mu)] + 1}\right) \\ &= \rho(a_f a^*_g).\end{aligned}$$

(If $f$ and $g$ are arbitrary functions in $L^2$, then in general $f_t$ and $\rho(a_f a^*_{g_t})$ have maximal analytic continuations only into the upper half-plane $\{z = t + iy \mid y > 0\}$, and $\rho(a^*_g a_{f_t})$ only into the region $\{z = t + iy \mid y < \beta\}$. However, if either $\tilde{f}$ or $\tilde{g}$ has compact support, for example, then the maximal analytic continuation of any of the expressions above is in fact an entire function.) The proof of the KMS property of $\rho$ for arbitrary elements of the algebra will not be given here, because of the amount of combinatorics it requires. The gauge transformation makes an appearance because of the extension of the state to the whole field algebra. If one deals only with the gauge-invariant algebra of observables $\mathscr{A}''_E$ (1.3.14), then the automorphism $\tau$ does not depend on $\mu$, so it is identical to the free time-evolution.

2. Free bosons. Let $\omega_\varphi$ be the equilibrium state of the field algebra of the free Bose gas at temperature $1/\beta$ and density $\rho$ (see (2.5.51; 4)), which appears as the integrand in the decomposition of the canonical limiting state in (2.5.51; 1). (The decomposition is nontrivial iff $\rho > \rho_c(\beta)$—see also (2.5.33; 3).) The field algebra of the bosons is generated by the operators

$$W_f \equiv \exp[i(a_f^* + a_f)]; \qquad W_f W_g = \exp[-i\,\mathrm{Im}(f\,|\,g)]W_{f+g},$$

and the free time-evolution of the observables will be extended to the field algebra by $W_f \to W_{f_t}$,

$$\tilde{f}_t(\mathbf{k}) = \exp[it(|\mathbf{k}|^2 - \mu)]\tilde{f}(\mathbf{k}).$$

(The quantity $\mu = \mu(\rho)$ is a unique but not invertible function.) Then $A(f, g, t) \equiv \omega_\varphi(W_f W_{g_t})$ is the continuous boundary value of an analytic function of $z = t + iy$ on the strip $0 < y < \beta$, $t \in \mathbb{R}$, *viz.*,

$$\begin{aligned}\hat{A}(f, g, z) \equiv \exp\Biggl\{-\int \frac{d^3k}{(2\pi)^3}\Biggl[(|\tilde{f}(\mathbf{k})|^2 + |\tilde{g}(\mathbf{k})|^2)\left(\frac{1}{2} + \frac{1}{\exp[\beta(|\mathbf{k}|^2 - \mu)] - 1}\right)\\ + \tilde{f}^*(\mathbf{k})\tilde{g}(\mathbf{k})\exp[iz(|\mathbf{k}|^2 - \mu)]\left(1 + \frac{1}{\exp[\beta(|\mathbf{k}|^2 - \mu)] - 1}\right)\\ + \tilde{g}^*(\mathbf{k})\tilde{f}(\mathbf{k})\exp[-iz(|\mathbf{k}|^2 - \mu)]\left(\frac{1}{\exp[\beta(|\mathbf{k}|^2 - \mu)] - 1}\right)\Biggr]\Biggr\}\\ \times \exp\{2i\sqrt{\rho - \rho_c(\beta)}\,\Theta(\rho - \rho_c(\beta))\,\mathrm{Re}[(\tilde{f}(\mathbf{0}) + \tilde{g}(\mathbf{0}))\exp(i\varphi)]\},\end{aligned}$$

and the KMS condition is satisfied: $w(ab_{-t}) = w(a_t b) = w(ba_{t+i\beta})$,

$$\begin{aligned}\lim_{y\to+\beta} \hat{A}(f, g, t + iy) &= \omega_\varphi(W_{g_t} W_f) = \omega_\varphi(W_g W_{f_{-t}}) = A(g, f, -t)\\ &= \lim_{y\to+0} \hat{A}(g, f, -t + iy).\end{aligned}$$

It follows from $\rho < \rho_c(\beta)$ that $\mu(\rho) < 0$, so in this situation $f$ and $g$ can be arbitrary elements of $L^2$. However, $\mu(\rho) = 0$ for all $\rho \geq \rho_c(\beta)$, so $\omega_\varphi$ must be restricted, for example, to the algebra generated by the $W_f$ with $f \in L^1 \cap L^2$. For general $f$ and $g$ it is not possible to extend $\hat{A}(f, g, z)$ analytically beyond the strip described above. However, if the support of either $f$ or $g$ is compact, then $\hat{A}(f, g, z)$ is an entire function of $z$.

**Properties of a KMS state** $w$ (3.2.9)

1. *A* KMS *state w is invariant in time.*
2. *When extended to* $\pi_w(\mathscr{A})''$, *w remains* KMS.
3. *If w is faithful (as a positive functional), then* $\pi_w$ *is faithful, and vice versa.*
4. $\mathscr{Z} = \pi_w(\mathscr{A})' \cap \pi_w(\mathscr{A})''$ *consists of time-invariant elements.*
5. *The* KMS *states for any fixed* $\beta$ *form a weak-* compact, convex set.*
6. *If w is an extremal* KMS *state, then* $\pi_w$ *is a factor.*
7. *For any w, there exists a unique time-evolution under which w is a* KMS *state.*

**Remarks** (3.2.10)

1. According to (1.3.5), if $w$ is invariant in time, then on $\pi_w$ we can write $a_{-t} = U_t a U_t^{-1}$, and the time evolution, when extended to $\pi_w(\mathscr{A})''$, transforms this algebra into itself: $a_n \to a \Rightarrow a_n(-t) = U_t a_n U_t^{-1} \to U_t a U_t^{-1} \in \pi_w(\mathscr{A})''$.
2. Of course, the extension of $w$ to $\pi_w(\mathscr{A})''$ with cyclic vector $|\Omega\rangle$ is $w(a'') = \langle\Omega|a''|\Omega\rangle$ for all $a'' \in \pi_w(\mathscr{A})''$. Property 2 means that this state is KMS with respect to the time-evolution defined earlier on $\pi_w(\mathscr{A})''$.
3. According to (III: 2.3.10; 3),

$$\begin{aligned} \operatorname{Ker} w &= \{a \in \mathscr{A} : w(a) = 0\} \\ \supset \mathscr{N} &\equiv \{a \in \mathscr{A} : w(a^*a) = 0\} \\ \supset \operatorname{Ker} \pi_w &= \{a \in \mathscr{A} : w(b^*a^*ab) = 0 \quad \text{for all } b \in \mathscr{A}\}, \end{aligned}$$

   and the statement that $w$ is faithful means that $\mathscr{N} = \{0\}$. Property 3 thus means that if $\operatorname{Ker} \pi_w = \{0\}$, then $\mathscr{N} = \{0\}$, so $|\Omega\rangle$ is a separating vector for $\pi_w(\mathscr{A})$: $\pi_w(a)|\Omega\rangle \neq 0$ for all $\pi_w(a) \neq 0$. (Speaking field-theoretically, no operator annihilates the vacuum.) If the algebra is simple, and hence has only faithful representations, then all KMS states are also faithful.
4. If the system is asymptotically Abelian, then $\mathscr{R}' = \mathscr{Z}$. The center $\mathscr{Z}$ contains the macroscopic observables, which are therefore constant in time in this case.
5. By Property 5, convex combinations and weak limits of KMS states (at a given $\beta$) are KMS states.
6. In a finite system, with $\mathscr{A} = \mathscr{B}(\mathscr{H})$, $U_t = \exp(iHt)$, there is only one normal KMS state. At $t = 0$ the condition is that

$$\operatorname{Tr} \rho ab = \operatorname{Tr} \rho b \exp(-\beta H) a \exp(\beta H) = \operatorname{Tr} \exp(-\beta H) a \exp(\beta H) \rho b$$

   for all $b$, which means that $\rho a = \exp(-\beta H) a \exp(\beta H) \rho$ for all $a$, so $\exp(\beta H)\rho \in \mathscr{A}'$, and thus $\rho = \exp(-\beta H)$. Since the convex set of KMS states is compact, any KMS state may be decomposed into extremal KMS states. If the system is asymptotically Abelian, then according to Remark 6 a decomposition into extremal KMS states is the same as a decomposition into elements of the center (defined as a decomposition into factors (1.4.9)), which is the same as a decomposition into extremal invariant states. In the characterization of ergodic states (3.1.22; 2) we learned that a factor state is not decomposable into invariant states, and thus *a fortiori* not decomposable into KMS states. Conversely, it is now being claimed that it is always possible to decompose a KMS state $w$ further into other, extremal KMS states, if $\pi_w$ is not a factor. This means that the extremal KMS states are ergodic and, as factors, even mixing. Since the decomposition by the center is unique, so is the decomposition into extremal KMS states. Hence the set of extremal KMS states is a simplex.

7. If the time-evolution is given, then there can be one or more KMS states (see Problem 2). In contrast, by Property 7, if $w$ is given, then there is a unique time-evolution for which it is KMS.

**Proof of** (3.2.9)

1. Let $b = 1$; the function $\rho(a_t) = \rho(a_{t+i\beta})$ can be continued analytically to all of $\mathbb{C}$ and is periodic in Im $t$. Since it is bounded in a strip, it is bounded throughout $\mathbb{C}$ and therefore constant. It follows that $\rho$ is time-invariant.
2. This proposition follows from a more general one to be stated later (3.2.13).
3. If $a \in \mathcal{N}$, then $w(a^*a) = 0$, which implies that for all $b$, $w(ba) = 0$ (by Cauchy–Schwarz), which means that for all $b$ and $c$, $0 = w(c_{-i\beta}ba) = w(bac)$, and therefore $a \in \operatorname{Ker} \pi_w$.
4. Suppose $c \in \mathcal{Z}: w(a_t c) = w(a_{t+i\beta} c)$. As in Proposition 1, it can be concluded that $w(a_t c)$ is constant in $t$. If $a$ is replaced with $ab$, it follows that $w(a_t cb_t) = \langle\Omega | a U_t c U_{-t} b | \Omega\rangle$ is constant for all $a$ and $b$, so $c$ is constant.
5. Convexity is trivial. If $w_n$ converges in the weak-* sense to $w$, then for all $a \in \tilde{\mathcal{A}}$, $b \in \mathcal{A}$ and $t \in \mathbb{C}$, the quantities $w_n(a_t b)$ converge to $w(a_t b)$ and are dominated by $\pi\gamma \|a\| \, \|b\| \exp \alpha |\operatorname{Im} t|$. Consequently, the limit is holomorphic throughout $\mathbb{C}$ and satisfies $w(a_{t-i\beta} b) = w(ba_t)$. As in Problem 1, this relationship remains valid for norm-limits of $a$'s in the strip $0 \le \operatorname{Im} t \le \beta$, and can thus be extended to all of $\mathcal{A}$ (and, by Property 2, to all of $\mathcal{A}''$).
6. Unless $\pi_w$ is a factor, $\mathcal{Z}$ contains a nontrivial projection $P$. Therefore $w$ can be decomposed into a combination of $w_1(a) = w(Pa)/w(P)$ and $w_2(a) = w((1 - P)a)/w(1 - P)$, and both $w_i$ are KMS states: $w(Pa_t b) = w(a_t Pb) = w(Pba_{t+i\beta})$.
7. Suppose that $\tau_t$ and $\bar{\tau}_t$ are distinct automorphisms under which $w$ is a KMS state. Then if $a$ is entire with respect to $\tau$, and $b$ is entire with respect to $\bar{\tau}$, it follows that

$$F(t) \equiv w(\bar{\tau}_{-t}(\tau_t(a)) \cdot b) = w(\tau_t(a) \cdot \bar{\tau}_t(b)) = w(\bar{\tau}_t(b) \cdot \tau_{t+i\beta}(a))$$
$$= w(\tau_{t+i\beta}(a) \cdot \bar{\tau}_{t+i\beta}(b)) = F(t + i\beta).$$

This fact implies that $F$ is constant, so $\tau$ and $\bar{\tau}$ have the same action on $\tilde{\mathcal{A}}$ and hence on $\mathcal{A}$. □

The foregoing conclusions suggest an interpretation of the decomposition into extremal KMS states as a decomposition of an equilibrium state into its pure phases. Yet it will be apparent from examples that these pure phases are not necessarily identical to physical phases. Property 6 together with Remarks (3.1.26) ensures that these states have mixing properties, meaning that local perturbations eventually die out, and equilibrium gets re-established. The canonical states were characterized earlier as the states of greatest entropy at a given energy, and the evolution towards them can be thought of as a tendency toward greater entropy. On the other hand, if the system is infinite, it is not the total entropy that is finite, but rather the average entropy, which is unaffected by local perturbations. If a state is normal when restricted to a local algebra (1.3.3; 6), then it is possible to define the local entropy, which will then tend to its equilibrium value. It is not, however, claimed that it increases monotonically to that value.

The diagram in Figure 27 collects together the various properties of asymptotically Abelian systems in invariant states and shows their connection with the time-evolution. It will be shown later (3.3.17) that the spectrum of $H$ is ordinarily the whole real line $(-\infty, \infty)$. The spectral properties stated then include the supposition that the systems that we shall be concerned with have neither dense point spectrum nor singular continuous spectrum.

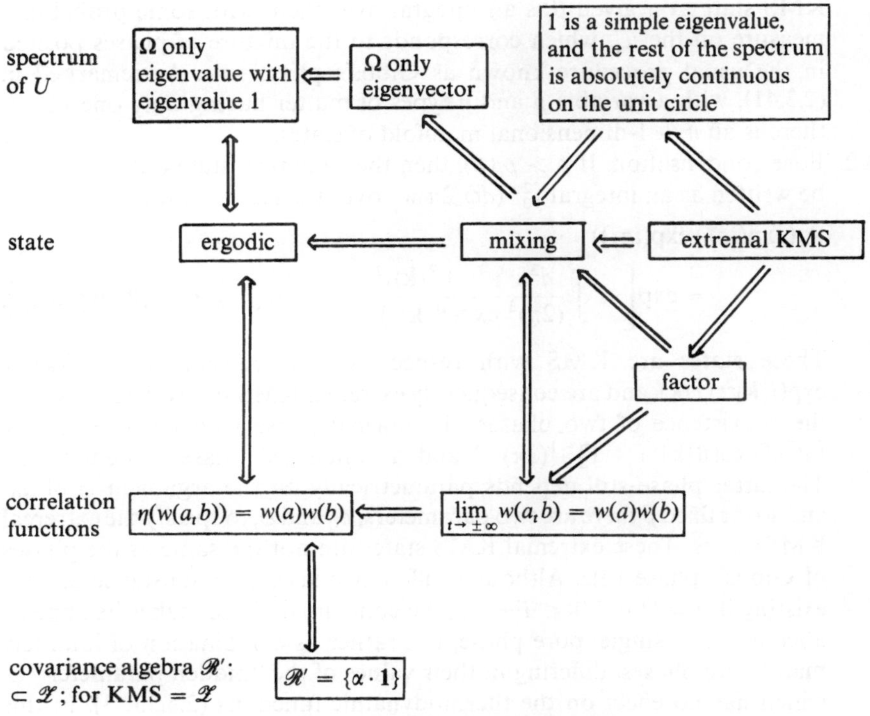


Figure 27 Implications among the ergodic properties.

**Examples** (3.2.11)

1. Free fermions. Consider a system of $n$ kinds of free fermions, described by the field operators $a_{\alpha,f}$, $\alpha = 1, \ldots, n$. The algebra $\mathscr{A}_E$ of observables will be taken to consist only of polynomials containing an equal number of $a_\alpha$ and $a_\alpha^*$ for any $\alpha$, in accordance with Definition (1.3.8). In other words it contains the densities and currents of the particles. The state is taken as the product of the grand canonical states (2.5.49), i.e.,

$$\langle a^*_{1,f_1^1} \cdots a^*_{1,f_{m_1}^1} a_{1,g_1^1} \cdots a_{1,g_{m_1}^1} a^*_{2,f_1^2} \cdots a^*_{2,f_{m_2}^2} a_{2,g_1^2} \cdots a_{2,g_{m_2}^2} a^*_{3,f_1^3} \cdots a_{n,g_{m_n}^n} \rangle = \prod_\alpha \mathrm{Det}\langle g_i^\alpha | \rho_1^\alpha f_j^\alpha \rangle,$$

$$\langle f | \rho_1^\alpha g \rangle = \int \frac{d^3k}{(2\pi)^3} \frac{\tilde{f}^*(\mathbf{k})\tilde{g}(\mathbf{k})}{\exp[\beta((|\mathbf{k}|^2/2m_\alpha) - \mu_\alpha)] + 1}.$$

It is KMS with respect to the automorphism $a_{\alpha, f_\alpha} \to a_{\alpha, f_\alpha(t)}$,

$$\tilde{f}_\alpha(t) = \exp\left(\frac{it|\mathbf{k}|^2}{2m_\alpha}\right)\tilde{f}_\alpha.$$

Observe that for this automorphism of the algebra of observables there is an $n$-parameter family of KMS states. They can be parametrized by the chemical potentials $\mu_\alpha$, and, as factor states, they are extremal. A general KMS state at a given $\beta$ is an integral over them with some probability measure on the $\mu_\alpha$, which corresponds to the mixture of phases posited in the usual procedure known as Gibbs's phase rule. As remarked in (2.3.41), with a variable $\beta$ and $n$ types of matter having only one phase, there is an $n + 1$-dimensional manifold of states.

2. Bose condensation. If $\rho > \rho_c(\beta)$, then the canonical state (2.5.51; 1) may be written as an integral $\int_0^{2\pi} (d\varphi/2\pi) w_\varphi$ over the factor states

$$w_\varphi(\exp(ia_f^*)\exp(ia_f)) = \exp\left[-\int \frac{d^3k}{(2\pi)^3} \frac{|\tilde{f}(\mathbf{k})|^2}{\exp(\beta|\mathbf{k}|^2) - 1} + 2i\sqrt{\rho_0}\,\mathrm{Re}(\tilde{f}(\mathbf{0})\exp(i\varphi))\right].$$

These states are KMS with respect to the transformation $\tilde{f}(\mathbf{k}) \to \exp(i|\mathbf{k}|^2 t)\tilde{f}(\mathbf{k})$, and are consequently extremal KMS states. They describe the coexistence of two phases, the normal phase with particle density $\int d^3k[\exp(\beta|\mathbf{k}|^2) - 1]^{-1}(2\pi)^{-3}$ and a condensed phase of density $\rho_0$. The latter phase still depends parametrically on the argument $\varphi$ of $a_0$, and so for fixed $\beta$ there are two parameters, $\rho_0$ and $\varphi$, to specify the extremal KMS states. These extremal KMS states are not the same as the phases of Gibbs's phase rule. Although different phases of a substance are coexisting if $\mu = 0$ and $0 < T < T_c$, the condensed phase makes its appearance not as a single, pure phase, but rather as combination of infinitely many pure phases, differing in their values of the "hidden parameter" $\varphi$, which has no effect on the thermodynamic functions (2.5.33; 3). In this way the decomposition into extremal KMS states is finer than the phase decomposition of (2.3.39) into extremal points of the concave function $\sigma(\varepsilon, \rho)$. If the field algebra is confined to its even part $\mathscr{A}_E$ (in the Fock representation, $\mathscr{A}_E = \mathscr{A}_B \cap \{N\}'$), then all the $w_\varphi$ become the same state. This is apparent when it is observed that gauge transformations $\tau_\varphi$: $W_f \to W_{\exp(i\varphi)f}$ transform the $w_\varphi$ into one another: $(w_\varphi \circ \tau_{\varphi'})(W_f) = w_{\varphi+\varphi'}(W_f)$. The restriction to $\mathscr{A}_E$ makes $\tau_{\varphi'}$ the identity, so $w_\varphi = w_{\varphi+\varphi'}$. Recall that for asymptotically Abelian systems the decomposition into extremal KMS states is unique according to (3.2.9; 6); the extremal states form a simplex. In contrast, we were not able to adduce any theoretical reasons for why the flat pieces of $\sigma(\varepsilon, \rho)$ had the structure of a simplex.

3. A model of a ferromagnet. The time-evolution of Example (2.3.33; 2) was investigated in (3.1.1; 4). We found that if $B = 0$ and $T < 2$, it was no longer an automorphism of the spin algebra $\mathscr{A} = \{\boldsymbol{\sigma}_i\}$, but rather of the strong closure $\pi(\mathscr{A})''$. The state

$$\langle \sigma_1^{\alpha_1} \cdots \sigma_m^{\alpha_m} \rangle = \int_{S_2} d\mathbf{n} s^m n_{\alpha_1} \cdots n_{\alpha_m}, \qquad s = \tanh(2\beta s),$$

is KMS with respect to this time-evolution. In each of the factors $\pi_\mathbf{n}$ it is a rotation about the axis $\mathbf{n}$ at angular velocity $4s$. For example, if $\mathbf{n}$ points in the $z$-direction, then $\sigma^+(t) = \exp(-4ist)\sigma^+$ and

$$\langle \sigma^+ \sigma^- \rangle = \frac{\langle 1 + \sigma \rangle}{2} = \frac{1+s}{2} = \langle \sigma^- \sigma^+_{i\beta} \rangle = \exp(4\beta s) \langle \sigma^- \sigma^+ \rangle$$

$$= \exp(4\beta s)\frac{1-s}{2},$$

because $s(1 + \exp(4\beta s)) = \exp(4\beta s) - 1$. The individual factors $\pi_\mathbf{n}$ thus give rise to extremal KMS states, corresponding to spontaneous magnetization in the direction $\mathbf{n}$. Again, from the physical point of view this model would be described as having one magnetized phase, whereas the decomposition into extremal KMS states would distinguish among different directions of $\mathbf{n}$, and treat magnetization in each direction as a distinct phase. Notice that the phase transition at $T = 2$ is connected with a change of the type of factor; if $T < 2$ the integral runs over factors of type III, while if $T > 2$, the factors are of type $\mathrm{II}_1$.

**Remarks** (3.2.12)

1. There are many different possible reasons for the existence of several KMS states. One is that the center of the algebra of observables $\mathscr{A}$ might be nontrivial. Unitary elements of the center generate transformations, which, like gauge transformations, leave each element of the algebra invariant. Therefore it is possible to combine the action of these transformations with that of time-evolution $\tau$ and study the KMS states with

respect to the resulting automorphisms. When restricted to $\mathscr{A}$, these automorphisms are identical to the time-evolution, so all such states are also $\tau$-KMS for $\mathscr{A}$ (cf Problem 2).

2. Many "degeneracies" of KMS states go away upon enlargement of the algebra of observables. If in Example 1 the particle number is also allowed to vary, for instance by a chemical reaction $(1) \rightleftharpoons (2) + (2)$, then noneven elements like $a_1^* a_2 a_2$ are introduced into the algebra of observables. They are not separately invariant under gauge transformations of the different types of particles, but are invariant only under certain combinations, e.g., if the generator of the transformation has the form $2N_1 + N_2$ in the Fock representation. Consequently, the KMS condition with the free time-evolution makes the chemical potentials satisfy a linear equation such as $\mu_1 - 2\mu_2 = 0$. Similarly, if two condensed Bose systems as in Example (3.2.11; 2) are coupled, the relative phase $\varphi$ becomes observable (the Josephson effect).
3. It is possible that a symmetry is broken, which means that the extremal KMS states $w$ are not invariant under some group $\sigma$ of automorphisms that commute with $\tau$. This is illustrated in Example (3.2.11; 2) with the gauge transformations and in (3.2.11; 3) with the rotations. If the symmetry is broken, then $w \circ \sigma_s$ is once again $\tau$-KMS; thus with continuous groups there are even infinitely many KMS states.
4. The theoretical justification of Gibb's phase rule for continuous systems is still an open problem (cf. [20]).
5. So far we have been considering $\beta$ as fixed. KMS states with different $\beta$'s are disjoint, i.e., if $w = (w_{\beta_1} + w_{\beta_2})/2$, then $\pi_w = \pi_{\beta_1} \oplus \pi_{\beta_2}$. In this case the temperature $\beta^{-1}$ becomes an observable belonging to the center of $\pi_w(\mathscr{A})$.

As discussed in §1.1, the ergodic property of a system has been an important ingredient of the justification of statistical mechanics throughout its history. Even though today ergodicity is no longer viewed as the central requirement, it can still be a noteworthy property of realistic systems, so it can still be valuable to have a formulation of ergodicity for infinite quantum systems. In a classical system, if there existed additional constants of the motion beyond $H$, it would be impossible for the trajectory of almost every point to wind densely throughout the energy shell. However, constants such as momentum or angular momentum are infinite for infinite systems, so ergodicity can not be defined as the absence of additional constants of the motion. But recall that classically constants of the motion also generate diffeomorphisms that commute with the flow of time (see I, §3.3). This property carries over to infinite systems, and even the notions of indecomposable time-invariant surfaces and of dense trajectories have analogies.

In order to characterize ergodic systems, it is only necessary to generalize (3.2.5) to infinite systems.

**Modular Automorphisms of a von Neumann Algebra** (3.2.13)

*Let $\mathscr{M}$ be a von Neumann algebra of operators on a Hilbert space $\mathscr{H}$. For every vector $|\Omega\rangle$ that is both cyclic and separating (i.e., $\mathscr{M}|\Omega\rangle = \mathscr{H}$, and if $a|\Omega\rangle = 0$ for any $a \in \mathscr{M}$, then $a = 0$), there exists a unique one-parameter group of automorphisms $a \to \tau_t(a)$ and a conjugate-linear operator $J$ such that*

(i) $w(a) \equiv \langle\Omega|a|\Omega\rangle$ *is* $\tau$-KMS (*with* $\beta = 1$);
(ii) $J^2 = \mathbf{1}$, $J\mathscr{M}J = \mathscr{M}'$; *and*
(iii) $U_{-i/2}a|\Omega\rangle = Ja^*|\Omega\rangle$, *where* $\tau_t(a) = U_{-t}aU_t$.

**Remarks** (3.2.14)

1. The idea of the proof follows that of (3.2.3), but with additional technical complications, for which reason the reader is referred to [21].
2. Properties (3.2.6) of the correlation functions hold also in the general case. Specifically, (iii) means that $\mathscr{A}|\Omega\rangle \subset D(\exp(-H/2))$, where $U_t = \exp(-iHt)$, from which it follows that $\mathscr{A}|\Omega\rangle \subset D(\exp(-yH)$ for $0 \le y \le \frac{1}{2}$, and $w(a^* \exp(-H)a) = w(aJ^2a^*) \le \|a\|^2$. The proofs of the other properties can be repeated verbatim.
3. It is clear that a further generalization to arbitrary $C^*$ algebras will not work. The state in Example (3.2.11; 3) is obviously faithful on the $\sigma$'s, so it is a candidate for $w$. However, we have found that the related automorphism under which $w$ is a KMS state maps the $C^*$ algebra generated by the $\sigma$'s out of itself, leaving only the von Neumann algebra $\pi_w(\mathscr{A})''$ invariant.
4. Suppose that $w$ is a KMS state on the algebra $\mathscr{A}$ with respect to the time-evolution $\tau_t$. By Property (3.2.9; 3) the vector $|\Omega\rangle$ given in the GNS representation $\pi_w$ is cyclic and separates $\pi_w(\mathscr{A})''$, even if $w$ fails to be faithful, and the representation of $\tau_t$ is identical to the modular automorphism.

**Ergodic Quantum Systems** (3.2.15)

*Let $\tau$ be the time-evolution under which the $C^*$ algebra $\mathscr{A}$ of observables is asymptotically Abelian, and let $\mathscr{T}$ be the set of faithful states $w$ with the property that the normal extension of $w$ to $\pi_w(\mathscr{A})''$ is also faithful. Then the following two properties are equivalent:*

(i) *A state $w \in \mathscr{T}$ is ergodic if and only if it is an extremal* KMS *state; and*
(ii) *There is no $w \in \mathscr{T}$ such that its modular automorphism $\sigma$ differs from $\tau$, but $[\sigma, \tau] = 0$.*

If a system has these properties, we shall call it **ergodic**.

**Proof that (i) ⇔ (ii)**

Not (ii) ⇒ not (i). Let $w$ be the $\sigma$-KMS state. Since $\sigma$ and $\tau$ commute, $\rho \equiv \eta_t(w \circ \tau_t)$ is also $\sigma$-KMS, so our strategy will be to use it to construct a $\tau$-ergodic state. Think of $\rho$ as decomposed in two separate ways, on the one hand into $\tau$-ergodic states and on the other into extremal $\sigma$-KMS states. By Remark (3.2.10; 6) the latter decomposition is the same as the decomposition into factors, whereas according to Remark (3.1.21; 3) the $\tau$-ergodic decomposition is coarser than the factor decomposition. This means that the $\tau$-ergodic components of $\rho$ are combinations of extremal $\sigma$-KMS states, but not vice versa. Hence any such component is $\tau$-ergodic but not $\tau$-KMS, since it is not possible for it to be KMS with respect to $\sigma$ and $\tau \neq \sigma$ at the same time.

Not (i) ⇒ not (ii). Suppose that $w(a) = \langle\Omega|a|\Omega\rangle$ is $\tau$-ergodic, and let $\sigma$ denote the modular automorphism of $\pi_w(\mathscr{A})''$. Since $w$ is invariant under $\tau$ and $\sigma$, both groups have unitary representations on $\pi_w$. Let $\exp(iHt)$ and $\exp(iGs)$ denote their representations. Since $w$ is also $\sigma$-KMS, given any $a$ and $b \in \mathscr{A}$,

$$\langle\Omega|\tau_t(a)\sigma_i(b)|\Omega\rangle = \langle\Omega|b\tau_t(a)|\Omega\rangle = \langle\Omega|\tau_{-t}(b)a|\Omega\rangle = \langle\Omega|\sigma_{-i}(a)\tau_{-t}(b)|\Omega\rangle,$$

so

$$\langle\Omega|a \exp(-iHt)\exp(-G)b|\Omega\rangle = \langle\Omega|a \exp(-G)\exp(-iHt)b|\Omega\rangle.$$

Since the vectors of the form $a|\Omega\rangle$ are dense, it follows that $[\exp(-G), \exp(-iHt)] = 0$, so $[\tau, \sigma] = 0$. However, if $w$ is not KMS with respect to $\tau$, then the groups of automorphisms must be different, since $w$ is KMS with respect to $\sigma$. □

**Remarks** (3.2.16)

1. Unfortunately, no examples of ergodic quantum systems are known. Although the grand canonical state (2.5.49) of free particles is mixing, there are ergodic states that fail to be KMS: The momentum distribution $[\exp(\beta(|\mathbf{k}|^2 - \mu)) \pm 1]^{-1}$ would just have to be replaced with some other positive, integrable function. The state would then be time invariant and, as a factor state, ergodic, but not KMS. The hope is that when interactions are switched on, states of this kind will turn into equilibrium states (see §3.3).
2. Property (3.2.15(ii)) forbids the existence of additional constants of the motion. In finite quantum systems, in addition to the Hamiltonian $H$ there are also the constants of the form $f(H)$. If $H$ is nondegenerate, then this accounts for all the constants, because $\{H\}'$ is generated by $f(H)$ and the unitary transformations of the degeneracy space. If the system is infinite, then $H$ exists only in representations $\pi_w$ of invariant states $w$, and does not belong to $\pi_w(\mathscr{A})$. It can be shown [22] that only linear functions $f(H)$ produce automorphisms of $\pi_w(\mathscr{A})$. However, the function

$H \to cH$ does nothing more than change the scale of time, and we consider scaled time-evolutions as identical.

3. If particle numbers are conserved, then gauge transformations $a_f \to \exp(i\alpha)a_f$, $\alpha \in \mathbb{R}$, certainly commute with time-evolution, and the system is not ergodic as defined by (3.2.15). Yet the corresponding KMS states $w$ are of the form (2.5.49) with infinite temperature but $\beta\mu = 1$,

$$w(a_f a_g^*) = \int \frac{d^3k}{(2\pi)^3} \frac{\tilde{f}^*(\mathbf{k})\tilde{g}(\mathbf{k})}{e + 1}.$$

   The particle density in this state is infinite, $w(a(\mathbf{x})a^*(\mathbf{x})) = \delta(\mathbf{0})/(1 + e)$, however, so it is not of physical interest. This shows that in a nonergodic infinite system it may happen that the states that are ergodic but not KMS never actually occur, so the system behaves ergodically anyway. On the other hand, there is no similar objection to this state on a lattice system, for which $\mathbf{k}$ varies only over a compact region.
4. If an infinite system is homogeneous and isotropic, then translations and rotations commute with $\tau$. The KMS states of these automorphisms have the same defect as that of Remark 3, that the local particle density is infinite.
5. Since under the measurability assumptions of (3.1.22; 3) ergodic states are time-averages of a pure state, the same will be true of the extremal KMS states of ergodic systems. This is the fulfillment of the hope of classical ergodic theory that the equilibrium state can be obtained as the closure of a single trajectory.

| state \ system | Finite, classical<br>There are no additional constants of the motion | Finite, quantum-mechanical<br>$H$ is nondegenerate | Infinite, quantum-mechanical<br>There exists no KMS $\sigma$ such that $\sigma \neq \tau$, $[\sigma, \tau] = 0$ |
|---|---|---|---|
| Microcanonical | Ergodic<br>Time-average of pure states<br>Not faithful | Ergodic<br>Time-average of pure states<br>Not faithful | |
| Canonical | Not ergodic<br><br>Faithful | Not ergodic<br>Time-average of pure states<br>Faithful | |
| Extremal KMS | | | Ergodic<br>Time-average of pure states<br>Faithful |

If we wish to conceive of ergodicity roughly as the absence of constants of motion other than $f(H)$, then it is useful to make a table of the implications of this for equilibrium states of systems of various types. As can be seen below, the KMS states of infinite quantum systems inherit the good properties of the canonical and microcanonical states of finite systems.

**Problems** (3.2.17)

1. Consider a sequence of states $w_N$ on a $C^*$ algebra $\mathscr{A}$ converging to $w$ (in the weak-* sense). Show that if the modular automorphism $\tau_{N,t}(a)$ is a norm-convergent sequence in $\mathscr{A}$ for all $a \in \mathscr{A}$ and $t \in \mathbb{R}$, then the $\tau_{N,t}$ converge to the modular automorphism belonging to $w$.
2. Find an example of an algebra $\mathscr{A} \subset \mathscr{B}(\mathbb{C}^4)$ such that some nontrivial automorphism has many KMS states.
3. Construct the KMS states for translation and rotation of a system of free fermions.
4. In both classical and quantum mechanics, study the automorphisms of the anisotropic oscillator $H = \frac{1}{2}(p_1^2 + p_2^2 + \omega_1^2 q_1^2 + \omega_2^2 q_2^2)$, with $\omega_1/\omega_2$ irrational, that commute with the time-evolution. Is the system ergodic?

**Solutions** (3.2.18)

1. Consider the limits of the correlation functions $w_N(\tau_{N,t}(a(N, f))b)$, where

$$a(N, f) \equiv \int dt \tau_{N,t}(a) f(t),$$

and $f$ is as in (3.2.6(v)), and let $\tau_t(a) = \lim \tau_{N,t}(a)$. The norm-limit of $\tau_{N,t}(a(N, f))$ is $\tau_t(a(f))$ by the dominated convergence theorem, even for complex $t$, since $\int |f(t + iy)| dt \leq \pi\gamma \exp(\alpha|y|)$. The first term of $[w(\tau_t(a(f))b) - w_N(\tau_t(a(f))b)] + w_N(\tau_t(a(f)) - \tau_{N,t}(a(N, f))b)$ goes to zero because of the weak-* convergence $w_N \to w$, and the second term goes to zero as a consequence of the norm-convergence of $a(N, f)$ to zero. Therefore, for all $a \in \mathscr{A}$ and $t \in \mathbb{C}$,

$$w_N(\tau_{N,t}(a(N, f))b) \to w(\tau_t(a(f))b).$$

These holomorphic functions converge pointwise and are uniformly bounded on every compact set in $\mathbb{C}$, because they are $\leq \|a\| \, \|b\| \pi\gamma \exp(\alpha|y|)$; the limit is therefore holomorphic and identical to $w(b\tau_{t+i}(a(f)))$.

This means that the KMS condition holds for all $a \in \tilde{\mathscr{A}}$, and of course boundedness in the strip (3.2.6(ii)) is preserved in limits. Passing by norm-limits $a_n \to a$ to general $a \in \mathscr{A}$, if $-1 \leq \operatorname{Im} t \leq 0$, then $w(\tau_t(a_n)b)$ converges uniformly to $w(\tau_t(a)b)$, which is consequently continuous on the strip and holomorphic in its interior.

It is trivial to see that the identity $w(\tau_t(a)b) = w(b\tau_{t+i}(a))$ continues to hold for limits, as do the group property $\tau_{t+s} = \tau_t \circ \tau_s$ and the invariance of $w$: $w \circ \tau_t = w$. The GNS construction can now be carried out, so that $\tau_t$ is represented unitarily on $\pi_w$ as $U_t$. If $\pi(a_n)$ converges weakly to $b \in \pi(\mathscr{A})''$, then $U_{-t}\pi(a_n)U_t$ converges weakly to $U_{-t}bU_t \equiv \tau_t(b)$. Therefore $\tau_t$ maps $\pi(\mathscr{A})''$ into itself, and is identical to the modular automorphism according to (3.2.9; 7) and (3.2.14; 4).

2. Let $\mathscr{A}$ be spanned by $(\mathbf{1}, \boldsymbol{\tau}) \otimes (\mathbf{1}, \sigma_3)$, and let the time-evolution be $\tau^{\pm}(t) = \exp(\pm i\omega t)\tau^{\pm}(0)$, with $\tau_3$ and $\sigma_3$ constant. For a given $\beta$ the density matrices of the form
$$\rho = \frac{\exp(-\beta\tau_3 - \alpha\sigma_3)}{\operatorname{Tr} \exp(-\beta\tau_3 - \alpha\sigma_3)}$$
yield KMS states for all real $\alpha$.

3. They have the same structure as in (2.5.49), with
$$\langle f \,|\, \rho, g \rangle = \int \frac{d^3k}{(2\pi)^3} \frac{\tilde{f}^*(\mathbf{k})\tilde{g}(\mathbf{k})}{1 + \exp(k_1)}$$
for translations in the 1-direction, and
$$\int_0^\infty r^2 \, dr \sum_{l,m} \frac{\hat{f}^*_{lm}(r)\hat{g}_{lm}(r)}{1 + \exp m}$$
for rotations about the 3-axis, where $\hat{f}_{lm}$ denote the expansion coefficients of $f$ in spherical harmonics.

4. Classically, $H_i = \frac{1}{2}(p_i^2 + \omega_i^2 q_i^2)$ are two independent constants of the motion, and generate flows that commute with time-evolution. The system is not ergodic in the sense of Table I. Quantum mechanically, $H$ has the eigenvalues $(n_1 + \frac{1}{2})\omega_1 + (n_2 + \frac{1}{2})\omega_2$ and is thus nondegenerate. All constants are of the form $f(H)$, and the system is ergodic in the sense of Table I.

## 3.3 Stability and Passivity

*The distinguishing feature of the equilibrium state is that it does not change abruptly when subjected to a local perturbation. The second law of thermodynamics can be proved in a version stating that a system prevents energy from being extracted by a cyclic perturbation only if it is in equilibrium.*

The final part of the general theory that will be investigated will be the influence of local perturbations on equilibrium. In the mathematical treatment local perturbations play the role of the speck of dust invoked in the traditional theory of statistical mechanics to convert stationary states, not yet in equilibrium, into equilibrium states. As a matter of fact, what makes the KMS states special in the mathematical theory is that they have certain stability properties—they change continuously when the Hamiltonian is perturbed slightly. This is certainly not true of all stationary states, and can even be used to characterize the extremal KMS states of an infinite system; they are precisely the set of states that turn continuously into the unperturbed states as a certain family of perturbations tends to zero. Mixed KMS states represent quantum-mechanical mixtures of phases, and lead to a nontrivial center of the algebra. If an observable from the center is added to $H$, the time-automorphism is unchanged, but the KMS states do change. Hence mixtures of KMS states exhibit a kind of instability in that they do not remain unchanged under the influence of a family of perturbations moving spatially off to infinity, and hence entering the center of the algebra.

A second important characteristic of KMS states is their passivity, which is the requirement that the energy of the system at time $t$ can only have increased if the Hamiltonian depends on time and has returned to its initial form at time $t$. This condition also fixes the sign of $\beta$ and means that no energy can be removed from a KMS state having $\beta > 0$, just as a periodic process can extract no energy from the ground state. This property does not constitute a kind of stability, and sheds no light on why Nature chiefly produces KMS states. However, it does show the most important empirically familiar feature of equilibrium.

As usual, the study of a finite system will provide us with a first exposure to the effects of perturbations. Its time-evolution will be caused by a self-adjoint operator, which also determines the equilibrium state $w$ by $a_t = \exp(iHt)a\exp(-iHt)$. $w(a) = \operatorname{Tr}\exp(-\beta H)a/\operatorname{Tr}\exp(-\beta H)$. If $H$ is subjected to a bounded, self-adjoint perturbation $h$, the effects can be written down as norm-convergent series. A simple generalization of (III: 3.4.10; 3) shows that

$$\exp(i(H+h)t)a\exp(-i(H+h)t) = a_t + \sum_{n\geq 1} i^n \int_{0\leq t_1\leq t_2\cdots\leq t_n\leq t} dt_1\, dt_2 \cdots dt_n [h_{t_1}, [h_{t_2}, \ldots, [h_{t_n}, a_t]\cdots]], \tag{3.3.1}$$

$$\exp(-H-h) = R_h \exp(-H), \qquad \exp(-(H+h)/2) = S_h \exp(-H/2),$$

$$R_h \equiv \mathbf{1} + \sum_{n\geq 1} (-1)^n \int_{0\leq s_1\leq\cdots\leq s_n\leq 1} ds_1 \cdots ds_n h_{is_1} \cdots h_{is_n},$$

$$S_h \equiv \sum_{n\geq 0} (-1)^n \int_{0\leq s_1\leq\cdots\leq s_n\leq 1/2} ds_1 \cdots ds_n h_{is_1} \cdots h_{is_n}. \tag{3.3.2}$$

**Remarks** (3.3.3)

1. Initially, $h_{is}$ is well defined only if $h$ is analytic in time (3.2.6(v)), but since such operators are dense in $\mathscr{A}$ in norm, the formulas it appears in extend to $\mathscr{A}$ by continuity.
2. Inequalities (2.1.8: 3) and (2.1.8; 7) yield the estimates

$$\exp(-\|h\|) \leq \exp\left(\frac{-\operatorname{Tr}\exp(-H)h}{\operatorname{Tr}\exp(-H)}\right) \leq \frac{\operatorname{Tr}\exp(-H-h)}{\operatorname{Tr}\exp(-H)}$$

$$= \frac{\operatorname{Tr} R_h \exp(-H)}{\operatorname{Tr}\exp(-H)} \leq \min\{\|R_h\|, \|\exp(-h)\|\}.$$

Equation (3.3.1) can now be extended to cover infinite systems, for which $H$ has continuous spectrum, as follows.

**Perturbation of the Time–Evolution and KMS State** (3.3.4)

Let $a \to a_t$ be an automorphism of a $C^*$ algebra $\mathscr{A}$, and let $\tilde{\mathscr{A}}$ be the subalgebra that is analytic in time and $w$ be a KMS state. Assume $\beta = 1$. If $h \in \tilde{\mathscr{A}}$ is self-adjoint, then a perturbed automorphism $a \to \tau_t^h(a)$ and perturbed state are defined by

$$\tau_t^h(a) = a_t + \sum_{n \geq 1} i^n \int_{0 \leq t_1 \leq \cdots \leq t_n \leq t} dt_1\, dt_2 \cdots dt_n [h_{t_1}, [h_{t_2}, \ldots, [h_{t_n}, a_t] \cdots]],$$

$$w_h(a) = \frac{w(aR_h)}{w(R_h)} = \frac{w(R_h^* a)}{w(R_h)} = \frac{w(S_h^* a S_h)}{w(R_h)},$$

where $R_h$ and $S_h$ are defined as in (3.3.2).

**Remarks** (3.3.5)

1. The operator $h$ exists as a local perturbation on a purely algebraic level, whereas $H$ exists only in certain representations. For that reason it is not possible to define $\tau_t^h(a)$ simply as $\exp(i(H + h)t)a \exp(-i(H + h)t)$. As in (3.3.2), for finite times the sums converge in norm.
2. If the system is asymptotically Abelian sufficiently strongly, then the limits as $t \to \pm\infty$ of $\tau_t^h \circ \tau_{-t}^0$ exist. However, such a limit may fail to be an automorphism; like the Møller transformations it might not be surjective. If it is surjective, its inverse transforms $w$ into the perturbed state

$$w_h = \lim_{t \to \pm\infty} w \circ \tau_{-t}^0 \circ \tau_t^h.$$

3. See Problem 1 for the equivalence of the definitions of $w_h$.
4. $(\partial/\partial t)\tau_t^h(a) = \tau_t^h((\partial/\partial s)a_s|_{s=0}) + i\tau_t^h([h, a])$.
5. The function $\mathscr{A} \to \mathscr{A} : h \to \tau_t^h(a)$ is continuous for all $t \in \mathbb{R}$ and $a \in \mathscr{A}$, if $\mathscr{A}$ has either the strong or the norm topology.
6. The state $w_h$ is KMS with respect to $\tau_t^h$ for $\beta = 1$: As shown by (3.3.1), $D(\exp(-H - h)) = D(\exp(-H))$ in the representation using $\pi_w$, and because $\exp(H) = \exp(H + h)R_h$, the domains of definition of $\exp(H + h)$ and $\exp(H)$ are also identical. Hence for all $a$ and $b \in \mathscr{A}$,

$$w_h(\tau_{-i}^h(a)b) = \frac{w(R_h^* \exp(H + h)a \exp(-H - h)b)}{w(R_h)}$$

is well defined. From (3.3.1) and the KMS condition for $w$,

$$w_h(\tau_{-i}^h(a)b) = \frac{w(\tau_{-i}(aR_h)b)}{w(R_h)} = w_h(ba).$$

7. There is an analogue of the variational principle for the free energy, which generalizes (2.1.8; 3) for infinite systems. It is a consequence of the convexity of the function $h \to \ln w(R_h)$, which can be proved as follows: From Duhamel's formula (cf. the proof of (III: 3.3.15)),

$$\frac{a}{d\lambda} \exp(-(H + \lambda a))$$
$$= -\int_0^1 ds \exp(-s(H + \lambda a))a \exp(-(1 - s)(H + \lambda a)),$$

it can be calculated that

$$\frac{d}{d\lambda} w(R_{h+\lambda a})|_{\lambda=0} = \int_0^1 w(\tau_{is}^h(a)R_h)\, ds = w(aR_h).$$

The second part of the equality makes use of the invariance of $w_h$ under $\tau^h$, which follows from the KMS condition shown above. Likewise,

$$\frac{d^2}{d\lambda^2} w(R_{h+\lambda a})|_{\lambda=0} = \int_0^1 dsw(a\tau_{is}^h(a)R_h),$$

and

$$\frac{d^2}{d\lambda^2} \log w(R_{h+\lambda a})|_{\lambda=0} = \frac{w(R_{h+\lambda a})''}{w(R_h)} - \left(\frac{w(R_{h+\lambda a})'}{w(R_h)}\right)^2$$
$$= \int_0^1 dsw_h((a - w_h(a))\tau_{is}^h(a - w_h(a))).$$

In (3.2.6(ii)) it was seen that the integrands are positive. As in (3.3.3; 2) this fact can be used to show that $w(R_h) \geq \exp(-w(h)) \geq \exp(-\|h\|)$.

If there is a bounded sequence of perturbations $h^{(n)}$ all the commutators of which with $\mathscr{A}$ tend to zero as $n \to \infty$, then the automorphism $\tau_t^{h^{(n)}}$ converges to the unperturbed automorphism because

$$\|\tau_t^h(a) - a_t\| \leq \exp(2\|h\|t) \int_0^t \|[h, a_{t-s}]\|\, ds.$$

This state of affairs can arise, for instance, if the algebra is asymptotically Abelian with respect to spatial translations. If $\Lambda_n$ denotes the region $\Lambda$ translated by $n\mathbf{a}$, $\mathbf{a} \in \mathbb{R}^3$, and $h^{(n)} \in \mathscr{A}_{\Lambda_n}$ is the corresponding translate of the operator $h$, then $\|[h^{(n)}, a]\| \to 0$, and consequently $\tau_t^h(a) \to a_t$. The question of whether the associated KMS states $w_{h(y)}$ likewise converge to the unperturbed $w$ depends on whether the KMS states are extremal. This is illustrated even in the finite-dimensional case by

**Example** (3.3.6)

With the notation of (1.1.1), let $\mathscr{A}$ be generated by $\{\mathbf{1}, \sigma_1, \sigma_1^\pm, \sigma_2\}$, and suppose that these observables evolve in time into $\{\mathbf{1}, \sigma_1, \exp(\mp 2it)\sigma_1^\pm, \sigma_2\}$. This time-evolution has a unitary representation as $U_t = \exp(it(\sigma_1 + c\sigma_2))$ for all $c \in \mathbb{R}$, so there is a one-parameter family of KMS states with density matrix $\rho = \exp(-\beta(\sigma_1 + \mu\sigma_2))$, which is not extremal, because

$$\exp(-\beta\mu\sigma_2) = \exp(-\beta\mu)\mathbf{1} \otimes \begin{vmatrix} 1 & 0 \\ 0 & 0 \end{vmatrix} + \exp(\beta\mu)\mathbf{1} \otimes \begin{vmatrix} 0 & 0 \\ 0 & 1 \end{vmatrix},$$

and $\exp(-\beta\sigma_1) \otimes \left(\begin{smallmatrix} 1 & 0 \\ 0 & 0 \end{smallmatrix}\right)$ provides a KMS state.

Although adding $h^{(n)} = (1/n)\sigma_1 + c'\sigma_2$ to the Hamiltonian leads to the same time-evolution as $n \to \infty$, the KMS state is different. Only the extremal KMS states provide two-dimensional representations, for which this can not happen.

Infinite systems generically have the property known as

**Spatially Asymptotic Dynamical Stability** (3.3.7)

*Let $\mathscr{A}$ be a quasilocal algebra and $w$ be a locally normal* KMS *state on $\mathscr{A}$. The state $w$ is an extremal* KMS *state iff for each sequence $h^{(n)}$ of perturbations such that $\|h^{(n)}\|$ and $\|h_i^{(n)}\|$ are bounded in $n$ and $\tau^{h^{(n)}}(a) \to a_t$ for all $a \in \mathscr{A}$, the sequence $w^{(n)} \equiv w_{h^{(n)}} \to w$ converges in the weak-* sense to $w$.*

**Remarks** (3.3.8)

1. The assumption that $\mathscr{A}$ is quasilocal (1.3.3; 8) serves to guarantee the existence of suitable sequences $h^{(n)}$.
2. If $\mathscr{A}$ is also asymptotically Abelian in time, then the following propositions are equivalent for KMS states (recall Figure 27):

   (a) $w$ is an extremal KMS state;
   (b) $\pi_w$ is a factor;
   (c) $\lim_{t\to\infty} w(ab_t) = w(a)w(b)$;
   (d) $w_{h^{(n)}} \to w$ for all $h^{(n)}$ as described in (3.3.7).

**Proof**

1. If $w$ is extremal, then $w^{(n)} \to w$: By assumption $\|h_i^{(n)}\|$ are bounded uniformly in $n$, so the same is true of the norms of $R_{h^{(n)}}$. Since, moreover, $w(R_{h^{(n)}}) \geq \exp(-\|h^{(n)}\|)$,

$$\rho_n = \frac{R_{h^{(n)}}}{w(R_{h^{(n)}})}$$

is a bounded sequence of operators. Bounded sequences of operators are weakly relatively compact ([33], VI; 9.6), and the set of states is weak-* compact (III: 2.1.23; 2), so there is a subsequence $h^{(k)}$, $k \in \mathbb{I} \subset \mathbb{N}$, such that $\bar{w} = \lim w^{(k)}$ and $\rho = \lim \rho_k$ exist, and $\bar{w}(a) = w(a\rho)$.
The automorphisms converge by assumption, and by Problem (3.2.17; 1) $\bar{w}$ is $\tau$-KMS. But this means that $\rho$ belongs not only to $\pi_w(\mathscr{A})''$ (by construction), but also to $\pi_w(\mathscr{A})'$ and thus belongs to the center:

$$w(a\rho b) = w(b_{-i}a\rho) = \bar{w}(b_{-i}a) = \bar{w}(ab) = w(ab\rho)$$

and

$$w(a\rho bc) = w(ab\rho c).$$

However, $\pi_w$ is a factor, so $\rho = 1$, and since $\bar{w} = w$ is the only point of accumulation it is the limit of $w^{(n)}$.

2. Suppose now that $w$ is not extremal. There is a nontrivial invariant element $z = z^*$ in the center of $\pi_w(\mathscr{A})''$. By Kaplansky's theorem [4] the unit ball of $\mathscr{A}$ is strongly dense in the unit ball of $\mathscr{A}''$, so $z$ belongs to the closure of a bounded set of self-adjoint operators $h$ of $\mathscr{A}$. Because of the locality assumption the closure of $\mathscr{A}_\Lambda|\Omega\rangle$ is a separable subspace of

$$\mathscr{H} = \overline{\mathscr{A}|\Omega\rangle} = \overline{\bigcup_n \mathscr{A}_{\Lambda(n)}|\Omega\rangle} \qquad (\Lambda(n) \to \mathbb{R}^3),$$

so $\mathscr{H}$ is also separable. As a consequence the strong topology on bounded sets of operators is metrizable, so $z$ is actually the limit of some sequence $h^{(n)}$ in $\bigcup_n \mathscr{A}$. According to (3.3.4) $\tau_t^h$ converges to $\tau_t^z = \tau_t^0$. As in (3.2.6(v)) $\rho_n$ can be constructed with the $h^{(n)}(f)$, as they converge to $z_t = z(f) = z$, just like $h_t^{(n)}(f)$ and $h_{is}^{(n)}(f)$. By the dominated convergence theorem it follows that

$$\lim_{n\to\infty} R_{h^{(n)}(f)} = R_z = \exp(-z),$$

and therefore

$$\lim w_{h^{(n)}(f)}(a) = \frac{w(\exp(-z)a)}{w(\exp(-z))}$$

is a KMS state different from $w$. □

The next topic is that of stability properties that can distinguish the extremal KMS states from other stationary states giving rise to factors. As shown by (3.3.4), if there is an extremal KMS state, then for all $h \in \mathscr{A}$ there exists a state that is stationary under the time-evolution including $h$ as a perturbation, and which transforms continuously into the unperturbed state as $h \to 0$. It is not obvious that such a "linear-response theory" is possible. In fact, we learned (I, §3.3) that even in classical physics there are constants of motion that are not continuous in a parameter of the Hamiltonian. A density in phase space that is a function of such a constant will be unstable when perturbed, no matter by how little. This phenomenon is illustrated in quantum mechanics by the trivial

**Example** (3.3.9)

$\mathscr{H} = \mathbb{C}^2, H = 0 \in \mathscr{B}(\mathbb{C}^2)$. Every density matrix $\rho$ corresponds to a stationary state, but with the perturbation $h = \mathbf{n} \cdot \boldsymbol{\sigma}$ the only stationary density matrices are $\rho = \mathbf{1}/2 + \lambda \mathbf{n} \cdot \boldsymbol{\sigma}$, $\lambda < |\mathbf{n}|/2$. This shows that only the density matrix $\rho = \mathbf{1}/2$ goes continuously into a density matrix that is stationary under all possible perturbed time-evolutions.

The example illustrates that only density matrices of the form $f(H)$, which are proportional to the identity in each degeneracy space of $H$, adapt themselves well to arbitrary perturbations. Despite the possibility of diagonalizing any stationary density matrix simultaneously with $H$, there is no telling from stationariness alone how it might vary within a degeneracy space. A requirement that two independent systems be stable would impose an additional restriction on the function $f$ such that $w = f(H)$. The existence of two subsystems shows up mathematically as a tensor product, so if $H = H_1 \otimes \mathbf{1} + \mathbf{1} \otimes H_2$, then we would require that $f(H_1 \otimes \mathbf{1} + \mathbf{1} \otimes H_2) = f(H_1) \otimes f(H_2)$. Since $H_1$ and $H_2$ commute, both $H_i$ may be regarded as ordinary numbers in their common spectral representation. Since the only reasonable functions satisfying $f(x + y) = f(x)f(y)$ are of the form $f(x) = \exp(-\beta x)$, we are led to the canonical density matrix, if the $H_i$ may have arbitrary real spectral values. Since our infinite systems are asymptotically Abelian with respect to translations, and thus come to resemble tensor products of independent systems, it is a reasonable expectation that the condition of stability for such systems characterizes the KMS states. It will now be seen that this is the case, given some assumptions.

**Local Dynamical Stability** (3.3.10)

Suppose that the algebra $\mathscr{A}$ is asymptotically Abelian with respect to $\tau^0$, and let $w$ be a stationary factor state, and hence mixing. The question is whether for any perturbed automorphism $\tau^h$ it is possible for there to be a unique state $w_h$ that is invariant under $\tau^h$ and turns into $w$ as $h \to 0$. The states

$$w_{\pm} = \lim_{t \to \pm\infty} w \circ \tau_t^h$$

are reasonable candidates for $w_h$. If the limits exist, they would be invariant under $\tau^h$, and the uniqueness of $w_h$ means that the limits are equal. If $\tau^h$ is expanded as in (3.3.4) and we use the invariance of $w$ under $\tau^0$, we obtain the

**Stability Condition to First Order in** $h$ (3.3.11)

If an invariant factor state $w$ on an algebra $\mathscr{A}$ asymptotically Abelian in time is stable against arbitrary perturbations in the sense stated above, then for all $h$ and $a \in \mathscr{A}$,

$$\int_{-\infty}^{\infty} dt w([h, a_t]) = 0.$$

**Remarks** (3.3.12)

1. The assumption that $h \in \mathscr{A}$ means that we consider only local perturbations. The requirement that $\mathscr{A}$ be asymptotically Abelian makes the commutator $[h, a_t]$ vanish as $t \to \pm\infty$. Condition (3.3.11) requires, roughly speaking, that $w(i[h, a_t])$ is equally often positive and negative.
2. The physical significance of (3.3.11) is that to first order in $h$ the scattering transformation is the identity in the representation $\pi_w$. This can be interpreted as meaning that $w$ is a locally perturbed equilibrium state with respect to the time-automorphism $\tau^h$ and should become the equilibrium state as $t \to \pm\infty$, so there is no net change between $t = -\infty$ and $t = +\infty$. In the kinetic theory of gases this is reflected in the argument that collisions do not alter the equilibrium distribution.

Let us introduce the abbreviations

$$F_{ab}(t) = w(ba_t) - w(a)w(b)$$

and

$$G_{ab}(t) = w(a_t b) - w(a)w(b) \tag{3.3.13}$$

in order to exploit (3.3.11) more fully.

**Consequences for the Correlation Functions**

Condition (3.3.11) makes

$$\int_{-\infty}^{\infty} dt(F_{ab}(t) - G_{ab}(t)) = 0.$$

Under the assumptions of (3.3.10) we know that $F$ and $G$ tend to zero as $t \to \pm\infty$. In order to ensure that this integral and others to follow make sense, it will be assumed that the correlation functions $F$ and $G$ are integrable in time from $-\infty$ to $+\infty$, at least for a dense set $\mathscr{S} \subset \mathscr{A}$. Since they are bounded, they belong to all $L^p(\mathbb{R})$ for $1 \leq p \leq \infty$. The assumption holds, for example, for free fermions. It will also be assumed that the higher correlation functions decrease rapidly enough for elements of $\mathscr{S}$ that integrals and limits may be interchanged.

If the state is a factor state, then as $u \to \pm\infty$, $w(ab_u c_t d_{t+u} - c_t d_{t+u} ab_u)$ tends to $w(ac_t)w(b d_t) - w(c_t a)w(d_t b)$. Therefore

$$\int_{-\infty}^{\infty} dt(F_{ca}(t)F_{db}(t) - G_{ca}(t)G_{db}(t)) = 0$$

for all $a, b, c,$ and $d \in \mathscr{S}$. Similarly, from considering what happens to $w([ab_u c_v, d_t e_{u+t} f_{v+t}])$ as $u \to \infty$ and as $v \to \infty$,

$$\int_{-\infty}^{\infty} dt(F_{da}(t)F_{cf}(t)F_{be}(t) - G_{da}(t)G_{cf}(t)G_{be}(t)) = 0$$

for all $a, b, c, d, e,$ and $f \in \mathscr{S}$. Because $F$ and $G$ belong to $L^1$, their Fourier transforms $\tilde{F}$ and $\tilde{G}$ exist and are continuous. Then if $a, b, c, d, e,$ and $f \in \mathscr{S}$, the last three equations imply that

$$\tilde{F}_{ab}(0) = \tilde{G}_{ab}(0),$$

$$\int dE\tilde{F}_{ab}(E)\tilde{F}_{cd}(-E) = \int dE\tilde{G}_{ab}(E)\tilde{G}_{cd}(-E),$$

and

$$\int dE_1\, dE_2\, \tilde{F}_{ab}(E)\tilde{F}_{cd}(E' - E)\tilde{F}_{ef}(-E') = \int dE_1\, dE_2\, \tilde{G}_{ab}(E)\tilde{G}_{cd}(E' - E)\tilde{G}_{ef}(-E'). \tag{3.3.14}$$

We shall now see that these equations imply the KMS condition.

In order to arrive at the KMS condition in Fourier-transformed space, $\tilde{F}_{ab}(E) = \exp(\beta E)\tilde{G}_{ab}(E)$, information about the supports of $\tilde{F}$ and $\tilde{G}$ is needed. It is at least clear that they are contained in the spectrum of $H$: Let $a_t = U_t^{-1} a U_t$, $U_t = \exp(-iHt)$, writing $H$ as in (1.3.5) in the representation determined by $w$. Then

$$w(ba_t) = \langle b^*\Omega | U_t^{-1} | a\Omega\rangle,$$

so if $E \neq 0$, then

$$\tilde{F}_{ab}(E) = \tilde{F}_{b^*a^*}(E)^* = \tilde{G}_{ba}(-E) = \langle b^*\Omega | \delta(E - H) a\Omega\rangle. \tag{3.3.15}$$

This expression is to be interpreted in the spectral representation of $H$, in which the functions depend continuously on $E$ when $a$ and $b \in \mathscr{S}$.

In order to draw more far-reaching conclusions from these relationships, more information is needed about the energy spectrum. It would simply be additive if the Harmiltonian were the tensor product of Hamiltonians of independent systems: If $H_1$ and $H_2$ have eigenvalues $e_n^{(1)}$ and $e_n^{(2)}$, then

$H^{(1)} \otimes \mathbf{1} + \mathbf{1} \otimes H^{(2)}$ has eigenvalues $e_n^{(1)} + e_m^{(2)}$. This fact generalizes to an infinite system provided that the system is asymptotically Abelian with respect to an automorphism, such as the translations, that commutes with the time-evolution.

**The Additivity of the Spectrum of** $H$ (3.3.16)

*Let $H$ generate a time-evolution $\tau$ on a factor state $w$, and suppose that the system is asymptotically Abelian with respect to an automorphism $\sigma$ such that $[\sigma, \tau] = 0$ and $w \circ \sigma = w$. If $H$ has the spectral values $E_1$ and $E_2$, then $E_1 + E_2$ also belongs to the spectrum of $H$.*

**Proof**

Given any neighborhoods $U_i$ of $E_i$, $i = 1, 2$, by assumption there exist $f_i$ such that

$$a_{f_i} \Omega\rangle \equiv \int_{-\infty}^{\infty} dt a_t f_i(t)|\Omega\rangle \neq 0,$$

where the Fourier transforms $\tilde{f}_i$ have their supports in $U_i$. Since by Property (3.1.18; 4) $\|\sigma_s(a_{f_1})a_{f_2}|\Omega\rangle\|^2$ approaches

$$\|a_{f_2}|\Omega\rangle\|^2 \|a_{f_1}|\Omega\rangle\|^2 \neq 0$$

as $s \to \infty$, there must be a sufficiently large $s$ that this vector is nonzero. Since the vector is supported in $E_1 + E_2 + U_1 + U_2$ in the spectral representation of $H$ for all $s$, there are spectral values in every neighborhood of $E_1 + E_2$. Since the spectrum is closed, $E_1 + E_2$ itself belongs to the spectrum. □

**Remark** (3.3.17)

If the system is asymptotically Abelian with respect to $\tau$, then of course it is possible to take $\tau = \sigma$. Since $w$ provides a factor, according to Table I in this case $|\Omega\rangle$ is the only eigenvector, and $H$ has no eigenvalues other than 0. Since the spectrum is additive, it is either $0 \cup [\pm c, \pm\infty)$ for some $c \geq 0$, or else $(-\infty, \infty)$. In the first case there is a ground state; we shall be concerned only with the second possibility.

**Derivation of the KMS Condition** (3.3.18)

Let $E_0$ be in the spectrum of $H$ and $f$ be a function of the kind described in (3.2.6(v)) with $\tilde{f}(E_0) = 1$, supp $\tilde{f} \subset I \supset E_0$. Then $U_f \equiv \int dt f(t) U_t \neq 0$, and there exists an $a \in \mathscr{S}$ such that $U_f a\Omega = a_f \Omega \neq 0$. The operator $a_f$ belongs to $\mathscr{S}$ whenever $a$ does, and the functions $\tilde{F}$ and $\tilde{G}$ constructed with $a_f$ are also supported in $I$, because

$$\tilde{F}_{a\,b}(E) = \tilde{f}(E)\tilde{F}_{ab}(E),$$
$$\tilde{G}_{a_f b}(E) = \tilde{f}(E)\tilde{G}_{ab}(E).$$

Let $b = a_f^*$ and shrink $I$ down to $E_0$; this makes $\tilde{F}$ and $\tilde{G}$ proportional to $\delta(E - E_0)$. If we normalize so that

$$\int_{-\infty}^{\infty} dE\tilde{F}_{a_f a_f^*}(E) = w(a_f^* a_f) - |w(a_f)|^2 = 1,$$

and if

$$\int_{-\infty}^{\infty} dE\tilde{G}_{a_f a_f^*}(E) = w(a_f a_f^*) - |w(a_f)|^2 \geq 0$$

converges to some $\Phi \in \mathbb{R}^+$ (possibly after passage to some subsequence), then, because of the continuity of $\tilde{F}$ and $\tilde{G}$, (3.3.14) yields

$$\tilde{F}_{cd}(E_0) = \Phi\tilde{G}_{cd}(E_0) \quad \text{for all } c \text{ and } d \in \mathscr{S}.$$

This also proves that $\Phi$ may not be either 0 or $\infty$. Since this is true for all $E_0 \in \operatorname{Sp} H = \mathbb{R}$, there exists a universal function $\Phi(E)$ such that

$$\tilde{F}_{cd}(E) = \Phi(E)\tilde{G}_{cd}(E).$$

It follows from (3.3.15) that

$$\Phi(-E) = \Phi(E)^{-1} = \Phi^*(-E),$$

and the functional form then follows from the last equation of (3.3.14):

$$\int dE\, dE'(1 - \Phi(E)\Phi(E' - E)\Phi(-E'))\tilde{G}_{ab}(E)\tilde{G}_{cd}(E' - E)\tilde{G}_{ef}(-E') = 0$$

implies that

$$\Phi(E)\Phi(E' - E)\Phi(-E') = 1 \quad \text{for all } E \text{ and } E' \in \mathbb{R}.$$

Because of the equation derived above this,

$$\Phi(E)\Phi(-E') = \Phi(E - E'),$$

and since $\Phi$ is continuous it therefore has the functional form

$$\Phi(E) = \exp(\beta E) \quad \text{for some } \beta \in \mathbb{R}.$$

This shows the KMS condition for the dense set $\mathscr{S}$. However, since it can be written with the aid of (3.3.15) in the form

$$\langle b^*\Omega | f(-H)a\Omega\rangle = \langle a^*\Omega | f(H)\exp(-\beta H)b\Omega\rangle$$

for any bounded, continuous $f(H)$, it clearly suffices to derive it on a dense set.

In sum, the foregoing argument has shown the

**Equivalence of Dynamical Stability and the KMS Condition** (3.3.19)

*Suppose that the algebra $\mathscr{A}$ is asymptotically Abelian with respect to the time-evolution and that w is a stationary state creating a factor representation. If for all $h \in \mathscr{A}$ there exists a normal state $w_1$ for $\pi_w(\mathscr{A})''$ to first order in h, such that w and $w_1$ are both stationary to first order under the perturbed time-evolution, and if w has an absolutely integrable correlation function, then either w is a* KMS *state, or else the spectrum of H is $\{0\} \cup [\pm c, \pm\infty)$, in which case w is the ground state.*

**Remarks** (3.3.20)

1. It does not follow from this argument that $\beta > 0$. This fact did not even emerge from our argument with the tensor product of finite systems.
2. It is hard to tell how much the result suffers from the sharpening of the hypothesis of asymptotic commutativity. All the hypotheses are satisfied by a system of free fermions, but with a Coulomb interaction it is not even known if they hold in weakened forms. To a certain extent our assumptions about decrease at infinity and the interchangeability of limits belong to the realm of unproven hopes.
3. This shows that stability to first order in $h$ implies KMS. Conversely, we have seen that KMS implies stability to every order in $h$, which means that the higher orders contribute no new information in this respect.

Whereas all the perturbations considered until now have been independent of time, we shall now turn our attention to perturbations $h(t)$ depending explicitly on time; they would be due to interference from outside the system. The time-evolution will not have the group property, but it will still be a one-parameter family of automorphisms. Let us, as usual, start by studying finite systems, for which the automorphisms are implemented by the unitary transformations

$$U_t = T\exp\left[-i\int_0^t dt'(H + h(t'))\right] \tag{3.3.21}$$

(cf. (III: 3.3.6)).

The most important quality of a passive state for our purposes will be that a system in a passive state will have gained energy when the perturbation has been switched off.

**The Passivity of a State** (3.3.22)

Let us suppose that a finite system evolves under the influence of $H + h(t)$, where by definition $h(0) = h(\tau) = 0$. The Hamiltonian generates a unitary time-evolution (3.3.21), so the change in energy from $t = 0$ to $t = \tau$ in the state $w$ is given by $\mathrm{Tr}\,\rho(U_\tau H U_\tau^{-1} - H)$. A state is said to be **passive** if the change in energy is positive for all self-adjoint $h \in \mathscr{B}(\mathscr{H})$, in which case $\mathrm{Tr}\,\rho U H U^{-1} \geq \mathrm{Tr}\,\rho H$ for all $U = U^{*-1} \in \mathscr{B}(\mathscr{H})$.

**Examples** (3.3.23)

1. The canonical density matrix. Let $\rho = \exp(-\beta H)/\mathrm{Tr}\exp(-\beta H)$ and $\sigma = U^{-1}\rho U$. From (2.2.22(ii)) we know that
$$0 \leq \mathrm{Tr}\,\sigma(\ln \sigma - \ln \rho) = \mathrm{Tr}(\rho - \sigma)\ln \rho = -\beta\,\mathrm{Tr}(\rho - U^{-1}\rho U)H,$$
so the system is passive.
2. Negative temperatures. Let $\rho$ be as above, but $\beta < 0$. In order for $\mathrm{Tr}\exp(-\beta H)$ to be finite, $H$ must be bounded from above; this would be realistic for a spin system. The inequality is then reversed, $\mathrm{Tr}(\rho - U^{-1}\rho U)H > 0$, so the system is not passive.

**Remarks** (3.3.24)

1. If it is desired to keep the energy $E = F + TS$ from increasing, the best tactic is to keep $S$ constant (when $T > 0$). Our unitary time-evolution manages this automatically, and so the change in the energy $E$ equals the change in the free energy $F$. Since the free energy is minimized with the canonical density matrix $\rho$, in the state $\rho$ the only possibility is for $E$ to increase, so $\rho$ is passive.
2. Obviously, passivity requires the states of lower energy to be more densely occupied, so that the system is ready to gain energy. This is not the case when $\beta < 0$, in which circumstances the system would prefer to give energy away.

**The General Form of Passive Density Matrices for Finite Systems** (3.3.25)

*A density matrix $\rho$ on a finite system corresponds to a passive state if and only if*

(i) $[\rho, H] = 0$; *and*
(ii) *if $\rho_i$ and $e_i$ designate respectively the ordered eigenvalues of $\rho$ and $H$, then*
$$(e_i - e_k)(\rho_i - \rho_k) \leq 0.$$

**Remarks** (3.3.26)

1. The condition on the eigenvalues means that if the $k$th eigenvalue of $H$ is greater than the $i$th, then the $k$th eigenvalue of $\rho$ must be less than or equal to the $i$th. However, it is not necessary for $\rho$ to be simply a function

of $H$, since in a degeneracy space for which $e_i = e_k$ it may happen that $\rho_i \neq \rho_k$.

2. The physical implication of the monotony is that lower-lying states are more densely occupied. On the other hand it implies nothing for the values of $\rho$ where $H$ does not vary:

$$H = \begin{pmatrix} 0 & & \\ & 0 & \\ & & 1 \end{pmatrix}, \quad \rho = \begin{pmatrix} \frac{1}{4} & & \\ & \frac{1}{2} & \\ & & \frac{1}{4} \end{pmatrix},$$

is passive.

**Proof**

(i) and (ii) $\Rightarrow$ passive $\Leftrightarrow \operatorname{Tr} \rho H \leq \operatorname{Tr} \rho U H U^{-1}$.
Let $U$ be given in a matrix representation in the common eigenvectors of $H$ and $\rho$ as $U_{ik}$. The matrix $|U_{ik}|^2$ is doubly stochastic and therefore a convex combination of permutation matrices or a limit of such matrices (cf. (2.1.11; 4)). For any such matrix,

$$\operatorname{Tr} \rho U H U^{-1} = \sum_{i,k} e_i \rho_k \|U_{ik}\|^2 = \sum_P c_P \sum_i e_i \rho_{P_i},$$

where $\sum_P c_P = 1$, $c_P \geq 0$, and $\{P_i\}$ is a permutation of the $i \in \mathbb{Z}^+$. If $e_i < e_k$ implies that $\rho_i \geq \rho_k$, then for any permutation, $\sum_i e_i \rho_{P_i} \geq \sum_i e_i \rho_i = \operatorname{Tr} \rho H$.
Passive $\Rightarrow$ (i) and (ii). Suppose that $\operatorname{Tr} \rho U H U^{-1}$ has its minimum at $U = \mathbf{1}$, and write $U = \mathbf{1} + M_1 + M_2 + \cdots$, where $\|M_k\| < \varepsilon^k$ for sufficiently small $\varepsilon$. Then $\operatorname{Tr} \rho U H U^{-1} = \operatorname{Tr} \rho H + \operatorname{Tr}([H, \rho] M_1) + O(\varepsilon^2)$. The operator $M_1$ only needs to satisfy the condition that $M_1^* = -M_1$, and since $[\rho, H]$ is anti-Hermitian, it must equal zero, as otherwise the energy could be lowered. In order to prove (ii), choose $U$ to have the form

$$U = \begin{pmatrix} \cos \varphi & \sin \varphi \\ -\sin \varphi & \cos \varphi \end{pmatrix}$$

on the subspace spanned by $v_i$ and $v_k$, the eigenvectors with eigenvalues $e_i, \rho_i$ and $e_k, \rho_k$. Then

$$\operatorname{Tr} \rho U H U^{-1} - \operatorname{Tr} \rho H = -(e_i - e_k)(\rho_i - \rho_k) \sin^2 \varphi,$$

which is positive only if $(e_i - e_k)(\rho_i - \rho_k) \leq 0$. □

In order to progress beyond the monotonic property to the statement that the function is exponential we must investigate infinite systems. We may either construct the infinite system by taking tensor products of copies of finite systems or go directly to the analysis of some asymptotically Abelian system. As before, the limiting case $\beta = \infty$, i.e., the ground state, would require a special treatment, which we shall not go into. Assuming therefore that $\beta$ is finite, we can state the main proposition on the

**Passivity of Infinite Systems** (3.3.27)

*Within the set of faithful factor states w on a $C^*$ algebra with a time-automorphism $\tau$ and another automorphism commuting with $\tau$ and under which w is invariant and asymptotically Abelian, the passive states are precisely the KMS states, for any $\beta \geq 0$.*

**Remarks** (3.3.28)

1. Translations of a homogeneous infinite system commute with the time-evolution. Since the local field algebra is asymptotically Abelian with respect to translations, this theorem can be used even if it is not known whether the time-evolution is asymptotically Abelian.
2. The sign of $\beta$ is fixed by passivity, though of course its value is not.
3. To ensure that $H$ is well-defined, assume that the time-evolution can be represented unitarily; then passivity is equivalent to the property that $w(U^{-1}HU - H) \geq 0$ for all unitary $U \in \mathscr{A}$.
4. Since the condition for passivity is linear in $w$, the passive states form a convex set. Passivity does not single out the extremal KMS states. We shall consider only factor states, which can not be decomposed further, as shown in §3.1.

**Proof**

Passive ⇒ KMS. If the condition of passivity for an infinite system is written as $w(UHU^{-1}) \geq w(H)$, and we choose $U = \exp(i\varepsilon a)$ for $a$ self-adjoint, then the first two terms of the expansion in powers of $\varepsilon$ lead to

(i) $w([a, H]) = 0$ for all $a \in \mathscr{A}$, and
(ii) $w([a, [H, a]]) \geq 0$ for all $a \in \mathscr{A}$.

Equation (i) means that $(\partial/\partial t)w(a_t) = 0$, so $w$ is stationary. In order to deduce the KMS condition from (ii) we employ the modular automorphism of $w$—call its generator $\bar{H}$. The KMS condition with respect to $\bar{H}$ can be used to write (ii) as

$$\begin{aligned} 0 &\leq \langle \Omega | 2aHa - Ha^2 - a^2H | \Omega \rangle \\ &= \langle \Omega | 2aHa - a\exp(-\bar{H})Ha - aH\exp(-\bar{H})a | \Omega \rangle \\ &= 2\langle \Omega | aH(\mathbf{1} - \exp(-\bar{H}))a | \Omega \rangle. \end{aligned}$$

In the last step we used the fact that $[H, \bar{H}] = 0$, in accordance with our assumption. Since the inequality holds for all $a = a^* \in \mathscr{A}$, it follows that $H(\mathbf{1} - \exp(-\bar{H})) \geq 0$. This means that in the common spectral representation of $H$ and $\bar{H}$ the spectrum is restricted to the hatched region of the $(H, \bar{H})$-plane shown in Figure 28. Now the existence of the commuting,

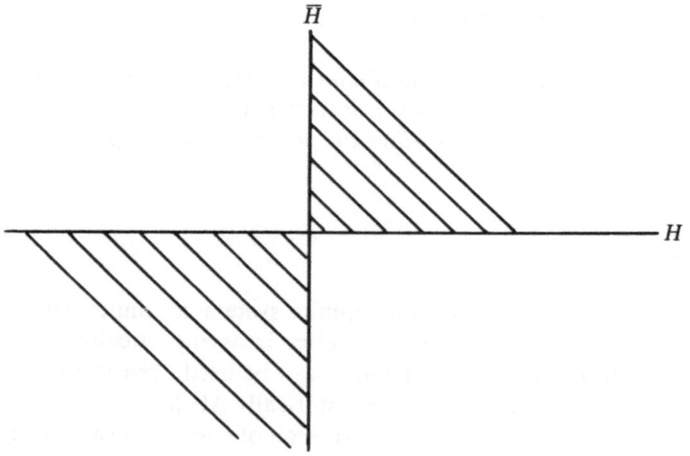


Figure 28 Possible location of the spectra of $H$ and $\bar{H}$.

asymptotically Abelian automorphisms comes into play. According to (3.3.16), this implies that the spectrum is additive, i.e., if $(h_1, \bar{h}_1)$ and $(h_2, \bar{h}_2)$ are in the spectrum, then so is $(h_1 + h_2, \bar{h}_1 + \bar{h}_2)$. As a consequence the spectrum can at most be on a line through (0, 0), so $\bar{H} = \beta H$ for some $\beta > 0$.
KMS $\Rightarrow$ passive. Since $x \geq 1 - \exp(-x)$,

$$w(UHU^{-1}) \geq w(UU^{-1}) - w(U\exp(-H)U^{-1})$$
$$= w(UU^{-1}) - w(U^{-1}U) = 0. \qquad \square$$

**Remarks** (3.3.29)

1. The last inequality proved above is only the first of a whole family of inequalities that the expectation values in KMS states satisfy, and which completely characterize the KMS states [24]. They generalize trace inequalities, which are not directly applicable to infinite systems, since $\exp(-\beta H)$ is not trace-class.
2. Example (3.3.23; 1) showed that for finite systems, passivity follows from thermodynamic stability, or, in other words, from the minimum property of the free energy. This fact generalizes to infinite systems, for many of which the implication goes both ways, KMS $\Leftrightarrow$ thermodynamic stability, for instance for lattice systems with finite-range interactions. For these systems KMS is equivalent to global thermodynamic stability, provided that only translation-invariant states are considered, and that the free energy is interpreted as the free-energy density. However, for systems with long-range forces there exist KMS states that do not minimize the free energy; they are instead metastable, minimizing the free energy only one some reduced set of comparison states. Since the free energy is a convex functional on the states, it can not have a relative minimum on the set of all states that fails to be absolute.

3. The state $w_{\beta_1} \otimes w_{\beta_2}$ of two independent systems at different temperatures $T_1 > T_2$ is KMS with respect to the automorphism generated by $\beta_1 H_1 + \beta_2 H_2$. A perturbation $h(t)$ can cause the temperatures to equalize, and it may happen that the first system will have given up a positive amount of energy $\Delta E_1 \equiv E_1(0) - E_1(\tau) > 0$ by the end of the period. However, because the state is passive, $\beta_1 \Delta E_1 + \beta_2 \Delta E_2 \leq 0$, and the change in the total energy $\Delta E = \Delta E_1 + \Delta E_2$ is bounded by $\Delta E/\Delta E_1 \leq (T_1 - T_2)/T_1$. Since the total entropy remains constant under the unitary time-evolution, $\Delta E$ is the amount of energy provided by the total system, and this inequality is Carnot's classical bound on the thermal efficiency.

Another way to characterize the KMS states of an infinite system is known as reservoir stability, and it further justifies the physical interpretation of $\beta$ as the reciprocal of the temperature. In outline it means that the KMS states are precisely the states that are suitable for thermal reservoirs, allowing the temperature $1/\beta$ to be defined. A more careful formulation states that if the reservoir is coupled to a finite system in the canonical state $w$, then in the weak-coupling limit $w$ is invariant under the resulting semigroups (cf. (3.1.12)) for a reasonable class of couplings iff the reservoir is in a KMS state [24].

**Problems** (3.3.30)

1. Show that $w(R_h^* a) = w(aR_h) = w(S_h^* a S_h)$.
2. Estimate the length of time for which the "linear-response theory" remains valid; i.e., estimate

$$\left\| \tau_t^h(a) - a_t - i \int_0^t dt_1 [h_{t_1}, a_t] \right\|$$

3. Use the methods of §2.1 to conclude from $e_i > e_j \Rightarrow \rho_i \leq \rho_j$ that

$$\sum_i e_i \rho_i \leq \sum_i e_i \rho_{P_i}$$

for every permutation $P$.

**Solutions** (3.3.31)

1. Since $H$ exists in the GNS representation with $w$, Equations (3.3.1) are applicable. The invariance of $\Omega$ holds also for complex $z$,

$$\exp(zH)|\Omega\rangle = |\Omega\rangle, \qquad R_h|\Omega\rangle = \exp(-H - h)|\Omega\rangle.$$

Now use the KMS condition for $w$ in the form $w(ab) = \langle \Omega | b \exp(-H) a | \Omega \rangle$:

$$w(aR_h) = \langle \Omega | R_h \exp(-H) a | \Omega \rangle = \langle \Omega | \exp(-H - h) a | \Omega \rangle = w(R_h^* a).$$

It is also true in this representation that $S_h \exp(-H) S_h^* = \exp(-H - h)$, so

$$w(R_h^* a) = \langle \Omega | S_h \exp(-H) S_h^* a | \Omega \rangle = w(S_h^* a S_h).$$

2. Apply Taylor's formula $\|f(\alpha) - f(0) - \alpha f'(0)\| \leq \|\int_0^1 d\zeta(1-\zeta)f''(\alpha\zeta)a^2\|$ to $f$: $[0, 1] \to \mathscr{B}(\mathscr{H})$, $\alpha \to \tau_t^{\alpha h}(a)$. According to (3.3.4),

$$\left\| \frac{\partial^2}{\partial\alpha^2} \tau_t^{\alpha h}(a) \right|_{\alpha=0} \left\| = \left\| 2 \int_0^t dt_2 \int_0^{t_2} dt_1 [h_{t_1}, [h_{t_2}, a_t]] \right\| \leq 4t^2 \|h\|^2 \|a\|.$$

This is also true when $\alpha \in [0, 1]$; the only change when $\alpha > 0$ is that the time-evolution $a, h \to a_t, h_t$ becomes $a, h \to \tau_t^{\alpha h}(a), \tau_t^{\alpha h}(h)$, which does not affect the norms. Consequently the answer is that $\|\cdots\| \leq (2t\|h\|)^2 \|a\|/2$. Recall that if $\|h\|$ is on the order of a Rydberg, then $t\|h\| \ll 1$ when $t \ll 10^{-15}$ sec. Therefore this *a priori* estimate guarantees only that the linear approximation remains valid for times on the atomic scale, and not for times measured in seconds. To go further would require knowing that the commutators go to zero for longer than macroscopic times.

3. Order $e_i$ and $\rho_i$; then

$$e_1\rho_1 + e_2\rho_2 + e_3\rho_3 + \cdots = (e_1 - e_2)\rho_1 + (e_2 - e_3)(\rho_1 + \rho_2) + (e_3 - e_4)(\rho_1 + \rho_2 + \rho_3) + \cdots.$$

All the summands are positive, and permuting the $\rho_i$ can at most make the summands larger.

# Physical Systems 4

## 4.1 Thomas–Fermi Theory

*Among the best examples of large quantum systems are atoms and molecules with highly charged nuclei. Classical features arise in the limit $Z \to \infty$, $N \to \infty$, except that the Fermi statistics continue to have an important effect.*

Matter around us and within us consists of electrons and atomic nuclei, which are governed by the laws of quantum mechanics. Relativistic effects arise only in the fine details (cf. III, §1), so the forces of primary relevance are electrostatic and, for cosmic bodies, gravistatic (nonrelativistic). Moreover, the precise nature of the atomic nuclei is of little consequence on the macroscopic scale, so they can be considered as point charges. In order to understand the gross features of matter we shall study a Hamiltonian

$$H_{\text{mat}} = \sum_{i=1}^{M} \frac{|\mathbf{p}_i|^2}{2m_i} + \sum_{i>j} \frac{(e_i e_j - \kappa m_i m_j)}{|\mathbf{x}_i - \mathbf{x}_j|} \tag{4.1.1}$$

for ordinary matter. The first important issue to confront is that of why macroscopic bodies behave classically; in what sense is the thermodynamic limit $N \to \infty$ equivalent to the classical limit $\hbar \to 0$? There are a variety of ways to pass to the limit $N \to \infty$. In this section we begin by letting the nuclear charge $Z$ and the nuclear masses both tend to infinity, while continuing to neglect gravity. This will permit a rather explicit mathematical treatment, as the action is determined by an average field, and the single-particle model becomes exact. The same will be true in §4.2 when we deal with cosmic bodies, for which gravitation predominates. However, macroscopic bodies on the scale of humans are far from these limits: nuclear charges are for the most part small, and yet gravitation is of little importance. In this intermediate range of normal matter it would be too much to hope for an explicit solution. Section 4.3 will discuss this case, but the results will be confined to general existence theorems and rather crude bounds on the values of observables of physical interest.

Let us consider now what happens to electrons in the field of fixed point charges. In order not to be distracted from the most important facts by physical constants, we shall use units in which $\hbar = 2m = e = k = 1$, so that (4.1.1) becomes

**The Hamiltonian for Normal Matter** (4.1.2)

$$H_N = \sum_{i=1}^{N} |\mathbf{p}_i|^2 - \sum_{i=1}^{N} \sum_{k=1}^{M} \frac{Z_k}{|\mathbf{x}_i - \mathbf{X}_k|} + \sum_{i>j} \frac{1}{|\mathbf{x}_i - \mathbf{x}_j|} + \sum_{k>l} \frac{Z_k Z_l}{|\mathbf{X}_k - \mathbf{X}_l|} + \sum_{i=1}^{N} W(\mathbf{x}_i).$$

**Remarks** (4.1.3)

1. The notation follows that of (III: 4.6.9), that is, $\mathbf{x}_i$ and $\mathbf{p}_i$ are the position and momentum of the $i$th electron, $\mathbf{X}_k$ and $Z_k$ are the position and charge of the $k$th fixed nucleus, $N$ is the number of electrons, and $M$ is the number of nuclei.
2. The Hamiltonian $H$ operates on an $n$-fold antisymmetrized tensor product of $L^2(\mathbb{R}^3) \otimes \mathbb{C}^2$ = configuration space $\otimes$ spin of a given electron. The nuclear coordinates $\mathbf{X}_i$ commute with everything, and are to be regarded as ordinary 3-vectors of numbers.
3. It is usually most convenient to study the many-particle system in the framework of the field algebra (1.3.2). If $a_\alpha(\mathbf{x})$, $\alpha = 1, 2$, denote the annihilation operators of electrons with spin up ($\alpha = 1$) and spin down ($\alpha = 2$), then (4.1.2) reads

$$H = \sum_\alpha \int d^3x \left[ \nabla a_\alpha^*(\mathbf{x}) \cdot \nabla a_\alpha(\mathbf{x}) + \left( \sum_{k=1}^{M} \frac{-Z_k}{|\mathbf{x} - \mathbf{X}_k|} + W(\mathbf{x}) \right) a_\alpha^*(\mathbf{x}) a_\alpha(\mathbf{x}) \right] + \sum_{\alpha,\beta} \frac{1}{2} \int d^3x \, d^3x' \, \frac{a_\alpha^*(\mathbf{x}) a_\beta^*(\mathbf{x}') a_\beta(\mathbf{x}') a_\alpha(\mathbf{x})}{|\mathbf{x} - \mathbf{x}'|} + \sum_{k>l} \frac{Z_k Z_l}{|\mathbf{X}_k - \mathbf{X}_l|}.$$

4. If the temperature is finite, then the attraction of the nuclei is not strong enough to prevent the electrons from escaping to infinity, and the system must be imagined confined to a box. The box can be represented by a potential $W$, adding a term $\sum_\alpha \int d^3x W(\mathbf{x}) a_\alpha^*(\mathbf{x}) a_\alpha(\mathbf{x})$ to $H$. The wall potential $W$ will be chosen to be the $v_L$ of (2.5.23).

Most interesting systems are approximately neutral, so $N$ is assumed to be about $\sum_{k=1}^{M} Z_k$. The thermodynamic limit $N \to \infty$ can consist either in

$M \to \infty$ or $Z_k \to \infty$. For the moment consider the latter case; the limit $M \to \infty$ will be studied in §4.3. The first step is to bound the grand canonical partition function in terms of the grand canonical partition function of a theory with free electrons in an external field. This means that the bounds of (III, §4.5) for the energies have to be generalized for arbitrarily complex systems at nonzero temperatures. After that we shall show that the upper and lower bounds coalesce (when properly scaled) as $Z_k \to \infty$, so the partition function can be calculated exactly in the thermodynamic limit. Finally, the limit of the grand canonical state will be analyzed.

**Upper Bounds for the Partition Function** (4.1.4)

These correspond to lower bounds for the Hamiltonian like those derived in (III: 4.5.20). The inequality (III: 4.5.24), though, is not well suited to our current purposes, and must be replaced with a variant, which will appear as a by-product of Thomas–Fermi theory in (4.1.46; 2). In it the Coulomb repulsion of the electrons is replaced by their energy in an external field:

$$\sum_{i>j=1}^{N} |\mathbf{x}_i - \mathbf{x}_j|^{-1} \geq \sum_{i=1}^{N} \int \frac{d^3x n(\mathbf{x})}{|\mathbf{x}_i - \mathbf{x}|} - \frac{1}{2} \int \frac{d^3x\, d^3x'}{|\mathbf{x} - \mathbf{x}'|} n(\mathbf{x})n(\mathbf{x}') - 3.68\gamma N$$
$$- \frac{3}{5\gamma} \int d^3x n^{5/3}(\mathbf{x})$$
$$\text{for all } \mathbf{x}_i \in \mathbb{R}^3, \quad \gamma > 0, \quad n \in L^1(\mathbb{R}^3) \cap L^{5/3}(\mathbb{R}^3). \qquad (4.1.5)$$

This yields a bound on the expression in (4.1.3; 3), which is quartic in the $a$'s, in terms of a quadratic expression,

$$\frac{1}{2} \sum_{\alpha, \beta} \int \frac{d^3x\, d^3x'}{|\mathbf{x} - \mathbf{x}'|} a_\alpha^*(\mathbf{x}) a_\beta^*(\mathbf{x}') a_\beta(\mathbf{x}') a_\alpha(\mathbf{x}) \geq \sum_\alpha \int d^3x a_\alpha^*(\mathbf{x}) a_\alpha(\mathbf{x})$$
$$\times \left[ \int \frac{d^3x' n(\mathbf{x}')}{|\mathbf{x} - \mathbf{x}'|} - 3.68\gamma \right] - \frac{1}{2} \int \frac{d^3x\, d^3x'}{|\mathbf{x} - \mathbf{x}'|} n(\mathbf{x}')n(\mathbf{x}) - \frac{3}{5\gamma} \int d^3x n^{5/3}(\mathbf{x}).$$

Consequently, $H$ is bounded by a

**Hamiltonian with an Effective Field** (4.1.6)

$$H - \mu N \geq H_n - C_n + \sum_{k>l} \frac{Z_k Z_l}{|\mathbf{X}_k - \mathbf{X}_l|},$$

where

$$H_n \equiv \sum_\alpha \int d^3x \Bigg\{ \nabla a_\alpha^*(\mathbf{x}) \cdot \nabla a_\alpha(\mathbf{x}) + a_\alpha^*(\mathbf{x}) a_\alpha(\mathbf{x})$$
$$\times \left[ - \sum_k \frac{Z_k}{|\mathbf{x} - \mathbf{X}_k|} + \int \frac{d^3x' n(\mathbf{x}')}{|\mathbf{x} - \mathbf{x}'|} + W(\mathbf{x}) - \mu - 3.68\gamma \right] \Bigg\}$$
$$- \frac{1}{2} \int \frac{d^3x\, d^3x'}{|\mathbf{x} - \mathbf{x}'|} n(\mathbf{x}) n(\mathbf{x}'),$$

and

$$C_n = \frac{3}{5\gamma} \int d^3x n^{5/3}(\mathbf{x}).$$

**Remarks** (4.1.7)

1. Although Inequality (4.1.5) holds for any $n(\mathbf{x})$, the optimal choice identifies it with the electron density. Thus the effective potential in the square brackets $[\cdots]$ consists of the attraction to the nuclei, the repulsion from other electrons, and the chemical potential. However, this interpretation counts the electron repulsion twice, as in $\sum_{i \neq k} |\mathbf{x}_i - \mathbf{x}_k|^{-1}$. The last term in $H_n$ corrects this overcounting.
2. The correlations among the electrons due to their Fermi statistics have the effect of reducing their repulsion. Also, $H_n$ contains the self-energy of the individual electrons. The constant $C_n$ and $-3.68\gamma N$ serve to control any possible effect from these corrections.

The monotonic property (2.1.7; 4) translates (4.1.6) into an inequality for the partition function. Then with the aid of the maximum principle of (2.5.16; 2) the inequality can be expressed as the supremum of an expression linear in $n$.

**The Partition Function with an Effective Field** (4.1.8)

$$\Xi(H - \mu N) \equiv T \ln \operatorname{Tr} \exp[-\beta(H - \mu N)] \leq \Xi\left(H_n - C_n + \sum_{k>l} \frac{Z_k Z_l}{|\mathbf{X}_k - \mathbf{X}_l|}\right)$$

$$\leq \Xi(H_n) + C_n - \sum_{k>l} \frac{Z_k Z_l}{|\mathbf{X}_k - \mathbf{X}_l|},$$

$$\Xi(H_n) = \operatorname{tr} 2 \ln(\mathbf{1} + \exp(-\beta h_n)) + \frac{1}{2} \int d^3x \, d^3x' \frac{n(\mathbf{x})n(\mathbf{x}')}{|\mathbf{x} - \mathbf{x}'|}$$

$$= \sup_{\rho_1} 2 \operatorname{tr}\{T(-\rho_1 \ln \rho_1 - (\mathbf{1} - \rho_1) \ln(\mathbf{1} - \rho_1)) - \rho_1 h_n\}$$

$$+ \frac{1}{2} \int d^3x \, d^3x' \frac{n(\mathbf{x})n(\mathbf{x}')}{|\mathbf{x} - \mathbf{x}'|},$$

$$h_n = |\mathbf{p}|^2 - \sum_k \frac{Z_k}{|\mathbf{x} - \mathbf{X}_k|} + \int \frac{d^3x \, n(\mathbf{x}')}{|\mathbf{x} - \mathbf{x}'|} + W(\mathbf{x}) - \mu - 3.68\gamma.$$

**Remarks** (4.1.9)

1. The Hamiltonian $h_n$ of one particle in the effective field acts on the space $\mathscr{H}_1 = L^2(\mathbb{R}^3)$. Spin is accounted for by the factor 2, and tr denotes the trace on $\mathscr{H}_1$.

2. As in Remark (2.5.16; 2), $\sup_{\rho_1}$ denotes the supremum over one-particle density matrices $\rho_1$ such that

$$0 \le \rho_1 \le 1,\ 2\,\mathrm{tr}\,\rho_1 = N \equiv \left\langle \sum_\alpha \int d^3x a_\alpha^*(\mathbf{x}) a_\alpha(\mathbf{x}) \right\rangle.$$

3. There exist $c_i \ge 0$ such that $h_n \ge c_1|\mathbf{p}|^2 + W(\mathbf{x}) - c_2$. This ensures that $\mathrm{tr}\ln(\mathbf{1} + \exp(-\beta h_n)) < \infty$.

The next task is to optimize the upper bound. The infimum over $n$ of $\Xi(H_n)$ is in fact achieved. This is a consequence of the

**Properties of the Functional** $\Xi(H_n)$ (4.1.10)

*The mapping $n \to \Xi(H_n)$ from $\mathcal{N}$ to $\mathbb{R}^+$, where $\mathcal{N}$ is the real Hilbert space of measurable functions $\mathbb{R}^3 \to \mathbb{R}$ finite in the norm*

$$\|n\|_c^2 = \langle n|n\rangle_c = \int \frac{d^3x\, d^3x' n(\mathbf{x})n(\mathbf{x}')}{|\mathbf{x} - \mathbf{x}'|}$$

is

(i) *weakly lower semicontinuous*;
(ii) *strictly convex*; *and*
(iii) *greater than* $\frac{1}{2}\|n\|_c^2$.

**Proof**

(i) In the second version $\Xi(H_n)$ depends on $n$ through $\mathrm{tr}\,\rho_1 h_n$ and $\|n\|_c$. The norm is $\sup_{n' \in \mathcal{N},\, \|n'\|_c \le 1} \langle n'|n\rangle_c$, and $\mathrm{tr}(\rho_1 \int n(\mathbf{x}')\, d^3x'/|\mathbf{x} - \mathbf{x}'|)$ is weakly continuous for

$$\rho_1 \in C_M \equiv \left\{\rho_1 \colon \int \frac{d^3x_1\, d^3x_2}{|\mathbf{x}_1 - \mathbf{x}_2|} \langle \mathbf{x}_1|\rho_1|\mathbf{x}_1\rangle\langle \mathbf{x}_2|\rho_1|\mathbf{x}_2\rangle < M \in \mathbb{R}^+\right\}.$$

The supremum is attained when $\rho_1 = (\exp(\beta h_n) + \mathbf{1})^{-1}$, which belongs to some $C_M$. Hence $\sup_{\rho_1}$ may be written as $\sup_{M \in \mathbb{R}^+} \sup_{\rho_1 \in C_M}$. In this way $\Xi(H_n)$ is expressed as the supremum over continuous functions, which is always lower semicontinuous.

(ii) This follows in the first version of $\Xi(H_n)$, when it is observed that $h \to \mathrm{tr}\ln(\mathbf{1} + \exp(-\beta h))$ is convex, $n \to h_n$ is linear, and $n \to \|n\|_c^2$ is strictly convex.

(iii) This follows in the first form of $\Xi(H_n)$, since $\mathrm{tr}\ln(\mathbf{1} + \exp(-\beta h)) \ge 0$. □

**Corollaries** (4.1.11)

1. Because of Property (iii), the infimum over $n$ lies in a compact region where $\|n\|_c < C$. Property (i) means that it is attained at some $n_0$, which is unique because of (ii).

2. Because of the convexity, we know that the function $\mathbb{R} \to \mathbb{R}^+ : t \to \Xi(H_{n_0+tn_1})$ has a right derivative everywhere, and the minimum is attained at $n_0$ if and only if

$$\lim_{t\downarrow 0} t^{-1}(\Xi(H_{n_0+tn_1}) - \Xi(H_{n_0})) \geq 0 \quad \text{for all } n_1 \in \mathcal{N}.$$

Although convexity does not imply the existence of a derivative, analyticity can be proved by a variant of Theorem (2.4.7). Granting that, the formal rules for differentiating tr ln(1 + $A$) are justified:

$$\frac{d}{dt} \operatorname{tr} \ln(\mathbf{1} + \exp(-\beta h_{n_0+tn_1}))|_{t=0} = -\operatorname{tr} \int \frac{d^3x' n_1(\mathbf{x}')}{|\mathbf{x}-\mathbf{x}'|} \frac{\beta}{\exp(\beta h_{n_0}) + \mathbf{1}}.$$

Therefore the minimum at $n_0$ is characterized by

$$\int \frac{d^3x'\, d^3x}{|\mathbf{x}-\mathbf{x}'|} n_0(\mathbf{x}) n_1(\mathbf{x}') = 2 \operatorname{tr} \int \frac{d^3x' n_1(\mathbf{x}')}{|\mathbf{x}-\mathbf{x}'|} \frac{1}{\exp(\beta h_{n_0}) + \mathbf{1}} \quad \text{for all } n_1 \in \mathcal{N}.$$

If $n_1$ is made to tend to $\Delta\delta(\mathbf{x} - \mathbf{x}_0)$, then there results an equation for $n_0(\mathbf{x}_0)$. Since the integral kernel $K(\mathbf{x}, \mathbf{x}')$ of $(\exp(\beta h_n) + \mathbf{1})^{-1}$ is analytic for $\mathbf{x}, \mathbf{x}' \neq \mathbf{X}_k$ even though $\Delta\delta$ does not belong to $\mathcal{N}$, we have the

**Existence of the Self-Consistent Field** (4.1.12)

*The equation*

$$n_o(\mathbf{x}) = 2\langle \mathbf{x} | (\exp(\beta h_{n_0}) + 1)^{-1} | \mathbf{x} \rangle$$

*has a unique solution, which minimizes* $\Xi(H_n)$.

**Remarks** (4.1.13)

1. Since $2\langle \mathbf{x} | (\exp(\beta h_{n_0}) + \mathbf{1})^{-1} | \mathbf{x} \rangle$ equals $\sum_\alpha \langle a_\alpha^*(\mathbf{x}) a_\alpha(\mathbf{x}) \rangle_{n_0}$, it is the mean electron density in the state determined by the one-particle Hamiltonian $h_{n_0}$.
2. The ease with which the existence of the solution of the generalized Hartree equation (4.1.12) was proved depended on the wall potential $W$. In an infinite space without $W$ there fails to be a solution when $N > \sum_k Z_k$, even at absolute zero temperature—the electrons escape to infinity, and the infimum is never attained. This is a reflection of the general mathematical fact that a strictly convex function need not achieve its infimum on a noncompact region; for example $1/x$ never reaches the value 0 on $[1, \infty)$.
3. A convex function on a finite-dimensional space is continuous on the interior of its domain of definition. This is not always the case when the dimension of the space is infinite, and $\|n\|_c^2$ is in fact not weakly continuous: The norms $\| \ \|_c$ of the charge distributions $n_R(\mathbf{x}) = R^{-5/2}\Theta(R - |\mathbf{x}|)$ are

all equal, but $\int d^3x n_R(\mathbf{x}) \to 0$ as $R \to 0$. Consequently $\langle n_R | n \rangle_c \to 0$ for all $n$, if

$$V_n(x) \equiv \int \frac{d^3x' n(\mathbf{x}')}{|\mathbf{x} - \mathbf{x}'|} \in L^\infty(\mathbb{R}^3).$$

Since the $n$'s such that $V_n \in L^\infty$ are dense in $\mathcal{N}$, $n_R \rightharpoonup 0$, even though $\|n_R\|_c \not\to 0$. There even exist convex functions that fail to be lower semicontinuous, for example the functional of (III: 2.1.15; 2). Of course the function $n \to \|n\|_c^2$ is continuous in norm, but this finer norm topology can not be used, because we need the compactness of bounded sets.

4. At the minimum (4.1.12), it is indeed true that $n(\mathbf{x}) > 0$ and $\int d^3x n(\mathbf{x}) = N$.

**Lower Bounds for $\Xi(H)$ (4.1.14)**

In (III, §4.5) upper bounds on the energy were provided by the min–max principle, the generalization of which for nonzero temperatures is the Peierls–Bogoliubov inequality (2.1.8; 3) with $\Xi = -F$. Because

$$\left\langle H - \bar{\mu}N - \sum_{k>l} \frac{Z_k Z_l}{|\mathbf{X}_k - \mathbf{X}_l|} - H_{no} \right\rangle_{no}$$

$$= \sum_{\alpha,\beta} \frac{1}{2} \int \frac{d^3x\, d^3x'}{|\mathbf{x} - \mathbf{x}'|} (\langle a_\alpha^*(\mathbf{x}) a_\beta^*(\mathbf{x}') a_\beta(\mathbf{x}') a_\alpha(\mathbf{x}) \rangle_{no} - n_0(\mathbf{x}) n_0(\mathbf{x}'))$$

$$= -\sum_{\alpha,\beta} \frac{1}{2} \int \frac{d^3x\, d^3x'}{|\mathbf{x} - \mathbf{x}'|} |\langle a_\alpha^*(\mathbf{x}) a_\beta(\mathbf{x}') \rangle_{no}|^2 \equiv -A(n_0) < 0,$$

where $\bar{\mu} = \mu - 3.68\gamma$, it implies that

$$\Xi(H - \bar{\mu}N) \geq \Xi(H_{no}) + A(n_0) - \sum_{k>l} \frac{Z_k Z_l}{|\mathbf{X}_k - \mathbf{X}_l|}.$$

When this is combined with (4.1.8), it yields

**Two-Sided Bounds for $\Xi$ (4.1.15)**

$$0 \leq A(n_0) \leq \Xi(H - \bar{\mu}N) + \sum_{k>l} \frac{Z_k Z_l}{|\mathbf{X}_k - \mathbf{X}_l|} - \Xi(H_{no}) \leq \frac{3}{5\gamma} \int d^3x\, n_0^{5/3}(\mathbf{x}).$$

**Remarks (4.1.16)**

1. This means that the true partition function exceeds the partition function with an effective field by more than $A$ but less than $C$.
2. In particular (4.1.15) states for the exchange energy that $0 \leq A \leq C$. If $Z$ is large, then $n_0$ approaches the electron density in Thomas–Fermi Theory, and we shall discover that $\int n^{5/3} \sim Z^{7/3}$. If $\gamma$ is chosen $\sim (Z^{7/3}/N)^{1/2}$, then $C$ and the additional term $3.68\gamma N$ in $\mu$ becomes $\sim N^{1/2} Z^{7/6}$. Since $H$ goes as $Z^{7/3}$, if $N \sim Z$, then the relative error is $O(Z^{-2/3})$.

**The Classical Limit** (4.1.17)

The next topic of study is the way in which $\Xi(H_n)$ approaches the classical phase-space integral (2.5.17) as $Z \to \infty$. According to the general considerations of (1.2.4) the interesting limit would be expected to be that in which the system shrinks as $Z^{-1/3}$. Consider, therefore, a sequence of Hamiltonians $H_Z$ in which not only do the nuclear charges increase as $Z_k = Zz_k$, $\sum_k z_k = 1$, $z_k$ fixed, but also the nuclear coordinates are scaled by changing $\mathbf{X}_k$ into $Z^{-1/3}\mathbf{X}_k$ and the wall potential varies at the same time:

$$H_Z \equiv \sum_\alpha \int d^3x \Bigg[ \nabla a_\alpha^*(\mathbf{x}) \cdot \nabla a_\alpha(\mathbf{x}) + a_\alpha^*(\mathbf{x}) a_\alpha(\mathbf{x}) \left( -Z \sum_{k=1}^{M} \frac{z_k}{|\mathbf{x} - \mathbf{X}_k Z^{-1/3}|} + Z^{4/3} W(Z^{1/3}\mathbf{x}) \right) \Bigg] + \frac{1}{2} \sum_{\alpha,\beta} \int a_\alpha^*(\mathbf{x}) a_\beta^*(\mathbf{x}) a_\beta(\mathbf{x}') a_\alpha(\mathbf{x}') \frac{d^3x\, d^3x'}{|\mathbf{x} - \mathbf{x}'|} + \sum_{k>l} \frac{z_k z_l}{|\mathbf{X}_k - \mathbf{X}_l|} Z^{7/3},$$

$$H_{Z,n} \equiv \sum_\alpha \int d^3x \Bigg\{ \nabla a_\alpha^*(\mathbf{x}) \cdot \nabla a_\alpha(\mathbf{x}) + a_\alpha^*(\mathbf{x}) a_\alpha(\mathbf{x}) \Bigg[ -Z \sum_{k=1}^{M} \frac{z_k}{|\mathbf{x} - \mathbf{X}_k Z^{-1/3}|} + \int \frac{d^3x' n^Z(\mathbf{x}')}{|\mathbf{x} - \mathbf{x}'|} + Z^{4/3}(W(\mathbf{x}Z^{1/3}) - \mu) - 3.68\gamma \Bigg] \Bigg\} - \frac{1}{2} \int \frac{d^3x\, d^3x'}{|\mathbf{x} - \mathbf{x}'|} n^Z(\mathbf{x}) n^Z(\mathbf{x}');$$

$$n^Z = Z^2 n(Z^{1/3}\mathbf{x}).$$

In order always to work in a fixed volume and see what happens in the limit $Z \to \infty$, use a canonical transformation to convert the electron coordinates $\mathbf{x}$ into $Z^{-1/3}\mathbf{x}$ and $\mathbf{p}$ into $Z^{1/3}\mathbf{p}$ at the same time—this entails $a(\mathbf{x}) \to Z^{-1/2} a(Z^{-1/3}\mathbf{x}))$ as well. Since the number of electrons also grows as $Z$, the mean momentum of the electrons grows as $Z^{2/3}$, and every kind of energy per particle, such as $T$ or $\mu$, will depend in the same way on $Z$. Thus if we calculate $\operatorname{Tr} \exp[-\beta_Z(H_Z - \mu_Z N)]$ with $\beta_Z = Z^{-4/3}\beta$, and $\mu_Z = Z^{4/3}\mu$, and scale $n$ appropriately, we are led to $\operatorname{tr} \ln(\mathbf{1} + \exp(-\beta h_n))$ with

$$h_n = Z^{-2/3}|\mathbf{p}|^2 - \sum_j \frac{z_j}{|\mathbf{x} - \mathbf{X}_j|} + \int \frac{d^3x' n(\mathbf{x}')}{|\mathbf{x} - \mathbf{x}'|} + W(\mathbf{x}) - \mu_\gamma$$

and

$$\mu_\gamma = \mu + Z^{-4/3} \cdot 3.68\gamma.$$

Observe that $Z^{-1/3}$ occurs in the position of $\hbar$, making the limit $Z \to \infty$ equivalent to the classical limit $\hbar \to 0$. Now use the bound (2.5.17) with

$$u^2(\mathbf{x}) = \frac{\kappa^3}{8\pi}\exp(-\kappa r).$$

The Fourier transform of this density is

$$\widetilde{u^2}(\mathbf{k}) = \frac{\kappa^4}{(|\mathbf{k}|^2 + \kappa^2)^2}.$$

Consequently $\int |\nabla u|^2\, d^3x = \kappa^2$, and if $v = 1/r$, then

$$v_u(q) \equiv \int d^3x \frac{1}{|\mathbf{x}|}|u(\mathbf{x} - \mathbf{q})|^2 = \frac{1}{q} - \frac{\exp(-\kappa q)}{q} - \frac{\kappa}{2}\exp(-\kappa q) \equiv \frac{1}{q} - v_s(q).$$

**The Classical Upper Bound** (4.1.18)

Since $1/r$ can not be represented as a smeared potential, $v^u$ makes no sense. Thus it is first necessary to remove $v_s$, the short-range, singular part of $1/r$, and handle it separately. It can be neglected as $\kappa \to \infty$, and if the smeared remainder is unsmeared, we recover $1/r$:

$$\frac{1}{r} = v_u + v_s, \qquad (v_u)^u = \frac{1}{r}.$$

Let $h_c$ be like the $h_n$ of (4.1.17), but with $v_u$ in place of $1/r$. Then

$$h_n = h_c + V_s,$$

$$h_c = \int \frac{d^3q\, d^3p}{(2\pi)^3}|\mathbf{q}, \mathbf{p}\rangle\langle\mathbf{q}, \mathbf{p}|\left(Z^{-2/3}(|\mathbf{p}|^2 - \kappa^2) - \sum_j \frac{z_j}{|\mathbf{q} - \mathbf{X}_j|}\right.$$
$$\left. + \int \frac{d^3x n(\mathbf{x})}{|\mathbf{q} - \mathbf{x}|} + W^u(\mathbf{q}) - \mu_\gamma\right),$$

$$V_s = -\sum_j z_j v_s(\mathbf{x} - \mathbf{X}_j) + \int d^3y n(\mathbf{y}) v_s(\mathbf{x} - \mathbf{y}).$$

In the $x$-representation, $|\mathbf{q}, \mathbf{p}\rangle$ is

$$\left(\frac{\kappa^3}{8\pi}\right)^{1/2} \exp(i\mathbf{p}\cdot\mathbf{x})\exp(-\kappa|\mathbf{x} - \mathbf{q}|),$$

and we let $W^u(\mathbf{x})$ be the unsmeared wall potential $W$ of (2.5.23). Convexity can be appealed to to bound the influence of $V_s$:

$$\begin{aligned} \operatorname{tr}\ln(\mathbf{1} + \exp(-\beta h_n)) &\le \operatorname{tr}\ln(\mathbf{1} + \exp(-\beta h_c)) \\ &\quad + \alpha^{-1}\operatorname{tr}[\ln(\mathbf{1} + \exp(-\beta(h_c + \alpha V_s))) \\ &\quad - \ln(\mathbf{1} + \exp(-\beta h_c))] \qquad \text{for all } \alpha \ge 1. \end{aligned}$$

The number $\alpha$ will be picked so large that the addition to the first term on the right side goes away in the limit $Z \to \infty$. By (2.5.17), the second term is bounded by

$$\operatorname{tr} \ln(\mathbf{1} + \exp(-\beta h_c)) \leq \int \frac{d^3q\, d^3p}{(2\pi)^3} \ln\left(\mathbf{1} + \exp\left[-\beta\left(Z^{-2/3}(|\mathbf{p}|^2 - \kappa^2)\right.\right.\right.$$

$$\left.\left.\left. - \sum_j \frac{z_j}{|\mathbf{q} - \mathbf{X}_j|} + \int \frac{d^3x n(\mathbf{x})}{|\mathbf{q} - \mathbf{x}|} + W^u(\mathbf{q}) - \mu_\gamma\right)\right]\right)$$

$$= Z \int \frac{d^3q\, d^3p}{(2\pi)^3} \ln\left(\mathbf{1} + \exp\left[-\beta\left(|\mathbf{p}|^2 - \sum_j \frac{z_j}{|\mathbf{q} - \mathbf{X}_j|}\right.\right.\right.$$

$$\left.\left.\left. + \int \frac{d^3x n(\mathbf{x})}{|\mathbf{q} - \mathbf{x}|} + W^u(\mathbf{q}) - \mu_\gamma - Z^{-2/3}\kappa^2\right)\right]\right).$$

The additional part containing $V_s$ can be taken care of because even for a singular potential $V(\mathbf{x}) \in L^{5/2}(\mathbb{R}^3)$ there is a bound of this form weakened by a factor $C \cong 7$:

$$\operatorname{tr} \ln(\mathbf{1} + \exp[-\beta(|\mathbf{p}|^2 + V(\mathbf{x}))])$$

$$\leq c \int \frac{d^3p\, d^3q}{(2\pi)^3} \ln(1 + \exp[-\beta(|\mathbf{p}|^2 + V(\mathbf{q}))]). \tag{4.1.19}$$

The derivation of this formula is left for Problems 1 and 2. In this case it leaves us with

$$\operatorname{tr} \ln(\mathbf{1} + \exp[-\beta(h_c + \alpha V_s)])$$

$$\leq cZ \int \frac{d^3q\, d^3p}{(2\pi)^3} \ln\left(1 + \exp\left[-\beta(|\mathbf{p}|^2 + W^u(\mathbf{q})\right.\right.$$

$$- \sum_j z_j\left(\frac{1}{|\mathbf{q} - \mathbf{X}_j|} + (\alpha - 1)v_s(\mathbf{q} - \mathbf{X}_j)\right)$$

$$\left.\left. + \int d^3y n(\mathbf{y})\left(\frac{1}{|\mathbf{q} - \mathbf{y}|} + (\alpha - 1)v_s(\mathbf{q} - \mathbf{y})\right) - \mu_\gamma - Z^{-2/3}\kappa^2)\right]\right).$$

It remains to be shown that $\alpha$ and $\kappa$ can be sent to infinity with $Z$ in such a way that the additions to the classical one-particle potential in the effective field become negligible. To this end assume that $W^u$ tends to infinity outside some compact set $K$ containing the $\mathbf{X}_i$ so rapidly that the contribution to the integral over the complement $CK$ is insignificant, that is, $\int_K d^3q \ln(\cdots) > \int_{CK} d^3q \ln(\cdots)$ for all $\alpha > 0$. Then it suffices to estimate the integral over $K$, which can be done in terms of the $L^p$ norms of the potential on $K$, i.e., $\|V\|_p = (\int_K d^3q |V(\mathbf{q})|^p)^{1/p}$. If $|x|_- \equiv |x|\Theta(-x)$, then

$$\ln(1 + \exp(-x)) = |x|_- + \ln(1 + \exp(-|x|)) \leq |x|_- + \exp(-|x|),$$

and if

$$\mathbf{q} \in K_- \equiv \{\mathbf{q} \in K \mid V(\mathbf{q}) < 0\},$$

then with $\varepsilon = |V(\mathbf{q})|\eta$,

$$I \equiv \int_0^\infty d\varepsilon\sqrt{\varepsilon}\ln(1 + \exp[-\beta(\varepsilon + V(\mathbf{q}))])$$

$$< \int_0^\infty d\eta\sqrt{\eta}(\beta|\eta - 1|_-|V(\mathbf{q})|^{5/2} + |V(\mathbf{q})|^{3/2}\exp[-\beta|V(\mathbf{q})||\eta - 1|]),$$

and if $\mathbf{q} \in K_+ \equiv \{\mathbf{q} \in K \,|\, V(\mathbf{q}) > 0\}$, then

$$I < \int_0^\infty d\eta\sqrt{\eta}\,V(\mathbf{q})^{3/2}\exp[-\beta V(\mathbf{q})(\eta + 1)].$$

Because $|\eta - 1| \le \eta + 1$ for all $\eta \ge 0$ and

$$\int_0^\infty d\varepsilon\sqrt{\varepsilon}\ln(1 + \exp(-\beta\varepsilon)) < \beta^{-3/2}\sqrt{\pi}/2,$$

if $K' = K_+ \cup K_-$, then

$$\int_K d^3q \int_0^\infty d\varepsilon\sqrt{\varepsilon}\ln(1 + \exp[-\beta(\varepsilon + V(\mathbf{q}))])$$

$$< \int_{K'} d^3q \int_0^\infty d\eta\sqrt{\eta}(\beta|V(\mathbf{q})|_-^{5/2}|\eta - 1|_-$$

$$+ |V(\mathbf{q})|^{3/2}\exp[-\beta|V(\mathbf{q})||\eta - 1|]) + \beta^{-3/2}\frac{\sqrt{\pi}}{2}.$$

The required bound now follows from

$$\int_0^\infty d\varepsilon\sqrt{\varepsilon}\exp(-\gamma|\varepsilon - 1|) \le \int_0^1 d\varepsilon\sqrt{\varepsilon} + \int_0^\infty d\varepsilon(\sqrt{\varepsilon} + 1)\exp(-\gamma\varepsilon)$$

$$= \tfrac{2}{3} + \gamma^{-1} + \frac{\sqrt{\pi}}{2}\gamma^{-3/2},$$

for

$$\int_0^\infty d\varepsilon\sqrt{\varepsilon}\int_K d^3q\ln(1 + \exp[-\beta(\varepsilon + V(\mathbf{q}))])$$

$$\le \int_K d^3q\left[\frac{4\beta}{15}|V|_-^{5/2} + \tfrac{2}{3}|V|^{3/2} + \beta^{-1}|V|^{1/2} + \sqrt{\pi}\beta^{-3/2}\right].$$

In the case at hand, since $\|V_s\|_p \sim \kappa^{1-3/p}$ and

$$\int d^3x|V + (\alpha - 1)V_s|^p \le (\|V\|_p + (\alpha - 1)\|V_s\|_p)^p, \qquad p = \tfrac{5}{2}, \tfrac{3}{2},$$

or, respectively,

$$\int d^3x|V + (\alpha - 1)V_s|^{1/2} \le \|V\|_{1/2}^{1/2} + \sqrt{\alpha - 1}\,\|V_s\|_{1/2}^{1/2},$$

it follows that $\mathbf{1} + \exp[-\beta(h_c + \alpha V_s)]$ remains bounded in the limit as $\alpha$ and $\kappa \to \infty$ when $\alpha \sim \kappa^{1/5}$. If $\kappa$ goes as $Z^{1/3-\varepsilon}$, $0 < \varepsilon < \frac{1}{3}$, then the correction $Z^{-2/3}\kappa^2$ to the kinetic energy tends to zero, and all corrections to the classical one-particle phase-space integral with the effective field are smaller than this quantity by a factor $Z^{-1/15+\varepsilon/5}$. The quantity $\mu_\gamma$ is no trouble at all, since it approaches $\mu$, provided that $\gamma Z^{-4/3} \to 0$. Likewise, $W^u(\mathbf{q})$ and $W_u(\mathbf{q})$ approach $W(\mathbf{q})$ in the limit $\kappa \to \infty$.

**The Classical Lower Bound** (4.1.20)

For the classical bound (2.5.17), the $1/r$ occurring in the classical phase-space integral has to be replaced with $v_u = 1/r - v_s$. As before, convexity is useful for estimating the influence of the $v_s$, except that this time the convexity of $f$ for $\alpha > 0$,

$$f(-1) \geq f(0) + \frac{f(0) - f(\alpha)}{\alpha},$$

is used for the other side of the equation. The result is

$$\begin{aligned}
&\operatorname{tr}\ln(\mathbf{1} + \exp(-\beta h_n))\\
&\geq Z\int\frac{d^3q\,d^3p}{(2\pi)^3}\ln\Bigg(1 + \exp\Bigg[-\beta\Bigg\{|\mathbf{p}|^2 + Z^{-2/3}\kappa^2 - \mu_\gamma + W_u(\mathbf{q})\\
&\quad - \sum_j z_j\left(\frac{1}{|\mathbf{q} - \mathbf{X}_j|} - v_s(\mathbf{q} - \mathbf{X}_j)\right)\\
&\quad + \int d^3x\,n(\mathbf{x})\left(\frac{1}{|\mathbf{q} - \mathbf{x}|} - v_s(\mathbf{q} - \mathbf{x})\right)\Bigg\}\Bigg]\Bigg)\\
&\geq Z\int\frac{d^3q\,d^3p}{(2\pi)^3}\Bigg[\ln\Bigg(1 + \exp\Bigg[-\beta\Bigg\{|\mathbf{p}|^2 + Z^{-2/3}\kappa^2 - \sum_j\frac{z_j}{|\mathbf{q} - \mathbf{X}_j|}\\
&\quad + \int\frac{d^3x\,n(\mathbf{x})}{|\mathbf{q} - \mathbf{x}|} + W(\mathbf{q}) - \mu_\gamma\Bigg\}\Bigg]\Bigg)\left(1 + \frac{1}{\alpha}\right)\\
&\quad - \frac{1}{\alpha}\ln\Bigg(1 + \exp\Bigg[-\beta\Bigg\{|\mathbf{p}|^2 + Z^{-2/3}\kappa^2\\
&\quad - \sum_j z_j(|\mathbf{q} - \mathbf{X}_j|^{-1} + \alpha v_s(\mathbf{q} - \mathbf{X}_j))\\
&\quad + \int d^3x\,n(\mathbf{x})(|\mathbf{q} - \mathbf{x}|^{-1} + \alpha v_s(\mathbf{q} - \mathbf{x})) + W_u(\mathbf{q}) - \mu_\gamma\Bigg\}\Bigg]\Bigg)\Bigg].
\end{aligned}$$

The integrals that show up are the same as for the upper bounds, so with $\alpha = \kappa^{1/5}$, $\kappa = Z^{1/3-\varepsilon}$, $0 < \varepsilon < \frac{1}{3}$, the corrections to the classical expression vanish as $Z \to \infty$. The $n(\mathbf{x})$ considered earlier was constant, while that defined by (4.1.12) depends on $Z$. However, it is shown in Problem 4 that the minimum values also converge, so our bounds prove the

**Classical Limit of the Partition Function** (4.1.21)

$$\lim_{Z\to\infty} Z^{-1} \ln \operatorname{Tr} \exp[-\beta(Z^{-4/3}H_Z - \mu N)]$$

$$= \lim_{Z\to\infty} Z^{-1} \ln \operatorname{Tr} \exp(-\beta Z^{-4/3} H_{Z,n}) - \beta \sum_{k>l} \frac{z_k z_l}{|\mathbf{X}_k - \mathbf{X}_l|}$$

$$= 2 \int \frac{d^3p\, d^3q}{(2\pi)^3} \ln\left(1 + \exp\left[-\beta\left(|\mathbf{p}|^2 - \sum_j \frac{z_j}{|\mathbf{q} - \mathbf{X}_j|} + \int d^3y \frac{n(\mathbf{y})}{|\mathbf{q} - \mathbf{y}|} + W(\mathbf{q}) - \mu\right)\right]\right)$$

$$- \beta \sum_{k>l} \frac{z_k z_l}{|\mathbf{X}_k - \mathbf{X}_l|} + \frac{\beta}{2} \int \frac{d^3x\, d^3y}{|\mathbf{x} - \mathbf{y}|} n(\mathbf{x}) n(\mathbf{y}).$$

According to Remark (2.5.18; 4), the optimal density for this formula satisfies

$$n(\mathbf{x}) = 2 \int \frac{d^3p}{(2\pi)^3} \left[\exp\left(\beta\left(|\mathbf{p}|^2 - \sum_j \frac{z_j}{|\mathbf{x} - \mathbf{X}_j|} + \int d^3y \frac{n(\mathbf{y})}{|\mathbf{q} - \mathbf{y}|} + W(\mathbf{x}) - \mu\right) + 1\right)\right]^{-1}. \quad (4.1.22)$$

**Remarks** (4.1.23)

1. The classical functional also has Properties (4.1.10), which ensure the existence and uniqueness of a solution of (4.1.22).
2. As yet unproved conjectures [11] imply that Equation (4.1.19) holds even with $c = 1$. If that turns out to be true, then many of the proofs given here can be simplified.

**The Density in Phase Space** (4.1.24)

Now that $\Xi$ has been shown to converge, we can study the limiting behavior of the expectation values of a suitable subalgebra of observables. The densities on classical phase space would seem to be an appropriate subalgebra, since in the classical limit $Z \to \infty$ it ought to make sense to speak of position and momentum simultaneously. As mentioned above (cf. (1.2.4)) position goes as $Z^{-1/3}$ while momentum goes as $Z^{2/3}$, so the product of their relative mean-square deviations would be expected to go as $Z^{-2/3}$, and as $Z \to \infty$ the physics should become classical. This rather airy argument can be made mathematically substantial, and we shall discover that in convenient units, fermions distribute themselves in phase space according to

$$\rho(\mathbf{q}, \mathbf{p}) = \left[\exp \beta\left\{|\mathbf{p}|^2 - \sum_j \frac{z_j}{|\mathbf{q} - \mathbf{X}_j|} + \int \frac{d^3x n(\mathbf{x})}{|\mathbf{x} - \mathbf{q}|} + W(\mathbf{q}) - \mu\right\} + 1\right]^{-1}.$$

Particularly interesting is the observation that fermions behave more classically than bosons. The latter have a $-1$ in the denominator, so $\rho(\mathbf{q}, \mathbf{p})$ becomes negative when $\mathbf{q} = \mathbf{X}_j$, and thus can not turn out to be a probability density on phase space.

To make the connection with (2.2.10; 5) we define creation and annihilation operators at the point $(\mathbf{q}, \mathbf{p})$ in phase space, and choose $u$ as a sufficiently smooth, decreasing function such that $\|u\|_2 = 1$, like the function of (4.1.17):

**The Field Algebra on Phase Space** (4.1.25)

The operators

$$a_{\mathbf{q},\mathbf{p};\alpha} = Z^{3\varepsilon/2} \int d^3x a_\alpha(\mathbf{x}) \exp(iZ^{2/3}\mathbf{p}\cdot\mathbf{x}) u(Z^\varepsilon(\mathbf{x} - Z^{-1/3}\mathbf{p})),$$

$$\tfrac{1}{3} < \varepsilon < \tfrac{2}{3}, \; u^* = u,$$

satisfy the commutation relations

$$[a_{\mathbf{q},\mathbf{p};\alpha}, a^*_{\mathbf{q}',\mathbf{p}';\beta}]_+ = \delta_{\alpha\beta} \int d^3x \exp(iZ^{2/3-\varepsilon}\mathbf{x}\cdot(\mathbf{p} - \mathbf{p}')) u(\mathbf{x} - Z^{\varepsilon-1/3}\mathbf{q}) \times u(\mathbf{x} - Z^{\varepsilon-1/3}\mathbf{q}').$$

If $\mathbf{q} = \mathbf{q}'$ and $\mathbf{p} = \mathbf{p}'$, then the right side is $\delta_{\alpha\beta}$, and otherwise it goes to zero as $Z \to \infty$. Hence $\rho_{\mathbf{q},\mathbf{p}} = \sum_\alpha a^*_{\mathbf{q},\mathbf{p};\alpha} a_{\mathbf{q},\mathbf{p};\alpha}$ are bounded above and below by $0 \le \rho_{\mathbf{q},\mathbf{p}} \le 2$, and generate an algebra that is Abelian in the limit $Z \to \infty$.

Defining $d\Omega \equiv d^3q\, d^3p/(2\pi)^3$, we calculate

$$\int d\Omega \rho_{\mathbf{q},\mathbf{p}} F(\mathbf{q}) = Z^{-1} \sum_\alpha \int d^3x a^*_\alpha(\mathbf{x}) a_\alpha(\mathbf{x}) Z^{3\varepsilon} |u(Z^3(\mathbf{x} - \mathbf{x}'))|^2 \times F(Z^{1/3}\mathbf{x}')\, d^3x',$$

$$\int d\Omega \rho_{\mathbf{q},\mathbf{p}} |\mathbf{p}|^2 = Z^{-7/3} \sum_\alpha \int d^3x (\nabla a^*_\alpha(\mathbf{x}) \cdot \nabla a_\alpha(\mathbf{x}) + a^*_\alpha(\mathbf{x}) a_\alpha(\mathbf{x}) Z^{2\varepsilon} \int d^3y |\nabla u(\mathbf{y})|^2)$$

$$\int d\Omega\, d\Omega' \frac{\rho_{\mathbf{q},\mathbf{p}} \rho_{\mathbf{q}',\mathbf{p}'}}{|\mathbf{q} - \mathbf{q}'|} = Z^{-7/3} \sum_{\alpha,\beta} \int d^3x\, d^3x' a^*_\alpha(\mathbf{x}) a^*_\beta(\mathbf{x}') a_\beta(\mathbf{x}') a_\alpha(\mathbf{x}) v_{uu}(\mathbf{x} - \mathbf{x}') + Z^{-7/3} \sum_\alpha \int d^3x a^*_\alpha(\mathbf{x}) a_\alpha(\mathbf{x}) v_{uu}(\mathbf{0}), \qquad (4.1.26)$$

where $F \in L^\infty(\mathbb{R}^3)$ and

$$0 \leq v_{uu}(\mathbf{x} - \mathbf{x}') \equiv \int d^3q\, d^3q' \frac{|u(Z^\varepsilon(\mathbf{x} - \mathbf{q}))|^2 |u(Z^\varepsilon(\mathbf{x}' - \mathbf{q}'))|^2}{|\mathbf{q} - \mathbf{q}'|} Z^{6\varepsilon} < \frac{1}{|\mathbf{x} - \mathbf{x}'|}.$$

**Remarks** (4.1.27)

1. As $Z \to \infty$, $Z^{3\varepsilon}|u(Z^\varepsilon(\mathbf{x} - \mathbf{x}'))|^2$ approaches $\delta(\mathbf{x} - \mathbf{x}')$. It is not hard to convince oneself that when the classical Hamiltonian with $\rho$ or $\rho \cdot \rho$ is integrated, the result is $H$ to order $Z^{7/3}$.
2. For neutral states, i.e., $\sum_\alpha \langle N_\alpha \rangle = Z$, it follows that $\langle \int d\Omega \rho_{\mathbf{q},\mathbf{p}} \rangle = 1$.

The convexity of the partition function (2.4.7) can be used to calculate an expectation value by allowing it to be written as the derivative of the partition function by a perturbation parameter. We shall show that the perturbed $\Xi$ still converges as $Z \to \infty$, which will simultaneously prove that the foregoing results are stable against small variations in $H$. The limit will turn out to be likewise convex and differentiable in the perturbation parameters, so by Problem (2.4.18; 3) the limit of the derivative is the derivative of the limit. Since our real aim is to prove that the expectation value of $\rho_{\mathbf{q},\mathbf{p}}$ approaches the Thomas–Fermi density and that the deviations of $\rho_{\mathbf{q},\mathbf{p}}$ vanish, we will perturb $H$ both linearly and quadratically in $\rho$. To an accuracy of $Z^{-2/3}$ we can by-pass the intermediate steps (4.1.15), so we shall not require the more refined inequality (4.1.5). Thus we get by with a somewhat simpler effective Hamiltonian.

**The Perturbed Hamiltonian** (4.1.28)

$$H_\lambda \equiv H_Z + \lambda_1 Z^{7/3} \int d\Omega \rho_{\mathbf{q},\mathbf{p}} f(\mathbf{q}, \mathbf{p}) + \lambda_2 \frac{Z^{7/3}}{2} \left( \int d\Omega \rho_{\mathbf{q},\mathbf{p}} f(\mathbf{q}, \mathbf{p}) \right)^2,$$

$$H_{\lambda,n} \equiv \int d^3x \sum_\alpha \Bigg\{ \nabla a_\alpha^*(\mathbf{x}) \cdot \nabla a_\alpha(\mathbf{x}) + a_\alpha^*(\mathbf{x}) a_\alpha(\mathbf{x})$$

$$\times \left[ -Z \sum_{k=1}^{M} \frac{z_k}{|\mathbf{x} - Z^{-1/3}\mathbf{X}_k|} + Z^{4/3}(W(Z^{1/3}\mathbf{x}) - \mu) \right] \Bigg\}$$

$$+ Z^{7/3} \int \frac{d\Omega\, d\Omega'}{|\mathbf{q} - \mathbf{q}'|} (\rho_{\mathbf{q},\mathbf{p}} - \tfrac{1}{2} n(\mathbf{q}, \mathbf{p})) n(\mathbf{q}', \mathbf{p}')$$

$$+ Z^{7/3}(\lambda_1 + \lambda_2 g) \int d\Omega \rho_{\mathbf{q},\mathbf{p}} f(\mathbf{q}, \mathbf{p}) - Z^{7/3} \lambda_2 g^2/2,$$

where $\lambda_i \in \mathbb{R}$ and $f \in C_0^\infty$. We shall choose $n(\mathbf{q}, \mathbf{p})$ as $\langle \rho_{\mathbf{q},\mathbf{p}} \rangle$ and let $g \equiv \int d\Omega n(\mathbf{q}, \mathbf{p}) f(\mathbf{q}, \mathbf{p})$. With the idea of (4.1.24), because $0 \leq v_{uu}(\mathbf{x}) \leq 1/|\mathbf{x}|$,

$$
\begin{aligned}
H_\lambda &- Z^{4/3}\mu N - Z^{7/3} \sum_{k>l} \frac{z_k z_l}{|\mathbf{X}_k - \mathbf{X}_l|} - H_{\lambda,n} \\
&= \frac{1}{2} \int \frac{d^3x\, d^3x'}{|\mathbf{x} - \mathbf{x}'|} a_\alpha^*(\mathbf{x}) a_\beta^*(\mathbf{x}') a_\beta(\mathbf{x}') a_\alpha(\mathbf{x}) \\
&\quad - Z^{7/3} \int \frac{d\Omega\, d\Omega'}{|\mathbf{q} - \mathbf{q}'|} (\rho_{\mathbf{q},\mathbf{p}} - \tfrac{1}{2} n(\mathbf{q}, \mathbf{p})) n(\mathbf{q}', \mathbf{p}') \\
&\quad + Z^{7/3} \frac{\lambda_2}{2} \left( \int d\Omega \rho_{\mathbf{q},\mathbf{p}} f(\mathbf{q}, \mathbf{p}) - g \right)^2 \\
&\geq \frac{Z^{7/3}}{2} \int d\Omega\, d\Omega' (\rho_{\mathbf{q},\mathbf{p}} - n(\mathbf{q}, \mathbf{p}))(\rho_{\mathbf{q}',\mathbf{p}'} - n(\mathbf{q}', \mathbf{p}')) \\
&\quad \times \left[ \frac{1}{|\mathbf{q} - \mathbf{q}'|} + \lambda_2 f(\mathbf{q}, \mathbf{p}) f(\mathbf{q}', \mathbf{p}') \right] - \frac{N}{2} v_{uu}(\mathbf{0}).
\end{aligned}
$$

**Remarks** (4.1.29)

1. Since the Fourier transform in the $q$ variables, $\tilde{f}(\mathbf{k}, \mathbf{p})$, decreases in $\mathbf{k}$ faster than any power, $|\mathbf{k}|^2 + \lambda_2 \tilde{f}(\mathbf{k}, \mathbf{p}) \tilde{f}(\mathbf{k}, \mathbf{p}')$ is positive for sufficiently small $|\lambda_2|$. The expression in square brackets $[\cdots]$ is then of positive type, and the inequality extends to the statement that

$$
H_\lambda - Z^{4/3}\mu N - Z^{7/3} \sum_{k>m} \frac{z_k z_m}{|\mathbf{X}_k - \mathbf{X}_m|} - H_{\lambda,n} \geq -\frac{N}{2} v_{uu}(\mathbf{0}).
$$

It is easy to calculate that $v_{uu}(\mathbf{0}) \sim Z^\varepsilon$, so the right side is dominated by $Z^{7/3}$, and in the limit as $Z \to \infty$,

$$
Z^{-1}\Xi(Z^{-4/3}H_\lambda - \mu N) \leq Z^{-1}\Xi(Z^{-4/3}H_{\lambda,n}) - \sum_{k>m} \frac{z_k z_m}{|\mathbf{X}_k - \mathbf{X}_m|}.
$$

2. According to (4.1.24),

$$
Z^{7/3} \int d\Omega\, d\Omega' \frac{\rho_{\mathbf{q},\mathbf{p}} n(\mathbf{q}', \mathbf{p}')}{|\mathbf{q} - \mathbf{q}'|} = \sum_\alpha \int d^3x\, d^3x' a_\alpha^*(\mathbf{x}) a_\alpha(\mathbf{x}) v_u(\mathbf{x} - \mathbf{x}') n(\mathbf{x}'),
$$

where

$$
n(\mathbf{x}) = \int \frac{d^3p}{(2\pi)^3} n(\mathbf{x}, \mathbf{p}).
$$

Therefore the Coulomb repulsion of the electrons in the Hamiltonian $H_{\lambda,n}$ of (4.1.28) is reduced by $v_s = 1/r - v_u$. As in (4.1.14) the Hamiltonian $H_{\lambda,n}$ with $v_u$ in place of $1/r$ furnishes a lower bound for $\Xi$. On the other hand, it was shown in (4.1.18) and (4.1.20) that the effect of $v_s$ on $\Xi(H_{\lambda,n})$ was negligible as $Z \to \infty$. Moreover, $\int d\Omega \rho_{\mathbf{q},\mathbf{p}} f(\mathbf{q}, \mathbf{p})$ is the second quantization of the one-particle operator $\int d\Omega |\mathbf{q}, \mathbf{p}\rangle\langle\mathbf{q}, \mathbf{p}| f(\mathbf{q}, \mathbf{p}), |\mathbf{q}, \mathbf{p}\rangle = \exp(i\mathbf{p}\cdot\mathbf{x})u(\mathbf{x} - \mathbf{q})$, the expectation value of which in the state $|\mathbf{q}', \mathbf{p}'\rangle$ reduces to $f(\mathbf{q}', \mathbf{p}')$ in the limit $Z \to \infty$. The generalization of (4.1.21) is consequently

$$\lim_{Z\to\infty} Z^{-1} \ln \operatorname{Tr} \exp[-\beta(Z^{-4/3}H_\lambda - \mu N)]$$

$$= 2\int \frac{d^3q\, d^3p}{(2\pi)^3} \ln\left\{1 + \exp\left[-\beta\left(|\mathbf{p}|^2 - \sum_k \frac{z_k}{|\mathbf{q} - \mathbf{X}_k|}\right.\right.\right.$$

$$\left.\left.\left. + \int \frac{d^3q' n_\lambda(\mathbf{q}')}{|\mathbf{q} - \mathbf{q}'|} + W(\mathbf{q}) + f(\mathbf{q}, \mathbf{p})(\lambda_1 + \lambda_2 g_\lambda) - \frac{\lambda_2}{2} g_\lambda^2 - \mu\right)\right]\right\}$$

$$+ \beta\left(\frac{1}{2}\int \frac{d^3q\, d^3q'}{|\mathbf{q} - \mathbf{q}'|} n_\lambda(\mathbf{q}) n_\lambda(\mathbf{q}') - \sum_{k>l} \frac{z_k z_l}{|\mathbf{X}_k - \mathbf{X}_l|}\right), \tag{4.1.30}$$

where

$$n_\lambda(\mathbf{q}) = \int \frac{d^3p}{(2\pi)^3} n_\lambda(\mathbf{q}, \mathbf{p}),$$

$$n_\lambda(\mathbf{q}, \mathbf{p}) = 2\left\{\exp \beta\left[|\mathbf{p}|^2 - \sum_k \frac{z_k}{|\mathbf{q} - \mathbf{X}_k|} + \int \frac{d^3q' n_\lambda(\mathbf{q}')}{|\mathbf{q} - \mathbf{q}'|}\right.\right.$$

$$\left.\left. + W(\mathbf{q}) + f(\mathbf{q}, \mathbf{p})(\lambda_1 + \lambda_2 g_\lambda) - \frac{\lambda_2}{2} g_\lambda^2 - \mu\right] + 1\right\}^{-1},$$

$$g_\lambda = \int d\Omega\, n_\lambda(\mathbf{q}, \mathbf{p}) f(\mathbf{q}, \mathbf{p}),$$

and $|\lambda_2|$ is sufficiently small.

Differentiation by $\lambda_1$ and $\lambda_2$ at $\lambda_1 = \lambda_2 = 0$ and an optimization of $f \in C_0^2(\mathbb{R}^6)$ reveal the

**Convergence of the Expectation Values** (4.1.31)

$$\lim_{Z\to\infty} \langle\rho_{\mathbf{q},\mathbf{p}}\rangle_Z \equiv \lim_{Z\to\infty} \frac{\operatorname{Tr}(\rho_{\mathbf{q},\mathbf{p}} \exp[-\beta(Z^{-4/3}H_Z - \mu N)])}{\operatorname{Tr} \exp[Z^{-4/3}H_Z - \mu N]}$$

$$= 2\left\{\exp\left[\beta\left(|\mathbf{p}|^2 - \sum_k \frac{z_k}{|\mathbf{q} - \mathbf{X}_k|}\right.\right.\right.$$

$$\left.\left.\left. + \int \frac{d^3q' n_0(\mathbf{q}')}{|\mathbf{q} - \mathbf{q}'|} + W(\mathbf{q}) - \mu\right)\right] + 1\right\}^{-1} = n_0(\mathbf{q}, \mathbf{p}),$$

$$\lim_{Z\to\infty} \langle\rho_{\mathbf{q},\mathbf{p}}, \rho_{\mathbf{q}',\mathbf{p}'}\rangle_Z = n_0(\mathbf{q}, \mathbf{p}) n_0(\mathbf{q}', \mathbf{p}').$$

**Remarks** (4.1.32)

1. Since $f$ is not arbitrary, but assumed in $C_0^2(\mathbb{R}^6)$, the limit converges only in the sense of distributions. The $C^*$ algebra $\mathscr{A}_Z$ generated by the "smeared" densities on phase space, $\rho_g \equiv \int d\Omega g(\mathbf{q}, \mathbf{p})\rho_{\mathbf{q},\mathbf{p}}$, together with the identity becomes Abelian in the "weak" limit $Z \to \infty$. Hence, according to the Gel'fand isomorphism (III: 2.2.28), if $Z = \infty$, then $\mathscr{A}_Z$ can be represented as the set of continuous functions on a compact Hausdorff space. The space of characters of an Abelian $C^*$ algebra $\mathscr{A}$, i.e., *-homomorphisms from $\mathscr{A}$ to $\mathbb{C}$, is the same as the set $\mathscr{E}$ of pure states and is a compact Hausdorff space in the (relative) weak-* topology. With the identification $[a](\omega) = \omega(a) \in \mathbb{C}$ for all $a \in \mathscr{A}$ and $\omega \in \mathscr{E}$, $\mathscr{A}$ is equivalent to the $C^*$ algebra of the continuous functions with the supremum norm on the set $\mathscr{E}$, given the weak-* topology. In our case, $\mathscr{E} = \{n \in L^\infty(\mathbb{R}^6) | n \geq 0 \text{ a.e.}, \|n\|_\infty \leq 2\}$, with the weak-* topology with respect to the linear functionals belonging to the predual $L^1(\mathbb{R}^6)$. (Since $C_0^2(\mathbb{R}^6)$ is dense in $L^1(\mathbb{R}^6)$ in norm, the corresponding weak-* topologies agree on $L^\infty(\mathbb{R}^6)$.) Since $\mathscr{E}$ is the intersection of the cone of the functions that are nonnegative a.e., which is a weak-* closed set, with a multiple of the unit cube of $L^\infty$, it is weak-* compact. The Gel'fand isomorphism correlates $\rho_g$ with the mapping $[\rho_g](n) = \int ng\, d\Omega$, and since $\|[\rho_g] - [\rho_{g'}]\|_\infty \leq 2\|g - g'\|_1$, the completion contains for instance all $\rho_g$ such that $g \in L^1(\mathbb{R}^6)$. The set of all states on the algebra is the weak-* closure of the convex combinations of characters and can be represented as a set of probability measures; pure states correspond to point measures. If the state is mixed, $\alpha\langle\ \rangle_{n_1} + (1 - \alpha)\langle\ \rangle_{n_2}$, then the two-point function can not be factorized:

$$\begin{aligned}
\alpha\langle\rho_{z_1}\rho_{z_2}\rangle_{n_1} + (1-\alpha)\langle\rho_{z_1}\rho_{z_2}\rangle_{n_2} &= \alpha n_1(z_1)n_1(z_2) + (1-\alpha)n_2(z_1)n_2(z_2) \\
&= (\alpha\langle\rho_{z_1}\rangle_{n_1} + (1-\alpha)\langle\rho_{z_1}\rangle_{n_2})(\alpha\langle\rho_{z_2}\rangle_{n_1} \\
&\quad + (1-\alpha)\langle\rho_{z_2}\rangle_{n_2}) \quad \text{for all } z_1, z_2 \\
&\Rightarrow n_1(z) = n_2(z) \quad \text{for all } z = (\mathbf{q}, \mathbf{p}).
\end{aligned}$$

   Hence it follows from (4.1.31) that the limiting state is a character, and consequently pure.
2. Although the system acts classically on a distance scale $\sim Z^{-1/3}$, it would be expected to behave like a free Fermi gas on the scale $Z^{-2/3} \sim$ the average distance between particles $\sim$ reciprocal of momentum. If the microscopic field operators

$$a_{\mathbf{q}}(\xi) = Z^{-1}a(Z^{-1/3}\mathbf{q} + Z^{-2/3}\xi), \qquad [a_q(\xi), a_q^*(\xi')]_+ = \delta(\xi - \xi')$$

are introduced, it can be seen from (4.1.31) that its expectation value for free Fermions is

$$\int \frac{d^3p}{(2\pi)^3} \exp(i\mathbf{p}\cdot\boldsymbol{\xi})\rho_{\mathbf{q},\mathbf{p}}$$

$$= \int \frac{d^3p}{(2\pi)^3} \exp[i\mathbf{p}\cdot(Z^{2/3}(\mathbf{x}-\mathbf{x}') + \boldsymbol{\xi})]\, a^*(\mathbf{x})a(\mathbf{x}')\frac{Z^{3/2}}{\pi^{3/2}}$$

$$\times \exp\left[-\frac{Z}{2}(|\mathbf{x}-Z^{-1/3}\mathbf{q}|^2 + |\mathbf{x}'-Z^{-1/3}\mathbf{q}|^2)\right] d^3x\, d^3x'$$

$$= \int d^3z a^*_{\mathbf{q}+\mathbf{x}}\left(-\frac{\boldsymbol{\xi}}{2}\right)a_{\mathbf{q}+\mathbf{x}}\left(\frac{\boldsymbol{\xi}}{2}\right)\exp\left[-Z^{1/3}|\mathbf{x}|^2 - \frac{Z^{-1/3}|\boldsymbol{\xi}|^2}{4}\right]\frac{Z^{1/2}}{\pi^{3/2}},$$

where the chemical potential is determined by the potential $V(\mathbf{q})$ at the point $\mathbf{q}$, and we set $\varepsilon = \frac{1}{2}$, $u = \pi^{-3/4}\exp(-|\mathbf{x}|^2/2)$,

$$\int \frac{d^3p}{(2\pi)^3} \exp(i\mathbf{p}\cdot\boldsymbol{\xi})\rho_{\mathbf{q},\mathbf{p}}$$

$$= \int \frac{d^3p}{(2\pi)^3} \exp[i\mathbf{p}\cdot(Z^{2/3}(\mathbf{x}-\mathbf{x}') + \boldsymbol{\xi})]a^*(\mathbf{x})a(\mathbf{x}')\frac{Z^{3/2}}{\pi^{3/2}}$$

$$\times \exp\left[-\frac{Z}{2}(|\mathbf{x}-Z^{-1/3}\mathbf{q}|^2 + |\mathbf{x}'-Z^{-1/3}\mathbf{q}|^2)\right] d^3x\, d^3x'$$

$$= \int d^3x a^*_{\mathbf{q}+\mathbf{x}}\left(-\frac{\boldsymbol{\xi}}{2}\right)a_{\mathbf{q}+\mathbf{x}}\left(\frac{\boldsymbol{\xi}}{2}\right)\exp\left[-Z^{1/3}|\mathbf{x}|^2 - \frac{Z^{-1/3}|\boldsymbol{\xi}|^2}{4}\right]\frac{Z^{1/2}}{\pi^{3/2}}.$$

Therefore

$$\int \frac{d^3p\, \exp(i\mathbf{p}\cdot\boldsymbol{\xi})(2\pi)^{-3}}{\exp[\beta(|\mathbf{p}|^2 - V(\mathbf{q}))] + 1}$$

$$= \lim_{Z\to\infty} \int d^3x \frac{Z^{1/2}}{\pi^{2/3}} \exp\left[-Z^{1/3}|\mathbf{x}|^2 - \frac{Z^{-1/3}|\boldsymbol{\xi}|^2}{2}\right]$$

$$\times \left\langle a^*_{\mathbf{q}+\mathbf{x}}\left(-\frac{\boldsymbol{\xi}}{2}\right)a_{\mathbf{q}+\mathbf{x}}\left(\frac{\boldsymbol{\xi}}{2}\right)\right\rangle_{n_0} \equiv \left\langle a^*_{\mathbf{q}}\left(-\frac{\boldsymbol{\xi}}{2}\right)a_{\mathbf{q}}\left(\frac{\boldsymbol{\xi}}{2}\right)\right\rangle_{\infty},$$

$$V(\mathbf{q}) = -\sum_k \frac{z_k}{|\mathbf{q}-\mathbf{X}_k|} + \int \frac{d^3x n_0(\mathbf{x})}{|\mathbf{q}-\mathbf{x}|} + W(\mathbf{q}) - \mu.$$

3. Results have also been obtained concerning the time-evolution in the limit $Z \to \infty$ [26], but they have only been proved for regularized potentials $v_u$ and not for $1/r$, so they will not be presented here. At any rate the time-evolution of $\omega(a_t)$, where the nonstationary state $\omega$ has the

same scaling properties as the grand canonical state $\rho$, is the free time-evolution, as is that of $\rho(a_t b)$, when only the microscopic observables (4.1.32; 2) are considered. The equation for the expectation values of the macroscopic observables $\rho_{\mathbf{q},\mathbf{p}}$ is known as the **Vlasov equation**; it describes a classical time-evolution according to

$$\frac{dn}{dt} = \mathbf{p} \cdot \frac{\partial n}{\partial \mathbf{q}} - \frac{\partial V}{\partial \mathbf{q}} \cdot \frac{\partial n}{\partial \mathbf{p}},$$

where the potential itself depends on the particle density,

$$V(\mathbf{q}) = -\sum_k \frac{Z_k}{|\mathbf{q} - \mathbf{X}_k|} + \int \frac{d^3q'\, d^3p'}{(2\pi)^3} \frac{n(\mathbf{q}', \mathbf{p}')}{|\mathbf{q} - \mathbf{q}'|}.$$

Thomas–Fermi theory thus reduces the quantum-mechanical many-body problem to the solution of the integral equation (4.1.22). Although (4.1.22) is much simpler than the original Schrödinger equation, it can still be solved with reasonable numerical effort and skill only in the radially symmetric case. Despite that, some valuable relationships and properties can be obtained just from the maximum property.

### The Relationships among the Contributions to $\Xi$ (4.1.33)

*Write*

$$-\lim Z^{-1}\Xi(Z^{-4/3}H_Z - \mu N) - \sum_{k>l} \frac{z_k z_l}{|\mathbf{X}_k - \mathbf{X}_l|}$$

$$= \inf_{0 \le n \le 2} \int \frac{d^3q\, d^3p}{(2\pi)^3} \left\{ 2T\left[\frac{n(\mathbf{q}, \mathbf{p})}{2} \ln \frac{n(\mathbf{q}, \mathbf{p})}{2} + \left(1 - \frac{n(\mathbf{q}, \mathbf{p})}{2}\right) \ln\left(1 - \frac{n(\mathbf{q}, \mathbf{p})}{2}\right)\right] \right.$$

$$+ n(\mathbf{q}, \mathbf{p}) \left( -\mu + |\mathbf{p}|^2 - \sum_{j=1}^{M} \frac{z_j}{|\mathbf{q} - \mathbf{X}_j|} \right.$$

$$\left. \left. + \frac{1}{2} \int \frac{d^3q'\, d^3p'}{2(\pi)^3} \frac{n(\mathbf{q}', \mathbf{p}')}{|\mathbf{q} - \mathbf{q}'|} + W(\mathbf{q}) \right) \right\}$$

$$= -TS - \mu\lambda + K - A + R + W,$$

*where*

$$\lambda = \int \frac{d^3q\, d^3p}{(2\pi)^3} n(\mathbf{q}, \mathbf{p}) = \lim_{Z \to \infty} \frac{N}{Z},$$

*K is the kinetic energy of the electrons, A is the potential attracting the electrons to the nuclei, and R is the interelectronic repulsion. Then for the values of* $\mu$

*at which the infimum is attained as a minimum (at a given phase-space density* $n_0$),

(i) $-3(TS + \mu\lambda) + 5K - 3A + 6R + 3W = 0$; *and*
(ii) *if an atom is isolated and in the ground state, i.e.,* $M = 1, X_1 = 0, W = 0, T = 0$, *then*

$$-3\mu\lambda + 3K - 2A + 5R = 0.$$

**Proof**

(i) Take the infimum over $n'$ of the form $n_0(q, \gamma^{-1}p)$. A change of the variables of integration $\mathbf{p} \to \gamma_1 \mathbf{p}$ converts (4.1.33) into

$$-\gamma_1^3(TS + \mu\lambda + A - W) + \gamma_1^5 K + \gamma_1^6 R.$$

This has its minimum at $\gamma_1 = 1$ when condition (i) holds.
(ii) Now dilate $\mathbf{q}$ so that $n(\mathbf{q}, \mathbf{p}) = n_0(\gamma_2^{-1}\mathbf{q}, \mathbf{p})$, and proceed as before; then

$$\frac{d}{d\gamma_2}[\gamma_2^3(K - \mu\lambda) - \gamma_2^2 A + \gamma_2^5 R]|_{\gamma_2=1} = 0$$

yields Relationship (ii). □

**Corollary** (4.1.34)

In case (ii) with $\mu = 0$, the three contributions to the energy stand in the ratio

$$K:A:R = 3:7:1.$$

**Remarks** (4.1.35)

1. The dilatation required for (ii) affects the nuclear coordinates (other than $\mathbf{X}_1 = \mathbf{0}$) and the wall. The reason for setting $T = 0$ was to avoid problems connected with the latter.
2. Since $A$, $K$, and $R$ are positive, the second derivatives at $\gamma = 1$ are automatically positive.
3. If $T = \mu = 0$, then $-\Xi$ becomes the minimum of the energy without fixed particle number. We shall learn that the minimum is achieved by a neutral system in Thomas–Fermi theory, and that in case (ii)

$$\int \frac{d^3q\, d^3p}{(2\pi)^3} n_0(\mathbf{q}, \mathbf{p}) = z_1.$$

The comparison densities $n(\gamma^{-1}\mathbf{q}, \mathbf{p})$ and $n(\mathbf{q}, \gamma^{-1}\mathbf{p})$ correspond to different numbers of particles, and the mystical numbers in (4.1.34) reflect the stability of neutral atoms against spontaneous ionization.

In the limit $T \to 0$, the quantity $(\exp[\beta(\varepsilon - a)] + 1)^{-1}$ approaches $\Theta(a - \varepsilon)$. In that case $W$ may be chosen identically zero, and the integration over $\mathbf{p}$ becomes elementary. The computation yields

**The Electron Density in Configuration Space** (4.1.36)

$$\rho(\mathbf{x}) \equiv \int \frac{d^3p}{(2\pi)^3} n_0(\mathbf{x}, \mathbf{p}) = \frac{1}{3\pi^2} |\Phi(\mathbf{x}) + \mu|_+^{3/2}, \qquad |z|_+ \equiv |z|\Theta(z),$$

$$\Phi(\mathbf{x}) \equiv \sum_j \frac{z_j}{|\mathbf{x} - \mathbf{X}_j|} - \int \frac{d^3x' \, \rho(\mathbf{x}')}{|\mathbf{x} - \mathbf{x}'|}.$$

The kinetic-energy density is

$$\int \frac{d^3p}{(2\pi)^3} |\mathbf{p}|^2 n_0(\mathbf{x}, \mathbf{p}) = \tfrac{3}{5}(3\pi^2)^{2/3} \rho^{5/3}(\mathbf{x}).$$

(Since the particles have spin 1/2, the factor $(6\pi^2)^{2/3}$ of (2.5.32) has become $(3\pi^2)^{2/3}$.)

This reveals

**The Range of Values of $\mu$ and $\Phi(x)$** (4.1.37)

(i) *$\mu$ takes on the values $-\infty < \mu \leq 0$; and*
(ii) *$\Phi$ takes on the values $0 \leq \Phi < \infty$.*

**Proof**

We shall only demonstrate the impossibility of $\mu > 0$ and $\Phi < 0$; Problem 3 will assure us that a minimizing $\rho$ exists for all $\mu \leq 0$, and it can be seen directly that $\Phi(\mathbf{x})$ ranges over $[0, \infty)$ as $\mathbf{x}$ ranges over $\mathbb{R}^3$.

(i) Since $\rho(\mathbf{x})$ must be integrable, $\Phi(\mathbf{x}) \to 0$ as $|\mathbf{x}| \to \infty$. If $\mu > 0$, then $\rho(\mathbf{x})$ would have to approach $\mu^{3/2}/3\pi^2$ as $|\mathbf{x}| \to \infty$, which would contradict integrability.
(ii) The set $A \equiv \{\mathbf{x} \in \mathbb{R}^3 : \Phi(\mathbf{x}) < 0\}$ is open and does not contain $\mathbf{x}_i$. Because $\mu \leq 0$, the density $\rho$ vanishes identically on $A$, so $\Delta\Phi(\mathbf{x}) = 0$ holds throughout $A$. Since $\Phi$ equals zero on the boundary of $A$ and at infinity and is harmonic, it would have to equal zero on $A$, because its maximum would be attained either on $\partial A$ or at infinity. However, this contradicts the definition of $A$, so $A$ must be empty. □

The quantity $\lambda \equiv \int d^3x \rho(\mathbf{x}) = \lim_{Z \to \infty} N/Z$, where $N$ is the number of electrons and $Z$ is the sum of the nuclear charges, is more intuitively understandable than $\mu$. By expressing the energy as a function of $\lambda$, we can find the limits of the observables studied in (III: §4.5).

**Properties of the Thomas–Fermi Functional at $T = 0$ (4.1.38)**

*Let*

$$K(\rho) = \tfrac{3}{5}(3\pi^2)^{2/3} \int d^3x \rho^{5/3}(\mathbf{x}),$$

$$A(\rho) = \sum_j z_j \int \frac{d^3x \rho(\mathbf{x})}{|\mathbf{x} - \mathbf{X}_j|},$$

$$R(\rho) = \frac{1}{2} \int \frac{d^3x\, d^3x'}{|\mathbf{x} - \mathbf{x}'|} \rho(\mathbf{x})\rho(\mathbf{x}'),$$

$$E(\rho) \equiv K(\rho) - A(\rho) + R(\rho),$$

*and*

$$S_\lambda = \left\{\rho: \rho(\mathbf{x}) \geq 0, \int d^3x \rho(\mathbf{x}) = \lambda\right\}, \qquad \sum_j z_j = 1.$$

*Then* $E[\lambda] = \inf_{\rho \in S_\lambda} E(\rho)$ *satisfies*

(i) $$E[\lambda] = -\inf_\mu \left(\Xi_\infty(\mu, 0) - \mu\lambda + \sum_{k>l} \frac{z_k z_l}{|\mathbf{X}_k - \mathbf{X}_l|}\right),$$

$$\Xi_\infty(\mu, T) = \lim_{Z \to \infty} Z^{-1} T \ln \operatorname{Tr} \exp[-\beta(Z^{-4/3}H - \mu N)];$$

(ii) $\partial E/\partial\lambda = \mu$ *if* $\lambda \leq 1$, *and* $= 0$ *if* $\lambda > 1$;

(iii) $E[\lambda]$ *is a nonpositive, convex, decreasing function of* $\lambda$; *and*

(iv) *in the atomic case* $z_1 = 1$, *all other* $z_i = 0$, $-\lambda^{-1/6}(-E[\lambda])^{1/2}$ *is a concave, increasing function of* $\lambda$

**Proof**

(i) Observe first that $E[\lambda]$ is convex, since the convexity of $E(\rho)$ as a function of $\rho$ means that $E[\alpha\lambda_1 + (1 - \alpha)\lambda_2] \leq E(\alpha\rho_1 + (1 - \alpha)\rho_2) \leq \alpha E[\lambda_1] + (1 - \alpha)E[\lambda_2]$, in which $E[\lambda_i] = E(\rho_i)$, because $\alpha\rho_1 + (1 - \alpha)\rho_2 \in S_{\alpha\lambda_1 + (1-\alpha)\lambda_2}$. As remarked in (2.4.15; 2(i)), the Legendre transformation

$$-\Xi_\infty(\mu, 0) = \inf_\lambda \inf_{\rho \in S_\lambda} \left(E(\rho) - \mu\lambda + \sum_{k>m} \frac{z_k z_m}{|\mathbf{X}_k - \mathbf{X}_m|}\right)$$

can be inverted for the concave function $-E[\lambda]$, yielding (i).

(ii) The formula $dE/d\lambda = \mu$ will follow from Property (i) once $E[\lambda]$ has been shown to be differentiable. Let $\rho_\lambda$ denote the minimizing $\rho$ (4.1.36). A calculation shows that

$$\frac{\partial}{\partial t} E(\rho_\lambda(1 + t))|_{t=0} = \mu\lambda,$$

so $E[(1+t)\lambda] - E[\lambda] \leq t\mu\lambda + o(t)$ and $E[(1-t)\lambda] - E[\lambda] \leq -t\mu\lambda + o(t)$. In the limit $t \to 0$, this becomes $dE/d\lambda = \mu$. It remains to show that $\lambda < 1 \Leftrightarrow \mu < 0$ and $\lambda = 1 \Rightarrow \mu = 0$, which $\Rightarrow \lambda \geq 1$. Note that $\Phi$ goes asymptotically as $(1-\lambda)/r$. If $\mu$ were 0, then

$$\rho \overset{r \to \infty}{\sim} \left(\frac{1-\lambda}{r}\right)^{3/2},$$

which would not be integrable; thus $\mu$ must be negative when $\lambda < 1$. When $\lambda > 1$, there is no minimum, since if there were, then $\Phi$ would be negative as $r \to \infty$, which is impossible because of (4.1.37). However, the infimum has to be $E(1)$, since for $\lambda > 1$ and for any $\varepsilon > 0$ a $\rho$ can be constructed such that $E(\rho) < E(1) + \varepsilon$; start with a $\rho_1$ with $\lambda = 1$ and compact support, and such that $E(\rho_1) \leq E(1) + \varepsilon/2$, and then let

$$\rho_k = \rho_1 + \frac{1}{k}\chi_k, \qquad k \in \mathbb{N},$$

where the characteristic functions $\chi_k$ satisfy $\chi_k \rho_1 \equiv 0$ and $\|\chi_k\|_1 = k(\lambda - 1)$ to ensure that $\rho_k \in S_\lambda$. Then $\|\rho_1 - \rho_k\|_p \to 0$ for all $p > 1$, and it is easy to verify that $E(\rho_k) \to E(\rho_1)$. This accords with the intuitive feeling that a thin electron cloud at a great distance affects the energy only slightly. It means that $E[\lambda]$ decreases while $0 < \lambda < 1$, and becomes constant thereafter.

(iii) This follows from the proofs of (i) and (ii), since $\mu \leq 0$.

(iv) Make both of the scaling transformations of (4.1.33) simultaneously and define

$$\inf_{\rho \in S_1}(K(\rho) - ZA(\rho) + \alpha R(\rho)) = Z^2 \inf_{\rho \in S_1}\left(K(\rho) - A(\rho) + \frac{\alpha}{Z} R(\rho)\right)$$
$$\equiv Z^2 f\left(\frac{\alpha}{Z}\right).$$

This is the infimum of a set of linear functions and consequently concave in $(Z, \alpha)$. The condition that

$$\frac{\partial^2}{\partial Z^2}\frac{\partial^2}{\partial \alpha^2} \leq \left(\frac{\partial^2}{\partial Z \alpha}\right)^2$$

implies that $2f'' \leq f'^2/f$, so $-\sqrt{-f}$ is concave. Because $f' = R(\rho) > 0$, the function $f$ is increasing. With still another scaling transformation, with $\rho(\mathbf{x}) = \lambda\bar{\rho}(\lambda^{2/3}\mathbf{x})$,

$$f(\lambda) = \inf_{\rho \in S_1}(K(\rho) - A(\rho) + \lambda R(\rho)) = \lambda^{-1/3} \inf_{\bar{\rho} \in S^\lambda}(K(\bar{\rho}) - A(\bar{\rho}) + R(\bar{\rho}))$$
$$= \lambda^{-1/3}E[\lambda]. \qquad \square$$

The at first sight contradictory properties (iii) and (iv) determine the form of $E[\lambda]$ rather narrowly for an atom, making it almost linear:

**Properties of $f(\lambda) = \lambda^{-1/3}E[\lambda]$ for an Atom** (4.1.39)

(i) $0 \leq f' \leq -\dfrac{1}{3\lambda}f$;

(ii) $\dfrac{2f}{9\lambda^2} - \dfrac{2f'}{3\lambda} \leq f'' \leq \dfrac{f'^2}{2f}$.

**Proof**

(i) This follows from $E' < 0$ and $f' = \lambda^{-4/3}R(\rho_\lambda) = R(\rho_1) > 0$, where $\rho_\lambda$ and $\rho_1$ are the minimizing densities of $S_\lambda$ and $S_1$.
(ii) This follows from $E'' \geq 0$ and the concavity of $-\sqrt{-f}$. □

**Consequence** (4.1.40)

1. With the aid of the virial theorem, $2K = A - R$, which follows from (4.1.33) for any $\mu$, Property (i) may be rewritten as $7R(\rho_\lambda) < A(\rho_\lambda)$, $0 \leq \lambda < 1$. This generalizes Corollary (4.1.34), which held for $\lambda = 1 \Rightarrow \mu = 0$, to the statement that $7R = A$, provided that $0 \leq \lambda < 1$.
2. It is not hard to calculate analytically that $f(0) = -0.572$ and $f'(0) = 0.2424$ (Problem 4); computer analysis of the Thomas–Fermi equation has shown that $f(1)$ is $-0.384$, and by (4.1.38(ii)) and (4.1.34), $f'(1) = -f(1)/3$. Integrating Property (ii) leads to the bounds

$$\max\{-\lambda^{-1/6}|f(1)|^{1/2}, -\lambda|f(1)|^{1/2} - (1-\lambda)|f(0)|^{1/2}\} \leq -|f(\lambda)|^{1/2}$$
$$\leq \min\left\{-|f(0)|^{1/2}\left(1 + \frac{\lambda}{2}\frac{f'(0)}{f(0)}\right), -|f(1)|^{1/2}\frac{7-\lambda}{6}\right\}$$

(cf. (III: 4.3.21)). The concave hull of the left side can be taken, in which case the greatest difference between the bounds is <2% (see Figure 29). Since this is already better accuracy than that of the Thomas–Fermi theory itself, there is no point in making fancy numerical calculations of $E[\lambda]$.

If from (4.1.36) we now deduce

**The Thomas–Fermi Equation** (4.1.41)

in the form

$$\Delta\Phi(\mathbf{x}) = -4\pi\delta^3(\mathbf{x}) + 4\pi\rho(\mathbf{x}) = -4\pi\delta^3(\mathbf{x}) + \frac{4}{3\pi}|\mu + \Phi(\mathbf{x})|_+^{3/2},$$

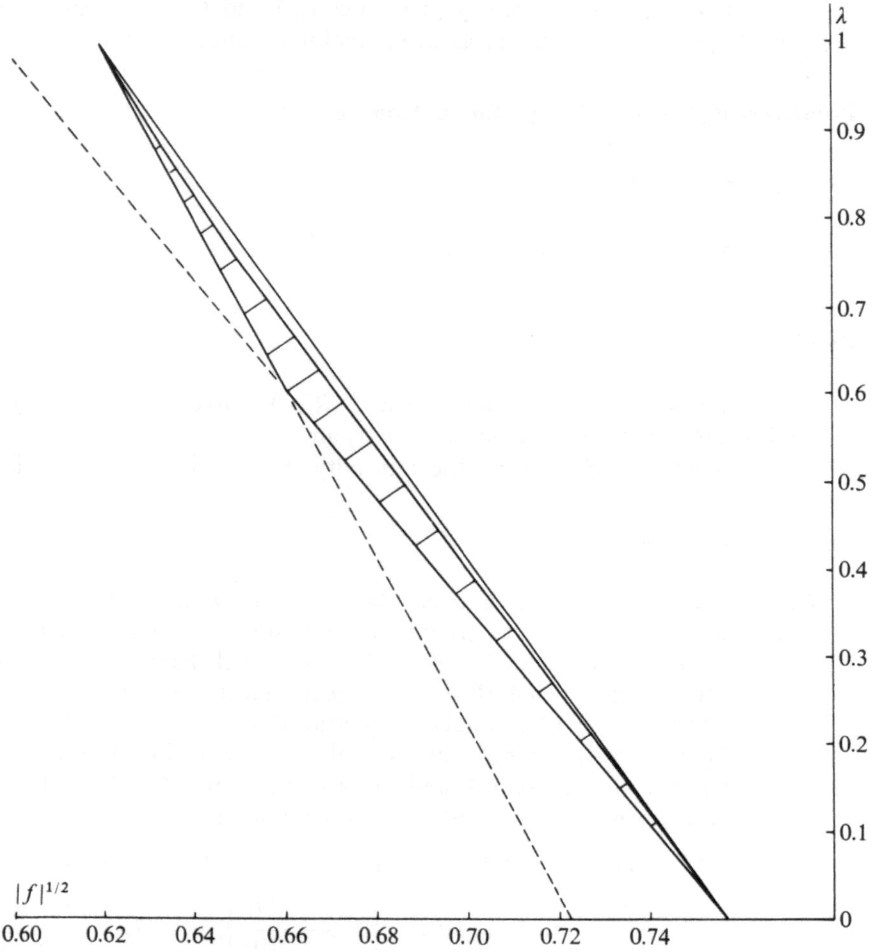


Figure 29 The bounds (4.1.40; 2) from the concavity of $f(\lambda) = \lambda^{-1/3}E(\lambda)$. The hatched region is allowed.

then it reduces to $\sqrt{\zeta}\chi'' = \chi^{3/2}\Theta(\chi)$ for spherically symmetric densities, with the substitution $|\mathbf{x}| = r = \zeta(3\pi/4)^{2/3}$, $\Phi(\mathbf{x}) + \mu = \chi(\zeta)/r$. The delta function is taken care of by the boundary condition $\chi(0) = 1$. The second boundary condition, required to make the solution unique, is $\chi'(\infty) = \mu$, which follows from $\int \rho \leq 1$ with Gauss's theorem. The function $\chi$ is concave and decreasing, and has the limiting forms

$$\chi(\zeta) \begin{array}{l} \xrightarrow{\zeta \to 0} 1 - 1.59\zeta \\ \xrightarrow[\zeta \to \infty]{} 144/\zeta^3 \end{array}$$

for $\mu = 0$. This means that for neutral atoms $\rho$ behaves like $r^{-3/2}$ at small $r$, and like $r^{-6}$ at large $r$. A numerical solution is shown in Figure 30. A compu-

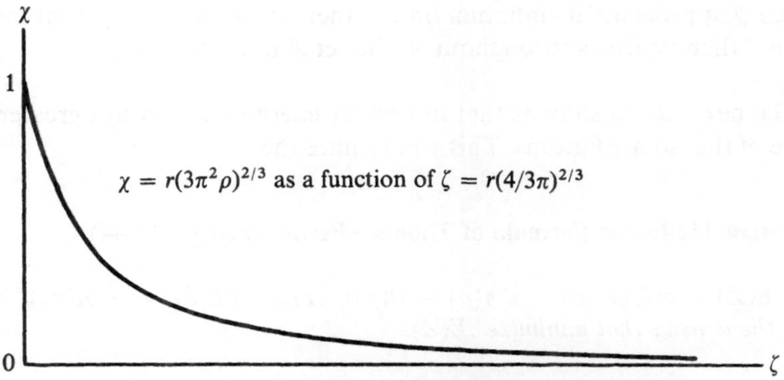


Figure 30 The Thomas–Fermi density of an atom.

tation of the energy of the solution yields the value $E(1) = -0.384$, i.e., $-0.77$ atomic units, or $-20.7$ eV.

The final proposition deduced from Thomas–Fermi theory will be that there is no chemical binding, which means that actual chemical binding energies must be smaller than the errors in the theory. In §4.3 it will be learned that this theory with some constants changed gives a lower bound for quantum-mechanical energies even for finite $Z$, and thereby leads to a simple proof of the stability of matter. Finally, we shall obtain the long-deferred proof of Inequality (4.1.5).

**Monotony of the Thomas–Fermi Potential with Respect to the Nuclear Charges** (4.1.42)

*Let $\Phi_{1,2}$ and $\rho_{1,2}$ be the solutions of the Thomas–Fermi equation with $\mu = 0$ and nuclear charges $z_k^{(1,2)}$. If $z_k^{(1)} \geq z_k^{(2)}$ for all k, then $\Phi_1(\mathbf{x}) \geq \Phi_2(\mathbf{x})$ and $\rho_1(\mathbf{x}) \geq \rho_2(\mathbf{x})$ for all* **x**.

**Remarks** (4.1.43)

1. The normalization $\sum_k z_k^{(i)} = 1$ has of course been dropped.
2. The condition $\mu = 0$ means $\int d^3x\rho_1(\mathbf{x}) = \sum_k z_k^{(1)} \geq \int d^3x\rho_2(\mathbf{x}) = \sum_k z_k^{(2)}$.
3. This can be interpreted as showing how increasing all the nuclear charges causes the configuration with lower energy to have a higher electron density.

**Proof**

As in the proof of (4.1.37(ii)), let $A \equiv \{\mathbf{x} \in \mathbb{R}^3 : \Phi_1(\mathbf{x}) < \Phi_2(\mathbf{x})\}$. Then $A$ is open and contains none of the $\mathbf{X}_k$, and on it $\psi(\mathbf{x}) \equiv \Phi_1(\mathbf{x}) - \Phi_2(\mathbf{x})$ is negative, continuous, and satisfies

$$\Delta\psi|_A = (\Phi_1^{3/2} - \Phi_2^{3/2})|_A < 0.$$

Hence $\psi$ approaches its infimum on $A$ either on the boundary or at infinity. Since it then vanishes throughout $A$, the set $A$ must be empty. □

The next fact to show is that molecular energies are always greater than those of the isolated atoms. This will require the

**Feynman–Hellmann Formula of Thomas–Fermi Theory** (4.1.44)

*Let* $E(Z) = \inf_\rho (K(\rho) - ZA(\rho) + R(\rho))$. *Then* $\partial E/\partial Z = -A(\rho_Z)$, *where* $\rho_Z$ *is the density that minimizes* $E(Z)$.

**Proof**

The function $E(Z)$ is concave, and its right and left derivatives are $\lim_{\varepsilon \downarrow 0}(-A(\rho_{Z\pm\varepsilon}))$, another consequence of the interplay between the concavity of $E(Z)$ and the convexity of the functional in the variable $\rho$ as in (4.1.38(ii)). Since, as shown in Problem 3, for any $Z$ there exists a unique minimizing $\rho_Z$ on a certain compact set, the densities $\rho_Z$ depend continuously on $Z$. In fact the individual contributions to $E(Z)$ are continuous in $Z$ as well as $E(Z)$ itself. Therefore both the right and the left derivative coincide with $-A(\rho_Z)$. □

Let us now start treating $E$ as a function of each of the nuclear charges, so

$$E(z_1, \ldots, z_M) = \inf_{\rho \in S}\left\{\tfrac{3}{5}(3\pi^2)^{2/3}\int \rho^{5/3} - \sum_{k=1}^{M} z_k \int \frac{\rho(\mathbf{x})}{|\mathbf{x} - \mathbf{X}_k|}\right.$$
$$\left. + \frac{1}{2}\int \frac{\rho(\mathbf{x})\rho(\mathbf{y})}{|\mathbf{x} - \mathbf{y}|} + \sum_{k>i} \frac{z_k z_i}{|\mathbf{X}_k - \mathbf{X}_i|}\right\},$$

and define

$$\begin{aligned} E(Z) &= E(Zz_1, Zz_2, \ldots, Zz_M), \\ E_1(Z) &= E(Zz_1, \ldots, Zz_j, 0, 0, \ldots), \\ E_2(Z) &= E(0, \ldots, 0, Zz_{j+1}, \ldots, Zz_M). \end{aligned}$$

Let $\rho \geq \rho_{1,2}$ and $\Phi \geq \Phi_{1,2}$ be the solutions of the appropriately subscripted Thomas–Fermi equations. Then

$$\begin{aligned} \frac{\partial E_1}{\partial Z} &= \sum_{k=1}^{j} z_k\left\{Z \sum_{i \neq k} \frac{z_i}{|\mathbf{X}_i - \mathbf{X}_k|} - \int \frac{d^3x\rho(\mathbf{x})}{|\mathbf{x} - \mathbf{X}_k|}\right\} \\ &= \sum_{k=1}^{j} z_k \lim_{\mathbf{x} \to \mathbf{X}_k}\left(\Phi_1(\mathbf{x}) - \frac{Zz_k}{|\mathbf{x} - \mathbf{X}_k|}\right), \end{aligned}$$

and likewise for $E_2$. The difference between the energy of the total system and the sum of the energies of the subsystems is easily found to satisfy

$$\frac{\partial E}{\partial Z} - \frac{\partial E_1}{\partial Z} - \frac{\partial E_2}{\partial Z} = \sum_{k=1}^{j} z_k(\Phi(\mathbf{x}_k) - \Phi_1(\mathbf{x}_k)) + \sum_{k=j+1}^{M} z_k(\Phi(\mathbf{x}_k) - \Phi_2(\mathbf{x}_k)) \geq 0.$$

Since $E$ and $E_{1,2}$ become zero when $Z = 0$, this calculation proves the

**Instability of Molecules in Thomas–Fermi Theory** (4.1.45)

$$E(z_1, \ldots, z_M) \geq E(z_1, \ldots, z_j) + E(z_{j+1}, \ldots, z_M).$$

**Remarks** (4.1.46)

1. In the absence of nuclear repulsion the inequality is reversed; in that case
$$\frac{\partial E}{\partial Z} - \frac{\partial E_1}{\partial Z} - \frac{\partial E_2}{\partial Z} = \sum_{k=1}^{j} z_k \int \frac{\rho_1(\mathbf{x}) - \rho(\mathbf{x})}{|\mathbf{x} - \mathbf{X}_k|} + \sum_{k=j+1}^{M} \int \frac{\rho_2(\mathbf{x}) - \rho(\mathbf{x})}{|\mathbf{x} - \mathbf{X}_k|} \leq 0.$$
Although Thomas–Fermi theory predicts some attraction between the nuclei, it is weaker than their Coulomb repulsion. It can even be shown that if the nuclear coordinates are scaled by $X_k \to RX_k$, then $E$ is a convex, decreasing function of $R$. Thus Thomas–Fermi theory predicts positive pressure and compressibility. However, the molecular energy is not a sum of pair potentials, but contains many-body potentials with alternating signs [34].
2. An alternative version of this theorem reads
$$\sum_{k>i} \frac{z_k z_i}{|\mathbf{X}_k - \mathbf{X}_i|} \geq \sum_{k=1}^{M} z_k \int \frac{\rho(\mathbf{x})}{|\mathbf{x} - \mathbf{X}_k|} - \frac{1}{2} \int \frac{\rho(\mathbf{x})\rho(\mathbf{y})}{|\mathbf{x} - \mathbf{y}|} - \frac{3}{5}(3\pi^2)^{2/3} \int \rho^{5/3} + \sum_k E(z_k)$$
for all $\mathbf{X}_k \in \mathbb{R}^3$ and $\rho \in S$. If $K(\rho)$ is replaced with $(1/\gamma)K(\rho)$, then, because of the way dilatations affect single atoms, $E(z_k)$ becomes $\gamma E(z_k)$. The computed value $E(1) = -0.384$ then leads to Equation (4.1.5), provided that $\mathbf{X}_k$ are interpreted as the coordinates of the electrons.
3. The proof of (4.1.45) works the same way for a Yukawa potential $\exp(-\mu r)/r$ in place of $1/r$. Because $\Delta \exp(-\mu r)/r + 4\pi\delta^3(\mathbf{x}) = \mu^2 \exp(-\mu r)/r > 0$, the argument with subharmonicity likewise works: $\Delta\psi|_A = \Phi_1^{3/2} - \Phi_2^{3/2} + \mu^2(\Phi_1 - \Phi_2) < 0$, which implies that $A$ must be empty.

**Problems** (4.1.47)

1. Let $H = |\mathbf{p}|^2 + V(\mathbf{x})$ act on $L^2(\mathbb{R}^3)$, and assume that $|V|_- \in L^{5/2}(\mathbb{R}^3)$ and let $e_i$ be the negative eigenvalues of $H$. Use the bound of Ghirardi and Rimini (III: 3.5.37; 2) to show that

$$\sum_i |e_i| \leq \frac{4}{15\pi} \int d^3x |V(\mathbf{x})|_-^{5/2}$$

and derive Inequality (4.1.19) from this fact.

2. Use Problem 1 to prove the inequality

$$T \equiv \langle\psi| - \sum_{i=1}^{N} \Delta_i |\psi\rangle \geq \frac{3}{5}\left(\frac{3\pi}{4}\right)^{2/3} \int d^3x \rho^{5/3}(\mathbf{x})$$

for spin $-\frac{1}{2}$ fermions, where

$$\rho(\mathbf{x}_1) = N \sum_{\alpha_i} \int d^3x_2 \cdots d^3x_N |\psi(\mathbf{x}_1, \mathbf{x}_2, \ldots, \mathbf{x}_N; \alpha_2, \ldots, \alpha_N)|^2,$$

$\alpha$ being the spin index. (Hint: use $\rho^{2/3}$ as the potential in Problem 1.)

3. Show that the sets $S \equiv \{\rho \in L^1 \cap L^{5/3}: \rho \geq 0, \|\rho\|_1 \leq N, \|\rho\|_{5/3} \leq K\}$ are compact in the weak $L^{5/3}$ topology, and that the functional $S \to \mathbb{R}$:

$$\varepsilon(\rho) = \tfrac{3}{5}(3\pi^2)^{2/3} \int d^3x \rho^{5/3}(\mathbf{x}) - \int d^3x \rho(\mathbf{x}) \left(\sum_k \frac{z_k}{|\mathbf{x} - \mathbf{X}_k|} + \mu\right)$$
$$+ \frac{1}{2} \int \frac{d^3x\, d^3x'}{|\mathbf{x} - \mathbf{x}'|} \rho(\mathbf{x})\rho(\mathbf{x}') + \sum_{k>j} \frac{z_k z_j}{|\mathbf{X}_k - \mathbf{X}_j|}$$

has Properties (4.1.10) if $\mu \leq 0$: It is

(i) weakly $L^{5/3}$ lower semicontinuous;
(ii) strictly convex; and
(iii) $\geq \frac{3}{5}(3\pi^2)^{2/3} \|\rho\|_{5/3}^{5/3} - 3(\frac{2}{3})^{5/6}(8\pi)^{1/3} \|\rho\|_{5/3}^{5/6} + |\mu| \|\rho\|_1$.

Conclude that the infimum is attained, and in fact precisely with the $\rho$ of (4.1.36).

4. Solve the Thomas–Fermi equation without Coulomb repulsion, compare with (III: 4.5.9), and conclude that the next correction is $0(N^{6/3})$. Use the solution to calculate $f(0)$ and $f'(0)$ of (4.1.40; 2).

5. Minimize the functional

$$E(\rho) = \int d^3x \left(\frac{\rho^2(\mathbf{x})2\pi}{\mu^2} - \sum_i \frac{z_i}{|\mathbf{x} - \mathbf{X}_i|} \rho(\mathbf{x}) + \frac{1}{2} \int \frac{d^3y \rho(\mathbf{y})}{|\mathbf{x} - \mathbf{y}|}\right) + \sum_{i>j} \frac{z_i z_j}{|\mathbf{X}_i - \mathbf{X}_j|},$$

and use the result for a new derivation of (III: 4.5.24):

$$\sum_{i>j} |\mathbf{X}_i - \mathbf{X}_j|^{-1} \geq \sum_{j=1}^{N} \int \frac{\rho(\mathbf{x})}{|\mathbf{x} - \mathbf{X}_i|} - \frac{1}{2} \int \frac{\rho(\mathbf{x})\rho(\mathbf{y})}{|\mathbf{x} - \mathbf{y}|} - \frac{2\pi}{\mu^2} \int \rho^2 - \frac{\mu N}{2}$$

for all $\mathbf{X}_i \in \mathbb{R}^3$, $\rho \in L^1 \cap L^2$.

**Solutions** (4.1.48)

1. Let $N_E(V)$ denote the number of eigenvalues less than or equal to $E$. According to (III: 3.5.37; 2), for all $\alpha > 0$,

$$N_{-\alpha}(V) \leq N_{-\alpha/2}\left(\left|V + \frac{\alpha}{2}\right|_-\right) \leq \operatorname{tr}\left[\left(|\mathbf{p}|^2 + \frac{\alpha}{2}\right)^{-1/2} \left|V + \frac{\alpha}{2}\right|_- \left(|\mathbf{p}|^2 + \frac{\alpha}{2}\right)^{-1/2}\right]^2$$

$$= \frac{1}{(4\pi)^2}\int d^3x\, d^3y \left|V(\mathbf{x}) + \frac{\alpha}{2}\right|_- \frac{\exp(-\sqrt{2\alpha}|\mathbf{x}-\mathbf{y}|)}{|\mathbf{x}-\mathbf{y}|^2}\left|V(\mathbf{y}) + \frac{\alpha}{2}\right|_-$$

$$\leq \frac{1}{4\pi\sqrt{2\alpha}}\int d^3x \left|V(\mathbf{x}) + \frac{\alpha}{2}\right|_-^2 .$$

The last step used Young's inequality, $\|f\cdot(v * g)\|_1 \leq \|v\|_1 \|f\|_2 \|g\|_2$. Now simply think about what $N_E(V)$ means (see Figure 31).

$$\sum_j |e_j(V)| = \int_0^\infty d\alpha N_{-\alpha}(V) \leq \frac{\sqrt{2}}{8\pi}\int d^3x \int_0^{2|V(\mathbf{x})|_-} \frac{d\alpha}{\sqrt{\alpha}}\left(V(\mathbf{x}) + \frac{\alpha}{2}\right)^2$$

$$= \frac{4}{15\pi}\int d^3x |V(\mathbf{x})|_-^{5/2}.$$

If $|V|_- \in L^{5/2}$, then the negative part of the spectrum of $H$ is discrete, and we may also write

$$\operatorname{Tr}||\mathbf{p}|^2 + V(\mathbf{x})|_- \leq 4\pi \int \frac{d^3x\, d^3p}{(2\pi)^3} ||\mathbf{p}|^2 + V(\mathbf{x})|_- .$$

The partition function can be bounded with the observation that

$$\ln(1 + \exp(-\beta H)) = \int_{-\infty}^{\infty} dE |H - E|_- \beta(1 + \exp(\beta E))^{-1} \Rightarrow$$

$$\operatorname{tr}\ln(1 + \exp[-\beta(|\mathbf{p}|^2 + V(\mathbf{x}))]) \leq 4\pi \int \frac{d^3x\, d^3p}{(2\pi)^3} \ln(1 + \exp[-\beta(|\mathbf{p}|^2 + V(\mathbf{x}))]).$$

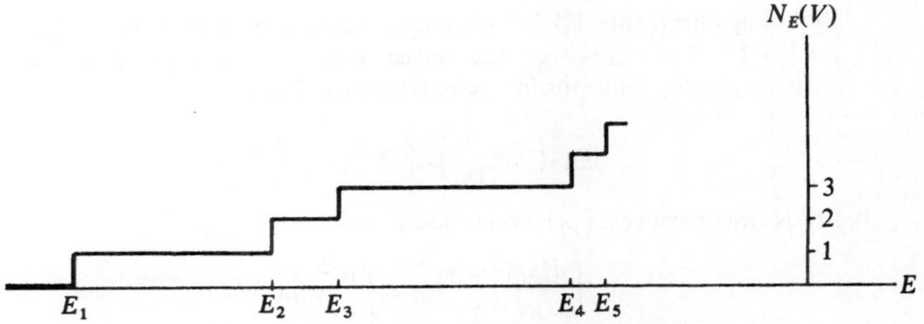


Figure 31 The dependence of the function $N_E(V)$ on $E$.

2. Let

$$\rho_\pm(\mathbf{x}_1) = N \sum_{\alpha_2,\ldots,\alpha_N} \int d^3x_2 \cdots d^3x_N |\psi(\mathbf{x}_1, \mathbf{x}_2, \ldots, \mathbf{x}_N; \pm, \alpha_2, \ldots, \alpha_N)|^2$$

be the densities of the electrons with spin $\pm\frac{1}{2}$ and $K = T(\int \rho_+^{5/3} + \int \rho_-^{5/3})^{-1}$. Because of Problem 1 and the min–max principle the lowest energy $E_0$ of the Hamiltonian

$$H = \sum_i \left( |\mathbf{p}_i|^2 - \frac{5K}{3} [\pi_{i,+} \rho_+^{2/3}(\mathbf{x}_i) + \pi_{i,-} \rho_-^{2/3}(\mathbf{x}_i)] \right),$$

where $\pi_{i,\pm}$ are the spin projections, satisfies the inequalities

$$-\frac{4}{15\pi}\left(\frac{5K}{3}\right)^{5/2}\left(\int \rho_+^{5/3} + \int \rho_-^{5/3}\right) \le E_0 \le \langle\psi|H|\psi\rangle = T - \frac{5K}{3}\left(\int \rho_+^{5/3} + \int \rho_-^{5/3}\right).$$

This implies that $K \ge \frac{3}{5}(3\pi/2)^{2/3}$, and then the convexity of the function $x \to x^{5/3}$ yields the inequality for $\rho = \rho_+ + \rho_-$.

3. Since

$$\|\rho\|_{5/3} = \sup_{\|\rho'\|_{5/2}=1} |\langle\rho'|\rho\rangle| \quad \text{and} \quad \|\rho\|_1 = \sup_{\substack{\rho' \in L^{5/2} \cap L^\infty \\ \|\rho'\|_\infty = 1}} |\langle\rho'|\rho\rangle|$$

are suprema over weakly continuous functions, they are weakly lower semicontinuous, so $S$ is weakly compact.

(i) This proposition is equivalent to the statement that $\rho_n \rightharpoonup \rho \Rightarrow \underline{\lim}\, \varepsilon(\rho_n) \ge \varepsilon(\rho)$. First note that $\|\rho\|_{5/3}$ is weakly lower semicontinuous, i.e., $\underline{\lim}\|\rho_n\|_{5/3} \ge \|\rho\|_{5/3}$. Moreover, $\lim_{n\to\infty} \int \rho_n(1/|\mathbf{x}|) = \int \rho(1/|\mathbf{x}|)$. If the potential $1/|\mathbf{x}|$ is broken up as $1/|\mathbf{x}| = V_1 + V_2$, where $V_1 \in L^{5/2}$, $V_2 \in L^p$, $3 < p \le \infty$, then by assumption $\int \rho_n V_1$ converges to $\int \rho V_1$. Since $\sup_n \|\rho_n\|_1 < \infty$ (by assumption $\{\rho_n\}$ is bounded in $L^1$), $\rho_n \to \rho$ in the weak topologies of all $L^q$ spaces with $1 < q \le \frac{5}{3}$. This follows from the density of $L^{5/2} \cap L^s$ in $L^s$ for $s \ge \frac{5}{2}$, $1/s + 1/q = 1$, and $\sup_n \|\rho_n\|_q < \infty$, because $\|\rho\|_q \le \|\rho\|_p^\alpha \|\rho\|_r^{1-\alpha}$ for $1/q = \alpha/p + (1-\alpha)/r$. Hence also $\int \rho_n V_2 \to \int \rho V_2$, proving the convergence of the nuclear attraction. Finally, for the repulsion of the electrons we can write

$$\left\|\left(\rho_n * \frac{1}{|\mathbf{x}|}\right)\rho_n\right\|_1 = \left\|\left((\rho_n - \rho) * \frac{1}{|\mathbf{x}|}\right)(\rho_n - \rho)\right\|_1 + 2\left\|\left(\rho * \frac{1}{|\mathbf{x}|}\right)\rho_n\right\|_1 - \left\|\left(\rho * \frac{1}{|\mathbf{x}|}\right)\rho\right\|_1$$

By Young's inequality, if $V$ is broken up as above and $\rho \in L^1$, then $\rho * V_1 \in L^{5/2}$, $\rho * V_2 \in L^p$, $3 < p \le \infty$, so the mixed term on the right converges to $2\|(\rho * 1/|\mathbf{x}|)\rho\|_1$, while the first term is positive. Therefore

$$\underline{\lim}\left\|\left(\rho_n * \frac{1}{|\mathbf{x}|}\right)\rho_n\right\| \ge \left\|\left(\rho * \frac{1}{|\mathbf{x}|}\right)\rho\right\|.$$

(ii) $\rho^{5/3}$ is strictly convex, $\int \rho(1/|\mathbf{x}|)$ is linear, and

$$\int \frac{d^3x\rho(\mathbf{x})\rho(\mathbf{y})}{|\mathbf{x} - \mathbf{y}|} = c \int \frac{d^3k}{|\mathbf{k}|^2} |\tilde{\rho}(\mathbf{k})|^2,$$

$c > 0$, is strictly convex.

(iii) The proof of semiboundedness on $S$ will require the following refinements of our earlier estimates. Let $R > 0$,

$$I_+ \equiv \int_{|\mathbf{x}| \geq R} d^3x \frac{\rho(\mathbf{x})}{|\mathbf{x}|}, \qquad I_- \equiv \int_{|\mathbf{x}| \leq R} d^3x \frac{\rho(\mathbf{x})}{|\mathbf{x}|} \quad \text{and} \quad f(\mathbf{x}) = \frac{\delta(|\mathbf{x}| - R)}{4\pi}.$$

It follows from

$$\int \frac{d^3x\, d^3y}{|\mathbf{x} - \mathbf{y}|} (\rho(\mathbf{x})\Theta(|\mathbf{x}| - R) - f(\mathbf{x}))(\rho(\mathbf{y})\Theta(|\mathbf{y}| - R) - f(\mathbf{y})) \geq 0,$$

$$I_+ = \int_{\|\mathbf{y}\| \geq R} \frac{d^3x\, d^3y}{|\mathbf{x} - \mathbf{y}|} f(\mathbf{x})\rho(\mathbf{y}),$$

and

$$\int \frac{d^3x\, d^3y}{|\mathbf{x} - \mathbf{y}|} f(\mathbf{x}) f(\mathbf{y}) = \frac{1}{R}$$

that

$$\frac{1}{2} \int_{\substack{|\mathbf{x}| \geq R \\ |\mathbf{y}| \geq R}} \frac{d^3x\, d^3y}{|\mathbf{x} - \mathbf{y}|} \rho(\mathbf{x})\rho(\mathbf{y}) - I_+ \geq -\frac{1}{2R},$$

and by Hölder's inequality,

$$|I_-| \leq \left\| \frac{\Theta(R - |\mathbf{x}|)}{|\mathbf{x}|} \right\|_{5/2} \|\rho\|_{5/3} = (64\pi^2 R)^{1/5} \|\rho\|_{5/3}.$$

If $R$ is chosen as $R = \frac{5}{2}((8\pi)^{2/5} \|\rho\|_{5/3})^{-1}$, then with $\sum_k z_k \leq 1$,

$$\varepsilon(\rho) \geq \tfrac{3}{5}(3\pi^2)^{2/3} \|\rho\|_{5/3}^{5/3} - 3(\tfrac{2}{3})^{5/6}(8\pi)^{1/3} \|\rho\|_{5/3}^{5/6} + \sum_{k > j} \frac{z_k z_j}{|\mathbf{X}_k - \mathbf{X}_j|},$$

and the function $ax^2 - bx + c$ is bounded below on $\mathbb{R}$ for non-negative $a$, $b$, and $c$.

If $\mu < 0$, then because of (iii) the infimum is attained for a $\rho$ in the interior of one of the compact sets $S$, and $\rho$ must satisfy the Thomas–Fermi equation (4.1.36) by the same argument as in (4.1.12). If $\mu = 0$ then there is also the possibility that the infimum is attained on the boundary $\|\rho\|_1 = N$ of every set $S$. In that event it would still satisfy the Thomas–Fermi equation with some $\mu$ as the Lagrange multiplier for the constraint $\|\rho\|_1 = N$. However, if $N > \sum_i z_i \leq 1$, then there is no such solution, as otherwise $\Phi(\mathbf{x})$ would be negative for large $|\mathbf{x}|$, contradicting (4.1.37(ii)). Therefore, if $\mu = 0$, then the infimum still lies in the interior of some set $S$.

4. Use units such that $e = \hbar = 2m = 1$, and suppose there is spin. Then

$$E = \int d^3x \left[ \frac{3}{5}(3\pi^2)^{2/3} \rho^{5/3} - \frac{Z}{r} \rho \right].$$

From the Thomas–Fermi equations,

$$(3\pi^2\rho)^{2/3} - \frac{Z}{r} + \mu = 0 \Rightarrow \rho = \frac{Z^{3/2}}{3\pi^2}\left(\frac{1}{r} - \frac{1}{R}\right)^{3/2}, \qquad \mu = Z/R.$$

$$N = \frac{Z^{3/2}}{3\pi^2} 4\pi \int_0^R r^2\, dr\left(\frac{1}{r} - \frac{1}{R}\right)^{3/2} = \frac{(ZR)^{3/2}}{12} \Rightarrow R = N^{-1/3}\frac{N}{Z}4(\tfrac{3}{2})^{2/3},$$

$$-V = Z^{5/2}\frac{4\pi}{3\pi^2}\int_0^R r\, dr\left(\frac{1}{r} - \frac{1}{R}\right)^{3/2} = \frac{6NZ}{R},$$

$$T = \tfrac{3}{5}Z^{5/3}\frac{4\pi}{3\pi^2}\int_0^R r^2\, dr\left(\frac{1}{r} - \frac{1}{R}\right)^{5/2} = -\frac{V}{2},$$

and

$$E = -T = \frac{3NZ}{R} = -\tfrac{1}{2}(\tfrac{3}{2})^{1/3}Z^2N^{1/3} \quad \text{(in units with } 2m = 1\text{; twice this if } m = 1\text{)},$$

so

$$f(0) = -\tfrac{1}{2}(\tfrac{3}{2})^{1/3} = -0.572,$$

$$f'(0) = \frac{1}{2}\int \frac{d^3x\, d^3y}{|\mathbf{x} - \mathbf{y}|}\rho(\mathbf{x})\rho(\mathbf{y}) = \left(\frac{4}{3\pi}\right)^2 \int_0^1 \frac{dr}{\sqrt{r}}(1 - r)^{3/2}\int_0^r dr'\sqrt{r'}(1 - r')^{3/2}\frac{(ZR)^3}{R}$$

$$= 0.24244, \quad \text{if } Z = N = 1.$$

If we read the exact ground-state energy off from (III: 4.5.15), then to $o(N^{-1/3})$,

$$E_{\text{exact}}/E_{\text{Thomas–Fermi}} = 1 - N^{-1/3}\tfrac{1}{2}(\tfrac{3}{2})^{1/3}.$$

Thus the Thomas–Fermi energy is below the actual ground-state energy.

5. The density that minimizes $E$ is

$$\rho_0(\mathbf{x}) = \frac{\mu^2}{4\pi}\sum_j z_j \frac{\exp(-\mu|\mathbf{x} - \mathbf{X}_j|)}{|\mathbf{x} - \mathbf{X}_j|},$$

with which

$$E(\rho_0) = -\frac{N\mu}{2} + \sum_{i>j} z_i z_j\left(\frac{1}{|\mathbf{X}_i - \mathbf{X}_j|} - \mu\exp[-\mu|\mathbf{X}_i - \mathbf{X}_j|]\right) > -\frac{N\mu}{2}$$

for all $\mathbf{X}_i \in \mathbb{R}^3$.

If $z_i = 1$, this reduces to (III: 4.5.24). In this variant of Thomas–Fermi theory the electron cloud creates an attractive potential $-\mu\exp(-\mu r)$ between the nuclei, which is also weaker than their $1/r$ Coulomb repulsion.

## 4.2 Cosmic Bodies

*The Thomas–Fermi theory of stars is thermodynamically more interesting than that of atoms, since it predicts an unusual phase transition*

In the year 1926 great discoveries about the laws of matter appeared in rapid succession. Shortly after E. Schrödinger published the equation named after him, E. Fermi discovered the distribution law (2.5.22; 1) governing particles that satisfy the exclusion principle. This inspired L. Thomas's ingenious idea that the electron cloud of a large atom should satisfy equation (4.1.36) at $T = 0$. Then, still in the year 1926, R. Fowler realized that the stability of cosmic matter is ensured by the zero-point energy of the electrons, and that a cosmic body is closely analogous to a "gigantic molecule in the ground state." Yet it has taken considerably longer to found this vision in mathematics and derive everything from the Schrödinger equation. Today, however, the derivation goes through without gaps, and the Thomas–Fermi theory of atoms and stars is the only many-body problem with realistic forces to have succumbed, in the appropriate thermodynamic limit, to mankind's attempts at calculation.

Yet the zero-point energy guarantees stability only in so far as the speeds of the electrons remain slow in comparison with light. If they enter the regime of relativistic kinematics, for which the kinetic energy $\sim c|\mathbf{p}|$, then the zero-point energy goes as $N(N/V)^{1/3}$, whereas the gravitational energy goes as $-\kappa N^2/V^{1/3}$. If $N > (\kappa m_p^2)^{-3/2} \sim 10^{57}$, then the latter predominates, and as $V$ becomes smaller and smaller, the total energy goes to $-\infty$. We shall avoid this instability by remaining within the framework of nonrelativistic kinematics, considering only stars of masses somewhat smaller than that of the sun. Then, according to the estimates (1.2.23; 3), if $N > 10^{54}$, the minimum energy occurs when $V \sim N^{-1}$. The situation is again like that of Thomas–Fermi theory, which leads to the hope that the many-body problem can be solved in the limit $N \to \infty$ with the Hamiltonian

$$H = \sum_{i=1}^{N} \frac{|\mathbf{p}_i|^2}{2m_i} + \sum_{i>j} \frac{e_i e_j - \kappa m_i m_j}{|\mathbf{x}_i - \mathbf{x}_j|}. \tag{4.2.1}$$

In this limit the system becomes a highly compressed plasma, so the average gravitational field would be expected to be so dominant that the Thomas–Fermi equation is valid. Of course, the total charge of the system must be zero, or, more exactly, the possible excess charge $\Delta Q$ is bounded by $(\Delta Q)^2 \leq \kappa m_p^2 N_p^2$, so for gravity to predominate, $\Delta Q < 10^{-19} N_p$. Indeed, these conjectures can be derived mathematically for all three ensembles:

**The Asymptotic Forms of the State Functions** (4.2.2)

*Let $H_{N,V}$ be the Hamiltonian (4.2.1) for $N_1$ positive and $N_2$ negative fermions of masses $M_{1,2}$, charges $e_1$ and $e_2 = -e_1$, and spin $\frac{1}{2}$ in a volume $V$. Let $N$ denote the pair $(N_1, N_2)$. Then the limits*

$$E(N, S, V) = \lim_{\lambda \to \infty} \lambda^{-7/3} \inf_{\mathscr{H}_{\lambda S}} \operatorname{Tr}_{\mathscr{H}_{\lambda S}} H_{\lambda N, \lambda^{-1} V},$$

$$F(N, \beta, V) = -\lim_{\lambda \to \infty} \beta^{-1} \lambda^{-1} \ln \operatorname{Tr} \exp(-\beta \lambda^{-4/3} H_{\lambda N, \lambda^{-1} V}). \tag{4.2.3}$$

*exist*. The grand-canonical function $\Xi$ is not defined as in (4.1.8), as now the finiteness of the sum $\sum_N$ requires a factor $N^{-2/3}$ in the interaction and $V \sim N$ [27] (see (4.2.10; 4)).

With the solution of the Thomas–Fermi equation

$$\rho_\alpha(\mathbf{x}) = 2\int \frac{d^3p}{(2\pi)^3}\left[1 + \exp\left(\beta\left(\frac{|\mathbf{p}|^2}{2M_\alpha} + W_\alpha(\mathbf{x}) - \mu_\alpha\right)\right)\right]^{-1}, \tag{4.2.4}$$

$$W_\alpha(\mathbf{x}) = \sum_\beta \int_V d^3x' \frac{e_\alpha e_\beta + \kappa M_\alpha M_\beta}{|\mathbf{x} - \mathbf{x}'|}\rho_\beta(\mathbf{x}'), \qquad \alpha, \beta = 1, 2, \tag{4.2.5}$$

$$\int_V d^3x \rho_\alpha(\mathbf{x}) = N_\alpha, \tag{4.2.6}$$

these quantities are found to be

$$E(N, S, V) = \sum_\alpha \int_V d^3x \Bigg\{\tfrac{1}{2}\rho_\alpha(\mathbf{x})W_\alpha(\mathbf{x}) + 2\int \frac{d^3p}{(2\pi)^3}\frac{|\mathbf{p}|^2/2M_\alpha}{1 + \exp[\beta(|\mathbf{p}|^2/2M_\alpha + W_\alpha(\mathbf{x}) - \mu_\alpha)]}\Bigg\}, \tag{4.2.7}$$

$$F(N, \beta, V) = \sum_\alpha \Bigg\{-\int_V d^3x \tfrac{1}{2}\rho_\alpha(\mathbf{x})W_\alpha(\mathbf{x}) + N_\alpha\mu_\alpha - 2T\int_V d^3x \int \frac{d^3p}{(2\pi)^3} \times \ln\left(1 + \exp\left[-\beta\left(\frac{|\mathbf{p}|^2}{2M_\alpha} + W_\alpha(\mathbf{x}) - \mu_\alpha\right)\right]\right)\Bigg\}, \tag{4.2.8}$$

and

$$\Xi(\mu_1, \mu_2, \beta, V) = \sum_\alpha \Bigg\{\int_V d^3x \tfrac{1}{2}\rho_\alpha(\mathbf{x})W_\alpha(\mathbf{x}) + 2T\int_V d^3x \int \frac{d^3p}{(2\pi)^3} \times \ln\left(1 + \exp\left[-\beta\left(\frac{|\mathbf{p}|^2}{2M_\alpha} + W_\alpha(\mathbf{x}) - \mu_\alpha\right)\right]\right)\Bigg\}. \tag{4.2.9}$$

**Gloss** (4.2.10)

1. For $\lambda S \in \ln \mathbb{Z}^+$, $\mathcal{H}_{\lambda S}$ is an $\exp(\lambda S)$-dimensional subspace of $\mathcal{H}$.
2. The thermodynamic limit has been taken in the sense discussed in (1.2.19), i.e., $E \sim N^{7/3}$, $V \sim N^{-1}$, $S \sim N$, and $T \sim E/S \sim N^{4/3}$. The energies $E$ and $F$ are accordingly neither per particle nor per volume; these specific energies and energy densities do not have thermodynamic limits.
3. The quantity $S = \beta(E - F)$ is extensive for $\beta \sim N^{-4/3}$ and $E - F \sim N^{7/3}$.
4. If one insists on the usual relationships $E \sim N$, $V \sim N$, $S \sim N$, with $T$ constant, then according to (1.2.19) the interaction has to be taken as

$$N^{-2/3}\sum_{i>j}\frac{e_i e_j - \kappa m_i m_j}{|\mathbf{x}_i - \mathbf{x}_j|}.$$

This means that the system is imagined as getting larger and larger with an ever weaker interaction; all such problems are mathematically equivalent because of the scaling law of (1.2.1). Physically relevant systems are large but finite and have weak, but still nonzero, gravitational interaction. The question of how reasonable the thermodynamic limit is depends only on whether the physical object is sufficiently like the limiting system. If so, the convergence of the thermodynamic quantities (4.2.2) guarantees that the relevant observables of the finite system will have values fairly near those of the infinite idealization.

5. Since $\rho_\alpha$ is a strictly monotonic function of $\mu_\alpha$, the normalization (4.2.6) is an implicit equation for $\mu_\alpha$.
6. We shall soon discover that for certain values of $\beta$, $N$, and $V$ there is more than one solution of the Thomas–Fermi equations. The question of which solutions are the correct limits (4.2.3) is decided by the minimum principles for the thermodynamic potentials (2.3.3; 4), (2.2.23; 1), and (2.5.3), which survive the limit $\lambda \to \infty$ in the following manner (cf. (4.1.21)): The functionals for energy, entropy, and the phase-space densities $n_\alpha$ are

$$E(n) = -\frac{1}{2}\sum_{\alpha,\beta}\int d^3x\, d^3x' \frac{d^3p\, d^3p'}{(2\pi)^6} n_\alpha(\mathbf{x}, \mathbf{p}) n_\beta(\mathbf{x}', \mathbf{p}') \frac{e_\alpha e_\beta - \kappa M_\alpha M_\beta}{|\mathbf{x} - \mathbf{x}'|}$$

$$+ \sum_\alpha \int d^3x \frac{d^3p}{(2\pi)^3} \frac{|\mathbf{p}|^2}{2M_\alpha} n_\alpha(\mathbf{x}, \mathbf{p}),$$

$$S(n) = -2\sum_\alpha \int d^3x \frac{d^3p}{(2\pi)^3}\left[\frac{n_\alpha}{2}\ln\frac{n_\alpha}{2} + \left(1 - \frac{n_\alpha}{2}\right)\ln\left(1 - \frac{n_\alpha}{2}\right)\right],$$

$$N_\alpha(n) = \int d^3x \frac{d^3p}{(2\pi)^3} n_\alpha(\mathbf{x}, \mathbf{p}).$$

The correct Thomas–Fermi densities are those that minimize the energy for given $N_\alpha$ and $S$. The variational derivative with $T$ and $\mu_\alpha$ as Lagrange multipliers leads to the Thomas–Fermi equations (4.2.4)–(4.2.7) again, with

$$\rho_\alpha(\mathbf{x}) = \int \frac{d^3p}{(2\pi)^3} n_\alpha(\mathbf{x}, \mathbf{p}),$$

for the solution of

$$\frac{\delta}{\delta n_\alpha(\mathbf{x}, \mathbf{p})}(E - TS + \mu_1 N_1 + \mu_2 N_2) = 0.$$

However, this equation is also satisfied by merely local extrema and by saddle points. At the minimizing density, $E(n) = E(N_1, N_2, S, V)$.

7. The ensembles are equivalent only in the region where the convex hull of the function $E(S)$ is the same as $E(S)$.

8. We have written $E$ and $F$ as functions of three variables, but it is clear from the definition that they depend on only two ratios. This is reflected in the Thomas–Fermi equation by its scaling behavior when $\mathbf{x} \to \lambda \mathbf{x}$.

**Proof**

If the only force is gravitation, as in a neutron star ($e = 0$), the methods of §4.1 are applicable; the lower bound for $\Xi$ is trivial, and Inequality (2.1.8; 3) can be used for the upper bound. However, since it requires knowledge of the expectation value of $H$, it is necessary to estimate the norm of the quantum fluctuations. If $e$ and $\kappa$ differ from zero the estimate is much more difficult than that of §4.1, and can not be given in detail here. The strategy is as follows.

1. Regularization of the potential
   Since one expects the motion of the particles to be determined by an average field, the singular part of the $1/r$ potential should first be cut off, so that the influence of a near-by particle will not be stronger than that of the average field. There are also good physical grounds to insist that the important part of the potential is its long range rather than the singularity, as in reality the singularity is smoothed out with some form factor. By "long range" is meant a length comparable to the diameter of the star, which shrinks to zero as $\lambda \to \infty$. Hence the cut-off length has to be reduced while $\lambda$ increases, or alternatively one can work in the scaled system (4.2.10; 4). It is thus useful to show that changing the potential by, say, $1/r \to (1 - \exp(-\lambda^{1/3} sr))/r$ makes little difference for large $s$ in comparison with the main contribution to the energy, which is $\sim N^{7/3}$. This fact can be shown by an argument similar to the estimate (1.2.21) and making use of the bound (III: 4.5.15) on the number of bound states of a short-range potential.
2. Replacing the potential with a step function
   Since Thomas–Fermi theory is oriented toward free particles in a box, it is useful to divide the volume $V$ into cells inside of which the potential is made constant. The proof that changing the potential to a step function has only a slight effect is trivial, since the continuous function $(1 - \exp(-sr))/r$ can be approximated uniformly on any compact set by a step function.
3. Insertion of walls
   In each of the cells of constant potential the Schrödinger equation reduces to the force-free equation, if they are separated by impenetrable walls. It is thus useful to show that inserting walls will not alter the result much. It is clear that the effect will be to raise all the energy levels. The min–max principle can be called upon to show that they do not rise by too much. The presence of the walls means that the wave-function vanishes at their positions, which costs kinetic energy. It is possible to patch together wave-functions for the system without walls so that they vanish at the

positions of the walls, and the expectation value of the kinetic energy in such a state is not increased by too much. It is important that the number of walls in this procedure remain constant in the limit $N \to \infty$ so that their effect can be neglected in comparison with $N^{7/3}$.

4. Filling the boxes
   The foregoing manipulations leave the particles in separated boxes moving in constant potentials, which, however, depend on the distribution of the particles. One now finds that the thermodynamic functions of (4.2.2) are dominated by the contribution from a certain distribution of the particles among the boxes, which is determined by a self-consistent equation, namely the Thomas–Fermi equation for the step potential with walls.
5. Continuity of the Thomas–Fermi equation
   Since we wish to end up with the Thomas–Fermi equation for a $1/r$ potential, we still need to show that the approximations made above for the $1/r$ potential do not change the energy of the solution much. Otherwise, if the solution depended discontinuously on the potential, it would be worthless; the Thomas–Fermi equations can not be solved analytically, and a numerical solution on a computer approximates the potential by a step function. It is thus indispensible, but fortunately also possible, to show that the Thomas–Fermi energy has the required continuity with respect to the potential. □

The structure of the Thomas–Fermi equation is different for stars than for atoms. The energy loses the properties of convexity and weak lower semicontinuity. Consequently the solution is not guaranteed to be unique and there is a possibility of a phase transition, which will be discussed at the conclusion of this section. Meanwhile, we prepare by proving a general

**Virial Theorem** (4.2.11)

*The pressure*

$$P \equiv -\frac{\partial}{\partial V} F(N, \beta, V),$$

*kinetic energy*

$$E_k = \sum_\alpha 2 \int_V d^3x \int \frac{d^3p}{(2\pi)^3} \frac{|\mathbf{p}|^2/2M_\alpha}{1 + \exp[\beta(|\mathbf{p}|^2/2M_\alpha + W_\alpha(\mathbf{x}) - \mu_\alpha)]},$$

*and potential energy*

$$E_p = \sum_\alpha \int d^3x \tfrac{1}{2}\rho_\alpha(\mathbf{x}) W_\alpha(\mathbf{x})$$

*are connected by*

$$3PV = 2E_k + E_p.$$

**Proof**

We start by convincing ourselves of the usual thermodynamic relationships

$$\frac{\partial F}{\partial N_\alpha} = \mu_\alpha \quad \text{and} \quad \beta \frac{\partial F}{\partial \beta} = E - F, \tag{4.2.12}$$

which follow directly from differentiating (4.2.8). For this purpose note that $\rho$ depends on $\beta$ and $N$, and thus implicitly so does $W$, but that this dependence does not show up when the Thomas–Fermi equations are satisfied. Next rewrite (4.2.8) by integrating by parts in the variable $\mathbf{p}$. Then

$$\frac{|\mathbf{p}|^2}{2M_\alpha} = \varepsilon, \qquad \int_0^\infty d\varepsilon \sqrt{\varepsilon} \ln(1 + \exp[-\beta(\varepsilon + c)]) = \tfrac{2}{3}\beta \int_0^\infty \frac{d\varepsilon\, \varepsilon^{3/2}}{1 + \exp[\beta(\varepsilon + c)]},$$

and we conclude that

$$F = \sum_\alpha N_\alpha \mu_\alpha - \tfrac{2}{3}E_k - E_p. \tag{4.2.13}$$

Finally, the dilatation relationship mentioned earlier implies that $F(N, \beta, V) = \lambda^{-7/3} F(\lambda N, \lambda^{-4/3}\beta, \lambda^{-1}V)$ for all $\lambda \in \mathbb{R}^+$.

With reference to (4.2.12), the derivative by $\lambda$ produces

$$0 = -\tfrac{7}{3}F + \sum_\alpha N_\alpha \mu_\alpha - \tfrac{4}{3}(E - F) + PV,$$

which concludes the proof of the theorem when combined with (4.2.13). □

The local densities in phase space,

$$n_\alpha(\mathbf{x}, \mathbf{p}) = 2\left[\exp\left(\beta\left(\frac{|\mathbf{p}|^2}{2m} + W_\alpha(\mathbf{x}) - \mu_\alpha\right)\right) + 1\right]^{-1},$$

have the same significance as in §4.1. They are stationary solutions of the Vlasov equation (4.1.32; 3),

$$\sum_j \frac{\mathbf{p}_j}{M_\alpha} \cdot \frac{\partial}{\partial \mathbf{x}_j} n_\alpha(\mathbf{x}, \mathbf{p}) - \frac{\partial}{\partial \mathbf{p}_j} n_\alpha(\mathbf{x}, \mathbf{p}) \cdot \frac{\partial}{\partial \mathbf{x}_j} W_\alpha(\mathbf{x}) = 0. \tag{4.2.14}$$

In this equation quantum mechanics enters only through the initial condition $|n_\alpha(\mathbf{x}, \mathbf{p})| \le 1$. In fact, as a classical equation it is the basis of stellar dynamics [35]. When reduced to configuration space, the local densities describe the hydrostatic equilibrium between the pressure of the matter and of gravitation, in the spherically symmetric case. Since the fermions behave like free particles on the microscopic level, one would expect from (2.5.32) that

$$P(\mathbf{x}) = \tfrac{2}{3}E_k(\mathbf{x}) \equiv \frac{2}{3}\sum_\alpha 2 \int \frac{d^3p}{(2\pi)^3} \frac{|\mathbf{p}|^2/2M_\alpha}{1 + \exp[\beta(|\mathbf{p}|^2/2M_\alpha + W_\alpha(\mathbf{x}) - \mu_\alpha)]} \tag{4.2.15}$$

functions as the pressure, and in fact if (4.2.14) is multiplied by $\mathbf{p}_i$, integrated by $d^3p$ by parts, and one replaces $\mathbf{p}_i \cdot \mathbf{p}_j \to (|\mathbf{p}|^2/3)\delta_{ij}$, then

$$\nabla P(\mathbf{x}) = -\sum_\alpha \rho_\alpha(\mathbf{x}) \nabla W_\alpha(\mathbf{x}), \tag{4.2.16}$$

which is the equilibrium condition mentioned above. If the geometry is spherically symmetric, i.e., $V$ is a sphere of radius $R$ and the local observables depend only on $|\mathbf{x}| = r$, then (4.2.16) can be written as the nonrelativistic **Tolman–Oppenheimer equation**

$$\frac{d}{dr}\tfrac{2}{3}E_k(r) = -\sum_\alpha \frac{\rho_\alpha(r) M_\alpha(r)}{r^2},$$

$$M_\alpha(r) = -\sum_\beta (e_\alpha e_\beta - \kappa M_\alpha M_\beta) \int_0^r dr' 4\pi r'^2 \rho_\beta(r') \tag{4.2.17}$$

(cf. (II: 4.5.11)). The electric and gravitational forces have been expressed in terms of the charges and masses within the sphere.

**The Connection between the Local and Global Pressures** (4.2.18)

By integrating (4.2.17) by $(4\pi/3)\int_0^R drr^3$ one gets

$$V\tfrac{2}{3}E_k(R) - \tfrac{2}{3}E_k = \frac{4\pi}{3}\int_0^R drr^3 \frac{d}{dr}\tfrac{2}{3}E_k(r)$$

$$= \sum_{\alpha,\beta} \frac{e_\alpha e_\beta - \kappa M_\alpha M_\beta}{3} \int_0^R drr 4\pi\rho_\alpha(r) \int_0^r dr' r'^2 4\pi \rho_\beta(r') = \frac{E_p}{3},$$

so with the virial theorem (4.2.11) the thermodynamic pressure becomes simply the local pressure at the boundary,

$$P = P(R).$$

We see that Thomas–Fermi theory, which begins with the Schrödinger equation, leads eventually to the concepts of classical physics.

A more accurate evaluation of the state functions (4.2.2) requires numerical solutions of Equations (4.2.4) through (4.2.6). In order to lend more physical plausibility to those numbers, let us extend the intuitive arguments of §1.2 to finite temperatures. Since the theory is only valid if gravity is the dominant force, let us simplify by considering only one type of neutral fermion such as neutrons (without nuclear forces). If there were protons and electrons, then the former would provide most of the gravitational force and the latter most of the pressure. This would increase all lengths compared with a system of neutrons by a factor of the ratio of the mass of the neutron to that of the electron, about 2000. Thus, if $10^{57}$ neutrons are found to have a radius of about 30 km, a similar system made of hydrogen would have a radius of about 6000 km, i.e., that of the earth or of a white dwarf. We begin with the observation that at a nonzero temperature there is a thermal contribution

$$N\tfrac{3}{2}T = N\frac{3}{2}\left(\frac{N}{V}\right)^{2/3}\exp\left(\frac{2S}{3N} - 1\right)$$

in addition to the zero-point energy $N(N/V)^{2/3}$. At high temperatures this is exactly the classical expression. In order to interpolate to intermediate temperatures, we shall simply combine the zero-point energy with the classical expression. This turns out to approximate the energy of free fermions (2.5.32) to within about 20%. It remains to add in the gravitational energy. If the mass of the particles is $\frac{1}{2}$, then up to geometric factors we get

$$\frac{E}{N} = \left(\frac{N}{V}\right)^{2/3}\left(1 + \frac{3}{2e}\exp\left(\frac{2S}{3N}\right)\right) - \frac{\kappa N}{V^{1/3}}$$

in natural units. In checking the properties (2.3.10) of the microcanonical energy density, it becomes readily apparent that, in agreement with (4.2.10; 4),

$$\rho^{-1}\varepsilon = \rho^{2/3}\left(1 + \frac{3}{2e}\exp(2\sigma/3\rho)\right) - \kappa N^{2/3}\rho^{1/3}$$

is independent of $N$ only if $\kappa \sim N^{-2/3}$. Although $\varepsilon$ increases as a function of $\sigma$, conditions (2.3.10(ii)) and (2.3.10(iii)) are not always satisfied; our ansatz does not do justice to the subadditivity (2.3.5). The reason becomes apparent when it is observed that the pressure

$$P = -\left.\frac{\partial E}{\partial V}\right|_{S,N} = \frac{2}{3}\left(\frac{N}{V}\right)^{5/3}\left(1 + \frac{3}{2e}\exp\left(\frac{2S}{3N}\right)\right) - \frac{\kappa N^2}{3V^{4/3}} = \frac{E - E_p/2}{3V/2},$$

$$E_p = -\frac{\kappa N^2}{V^{1/3}},$$

consists of three parts, from the zero-point, thermal, and gravitational energies. The first two are positive and the last one is negative, and may dominate in the intermediate regime of average densities. However, a negative pressure is physically impossible; the system does not adhere to the walls and pull them inward. What happens is that the system shrinks itself down to such a small radius, $V_0 = (\kappa N^2/-2E)^3$, that it reaches $P = 0$. A better ansatz consists in replacing $V$ with $V_0$ in $E$ when $P < 0$,

$$\frac{E}{N} = \left(\frac{N}{V}\right)^{2/3}\left(1 + \frac{3}{2e}\exp\left(\frac{2S}{3N}\right) - \frac{\kappa N}{V^{1/3}}\right)\Theta_+ - \frac{\kappa^2 N^{4/3}/2}{2 + 3\exp((2S/3N) - 1)}\Theta_-,$$

$$\Theta_\pm = \Theta\left(\pm\left(2 + 3\exp\left(\frac{2S}{3N} - 1\right) - \kappa(NV)^{1/3}\right)\right).$$

The function $\Theta_\pm$ is also equal to $\Theta(\pm(E + \kappa N^2/2V^{1/3}))$, implying that if the total energy is sufficiently negative, then the system condenses into a volume $V_0$. As in Example (2.3.32; 1) this brings about a phase transition with negative specific heat: The calculation

$$\frac{3}{2}T = \frac{3}{2}\frac{\partial E}{\partial S}\bigg|_{V,N} = \frac{3}{2e}\left(\frac{N}{V}\right)^{2/3} \exp\left(\frac{2S}{3N}\right)\Theta_+ + \frac{3}{2e}\frac{\kappa^2 N^{4/3}\exp(2S/3N)}{(2+3\exp(2S/3N-1))^2}\Theta_-$$

$$= \left(\frac{E}{N} - \left(\frac{N}{V}\right)^{2/3} + \frac{\kappa V}{N^{1/3}}\right)\Theta_+ + \left[-\frac{E}{N} - \left(\frac{2E}{\kappa N^{5/3}}\right)^2\right]\Theta_-$$

reveals that the classical linear dependence of $T$ on $E$ becomes parabolic in the condensation region (see Figure 32). The temperature begins to rise again as $E$ decreases, and afterwards, when the zero-point energy gets larger than the gravitational energy, it falls to zero. It is in fact observed by astrophysicists that large gaseous masses contract under the influence of gravity, thereby heating up and radiating the gravitational energy that has been set free. This activity, which indicates a range of values for which $S(E)$ is convex and hence microcanonically a negative specific heat, is a direct consequence of the virial theorem and the theorem of equipartition: energy = $-$kinetic energy = $-(3N/2)$ temperature. Yet this is true only in the intermediate region, since it ignores the external virial (the pressure) and the equipartition theorem is not valid for degenerate gases. This also becomes visible in the computer solution of the Thomas–Fermi equation, as shown in Figures 33 and 34. At the smaller radius $R = 30$ km the zero-point energy predominates and the star acts normally, whereas an intermediate region of negative specific heat shows up at $R = 100$ km.

This phenomenon can not arise in the canonical ensemble, so our next topic will be what the situation is like in that ensemble. In the transition region the Thomas–Fermi equation has many solutions for a given $\beta$, and the

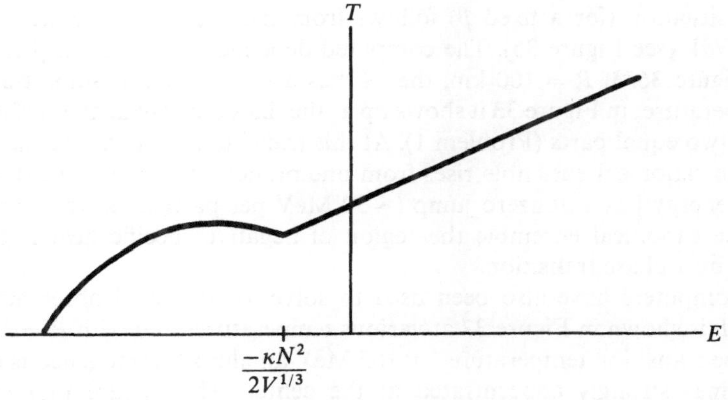


Figure 32 The function $T(E)$ for a conceptual model.

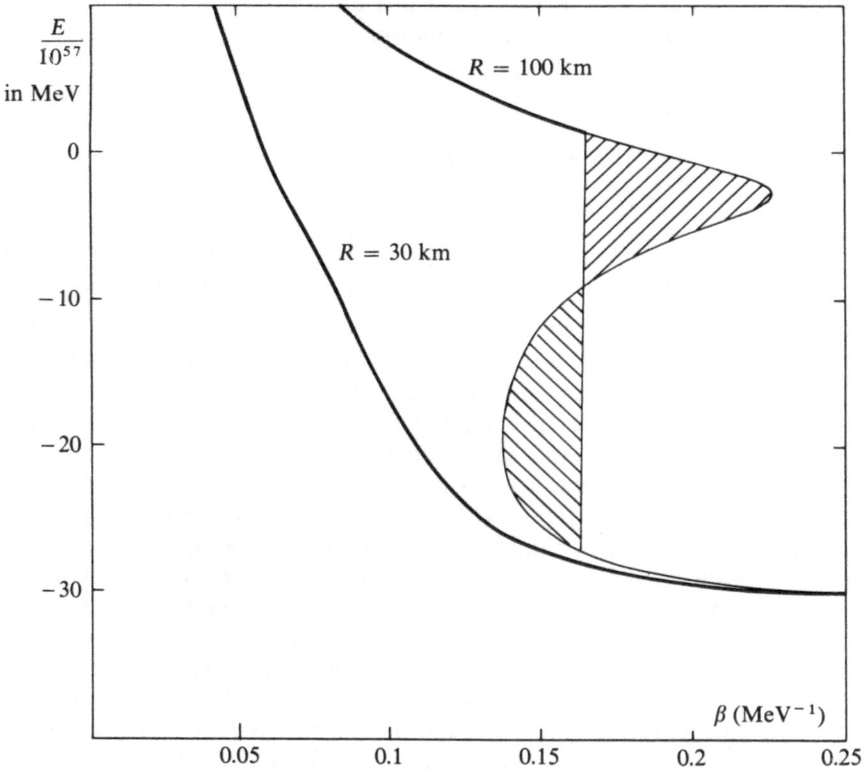


Figure 33 Phase transition in $E(\beta)$.

analysis leading to (4.2.2) shows that the right solution to choose is the one with the smallest value of $F$. The existence of many different values of $F$ in this situation (for a fixed $\beta$) follows from the change in the sign of $P = -\partial F/\partial V$ (see Figure 35). The computed dependence of $-F$ on $\beta$ is shown in Figure 36. If $R = 100$ km, then $F$ has a sharp bend at some transition temperature; in Figure 33 it shows up as the lines that divide the surface $E(\beta)$ into two equal parts (Problem 1). At this transition temperature the system in the canonical ensemble rises from one branch of the curve to the other. The energy has a nonzero jump ($\sim$30 MeV per particle) at the transition; in the canonical ensemble the region of negative specific heat is bridged over by a phase transition.

Computers have also been used to solve for the local observable $\rho(r)$, which is shown in Figure 37 at various temperatures and with $R = 100$ km. At the transition temperature 1/0.165 MeV an almost homogeneous density becomes strongly concentrated at the center. The picture that emerges is of a star with a rather definite surface and a central density about $10^6$

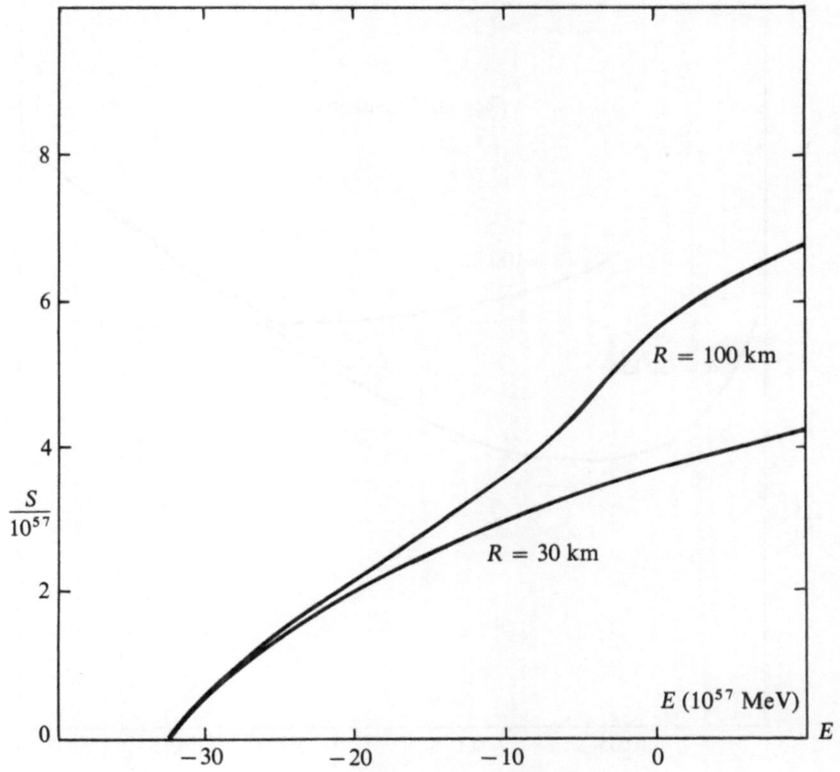


Figure 34 Convex region in $S(E)$.

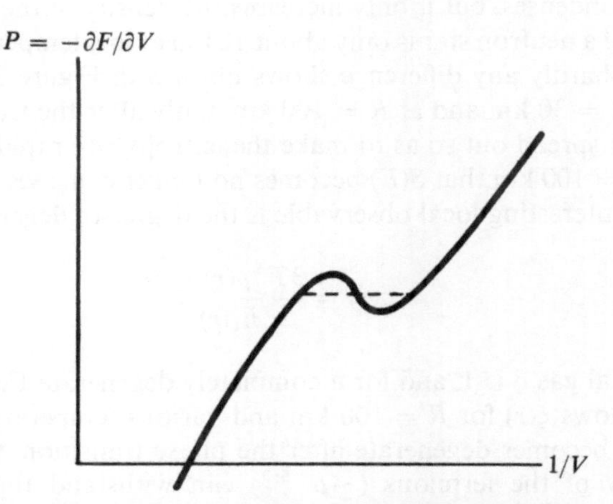


Figure 35 Phase transition with negative pressure.

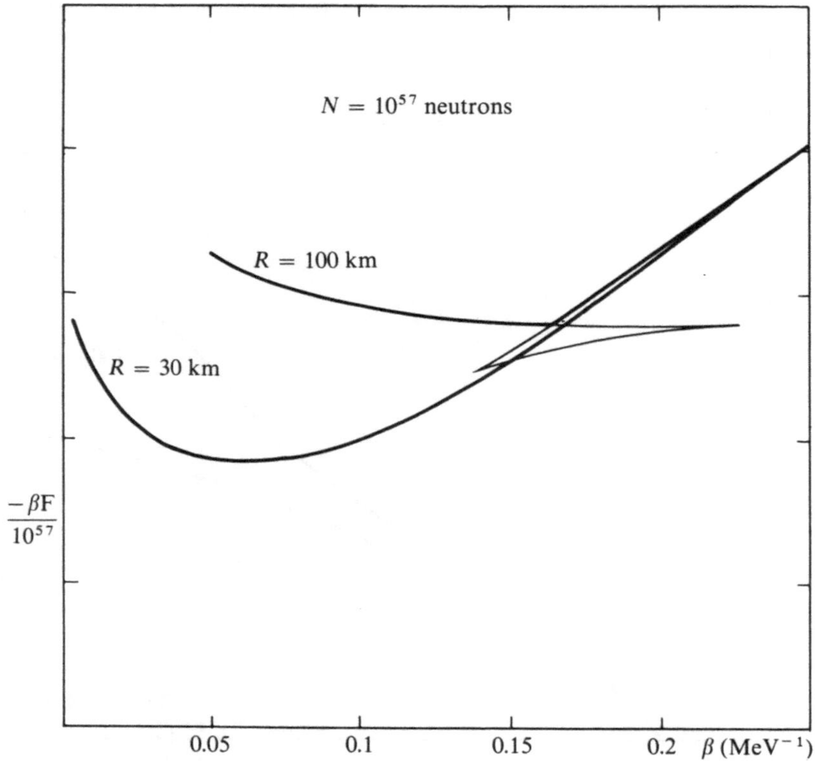


Figure 36 The negative free energy.

times the density of the atmosphere. At still lower temperatures the atmosphere also condenses, but it only increases the density of the star a tiny bit. The radius of a neutron star is only about 10 km at low temperature, which is why at first hardly any difference shows up in $S$ in Figure 34 between the systems at $R = 30$ km and at $R = 100$ km. Only after the transition energy does the star spread out so as to make the entropy rise rapidly enough in a box with $R = 100$ km that $S(E)$ becomes no longer concave.

Another interesting local observable is the degree of degeneracy

$$\xi(r) = \frac{3T}{2}\frac{\rho(r)}{E_k(r)}. \tag{4.2.19}$$

For a classical gas $\xi$ is 1, and for a completely degenerate Fermi gas it is 0. Figure 38 shows $\xi(r)$ for $R = 100$ km and various temperatures. It reveals that the gas becomes degenerate after the phase transition. Only the zero-point energy of the fermions ($\sim\rho^{-5/3}$) can withstand the gravitational pressure ($\sim\rho^{-4/3}$), while the classical pressure is weaker ($\sim\rho^{-1}$). This

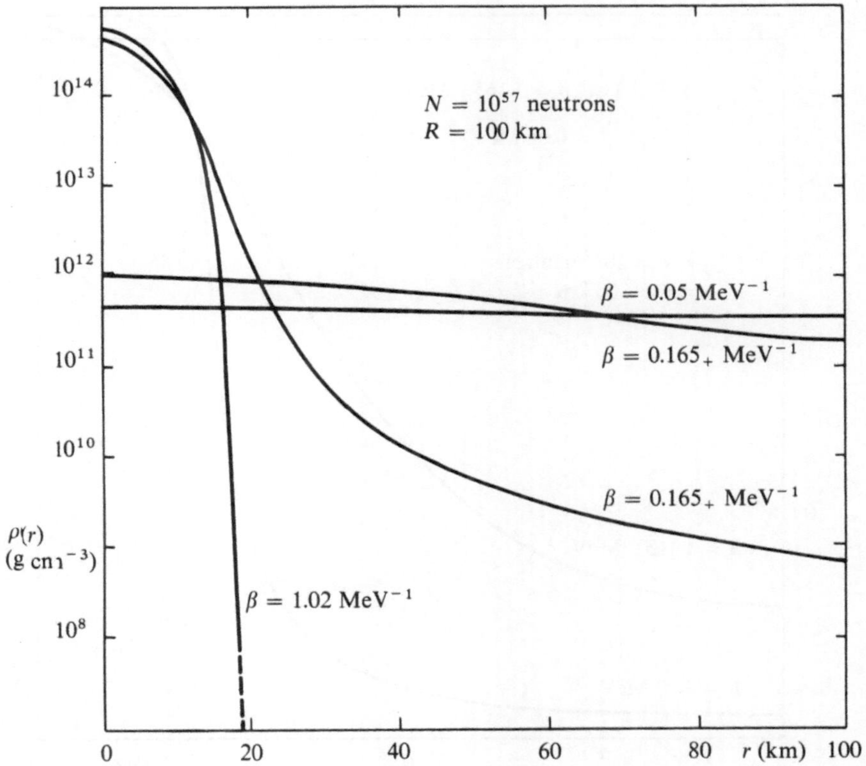


Figure 37 The change in the density at a phase transition.

means that the interior of the star is degenerate, while the atmosphere remains a classical gas.

**Problem** (4.2.20)

Show that the reciprocal $\beta_c$ of the transition temperature for the canonical ensemble is determined by

$$0 = \int_{E_2}^{E_1} dE(\beta(E) - \beta_c), \qquad \beta(E_1) = \beta(E_2) = \beta_c.$$

**Solution** (4.2.21)

Since $\beta = dS/dE$, the condition implies

$$S(E_1) - S(E_2) - \beta_c(E_1 - E_2) = \beta_c(F(E_2) - F(E_1)) = 0.$$

At $\beta_c$ the two branches of the curves $F(E)$ cross, and the canonical ensemble always selects the lower branch.

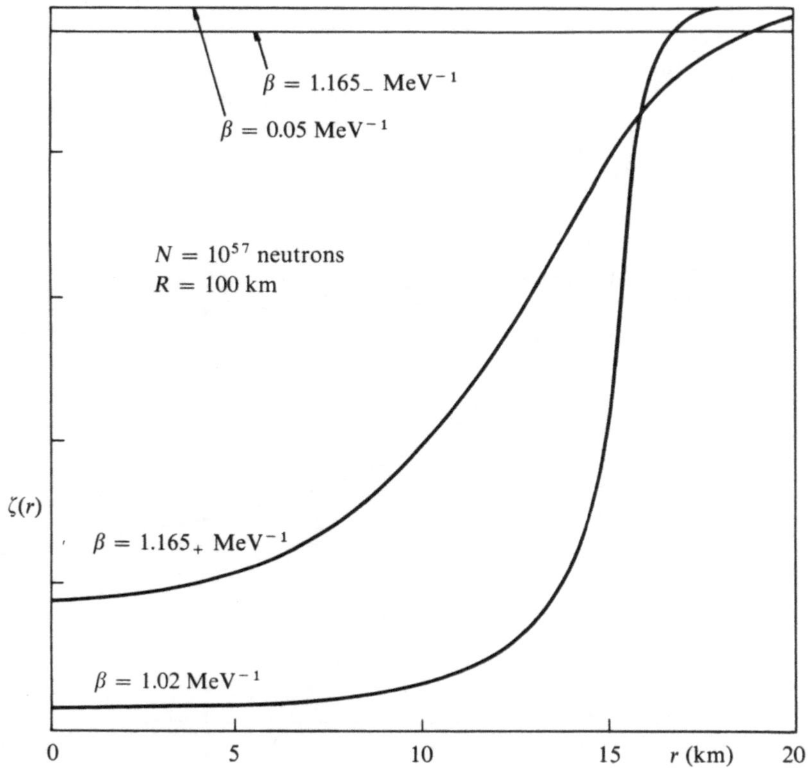


Figure 38 The change in the degree of degeneracy $\zeta$ at a phase transition.

## 4.3 Normal Matter

*Although matter consisting of electrons and atomic nuclei exhibits extremely varied and complicated phenomena, some of its essential features can be deduced from the fundamental physical laws.*

With the results of §4.1 we are now in a position to cope with a central problem, the stability of matter. As discussed in (1.2.17; 2), it is essential that the electrons follow Fermi statistics, though the statistics of the nuclei should not matter. Moreover, it is the mass of the electron rather than the nucleus that occurs in the basic Rydberg energy $e^4m^2/2$. We shall therefore assume that the nuclei are infinitely massive and use the Hamiltonian $H_N$ of (4.1.2); at any rate it provides a lower bound to (4.1.1) with $\kappa = 1$. The wall $W$ can then also be dispensed with. The question to be confronted is whether a bound $H_N > -AN$ can be found for fixed $Z_k$ but $M$ and $N \to \infty$. With this in mind, write (4.1.6) with $\mu = W = 0$ as

$$\begin{aligned} H_N &= \sum_{i=1}^{N} |\mathbf{p}_i|^2 - \sum_{i=1}^{N}\sum_{k=1}^{M} \frac{Z_k}{|\mathbf{x}_i - \mathbf{X}_k|} + \sum_{i>j} \frac{1}{|\mathbf{x}_i - \mathbf{x}_j|} + \sum_{k>l} \frac{Z_k Z_l}{|\mathbf{X}_k - \mathbf{X}_l|} \\ &\geq \sum_{i=1}^{N} |\mathbf{p}_i|^2 + \sum_{i=1}^{N} \left[ -\sum_k \frac{Z_k}{|\mathbf{x}_i - \mathbf{X}_k|} + \int \frac{d^3x' n(\mathbf{x}')}{|\mathbf{x}_i - \mathbf{x}'|} \right] \\ &\quad + \sum_{k>l} \frac{Z_k Z_l}{|\mathbf{X}_k - \mathbf{X}_l|} - 3.68\gamma N \\ &\quad - \frac{1}{2}\int \frac{d^3x\, d^3x'}{|\mathbf{x} - \mathbf{x}'|} n(\mathbf{x})n(\mathbf{x}') - \frac{3}{5\gamma}\int d^3x n(\mathbf{x}) \equiv H_n. \end{aligned} \tag{4.3.1}$$

The first step is to bound the kinetic energy by $\int \rho^{5/3}$ with the inequality of (4.1.47; 2) and set $n = \rho$. This is a bound for every expectation value with spin $-\frac{1}{2}$ fermions, so, again with the aid of (4.1.46; 2), we obtain

$$\begin{aligned} \langle \psi | H_N | \psi \rangle &\geq \frac{3}{5}\left(\left(\frac{3\pi}{4}\right)^{2/3} - \frac{1}{\gamma}\right) \int d^3x n^{5/3}(\mathbf{x}) - \int d^3x \sum_{k=1}^{M} \frac{Z_k n(\mathbf{x})}{|\mathbf{x} - \mathbf{X}_k|} \\ &\quad + \frac{1}{2}\int \frac{d^3x\, d^3x'}{|\mathbf{x} - \mathbf{x}'|} n(\mathbf{x})n(\mathbf{x}') + \frac{1}{2}\sum_k \frac{Z_k Z_m}{|\mathbf{X}_k - \mathbf{X}_m|} - 3.68\gamma N \\ &\geq -3.68\left(\gamma N + \frac{1}{(3\pi/4)^{2/3} - 1/\gamma} \sum_{k=1}^{M} Z_k^{7/3}\right) \quad \text{for } \|\psi\|^2 = 1. \end{aligned} \tag{4.3.2}$$

If this is optimized in $\gamma$, it shows the

**Stability of Matter** (4.3.3)

$$H_N \geq -2.08N\left[1 + \left(\sum_{k=1}^{M} \frac{Z_k^{7/3}}{N}\right)^{1/2}\right]^2.$$

**Remarks** (4.3.4)

1. If there were $q$ kinds of electrons instead of the two spin orientations, then the right side would be multiplied by $(q/2)^{2/3}$. Thus there is a bound $\sim N^{5/3}$ independently of the statistics of the electrons.
2. The solution of the Thomas–Fermi equation describes a neutral system, and accordingly the bound is $MZ^{7/3}$ if all $Z_k$ equal $Z = N/M$. The bound is certainly not optimal if $N \ll MZ$, for one would expect $\sim NZ^2$. However, (4.3.3) suffices for our purposes, as we are concerned only with the neutral case.

3. Inequality (4.1.47; 2) is presumably not optimal; on the right the constant should be increased by a factor $(4\pi)^{2/3}$ to $\frac{3}{5}(3\pi^2)^{2/3}$. If this conjecture were proved, then (4.3.3) would be improved by the same factor, reading
$$H \geq -0.385 \sum_{k=1}^{M} Z_k^{7/3}(1 + O(Z_k^{-7/6})).$$
If $Z_k \to \infty$ this approaches the sum of the Thomas–Fermi energies of the atoms. Such an optimal inequality can in fact be proved, although only in the form
$$H \geq -0.385 \sum_{k=1}^{M} Z_k^{7/3}(1 + O(Z_k^{-2/33}))$$
[28].
4. Inequality (4.3.3) holds *a fortiori* for a system in a finite volume.
5. Since the kinetic energy of the nuclei was not used, they may follow either Bose or Fermi statistics.
6. The important property of the Coulomb potential for stability is that $1/r$ is a function of positive type, i.e., $\tilde{v} \geq 0$. The Yukawa potential $v(r) = \exp(-\mu r)/r$ similarly satisfies $\tilde{v} > 0$, and stability can be proved analogously. In contrast the potential $v(r) = (a + br)\exp(-\mu r)$ with $b > a\mu > 0$, $\mu > 0$, which is even finite and of short range, does not lead to stability for the Hamiltonian $\sum_{i=1}^{N} |\mathbf{p}_i|^2 + \sum_{i>j} e_i e_j v(\mathbf{x}_i - \mathbf{x}_j)$, even for fermions: There is an $r_0 > 0$ such that $v(r_0) > v(0)$ (this would be impossible if $\tilde{v} > 0$), so let us confine $N/2$ positive and negative particles to separated balls of radius $r_0 \varepsilon$, $\varepsilon \ll 1$, arrayed at a distance $r_0$ from one another. Then the interaction between the balls, $-e^2 v(r_0)N^2/4$, wins out over the respulsive energy of the like-charged particles within the balls, $\sim e^2 v(0)N(N-2)/4$, and also wins out over the kinetic energy $\sim N^{5/3}(r_0\varepsilon)^2$ as $N \to \infty$. Thus the total energy goes to $-\infty$ as $-N^2$ when $N \to \infty$. This shows that the problem of the stability of matter has nothing to do with the long range of the Coulomb potential. The proof with the Yukawa potential is not any simpler; in a way it is more difficult, since stability with a Yukawa potential immediately implies stability with a Coulomb potential—as remarked in (1.2.17; 5) the difference produces stability—but not vice versa. However, as we have just seen, the $1/r$ singularity is not the only danger for stability; even regular potentials $v$ with energies $\sum_{i \leq j} e_i e_j v(\mathbf{x}_i - \mathbf{x}_j)$ that take on both signs can lead to instability. This shows the superficiality of the common opinion that stability is not a real physical problem, since actual potentials do not become singular.

**The Extensivity of the Volume** (4.3.5)

*If $H > -cN$ and the expectation value of $H$ in a state is nonpositive, $\langle H \rangle \leq 0$, then no volume $\Omega \leq \varepsilon N$ contains more than $N(\frac{20}{3}c)^{3/5}(4\varepsilon/3\pi)^{2/5}$ particles.*

**Proof**

Let $H = T + V$. Since the energy is proportional to the mass in a Coulomb system, $\frac{1}{2}\langle T\rangle \leq -\langle\frac{1}{2}T + V\rangle \leq 2cN$. Then it follows from

$$\langle T\rangle \geq \frac{3}{5}\left(\frac{3\pi}{4}\right)^{2/3}\int \rho^{5/3}$$

that

$$\int_\Omega \rho(\mathbf{x})\, d^3x \leq \left(\int_\Omega \rho^{5/3}\, d^3x\right)^{3/5}\left(\int_\Omega d^3x\right)^{2/5} \leq \left(\frac{5}{3}\left(\frac{4}{3\pi}\right)^{2/3} 4cN\right)^{3/5}(\varepsilon N)^{2/5}$$
$$\leq N\left(\frac{20}{3}c\right)^{3/5}\left(\frac{4\varepsilon}{3\pi}\right)^{2/5}. \qquad \square$$

**Remarks** (4.3.6)

1. If $\Omega$ is a ball, then it is possible to derive bounds of the form $\langle r^\nu\rangle \geq cN^{\nu/3}$, in analogy with (III: 4.5.28).
2. The material up to this point does not allow upper bounds of the form $r \sim N^{1/3}$ to be proved. Neutrality does not enter in an important way, and with an excess of electrons the Coulomb potential would cause the system to swell out to infinity. In other words, it has been proved that matter is stable in the sense that it does not implode, but it might still explode.

**The Existence of the Thermodynamic Functions** (4.3.7)

We are now faced with the question of how to define the energy density when $N \to \infty$ [30]. It clearly follows from (4.3.3) that $(1/V)E(V\sigma, V, \rho V) > -\rho\cdot$ constant for all $V$, and since it is easy to show that $E/V$ remains bounded above, $\underline{\lim}_{V\to\infty}(1/V)E(V\sigma, V, \rho V)$ could be regarded as $\varepsilon(\rho, \sigma)$ (by definition, $\underline{\lim}_{n\to\infty} a_n = \sup_{n'}\inf_{n>n'} a_n$). This cheap way out is physically unsatisfying, however; one would hope that the limit exists and that the energy density becomes independent of $V$ as the system is made infinitely large. This means that the sequence should be proved monotonic, as was done in (2.3.6). Unfortunately, the inductive procedure followed there, of imagining each cube to consist of smaller cubes, does not work in this case, since it is difficult to estimate the Coulomb interaction between cubes. Balls can be used instead of cubes, however, as their interactions are as if the charges were concentrated at their centers, according to a theorem dating from Newton. In particular, if they are overall neutral, then they do not interact with charges placed outside them. Of course, spheres do not fill space as densely as cubes, but by the use of spheres of different radii the unfilled volume can be made arbitrarily small. The convergence proof consequently proceeds by three steps.

(a) We must first show that the interaction between the spheres is not positive, in order to prove monotony.
(b) It must be shown that the radii of the balls can be chosen so that the fraction of volume outside them goes to zero in the limit.
(c) The distribution of particles in this procedure must lead to a homogeneous density in the limit.

**The Interaction between Balls** (4.3.8)

We consider

$$H = \sum_{i=1}^{N}\left(|\mathbf{p}_i|^2 - \sum_{k=1}^{M}\frac{Z_k}{|\mathbf{x}_i - \mathbf{X}_k|}\right) + \sum_{i>j}|\mathbf{x}_i - \mathbf{x}_j|^{-1} + \sum_{k>l}\frac{Z_k Z_l}{|\mathbf{X}_k - \mathbf{X}_l|} + \sum_{k=1}^{M}\frac{|\mathbf{p}_k|^2}{2M_k} \tag{4.3.9}$$

in a ball $B$, such that $\psi|_{\partial B} = 0$, and examine the neutral case with only one kind of nucleus: $N = MZ$, $N_t = N(1 + 1/Z) =$ the total number of particles. The eigenvalues $e_i(V, N_t)$, $i = 1, 2, \ldots$, of $H$ depend on the volume $V$ of $B$ and on $N_t$, and the microcanonical energy is given by

$$E(S, V, N_t) = \exp(-S)\sum_{i=1}^{\exp(S)} e_i(V, N_t),$$

where $E$ and $E_m$ have been identified in accordance with (2.3.13; 2). Now put $k$ disjoint balls $B_\alpha$ of volumes $V_\alpha$ into $B$,

$$B \supset \bigcup_{\alpha=1}^{k} B_\alpha,$$

and form a system of trial functions $\psi_i$ by taking tensor products of the eigenfunctions of $H_\alpha$, defined as $H$ for $N_\alpha$ particles in $B_\alpha$:

$$\psi_i = \psi_{i_1} \otimes \psi_{i_2} \otimes \cdots \otimes \psi_{i_k}$$

The trial functions then have to be antisymmetrized in the electron variables and either symmetrized or antisymmetrized in the nuclear coordinates, depending on the nuclear statistics. Yet since $\psi_{i_\alpha}$ and $\psi_{i_\beta}$ have disjoint support, there are no cross terms in their interaction, and the expectation values are the same as those with the unsymmetrized $\psi_i$. (The subscript $i$ is to be treated as a multi-index $i_1, \ldots, i_k$.) We always choose the first

$\exp(S_\alpha)$ eigenfunctions of the operators $H_\alpha$ (and denote the eigenvalues $e_{\alpha,i}$), so

$$\sum_{i=1}^{\exp(S)} = \sum_{i_1=1}^{\exp(S_1)} \cdots \sum_{i_k=1}^{\exp(S_k)},$$

where $S = \sum_{\alpha=1}^k S_\alpha$, $N = \sum_{\alpha=1}^k N_\alpha$, and $N_\alpha/Z + 1$ is an integer. Then each $B_\alpha$ can be filled with whole atoms, becoming neutral. As in (2.3.5), with the min–max principle (III: 3.5.21),

$$E(S, V, N) \le \exp(S) \sum_{i=1}^{\exp(S)} \langle \psi_i | H\psi_i \rangle = \sum_{\alpha=1}^{k} \exp(-S_\alpha) \sum_{i_\alpha=1}^{\exp(S_\alpha)} e_{\alpha,i_\alpha}(N_\alpha, N_\alpha) + U$$

$$= \sum_{\alpha=1}^{k} E_\alpha(S_\alpha, V_\alpha, N_\alpha) + U, \tag{4.3.10}$$

but this time there is an energy of the interaction between the balls,

$$U = \sum_{\alpha>\beta} \exp(-S_\alpha - S_\beta) \sum_{i_\alpha=1}^{\exp(S_\alpha)} \sum_{i_\beta=1}^{\exp(S_\beta)} U_{i_\alpha i_\beta},$$

$$U_{i_\alpha i_\beta} \equiv \sum_{j=1}^{N_\alpha} \sum_{m=1}^{N_\beta} e_j e_m \int \frac{d^{3N_\alpha}x\, d^{3N_\beta}y}{|\mathbf{x}_j - \mathbf{y}_m|} |\psi_{i_\alpha}(\mathbf{x}_1, \ldots, \mathbf{x}_{N_\alpha})|^2 |\psi_{i_\beta}(\mathbf{y}_1, \ldots, \mathbf{y}_{N_\beta})|^2.$$

Because of the spherical symmetry of $B_\alpha$ and $H_\alpha$, the functions $\psi_{i_\alpha}$ can be ordered according to the eigenvalues $l_\alpha$ of the total angular momentum $L_\alpha$ about the center of $B_\alpha$. The eigenvalues $e_{\alpha,i}$ do not depend on the $z$-component of the angular momentum (which has quantum numbers $m_\alpha$, $-l_\alpha \le m_\alpha \le l_\alpha$), and

$$\rho_\alpha(\mathbf{x}) = \sum_{i_\alpha} \int d^3x_2 \cdots d^3x_{N_\alpha} |\psi_i(\mathbf{x}, \mathbf{x}_2, \ldots, \mathbf{x}_{N_\alpha})|^2$$

will be spherically symmetric if the sum runs over a full $L$-shell. If the limits of summation $\exp(S_\alpha)$ corresponded exactly to full shells, then $U$ would equal zero by Newton's theorem. It will now be shown that the partially filled shells can be chosen to make $U$ negative. Let $\mu_\alpha$, and $\nu_\alpha$ be the indices nearest to $\exp(S_\alpha)$ corresponding to filled shells, such that $\mu_\alpha \le \exp(S_\alpha) \le \nu_\alpha$. Thus

$$\sum_{i_\alpha=1}^{\exp(S_\alpha)} \sum_{i_\beta=1}^{\exp(S_\beta)} U_{i_\alpha i_\beta} = \sum_{i_\alpha=\mu_\alpha}^{\exp(S_\alpha)} \sum_{i_\beta=\mu_\beta}^{\exp(S_\beta)} U_{i_\beta i_\alpha},$$

and the interaction energy can be written as

$$U = c \sum_{i_1=\mu_1}^{\exp(S_1)} \sum_{i_2=\mu_2}^{\exp(S_2)} \cdots \sum_{i_k=\mu_k}^{\exp(S_k)} U_{i_1,\ldots,i_k}, \qquad c > 0,$$

$$U_{i_1,\ldots,i_k} = \left\langle \psi_i \middle| \sum_{j>m} \frac{e_j e_m}{|\mathbf{x}_j - \mathbf{x}_m|} \psi_i \right\rangle.$$

We know that

$$\sum_{i_1=\mu_1}^{\nu_1} \sum_{i_2=\mu_2}^{\nu_2} \cdots \sum_{i_k=\mu_k}^{\nu_k} U_{i_1,\dots,i_k} = 0,$$

and since the eigenvalues $e_{i_\alpha}$ are degenerate if $\mu_\alpha \le i_\alpha \le \nu_\alpha$, it is possible to select $\exp(S_1) - \mu_1$ indices $i_1$ such that

$$\sum_{i_2=\mu_2}^{\nu_2} \cdots \sum_{i_k=\mu_k}^{\nu_k} \sum_{i_1=\mu_1}^{\exp(S_1)} U_{i_1,\dots,i_k} \le 0$$

without changing the first sum in (4.3.10). We now proceed inductively and choose $\exp(S_2) - \mu_2$ indices $i_2$ such that

$$\sum_{i_2=\mu_2}^{\exp(S_2)} \cdots \le 0$$

and so forth, until finally $U \le 0$. This proves the

**Monotony of the Energy** (4.3.11)

*If* $B \supset \bigcup_{\alpha=1}^{k} B_\alpha$, $N_t = \sum_{\alpha=1}^{k} N_\alpha$, *and* $N_\alpha/Z + 1$ *is integral*, $S = \sum_{\alpha=1}^{k} S_\alpha$, *and E is as defined in* (4.3.8), *then*

$$E(S, V, N_t) \le \sum_{\alpha=1}^{k} E_\alpha(S_\alpha, V_\alpha, N_\alpha).$$

**Remarks** (4.3.12)

1. The $B_\alpha$ are required only to be disjoint; how well they fill $B$ does not affect the validity of the equation.
2. All but one of the $B_\alpha$ have to be spherical and electrically neutral, but one of them need not be.
3. The theorem holds regardless of the statistics of the particles, which can affect it only by ensuring the existence of a bound on $E/N$.

The question of how completely $B$ can be filled by the $B_\alpha$ is a purely geometrical one. It is answered by the

**Swiss Cheese Theorem** (4.3.13)

*Let* $R_j = (1 + p)^j R_0$, $p \in \mathbb{Z}^+$, $1 + p \ge 27$, *be the radii of the balls of a given size indexed by j and let* $B_m$ *be a ball of size m. Then for all* $m > 0$, $B_m$ *contains*

*the union from $j = 1$ to $m - 1$ of $v_j$ disjoint balls of size $j$, where*

$$v_j = \frac{(1+p)^{3(m-j)}}{p}\left(\frac{p}{1+p}\right)^{m-j} \in \mathbb{Z}^+.$$

**Remarks** (4.3.14)

1. This theorem makes more precise the fact, clear at the intuitive level, that a large ball can be filled extremely well by smaller ones if their radii are chosen suitably. The total volume of the small balls is
$$\sum_{j=0}^{m-1} R_j^3 v_j = ((1+p)^m R_0)^3\left(1 - \left(\frac{p}{1+p}\right)^m\right),$$
so that the unfilled fraction is only $(p/(1+p))^m$, which tends to zero as $m \to \infty$.
2. Of course, the filling of a ball uses more small balls than large ones, but the fraction of volume filled by the balls of size $j$ is $(1/p)(p/(1+p))^{m-j}$, as the larger balls are much more voluminous.

**Proof**

See Problem 1. □

**The Homogeneity of the Density** (4.3.15)

The next step in §2.3 was to consider a sequence of larger and larger cubes, all of which had the same entropy and particle density. Nothing like that is possible in this situation, since to compensate for the gaps some of the balls will have greater densities than the average density overall. Since the unfilled volume gets smaller and smaller, however, it suffices to impose relatively large densities on the balls of size 0 and assign equal densities to all the others. Let us thus choose $N_\alpha/V_\alpha = \rho(p+1) \equiv \rho_0$ for $\alpha = 1, 2, \ldots, v_0$, so for the balls of size 0, $N_\alpha/V_\alpha = \rho$ for all $\alpha > v_0$. If $\rho_j$ is the density in a ball of size $j$, and we let $\rho_1, \ldots, \rho_m = \rho$, then the $\rho_j$ satisfy a recursion formula

$$\rho_m = \sum_{j=0}^{m-1} \rho_j v_j \left(\frac{R_j}{R_m}\right)^3 = \frac{\rho_0}{p}\left(\frac{p}{p+1}\right)^m + \frac{\rho}{p}\sum_{j=1}^{m-1}\left(\frac{p}{p+1}\right)^{m-j} = \rho$$

for all $m \geq 1$.

In the same way the entropy is distributed so that the entropy density $\sigma_j$ in the balls of size $j$ satisfies

$$\sigma_0 = \sigma(p+1), \qquad \sigma_1 = \sigma_2 = \cdots = \sigma_m = \sigma = \frac{1}{p}\sum_{j=0}^{m-1}\sigma_j\left(\frac{p}{p+1}\right)^{m-j}.$$

If $V_0 = 4\pi R_0^3/3$ and $E_j$ is the energy and $\varepsilon_j$ the energy density of the balls of size $j$, then Proposition (4.3.11) specializes for this particular filling to

$$
\begin{aligned}
E_k(S, N) &\leq \sum_{j=0}^{k-1} E_j(S_j, N_j)\nu_j, \\
\varepsilon_k(\sigma_k, \rho_k) &= [(1+p)^{3k}V_0]^{-1}E_k(S_k, N_k) \\
&\leq \frac{1}{V_0 p}\sum_{j=0}^{k-1}\left(\frac{p}{1+p}\right)^{k-j}(1+p)^{-3j}E_j(S_j, N_j) \\
&= \frac{1}{p}\sum_{j=0}^{k-1}\left(\frac{p}{1+p}\right)^{k-j}\varepsilon_j(\sigma_j, \rho_j).
\end{aligned}
$$

This is a modification of (2.3.6) and similarly allows the convergence of $\varepsilon_k \equiv \varepsilon_k(\sigma_k, \rho_k)$ to be demonstrated: There exist numbers $c_k \leq 0$ such that

$$
\varepsilon_k = c_k + \frac{1}{p}\sum_{j=0}^{k-1}\left(\frac{p}{1+p}\right)^{k-j}\varepsilon_j.
$$

The recursion formula has the solution

$$
\varepsilon_k = c_k + \frac{1}{1+p}\left(\varepsilon_0 + \sum_{j=0}^{k-1} c_j\right). \tag{4.3.16}
$$

Since the sequence $\{\varepsilon_k\}$ is bounded from below, $\sum_j c_j$ must converge, so $\lim_{k\to\infty} c_k = 0$. Since $\varepsilon_k - c_k$ decreases monotonically as a function of $k$ by (4.3.16), $\varepsilon_k$ must tend to a limit. If $k > 0$, then all the densities had the same values $(\sigma, \rho)$, and we arrive at the

**Existence of the Thermodynamic Limit** (4.3.17)

*For the H of* (4.3.19), *the limit* $\varepsilon(\sigma, \rho) \equiv \lim_{V\to\infty} (1/V)E(\sigma V, \rho V)$ *exists.*

**Remarks** (4.3.18)

1. The theorem has been proved for spherical volumes, but it generalizes to other shapes with a reasonable relationship between volume and surface area.
2. Although the theorem and proof are given here for strictly neutral systems, it is clear that a small excess charge $\Delta Q$ can be allowed as long as its electrostatic energy $\sim(\Delta Q)^2/V^{1/3}$ can be neglected in comparison with $E$.
3. Although we have assumed there was only one kind of nucleus, the case of any number of kinds of nucleus can be covered simply by generalizing the notation.

4. Since $\varepsilon_k - c_k$ is a monotonic sequence, Dini's theorem guarantees that $\varepsilon_k$ converges uniformly on compact sets in $(\sigma, \rho)$; to use this argument it is necessary to extend the definition of the function $\varepsilon_V$, which was initially defined for finite $V$ on a discrete set, to make it continuous. The continuity of $\varepsilon$ will follow from the convexity to be proved below.
5. The Hamiltonian (4.3.9) includes the kinetic energy of the nuclei. Strangely, the existence of the thermodynamic limit (4.3.17) has not been proved in the apparently simpler case where $M_k = \infty$.

The existence of the limit means that all systems characterized by $N$ have the same dependence on the averaged quantity $\varepsilon$ provided that they are large enough. But does the theory predict a reasonable dependence? The temperature, pressure, specific heat, and compressibility should at least be positive in accordance with our experience. The positivity of the temperature and pressure are ensured by our definition of entropy and by the boundary conditions. With the aid of (2.3.29), the positivity of the other observables is a consequence of the convexity of the function $(\sigma, \rho) \to \varepsilon(\sigma, \rho)$, which, however, does not follow directly from the definitions—recall that the preceding chapter illustrated this with a counter example. Yet it is possible to formulate a theorem on the

**Thermodynamic Stability of Coulomb Systems** (4.3.19)

*The mapping* $\mathbb{R} \times \mathbb{R}^+ \to \mathbb{R}: (\sigma, \rho) \to \varepsilon(\sigma, \rho)$ *is*

(i) *convex;*
(ii) *nondecreasing in* $\sigma$;
(iii) *bounded below by* $-c\rho$ *for* $c \in \mathbb{R}^+$;
(iv) *such that* $\rho^{-1}\varepsilon(\sigma\rho, \rho)$ *is an increasing function of* $\rho$.

**Proof**

(i) Let $p$ be an odd integer, so that $v_j = (1 + p)^{2(k-j)}p^{k-j-1}$ is even for $0 \leq j \leq k - 1$, and fill half of the balls of a given size with densities $\rho$, $\sigma$ (or, respectively, $\rho_0 = \rho(1 + p)$, $\sigma_0 = \sigma(1 + p)$) and the other half with $\rho'$, $\sigma'$ (or, respectively, $\rho'_0 = \rho'(1 + p)$, $\rho'_0 = \sigma'(1 + p)$). Then, since the energy is monotonic as in (4.3.11),

$$\varepsilon_k(\bar{\sigma}_k, \bar{\rho}_k) \leq \frac{1}{2p} \sum_{j=0}^{k-1} \left(\frac{p}{1 + p}\right)^{k-j} [\varepsilon_j(\sigma_j, \rho_j) + \varepsilon_j(\sigma'_j, \rho'_j)],$$

$$\bar{\sigma}_k = \frac{1}{2p} \sum_{j=0}^{k-1} \left(\frac{p}{1 + p}\right)^{k-j} (\sigma_j + \sigma'_j),$$

and

$$\bar{\rho}_k = \frac{1}{2p} \sum_{j=0}^{k-1} \left( \frac{p}{1+p} \right)^{k-j} (\rho_j + \rho_j'),$$

which implies that

$$\varepsilon(\tfrac{1}{2}(\sigma + \sigma'), \tfrac{1}{2}(\rho + \rho')) \le \tfrac{1}{2}(\varepsilon(\sigma, \rho) + \varepsilon(\sigma', \rho'))$$

as $k \to \infty$. Now note that $\varepsilon$ is monotonic in $\sigma$ and $\rho^{-1}\varepsilon(\sigma\rho, \rho)$ is monotonic in $\rho$, so according to (2.3.11; 1) $\varepsilon$ is convex not just with coefficient $\frac{1}{2}$ but with all $\alpha \in [0, 1]$. Hence it is continuous on the interior of $\mathbb{R}^+ \times \mathbb{R}^+$.

(ii) See Remark (2.3.3; 3).

(iii) This follows from the estimate (4.3.3) showing the stability of matter.

(iv) From the monotonic property (2.3.4) of the energy, $\partial E/\partial V|_{S, N = \text{const}} \le 0$. □

Since $\varepsilon$ has the right sort of convexity, one of the assumptions needed to prove the existence of the thermodynamic limit of the canonical ensemble is satisfied. More information about the function $\varepsilon(\sigma, \rho)$ is needed to verify the other hypotheses made in Theorem (2.4.14). In particular it needs to be shown that $\varepsilon$ increases rapidly enough with $\sigma$ that the $\sigma_0$ introduced in (2.3.11; 4) is finite, and $\lim_{\sigma \to \infty} \varepsilon/\sigma = \infty$. This is shown by the

**Lower Bound for the Energy Density** (4.3.20)

*If* $H = H_\alpha \equiv K + \alpha \sum_{i>j} e_i e_j |\mathbf{x}_i - \mathbf{x}_j|^{-1}$ *and* $\varepsilon_\alpha$ *are the corresponding energy densities, then*

$$\varepsilon_\alpha(\sigma, \rho) \ge \lambda \varepsilon_0(\sigma, \rho) - \frac{c\rho\alpha^2}{1 - \lambda} \quad \textit{for all } 0 \le \lambda < 1,$$

*where*

$$c = 2.08(1 + Z^{2/3})^2.$$

**Proof**

According to (2.3.3; 4), $\varepsilon_\alpha$ is concave in $\alpha$, and $\varepsilon_\alpha \ge \lambda\varepsilon_0 + (1 - \lambda)\varepsilon_{\alpha/(1-\lambda)}$. However, by (4.3.3), $-c\rho\alpha^2$ is a lower bound for all $\rho$ and $\sigma$. □

**Corollaries** (4.3.21)

1. Since it was shown in (2.5.23) that in the case of one kind of particle, $\varepsilon_0(\sigma, \rho) = c'\rho^{5/3} \exp(2\sigma/3\rho)$, $c' > 0$, is the limit as $\sigma \to \infty$, it follows that $\lim_{\sigma \to \infty\, \rho\, \text{fixed}} \varepsilon(\sigma, \rho)/\sigma = \infty$.

2. Even for a finite volume $-c\rho\alpha^2$ is a lower bound, which makes it easy to verify that there exists a function $s(\varepsilon, \rho)$ dominating $\sigma$ for all volumes, and satisfying $\lim_{\varepsilon \to \infty} s/\varepsilon = 0$.
3. In (4.3.43; 2) we shall find an upper bound on the ground-state energy density, of the form $c_1\rho^{5/3} - \alpha c_2 \rho^{4/3}$. When combined with (4.3.20) it yields an upper bound for the $\sigma_0$ of (2.3.11; 4) at which $\varepsilon(\sigma)$ starts to move up.

This fact is not yet enough to ensure that thermodynamics works perfectly. Let us write down a

**Thermodynamic Wish List** (4.3.22)

1. $\sigma_0 = 0$.
2. $\partial\varepsilon/\partial\sigma|_{\sigma=\sigma_0} = 0$.
3. $\lim_{\sigma \to \infty} (\partial\varepsilon/\partial\sigma) = \infty$.
4. The function $\varepsilon$ is continuously differentiable.
5. The function $\varepsilon$ is strictly convex for large $\sigma$ and is linear on certain intervals in $\sigma$ when $\sigma$ is small.

**Open Questions for the Wish List**

1. Statement 1 is a strong formulation of the third law of thermodynamics, and is unproved for Coulomb systems. Although there is an upper bound on $\sigma_0$ in (4.3.21; 3), it is not sharp enough to show that $\sigma_0 = 0$.
2. The second statement implies that the system does not fall into its ground state if the temperature is higher than absolute zero, and our bounds are likewise too crude to prove it.
3. The third statement means that there is no maximum temperature, and is proved by (4.3.21; 1).
4. Kinks in the graph of $\varepsilon$ would correspond to "anti-phase-transitions" at which either the temperature or the pressure shows a discontinuity while the energy remains continuous. The specific heat and the compressibility would be zero at such a point. Such things do not appear to happen in reality, though the arguments we have made do not exclude them.
5. It is known empirically that there are no phase transitions at high temperatures, only at low temperatures. However, this fact has not been proved in the theory.

The equivalence with the canonical ensemble requires only the positivity of the specific heat, which is guaranteed by (4.3.19). The assumptions of

Theorems (2.4.14) are fulfilled because of (4.3.18; 4), (4.3.19(i)), and (4.3.21; 2), so it leads to the

**Thermodynamic Limit of the Canonical Ensemble** (4.3.23)

*The limit*

$$\lim_{V\to\infty}\left(-\frac{T}{V}\ln \operatorname{Tr}\exp(-\beta H)\right) = \inf_{\varepsilon}(\varepsilon - T\sigma(\varepsilon, \rho)) = \varphi(T, \rho)$$

*exists.*

**Remarks** (4.3.24)

1. The properties of the free-energy density listed in (2.4.16) are also proved.
2. It is possible to prove the existence of the limit as $V \to \infty$ directly, but that is not enough to show the equivalence with the microcanonical $\varepsilon$. In particular it does not show that $\varepsilon$ is convex in $\sigma$.

Finally, consider the grand canonical ensemble, supposing there are $N_e$ electrons and $N_k$ nuclei with chemical potentials $\mu_e$ and $\mu_k$. The function to investigate is

$$P(T, \mu_e, \mu_k) \equiv \lim_{V\to\infty}\frac{T}{V}\ln \operatorname{Tr}\exp[-\beta(H - N_e\mu_e - N_k\mu_k)]. \quad (4.3.25)$$

One difficulty with (4.3.25) is that the trace contains the sum over all possible numbers of particles, and not only the neutral configuration for which $N_e = ZN_k$. Fortunately, it turns out that the non-neutral contributions have such large Coulomb energies that they play no role. Stated without proof [30], here is the resulting proposition on the

**Thermodynamic Limit of the Grand Canonical Ensemble** (4.3.26)

*The limit* (4.3.25) *exists, and*

$$P(T, \mu_e, \mu_k) = \sup_{\rho_e = Z\rho_k}(\mu_e\rho_e + \mu_k\rho_k - \varphi(T, \rho)),$$

$$\rho = \frac{N_e + N_k}{V} = \left(1 + \frac{1}{Z}\right)\rho_e.$$

**Remarks** (4.3.27)

1. Although the supremum is *a priori* over all density configurations, it is attained in the neutral sector.

2. Roughly speaking, to generalize this to cover arbitrarily many components it is only necessary to treat $\mu$ and $\rho$ as "isovectors."

**Bounds for** $\varepsilon(\sigma, \rho)$ (4.3.28)

The question that now arises is to what extent the qualitative propositions that have been derived about $\varepsilon(\sigma, \rho)$ can be sharpened and made quantitative. For instance, it would be desirable to find an upper bound to complement the lower bound (4.3.20); upper bounds are always easy to discover, since with the min–max principle it is only necessary to devise some good trial functions. In the limit $\rho \to 0$ an obvious upper bound for the ground-state energy is the sum of the energies of the individual atoms. If the density is finite, then one would think of using the ground state of the kinetic energy $K$ in the variational principle, and the result is the first-order perturbation-theoretic approximation to $H_\alpha = K + \alpha V$.

**Remarks** (4.3.29)

1. It is impossible for the expansion in powers of $\alpha$ to converge in the thermodynamic limit; if $\alpha < 0$, then the electrons would attract one another, as would the nuclei, whereas the nuclei would repel the electrons. The ground-state energy of fermions with an attractive $1/r$ potential goes as $-N^{7/3}$, and that of bosons goes as $-N^3$ (see (1.2.22) and (1.2.23; 3)). If a trial function is constructed with all the electrons on one side of the container and all the nuclei on the other, then the expectation value of the energy is greater than $-N^{7/3} + N^2/R \to -N^{7/3}$, so $E/N$ does not remain bounded from below. On the other hand, the convergence of a series in the limit $N \to \infty$ would imply that $\lim_{N\to\infty} E/N$ would be finite on the whole disc of convergence, which would include some negative values of $\alpha$. In fact the explicit calculation reveals that even the second-order contribution becomes infinite as $N \to \infty$. Even so, the first-order result is useful as an upper bound.
2. According to (III: 3.5.21) the min–max principle applies to finite $\sigma$ other than the ground state, but it is more difficult to calculate the microcanonical expectation values than the grand canonical ones. Hence, for nonzero temperatures it is better to use (2.1.8; 3) to bound the grand canonical partition function with $-P_\alpha \leq -P_0 + \text{Tr}\, V\rho_{GC}$.

**The Ground State** (4.3.30)

The simplest case is $T = 0$, so let us see how far we can get with the easiest methods. Take the expectation value of (4.3.1) in the ground state of the electrons; if they are confined in a box $\Lambda$ with periodic boundary conditions, the ground state is a plane wave, producing a constant electron density $\rho_e$. If the nuclear charges are all $Z$ and the nuclear masses are all $\mu$, that leaves

$$\langle H\rangle = \sum_{k=1}^{M} \frac{|\mathbf{p}_k|^2}{2\mu} + Z^2 \sum_{k>j} |\mathbf{X}_k - \mathbf{X}_j|^{-1} - \sum_k \int_\Lambda \frac{d^3x\rho_e Z}{|\mathbf{x} - \mathbf{X}_k|}$$
$$+ \frac{1}{2}\int_\Lambda \frac{d^3x\, d^3y\rho_e^2}{|\mathbf{x} - \mathbf{y}|}$$
$$+ \left\langle \sum_i |\mathbf{p}_i|^2 \right\rangle + \left\langle \sum_{i>k} |\mathbf{x}_i - \mathbf{x}_k|^{-1} - \frac{1}{2}\int_\Lambda \frac{d^3x\, d^3y\rho_e^2}{|\mathbf{x} - \mathbf{y}|}\right\rangle$$
$$+ \sum_k \left\langle \int \frac{d^3x\rho_e}{|\mathbf{x} - \mathbf{X}_k|} - \sum_j \frac{1}{|\mathbf{x}_j - \mathbf{X}_k|}\right\rangle. \tag{4.3.31}$$

The first line of this equation is the Hamiltonian $H_J$ of jellium (1.2.6) in the nuclear variables. If we therefore add the ground-state energy of jellium to the other expectation values, we get an upper bound on the ground-state energy of $H$, corresponding to a trial function consisting of the tensor product of the ground state of $H_J$ with the electron wave-function. The zero-point energy of the electrons is the next term in (4.3.31), followed by what is referred to as the exchange energy, and the final expectation value is zero. By (2.5.32), if the spin is $\frac{1}{2}$, the zero-point energy goes as

$$\left\langle \sum_{i=1}^{N} |\mathbf{p}_i|^2 \right\rangle = N\tfrac{3}{5}(3\pi^2\rho_e)^{2/3} = N\frac{2.2}{r_s^2}, \qquad r_s = \left(\frac{3}{4\pi\rho_e}\right)^{1/3}, \tag{4.3.32}$$

as $N \to \infty$, and with only a little difficulty the exchange energy can be calculated as

$$\left\langle \sum_{i>j} |\mathbf{x}_i - \mathbf{x}_j|^{-1} - \frac{1}{2}\int \frac{d^3x\, d^3x'\rho_e^2}{|\mathbf{x} - \mathbf{x}'|}\right\rangle = -0.458\frac{N}{r_s} \tag{4.3.33}$$

(Problem 3). It expresses the effect of the correlations among the electrons owing to their having to avoid each other to satisfy the exclusion principle. The result is to lower the Coulomb energy in comparison with that of a homogeneous charge distribution.

### **The Ground State of Jellium** (4.3.34)

As for $H_J$, an upper bound can be obtained by using plane waves as trial functions, for which $\langle H_J\rangle$ once again consists of zero-point energy and exchange energy. A lower bound comes from the sum of the zero-point energy and the minimum of the potential (1.2.10), and when combined they bound $E_J$ according to

$$\frac{2.2}{2\mu r_s^2} - \frac{0.9}{r_s} \le \frac{E_J}{N} \le \frac{2.2}{2\mu r_s^2} - \frac{0.458}{r_s} \tag{4.3.35}$$

if $Z = 1$ and the spin is $\frac{1}{2}$. If the density is large ($r_s \to 0$), then the bounds are close together, but they spread out if the density is small. At small densities it is better to array the nuclei on a lattice; give them wave-functions $\sim \sin(\pi r a)/r$, where $r$ is the distance from the lattice site if it is less than $a$ and otherwise let the wave-function be 0, and take $a$ small enough that the wave-functions will not overlap, and will thus be orthogonal. The most convenient configuration is a body-centered cubic lattice, which consists of two simple cubic lattices, one of which has been displaced along a diagonal so that its corners are at the centers of the other. If the density is 2, i.e., the lattice constant of the simple cubic lattice is 1, then $a$ must be less than $\sqrt{3}/4$ in order that the balls of radius $a$ do not intersect; in terms of $r_s$, the distance between nuclei,

$$a \leq \left(\frac{8\pi}{3}\right)^{1/3} \frac{\sqrt{3}}{4} r_s. \tag{4.3.36}$$

If the nuclei were concentrated at the points of the lattice, then the Coulomb energy per particle would be $-0.896/r_s$ according to (1.2.11; 2). Provided that they do not overlap, the repulsion between the nuclei will be the same even if they are somewhat spread out. On the other hand, their interaction (per particle) with the background would be affected by

$$\frac{\rho}{2} \int_0^a dr r^2 \sin^2 \frac{r\pi}{a} \Big/ \int_0^a dr \sin^2 \frac{r\pi}{a} = \frac{\rho}{2} a^2 \left(\frac{1}{3} - \frac{1}{2\pi^2}\right). \tag{4.3.37}$$

If this is added to the kinetic energy $(\pi/a)^2$ (for mass $\frac{1}{2}$), then the minimum

$$\frac{E}{N} = \left(\frac{\pi}{2} - \frac{3}{4\pi}\right)^{1/2} r_s^{-3/2} - \frac{0.896}{r_s} = 1.15 r_s^{-3/2} - \frac{0.896}{r_s}$$

is attained when

$$a = \left[\frac{\rho}{2\pi^2}\left(\frac{1}{3} - \frac{1}{2\pi^2}\right)\right]^{-1/4} = r_s^{3/4}\left[\frac{3}{8\pi^3}\left(\frac{1}{3} - \frac{1}{2\pi^2}\right)\right]^{-1/4}.$$

Condition (4.3.36) means that

$$r_s \geq \frac{8^3 \pi^4}{3(2\pi^2 - 3)} \left(\frac{3}{8\pi}\right)^{1/3} \cong 489. \tag{4.3.38}$$

If $r_s$ is smaller, then $a$ must be taken as $(8\pi/3)^{1/3}(\sqrt{3}/4)r_s$, which costs some kinetic energy, $12.75/r_s^2$, and raises the Coulomb interaction above that due to the background by $0.026/r_s$. The figures become more favorable, however, when it is recalled that wave-functions of nuclei with opposite spins do not need to be spatially orthogonal to avoid incurring exchange energy. Suppose that the nuclei have spin $\frac{1}{2}$, as with protons, and put nuclei with spin up on one of the simple cubic lattices and nuclei with spin down on the other. Then the spheres are required only not to overlap with other spheres on the same simple cubic lattice. This weakens the bound (4.3.36) to

$$a \leq \left(\frac{8\pi}{3}\right)^{1/3} \frac{r_s}{2},$$

which weakens the lower bound on $r_s$ (4.3.38) by a factor $\frac{9}{16}$, so

$$r_s \geq 275, \tag{4.3.39}$$

and also diminishes the zero-point energy by $\frac{3}{4}$ to $9.54/r_s^2$ and increases the interaction with the background by the same factor. The Coulomb repulsion between neighboring nuclei decreases, but only by an insignificant amount $10^{-3}/r_s$ The net effect is to produce

**Bounds on the Ground-State Energy of Spin $-\frac{1}{2}$ Jellium** (4.3.40)

$$\leq \frac{2.2}{r_s^2} - \frac{0.458}{r_s} \qquad \text{(i)}$$

$$\frac{2.2}{r_s^2} - \frac{0.9}{r_s} \leq \frac{E}{N} \leq \frac{9.58}{r_s^2} - \frac{0.85}{r_s} \qquad \text{(ii)}$$

$$\leq \frac{1.15}{r_s^{3/2}} - \frac{0.89}{r_s} \quad \text{if } r_s > 275, \qquad \text{(iii)}$$

where $e = 2\mu = 1$. (See Figure 39).

**Remarks** (4.3.41)

1. The distance between particles as measured in Bohr radii with the appropriate mass is $r_s$. If $H_J$ is the Hamiltonian of the nuclei, and the pressure is not too huge, then $r_s$ is on the order of the ratio of the mass of the nucleus to that of the elctron, which is at least 2000. This means that (4.3.40(i)) will be the best of the bounds. If jellium is taken as a model of electrons in a metal, then $r_s \sim 1$, and (4.3.40(i)) is best.
2. There are conjectures that the transition from homogeneity to a lattice structure as $r_s$ increases is accompanied by a phase transition. It is even believed that the exchange energy, which favors parallel spins, causes ferromagnetism. Despite the simple form of $H_J$ it has not been possible to prove these speculations.

If we focus attention again on real matter, we must add the contribution from the electrons to that of the protons. Observe first that for nuclei the parameter $r_s \sim \rho^{-1/3}$/Bohr radius is increased by a factor $\mu Z^2$, but at the same time the energies in (4.3.40) are multiplied by $\mu Z^2$. Since the zero-point energy obtains an extra factor $1/\mu$, it can be neglected. For the densities of interest, $r_s > 275/\mu Z^2$, so (4.3.40(iii)) applies to nuclei. Of course, the trial function with a homogeneous electron distribution is poor when $Z$ is large, and does not contribute the right dependence on $Z$. If $Z = 1$, our earlier results on the energy per electron are only

**Crude Bounds** (4.3.42)

$$-8.32 \le \frac{E}{N} \le \frac{2.2}{r_s^2} - \frac{1.34}{r_s}.$$

**Refinements** (4.3.43)

1. The lower bound. The Birman–Schwinger bound (III: 3.5.36) can be improved with the methods of functional integration [31], sharpening Inequality (4.1.47; 2) by a factor of 1.5. Then with (4.3.20), if the density is finite, $\lambda$ is chosen optimally, and $\varepsilon_0 = 5.74\rho^{5/3}$, or equivalently $E_0/N = 2.2/r_s^2$, there results
$$\frac{2.2}{r_s^2} - \frac{5.5}{r_s} \le E.$$

2. The upper bound. The ground-state energy in a box of volume $V$ is of the form
$$E = V^{-2/3} f(V^{1/3}\alpha).$$
The facts that $\partial E/\partial V \le 0$ and $\partial^2\varepsilon/\partial\rho^2 \ge 0$ and the convexity in $\alpha$ are expressed by the inequalities
$$f(x) \ge \frac{x}{2} f'(x) \quad \text{and} \quad 6xf'(x) - 10f(x) \le x^2 f''(x) \le 0.$$
Since $\partial E/\partial V \le 0$, a linear bound $f(x)/f(0) \le 1 - \gamma x$ for $x > 2/\gamma$ can be improved by a parabolic bound $f(x)/f(0) \le -x^2(\gamma/2)^2$. By (4.3.43; 1) $\gamma^{-1} = 2.2/1.34$, so if $r_s > 2\gamma^{-1} = 3.28$, then $f$ is less than $-f(0)x^2 \cdot 1.34/4(2.2)^2$. It follows that
$$\frac{E}{N} \le \begin{cases} 2.2/r_s^2 - 1.34/r_s, & \text{if } r_s < 3.28 \\ -0.204, & \text{if } r_s \ge 3.28. \end{cases}$$

These bounds are far from satisfactory. Not only do they fail to allow finer details to be discerned, but indeed they do not even prove that hydrogen holds together at $T = 0$ rather than breaking up into separated atoms. In these units the energy of a separated hydrogen atom is $-\frac{1}{4}$, i.e., less than the upper bound, which only shows how large a territory still remains open to exploration with exact methods in physics.

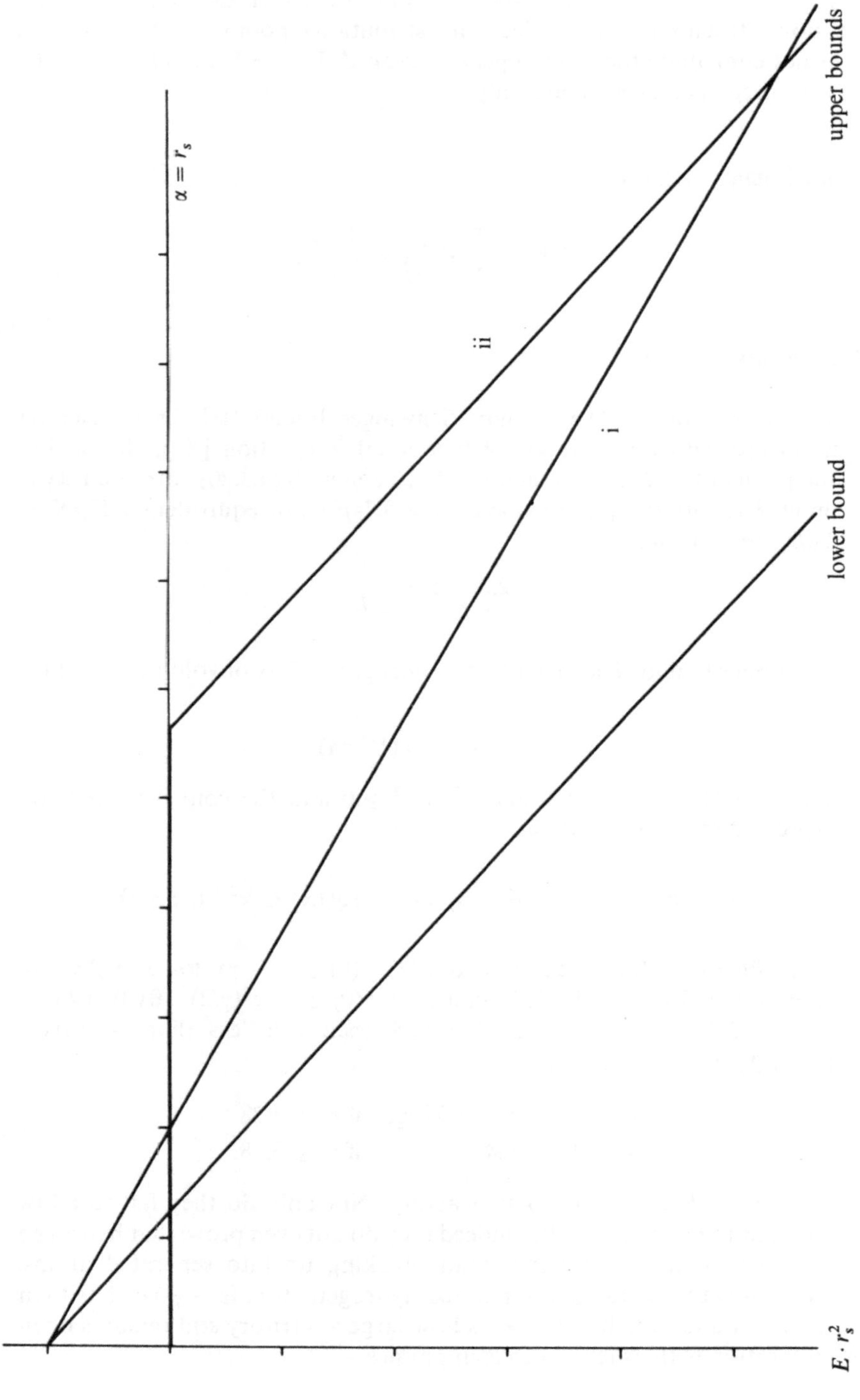


Figure 39 Bounds (4.3.40) for the energy of jellium.

## Problems (4.3.44)

1. Prove the Swiss cheese theorem (4.3.13): For any region $\Lambda \subset \mathbb{R}^3$ and any real number $h$ let $\Lambda_h = \{\mathbf{x} \in \Lambda: d(\mathbf{x}, \Lambda^c) < h\}$, if $h > 0$ and $\Lambda_h = \{\mathbf{x} \in \Lambda^c: d(\mathbf{x}, \Lambda) \leq -h\}$, if $h \leq 0$, and denote the volume of $\Lambda_h$ by $V(h, \Lambda)$.
Then prove the following two lemmas: (i) Suppose $\Lambda$ is covered by closed cubes of side $l$, the interiors of which do not intersect, and let $v$ be the number of cubes entirely contained in $\Lambda$. Then the volume of $\Lambda$ not covered by these $v$ cubes is at most $V(l\sqrt{3}, \Lambda)$. (ii) Let $B \subset \mathbb{R}^3$ be an open ball of radius $R$ and $y$ a number satisfying the inequality $R \geq 2\sqrt{3}\,y \geq 0$. Then $V(2\sqrt{3}\,y, B) \leq V(-2\sqrt{3}\,y, B) \leq 56\pi R^2 y/\sqrt{3}$. Finish the proof of the theorem by covering $B_0$ with a cubic lattice of spacing $2R_1$, and in each cube of the lattice place a ball of radius $R_1$, then cover the balls with a cubic lattice of spacing $R_2$, etc. Use the lemmas to estimate $v_j$ and the fraction of volume taken up by the balls of size $j$.

2. Use Inequalities (III: 4.5.24) and (4.1.5) to find a lower bound for the potential energy of jellium,

$$U = \sum_{j>k} |\mathbf{x}_j - \mathbf{x}_k|^{-1} - \sum_i \int d^3x \rho(\mathbf{x}) |\mathbf{x} - \mathbf{x}_i|^{-1} + \frac{1}{2} \int \frac{d^3x\, d^3y}{|\mathbf{x} - \mathbf{y}|} \rho(\mathbf{x})\rho(\mathbf{y})$$

and compare with (1.2.10). (Let $\rho$ be constant in any ball.)

3. Calculate

$$\lim_{V \to \infty} V^{-1} \langle \psi | \left( \sum_{i>k} |\mathbf{x}_i - \mathbf{x}_k|^{-1} - \frac{1}{2} \int_\Lambda \frac{d^3x\, d^3y}{|\mathbf{x} - \mathbf{y}|} \rho_e^2 \right) | \psi \rangle$$

if $\psi$ is the ground state of a system of free electrons in a box of volume $V$. (The momentum states in both spin orientations are occupied up to a maximum momentum $p$ such that $p^3/3\pi^2 = N/V = 3/4\pi r_s^3$.)

4. Verify that the concavity of $E$ as a function of $(1/m, \alpha)$ is no more severe a restriction than the concavity of $f$ in (4.3.43; 2).

## Solutions (4.3.45)

1. (i) If $\Lambda$ is covered by cubes of length $l$, but all cubes intersecting $\Lambda^c$ are removed, then the uncovered portion of $\Lambda$ is contained in $\Lambda_{l\sqrt{3}}$. (Hence the number $v_{2l}$ of cubes of length $2l$ that can be packed entirely into $\Lambda$ is at least $(2l)^{-3}[V(\Lambda) - V(2\sqrt{3}l, \Lambda)]$.)
   (ii) If $0 \leq h \leq R$, then

$$V(h, B) = \frac{4\pi}{3}[R^3 - (R-h)^3] \leq V(-h, B) = \frac{4\pi}{3}[(R+h)^3 - R^3].$$

The lemma is then a consequence of the convexity of the function $f(\varepsilon) \equiv (1+\varepsilon)^3 - 1$, which implies that $f(\varepsilon) \leq f(0) + \varepsilon[f(1) - f(0)] = \varepsilon[2^3 - 1] = 7\varepsilon$.

Proof of the packing estimates. For simplicity assume that $R_0 = 1$, and let $v_j = p^{j-1}(1+p)^{2j}$. If a unit ball is covered by cubes of length $2R_1 = 2(1+p)^{-1}$, then it contains $v_1$ cubes, as we shall show. If we then cover the unit ball with a lattice of length $2R_2$, then there are $v_2$ cubes contained in the unit ball and not intersecting the first $v_1$ balls of size 1. The general fact will follow by induction. Therefore it needs to be shown that when the ball has been filled with smaller balls up to size $j$, it is still possible to pack $v_{j+1}$ balls of radius $R_{j+1}$ into the remaining space $B - \bigcup_{k=1}^{j}$ (balls of size $k$) $\equiv \Omega_j$:

$$V(\Omega_j) = \frac{4\pi}{3}\left(1 - \sum v_k R_k^3\right) = \frac{4\pi}{3}\left(\frac{p}{p+1}\right)^j.$$

$V(2\sqrt{3}R_{j+1}, \Omega_j) \leq M_j$, defined as the sum of $V(-2\sqrt{3}R_{j+1}, B)$ for all balls of size $\leq j$ and $V(2\sqrt{3}R_{j+1}, B)$, where $B$ is the unit ball. Because of (ii) and the inequality $2\sqrt{3}R_{j+1} < R_j$,

$$\begin{aligned} V(2\sqrt{3}R_{j+1}, \Omega_j) \leq M_j &\leq \frac{56\pi}{\sqrt{3}} R_{j+1}\left[1 + \sum v_k R_k^2\right] \\ &= (p^j + p - 2)(p-1)^{-1}(1+p)^{-(j+1)}\frac{56\pi}{\sqrt{3}} \equiv \tilde{M}_j. \end{aligned}$$

Therefore it suffices to show that

$$(2R_{j+1})^3 v_{j+1} \leq [V_j - \tilde{M}_j] \leq [V(\Omega_j) - V(2\sqrt{3}R_{j+1}, \Omega_j)],$$

i.e.,

$$1 \leq \frac{\pi}{6}\left[p + 1 - 14\sqrt{3}\,\frac{1 + p^{-j}(p-2)}{p-1}\right].$$

Since $p^{-j}(p-2) \leq (p-2)$, this reduces to

$$1 \leq \frac{\pi}{6}\left[p + 1 - 14\sqrt{3}\right],$$

which is true when $p + 1 \geq 27$. The fraction of the volume taken up by the balls of radius $R_j$ is

$$\frac{p^{j-1}}{(1+p)^j},$$

which shows that the packing fills the original ball exponentially fast.

2. From (III; 4.5.24),

$$U \geq -\tfrac{3}{4}\left[8\pi N^2 \int d^3x\rho^2(\mathbf{x})\right]^{1/3} = -1.35N/r_s,$$

and from (4.1.5),

$$U \geq -2\left[3.68N\frac{3}{5}\int \rho^{5/3}\right]^{1/2} = -1.84N/r_s.$$

3. As $N$ and $V \to \infty$, make the replacements

$$\frac{1}{V}\sum_{|k|\leq p} \to \int_{|k|\leq p} \frac{d^3k}{(2\pi)^3},$$

$$v(\mathbf{k}) \equiv \frac{1}{V}\int \frac{d^3x\, d^3x'}{|\mathbf{x}-\mathbf{x}'|}\exp[i\mathbf{k}\cdot(\mathbf{x}-\mathbf{x}')] \to 4\pi/|\mathbf{k}|^2,$$

to find that

$$\begin{aligned}\sum_{\substack{|\mathbf{k}|\leq p\\ |\mathbf{k}'|\leq p}} v(\mathbf{k}-\mathbf{k}') &= \int \frac{d^3k}{(2\pi)^3}\frac{4\pi}{|\mathbf{k}|^2}\int \frac{d^3q}{(2\pi)^3}\Theta(p-|\mathbf{q}|)\Theta(p-|\mathbf{k}-\mathbf{q}|)\\ &= \frac{2}{\pi}\int_0^{2p} dk\,\frac{p^3}{6\pi^2}\left[1-\frac{3k}{4p}+\frac{k^3}{16p^3}\right] = \frac{p^3}{3\pi^2}\frac{p}{\pi}\frac{3}{4}\\ &= \frac{N}{V}\left(\frac{9\pi}{4}\right)^{1/3}\frac{1}{r_s}\frac{3}{4\pi} = \frac{N}{V}\frac{0.458}{r_s}.\end{aligned}$$

In order to justify this formal calculation, make a convolution so that

$$v(\mathbf{k}) = \frac{4\pi}{|\mathbf{k}|^2} * F(\mathbf{k}),$$

where

$$\begin{aligned}F(\mathbf{k}) &= \frac{1}{V}\int_{\mathbf{x}\in V,\,\mathbf{x}'\in V} d^3x\, d^3x' \exp[i\mathbf{k}\cdot(\mathbf{x}-\mathbf{x}')]\\ &= L^{-3}\left(\frac{\sin k_1L/2}{L/2}\right)^2\left(\frac{\sin k_2L/2}{L/2}\right)^2\left(\frac{\sin k_3L/2}{L/2}\right)^2\end{aligned}$$

is the Fourier transform of the characteristic function of the box, and use Lebesgue's dominated convergence theorem to show that the integrals have the limits given above.

4. With $1/m = v$: $E = vf(\alpha/v)$,

$$E_{,\alpha\alpha} = \frac{1}{v}f'', \qquad E_{,vv} = \frac{\alpha^2}{v^3}f'',$$

$$E_{,v\alpha} = -\frac{\alpha}{v^2}f'', \qquad E_{,\alpha\alpha}E_{,vv} - (E_{,v\alpha})^2 = 0.$$

# Bibliography

## Works Cited in the Text


[1] F. J. Dyson and A. Lenard. Stability of Matter, I. *J. Math. Phys.* **8**, 423–433, 1967; Stability of Matter, II. *Ibid.* **9**, 698–711, 1968.

[2] E. H. Lieb. The $N^{5/3}$ Law for Bosons. *Phys. Lett.* **70A**, 71–73, 1979.

[3] R. A. Goldwell–Horstall and A. A. Maradudin. Zero-Point Energy of an Electron Lattice. *J. Math. Phys.* **1**, 395–404, 1960.

[4] J. Dixmier. Les Algèbres d'Opérateurs dans l'Espace Hilbertien. Paris, Gauthier-Villars, 1969.

[5] A. Wehrl. General Properties of Entropy. *Rev. Mod. Phys.* **50**, 221–260, 1978. E. H. Lieb. Convex Trace Functions and the Wigner–Yanase–Dyson Conjecture. *Adv. Math.* **11**, 267–288, 1973. B. Simon. Trace Ideals and their Applications. London and New York, Cambridge Univ. Press, 1979. A. Uhlmann. Relative Entropy and the Wigner–Yanase–Dyson–Lieb Concavity in an Interpolation Theory. *Commun. Math. Phys.* **40**, 147–151, 1975; Sätze über Dichtematrizen. *Wiss. Z. Karl-Marx-Univ. Leipzig* **20**, 633, 1971; The Order Structure of States. In: Proc. Intl. Symp. on Selected Topics in Statistical Mechanics. JINR-Publ. D17-11490. Dubna USSR, 1978.

[6] M. B. Ruskai. Inequalities for Traces on von Neumann Algebras. *Commun. Math. Physics* **26**, 280–289, 1972. M. Breitenecker, H.-R. Grümm. Note on Trace Inequalities. *Commun. Math. Phys.* **26**, 276–279, 1972. K. Symanzik. Proof and Refinements of an Inequality of Feynman. *J. Math. Phys.* **6**, 1155–1156, 1965.

[7] J. Aczel, B. Forte, and C. T. Ng. Why the Shannon and Hartley Entropies are "Natural." *Adv. Appl. Prob.* **6**, 131–146, 1974.

[8] E. H. Lieb and M. B. Ruskai. Proof of the Strong Subadditivity of Quantum-Mechanical Entropy. *J. Math. Phys.* **14**, 1938–1941, 1973. H. Araki and E. H. Lieb. Entropy Inequalities. *Commun. Math. Phys.* **18**, 160–170, 1970.

[9] P. C. Martin and J. Schwinger. Theory of Many-Particle Systems, I. *Phys. Rev.* **115**, 1342–1373, 1959.

[10] R. Peierls. Surprises in Theoretical Physics. Princeton, Princeton Univ. Press. 1976.

[11] E. H. Lieb and W. Thirring. Inequalities for the Moments of the Eigenvalues of the Schrödinger Hamiltonian and their Relation to Sobolev Inequalities. In: Studies in Mathematical Physics, Essays in Honor of Valentine Bargmann, A. S. Wightman, E. H. Lieb, and B. Simon, eds. Princeton, Princeton Univ. Press, 1976.

[12] E. T. Whittaker and G. N. Watson. A Course of Modern Analysis. Cambridge, at the University Press, 1969.

[13] J. T. Cannon. Infinite Volume Limits of the Canonical Free Bose Gas States on the Weyl Algebra. *Commun. Math. Phys.* **29**, 89–104, 1973.

[14] G. Lindblad. On the Generators of Quantum Dynamical Semigroups. *Commun. Math. Phys.* **48**, 119–130, 1976.

[15] A. Kossakowski and E. C. G. Sudarshan. Completely Positive Dynamical Semigroups of $N$-Level Systems. *J. Math. Phys.* **17**, 821–825, 1976.

[16] T. L. Saaty and J. Bram. Nonlinear Mathematics. New York, McGraw-Hill, 1964.

[17] D. Ruelle. Statistical Mechanics, Rigorous Results. New York, Benjamin, 1969.

[18] A. Guichardet. Systèmes Dynamiques non Commutatifs. *Astérisque* **13-14**, 1974.

[19] I. M. Gel'fand, R. A. Minlos, and Z. Ya. Shapiro. Representations of the Rotation and Lorenz Group and their Applications. Oxford, Pergamon Press, 1963.

[20] R. B. Israel, ed. Convexity in the Theory of Lattice Gases. Princeton, Princeton Univ. Press, 1979.

[21] O. Bratteli and D. W. Robinson. Operator Algebras and Quantum Statistical Mechanics, in two volumes. New York, Springer, 1979, 1980.

[22] H. Narnhofer. Kommutative Automorphismen und Gleichgewichtszustände. *Acta Phys. Austriaca* **47**, 1–29, 1977.

[23] H. Narnhofer. Scattering Theory for Quasi-Free Time Automorphisms of $C^*$ Algebras and von Neumann Algebras. *Rep. Math. Phys.* **16**, 1–8, 1979.

[24] H. Araki and G. L. Sewell. KMS Conditions and Local Thermodynamic Stability of Quantum Lattice Systems. *Commun. Math. Phys.* **52**, 103–109, 1977.

[25] G. L. Sewell. Relaxation, Amplification, and the KMS Conditions. *Ann. Phys.* (N.Y.) **85**, 336–377, 1974.

[26] H. Narnhofer and G. L. Sewell. Vlasov Hydrodynamics of a Quantum Mechanical Model. *Commun. Math. Phys.* **79**, 9–24, 1981.

[27] J. Messer. The Pressure of Fermions with Gravitational Interaction. *Z. Phys.* **B33**, 313–316, 1979.

[28] W. Thirring. A Lower Bound with the Best Possible Constant for Coulomb Hamiltonians. *Commun. Math. Phys.* **79**, 1–7, 1981.

[29] E. H. Lieb. The Stability of Matter. *Rev. Mod. Phys.* **48**, 553–569, 1976.

[30] J. Lebowitz and E. H. Lieb. The Constitution of Matter: Existence of Thermodynamics for Systems Composed of Electrons and Nuclei. *Adv. Math.* **9**, 316–398, 1972.

[31] E. H. Lieb. The Number of Bound States of One-Body Schroedinger Operators and the Weyl Problem. *Proc. Symposia in Pure Math.* **36**, 241–252, 1980.

[32] E. H. Lieb. Proof of an Entropy Conjecture of Wehrl. *Commun. Math. Phys.* **62**, 35–41, 1978.

[33] N. Dunford and J. T. Schwartz. Linear Operators, part I. New York, Wiley-Interscience, 1967.

[34] E. H. Lieb. Thomas–Fermi and Related Theories of Atoms and Molecules. *Rev. Mod. Phys.* **53**, 603–641, 1981.

[35] S. Chandrasekhar. An Introduction to the Study of Stellar Structure. New York, Dover, 1967.

## Further Reading

Section 1.1

H. Wergeland. Irreversibility in Many-Body Systems. In: Irreversibility in the Many-Body Problem, J. Biel and J. Rae, eds. New York and London, Plenum, 1972.

(1.1.1)

G. Emch. Non-Markovian Model for the Approach to Equilibrium. *J. Math. Phys.* **7**, 1198–1206, 1966.

(1.1.13)

E. Schrödinger. Zur Dynamik elastisch gekoppelter Punktsysteme. *Ann. Phys.* (Leipzig) **44**, 916–934, 1914.

I. Prigogine and G. Klein. Sur la Mécanique Statistique des Phénomènes Irréversibles, III. *Physica* **19**, 1053–1071, 1953.

(1.2.10)

E. H. Lieb and H. Narnhofer. The Thermodynamic Limit for Jellium. *J. Stat. Phys.* **12**, 291–310, 1975.

Section 1.3

F. A. Berezin. Method of Second Quantization. New York, Academic Press, 1966.

M. Reed and B. Simon. Methods of Modern Mathematical Physics, vol. II: Fourier Analysis, Self-Adjointness. New York, Academic Press, 1975.

(1.4.2)

J. von Neumann. On Infinite Direct Products. *Compositio Math.* **6**, 1–77, 1939.

(1.4.9)

O. Bratteli and D. W. Robinson. Operator Algebras and Quantum Statistical Mechanics, vol. I. New York, Springer, 1979.

Section 2.1

A. Wehrl. How Chaotic is a State of a Quantum System? *Rep. Math. Phys.* **6**, 15–28, 1974.

A. Uhlmann. Sätze über Dichtematrizen. *Wiss. Z. Karl-Marx-Univ. Leipzig* **20**, 633, 1971; Endlichdimensionale Dichtematrizen, I. *Ibid.* **21**, 421; 1972; Endlichdimensionale Dichtematrizen, II. *Ibid.* **22**, 139, 1973.

(2.1.7)

E. H. Lieb. Some Convexity and Subadditivity Properties of Entropy. *Bull. Amer. Math. Soc.* **81**, 1–13, 1975.

B. Simon. Trace Ideals and their Applications. London and New York, Cambridge Univ. Press, 1979.

(2.2.9)

B. Baumgartner. Classical Bounds on Quantum Partition Functions. *Commun. Math. Phys.* **75**, 25–41, 1980.

(2.2.11)

F. A. Berezin. Wick and Anti-Wick Operator Symbols. *Math. USSR Sbornik* **15**, 577–606, 1971. (Translation of Vikovskie i antivikovskie simboly operatorov. *Mat. Sbornik.* **86(128)**, 578–610, 1971.)

(2.2.22)

G. Lindblad. Entropy, Information, and Quantum Measurements. *Commun. Math. Phys.* **33**, 305–322, 1973.

H. Umegaki. Conditional Expectation in an Operator Algebra. *Kodai Math. Seminar. Rep.* **14**, 59–85, 1962.

H. Araki. RIMS preprint 190, Kyoto, 1975.

Section 2.3

R. Griffiths. Microcanonical Ensemble in Quantum Statistical Mechanics. *J. Math. Phys.* **6**, 1447–1461, 1965.

(2.3.39)

A. S. Wightman. Convexity and the Notion of Equilibrium State in Thermodynamics and Statistical Mechanics. In: Convexity in the Theory of Lattice Gases, R. Israel, ed. Princeton, Princeton Univ. Press, 1979.

(2.4.7)

H. D. Maison. Analyticity of the Partition Function for Finite Quantum Systems. *Commun. Math. Phys.* **22**, 166–172, 1971.

(2.4.9)

E. H. Lieb. The Classical Limit of Quantum Spin Systems. *Commun. Math. Phys.* **31**, 326–341, 1973.

(2.4.15; 2)

A. S. Wightman, *op. cit.* in (2.3.39)

(2.5.15)

W. Thirring. Bounds on the Entropy in Terms of One Particle Distributions. *Lett. Math. Phys.* **4**, 67–70, 1980.

(2.5.26)

K. Huang. Statistical Mechanics. New York; Wiley, 1963.
F. Reif. Fundamentals of Statistical and Thermal Physics. New York, McGraw-Hill, 1965.

Section 3.1

D. Ruelle. Statistical Mechanics, Rigorous Results. New York, Benjamin, 1969.
G. Emch. Algebraic Methods in Statistical Mechanics and Quantum Field Theory. New York, Wiley, 1971.

(3.1.2)

E. B. Davies. Quantum Theory of Open Systems. New York, Academic Press, 1976.

(3.1.4)

V. Gorini, A. Frigerio, M. Verri, A. Kossakowski, and E. C. G. Sudarshan. Properties of Quantum Markovian Master Equations. *Rep. Math. Phys.* **13**, 149–173, 1978.
P. Martin. Modèles en Mécanique Statistique des Processus Irréversibles. New York and Berlin, Springer, 1979.

(3.1.12)

G. Lindblad. Completely Positive Maps and Entropy Inequalities. *Commun. Math. Phys.* **40**, 147–151, 1975.
A. Uhlmann. Relative Entropy and The Wigner–Yanase–Dyson–Lieb Concavity in an Interpolation Theory. *Commun. Math. Phys.* **54**, 21–32, 1970.

(3.1.14)

F. Greenleaf. Invariant Means on Topological Groups. New York, Van Nostrand, 1966.
A. Guichardet. Systèmes Dynamiques non Commutatifs. *Astérisque* **13–14**, 1974.

(3.1.25)

V. I. Arnol'd and A. Avez. Ergodic Problems of Classical Mechanics. New York, Benjamin, 1968.

(3.2.1)

N. M. Hugenholtz. Article in: Mathematics of Contemporary Physics, R. Streater, ed. New York and London, Academic Press, 1972.

(3.2.6)

R. Haag, N. M. Hugenholtz, and M. Winnink. On the Equilibrium States in Quantum Statistical Mechanics. *Commun. Math. Phys.* **5**, 215–236, 1967.

(3.2.13)

M. Takesaki. Tomita's Theory of Modular Hilbert Algebras and Its Applications, Lecture Notes in Mathematics, vol. 128. New York and Berlin, Springer, 1970.

(3.3.7)

H. Narnhofer and D. W. Robinson. Dynamical Stability and Pure Thermodynamic Phases. *Commun. Math. Phys.* **41**, 89–97, 1975.

(3.3.10)

R. Haag, D. Kastler, and E. B. Trych-Pohlmeyer. Stability and Equilibrium States. *Commun. Math. Phys.* **38**, 173–193, 1974.

(3.3.22)

W. Pusz and S. L. Woronovicz. Passive States and KMS States for General Quantum Systems. *Commun. Math. Phys.* **58**, 273–290, 1978.

A. Lenard. Thermodynamical Proof of the Gibbs Formula for Elementary Quantum Systems. *J. Stat. Phys.* **19**, 575–586, 1978.

Section 4.1

E. H. Lieb and B. Simon. The Thomas–Fermi Theory of Atoms, Molecules, and Solids. *Adv. Math.* **23**, 22–116, 1977.

H. Narnhofer and W. Thirring. Asymptotic Exactness of Finite Temperature Thomas–Fermi Theory. *Ann. Phys.* (N.Y.) **134**, 128–140, 1981.

B. Baumgartner. The Thomas–Fermi Theory as Result of a Strong-Coupling Limit. *Commun. Math. Phys.* **47**, 215–219, 1976.

Section 4.2

P. Hertel, H. Narnhofer, and W. Thirring. Thermodynamic Functions for Fermions with Gravostatic and Electrostatic Interactions. *Commun. Math. Phys.* **28**, 159–176, 1972.

P. Hertel and W. Thirring. Article in: Quanten und Felder, H. Dürr, ed. Brunswick, Vieweg, 1971.

J. Messer. Temperature Dependent Thomas–Fermi Theory, Lecture Notes in Physics, vol. 147. New York and Berlin, Springer, 1979.

B. Baumgartner. Thermodynamic Limit of Correlation Functions in a System of Gravitating Fermions. *Commun. Math. Phys.* **48**, 207–213, 1976.

Section 4.3

E. H. Lieb. The Stability of Matter. *Rev. Mod. Phys.* **48**, 553–569, 1976.
W. Thirring. Stability of Matter. In: Current Problems in Elementary Particle and Mathematical Physics, P. Urban, ed. *Acta Phys. Austriaca Suppl.* **XV**, 337–354, 1976.

(4.3.22)

R. Griffiths. Microanonical Ensemble in Quantum Statistical Mechanics. *J. Math. Phys.* **6**, 1447–1461, 1965.

(4.3.40)

H. Narnhofer and W. Thirring. Convexity Properties for Coulomb Systems. *Acta Phys. Austriaca.* **41**, 281–297, 1975.

# Index